Food Security, Poverty and Nutrition Policy Analysis

Food Security, Poverty and Nutrition Policy Analysis

Statistical Methods and Applications

THIRD EDITION

Suresh C. Babu
Int'l Food Policy Research Institute, Washington, DC, United States;
Department of Agricultural Economics, Extension and Rural Development, University of Pretoria, Pretoria, South Africa

Shailendra N. Gajanan
Departments of Economics, Business Management and Education, University of Pittsburgh, Bradford, PA, United States

ACADEMIC PRESS
An imprint of Elsevier

Academic Press is an imprint of Elsevier
125 London Wall, London EC2Y 5AS, United Kingdom
525 B Street, Suite 1650, San Diego, CA 92101, United States
50 Hampshire Street, 5th Floor, Cambridge, MA 02139, United States
The Boulevard, Langford Lane, Kidlington, Oxford OX5 1GB, United Kingdom

Copyright © 2022 Elsevier Inc. All rights reserved.

No part of this publication may be reproduced or transmitted in any form or by any means, electronic or mechanical, including photocopying, recording, or any information storage and retrieval system, without permission in writing from the publisher. Details on how to seek permission, further information about the Publisher's permissions policies and our arrangements with organizations such as the Copyright Clearance Center and the Copyright Licensing Agency, can be found at our website: www.elsevier.com/permissions.

This book and the individual contributions contained in it are protected under copyright by the Publisher (other than as may be noted herein).

Notices
Knowledge and best practice in this field are constantly changing. As new research and experience broaden our understanding, changes in research methods, professional practices, or medical treatment may become necessary.

Practitioners and researchers must always rely on their own experience and knowledge in evaluating and using any information, methods, compounds, or experiments described herein. In using such information or methods they should be mindful of their own safety and the safety of others, including parties for whom they have a professional responsibility.

To the fullest extent of the law, neither the Publisher nor the authors, contributors, or editors, assume any liability for any injury and/or damage to persons or property as a matter of products liability, negligence or otherwise, or from any use or operation of any methods, products, instructions, or ideas contained in the material herein.

Library of Congress Cataloging-in-Publication Data
A catalog record for this book is available from the Library of Congress

British Library Cataloguing-in-Publication Data
A catalogue record for this book is available from the British Library

ISBN: 978-0-12-820477-1

For information on all Academic Press publications visit our website at https://www.elsevier.com/books-and-journals

Publisher: Charlotte Cockle
Acquisitions Editor: Nancy Maragioglio
Editorial Project Manager: Allison Hill
Production Project Manager: Sruthi Satheesh
Cover Designer: Victoria Pearson

Typeset by TNQ Technologies

To
Per Pinstrup-Andersen
Joachim von Braun
Shenggen Fan
Johan Swinnen
For their past and present leadership of
the International Food Policy Research
Institute—its research outputs in the past
45 years are the prime motivators of the
contents of this book.

Praises from around the world for the third edition

The third edition of this important book makes seminal contributions on issues relating to food security and nutritional policy analysis and presents research on recent topics such as the importance of food systems, multidimensional poverty index, and the impact of covid-19 on food security. The book deepens the understanding of the reader through several examples using STATA and R, which will be beneficial for faculty, researchers, and students. The materials in the book could immensely benefit the teaching of advanced undergraduate and graduate students in Development Economics and Public Policy.
Dr. Devinder Malhotra, Chancellor, Minnesota State Colleges and Universities, USA.

In this third edition of *Food Security, Poverty and Nutrition Policy Analysis*, Babu and Gajanan bring their skills and substantial amount of evidence to bring together analytical concepts, statistical methods, and data where they are available to address some of the key issues developing countries face in these areas. It is a cornucopia of ideas and approaches for students and teachers of food, nutrition, and poverty.
Dr. Uma Lele, President, International Association of Agricultural Economists, Former Director, World Bank.

This comprehensive work by highly accomplished and experienced researchers could not be more timely. With hunger on the increase and only a decade to go to meet the SGD goals, policymakers are searching for evidence-based solutions to the intractable problems of poverty, hunger, and malnutrition (including obesity). This text provides a sound theoretical base and highly relevant practical examples for trainers, students, and practitioners in the essential elements to provide much-needed evidence for decision-making and program evaluation.
Prof. Sheryl L. Hendriks, Head of Department, University of Pretoria, South Africa.

This book provides an excellent introduction to the analysis of food security, poverty, and nutrition policy. It provides theoretical foundations for such analyses, backed up by specific examples, and is accompanied with STATA and R codes. It is an ideal textbook for any course on food and nutrition security and is a valuable addition to the tool kit of students of development studies, development practitioners, program evaluators, and policy analysts.

Prof. J.V. Meenakshi, Delhi School of Economics, India.

The launch of the third edition of the book *Food Security, Poverty and Nutrition Policy Analysis* takes place at a crucial moment, in which we see the worsening of the problem of hunger, especially in countries of the Global South. In a time when obscurantist interpretations have questioned the science, this book is an ideal tool for training students in our food security-related postgraduate courses.

Prof. Sonia Bergamasco and Prof. Ricardo Borsatto, University of Campinas, Brazil.

The contents of this edition and the previous editions of this book were taught in my course that addresses global food issues at the University of Maryland. The book's self-learning approach to wide ranging policy issues relating the food security, nutrition, and poverty provides practical applications that would help students, policy analysts, and development practitioners a quick start on public policy analysis.

Professor Prabhakar Tamboli, University of Maryland, USA.

The new edition of this important textbook on food security and nutrition policy analysis is very welcome and timely. It is updated to include impacts of covid-19 on food security, the application of food systems approaches, and guidance on the use of "R" open-source software in addition to the already very useful examples of applications of Stata. The book is comprehensive, rigorous but also very accessible, and structured in a systematic way to guide student learning. We use this book as a core textbook in our MSc Food Security Policy and Management, and it equips students with the holistic knowledge and technical skills required to be able to conduct good applied policy analysis, which is a major requirement to address current global food systems challenges. I strongly recommend this book both for academics and for practitioners working in this field.

Dr Nick Chisholm, Senior Lecturer in International Development, University College Cork, Ireland.

This book provides sound statistical foundation for policy analysis of food and agriculture sectors. African development crucially depends on such evidence-based analysis. I highly recommend this book for all development practitioners and teachers and students of agricultural economics, economic development, and public policy programs in African universities.

Dr. Guy Blaise Nkamleu, President, African Association of Agricultural Economists (AAAE), Lead Economist and Advisor, African Development Bank.

Contents

Preface xxi
Introduction xxv

Section I
Food security policy analysis

1. **Introduction to food security: concepts and measurement** 3

   Introduction 3
   Conceptual framework of food security 4
   Food security in the developed world 7
   Other policy issues in the United States 11
   Food security concerns in other countries 11
   Measurement of the determinants of food security 16
   Food availability 16
   Measuring food availability 17
   Measuring food access 17
   Food utilization 18
   Measuring food utilization 18
   Stability of availability 19
   Alternative approaches in measuring food security 19
   Conclusions 21
   A natural question is why is measuring food insecurity important for better program design in developing countries? 23
   Exercises 26

2. **Implications of technological change, postharvest technology, and technology adoption for improved food security—application of t-statistic**

   Introduction 28
   Review of selected studies 30
   Postharvest technology and implications for food security 33
   Food security issues and technology in the United States 36
   Biofuels—the Chinese experience 38

US Farm Policy and food security—background and
    current issues   40
GEO-5 and coping mechanisms for the future   44
Empirical analysis—a basic univariate approach   45
Data description and analysis   46
Two measures of household food security are
    computed   47
Consumption components of the food security index   49
Descriptive statistics   51
Threshold of food security by each individual component   52
Tests for equality of variances   52
Student's *t*-test for testing the equality of means   53
Policy implications   56
Technical appendices   58
Constructing the cutoff points for components of the food
    security index   58
Variable definitions   58
Using STATA for *t*-tests   59
Independent group *t*-test   59
Independent sample *t*-test assuming unequal variances   60
Exercises   63
STATA exercise   64

3. **Effects of commercialization of agriculture (shift from traditional crop to cash crop) on food consumption and nutrition—application of chi-square statistic**

Introduction   68
A few concepts   70
What is commercialization?   70
Review of selected studies   74
Organic farms and commercialization in the United States   79
Organic farming in a global context   81
Empirical analysis   82
Data description and analysis   83
Descriptive analysis: cross-tabulation results   84
Chi-square tests   87
Chi-square tests using STATA   89
Conclusion and policy implications   93
Technical appendices   95
Pearson's chi-square (χ^2) test of independence   95
Student's *t*-test versus Pearson's chi-square (χ^2) test   96
Limitations of the chi-square procedure   96
Descriptive analysis: cross-tabulation results   97
Chi-square tests using R   99
Exercises   101
STATA exercise   102

Contents **xi**

4. **Effects of technology adoption and gender of household head: the issue, its importance in food security—application of Cramer's V and phi coefficient**

Introduction	105
Review of selected studies	107
Female farm operators in Kenya and Ethiopia: recent evidence	109
Rights, norms, and institutions: beyond technology adoption	110
Female farm operators in the United States	112
Women in agriculture: the global scene	113
Uganda's coffee market: a case study	116
Empirical analysis	118
Data description and analysis	119
Descriptive analysis: cross-tabulation results	120
Cramer's V and phi tests	121
Conclusion and policy implications	123
Section highlights: Covid-19, women's burden, and the digital divide	123
Cramer's V in STATA	126
Technical appendices	128
Phi coefficient and Cramer's V	128
Applications in R	128
Exercises	131
STATA exercise	131

5. **Changes in food consumption patterns: its importance to food security—application of one-way ANOVA**

Introduction	135
Determinants of food consumption patterns and its importance to food security and nutritional status	138
Impact on food security	139
Review of selected studies	140
Food consumption patterns for developing countries	141
Nutritional and economic outcomes	144
Food consumption patterns in the United States	145
Food consumption patterns in India and China	147
Empirical analysis and main findings	150
Data description	150
Analysis method	150
Results	150
One-way ANOVA in STATA	153
Conclusion and policy implications	158
One-way ANOVA	159
Underlying assumptions in the ANOVA procedure	160
Decomposition of total variation	160
Number of degrees of freedom	161

F-test and distribution	161
Relation of *F* to *T*-distribution	161
STATA workout	162
Fruit intakes per week for three income groups (F_1, F_2, and F_3)	162
Compute mean square between and mean square within	164
Compute the calculated value of *F*	164
One-way ANOVA in R	165
Exercises	168

6. Impact of market access on food security—application of factor analysis

Introduction	170
Assessing the linkages of market reforms on food security and productivity	171
Review of selected studies	173
Food deserts in the United States	175
Access, information, and food security in Africa	178
The role of the informal sector and food security	180
Food security issues in Middle East and North Africa	180
Food insecurity in South Asia: case studies of India and Afghanistan	181
Empirical analysis	185
Technical concepts	186
Eigenvalues and eigenvectors	186
Properties of eigenvalues	187
Data description and methodology	188
Food indicators	189
Asset indicators	189
Yield/technology indicators	189
Market access indicators	189
Household level characteristics	190
Factor analysis by principal components	190
Step 1: Computing the observed correlation matrix	190
Step 2: Estimating the factors	192
Principal components analysis	193
Examining eigenvalues	193
Scree plot	193
Step 3: Making the factors easier to interpret: rotation procedure	197
Varimax orthogonal rotation	197
Step 4: Computing factor scores	198
Principal components analysis in STATA	201
Conclusion and policy implications	204
Technical appendices	206
Factor analysis decision process	206
Exercises	211
STATA workout	212

Section II
Nutrition policy analysis

7. **Impact of maternal education and care on preschoolers' nutrition—application of two-way ANOVA**

Introduction	222
Conceptual framework: linkages between maternal education, child care, and nutritional status of children	223
Possible linkages	223
Conceptual and measurement issues on child care	224
Measurement issues	225
Review of selected studies	227
Maternal education and nutrition status in the United States	**229**
Children's nutrition and maternal education in Africa	**232**
Kenya	232
Uganda and Nigeria	232
Substantive findings from Asia and Latin America	233
Empirical analysis	234
Data description	234
Cross-tabulation of weight for height with mothers' educational levels	**235**
Two-way ANOVA results	237
Definition of main effect	237
Partitioning sum of squares	237
Interpreting the interaction effect and post hoc tests	**241**
Two-way ANOVA in STATA	242
Conclusion	**248**
Technical appendices	**249**
Scoring system used to create the care index (Ruel et al., 1999) care index by age group	249
Post hoc procedures	**249**
ANOVA in R	250
Exercises	**253**
STATA workout	**254**

8. **Indicators and causal factors of nutrition—application of correlation analysis**

Introduction	259
Review of selected studies	261
Food insecurity and nutrition in the United States	264
Food insecurity in Brazil	269
Global Monitoring Report on Nutrition and Millennium Development Goals	269
Impact of food price spike and domestic violence in rural Bangladesh	270
Malnutrition and chronic disease in India	271

Malnutrition in Guatemala		272
Empirical analysis and main findings		273
Data description and methodology		274
Concepts in correlation analysis		275
Inference about population parameters in correlation		276
Descriptive analysis		277
Main results		278
Correlation analysis of the outcome variables		279
Estimating correlation using STATA		281
Conclusion and policy implications		283
Estimating correlation using R		286
Exercises		289
STATA workout		289

9. Effects of individual, household, and community indicators on child's nutritional status—application of simple linear regression

Introduction	296
Conceptual framework and indicators of nutritional status	297
Household utility maximization framework	297
Core indicators of nutritional status	299
Review of studies on the determinants of child nutritional status	303
Child's nutritional status in the United States	307
The economics of double burden	308
AIDS and double burden in Africa	309
Malnutrition and mortality in Pakistan and India	310
Social participation as social capital, women empowerment, and nutrition in Peru	312
Social capital and policy during the pandemic	313
Empirical analysis and main findings	316
Data description	316
Incidence of stunting and wasting	318
Normality tests and transformation of variables	319
Regression results	321
Simple regression in STATA	325
Conclusion	327
Simple regression in R	328
Exercises	330
STATA workout	331

10. Maternal education and community characteristics as indicators of nutritional status of children—application of multivariate regression

Introduction	336
Selected studies on the role of maternal education and community characteristics on child nutritional status	337
Community characteristics and Children's nutrition in the United States	342

Community characteristics and child nutrition in Kenya	345
Financial crisis and child nutrition in East Asia	345
Double burden within mother–child pairs: Asian case	346
Empirical analysis	347
Data description and methodology	350
Descriptive summary of independent variables	352
Main results	353
Step 1: Estimating the coefficients of the model	353
Step 2: Examining how good the model predicts	355
Step 3: hypotheses testing	357
Step 4: Checking for violations of regression assumptions	362
STATA Output	368
Conclusions	371
Multiple regression in *R*	372
Exercises	376

Section III
Special topics on poverty, nutrition, and food policy analysis

11. Predicting child nutritional status using related socioeconomic variables—application of discriminant function analysis

Introduction	382
Conceptual framework: linkages between women's status and child nutrition	384
Linkages between women's status and child nutrition	384
Review of selected studies	385
Direct linkage studies between women's status and children's nutritional status	385
Indirect linkages between women's status and child's nutritional status	388
USDA nutrition assistance programs: a case study from the United States	390
Case studies of women's status and child nutritional status from Africa, Asia, and Latin America	393
Food security and welfare in Africa: social customs, technology, and climate change	393
Childhood undernutrition and climate change in Asia	396
Adaptive strategies and sustainability lessons from Latin America	397
Can garden plots save Russia?	398
Empirical analysis and main findings	398
Data description and analysis	399
Descriptive statistics	401
Testing the assumptions underlying discriminant analysis model	403
Box's M Test	403
Tests of equality of group means	404

xvi   Contents

      Summary of main findings   405
      Relative impact of the predictor variables on ZWHNEW   407
      Correlation between the predictor variables and discriminant
          function   408
      Classification statistics   409
      Classification function based on equal and unequal prior
          probabilities   410
      Canonical discriminant analysis using STATA   412
      Conclusions   416
      Technical appendix: discriminant analysis   417
      Discriminant analysis decision process   417
      Canonical discriminant analysis using R   424
      Exercises   426
      STATA workout   426

12. **Measurement and determinants of poverty—application of logistic regression models**

      Introduction   434
      Dimensions and rationale for measuring poverty   435
      Defining and measuring poverty   435
      Monetary approach   435
      Basic needs approach   436
      Capability approach   436
      Participatory poverty approach   437
      Rationale for measuring poverty   439
      Indicators in measuring poverty   439
      Income measure   440
      Consumption expenditure   440
      Construction of poverty lines using food energy intake and cost of basic needs approaches   442
      Poverty lines in theory   442
      Absolute and relative poverty   443
      Referencing and identification problems   443
      Deriving a poverty line   444
      Food energy intake method   444
      Cost of basic needs method   446
      New measures of poverty based on the engel curve   449
      Measures of poverty   449
      Selected review of studies on determinants of poverty   454
      Poverty and welfare in the United States   458
      Agriculture and poverty in Laos and Cambodia   463
      Financial crisis and poverty in the Russian Federation   465
      Poverty in Europe   465
      Poverty in developing countries: China and India   467
      Determinants of poverty—binary logistic regression analysis   468

Dichotomous logistic regression model	469
An example with the Malawi dataset	469
Expected determinants of household welfare	470
Empirical results	472
Measuring model fit	472
Log-likelihood ratio	472
Hosmer–Lemeshow goodness-of-fit test	473
Generalized coefficient of determination	474
Classification table	475
Interpreting the logistic coefficients and discussion of results	476
Estimating logistic regression models in STATA	478
Example 1	478
Example 2	482
Conclusions and implications	485
Technical appendices	486
Technical notes on logistic regression model	486
Estimating logistic regression models in R	488
Exercises	489
STATA workout	490

13. Classifying households on food security and poverty dimensions—application of K-Means cluster analysis

Introduction	493
Food hardships and economic status in the United States	495
Food security, economic crisis, and poverty in India	499
Cluster analysis: various approaches	501
Hierarchical clustering method	501
Single linkage (nearest neighbor method)	502
Complete linkage (farthest neighbor method)	502
Average linkage method	503
K-means method	503
Review of selected studies using cluster analysis	503
Empirical analysis: *K*-Means clustering	506
Data description	507
Initial partitions and optimum number of clusters	507
Descriptive characteristics of the cluster of households	508
Cluster centers	510
Cluster analysis in STATA	513
Conclusion and implications	516
Cluster analysis in R	517
Exercises	519
STATA workout–1	519
STATA workout–2	523

14. Household care as a determinant of nutritional status—application of instrumental variable estimation

Introduction	527
Review of selected studies	529
Federal nutrition programs and children's health in United States	534
Parental unemployment and children's health in Germany	536
Food security using the Gallup World Poll	538
Empirical analysis	539
Stage 1: estimating child-care practices	540
Stage 2: estimating the determinants of child health (weight-for-age Z-Scores)	541
IV estimation using STATA	543
Conclusions	545
Instrumental variable estimation using R	546
Exercises	547
STATA workout 1	548
STATA workout 2	552

15. Achieving an ideal diet—modeling with linear programming

Introduction	557
Review of the literature	559
Linear programming model	563
Solution procedures	565
Graphical solution approach	565
Some qualifications about the optimum	567
Using solver in excel to obtain an LP solution	568
Step 1: Setting the problem in excel	568
Step 2: Solving the parameters of the model	570
Step 3: Deriving the results	570
Summary	572
Exercises	573

16. Food and nutrition program evaluation

Introduction	575
Recent developments	577
Randomization	577
Instrumental variables	579
Difference-in-difference	583
Regression discontinuity design	585
Propensity score matching and pipeline comparisons	587
Randomization and development policy: applying the methods	588
Summary and conclusions	591
Section highlights: nobelprize worthy	591
STATA workout 1	592
STATA workout 2	594

17. Multidimensional poverty and policy

Multidimensional child poverty and gender inequalities	602
Multidimensional energy poverty	603
Financial exclusion and Multidimensional Poverty Index	604
The Alkire–Foster method	604
STATA implementation	609
STATA workout	614

Section IV
Technical appendices

Appendix 1:	Introduction to software access and use	619
Appendix 2:	Software information	621
Appendix 3:	SPSS/PC+ environment and commands	623
Appendix 4:	Data handling	633
Appendix 5:	SPSS programming basics	641
Appendix 6:	STATA—a basic tutorial	655
Appendix 7:	Anthropometric indicators—computation and use	663
Appendix 8:	Elements of matrix algebra	669
Appendix 9:	Some preliminary statistical concepts	675
Appendix 10:	Instrumental variable estimation	679
Appendix 11:	Statistical tables	685

References	695
Index	735

Preface

Any applied problem-solving book will need regular updating, as concepts, theories, and methods are constantly changing. The contents of this book have been revised to keep up with the emerging issues and methods in the field. The motivation for the third edition of this book comes from four key sources. First, the issues and challenges related to food security, nutrition, and poverty, as development problems continue to be at the top of the policy makers' agenda. Second, levels of food insecurity, malnutrition, and poverty are still at unacceptable levels, and the COVID-19 pandemic is likely to increase these levels. Third, the approach to solving challenges of food insecurity, malnutrition, and rural poverty requires a wholistic food system perspective. The proposed UN Food System Summit in 2021 is a testimony to the global movement toward system approach to food security and nutrition challenges. Finally, there has been progress in conceptual and methodological issues, which require attention of the policy analysts. This edition attempts to meet all these emerging needs.

In the second edition of this book, we added the following. First, since the publication of the first edition of this book in 2009, there was increasing demand and reminders from the readers and the teachers of the content to incorporate the latest developments in the field of food security, nutrition, and poverty analysis. Second, there was an overwhelming response to the first edition from both researchers and practitioners in the field and the effectiveness of the contents in practical problem-solving. Third, the use of software for quantitative analysis had also evolved in the past 15 years, and there is a need to cast the problems and issues in this context to develop effective pedagogical methods. Finally, the second edition addressed the financial and food crisis that had pushed millions of people into food insecurity and poverty even in developed countries, which generated a need for teaching and addressing these issues in the European, North American, and Australian universities.

The first edition of this book had its conceptual origin from the lecture materials of the training courses taught by one of the authors in the early 1990s. It was during this period that in several developing nations, particularly in Africa, even when the signs of widespread hunger and abject poverty were visible, policy makers did not act for want of "empirical evidence." Some policy makers even dismissed the severity of the problem saying that the hunger reports prepared by government officials were not rigorous enough to take them seriously. Some decision-makers entirely rejected the reports prepared by the officials, stating that the analysis of data was "not statistically sound" to draw reliable inference and undertake the desirable public actions. The final result was inaction on the part of the policy makers. Little has changed since then as evidenced by the continuing food crises in several countries. Generating empirical evidence on causal factors and severity of food insecurity, nutrition, and poverty problems becomes more urgent also in the context of the recent sharp increases in global food prices.

The capacity to collect, process, and analyze data on food security, nutrition, and poverty problems continues to remain low in many developing countries. While students are trained adequately in their individual fields of specialization, such as nutrition, economics, sociology, political science, international development, anthropology, and geography, they are often ill-prepared for the task of policy analysts in the governments, academic and research institutions, civil society organizations, and the private sector. Developing applied policy analysis skills requires a combination of several related abilities in statistical data analysis, computer literacy, and using the results for developing policy alternatives. In addition, an understanding of issues, constraints, and challenges facing policy makers on particular hunger, malnutrition, and poverty problems is critical. Such capacity is also needed in developed countries to address the problems of hunger and poverty. These capacity challenges remain as we bring out the third edition of this book.

The first edition of this book was largely motivated by and based on three decades of food and nutrition policy research at the International Food Policy Research Institute. In the mid-1990s, the data-based statistical methods were combined with selected case studies from IFPRI research on food and nutrition security issues to form a training manual. It was well received among the training institutions and university departments teaching courses on food security and nutrition policy analysis both in the North and in the South. Selected contents of this manual were taught by the first author over the years at various institutions in many parts of the world including the University of Maryland, University of Sweden, University of Hohenheim, Tufts University, University of Malawi, University of Zimbabwe, Indian Agricultural Research Institute, Andhra Pradesh Agricultural University, Eduardo Mondlane University, China Academy of Agricultural Sciences, Ghana University of Development Studies, University of Pretoria, and Lamolina University. Since the publication of the first edition, we have been teaching the contents of the book in several other universities and national research systems in countries including India, China, Bangladesh, Myanmar, Malawi, Nigeria, South Africa, Ireland, Nepal, Bhutan, Sri Lanka, selected European countries, and North America. As shown by our correspondences with them, researchers, students, and academic faculties in many more countries are adopting this book for their research methods and policy analysis learning and curriculum.

The second edition of the book was a substantially revised version of the first edition and continues to impart the combined skills of statistical data analysis, computer literacy, and using the results for developing policy alternatives through a series of statistical methods applied to real-world food insecurity, malnutrition, and poverty problems. It continued to base its approach of combining case studies with data-based analysis for teaching policy applications of statistical methods from several training courses and class lectures taught in the previous 20 years. Thus, the second edition also had the benefit of the feedback and comments from the users of the earlier edition of the book and the participants of the aforementioned training courses offered by the authors since the first edition of the book. It contained new sections related to the problems of the developed countries and a new chapter on program evaluation. All the analytical chapters were also based on STATA programming.

The third edition of the book attempts to bring the emerging issues, challenges, opportunities, and methods in all chapters. We introduce new literature in the chapters and additional sections related to the new approaches to concepts and issues including the COVID-19 pandemic. We have taken the new food system approach as a cross-cutting theme in all the chapters. Finally, all the chapters now contain the programming codes and selected results in R programming language. As in the first and second editions, the contents of the third edition can be taught in a semester-long course of 15 weeks.

As in the first two editions, this book is primarily addressed to students with a bachelor's degree who have familiarity with food security, nutrition, and poverty issues and who have taken a beginner's course in statistics. It is ideally suited for first-year postgraduate courses in food sciences, nutrition, agriculture, development studies, economics, and international development. The book is self-contained with statistical appendices, computer programs, and interpretation of the results for policy applications. It could be used as course material both in face-to-face and distance learning programs or a combination of these approaches.

While preparing the contents of all the three editions, we were inspired and benefited from the following contents of courses offered by the following professors. The contents of this book should be of valuable addition to these course materials.

1. Prabhakar Tamboli, University of Maryland
2. Sudhanshu Handa, University of North Carolina
3. Rolf Klemm and Keith West, John Hopkins University
4. Ellen Messer, Boston University
5. Rosamond L. Naylor, Stanford University
6. James Tillotson, Tufts University
7. Partick Webb, Tufts University
8. William Masters, Tufts University
9. Constance Gewa, George Mason University
10. Robert Paarlberg, Harvard Kennedy School
11. Sue Horton, University of Waterloo

12. Esther Duflo, Economics, MIT
13. Abhijit Banerjee, Economics, MIT
14. Sendhil Mullainathan, Harvard University
15. Neha Khanna, Binghamton University, SUNY
16. Craig Gundersen, University of Illinois at Urbana-Champaign
17. Raghbendra Jha, Australian National University
18. Christopher Barrett, Cornell University
19. Per Pinstrup-Andersen, Cornell University
20. Joachim von Braun, University of Bonn
21. David Sahn, Cornell University
22. Christine Olson, Cornell University
23. Susan M Randolph, University of Connecticut
24. Timothy Dalton, Kansas State University
25. Katherine Cason, Clemson University
26. Kenneth A. Dahlberg, Western Michigan University
27. Jonathan Robinson, University of California
28. Helen H. Jensen, Iowa State University
29. Arne Hallam, Iowa State University
30. Bruce Meyer, University of Chicago
31. Marie-Claire Robitaille-Blanchet, University of Nottingham
32. Marion Nestle, New York University
33. Jane Kolodinsky, University of Vermont
34. Cynthia Donovan, Michigan State University
35. Prabhu Pingali, Cornell University

We hope that this new edition will be useful in developing a new generation of policy researchers and analysts who are well equipped to address the real-world problems of poverty, hunger, and malnutrition, whose policy recommendations will not be rejected for want of empirical evidence and will result in swift public and private action in both the developing and developed worlds.

SCB
SNG

Introduction

The nature and scope of food security, poverty, and nutrition policy analysis

Even before the advent of the recent pandemic COVID-19, in the year 2019, the issues of chronic food insecurity, poverty, and malnutrition remained fundamental human welfare challenges in developing and developed countries. Problems related to increasing food availability, feeding the population, improving their nutritional status, and reducing poverty levels continue to confront decision-makers. These challenges are further exacerbated by the emerging challenges of climate change, unsustainable food production practices, continued wastage of food, food safety, and inadequate inclusion of youth and women in the production process.

Program managers and policy makers who constantly deal with design, implementation, monitoring and evaluation of food security, nutrition, and poverty-related interventions have to make best decisions from a wide range of program and policy options. Information for making such policy and program decisions must be based on sound data-based analysis. Such analysis should be founded on statistical theory that provides an inferential basis for evaluating, refining, and, sometimes, rejecting the existing policy and program interventions.

This book deals with the application of statistical methods for analysis of food security, poverty, and nutrition policy and program options. A range of analytical tools are considered that could be used for analyzing various technological, institutional, and policy options and for developing policy and program interventions by making inferences from household-level socioeconomic data. It helps the students and researchers to explore policy analysis as a self-learning tool.

The objective of policy analysis is to identify, analyze, and recommend policy options and strategies that would achieve the specific goals of policy makers (Babu, 2013; Babu et al., 2000; Dunn, 1994). Issues related to increasing food security, reducing malnutrition, and alleviating poverty are high on the global development policy agenda as evidenced by recent unprecedented increases in food prices, resultant unrest in several developing countries, and a series of international summits convened to mitigate the effects of food price increase (UN Summit, 2008). This book addresses a wide range of policy and program options typically designed and implemented by government agencies, nongovernmental organizations, and communities to address the development challenges such as hunger, poverty, and malnutrition faced by households and communities.

Such policy and program options, for example, aim at increasing the availability of food, increasing the household entitlement, improving the efficiency of food distribution programs, enhancing the market availability for selling and buying food commodities, reducing malnutrition through the school feeding and nutrition programs, increasing technological options through introduction of high yielding varieties of seeds that farming communities in rural areas could grow to increase income, investing in technological advancements, implementing land reforms and distribution of land to poor households, increasing the education of mothers, improving child care and promoting changes in consumption patterns, and so on. Using such real-world policy options and interventions as case studies, the chapters of this book attempt to show how using the analysis of socioeconomic data sets can help in the development of policy and program interventions. The chapters also introduce various approaches to the collection of data, processing of collected data, and generation of various socioeconomic variables from the existing data sets. They also demonstrate applications of analysis of the relationship between causal policy variables and welfare indicators that reflect household and individual food security, nutrition, and poverty.

Why should a book that teaches statistical methods for analyzing socioeconomic data for generating policy and program options be important?

The goal of the decision-maker is to select the best option for intervention from a set of choices that are politically feasible and economically viable (Babu and Mthindi, 1995a, 1995b). Yet making such decisions requires a full understanding of the intended and unintended consequences of the proposed interventions. While the need for rigorous analysis—through assessment of the existing situation—is largely recognized by the policy decision-makers before taking necessary action, the needed capacity for undertaking such analysis is grossly lacking in many countries. Hence, much of the policy and program decisions related to food security, poverty, and nutrition continue to be made under the veil of ignorance.

Improved capacity for food security, poverty, and nutrition policy analysis is essential for achieving the Sustainable Development Goals (SDGs). The SDGs should be achieved by 2030. While some progress toward the SDGs has been made, recent COVID-19 pandemic is threatening to derail this progress. We have learned few lessons from the previous efforts as well. At the global level, the global community also develops a set of goals called Millennium

Development Goals (MDG) (UN, 2005). The major MDG of "reducing hunger, poverty, and malnutrition by half by the year 2015" remained unachievable in many parts of the world. It has been recognized that one of the major constraints in attaining the MDGs related to hunger and malnutrition is the lack of capacity for scaling up of food and nutrition interventions (World Bank, 2006). Scaling up requires capacity for monitoring, evaluation, and adoption of successful food and nutrition programs. Such capacity is severely lacking at the global, national, and local levels (Babu, 1997a, 1997b, 2001).These capacity challenges continue to thwart the efforts toward achieving the SDGs. Yet, least attention and investment goes to building local capacity for solving food insecurity, malnutrition, and poverty.

A good conceptual understanding of the issues related to food and nutrition, economic concepts, statistical techniques, and policy applications with case studies will help in understanding how quantitative analysis could be used for designing program and policy interventions. Students who take up jobs that involve designing, implementing, monitoring, and evaluation of development programs are often ill-prepared to undertake these tasks. Based on one statistical course, students take in the undergraduate program, and with their little exposure to food and nutrition issues, for example, they are expected to perform the role of policy and program analysts. Even if they are well trained in the individual discipline such as food and nutrition, statistics, monitoring and evaluation, or policy analysis, they are often not adequately trained to combine these disciplines to address real-world food and nutrition challenges (Babu and Mthindi, 1995b).

A book that brings together concepts and issues in food security, nutrition, and poverty policy analysis in a self-learning mode can serve thousands of policy analysts, program managers, and prospective students dealing with designing, implementing, monitoring, and evaluation of food security, nutrition, and poverty reduction programs.

Objectives of this book

The purpose of this book is to provide readers and practitioners with skills for specifying and using statistical tools that may be appropriate for analyzing socioeconomic data and enable them to develop various policy and program alternatives based on the inferences of data analysis.

The chapters of this book introduce a wide range of analytical methods through the following approaches:

- Review a broad set of studies that apply various statistical techniques and bring out inferences for policy applications.
- Demonstrate the application of the statistical tools using real-world data sets for policy analysis.

- Use the results of the analysis for deriving policy implications that provide useful learning for policy analysts in designing policy and program options.

Organization of this book

The 16 chapters of this book are organized into three broad sections. The first section deals with food security policy analysis, the second section addresses nutrition policy analysis, and the third section covers the special and advanced topics on food and nutrition policy analysis including measurement and determinants of poverty. This section also provides an introduction to modeling with linear programming methods and program evaluation.

To show the interconnectedness of the issues addressed by the chapters of this book to broad development goals, Fig. I.1 identifies the placement of the chapters, as they relate to specific policy challenges. The broad conceptual

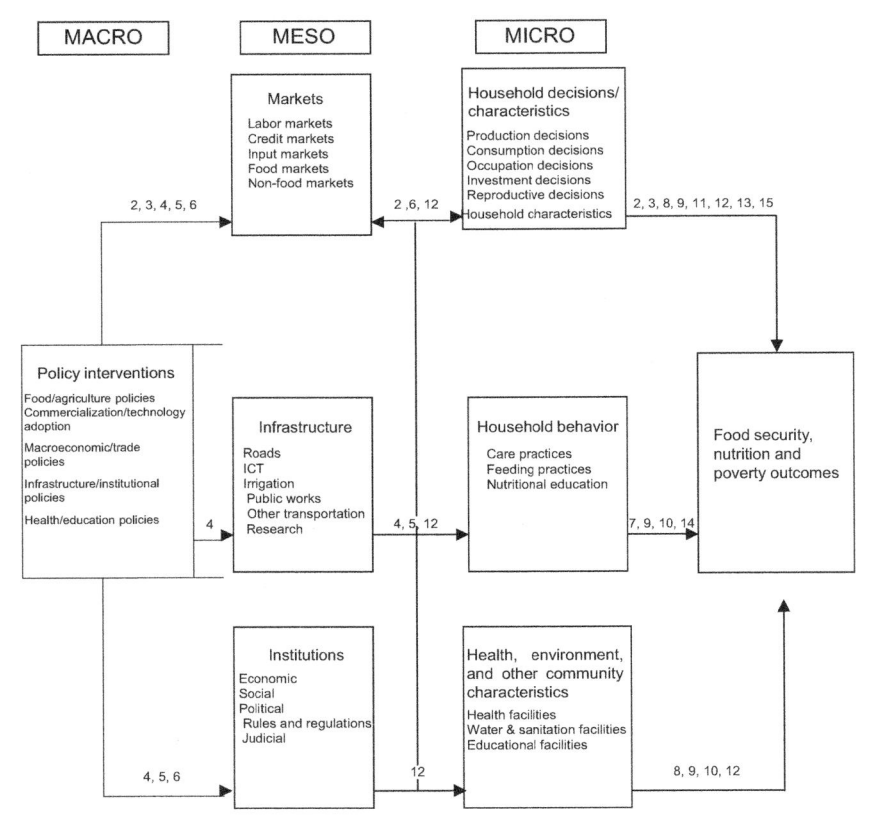

FIGURE I.1 Conceptual framework for designing food and nutrition security interventions. Numbers denote linkage across chapters in this book.

approach used throughout this book, explained later in greater detail, is also depicted in Fig. I.1.

The conceptual framework outlined in Fig. I.1 is a tool for analyzing the impacts of policies and programs on food and nutrition security outcomes at the household level. It links various policies at the macro-, meso- (markets), and micro- (household) levels (Metz, 2000). Economic changes induced by various macropolicies influence markets, which, in turn, affect food security at the household level. Food entitlements in terms of availability and access to food at the household level are affected by various policy interventions. Both macroeconomic (exchange rate, fiscal, and monetary policies) and sector-specific policies (agriculture, health, education, and other social services) affect markets, infrastructure, and institutions. The markets can be subclassified into food markets and other markets for essential consumer goods, production inputs, and credit. The main issues addressed in the chapters of this book relate to policy changes that affect food security through these markets. Infrastructure comprises the economic, social, as well as physical infrastructure; institutions are also affected by policy changes and affect household food security.

Changes induced by policies on different markets and on infrastructural factors affect household incomes, assets, human capital, and household behavioral changes. The aforementioned factors in turn determine household food security as well as household resources devoted to food production.

Income is one of the major determinants of household food security. Both the supply and the demand factors determine the level of household food entitlement. Household food security is achieved if subsistence production and household food purchases are sufficient to meet the household food requirements. Nutrition security, on the other hand, is determined by a complex set of interactions between food and nonfood determinants. For example, nonfood determinants, such as the quality of healthcare facilities and services, education, sanitation, clean water, caring practices, and effective mechanisms for delivering these services, are important in improving the nutritional situation (IFPRI, 1995).

The aforementioned conceptual framework could be used to illustrate the linkages of the chapters of this book. Chapter 1 presents an introduction to the concepts, indicators, and causal factors of household food security and nutritional outcomes.

In Chapter 2, we address the following issues:

1. To what extent does adoption of new technologies improve household or individual food consumption.
2. How does technology adoption in agriculture including postharvest technologies translate into improved food security?

From the arrows in the diagram, we see that agricultural policies, such as technology adoption or commercialization, have close linkages to food and nutrition security, through securing food production and supply. The linkages are given by arrows bearing number 2.

Similarly, for example, Chapter 6 addresses the issue of how market access plays an important role in the agricultural food markets and thus affects household food security. Since marketing and pricing policies are affected by both the supply and demand side of the food economy, it is important for national governments simultaneously to provide incentive prices to producers to increase their incomes and to protect consumers against rapid price fluctuations to ensure steady food supplies. One of the ways that government marketing and pricing policies can reduce price instability is by allowing the private sector to participate in the market along with state parastatals through alteration of the infrastructural and institutional policies that affect food markets. The linkages are given by arrows bearing number 6.

As another example, in Chapter 10, we address the pathways through which maternal education improves child health. These pathways help us in understanding the impact of community characteristics (such as presence of hospitals and water and sanitation conditions) on child nutritional status. Social infrastructure, such as the presence of medical centers and improved water and sanitation conditions, can be beneficial for certain subgroups of the population, such as the low-income and less educated households. The time saved by not traveling to a medical center can be reallocated to leisure, health production, and other agricultural activities, which can improve household productivity and child nutritional status. As indicated by arrows with number 10, health and education policies, through their effect on markets and social infrastructure, can not only alter to improved provision of services but also alter household behavior through better child care and hygienic practices, which can eventually improve child nutritional status.

Currently, there is increased interest among policy makers and the media on the food security situation of the poor households in the United States. In this second edition, we address the issues in every chapter in the context of the problems households face in the United States. We provide results from recent research in the United States and demonstrate how each topic is of relevance to the United States.

In the second edition, we have added Chapter 16 on program evaluation to show the recent developments in the field of development economics. In the past 5—10 years, the application of randomized control trials to development interventions has become increasingly popular. These approaches help us to save resources and to understand the impact of pilot interventions. Chapter 16 deals with these newer techniques and provides information from recent results in the field. In the third edition we added a new Chapter 17 on multidimensional poverty analysis.

Rationale for statistical methods illustrated in the book

Before launching into an analytical technique, it is important to have a clear understanding of the form and quality of the data. The form of the data refers to whether the data are categorical or continuous. The quality of the data refers to the distribution, i.e., to what extent it is normally distributed or not.

Additionally, it is important to understand the magnitude of missing values in observations and to determine whether to ignore them or impute values to the missing observations. Another data quality measure is outliers, and it is important to determine whether they should be removed.

Quantitative approaches in this book consist of descriptive, inferential, and noninferential statistics. Descriptive statistics organize and summarize information in a clear and effective way (for example, means and standard deviations). Inferential statistics analyze population differences, examine relationships between two or more variables, and examine the effect of one variable or variables on other variables. The key distinction for inferential and noninferential techniques is in whether hypotheses need to be specified beforehand. In the latter methods, normal distribution is not a prerequisite. For example, in cluster analysis, one can use continuous or categorical variables to create cluster memberships, and there is no need for a predefined outcome variable.

The choice and application of analytical tools is largely motivated by policy and program issues at hand, and the type of data that are collected, which, in turn, is related to the policy and program objectives. In inferential methods, users can draw inferences about the population from a sample because it provides a measure of precision or variation with regard to the sample data. Inferential methods generally focus on parameter estimation and its changes over time. The primary inferential procedures are confidence intervals and statistical tests. While confidence intervals can be used for both point and interval estimates, statistical tests are ways to determine the probability that a result occurs by chance alone.

Different objectives related to the question at hand and the types of data necessitate that the user choose an analysis from a number of possible approaches. The selection of a statistical procedure must consider the following key characteristics: independence of samples, type of data, equality of variances, and distribution assumptions. The conceptual diagram (Fig. I.2) illustrates how an analysis can be undertaken using different approaches for bivariate and multivariate statistical procedures.

The conceptual diagram can be understood with the following questions and answers that lead to the appropriate statistical technique:

1. How many variables does the problem involve? For example, are there two variables or more than two variables? A question related to the first one is how does one want to treat the variables with respect to the scale of measurement? For example, are they both categorical (which includes nominal and ordinal variables)? Nominal variables are unordered categorical variables, such as sex of the child, while ordinal variables are ordered ones. For example, height of a child can be converted into short, average, and tall.
2. What do we want to know about the distribution of the variables? For example, in the case of a continuous variable, is the distribution normal? One can test this condition by superimposing the normal density over the histogram of the variable or by drawing a Q–Q plot.

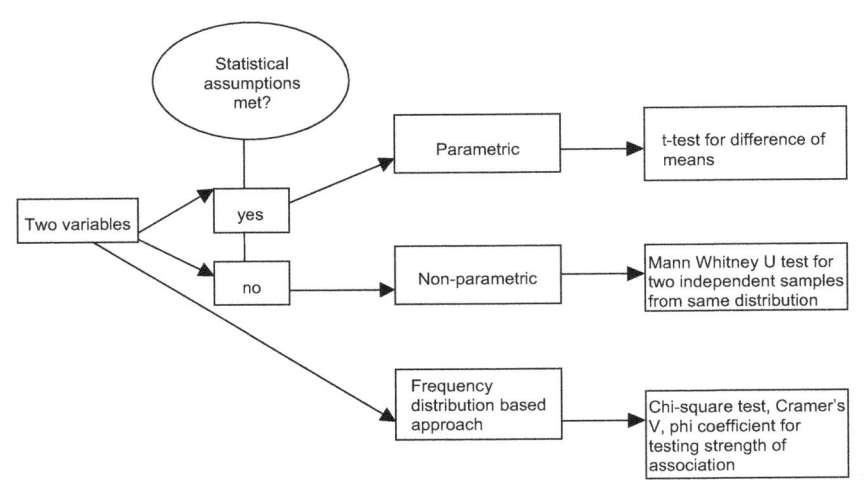

FIGURE I.2 Statistical procedures to test for determinants of food security, nutritional status, and poverty.

Examples of statistical tests used in this book

In the case of both the variables being nominal, with no distinction made between a dependent and an independent variable, one can measure association using a statistic based on the number of cases in each category. Various statistics based on the number of cases in each category are chi-square, Cramer's V, and phi or the contingency coefficient as illustrated in Chapters 3 and 4.

In contrast, in the case of two variables being continuous and no distinction being made between a dependent and an independent variable, one can test whether the means on the two variables are equal (for example, in Chapter 2, we address whether food security differs between the hybrid maize growers versus nongrowers). The difference of the means can be inferred using the t-test.

In the case of two variables, with one being nominal and the other continuous (the continuous variable being dependent), one can test the null hypothesis of statistical significance of differences between groups. By assuming homoscedasticity across levels of the independent variable, one can undertake an analysis of variance (ANOVA)/F-test. In Chapter 5, we address the issue of whether the share of calories from various food groups differs across households classified by different expenditure brackets. Since the per capita expenditure of different food groups is continuous and the expenditure brackets are nominal, this approach is appropriate.

It is important to mention here by way of digression that while t- and F-tests are based on assumptions such as equal variances and normality, data are rarely examined prior to execution of the desired tests (we do not undertake nonparametric analysis in this book). There are instances when these assumptions may not be met. These include small samples and a nonnormal distribution. In such cases, nonparametric tests may be appropriate. Also referred to as distribution-free methods, nonparametric tests are not concerned with specific parameters, such as mean in an ANOVAs, but with the distribution of the variates (Sokal and Rohlf, 1981). Nonparametric ANOVA is easy to compute and permits freedom from the distribution assumptions of an ANOVA. These tests are less powerful than parametric tests when the data are normally distributed. Under those circumstances, there is a greater likelihood of committing type II error using nonparametric tests. Some of the guidelines for deciding when to apply a nonparametric test are as follows:

1. fewer than 12 cases,
2. the sample is clearly not normally distributed, and
3. some values are excessively high or low.

However, it is important to bear in mind that nonparametric tests are counterparts to the parametric tests.

If the primary focus is to measure covariation (with no distinction made between dependent and independent variables), one can assign interval-scaled values to the categories of the variable to compute the product moment correlation coefficient. The main question addressed here: How much do the variables vary together (Sokal and Rohlf, 1981)? In Chapter 8, we illustrate this method with the different indicators of nutritional status such as height for age, weight for age, and weight for height.

In contrast to correlation, in a regression analysis, a distinction is made between an independent and a dependent variable. If the dependent variable is continuous and one treats the relationship between the variables as linear, then coefficients from the linear regression can predict how much the dependent variable changes with respect to changes in the independent variables. In Chapter 9, we use this method to predict the values of child nutritional status from the values of individual/household and community characteristics.

We then proceed to multivariate analysis of data, which allows the user to examine multiple variables using a single technique. While traditional univariate methods such as *t*-tests and chi-square tests can be very powerful, one can interpret the results based on the analysis of one manipulation variable. Multivariate techniques allow for the examination of many variables at once. There are different types of multivariate techniques that can be used to analyze food security, nutritional status, and poverty analysis. Some of these techniques such as multivariate regression, logistic regression, discriminant analysis, K-mean cluster analysis, and factor analysis are used in this book. While these techniques can be very powerful, their results should be interpreted with care. Some techniques are sensitive to particular data types and require that data be distributed normally. Others cannot be used with nonlinear variables (for example, classification). Thus, while using these techniques, it is important to understand their respective intended uses, strengths, and limitations.

Continuing with our examples, with more than two variables, we have the following: if there are more than two variables with a distinction being made between dependent (continuous) and independent variables (and relationship among the variables treated as additive and linear), the coefficients of multiple linear regression with their t-statistic will assign to each independent variable some of the explained variance in the dependent variable that the dependent variables share with other independent variables. This method has been used in examining the role of maternal education and community characteristics on child nutritional status in Chapter 10.

In contrast to multivariate regression, when the dependent variable is categorical (either nominal or ordinal), the coefficients from the ordinal logit regression accompanied with the Wald statistic can tell us the probability associated with being in a particular category of the dependent variable. The idea can be illustrated with our example of determinants of poverty as in Chapter 12 as follows: Suppose we want to examine the relationship between assets held by the household and probability of being poor. When the household has a very low level of assets, the probability of getting out of poverty is small and rises only slightly with increasing assets. But, at a certain point, the change of owning more assets begins to increase in an almost linear fashion, until eventually many households hold more assets, at which point the function levels off again. Thus, the outcome variable (in this case, the probability of being poor) varies from 0 to 1 since it is measured in probability.

Discriminant analysis, as introduced in Chapter 11, is used to determine which continuous variables discriminate between two or more naturally occurring groups. In this chapter, we investigate which variables discriminate between various levels of child nutritional status. This approach is particularly suitable, since it answers the questions: Can a combination of variables be used to predict group membership (e.g., differentiating between low wasting from severe wasting) and which variables contribute to the discrimination between groups?

However, this method is more restrictive than logistic models, since the key assumption required is multivariate normality of the independent variables and equal covariance structure for the groups as defined by the dependent variable. If the sample sizes are small and the covariance matrices are unequal, then the estimation process can be adversely affected.

The method builds a linear discriminant function that can be used to classify the households. The overall fit is assessed by looking at the degree to which the group means differ (Wilks' lambda) and how well the model classifies. By looking at the correlation between the predictor variables and the discriminant function, one can determine the discriminatory impact. This tool can help categorize a wasted child from a normal child.

We also explore data reduction and exploratory methods in the chapters of this book. In a cluster analysis, the main purpose is to reduce a large data set to meaningful subgroups of objects or households. The division is accomplished on the basis of similarity of the objects across a set of dimensions. The main problem with this method is outliers, which are often caused by including too many irrelevant variables. Secondly, it is also desirable to have uncorrelated factors. The analysis is especially important for exploring households that can be vulnerable in food insecurity and poverty dimensions. For example, this method can allow the researcher to identify households that are vulnerable in food insecurity dimension alone, households that are vulnerable in dimensions of poverty (such as lack of productive assets), and households that are vulnerable in both dimensions. The rules for developing clusters are that they should be different and measurable.

Finally, when there are many variables in a research design, it is often useful to reduce a large number of variables to a smaller number of factors. There is no distinction between dependent and independent variables and the relationships among variables are treated as linear. In this method, the researcher wants to explore the relationships among the set of variables by looking at the underlying structure of the data matrix. Multicollinearity is generally preferred between the variables, as the correlations are the key to data reduction. The "KMO-Bartlett test" is a measure of the degree to which every variable can be predicted by all other variables. This approach is suitable for constructing a food security index, since a large number of variables, which are the main determinants of food security, can be reduced to a smaller set of underlying components or factors that summarize the essential information in the variables. We use the principal component analysis to find the fewest number of variables that explain most of the variance. The new set of variables is created as linear combinations of the original set. In this procedure, if there were originally 15 variables that affected food security, the procedure can tell us which components explain a substantial percent of variability of the original set of 15 variables and thus reduce the number of factors to say 3. In essence, then, the number of variables to be analyzed has been reduced from 15 to 3.

In the second new edition, we introduced the application of STATA, which is a very popular and powerful statistical software package. We have used illustrative data and STATA to explain the issues identified in each of the chapters. The application of STATA to illustrative data is in addition to applications in SPSS that are still retained from the previous edition. Thus, readers familiar with either of SPSS or STATA will be able to use the book, to learn the same thematic and policy issues. We provide several "hands-on" tutorials to assist policy makers with STATA, and the associated statistical analysis, interpretation, and examples. We have also developed a technical appendix that discusses the application of STATA along with resources to guide researchers and students. This is kept in this third edition for readers familiar only with STATA.

In this new edition, we invite the readers to explore the programming language and the statistical package "R," which is becoming popular among the students throughout the world. Given it is open source, its adoption is becoming ubiquitous. Exposure to "R" will enable readers to adopt this programming language even when they are not affiliated to an institution that has license for using STATA. This is an important feature for self-learning when the readers do not have access to STATA and can be using updated versions of "R" generated by its community of users.

Learning objectives

Each of the analytical chapters in this book addresses four sets of learning objectives. First, each chapter is theme based. A thematic policy issue is chosen and introduced to provide motivation and discussion for policy analysis. As part of this introduction, students are introduced to selected case studies of policy analysis and research that addresses the chosen theme from various geographical, eco-regional, and policy contexts. Additional literature relevant to the theme is also reviewed.

Second, an appropriate empirical analytical technique to address policy issues of the chosen theme is demonstrated. The learning objective of this part of the chapter includes application of the statistical technique to the real-world data by describing the variables, calculation of new variables, development of welfare indicators, and applying a statistical model to the data to derive empirical results.

Third, each chapter has its own specific technical appendix that describes in detail the analytical method used in the chapter for implementing the statistical method using the software. Finally, the translation of analytical results into implications for policy and program development is shown relating the results back to the thematic issue introduced in the beginning of the chapter.

In addition, each chapter has its own set of exercises that tests readers' understanding of the issues, concepts, and analytical techniques and allows them to explore further the literature. All of the chapters use a single household data set (the Malawi household data set) that contains socioeconomic data on several causal factors and indicators of food security, poverty, and nutrition. The links to several publicly available data sets are provided in the publishing company's website.

Readers must note that our STATA results should be taken as illustrative examples and as helpful guide for research and analysis. In many instances, the illustrative data we use may not confirm to certain statistical assumptions and requirements, such as sample points, sampling adequacy, and normality of distributions. Also the results may differ depending on the software used for analysis and the nature of the data sets used. However, we draw the readers' attention to these issues whenever they arise, and use these instances as learning tools, and allow the readers to explore the assumptions further in the exercises sections.

Section I

Food security policy analysis

Chapter 1

Introduction to food security: concepts and measurement

Chapter outline

Introduction	3	Measuring food access	17
Conceptual framework of food security	4	Food utilization	18
		Measuring food utilization	18
Food security in the developed world	7	Stability of availability	19
Other policy issues in the United States	11	Alternative approaches in measuring food security	19
Food security concerns in other countries	11	Conclusions	21
Measurement of the determinants of food security	16	A natural question is why is measuring food insecurity important for better program design in developing countries?	23
Food availability	16		
Measuring food availability	17	Exercises	26
Food access	17		

We are not only not on track to eradicate hunger, food insecurity, and all forms of malnutrition by 2030, but also we need to redouble our efforts given the challenges brought about by COVID-19.

FAO Director-General Qu Dongyu.

Introduction

According to the latest report on the State of Food Security and Nutrition in the World (SOFI) (FAO, 2020), close to 690 million people in the world faced hunger in 2019. This figure is 10 million more compared with that in the year 2018. In addition, the report also suggested an additional 83 million-132 million people could face chronic hunger in 2020, due to COVID-19 pandemic. The report also predicts that we are not on track to achieve Zero Hunger by 2030 and that the number of people affected by hunger would surpass 840 million by 2030 (FAO, 2020). Why does chronic food insecurity exist even when the world has been producing enough food to feed it population at the aggregate level? This is a serious development concern and is

Food Security, Poverty and Nutrition Policy Analysis. https://doi.org/10.1016/B978-0-12-820477-1.00009-7
Copyright © 2022 Elsevier Inc. All rights reserved.

identified as the Sustainable Development Goal 2. In this chapter, we begin with the study of concepts, indicators, and measurements of food security. A common acceptable definition of food security exists. Yet, the concept of food security is understood and used differently depending on the context, time frame, and geographical region in question. In this chapter, we explore the definition and measurement of food security to provide a conceptual foundation to food security policy analysis. First, we introduce a widely used and well-accepted definition along with three core determinants of food security. Second, we explain the measurement of these determinants with examples of global, national, and regional data sets that provide information on these determinants. We also examine the implications of food security in the United States and in other countries. Finally, we explore some alternative approaches to measuring food security indicators.

Conceptual framework of food security

Before examining the determinants of food security, understanding several concepts associated with the definition of food security is necessary. This is because many developing countries continue to suffer from chronic food insecurity and high levels of malnutrition and they are under constant threats of hunger caused by economic crises and natural disasters. Designing policies and programs to improve nutritional status requires an understanding of the factors that cause malnutrition, knowledge of the pathways in which these factors affect vulnerable groups and households, and an awareness of policy options available to reduce the impact of these factors on hunger and malnutrition.

A multitude and complex set of factors determine nutritional outcomes. These factors have been identified, and Smith and Haddad (2000) elaborate on their linkages to nutrition.

The food and nutrition policy-focused conceptual framework presented in Fig. 1.1 identifies the causal factors of nutrition security and the food policy linkages to them. It also identifies the points of entry for direct and indirect nutrition programs and policy interventions as well as the capacity gaps for analysis and evaluation of food and nutrition policies and programs.

The framework was originally developed and successfully used for explaining child malnutrition (UNICEF, 1998; Haddad, 1999; Smith and Haddad, 2000). It was revised further to incorporate policy and program dimensions (Babu, 2001; Babu, 2009). Given the role of nutrition in the human life cycle, this framework attempts to encompass the life cycle approach to nutrition. In addition, it includes the causes of nutrition security at both the macro- and microlevels. As seen in Fig. 1.1, achieving food security at the macrolevel requires economic growth, resulting in poverty alleviation and increased equity in the distribution of income among the population. In a

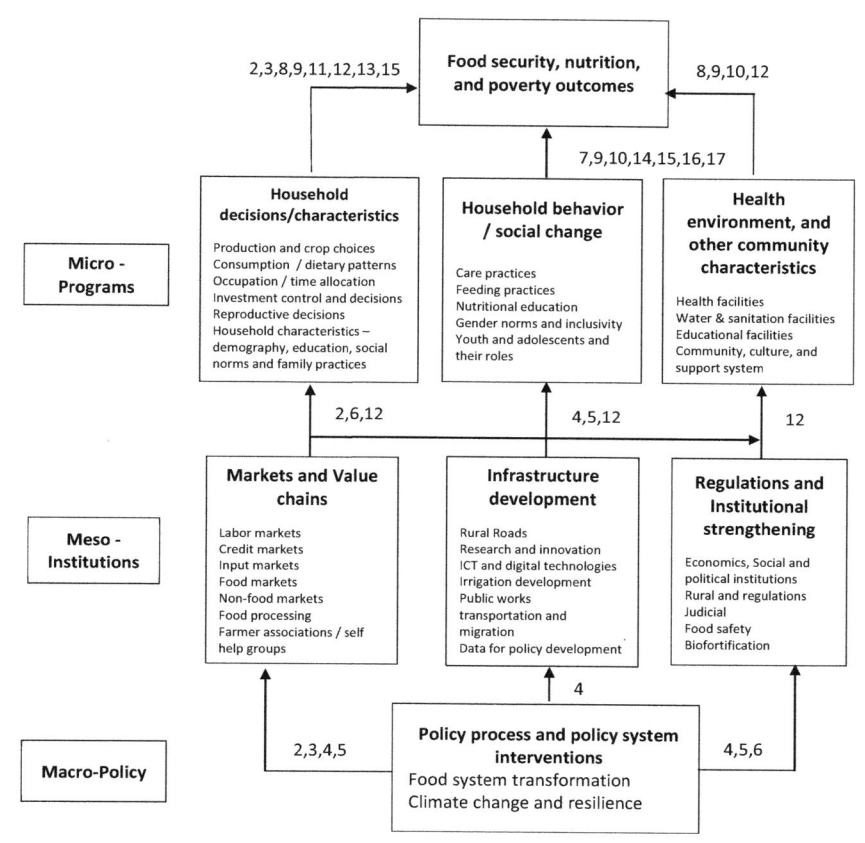

Introduction to food security: concepts and measurement **Chapter | 1** 5

```
                                  Food security, nutrition,
         2,3,8,9,11,12,13,15       and poverty outcomes        8,9,10,12

                                      7,9,10,14,15,16,17

              Household              Household behavior         Health
         decisions/characteristics    / social change        environment, and
                                                             other community
         Production and crop choices                          characteristics
Micro -  Consumption / dietary patterns   Care practices
Programs Occupation / time allocation     Feeding practices   Health facilities
         Investment control and decisions Nutritional education Water & sanitation facilities
         Reproductive decisions           Gender norms and inclusivity Educational facilities
         Household characteristics –      Youth and adolescents and Community, culture, and
         demography, education, social    their roles         support system
         norms and family practices

                   2,6,12                    4,5,12                  12

              Markets and Value          Infrastructure         Regulations and
                  chains                  development             Institutional
                                                                  strengthening
Meso -   Labor markets                 Rural Roads
Institutions Credit markets            Research and innovation Economics, Social and
         Input markets                 ICT and digital technologies political institutions
         Food markets                  Irrigation development   Rural and regulations
         Non-food markets              Public works             Judicial
         Food processing               transportation and       Food safety
         Farmer associations / self    migration                Biofortification
         help groups                   Data for policy development

                                             4
                                  Policy process and policy system
                 2,3,4,5                interventions              4,5,6
Macro-Policy                     Food system transformation
                                 Climate change and resilience
```

FIGURE 1.1 Conceptual framework for designing food security, nutrition, and poverty interventions. Numbers denote linkage across chapters in the book. *From: Smith, L.C., Haddad, L., 2000. Explaining child malnutrition in developing countries: A cross-country analysis. IFPRI Research Report 111, Figure 1. Washington, DC: International Food Policy Research Institute. Reproduced with permission from the International Food Policy Research Institute (IFPRI) www.ifpri.org. The original figure is available online at https://ebrary.ifpri.org/digital/collection/p15738coll2/id/125371.*

predominantly agrarian economy, economic growth is driven by increases in agricultural productivity and, therefore, depends on the availability of natural resources, agricultural technology, and human resources. These are depicted as potential resources at the bottom of Fig. 1.1. Recently, several authors have attempted to provide their own versions of conceptual frameworks linking food security, agriculture, and nutrition variables (see Fan and Brzeska, 2011 and Pinstrup-Andersen, 2012).

Agricultural technology and natural resources are necessary but, by themselves, are not sufficient to generate dynamic agricultural growth. Both policies that appropriately price the resources and allocate them efficiently along with stable investment in human and natural resources through political and legal institutions are necessary. These basic factors determine a set of underlying causes of nutrition security, i.e., food security, care, and health. These three underlying causes are associated with a set of resources necessary for this achievement. Attaining food security is shown to be one of the key determinants of nutritional status of individuals. Food security is attained when all people have physical and economic access to sufficient food at all times to meet their dietary needs for a productive and healthy life (World Bank, 1986).[1] While this definition is frequently applied at different levels, such as national, subnational, and household levels, it is more meaningful to use this concept at the household level. Resources for achieving food security are influenced by both policies and programs that increase food production, provide income for food purchases, and establish in-kind transfer of food through formal or informal supporting mechanisms.

Resources for the provision of care depend on policies and programs that increase the caregivers' access to income, strengthen their control of income use, and improve their knowledge, adoption, and practice of care. Care is the provision by households and communities of "time, attention, and support to meet the physical, mental, and social needs of a growing child and other household members" (ICN, 1992). Child feeding, health-seeking behavior, caring, and supporting of mothers during pregnancy and breastfeeding are some examples of caring practices. Resources for health could be improved through policies and programs that increase the availability of safe water, sanitation, healthcare, and environmental safety.

As mentioned earlier, food security that ensures a nutritionally adequate diet at all times and a care and health environment that ensures the biological utilization of food jointly determines the nutrition security of individuals. Thus, the immediate causes of nutrition security are dietary intake of macronutrients (energy, protein, and fat), micronutrients, and the health status of individuals. Adequate nutrition security for children results in the development of healthy adolescents and adults and contributes to the quality of human capital. Healthy female adults with continued nutrition security during pregnancy contribute to fewer incidences of low birth weight babies, thereby minimizing the probability of the babies becoming malnourished. In the case of adults, improved nutrition security, in terms of timely nutrient intakes, increases labor productivity (given opportunities for productive employment),

---

1. A thorough review of the food security concept and the conceptual frameworks used in the literature for analyzing food security is beyond the scope of this chapter. For such reviews, see Maxwell and Frankenberger (1992), Clay (1997) and Von Braun et al. (1992), Pinstrup-Andersen (2012), Dorward (2013).

thus resulting in reduced poverty. Lower prevalence of poverty increases the potential resources needed for attaining nutrition security. This conceptual framework has been applied by several authors to study the determinants of food security and nutritional status in the African context (Sahn and Alderman, 1997), Latin American context (Ruel et al., 1999), and the South Asian context (Babu, 2006).

Food security in the developed world

Food insecurity is not just the problem for poor countries. It affects more advanced economies in Europe and the United States. For example, to many, the idea of a significant group of people in the United States facing food insecurity and poverty will be very surprising. However, these are serious issues in the United States, and further, the policy makers in the United States actively pursue measures to mitigate the effects of food insecurity. For example, the United States Department of Agriculture (USDA) regularly reports on food security and hunger in US households and communities. Provision of such information facilitates informed public debate regarding food insecurity, its impact on the well-being of children, adults, families, and communities, as well as its relationship to public policies, public assistance programs, and the economy.[2]

Recent estimates from the USDA tell us that nearly 45 million or about 15% of US households were found "food insecure" in 2011.[3] Included in this are almost 170,000 households with "very low food security," which the report indicates is an increase from previous years. For example, the USDA reports indicate that, in 2010, 85.5% of US households were food secure throughout the entire year, and 14.5% of households were food insecure at least some time during that year. The USDA study also reports that about 57% of all food-insecure households participated in one or more of the three largest federal food and nutrition assistance programs during the month prior to the 2011 survey. The extent of food insecurity also strains the fiscal budgets, as dependence of food-insecure households on government assistance increases.

---

2. The USDA administers surveys and questionnaires and arrives at a measure that indicates the level of food insecurity. A household is considered "food insecure" if it answers affirmatively to three or more food insecurity questions that describe its ability to acquire enough food. Households indicating low levels of food insecurity (one or two affirmative responses) are considered food secure. For a critical evaluation of the USDA's approach to measure food security in the United States, see Opsomer et al. (2002) and Coleman-Jensen (2011).
3. According to the USDA, a household is food secure if all the members had access at all times to enough food for an active, healthy life for all household members. The term "food insecure" implies that the food intake of one or more household members was reduced and their eating patterns were disrupted at times during the year because the household lacked money and other resources for obtaining food (see Coleman-Jensen et al., 2011).

The USDA has produced these reports over the years to inform the policy makers in the United States about the continuous severity of the problem. For example, in 2010, 5.4% of households experienced food insecurity in the more severe range.[4]

There are many ways in which the food insecurity problem surfaces across the United States. In an interesting study, Olson (2004) links the issues around hunger to the broader context of poverty in rural America. Using a sample of over 300 poor rural households from 14 states, Olson (2004) examines how the level of human resources and the diversion of financial resources away from food are related to the food security status. Olson (2004) reports that food security status is determined by the mothers' ability to juggle many jobs and the skills required to perform them. For instance, mothers who used a greater number of food and financial management skills such as managing bills, making a budget, stretching groceries, and preparing meals were more likely to have food-secure households, compared with the mothers who were less resourceful. Most importantly, mental health and medical care costs were also closely linked to food security issues. Naturally, the results must appeal to US policy makers, to readjust their programs in rural areas.

Interestingly, Olson's concerns are also voiced by Blaylock and Blisard (1995), who use a production function approach and find that the food security situation has a significant influence on woman's self-evaluated health status. The most food-insecure women are also poor and tend to be less educated, and as a consequence lead unhealthy lives, which continue to add to the problem. The researchers suggest that the improvement of woman's health crucially depends upon food availability and that policy makers must also concentrate on educating the women on the usefulness of tobacco abstinence, weight reduction, and an increase in physical activity.

Section highlight

Who is food insecure?

A natural question arises as to how one determines whether a household is food secure or not. In the United States, a need for a reliable measure of hunger and food insecurity has been recognized since the early 1980s. Many researchers have contributed to this topic, and obviously the measure is a complicated procedure involving surveys, questionnaire design and implementation, development of a useful scale of severity, etc., to identify food security status. In this section, we examine the list of measures and a sample questionnaire that the USDA

---

4. For this and other effects of Clinton's welfare reform, and Obama's stimulus on food security in the United States, see http://www.ers.usda.gov/Briefing/FoodSecurity/. For a critical evaluation of the USDA's model, see Opsomer et al. (2002).

Section highlight—cont'd

implements to collect its data. This will be very useful for those researchers who wish to conduct similar studies.

The first task is to arrive at a measure of "What Is Food Security?" For the USDA research center, food security for a household means access by all members at all times to enough food for an active, healthy life. Food security includes, at a minimum, the following:

1. The ready availability of nutritionally adequate and safe foods
2. Assured ability to acquire acceptable foods in socially acceptable ways (that is, without resorting to emergency food supplies, scavenging, stealing, or other coping strategies)

Likewise, food insecurity is limited or uncertain availability of nutritionally adequate and safe foods or limited or uncertain ability to acquire acceptable foods in socially acceptable ways.

The next logical step is to develop an index to capture the variation in the food security status within the data. The food security status of each household is divided into four ranges, characterized as follows:

1. High food security—Households had no problems, or anxiety about, consistently accessing adequate food.
2. Marginal food security—Households had problems at times, or anxiety about, accessing adequate food, but the quality, variety, and quantity of their food intake were not substantially reduced.
3. Low food security—Households reduced the quality, variety, and desirability of their diets, but the quantity of food intake and normal eating patterns were not substantially disrupted.
4. Very low food security—At times during the year, eating patterns of one or more household members were disrupted and food intake reduced because the household lacked money and other resources for food.

The third step would be to develop questions and use the responses to place a household into the correct range. Many questions are developed to elicit responses about behaviors and experiences associated with difficulty in meeting food needs. The questions cover a wide range of severity of food insecurity. The responses are classified as "least severe," "somewhat more severe," "midrange severity," and "most severe."

The responses obtained from the aforementioned survey are then used to identify the households into different groups. For example, households that report three or more conditions that indicate food insecurity are classified as "food insecure." The three least severe conditions that would result in a household being classified as food insecure are as follows:

- They worried whether their food would run out before they got money to buy more.
- The food they bought did not last, and they did not have money to get more.
- They could not afford to eat balanced meals.

Households are also classified as food insecure if they report any combination of three or more conditions. Other instances as in households with no children

Continued

Section highlight—cont'd

present must report at least the three conditions listed before and also the following:
- Adults ate less than they felt they should.
- Adults cut the size of meals or skipped meals and did so in 3 or more months.

Here is a sample of a few survey questions used by the USDA to assess household food security:

1. "We worried whether our food would run out before we got money to buy more." Was that often, sometimes, or never true for you in the last 12 months?
2. "The food that we bought just didn't last and we didn't have money to get more." Was that often, sometimes, or never true for you in the last 12 months?
3. "We couldn't afford to eat balanced meals." Was that often, sometimes, or never true for you in the last 12 months?
4. In the last 12 months, did you or other adults in the household ever cut the size of your meals or skip meals because there wasn't enough money for food? (Yes/No)
5. (If yes to question 4) How often did this happen-almost every month, some months but not every month, or in only 1 or 2 months?
6. In the last 12 months, did you ever eat less than you felt you should because there wasn't enough money for food? (Yes/No)
7. In the last 12 months, were you ever hungry, but didn't eat, because there wasn't enough money for food? (Yes/No)
8. In the last 12 months, did you lose weight because there wasn't enough money for food? (Yes/No)
9. In the last 12 months did you or other adults in your household ever not eat for a whole day because there wasn't enough money for food? (Yes/No)
10. (If yes to question 9) How often did this happen almost every month, some months but not every month, or in only 1 or 2 months?

The USDA's food security statistics are based on a national food security survey conducted as an annual supplement to the monthly Current Population Survey (CPS). The CPS is a nationally representative survey conducted by the Census Bureau for the Bureau of Labor Statistics. The CPS provides data for the Nation's monthly unemployment statistics and annual income and poverty statistics.[5] Many researchers use these measures and derive implications

---

5. USDA's food security statistics are based on a national food security survey conducted annually. In December of each year, after completing the labor force interview, about 45,000 households respond to the food security questions and to questions about food spending and about the use of federal and community food assistance programs (http://www.ers.usda.gov/topics/food-nutrition-assistance/food-security-in-the-us/measurement.aspx).

for policy. For example, Andrews et al. (1998a,b) used a version of the scaling technique to derive implications for food security and find the rate of extreme food insecurity to be roughly 4% of the population.

Other policy issues in the United States

Finally, a whole range of issues surrounding the welfare system and the transfer programs in the United States have attracted the attention of many researchers to examine these concerns within the broader context of poverty, immigration, and technical change (Moffitt, 2003; DePolt et al., 2009; Blaylock and Blisard, 1995; Gundersen et al., 2011). DePolt et al. (2009) examine how participation in the Food Stamp and Temporary Assistance for Needy Families Programs is associated with self-reported household food hardships, using data from low-income families living in Boston, Chicago, and San Antonio. In addition to the measures of hardships and program participation, they also include measures of income, wealth, social resources, disability, physical health, and family structure. These measures help us to account for selection between recipient and nonrecipient households. They show that participation in the Food Stamp Program is associated with fewer food hardships, while participation in the Temporary Assistance for Needy Families program has no detectable association with hardships.

Gundersen et al. (2011) provide the important conclusions from recent research on US food insecurity. First, evidence that the Supplemental Nutrition Assistance Program (SNAP) reduces food insecurity has grown in recent years. Second, National School Lunch Program (NSLP) also reduces food insecurity. Third, potential benefits from increased health and productivity should be incorporated to correctly compute the cost—benefits of both SNAP and NSLP. Finally, idiosyncratic price shocks in assets and income can also increase food insecurity, and polices have to be designed to include the middle-income households who might be the first victims of such shocks.

Food security concerns in other countries

A key statement that highlights the findings of the study from the FAO, United Nations, is that economic growth is necessary but not sufficient to accelerate reduction of hunger and malnutrition.[6] Countries experience a reduction in food security for a variety of reasons. Food security concerns could arise due to poor macroeconomic performance, or due to threats from foreign policy or population pressures. The importance of food security is a growing concern for governments across all countries. But it is also important to note that most countries try different strategies and attempt many policy responses to understand and solve this problem. For instance, Motoyuki (2008) relates the

---

6. See the FAO study *The State of Food Insecurity in the World* (2012). Also see FAO (2013).

Multifunctionality of Japanese agriculture and derives policy implications for food security, rural development, and environment. Motoyuki (2008) indicates that the social and geographical nature of Japan has generated a collaborative farming system that is conducive to raising productivity. The farmland area, the irrigation system, and the daily life in Japan add to the multifunctionality aspect of Japanese rice production system. Policy makers in Japan have developed appropriate policies that exploit jointness in Japanese agriculture to combat food security.[7]

Additionally, fisheries provide an important source of food in Japan, and as a consequence, over the years, access to newer fishing grounds and international waters has become national priority. However, this concern over fisheries and food security also impinges on foreign policy and national security. Consequently, Japan provides an interesting case study where there is a dynamic interaction between international fisheries, food security, self-sufficiency, border security, and foreign relations. Policy makers in Japan have to constantly develop new strategies to cope with social structures and incentives to deal with food security and foreign policy.[8]

Rhoe et al. (2008) and Babu and Rhoe (2006) indicate the experience in Central Asia, during the time period of transition from planned to market economies. Interestingly, Rhoe et al. (2008) point out that the Republics within Central Asia experienced rising poverty, food insecurity, and malnutrition, as an indirect result of a poor macroeconomic environment, falling GDP, and high inflation. During the transition, the welfare of the Central Asian population as a whole clearly deteriorated at the household level, brought about mainly because of increased poverty and decreased food entitlements. Rhoe et al. (2008) examine these issues in detail for Kazakhstan, identify the sectors of the population suffering from food insecurity and their location, and provide policy reforms to reduce the informational barriers within the poor households. Food sector reforms, in Kazakhstan, particularly the removal of school and preschool nutrition programs and the elimination of food subsidies, exposed poor households to food insecurity and malnutrition. Therefore, the researchers conclude that land markets need to be further developed and a clear understanding of these markets by the people needs to be formed. They also stress domestic market reforms to increase food production, stronger trade arrangements between Central Asian Republics, and improvements in rural labor and credit markets. These policies must accompany investment in agriculture research, increases in knowledge among the head of the households, and increase in school enrollment, training, and entrepreneurship.

In the context of South Asia, the economic reforms, technological change, and regional trade play an important role in household food security

---

7. For details, see Motoyuki (2008), OECD (2013) and the related papers in this study.
8. For an interesting account of the development of the United States—Japan postwar treaties and food security, see Smith (2008).

(Babu, 2006; Babu and Gulati, 2005). For example, the idea that food insecurity and lack of access to food grains is driven by the lack of food supply and production can also be misleading, as shown in several of the writings by Sen (1999). An adequate amount of food production almost invariably accompanies frustrating questions about distribution and management. Nowhere is this more pronounced than in India, which has seen a phenomenal increase in food production. For example, Chandrasekhar (2012) notes that India is one of the world's largest rice exporters with exports at a record level of 10.4 million tons. However, in the same study, Chandrasekhar (2012) goes on to point out that the Indian government has failed to deliver a minimum amount of access to grains at affordable prices to the needy. Consequently, one can argue that the gains in exports come at the expense of the poor and needy.

Indeed, policy proposals aiming for food security in India have generated a lot of debate. For instance, the Parliamentary Standing Committee on Food, which drafts the food security bill, has argued for mandatory coverage of 67% of the population based on multiple criteria. The bill has scaled down the monthly entitlement of subsidized grain to a uniform 5 kg per month for every person covered under the act. However, the left-wing groups, civil society groups like the Right to Food Campaign, and many analysts argue for universal coverage on the grounds that targeting would in many ways be self-defeating.

The government's draft bill and the recommendations from the National Advisory Council (NAC) and from the PM's Expert Committee are all at odds with each other when it comes to defining who the "needy" population is and what the "needed" amounts are to ensure guaranteed access to wheat and rice at the "right support prices."[9]

The government favors targeting as a cost-saving effort, since universal coverage is usually infeasible, carrying excessive fiscal burden. However, critics are quick to point out that the government always celebrates India's recent growth in food production and note the fiscal preference given to the India's Vodafone, the IT sector, the Telecom industry, etc., implying a failure in resource mobilization needed for the critical delivery of food to the citizens. Indeed, as the debate continues, the points raised by the critics deserve special attention. Namely, the ratio of actual food subsidies over the past decade relative to GDP has amounted to between 0.6% and 0.8% of GDP. Raising this figure to more than 1% is a reasonable demand given the fact that the World Food Program estimates that, despite high growth over two decades and more, a quarter of the world's hungry population resides in India and around 43% of children under the age of 5 years are malnourished.

---

9. For an interesting debate on the potpourri of terms and schemes, see Chandrasekhar (2013).

Furthermore, the 2000s saw soaring corporate profits, increased tax concessions to the corporate sector, and a government mired in controversies over spectrum sale, coal blocks, and gas prices.

There is no reason, critics argue, as to why the rise in prosperity cannot trickle down in the form of adequate food security, universal access to food grains, and allocation of surplus to where it is needed most. Thompson (2012) also makes similar observations about food security and food production in India.

The State of Food Insecurity in the World 2012 indicates that almost 870 million people are chronically undernourished in 2010−12 and that this number of hungry people in the world remains unacceptably high.[10] The report presents these estimates based in terms of the distribution of dietary energy supply.

We get a good sense of the priority of these issues from the actions listed in the most recent G8 summit held in Lough Erne, United Kingdom. In June 2013, the European Union set out to establish regulations concerning trade, taxation, and transparency (the three T's). Agriculture and food security are top priority of the European Union's development policy. Indeed, every year around €1 billion is invested to that end. In 2010−11 alone, the Commission allocated nearly €5 billion to improve food security. A recent report on the European Union's Food Facility—the €1 billion facility set up in 2008 on the initiative of President Barroso to counter the negative effects of the food crisis—shows that in three years, the EU food facility has improved the lives of over 59 million people in 49 countries and provided indirect support for another 93 million others, particularly farmers (Europa, 2013).

A recent report from Europa (2013) points out that currently, 70 million people are still going hungry, and malnutrition is responsible for over 3 million child deaths annually. The report indicates that at the G8 summit, the European Union announced that it will spend an unprecedented €3.5 billion between 2014 and 2020 on improving nutrition in some of the world's poorest countries. The policy framework seeks political commitment for nutrition at country and international levels (Europa, 2013). The G8 summit also emphasizes the fight against climate change and provides the global negotiations toward an agreement on this issue in 2015.

On a positive side, the recent estimates also show that progress in reducing hunger has been more pronounced than previously believed. Most of the progress was achieved before 2007−08, and since that time, global progress in reducing hunger has slowed. The report also concludes that further improvements and better data are needed to capture the effects of food price and other

---

10. The undernourishment estimates do not fully reflect the effects on hunger of the 2007−08 price spikes or the economic slowdown experienced by some countries since 2009. For more recent statistics, see FAO Statistical Yearbook (2013).

TABLE 1.1 Global undernourishment 1990−92, 2010−12, 2020. Number (millions) and prevalence (%) of undernourishment.

Regions	1990−92	2010−12	2019*
World	1000 (19%)	868 (13%)	688 (9.8%)
Developed regions	20 (2%)	16 (1.4%)	(<2.5%)
Developing regions	980 (23%)	852 (15%)	N/A
Africa	175 (27%)	239 (23%)	250 (19.1%)
Asia	739 (24%)	563 (14%)	381 (8.3%)
Latin America	65 (15%)	49 (8%)	48 (7.4%)
Oceania	1 (14%)	1 (12%)	2.4 (5.8%)

2019*, projected by FAO (2020)s.

economic shocks. Newer indicators are also needed to provide a more holistic assessment of undernourishment and food security. Table 1.1 indicates the progress in major geographic regions:

In addition to these measures, the World Food Programme (http://www.wfp.org) has provided a host of studies for all the countries and has provided vital statistics of hunger, malnutrition, and related measures. Some of the insights that we can glean from WFP are as follows:

1. Eight hundred and seventy million people in the world do not have enough to eat. This number has fallen by 130 million since 1990, but progress slowed after 2008.
2. The vast majority of hungry people (98%) live in developing countries, where almost 15% of the population is undernourished.
3. Asia and the Pacific have the largest share of the world's hungry people (some 563 million), but the trend is downward.
4. If women farmers had the same access to resources as men, the number of hungry in the world could be reduced by up to 150 million.
5. Poor nutrition causes nearly half (45%) of deaths in children under 5—3.1 million children each year.
6. One out of six children—roughly 100 million—in developing countries is underweight.
7. One in four of the world's children is stunted. In developing countries, the proportion can rise to one in three.
8. 80% of the world's stunted children live in just 20 countries.
9. Sixty six million primary school-age children attend classes hungry across the developing world, with 23 million in Africa alone.
10. WFP calculates that US$3.2 billion is needed per year to reach all 66 million hungry school-age children.

The next section examines the measurement and determinants of food security based on the aforementioned conceptual framework.

Measurement of the determinants of food security

"Food security" is a flexible concept and is usually applied at three levels of aggregation: national, regional, and household or individual. At the 1996 World Food Summit, food security was defined as follows: "Food security exists when all people, at all times, have physical, social and economic access to sufficient food which meets their dietary needs and food preferences for an active and healthy life" (FAO, 1996). This definition is well accepted and widely used.

The three core original determinants of food security are

1. food availability,
2. food access, and
3. food utilization.

However, recently, the Committee on World Food Security (2012) adds the stability dimension to the aforementioned determinants and defines food security as follows:

"Food security exists when all people, at all times, have physical, social and economic access to sufficient, safe and nutritious food that meets their dietary needs and food preferences for an active and healthy life. The four pillars of food security are availability, access, utilization and stability. The nutritional dimension is integral to the concept of food security."

The measurement of various indicators of food security is a first step in quantifying food security of the population. Various approaches are used to collect and document data on food security indicators. We provide a brief introduction to these measures and their data sources.

Food availability

Information on food availability usually comes from national, regional, and subregional food balance sheets. This is obtained from the FAO food balance sheet database for individual countries and regions (http://faostat.fao.org/site/502/default.aspx). However, food balance sheets provide no information on consumption patterns and relate only to the supply or availability of food at the national level (Becker and Helsing, 1991). They depict annual production of food, changes in food stocks, and imports and exports and describe national dietary patterns in terms of the major food commodities. While they are useful to understand, aggregate indicators (such as macroeconomic and demographic factors) on food consumption, using the national food balance data, do not provide information on food security at the household level.

Measuring food availability

There are a variety of methods for measuring food availability. They are as diverse as participatory poverty profiles, principal component analysis, and spatial econometric tools. The small-area estimation method developed by Hentschel et al. (2000) and Elbers et al. (2001) is one of the most common methods in measuring household food availability. It is a statistical tool that combines survey and census data to estimate welfare or other indicators for disaggregated geographical units (such as rural regions and municipalities). In this method, the first step is to estimate a model of household welfare using the household survey data. In the second step, the parameter estimates are applied to the census data assuming that the relationship holds for the entire population. The household level results are then aggregated by a larger geographical region or area by taking the mean of the probabilities for the area. This allows the researcher to construct maps for different levels of food insecurity disaggregated across geographic units.

Food access

What do we mean by food access? It could be physical access to food in the market or economic access to food at the household level. While food availability at the national and regional levels and the associated infrastructure such as roads and market outlets to buy food determine physical access to food, economic access depends on the purchasing power of the household and the existing level of food prices, which could depend on the physical access to food (Thomson and Metz, 1998). A household's ability to spend on food is a good indicator of food access at the household level.

Measuring food access

Household food access is measured through food or nutrient intake at the household level. This is usually reported in "adult equivalent" units to facilitate comparison among individuals within a household as well as among households. The adult equivalent unit is a system of weighting household members according to the calorie requirements for different age and sex groups. Household income and expenditure surveys that collect information on household composition, household expenditure patterns with a focus on food and nonfood items, calorie intake, consumption of major products, and socioeconomic characteristics (such as head of the household, household education level, etc.) can be used to assess food access over time, by estimating amounts of food consumed, composition of the diet, and nutrient availability at the household and individual levels.

Food utilization

Food utilization relates to how food consumed is translated into nutritional and health benefits to the individuals. In this approach, the consumption of foods both in quantity and in quality that is sufficient to meet energy and nutrient requirements is a basic measure of food utilization.

The relationship between food security and nutrition security is depicted in Fig. 1.1. It shows links between nutritional status and other determinants at the household level. In this framework, the nutritional status is an outcome of food intake and health status. However, the underlying causes of health (namely environmental conditions, health services, and caring activities) are shown in different boxes due to their different underlying characteristics and features. A reduced state of health can be due to poor access to healthcare and poor housing and is possibly worsened by malnutrition, which makes individuals vulnerable to diseases. Thus, distinguishing between health services, caring activities, and environmental factors is crucial in selecting appropriate intervention strategies to improve food utilization.

Measuring food utilization

Food intake data, following conversion to nutrient composition, are evaluated by comparing them with recommended intakes of energy and other nutrients. Two terminologies are essential in understanding this approach. *Nutrient requirements* are the levels of particular nutrients in the lowest amount that is necessary to maintain a person in good health. They vary between individuals, although the requirements of a group of similar individuals (age, sex, body size, and physical activity) will fall within a certain range. *Recommended intakes* are the levels of nutrients that are thought to be high enough to meet the needs of all individuals within a similar group. The WHO and FAO set this recommendation by taking the mean minimum requirement for a nutrient plus two standard deviations. *Dietary guidelines* are the linkages for the general public between recommended nutrient intakes and the translation of these recommendations to food-based guidelines.

There is no method for establishing the minimum requirement levels for nutrients, and methods differ depending on the nutrient. Similarly, for the recommended intake levels, the usual guidelines are based on the estimates of the minimum requirements for a nutrient plus a standard additional amount. This amount is usually either two standard deviations or a fixed percentage increment of the mean requirement for the group. Since food balance sheet data are not very useful in describing dietary intake adequacy of a population and household surveys can provide limited information on the dietary adequacy of the household as a whole, the dietary intake approach yields precise application of standards or requirements to individual intake data.

Although food intake includes protein and other nutrients, *energy intake* is one of the main parameters and is extremely important in improving food utilization. Energy requirement for an individual is the amount of dietary energy (through food) needed to maintain health, growth, and an appropriate level of physical activity (Torun, 1996). Since energy requirements are derived from data originating in healthy populations, they need to be adjusted in communities that suffer from malnutrition and other debilitating diseases. Estimates of energy requirements are usually based on energy expenditure data, although it is possible to obtain rough estimates on the basis of energy intake data from dietary surveys. For children, there is an additional allowance for growth.

In food security assessment, the group distribution of the individuals' energy and nutrient requirements is assumed to be normally distributed. The determinants of energy requirements include basal metabolic rate (BMR) (constituting between 60% and 70% of total energy expenditure); physical activity; body size and composition; age; and climate and ecological factors.

In the basal metabolic factor approach, energy requirement is computed as the product of the BMR and physical activity level. The BMR is the minimal rate of energy expenditure required to maintain life. To calculate BMR, first individual oxygen consumption is measured and then converted into heat or energy output. Physical activity levels have been calculated for various occupational categories. A physical activity level of 1.55–1.65 is an average for most developed countries (Shetty et al., 1996).

The estimates of mean per capita energy requirement are thus dependent on the BMRs, physical activity levels, lactation, pregnancy, climate, and the degree of malnutrition. Scientifically, the range varies from 1900 to 2500 kcal per day. The National Academy of Sciences (1995) has arrived at a figure of 2100 kcal per day for use in food emergency situations, which is based on an assumption of light activity.

Stability of availability

The stability of all indicators of food availability, accessibility, and utilization is seen as an additional factor in determining food security. The resilience with which countries, regions within the countries, communities, and households attain food security needs to be taken into consideration in analysis of food security.

Alternative approaches in measuring food security

Although the aforementioned approaches are the most common ways of measuring food security, some recent alternative approaches are also in vogue in measuring food security depending on context specificity. They are as follows:

1. interaction approach,
2. coping strategy/chronic vulnerability approach, and
3. scaling approach.

The *interaction approach* developed by Haddad et al. (1994) is an overlap technique that seeks to determine to what extent a proportion of households are insecure on a particular dimension given that they are insecure on another dimension. For example, with the aforementioned approach, one can address the following: for all individuals who have inadequate drinking water facilities, what percentage is also food insecure. Thus, a combination of various indicators can be an important predictor of food insecurity. While the aforementioned approach is useful in combining various indirect indicators to determine household level food insecurity, its main limitation is that such combinations are endless. Thus, the analysis is purely suggestive.

Maxwell (1996) developed a *coping strategy* approach for households in the face of insufficient food consumption. The cumulative food security index is based on six food coping strategies. A scale was developed for the frequency of each individual strategy and was multiplied by the severity weighting factor based on ordinal ranking to derive the food security score. The advantage of this approach is to understand short-term food insufficiency. This approach does not require specialized enumerators or any complex statistical procedures. However, a major disadvantage of this approach is that it cannot differentiate between short-term food insecurity from long-term vulnerability indicators.

A natural way of extending the "coping strategy" approach is to bring in temporal dimensions of food insecurity. The *chronic vulnerability approach* to food security, originally developed by Sen (1981), seeks to identify why households become vulnerable in particular dimensions and thus characterizes a dynamic relationship. It is defined by Riely (1999) as "the probability of an acute decline in food access or consumption levels below minimum survival needs." It can result from both exposure to risk factors—drought, conflict, or extreme price fluctuations—or it can result from households having lower ability to cope due to various socioeconomic constraints. According to Riely, vulnerability can be viewed as the sum of exposure to risk and the inability to cope. Vulnerability tends to be higher when the risk of natural disasters increases, when adverse government policies result in chronically deficit household consumption, or when poorer households rely on a risky source of consumption or income (Scaramozzino, 2006).

Finally, the *scaling approach* assesses how households go through different experiential and behavioral stages and thus become more food insecure over time. This approach is widely used to measure household food security in the United States (Bickel et al., 2000). A core six item set of questions is used to determine a single overall food security scale, with greater values of the index indicating that households are more food insecure. While the food security scale shows that some member or members of the household are experiencing food insecurity, it does not capture other dimensions such as the nutritional status of children.

Conclusions

Measuring food security at the national, regional, community, and household levels is important for developing appropriate policy and program options. At the national level, measuring food available for consumption is based on food balance sheets. Food balance sheets provide a comprehensive picture of food supply during a particular reference period (usually one year) and is computed from the annual production of food, change in stocks, and imports and exports (FAO, 2001). While the food balance sheet is extremely useful in formulating agricultural policies related to production, consumption, and distribution of food, they can also be used in developing appropriate agricultural trade policies (for example, when a country faces a chronic deficit in food). The trends in food consumption over a longer time period at the national level can provide useful information on nutrient intake of the population. However, food balance sheets do not provide general information on nutrient intake within a country or among groups of households and thus should not be usually used in estimating nutritional inadequacy (Jacobs and Sumner, 2002).

At the regional level, targeting through small area estimation in smaller administrative areas improves the cost-effectiveness of development spending and reduces geographical disparity of food-insecure households. The aforementioned food security mapping exercise can be useful in various policy interventions, such as transfer of food aid throughout a country (as in Sri Lanka) or testing new technologies in a particular food-deficient area (such as in Mexico) (Hyman et al., 2005).

Community food security is a natural extension of the food security concept at the community level (Anderson and Cook, 1999). It is defined as "all persons in a community having access to culturally acceptable, nutritionally adequate food through local nonemergency sources at all times" (Winne et al., 1997, p. 1). However, the lack of a consensus of a general definition of a community among researchers and practitioners has hindered the measurement of a "food-insecure community" and its relationship to household and individual food security. It is thus important conceptually and operationally to define a "food-insecure community" for the purpose of survey design in various developing countries. While both contextual and global community factors are critical elements of a community survey, collecting and integrating all these data remains a major challenge in terms of cost-effectiveness.

We also examined the food insecurity situation in the United States, which surprisingly indicates that about 15% of US households were found "food insecure" in 2011. We examined the issues surrounding this aspect, including the relevant measures adopted by the USDA to compute food insecurity among

US households. The literature on food insecurity in the United States has given policy makers and program administrators several insights into the causes and consequences of food insecurity. Newer findings and techniques are constantly being adopted in the United States and elsewhere to get a handle on this critical issue.

The results from the FAO report imply that the Millennium Development Goal (MDG) target of halving the prevalence of undernourishment in the developing world by 2015 is within reach, if appropriate actions are taken to reverse the slowdown since 2007–08. Some of the policy recommendations to reach the MDG are as follows:

a. Growth needs to involve and reach the poor. For economic growth to enhance the nutrition of the most needy, the poor must participate in the growth process and its benefits.
b. The poor need to use the additional income for improving the quantity and quality of their diets and for improved health services.
c. Governments need to use additional public resources for public goods and services to benefit the poor and hungry.
d. Agricultural growth is vital in reducing hunger and malnutrition. Since most of the poor depend on agriculture, it is important that agricultural growth involving smallholders, especially women, be carried out most effectively.
e. Economic and agricultural growth should be "nutrition sensitive." Growth needs to result in better nutritional outcomes through enhanced opportunities for the poor to diversify their diets. This includes improved access to safe drinking water, sanitation, health services, better consumer awareness, and targeted distribution of supplements in situations of acute micronutrient deficiencies.
f. Social protection is crucial for accelerating hunger reduction. Social protection must be properly structured and must be effectively targeted toward smallholders, with an emphasis toward newer technology adoption.
g. To accelerate hunger reduction, economic growth needs to be accompanied by purposeful and decisive public action. Public policies and programs must be propoor oriented. The programs must include provision of public goods and services for the development of the productive sectors, equitable access to resources by the poor, empowerment of women, and design and implementation of social protection systems. An improved governance system, based on transparency, participation, accountability, rule of law, and human rights, is essential for the effectiveness of such policies. Finally, household level surveys in conjunction with individual level measures of dietary intake are another set of instruments in assessing food security at the household level. These surveys enable comparison of household food

security status by analyzing expenditure patterns on food and nonfood items and yield dietary intake patterns of individual members of the household in the context of resource constraints.

Assessment of food security at a community, region, or national level should be context specific and will depend on the purpose for which the data are collected. For example, emergency interventions may use data from rapid appraisal surveys while a long-term planning exercise will demand comprehensive household surveys. Similarly, monitoring and evaluation of food security interventions may collect a different set of indicators with varying levels of intensity and accuracy. Nevertheless, analysis and use of data for informing policy and program options require effective conversion of data into useful information for decision-making.

These ideas and conclusions are strongly emphasized in a most recent (OECD, 2013) report. In particular, the report concludes that (1) the challenge of eliminating global hunger is more about raising the incomes of the poor than an issue of food prices; (2) agricultural development has a key role to play in ensuring food security; (3) there is a need for increased investment in rural areas, which offers higher returns than agricultural subsidies; (4) efforts to raise incomes need to be complemented by other policies to improve nutritional outcomes; (5) sustainable agricultural productivity is central to ensuring that food will be available at prices people can afford; (6) policies that subsidize or mandate the use of biofuels should be removed; (7) public supported by development aid can complement and attract private investment although there is enough evidence to suggest that public investment and development aid can crowd out private investment; and (8) trade has an important role to play in ensuring food security. Reforming countries may need to put in place parallel measures to maximize the benefits and reduce the costs. We examine many of these issues in the remaining chapters.

A natural question is why is measuring food insecurity important for better program design in developing countries?

The search for better measures of food security still remains a major challenge due to the complex and multidimensional nature of food security. However, the issue remains important, as hundreds of millions of individuals and households are affected on a daily basis in both the developing and developed world (see Table 1.2). The recent food crises in Haiti due to a substantial hike in food prices, the chronic vulnerability of Ethiopian population to famine, and the food insecurity of households in the northern region of Malawi due to

TABLE 1.2 Nationally representative household surveys containing food expenditure data (1990−2013).

Region/country	Year	Sample size	Type
Sub-Saharan Africa			
Ghana	1998/99	5998	World Bank LSMS
South Africa	1993	9000	World Bank LSMS
Tanzania	1993	5200	World Bank LSMS
Ethiopia	1989, 2004, 2009	1477	IFPRI rural household survey
Malawi	2000−02	758	IFPRI complementary panel survey
South Asia			
India (Uttar Pradesh and Bihar)	1997−98	2250	World Bank LSMS
Nepal	1996	3373	World Bank LSMS
Bangladesh	2000, 2013	1120	IFPRI SHAHAR baseline survey
		6500 (Bangladesh Integrated Household Survey (BIHS) 2011−12)	
Pakistan	1991	4800	World Bank LSMS
East Asia			
China	1995 and 1997	780	World Bank LSMS
Cambodia	1999	6000	World Bank LSMS
Vietnam	1997/1998	5994	World Bank LSMS
Middle East and North Africa			
Morocco	1991	3323	World Bank LSMS
Egypt	1997	2500	IFPRI
Newly industrializing countries (NIC)			
Albania	2005	3638	World Bank LSMS
Armenia	1996	4920	World Bank LSMS
Azerbaijan	1995	2016	World Bank LSMS
Bosnia and Herzegovina	2001	5402	World Bank LSMS

higher maize prices all testify to the fact that constructing better measures will remain critical in the coming decades for addressing the substantial challenges posed.

Section highlights: food insecurity during a pandemic

The recent COVID-19 crisis has intensified the existing issues surrounding food insecurity among vulnerable populations across all countries. Resnick (2020) and Arndt et al. (2020) demonstrate how the lockdown has decreased South Africa's GDP by 5%, and recovery back to normal is predicted to be painfully slow and long. Dorosh et al. (2020) examines the impact of COVID-19 on Bangladesh and notes that the fall in incomes has produced a demand shock, which has induced a drastic decline in nutritional outcomes. Ray and Subramanian (2020) examine India's informal wage sector, a large portion of the economy's workforce that has taken a big hit in terms of job loss and wage reductions.

Although the Indian government has provided about 700 million rupees, equivalent to 100 million dollars, in some form of cash transfers, the situation is still far from settled. About 84% of people have experienced a decline in incomes.

Limaye et al. (2020) and Varshney et al. (2020) also note that about 30 million migrant urban workers have returned to their rural homes in the following states: Assam, MP, UP, WB, Jharkhand, Bihar, Odisha, Chhattisgarh, Gujarat, Maharashtra, Karnataka, and Rajasthan.

Households with migrant-returning workers face severe issues with food adequacy. Almost all the families have been forced to change food habits and report a decline in consumption, with respect to the number of food items and meals. The number of times families have had to borrow food from friends and relatives has increased substantially. Furthermore, there have been reports of increasing mortgaging of household items, which foreshadows long-run reduction in wealth and persistent poverty.

Researchers have also noted some of the reasons for the increase in food insecurity, in the ASEAN context:
- About 80% of rural households depend on public distribution system or PDS (which was equally stressed due to limited supplies) in India.
- Important in-school programs and other civic centers (known as *Poshan Abhiyaan*, and *Anganwadi*) in India have had to close.
- Although the government has increased funding, a very low percent of people have actually received the benefits, indicating implementation issues.

All of the aforementioned factors have contributed to reduced availability, accessibility, affordability, and quality of food. Reports indicate that about 50% of households face shortage of food items, and about 60% of households experience reduced food intakes.

The incidence of food insecurity remains high among rural households, whose livelihoods depend on sales of farm produce, dairy, and poultry (Limaye et al., 2020).

Most importantly, food insecurity and nutrient deficiency are high among pregnant women, which have direct implications on future difficulties with massive malnutrition among children.

Continued

> **Section highlights: food insecurity during a pandemic—cont'd**
>
> We highlight some of the issues in food security measurement that are relevant for policy analysis. First, the severity of food insecurity cannot be ascertained only from the national food balance sheet data. Additional household level and dietary intake surveys will be necessary to determine which segments of the population are particularly vulnerable. Second, there is a need to improve the tools and frameworks for targeting various interventions (especially for the vulnerable segments of a population) for achieving optimum resource allocation. This will require precise measures for locating the food-insecure households. Third, food availability, accessibility, utilization, and stability measures have to be addressed in a holistic manner to develop a gamut of policy and program interventions. Finally, both quantitative and qualitative measures of food security need to be identified in the context of a given resource base, agro-ecological constraints, and production and employment opportunities of the communities and households.

Exercises

1. Is there a single definition of food security? What are the core determinants of food security? Define each of these. How are these determinants measured?
2. Is food availability in a country the key determinant of food security? Explain.
3. Choose a developing country for understanding the concepts of food security. Using library and web-based resources, prepare a food balance sheet of the country for the last year (or latest year for which data are available) and for five years ago. Explain what you infer from the data about food availability in the country including trends in food production, food stocks, imports, and exports.
4. Choose a developed country, preferably a European country to describe the status of food security of its population and provide an analysis of the recent food policy changes in that country.
5. What are the advantages and disadvantages of using the coping strategy/vulnerability approaches in measuring food security?
6. Visit the website http://www.fao.org/docrep/016/i3027e/i3027e.pdf and use this document to see the definition of food insecurity that is adopted. Do you agree with this view?
7. Use the data in http://www.fao.org/docrep/016/i3027e/i3027e.pdf *and compare the food insecurity situation in a developed country with that of a developing country. What are their country classifications? Do these classifications make sense?*

Chapter 2

Implications of technological change, postharvest technology, and technology adoption for improved food security—application of t-statistic

Chapter outline

Introduction	**28**	Threshold of food security by each	
Review of selected studies	**30**	individual component	52
Postharvest technology and		Tests for equality of variances	52
implications for food security	33	**Student's *t*-test for testing the equality**	
The postharvest loss		**of means**	**53**
footprint	34	**Policy implications**	**56**
Coping strategies for		**Technical appendices**	**58**
postharvest loss	35	Constructing the cutoff points for	
Food security issues and technology		components of the food security	
in the United States	**36**	index	58
Biofuels—the Chinese experience	**38**	Variable definitions	58
US Farm Policy and food		Dichotomous variable	58
security—background and current		Interval variable	59
issues	**40**	**Using STATA for *t*-tests**	**59**
GEO-5 and coping mechanisms for the		Independent group *t*-test	59
future	**44**	Independent sample *t*-test	
Empirical analysis—a basic univariate		assuming unequal variances	60
approach	**45**	Using R for *t*-tests	60
Data description and analysis	**46**	Independent sample *t*-test	
Two measures of household food		assuming unequal variance	62
security are computed	**47**	Independent group *t*-test	
Consumption components of the		assuming equal variance	62
food security index	49	**Exercises**	**63**
Descriptive statistics	51	**STATA exercise**	**64**

Food Security, Poverty and Nutrition Policy Analysis. https://doi.org/10.1016/B978-0-12-820477-1.00029-2
Copyright © 2022 Elsevier Inc. All rights reserved.

It is unacceptable that hunger is on the rise at a time when the world wastes more than 1 billion tons of food every year. It is time to change how we produce and consume, including to reduce greenhouse emissions. Transforming food systems is crucial for delivering all the Sustainable Development Goals. As a human family, a world free of hunger is our imperative.

UN Secretary-General António Guterres 2020

Introduction

Food security of communities and countries depends on the productivity levels of their faming systems. Technological change in agriculture has long been accepted as a necessary condition for accelerating growth in food production. Adopters of yield increasing, or postharvest technology, are more likely to experience higher production per unit of land and the associated income benefits at the household level compared with nonadopters. The desired benefits of technological change such as increased agricultural or food production and income are expected to have a positive influence on household food consumption and nutritional adequacy. It is typically assumed that this income-mediated effect on food security and nutritional improvement operates through two main ways (Babu, 2002a; Babu and Rhoe, 2002). First, increased income can be used for greater food expenditures that directly increase food consumption, which in turn, may improve nutritional status by higher intake of energy and other nutrients. Second, increased income can result in higher nonfood expenditures such as health and sanitation that, along with food consumption, could indirectly have positive nutrition and health effects. Thus, to understand the relationship between technological adoption and food security and nutrition, it is important to answer the following two questions (Babu, 1999; Babu and Rajasekaran, 1991a,b):

1. To what extent adoption of new technologies improves household or individual food consumption and through what mechanisms is such an improvement, if any, achieved?
2. How does technology adoption in agriculture translate into measurable nutritional improvement?

The significance of questions such as the aforementioned can be understood in the context of the technological change that was christened as the "Green Revolution," which occurred in many parts of the world—first in the United States and Europe during the 1940 and 1950s (Griliches, 1957) and, later, in Asia and Latin America beginning in the 1960s. As pointed out by Conway (2003), "the first Green Revolution offered farmers new crop varieties that allowed them to improve agricultural yields." The new varieties were widely accepted and adopted by farmers in various countries such as India, Pakistan, Indonesia, Mexico, and the Philippines during the 1960s and the

1970s. To foster such technological change, the Ford and Rockefeller foundations, along with bilateral aid agencies such as the USAID, helped to fund the International Agricultural Research Centers in various parts of the world (see www.cgiar.org). These centers, such as the International Rice Research Institute (IRRI) in the Philippines and the International Maize and Wheat Improvement Center (CIMMYT) in Mexico, bred new varieties of rice and wheat, respectively, and developed new production and postharvest technologies to accompany them.

The impending food crisis and the need to avert a massive famine in several Asian countries motivated such global action (Babu, 2009). For example, before the Green Revolution, almost two-thirds of South Asia's rural population was food insecure and hungry, and the region depended on food aid for feeding its population. The Green Revolution brought the South Asian region and other Asian countries close to food self-sufficiency with surplus grain stocks available to the vulnerable regions, which could otherwise be affected by famines. Yet, many countries in sub-Saharan Africa continue to struggle to meet the food needs of their populations. Technological change requires increased investment for agricultural research and development, which has been declining lately in many developing countries (World Bank, 2008). Recent food price increases, caused by short-sighted policies that encourage diversion of food crops to biofuel production, have raised alarm bells and encouraged world leaders to recommit themselves to agricultural development (Food and Agriculture Organization (FAO), 2008; IFPRI, 2008; World Food Program (WFP), 2008).

Adoption of existing technologies that could increase food security depends on supporting programs and institutions. Such support increasingly requires convincing policy makers with empirical evidence on the benefits of technology adoption on human welfare. There is also an emerging international consensus that the adoption of agricultural biotechnology has the tremendous potential for making a substantial impact on many aspects of agriculture—crop productivity, yield sustainability, and environmental sustainability, thereby improving household food security in the developing world (World Bank, 2008). Recent advances in molecular biology and genomics can greatly enhance the plant breeder's capacity and introduce new traits in plants. The commercial applications of agricultural biotechnology have already produced crops such as Bt-maize, rice, potatoes, and sweet corn that can protect themselves against insects and herbicide-tolerant crops such as wheat, maize, rice, and onions, which allow for better weed management practices (Ozor and Igbokwe, 2007). At the same time, there is growing concern that current investments are increasingly driven by the private sector, which does not address the needs of the poor. There is an urgent need to increase public investments (both at the international and national levels) along with supporting programs and institutions so that the benefits of these technologies do not miss the poor households in developing countries.

This chapter, using the household level data from Malawi, shows how to analyze the impact of adoption of hybrid maize technology on household food security and nutritional situation. Maize remains an important food crop in Malawi. It is the main staple for Malawians and provides over 85% of the total calorie intake (see, for example, Kadzandira, 2003, p. 14).

Over the years, research on high- and early-yielding varieties of maize has achieved remarkable results in terms of yield gains and achieving food security (Smale and Jayne, 2003). This is important for Malawian and other maize-based farming systems, since maize (especially hybrid maize) will remain a crucial component of the food security in two ways: first, by satisfying the basic food requirements of a more diversified rural economy and, second, as a cash crop in areas where it is agroecologically suited to provide higher returns (Babu and Sanyal, 2008).

The relationship between technological change and food security is complex, and there are indirect and partial effects of new technology on food security, so that a focused approach has to be taken to disentangle the complexities of the relationship. We extend the analysis to cover food security issues in the United States and China. We provide examples from recent approaches in China that deals with technological transformation toward biofuels.

In this chapter, we introduce a statistical analysis using t-test to examine whether hybrid maize adopters and nonadopters are different with respect to their food security status. In other words, we examine whether food security differs between these two groups (adopters and nonadopters) and whether this difference is statistically significant. This test is most commonly used for assessing group differences. However, it is also one of the most restrictive tests in its assumptions concerning the underlying data. In general, the data need to be normally distributed, and the group variances need to be homoscedastic.[1] We provide examples that operationalize these concepts using STATA.

In what follows, we present selected case studies that analyze the role of technology adoption in achieving greater food security and a higher nutritional status. A discussion of the issues, data set, methods, and results of these case studies serves as motivation for the analytical method demonstrated and policy conclusions drawn from the analysis.

Review of selected studies

The relationship between technology adoption and food security continues to receive wide interest among food policy researchers. This is particularly true in

---

1. For testing the equality of group variances, SPSS provides Levene's test, which is a homogeneity of variance test. This test is less dependent on the assumption of normality than most other tests. For each observation, the program computes the absolute difference between the value of that observation, and its cell mean and performs a one-way analysis of variance (ANOVA) on those differences.

many African countries, where the threat of famine continues to be real. The case study of Zambia reviewed here is useful in developing and testing some of the maintained hypotheses about technology adoption and food security.

Kumar (1994) examined the nature and effects of technological change in maize production on food consumption and nutrition in the Eastern Province of Zambia and suggested a few policy implications. In Zambia, maize is the staple food. To achieve food production growth, the traditional approach has been extensive cultivation—expanding the land under cultivation of maize given the abundant supply of land. However, land expansion (extensive cultivation) alone is not sufficient for a sustained growth in maize production due to diminishing returns from land for a given level of labor supply. Therefore, it is important to increase yield per hectare of land (intensive cultivation). An effective way to do this is through adoption of improved technology. The technology adoption considered in this study is the use of high-yielding varieties of seeds—the hybrid maize.

Given the improvement in agricultural productivity through modern technological methods as a worthwhile food security intervention, the study generates some significant policy implications. It is observed that the majority of farmers in the Eastern Province of Zambia grow traditional maize for self-consumption and hybrid maize as a cash crop. The local maize can be easily stored and processed at home, while hybrid maize does not store well and requires processing at mills. The study also observes low adoption of hybrid maize in several areas due to limited availability and poor distribution channels of hybrid seeds and fertilizers. The government thus must encourage and invest in market infrastructure and distribution channels, including the construction of roads, processing and storage facilities, and improvement of marketing channels. Government incentives and support to improve on-farm storage capacity and village-level access to milling facilities will encourage the use of hybrid maize for households' own consumption. Policies that offer innovative extension and credit systems will also promote higher production of hybrid maize.

The results indicate that hybrid maize production is more profitable for smaller farms. Also, the positive effect of technology adoption is more pronounced on food consumption of households with smaller farms than larger ones. This can be attributed to larger farms requiring more labor. Since labor costs are high, investing in labor-saving technologies can fulfill the additional labor requirement. This substitution process, however, results in smaller gains to large farms compared with small farms, which are usually managed by a family. Incentives to encourage women's involvement in maize production are thus critical.

The study also finds that the adoption of hybrid maize decreases women's share of income particularly in larger farms. The reduction in income affects both production efficiency and family welfare adversely. The government should therefore offer women easier access to information about farming and agricultural production that results from technological change. Policies that provide equal access to inputs and credit to women farmers should be brought into effect, since increasing women's income share is associated with better food security and child nutrition.

Technological change or technological adoption and commercialization of agriculture are virtually synonymous in many cases (von Braun et al., 1994). An export-producing cooperative in Guatemala (von Braun and Immink, 1994) enabled its household members to have 18% more expenditure on food per capita on average and significantly improved their calorie consumption. The commercialization scheme under the cooperative resulted in an increase in income and affected health and nutrition positively in the form of decreased stunting and weight deficiency among the children of the households.

Potato production in the Gishwati forest area of Rwanda (Blanken et al., 1994) resulted in more expensive calories being acquired and made adopters surplus-calorie producers. It also resulted in reduction of malnutrition among adopting households.

Technological adoption for cultivation of tobacco and maize in Malawi (Peters and Herrera, 1994) resulted in a significant increase in calorie intake, especially for those in the top third of income distribution. The study did not find a significant difference in nutritional status between children of adopters and nonadopters.

Bouis (2000) examined the impact of three programs that provided credit and training to women in Bangladesh for the production of polyculture fish and commercial vegetables on micronutrient status of households. The study found a modest increase in incomes for adopting households compared with non-adopting ones. The adoption of polyculture technology did not improve the micronutrient status of members of adopting households.

A report by the International Food Policy Research Institute and International Center for Tropical Agriculture on the use of biofortification for health improvement of poor (IFPRI and CIAT, 2002) states that biofortified crops (crops that are bred for increased nutrient content) are one of the most promising new tools to fight and end malnutrition. However, lack of infrastructure, poor policies, lack of delivery systems for new varieties, low level of investment in research, and less demand for such crops in poorest countries make it difficult for commercial application and supplementation of such technology.

In a comprehensive study, Minten and Barrett (2008) examine how agricultural technology adoption and crop yields affect food prices, real wages for unskilled workers, and key welfare indicators for Madagascar. The novelty of the paper is twofold:

1. It is one of the few empirical studies that examines the linkages between agriculture and poverty in sub-Saharan Africa.
2. It relies on spatially explicit data from a complete census of Madagascar's communes—the smallest administrative unit—to under meso analysis.

The study examines three distinct pathways through which productivity enhancing technical change affects welfare measures—(1) lower real food prices, thus benefiting net food consumers; (2) output increases that surpass price declines, thereby benefiting net food suppliers; and (3) increase in real wages, which benefits unskilled workers.

The data for this study originate from three sources: a commune-level census conducted in 2001, the national population census of 1993, and geographical data from secondary sources. The unit of analysis is the commune, which is the smallest administrative unit with direct representation and funding from the central or provincial government. The analysis was undertaken using multivariate regression techniques.

Overall, the results of the study clearly demonstrate that better agricultural performance (as proxied by higher rice yields) is strongly correlated with real wages, as well as rice profitability and prices of staple food. The aforementioned results strengthen the conclusion that greater rice productivity reduces food insecurity in Madagascar for all the major subpopulations.

The results of the study show that increased agricultural yields are strongly associated with gains for each of the three subpopulations (net sellers, net buyers, and wage laborers) in the rural areas. While greater rice productivity outpaces local market price declines and thus benefits net sellers, higher rice yields benefit the other two subpopulations by driving down food prices and improving unskilled laborers' real wages. The net effect is the presence of fewer food-insecure households and shorter lean periods.

Second, cash crop production, but not mining activities, was associated with improvement in welfare outcomes. Finally, the results indicate that no single intervention is effective in improving agricultural productivity and reducing poverty and food insecurity in rural Madagascar. While technology diffusion remains important, equally important are improved rural transport infrastructure, increased literacy rates, secure land tenure, and access to extension services.

Postharvest technology and implications for food security

In many tropical developing countries, agricultural commodities can suffer significant losses after they are harvested and stored. This is often referred to as "postharvest loss (PHL)." Reducing food losses, especially in developing countries, is considered to be a major constraint in achieving food security (Toma et al., 1991). Crop losses can occur during the postharvest system at all levels including preprocessing, storage, packaging, and marketing. The final level of production is thus adversely affected. Adopting postharvest technologies can increase better quality of products and extend market opportunity.

Poor grain storage remains one of the most common problems in developing countries and estimates of grain losses range from 33% to 50%

(Kader, 2003). The inadequacy of storage accompanied with vulnerability of crops to damage makes middlemen and traders unwilling to store stocks beyond the minimum turnover period (Gabriel and Hundie, 2004). Due to the rapid perishability of food grains, the risk of loss could be quite high in magnitude. Thus, the role of postharvest management practices becomes critical.

Postharvest grain management practices can affect household food security through the following channels:

1. Output reduction in food grain availability due to physical losses.
2. Lower income due to lower prices when grains are sold immediately after harvest.

The postharvest loss footprint

Gustavasson et al. (2011), Sternmarck et al. (2016) and many other studies have concluded the following:

- Around 1/3rd of all food produced globally is wasted, amounting to roughly 1.3 billion tons annually.
- The loss in Europe is estimated at 88 million tons, worth around 143 billion euros.
- PHL in developed countries is 12 times more than in developing countries.
- The nutritional energy lost is equivalent to feeding 1.9 billion people.
- More than half of the PHL is preventable.

The aforementioned list must be evaluated in a context where, by 2050, there will be about 9.1 billion people waiting to be fed. Heavy PHL footprint, in recent years, is identified in Africa. Abass et al. (2014) shows that PHL in Eastern and Southern Africa amounts to $ 1.6 billion annually and constitutes about 13.5% of total grain production. The World Bank (2011) and related literature indicate that PHL in Ethiopia ranges from 5% to 26% of all production, while in sub-Saharan Africa, the PHL in grain production amounts to $ 4 billion, and about the PHL associated with maize is around 11.7% (see Chegere (2018), Sheahan and Barrett (2017)).

An overwhelming evidence of PHL across all links in the production—supply value chain has been noted in India. PHL footprints are evidenced in the production and supply of fruits in Punjab, pulses in Assam and Orissa (Dutta et al., 2018; Devi et al., 2017), drumsticks and fisheries in Gujarat (Singh et al., 2018; Sharma et al., 2017) and onions in Rajasthan (Sharma and Shukla, 2012).

It is important to emphasize that in the aforementioned channel, farmers' perception of risk of PHLs and other liquidity constraints can affect the marketing behavior and can produce suboptimal outcomes, which result in lower levels of household food security.

As pointed out by Goletti and Wolff (1999), the postharvest sector can play an important role in achieving higher agricultural growth and improved food security for the following reasons. First, the sector has high internal rates of return.[2] On average, the rate of return of the postharvest sector is comparable with that from production research and thus makes an almost equal contribution to income growth. Second, postharvest research has public good like characteristics and thus will be underfunded by the private sector. In the area of postharvest research, the International Center for Tropical Agriculture (CIAT) cassava project and the International Rice Research Institute's (IRRI) rice drying technology are examples that can be replicated in many countries. For cassava, the rapid deterioration and perishability of roots increases costs and risks. This leads to considerable losses to wholesale merchants, retailers, processors, and consumers. Thus, techniques for storage and processing should be adopted to prolong the root's useful life or in generating other products. If cassava roots can be stored for more than 3 days, the two advantages would be as follows:

1. Losses and marketing risks would be fewer, making cassava more acceptable to markets.
2. Managing the possibilities for the processor and consumers will be greater.

Finally, postharvest research contributes to food security in several ways. Improvement in storage technology reduces the losses and thus increases the amount of food available for consumption. Similarly, reduction of cyanide potential in crops such as cassava has an important effect on food safety and can improve the nutrition situation of a significant proportion of the population in many countries in Africa.

As evident from the aforementioned studies, adoption of modern technology can improve both household income and access to resources and, therefore, improve household food security, while its impact on household nutritional status is at best ambiguous. This may be due to gender biased effects, as there is a shift in the control over the crop from women to men with the adoption of new technology. To generate similar policy recommendations, this chapter undertakes an empirical analysis based on univariate t-test framework. This is done to understand the impact on food security from the adoption of new technology (hybrid maize adoption) using the socioeconomic household survey data of Malawi.

Coping strategies for postharvest loss

Several global policy action plans have been prescribed in recent years to build awareness and cope with PHL. Some of the well-known examples in this

---

2. The internal rate of return is the interest rate that makes the present value of an investment's income stream equal to zero.

context are the "Save Food" campaign by the FAO in 2011, the Sustainability Development Goals from the United Nations in 2015, the AgroCycle project, the FUSIONS project from the European Union, the WARP program in the United Kingdom, and the Milan Protocol from the Barilla Center for Food and Nutrition.

Garcia-Herreroa et al. (2018) present a unified methodology to estimate the Nutritional Food Loss Footprint (NFLF) for Spain. The main take-away message from all of the studies is clear: Governments must invest in identifying PHL hot spots at all stages of the food supply chain: production, postharvest storage, processing, distribution, and consumption.

A natural question arises as to whether some of the proposed strategies (namely, adopting modern storage methods, and improved seed varieties) actually work. Once again, in recent years, there has been a bulk of evidence stemming from Africa, which support such measures.

For instance, Wodimagegn and Nyasha (2018) show how in Ethiopia, improved storage technology increases the quality of diet and reduces child malnutrition and food insecurity. In a counterfactual simulation, the researchers also show that the nonusers of the technology would have also benefitted, had they adopted the same measures, such as metal silos, airtight drums, and modern storage structures.

Likewise, Ndiritu and Ruhinduka (2019) show that farmers in rural Tanzania significantly reduce their operational costs through modern storage facilities. Ainembabazi et al. (2018) note the value of PHL storage technologies in poverty reduction in Central Africa. Ricker-Gilbert and Jones (2016) for Malawi, and Omotilewa et al. (2018) for Uganda have also indicated the importance of considering PHL, while pursuing other complementary policies, such as chemical and input subsidies.

Food security issues and technology in the United States

The issue of long-term food security has attracted a lot of attention in the United States. For example, Dahlberg (2008) identifies the three main types of agriculture prevalent in the United States and lists the different types of risks that these types face. For instance, in the United States, the three main types of agriculture are agribusiness, agriculture in the middle, and alternative agriculture. Agribusiness is the high-tech version of the food system, which is the most homogeneous of all the systems. Agriculture in the middle refers to farms, which are spread through the rural areas and are independent agricultural communities. Alternative agriculture is the most heterogeneous system and includes a variety of systems including organic, ecological, biodynamic, and sustainable food systems.

Dahlberg (2008) points out that the types of threats and risks each system faces are different. High-tech agribusiness is the most vulnerable system, which can face losses from commodity and export subsidies, GMO (genetically modified organism) labeling issues, the large processing units with inherent economies of scale, climate changes, and bioterrorism. Agriculture in the middle is also exposed to the same risks, if the system aligns itself closely to high-tech agribusiness. This is already evident with the decline in the number of farms, farm families, and rural communities. Furthermore, this has also threatened the health of rural America, its landscapes, and the general biodiversity. Consequently, it is important for the agriculture in the middle to combine what was best from its past, which is a very challenging prospect. Alternative agriculture faces pressures from structural simplification and standardization brought about by commercialization, corporate competition, and governmental regulation. This sector also does not have any preemption from trade treaties and, most importantly, are excluded in health, safety, and environmental standards. Consequently, it is not possible to infer the long-term viability of this sector.

What are some of the risks concerning food security that are most visible in the United States and in these three systems? The three most glaring risks are (1) national security risks, (2) public health risks, and (3) loss of biodiversity. National security risks, in the name of energy independence, can force the United States to divert agriculture to the production of biofuels. Farmers will be forced to become "miners of soil" and soon farming will become a supplier to industrial by-products such as plastics and pharmaceuticals. Public health risks are seen in the rise of obesity due to the "supersize" effect. Furthermore, public health risks also arise due to the growing number of antibiotic resistant strains of disease that are caused by genetically modified feed with growth hormones and centralized confined feeding operations. Finally, the loss of the regenerative base of agriculture due to monocultural and a confinement production practice has reduced the diversity in varieties of crop and livestock.

What are some of the coping mechanisms and strategies available to the policy makers in the United States, in the light of the risks and threats to food security? Currently, there is need in the United States to move toward adaptive strategies in agriculture such as decentralization, diversification, and democratization. For example, removing oversized dams, and replacing them with small and midsized ones that suit the topography and climatic conditions, building of new canals, networks of ditches, water conservation, and flexible distribution of food to establish ecological and social systems are important steps in reducing the threats to food security. Subsidies to farmers should be scale- and commodity-neutral. Furthermore, infrastructure reforms must accompany legal reforms in terms of corporate reforms that reduce the oligopoly power in agriculture and food retail. Legal reforms must also make livestock owners responsible for health and environmental problems. A larger emphasis should be placed on rebuilding livestock infrastructures in milk processors, abattoirs, distributive centers, etc. There are many other strategies in terms of diversifying food systems including strengthening enforcement of air, water, wetlands, forest, and endangered species. Benbrook (2003) has also proposed that federal farm program payments be based on nitrogen uptake and diversity rotations. Current biofuels policy has concentrated on mining the soil for energy and has redefined agriculture within the framework of national security. This has reduced the cropland for food and has increased food prices. Rather, policy should be focused toward sustainability with a focus on building local self-reliance, than on competition for cropland for food production. The Chinese experience with their biofuels policy provides an interesting insight into this problem.

Biofuels—the Chinese experience

Dong (2007) has provided an interesting case study that traces the origins and the results of biofuel production in China. Dong (2007) analyzes the background, history, and current situation of biofuels development in China. The Chinese experience with biofuel development provides some key policy implications for developing countries. As China is a developing country with rapid economic growth, population growth, significant demand for fuels, and food security concerns, it serves as a good example for studying the opportunities and challenges faced by developing countries.

As mentioned in the previous section, biofuels production is expanding rapidly all over the world, driven by rising crude oil prices, the desire of countries to be energy independent, and concerns about climate change. As developed countries, especially the United States, expand biofuels production, developing countries have also started to expand their biofuels industries as well, to power their growing economies. However, developing countries, unlike food-rich nations, must address the food security issue when they develop biofuels.

Biofuels are usually proclaimed as the alternative fuel of the future, offering cleaner energy and new opportunities for farmers in developing countries. The large and rapid expansion of ethanol worldwide has affected virtually every aspect of food markets and prices in both domestic and international markets, ranging from the allocation of acreage among crops to exports and imports. As more food grains are used to produce biofuels, food grain carryover stocks will remain tight, and average grain prices will increase. Moreover, these price increases also increase the feed cost for livestock.

The same problems are also present in China, which is currently the third-largest producer and consumer of ethanol, behind Brazil and the United States. However, unlike Brazil, the United States, and some other developing countries, China's population is the largest population in the world. With limited arable land, food security is a major issue in China. Furthermore, the Chinese government does not encourage imports of crops. Rather, government policy supports food self-sufficiency for the sake of national security. Finally, China's incentives for developing fuel ethanol production include not only those common to other countries, such as rising oil prices, energy independency, and environmental issues, but also some unique concerns, which are listed in the following.

For the sake of national security, China's government advocates food self-sufficiency and has a national grain reserve system. However, in recent years, the grain reserves have exceeded expectations, due to bountiful harvests. Unfortunately, this has created two problems. The first is the massive increase in the government's administrative and maintenance costs of the excess reserve, thereby creating an unnecessary fiscal burden on the budget. The second problem is the consequent fall in grain prices, which has in turn produced a big income gap between urban and rural residents. The Chinese government has been forced to find a way to stabilize grain prices and increase farmers' incomes. Consequently, promotion of the fuel ethanol industry has been a good option for the use of the excess grains, for increased grain consumption and farmers' welfare.

Alongside excess supply, the increase in population, income levels, and economic growth has also forced China to import a lot of oil, which in the face of increasing energy prices has created a balance-of-payments crisis for the economy. Furthermore, emissions from automobiles and related activities have deteriorated the air quality in China considerably. Consequently, the Chinese government has seen fit to develop alternative energy sources, particularly biofuels, to deal with excessive imports and the environment.

As a result of the aforementioned efforts, China currently is the third-largest ethanol producer and consumer, following Brazil and the United States. However, with the expansion of biofuels, we have serious issues with food security in China. First, corn and wheat are the major inputs needed to produce fuel ethanol. As the demand for corn increased, the price of corn has risen significantly. Higher corn prices have changed the allocation of acreage among crops and have simultaneously increased the price of pork, since the main feed for hogs is corn. With higher food prices, the net incomes of most families have fallen, with the poor in the rural areas being the worst affected.

Surging food prices and related concerns have forced the government to institute a wide range of policies that now prevent biofuels production in arable lands. Furthermore, the government has also suspended all newer proposals for ethanol production and has decided that food security is more important than energy.

However, since biofuels development cannot be completely abandoned, the government has shifted its strategies from using main staple grains to using other nongrain crops or crop by-product sources. Crops currently receiving the most attention include cassava, sugarcane, sweet sorghum, and sweet potatoes. Even though these sources have been identified, several obstacles still remain:

1. Policies on ethanol production should be reformed to support fuel ethanol production from nongrain feedstock.
2. There are many difficulties in the collection and transportation of the feedstock from the field to the ethanol plant. Large investments in infrastructure and delivery mechanisms are necessary.

3. There are many logistical problems associated with the storage of nongrain feedstock, which makes the whole option infeasible.
4. The yield from Chinese nongrain feedstock is very low, which restricts the supply and the potential capacity of ethanol plants. Better seeds and technology are needed to increase the yields of nongrain feedstock for any future expansion of ethanol production.
5. Most crucially, biofuel production requires intensive use of natural resources particularly water. Besides affecting the location of biofuel plants, the production of biofuels is severely limited, given the acute water shortage in China.

What are the lessons one can learn from the Chinese experience? Although China has some unique characteristics, its experiences and lessons in biofuels development are still instructive for developing countries. Firstly, developing countries should seriously reevaluate the relationship between food security and biofuels expansion. These countries should develop an analytical framework that takes into account their local conditions and their long-term needs. Secondly, policy makers must ensure that biofuels development does not affect poor households, which are vulnerable to rising food prices. Thirdly, government support in the form of subsidies, tax benefits, infrastructure development, and related facilities is needed. Fourth, government should also allocate research funds for development of second-generation technologies based on feedstock such as grasses, wood, crop and forest residues, and municipal wastes. If these inputs become economically viable, then they could reduce the demand for food and feed crops for the production of biofuels and mitigate the competition of food and energy.

Policy makers must also initiate incentives to grow crops on degraded or barren land not suitable for food production. The expansion of biofuels production could contribute significantly to higher incomes for farmers through higher feedstock prices and new employment opportunities.

US Farm Policy and food security—background and current issues

The Food Security Act of 1985 has received a lot of attention among researchers and can serve as a good background toward evaluating the current situation in US Farm Policy. Erdman and Runge (1990) examine the Act and its 1990 version very comprehensively. Erdman and Runge (1990) point out that the goal of the Act under the Reagan administration was to reduce the role of the state and increase the market participation in agriculture. The "marketing loan," which let farmers repay their loans at a rate below the loan rate when world prices fall, is an example of a program under the Act.

The Export Enhancement Program is another example, which allowed farmers to export to specified countries at prices below those of the domestic market. The sod-swamp-buster programs were also established under the 1985 bill.

The programs were set up to discourage conversion of highly erodible land and wetlands. Although these programs were intended to improve the environment, reduce farm support costs, and increase farmers' incomes, Erdman and Runge (1990) conclude that the 1985 bill was not very effective in achieving many of these goals. The environmental regulations for instance were often misguided and applied ineffectively. The penalties following violations were not strictly enforced and were not properly set up to create incentives to improve the environment. There was substantial misallocation of costs associated with the marketing loan program, and the EEP program has been criticized for not being effective in terms of its destructive impact on world markets, with improving farmers' incomes. Over the years, the EEP program cost the taxpayers a lot, without augmenting exports or farm income, with the real victims being Australia, Canada, and other grain exporters.

Consequently, in addition to the aforementioned concerns, the 1990 version of the farm bill had to reckon with the huge budget deficits, which forced reduction of farm supports, and develop a consensus surrounding the "flexibility debate."[3] Basically the flexibility debate had to resolve the question, "how low can farm supports go without impinging too much on consumer prices." Obviously, a whole range of proposals dealing with this wedge came from different interest groups, and ultimately a compromise called the "triple-base" emerged, in which the government and farmers agree to leave some land idle, so as to afford reasonable supports. The 1990 farm bill also had to deal with the environmental movement of the 1990s, with particular emphasis on chemicals, fertilizers, and soil conservation practices. The Uruguay Round and the GATT negotiations were also key drivers in shaping the 1990 provisions. Overall, the 1990 provisions did make progress toward cost reductions, environmental standards and enforcement, establishment of the wetlands program, water quality, and integrated farm management. However, by the 1990s, it was clear that US Farm Policy continued to wrestle issues with the budget deficit and accumulated debt, environmental concerns, farming flexibility, and multilateral trade negotiations.

Sumner (2003) summarizes the effects of the US Farm Bill of 2002, which was cast as the Farm Security and Rural Investment Act of 2002 (FSRIA). Government payments remained the primary focus of commodity programs in the 2002 Bill. Even around 2002, the government payments were massive toward growers of program crops. Along with budgetary issues, Sumner (2003) points out that the 2002 version had to deal seriously with international trade negotiations, including at the World Trade Organization (WTO). Because WTO provisions affect trade and because trade agreements also discipline US policy, the 2002 Bill affected the negotiating positions of the United States.

---

3. See Erdman and Runge (1990) for the details and policy dilemmas surrounding the "flexibility debate."

Moreover, some of the 2002 provisions were also seen as a conflict between US compliance with existing international agreements, with a potential to lead to international disputes. However, Sumner (2003) does indicate that the 2002 provisions were more consistent with market forces and trade liberalization. By 2002, US policy makers were resistant to lower domestic supports in exchange for additional markets or lower export subsidies. Furthermore, by this time, other countries perceived the United States as a major source of distortion in the world markets. Developing countries were also threatening to place import barriers and production-distorting policies. For example, the large US farm subsidies were perceived to discourage China from enforcing its commitments to the WTO. Similarly, the European Union also decided to become lax with its compliance to international agreements. Hence, current US Farm policies continue to evolve with the same threats that have plagued the economy since the 1980s.[4]

Section highlights: the information age and precision agriculture

How is it possible to find out the optimal application of fertilizers for a crop in a field? At one point of time, the application of fertilizers was based on obtaining enough nutrients to match the requirements of a particular crop. However, this uniform management of fertilizers becomes problematic when fertilizer prices start increasing, as they have done in the past decade. A huge amount of farming budget, or about $ 120 billion, goes into the purchase of fertilizers worldwide. Obviously, it would be very useful in terms of savings to know how much fertilizers to apply based on information on soil conditions, nutrient needs, pest problems, and weather.

Such information can also address the environmental concerns from too much fertilizer use. In particular, poorly timed fertilizer application produces substantial runoffs into wells, waterways, wetlands, and estuaries. For example, during heavy rainfall, the nutrient delivery to the Gulf of Mexico enlarges the size of the hypoxic "dead zone" at the mouth of the Mississippi River. In 2009, the delivery of nutrients from agricultural runoffs to the Gulf was the highest on record. The Intergovernmental Panel on Climate Change has focused a lot of attention on reducing N_2O emissions through improved agricultural fertilizer application techniques.

Consequently, it will be very useful to incorporate technological advances to assist farmers and government policy makers to assist and guide with important information, so as to realize the cost savings and reduce greenhouse gasses. This is where precision agriculture comes in.

Precision agriculture—a suite of information technologies used as management tools in agricultural production—has recently become more accessible to

---

4. See Atkins (2004) for the negative effects of the 2002 Farm Bill provisions of $180 billion supports to US farmers on the Caribbean Community exports. For an Australian perspective, see Roberts and Jotzo (2002). For the positive effects of price stabilizing instruments, and the deficiency payment support programs under the 2002 provisions, see Ingersent (2003).

> **Section highlights: the information age and precision agriculture—cont'd**
>
> farmers. The adoption of precision agriculture can improve the efficiency of input use and reduce environmental harm from the overapplication of inputs such as fertilizers and pesticides.
>
> Schimmelpfennig and Ebel (2011) have documented the use of precision agriculture for major crops in the United States and find that efficient input use in agriculture has increasingly become a priority for producers, the public, and policy makers. One way to increase efficiency in agriculture is through the adoption of precision technologies, which use information gathered during field operations, from planting to harvest, to calibrate the application of inputs and economize on fuel use.
>
> Adoption of precision agriculture holds promise for improving the efficiency of input use. Schimmelpfennig and Ebel (2011) have documented the use of precision agriculture using yield monitors, variable-rate application technologies, and guidance systems with GPS.
>
> Yield monitoring is the most popular component of precision agriculture, with soybean producers, corn producers, and winter wheat growers. Yield monitors have allowed farmers to use global positioning system (GPS) maps to pinpoint yield variation within their fields. Yield monitoring is sometimes used to store plot-level information to roughly monitor crop moisture so as to minimize the cost of drying harvested grain. The technology also allows farmers to see if plot-level yields would benefit from a change in management practices within a single field.
>
> Variable-rate technology (VRT) allows producers to make use of factors that influence yields by adapting their practices "on the fly." VRTs are seeders, sprayers, and other fertilizer and pesticide application equipment that can be continually adjusted during field operations to optimize the application of inputs depending on field conditions. VRT use is prevalent among producers of corn, soybeans, and winter wheat.
>
> Guidance systems for field equipment make use of GPS readings to alert equipment operators as to their field position coordinates. GPS can improve the accuracy of variable-rate applicators and help operators reduce the incidence of overlapping or missed sections in their field operations.
>
> Schimmelpfennig and Ebel (2011) also indicate how precision agriculture can be put into practice. They identify a three-step approach to operationalize the concept:
> - In the first step, yield monitors provide and electronically store all the data on crop yields including bushels per acre.
> - In the second step, the data from the first step are integrated with the information on site-specific farming through soil maps and soils data, including soil properties, soil testing, organic matter, pH levels, topology, catchment area, stream power index, watershed boundary, cropland layers, temperature averages, precipitation averages, and nutrient availability. This information can be gathered from many data libraries such as USDA Natural Resources Conservation Service's National Cartography and Geospatial Center, National Soil Information System, and the Agricultural Resource Management Survey.

Continued

> **Section highlights: the information age and precision agriculture—cont'd**
>
> - In the third step, all the information is combined with decisions regarding investment in field equipment and machinery. This step enables the farmers to continuously change the rates of seeding, fertilizer, and pesticide applications.
>
>   The research by Schimmelpfennig and Ebel (2011) tells us that yield monitors are most popular among corn and soybean farmers practicing conservation tillage. Interestingly, they find that average fuel expenses per acre, for both corn and soybean farmers, are lower for farmers who used yield monitors and VRTs. They also indicate that adopters of yield monitors and VRTs and GPS had higher corn and soybean yields than nonadopters.
>
>   Although precision agriculture has obvious advantages, its adoption has been mixed among US farmers. Some of the reasons are farm operator education, technical sophistication, and farm management acumen. Schimmelpfennig and Ebel (2011) conclude that precision agriculture will be more popular in the future, as technologies become less expensive to buy and maintain, and as conservation tillage becomes widespread. Most importantly, relative prices of fuel and fertilizers may also be primary drivers of precision agriculture.

As the most recent Farm Bill of 2012 continues to evolve under this background, heated debates have started, with similar issues: support farm programs to provide a sufficient farm safety net versus reduction of federal spending. Paulson and Schnitkey (2011) point out that there are serious concerns raised by the Midwestern Grain Producers regarding certain provisions in the 2012 bill dealing with reduction in crop insurance reductions. Policy concerns of midwestern producers directly reflect the region's heavy reliance on corn, soybeans, and wheat production. Crop insurance plays a significant role in providing a safety net, with revenue insurance products at high coverage levels being the predominant crop insurance choice. Midwestern farm groups have championed an increasing emphasis on providing a revenue-based safety net. Policy makers in the United States are currently generating policies that tie direct payment program with ACRE and crop insurance programs. Policy reform is currently under way to integrate the ACRE program with insurance policies to derive the best potential from reduced spending with sufficient protection of risks. The 2012 Farm Policy also has ethanol and biofuels policy that are in line with Renewable Fuel Standards. Thus, US Farm Policy in recent years has evolved through issues surrounding budget constraints, international trade agreements, environment, and energy security.

GEO-5 and coping mechanisms for the future

UNEP (2013) provides a comprehensive report that details the impact of environmental changes on the corporate sector. The report also provides strategies and policies that the corporate sector must undertake to remain competitive and successful. Many of the policies concern food security and sustainability.

The UNEP (2013) report for the GEO-5 indicates that environmental trends indicate increases in greenhouse gases, severe weather patterns, lack of available land, water, and biodiversity, and an increase in water pollution, chemical exposure, and waste. All these issues affect food availability. The report provides implications for business and suggests several strategies to shift production and distribution, technical change, and market opportunities to combat these difficulties.

Empirical analysis—a basic univariate approach

As we have seen from the aforementioned case studies, policy makers and program managers are keen to know the food security impact of the adoption of new technology. Thus, the main question frequently asked is as follows: Do technology adopters and nonadopters have different levels of food security? To answer this question in a comprehensive manner, one needs to know information on household characteristics, such as age and sex, household income, and expenditure patterns on food and nonfood items and food intakes by the members of the family. The data can be collected in a panel form where the same households are surveyed over time (e.g., before and after technology adoption) or can be gathered from a cross section of households for a single time period from technology adopters and nonadopters.

In this section, we introduce a statistical technique to study the difference in the food security status of technology adopters and nonadopters. An independent sample t-test is undertaken to answer the aforementioned question. The key objective of this test is to determine the statistical significance of the observed differences in food security across the two groups, namely technology adopters versus nonadopters. The t-test computes sample means for each of two subgroups of observations and tests the hypothesis that the population means are the same between the subgroups. Thus, the hypothesis being tested is that, on average, adopters and nonadopters of new technology have the same level of food security.

In our analysis, the t-statistic for testing the aforementioned hypothesis (equality of means) is calculated under two different assumptions—equal and unequal variances. These two assumptions imply that either the food security of adopters will vary the same way within themselves as that of nonadopters or adopters and nonadopters will have different within group variances.

In the rest of this section, we use the data set from Malawi to demonstrate the following:

1. Data description and analysis
2. Descriptive statistics

3. Threshold of food insecurity by each individual component
4. Tests for equality of variances
5. *t*-test

Data description and analysis

In the first stage, we chose 604 households from regions Mzuzu, Salima, and Ngabu out of 5069 households based on whether the household had at least one child as member below the age of 5 (Fig. 2.1). These regions were chosen, since detailed data on food consumption patterns for the household and nutritional status of the children are available. Furthermore, they represent varied agroecological zones, cropping and livestock rearing patterns, consumption patterns, and geographical (northern, central and lakeshore and southern) locations within the country. Out of the 604 households, 197 had information on 304 children (below the age of 5) related to nutritional status and general health conditions. All the households had information pertaining to food intake, quantity harvested for various crops, and other socioeconomic information. The aforesaid sampling strategy was adopted to understand which households (who had at least one child below the age of 5) suffer from a nutrition insecurity problem. These household level data are rich in content, since they contain not only information on household characteristics such as age, education, and sex of the household head, but also expenditure on and share of different food and nonfood items consumed. Additionally, the data also contain information on the number of meals consumed by the household

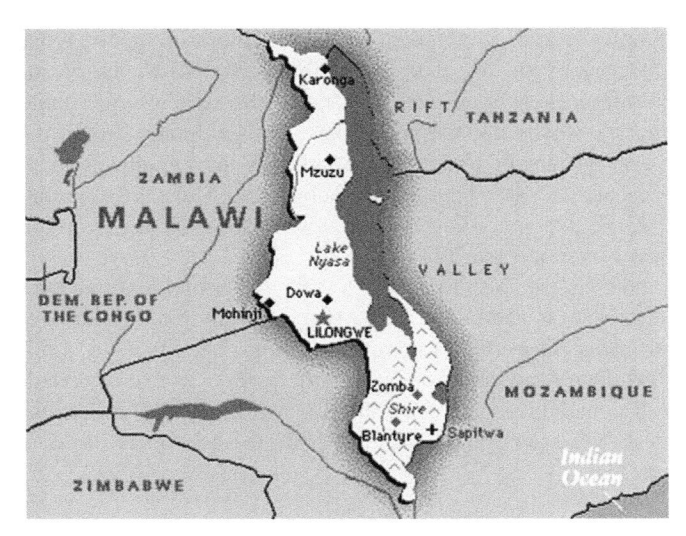

FIGURE 2.1 Map of Malawi. *From CIA, The World Factbook, 2004, https://commons.wikimedia.org/wiki/File:Malawi-CIA_WFB_Map.png.*

on a daily basis (this variable in combination with other variables is used as an indicator of food security) and the time after harvest when the household stock of food runs out.

In the second stage, we sort the data by add-code (agricultural district), epacode (village cluster), household number, and enumerator number. The add-code signifies the main region, whereas the epacode denotes a subregion within the main region. This is done to determine if there are regional variations (for example, between the northern and southern regions) in food security between technology adopters and nonadopters. The main variables used in the analysis are as follows:

HYBRID: This variable denotes technology adoption by a household and assumes two values: adoption of hybrid maize (a value of 1) and nonadoption (a value of 0).

Two measures of household food security are computed

1. The first measure of "food security" is a combination of household dependency ratio and the number of meals that a household consumes. It is thus an interaction variable on the lines of Haddad et al. (1994). The basic idea is that if a household has a large number of dependents (as measured by variable Depratio, the ratio of dependents to total household members, being above 0.5) and consumes less number of meals (NBR), say below three per day, then the household is relatively more food insecure (a value of 3). On the other hand, if the household has a less number of dependents and consumes more meals, then the household is relatively more food secure (a value of 0). This variable is coined as INSECURE and is computed on a 0–3 scale to indicate the different degrees of food security. Higher values of this indicator will thus denote that the household is more food insecure. This is computed as follows:

$$\begin{aligned} &\text{If Depratio} \geq 0.5 \text{ and NBR} \leq 2 \text{ then INSECURE} = 3 \\ &\text{If Depratio} < 0.5 \text{ and NBR} \leq 2 \text{ then INSECURE} = 2 \\ &\text{If Depratio} \geq 0.5 \text{ and NBR} > 2 \text{ then INSECURE} = 1 \\ &\text{If Depratio} < 0.5 \text{ and NBR} > 2 \text{ then INSECURE} = 0 \end{aligned} \quad (2.1)$$

2. The second measure of food security combines the income and consumption components into an overall index of food security. The "income component" is determined by total livestock ownership (LIVSTOCKSCALE) and measured in tropical livestock units (TLUs). This is an equivalence scale based on an animal's average biomass consumption. Livestock ownership can be a critical component of food security if assets ensure that household

consumption does not fall below a critical level even when incomes are insufficient. During periods of hardship, these assets can be sold to fill in income gaps and, hence, this indicator can serve as a proxy for income levels. The relevant conversion for the types of livestock reported in the questionnaire is given in Table 2.1.

Market prices come as a natural candidate for aggregating livestock resources. However, for a poor country such as Malawi, livestock prices may be highly seasonal and extremely variable from year to year. Additionally, spatial integration of markets is poor in remote regions, and thus, under these conditions, even accurate price data are of unreliable quality. Thus, the biophysical scale of TLU is used in aggregating the value of livestock and is consistent with physical measures applied to other assets (such as land for example). For scaling this indicator, we use the minimum value of 0 (for no animals owned), while the maximum is truncated at six TLU (i.e., all households with six or more have TLUs score 1 on this indicator). Like other assets, livestock ownership is a continuous variable with many possible values.

Table 2.2 provides an example of the possible values of livestock owned and their scaled equivalents. The scaling is done so that different indicators can be combined on a scale from 0 to 1. For example, food shortage in months and total livestock units measured in TLU units can be compared and combined. The basic formula for scaling a variable is given by this, so if the actual value of a variable is 5, the minimum value is 0, and the maximum value is 6, then using the aforementioned formula, we obtain the scaled or normalized value of the variable to be 0.83. This measure of food security can be used across time and space (e.g., between years and between different regions if data are available). Since this variable is continuous, a large number of intermediate values are feasible. The scaled values for livestock owned are given in Table 2.2 using the aforementioned approach.

TABLE 2.1 Tropical livestock unit values for different animals.

Animal type	TLU value
Cattle	0.8
Goat	0.1
Sheep	0.1
Pigs	0.2
Chicken, ducks, and doves	0.01

Courtesy: International Livestock Research Institute, 1999. ILRI (International Livestock Research Institute), Nairobi, Kenya.

TABLE 2.2 Scaled values for livestock owned.

Data value of livestock units (TLUs)	Scaled value
6+	1.00
5	0.83
4	0.67
3	0.5
2	0.33
1	0.17
0	0.00

Consumption components of the food security index

The "consumption components" of the food security index are number of meals (NBR) that the household consumes during a given day and the months when the stock of food runs out (RUNDUM). The number of meals that the household consumes is a strong indicator of household strategies to cope with short-run food insecurity and thus may be a potential measure of vulnerability. However, this measure is less sensitive to changes in situations of chronic food insecurity.

The second component provides an estimate of the number of months per year that households are able to meet their food needs, through either production or purchase. Months of adequate food provision is a relatively simple indicator and, when used in conjunction with the number of meals that the household consumes, provides a clear picture of the vulnerability of the household. Thus, these variables along with the income indicator such as livestock owned could show why particular households are vulnerable and provide a measure of the severity of vulnerability. The number of months that households are unable to eat enough to satisfy their hunger can be used as an indicator of vulnerability. We measure the number of meals on a 0–3 scale, with 0 denoting that the household went sometimes without eating in a day, while the maximum value of 3 denoting that the household is food secure (Table 2.3).

Months of adequate provisions or stock of food running out (RUNDUM) was also measured on a 0–3 scale, with the truncation being at the minimum value of 0. In other words, if the stock of food runs out before the enumerating period, the variable assumes a value of 0. For stock of food lasting between 1 and less than 3 months, the variable attains a value of 1. For stock of food lasting between 3 and less than 6 months, the variable assumes a value of 2 while, and for the food stock lasting more than 6 months, the variable attains

TABLE 2.3 Scaled values for number of meals per day.

Number of meals per day	Scaled value
3	1
2	0.67
1	0.33
0	0.00

the maximum value of 3. Thus, a higher value of this component indicates that the household is relatively more food secure. Finally, the overall food security index is a weighted average of these three components, namely (1) the number of livestock owned (LIVSTOCKSCALE), (2) the number of meals consumed per day (NBR), and (3) stocks of food running out (RUNDUM). The weights are chosen in proportion to the variance of each component. Thus, if one component has a higher variance relative to another component, it gets a lower weight in the food security index (FOODSEC) (explained in note 3).[5] The rationale for choosing the weights in such a way is to ensure that none of the components dominates the overall "food security index." The overall index is thus computed as where w_i are the respective weights and c_i are the three individual components.

$$\text{FOODSEC} = \sum_{i=1}^{3} w_i c_i \qquad (2.2)$$

For example, with $w_1 = 0.1$, $w_2 = 0.4$, and $w_3 = 0.5$ and with values of NBR $= 0.33$, RUNDUM $= 0.33$, and LIVSTOCKSCALE $= 0.67$, the value of the food security index would be 0.5. The aforementioned measure of food security is a continuous variable with higher values of index denoting food secure households. The index ranges between 0 and 1 with 0 denoting completely food-insecure households and 1 denoting fully food-secure households.

A household's score of this index has no easy explanation by itself, since it is a mathematical composite of three different factors. Following a common procedure adopted by Filmer and Pritchett (1999), by categorizing the bottom 40% of the population as food insecure, the cutoff value for this index was determined at 0.3. It should be emphasized that the aforementioned measure of

---

5. In other words, the costs and benefits add up to zero. The first step in measuring dietary diversity is to collect information on local consumption patterns to identify a diet that signifies food security. When this indicator is used as a measure of food security, foods are grouped not by economic value but by nutritional composition.

food security does not capture dietary diversity[6] or household's perception of food needs such as the food security measurement scale developed in the United States (Staatz et al., 1990).

Descriptive statistics

The independent variable HYBRID is nominal dichotomous variable and FOODSEC; the measure of food security is a continuous variable. From the analysis, the food security measure is given by the following:

$$\text{FOODSEC} = 0.2798 \text{ NBR} + 0.4821 \text{ RUNDUM} + 0.2381 \text{ LIVSTOCKSALE} \tag{2.3}$$

where 0.2798, 0.4821, and 0.2381 are, respectively, the variance[7] of the components NBR, RUNDUM, and LIVSTOCKSCALE.

The two groups—hybrid maize adopters and nonadopters—are compared in reference to food security. Let us first compare the basic descriptive statistics of food security for adopters and nonadopters (Table 2.4).

From Table 2.4, it is evident that the hybrid maize adopters have a higher mean for food security compared to nonadopters. This suggests that adoption of new technology improves food security given by Eq. (2.2). Next, we want to investigate whether these differences of mean and variance are statistically significant. In other words, we want to determine if the differences among the sample of technology adopters and nonadopters on food security are relevant for the population too.

TABLE 2.4 Group distribution of FOODSEC.

	Hybrid maize adoption	N	Mean	Standard deviation	Standard error mean
FOODSEC	Nonadopters	131	0.3439	0.144	0.01261
	Adopters	43	0.3970	0.152	0.02318

---

6. The first step in measuring dietary diversity is to collect information on local consumption patterns to identify a diet that signifies food security. When this indicator is used as a measure of food security, foods are grouped not by economic value but by nutritional composition.
7. The individual weights were computed from the variances as follows: suppose, that for any three series, X, Y, and Z the variances are $V(X)$, $V(Y)$, and $V(Z)$, with $V(X) > V(Z)$. Then, the weight of X is given by si4.gif, while the weight of Z is given by si5.gif. In other words, we give a higher weight to the series with a lower variance and a lower weight to the series with a higher variance. This procedure ensures that none of the series dominates the overall index.

Threshold of food security by each individual component

As mentioned before, the problem with a continuous indicator of food insecurity (FOODSEC) is that it does not contain rules or information to identify the food-insecure households from the rest. To fully understand the households that are food insecure in each of the aforementioned components (namely livestock ownership, number of meals consumed per day, and the month when the stock of food runs out), it is important to determine the cutoff point for each of the aforementioned components. This procedure is explained in the technical appendix of this chapter.

This is achieved by looking at the cumulative distribution of each component and identifying the cutoff point. Table 2.5 provides the cumulative distribution of households by the cutoff points for the individual components of food security.

Table 2.5 provides some additional interesting insights into the nature of food insecurity in the regions. From the number of meals consumed per day set at a threshold level of 1 or below, we find that about 13% of the population is food insecure. However, looking at the variable when food stock runs out (choosing the threshold value of 0.33, i.e., for RUNDUM <0.33), we find that almost 70% of the population is food insecure. Additionally, from the number of livestock owned as a measure of asset owned, we find that almost 75% of the population does not own any livestock. As the UN points out, "households with limited assets are vulnerable, not only because of their relative poverty, but also because they have few items to divest should they be forced to spend money on food or emergencies." (See, for example, the document of the Southern African Regional Poverty Network at http://www.sarpn.org.za/documents/d0000522/exec.php.)

Tests for equality of variances

A basic assumption underlying the use of parametric tests such as Student's t-test and analysis of variance (ANOVA) is that the variance of variables under study must be roughly equal for multiple samples (groups). Levene's test is used to test the null hypothesis that multiple population variances (corresponding to multiple samples or groups) are equal. The importance of the

TABLE 2.5 Threshold of food security components.

Indicator	Cutoff point	Cumulative percentage
NBR	0.33	13.4
RUNDUM	0.33	69.8
LIVSTOCKSCALE	0.16	74.7

assumption of equal variances in all groups depends on whether the groups have roughly equal sample sizes. If they do, violation of the equality of variance assumption has little effect on the observed significance levels. If the groups have very different sample sizes and the smaller groups have larger variances, then the chance of rejecting the null hypothesis when it is actually true increases (probability of type I error). On the other hand, if the larger groups have smaller variances, the chances of not rejecting the null hypothesis when it is actually false increase (probability of type II error).

Levene's test follows an F-distribution. The F ratio test of variance equality is computed by the ratio of the largest to the smallest variance. Assuming a normally distributed population, F follows the following distribution: $F\ (n_{max} - 1, n_{min} - 1)$, where n_{max} is the sample size of the group with the larger variance and n_{min} is the sample size of the group with the smaller variance. If the actual F-value is greater than the critical value, then the variance of the relevant variable is significantly different across groups. If the F-value is nonsignificant, then the variances do not differ statistically. If Levene's test produces a nonsignificant result (for example, a P-value greater than 0.05), then the t-test that assumes equal variances must be used. However, if Levene's test produces a significant result (e.g., a P-value less than 0.05), then the t-test that does not assume equal variances must be used. In the latter case, the computation includes a correction for the lack of homogeneity of variance (Table 2.6). When the variances are homogeneous, the standard error of the difference is computed by summing the standard deviations and dividing by the square root of the sum of the number of observations in each group. This is also called the standard error of the difference with pooled variance estimates. This is given by Eq. (2.5).

The hypothesis of equal variance of the interaction measures of food insecurity (INSECURE) and the weighted average of the income and consumption components of food security (FOODSEC) across the two groups, namely the hybrid maize growers and nongrowers, is not rejected at the 1% or 5% level of significance.

Student's *t*-test for testing the equality of means

We first introduce a few notations to show how the t-test is undertaken for testing difference of means of variables. Let μ_1 and μ_2 and σ_1 and σ_2 denote

TABLE 2.6 Levene's test of equality of variances.

Variables	F-statistic	P-value
INSECURE	0.566	0.452
FOODSEC	0.174	0.677

the means and standard deviations of technology adopters and nonadopters, respectively. These are the population parameters corresponding to the sample statistics. We want to test the hypothesis that

$$H_0: \mu_1 - \mu_2 = 0$$
$$H_1: \mu_1 - \mu_2 \neq 0 \quad (2.4)$$

In other words, the null hypothesis (H_0) asserts that the population parameters are equal. The statistic $(\overline{X}_1 - \overline{X}_2)$ is the difference between the sample means. If $(\overline{X}_1 - \overline{X}_2)$ differs significantly from zero, we will reject the null hypothesis and conclude that the population parameters are indeed different. Since the two random samples are independent, i.e., probabilities of selection of the elements in one sample are not affected by the selection of the other sample, we want to look at the sampling distribution of the variable $(\overline{X}_1 - \overline{X}_2)$. The standard error of the difference between the two means is given as follows:

$$S_{\overline{X}_1 - \overline{X}_2} = \sqrt{\left(\frac{s^2_{pooled}}{n_1} + \frac{s^2_{pooled}}{n_2}\right)} \quad (2.5)$$

where s^2_{pooled} = pooled variance estimate = $\frac{(n_1-1)s_1^2 + (n_2-1)s_2^2}{n_1+n_2-2}$. s_1^2 and s_2^2 are the estimates of the within group variability of the first and second groups, respectively. The t-test statistic is thus given by

$$t = \frac{\left(\overline{X}_1 - \overline{X}_2\right) - (\mu_1 - \mu_2)}{S_{\overline{X}_1 - \overline{X}_2}} \quad (2.6)$$

We undertake the t-test of equality of means assuming equal variances. However, we present the results assuming both equal and unequal variances across the technology adopters and nonadopters.

An independent sample t-test is used since different subjects (technology adopters and nonadopters) have been used in each condition. As explained before, Student's t-test for independent samples is used to determine whether the two samples were drawn from populations with different means. The t-test provides a t-statistic along with a corresponding level of significance. If the actual t-value exceeds the critical value at the 1% or 5% level of significance, we reject the null hypothesis of equal means of food security (as measured by either INSECURE or FOODSEC) and conclude that technology adopters and nonadopters have differing levels of food security. On the other hand, if the computed t-value is less than the critical value at the relevant level of significance (1% or 5%), then we do not reject the null hypothesis and conclude that technology adopters (maize growers) and nonadopters (maize nongrowers) have the same level of food security (Table 2.7).

TABLE 2.7 Student's *t*-test for equality of means.

Variables	Assumptions	*t*-Statistic	Attained significance (2-tailed)
INSECURE	Equal variance assumed	2.33	0.02
	Equal variance not assumed	2.363	0.019
FOODSEC	Equal variance assumed	−2.064	0.04
	Equal variance not assumed	−2.011	0.04

The validity of the *t*-test requires that the samples be drawn from normally distributed populations with equal (population) standard deviations. Levene's test can be used to test the hypothesis of equal variances across populations. If both samples are large, then the assumption of equal or unequal variance is not very relevant since the difference in results due to either becomes negligible.

It is important to note that if X_{ij} denotes the original score on a response variable X for case i in group j, Levene's test undertakes a minor transformation of the original scores as $X_{ij} = \left| X_{ij} - \overline{X}_j \right|$ before undertaking the hypothesis test of equal group variances. However, as pointed out earlier, Levene's test is less robust. This is because if the groups have very different sample sizes and the smaller groups have larger variances, then the chance of rejecting the null hypothesis when it is true increases. In other words, the probability of type I error increases. Under such circumstances (when the variances of the two groups are not equal), it is better to improve the robustness of equality of variance test by applying the Brown–Forsythe test. The Brown–Forsythe test is identical to Levene's test but uses group *j*'s sample median in transforming the original data.

From Table 2.7, considering the continuous measure of food security (FOODSEC), using Eqs. (2.5) and (2.6), respectively, the mean difference is −0.05307 and the mean standard error difference $(S_{\overline{X}_1 - \overline{X}_2})$ is 0.02571. Thus, the computed *t*-statistic is given by −2.064. For 172 degrees of freedom, the critical *t*-value that corresponds to an area of 0.05 (*P*-value) in both tails is 1.96. Since the actual value is greater than the critical value, the null hypothesis of equal means of food security for technology adopters and nonadopters is rejected. Thus, food security between these two groups is statistically different, and we conclude that hybrid maize adopters are better off in terms of food security. It is important to note here that the degrees of freedom are substantially reduced when the analysis is undertaken with the weighted average measure of food security (FOODSEC). This is because livestock information is available only for 174 households rather than the total sample of 604 households. Since one of the components of the food security index consists of owning livestock, we have only $(131 + 43 - 2) = 172$ degrees of freedom.

Policy implications

The impact of technology adoption on food security and nutrition is still under debate. In many cases cited earlier, adoption of improved technology did not always increase calorie consumption (food security) and/or nutrition. But, in general, there is a consensus that technological change does result in positive production and income effects. Now, it is not always the case that such an income effect is translated into either an increased expenditure on foods giving higher energy intake or an increased expenditure on nonfood expenditures related to health and sanitation or both. Such an effect is desirable, since it leads to better food security and nutrition for members of the household.

In addition to increased income, we know from the earlier discussion that here are other determinants that significantly affect the transition from technology adoption to improved nutrition. They are intrahousehold labor allocations; equal access for both men and women to credit and technology training; and agricultural policies that improve credit markets, storage and processing facilities, transport infrastructure, research and training, and distribution channels.

From the program design and policy intervention perspective, policy makers and program managers need research-based evidence on the role of various factors, such as aforementioned, which hinders translation of benefits from technological change into food security and nutrition outcomes. Since much of these outcomes are context specific, the results of the analysis of technological change and adoption should be viewed in a broader policy context.

Reardon et al. (1997) summarize the crucial policy actions that need to be undertaken for a successful implementation of technology adoption:

1. Improvement in input access and reduction of unit costs to farmers through infrastructure investment.
2. Increase in productivity of fertilizer and seed variety by encouraging complementary farm-level investments.
3. Improvement in distribution and marketing channels of inputs and outputs accompanied by incentives for private sector investment.
4. Facilitating easy and equal access to credit to both men and women to buy inputs.
5. Introduction of innovative credit schemes to reduce the farmers' financial risks of investment in inputs.
6. Evaluation of net economic and social benefits of agricultural support programs, including input subsidies.

The empirical analysis presented from the Malawi data set is a univariate approach that assesses the role of technology adoption (in the form of growing hybrid maize) on food security. Two measures of food security were constructed: the first being an interaction measure of the combination of the number of meals consumed per day and the proportion of dependents in the household, while the second measure (a broader one) was a weighted average of the income and consumption indicators to understand food access.

Since the aforementioned analysis indicates that hybrid maize adopters are better off in terms of food security, policies to encourage technology adoption by growing hybrid maize should be implemented by the government to increase household food security of the rural population. However, the results of such an empirical analysis should be taken cautiously. Though such an exercise is an effective immediate approach, policy makers should recognize that the effects of technology adoption can increase overall household food consumption, thereby satisfying the calorie requirement for the vulnerable sections of the population.

However, it can have an ambiguous effect on the nutritional status of adults (especially women) and children. For example, an increased yield of maize could possibly result in increased consumption by the adopting household but could substitute for other purchased foods. This may result in decrease of dietary diversity, which could then lead to micronutrient deficiencies.

Interestingly, the analysis also found that the nature of food insecurity in Malawi may be chronic and long term in duration. This is because, although most of the households are not food insecure as measured by the number of meals consumed per day, they are asset poor and do not have adequate food provisions. Thus, long-term factors and policies encouraging technology adoption in improving food security must take into account income effects, women's welfare, input and output subsidies, equal access to credit, infrastructural development, efficient marketing and distribution channels, and research and training.

A study that was reviewed earlier, on the effects of hybrid maize adoption on food consumption and nutrition in Zambia by Kumar (1994), indicates many pros and cons of technology adoption and outlined different policy implications. The Malawi data used in this chapter come from a very similar agroecological region as the Eastern region of Zambia. Thus, the policy implications could be similar to that study. To encourage higher production of hybrid maize by rural households, the study finds it important to promote the development of maize not only as a cash crop but also for self-consumption. The difficulty involved in processing and storing hybrid maize hinders such promotion. Hence, governments must invest and provide easy credit for private investment in storage facilities and processing mills. The limited availability and poor distribution channels of hybrid seeds and fertilizers constrain the yield levels. To increase the efficiency of yields, governments must invest in infrastructure and transport and communications, in addition to improving marketing channels.

Technical appendices

Constructing the cutoff points for components of the food security index

As discussed in this chapter, the problem with a continuous indicator such as FOODSEC is that it does not contain any principle of isolating the food insecure households from the rest, since it is a mathematical composite of three different components. Thus, to estimate the number of food insecure households, it is necessary to determine a cutoff point between food-secure and food-insecure households. Since it is reasonable to assume that households having less than one meal per day (NBR ≤ 0.33) and the stock of food lasting less than 3 months (RUNDUM ≤ 0.33) are relatively more food insecure, we look at the frequency distribution of these variables and the cumulative distribution function of the number of households falling below these threshold values. Since the livestock ownership by the households (LIVSTOCKSCALE) is a continuous variable with many possible values, we first looked at the mean of this variable. The mean was found to be 0.164. Thus, the number of households, which fell below this threshold value, was considered to be food insecure. In other words, we wanted to determine the probability that LIVSTOCKSCALE ≤ 0.164. By looking at the cumulative distribution function of this variable, we found the percentage of households below this threshold value to be 74.7.

The advantage of this approach is that a researcher can look at the various components of food insecurity and isolate the components from each other. Thus, although the households may not be food insecure by the number of meals consumed per day, they may be asset poor and may not have adequate food provision to sustain themselves throughout the year. The nature of food insecurity can be long term and chronic and should be captured for policy analysis.

Variable definitions

Dichotomous variable

A variable that categorizes data into two groups is a dichotomous variable. Generally, such a variable is indicated by values 0 and 1 with each representing a different category. For example, sex is a dichotomous variable and can be defined as 1, if female and 0, if male. A dichotomous variable is also called a dummy variable or indicator variable. Most importantly, one must always understand that, in the case of a dichotomous variable, each of the two numbers (e.g., 0 and 1) represents a qualitative category and in no way can they be used as quantities.

In general, a variable that categorizes data into two or more groups is a *nominal* or *categorical* variable. In the present chapter, HYBRID is a dichotomous variable, while INSECURE is a nominal or categorical variable. Thus, a dichotomous variable is just a special case of a set of nominal or categorical variables.

Interval variable

These variables are also known as continuous variables. They are measured on a scale that changes values smoothly rather than in stepwise fashion. They can be ordered in ascending or descending fashion. For such a class of variables, distances between the values are known, and these distances have meanings. In the present chapter, FOODSEC is an example of a continuous variable.

Using STATA for *t*-tests

We motivate this section with an illustrative example using 200 observations. As mentioned in the previous section, the *t*-test is designed to compare means of the same variable between two groups. Assume we have information about FOODSEC for 200 farmers, some of whom are ADOPTERS of new hybrid varieties, while others are NON-ADOPTERS. We first use the summarize command in STATA and examine the descriptive statistics:

In our sample we have 200 observations on the status of technology adoption (0 = NON-ADOPTERS and 1 = ADOPTERS), with relevant index of food security (foodsec) for each observation. We can further obtain specific descriptive statistics for each status using the tabulate, summarize command in STATA:

Note in our hypothetical example we have rejected the null hypothesis, which does not match our results using real-world data from Malawi. Even though our STATA example is illustrated using made-up data, it serves as a good lesson, reminding us that in research, not all our results will always work out the way we expect.

Independent group *t*-test

As mentioned earlier, in this example, we compare the mean food security index between the group of adopters and nonadopters of hybrid seed varieties. Ideally, these subjects are randomly selected from a larger population of subjects. The test assumes that variances for the two populations are the same. We generate the results using the following command in STATA:

Note that in this example, STATA computes the t-value as follows:

$$t = \frac{\overline{x_1} - \overline{x_2}}{\left[\frac{(n_1-1)s_1^2 + (n_2-1)s_2^2}{n_1+n_2-2}\right]^{1/2} \left[\frac{1}{n_1} + \frac{1}{n_2}\right]^{1/2}}$$

$$= \frac{52.82 - 51.73}{\left[\frac{(91-1)10.5^2 + (109-1)10.05^2}{91+109-2}\right]^{1/2} \left[\frac{1}{91} + \frac{1}{109}\right]^{1/2}} = \frac{1.09}{1.456} = 0.748$$

For details, see https://www.stata.com/manuals13/rttest.pdf page 9.

In this example, the *t*-statistic is 0.748 with 198 degrees of freedom. The corresponding two-tailed *P*-value is 0.4553, which is less than 0.05. We conclude that the difference of means in food security between adopters and nonadopters of hybrid varieties is not different from 0. In other words, for 198 degrees of freedom, the critical *t*-value that corresponds to an area of 0.05 (*P*-value) in both tails is 1.96. Since the calculated value is smaller than the critical value, the null hypothesis of equal means of food security for technology adopters and nonadopters is not rejected. Thus, food security between these two groups is not statistically different, and we conclude that hybrid maize adopters are not better off in terms of food security.

Independent sample *t*-test assuming unequal variances

We again compare means of the same variable between the two groups. In our example, we compare the mean food security index between the group of adopters and nonadopters of hybrid seed varieties. Ideally, these subjects are randomly selected from a larger population of subjects. We previously assumed that the variances for the two populations are the same. Here, we will allow for unequal variances in our samples. The interpretation for *P*-value is the same as in other type of *t*-tests. We generate the results in STATA using the following line command:

Note that in this example, STATA computes the t-value as follows:

$$t = \frac{\overline{x_1} - \overline{x_2}}{\left[\frac{s_1^2}{n_1} + \frac{s_2^2}{n_2}\right]^{1/2}} = \frac{52.82 - 51.73}{\left[\frac{10.5^2}{91} + \frac{10.05^2}{109}\right]^{1/2}} = \frac{1.09}{1.462} = 0.745$$

In this example, the *t*-statistic is 0.7451 with 188.463 degrees of freedom. The corresponding two-tailed *P*-value is 0.4572, which is less than 0.05. So we cannot reject the null that the difference in means between the two groups is zero. In other words, for 188.5 degrees of freedom, the critical *t*-value that corresponds to an area of 0.05 (*P*-value) in both tails is 1.96. Since the calculated value is smaller than the critical value, the null hypothesis of equal means of food security for technology adopters and nonadopters is not rejected. Thus, food security between these two groups is not statistically different, and we conclude that hybrid maize adopters are not better off in terms of food security.

Using R for t-*tests*

We motivate this section with an illustrative example using 200 observations. As mentioned in the previous section, the *t*-test is designed to compare means of the same variable between the two groups. Assume we have random information on FOODSEC for 200 farmers, some of whom are adopters of new hybrid varieties, while others are nonadopter. We first use the *summary* command in R and examine descriptive statistics. With the summary command, we can identify min, each quarter, and max values for each variable. Note that we often are going to use the *summary* command to investigate more information about variables and analysis throughout the book.

```
> summary(data_chapter2)
    adopters            foodsec
 Min.   :0.000      Min.    :28.00
 1st Qu.:0.000      1st Qu.:39.75
 Median :1.000      Median :52.00
 Mean   :0.545      Mean    :51.50
 3rd Qu.:1.000      3rd Qu.:63.00
 Max.   :1.000      Max.    :76.00
```

In our sample, we have 200 observations on the status of technology adoption (0 = NON-ADOPTERS and 1 = ADOPTERS), with a relevant index of food security (foodsec) for each observation. We can further obtain frequency statistics for a variable by using the *table* command in R.

```
> table(adopters)
adopters
  0   1
 91 109
```

When one wants to list more than two results together, we can use the *list* command in R. The *list* command allows us to enumerate information. In this case, we list the mean of food security index of two groups: adopters and nonadopters. As a result, those who adopt the hybrid seed varieties have a higher food security index than nonadopters.

```
> list(mean(foodsec[adopters==1]),mean(foodsec[adopters==0]))
[[1]]
[1] 54.14679

[[2]]
[1] 48.32967
```

To describe the variables' information graphically, we can draw plots to examine the relationship between two variables. The *boxplot* command produces box-and-whisker plots of the given values. We compare the mean of food security index between the groups, which are adopters and nonadopters of hybrid seed varieties.

```
> boxplot(foodsec ~ adopters)
```

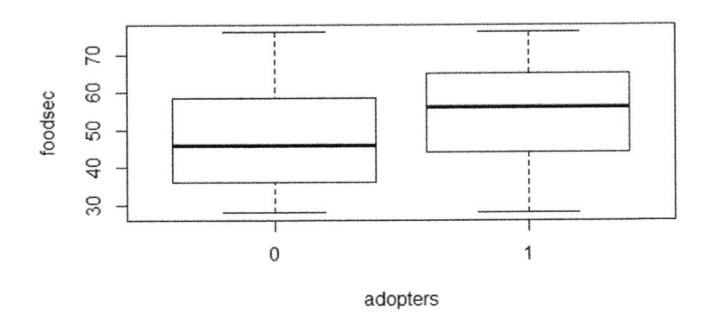

[Boxplot showing foodsec by adopters (0 and 1), y-axis ranging from 30 to 70]

Independent sample t-test assuming unequal variance

As mentioned earlier, we compare the mean food security index between the group of adopters and nonadopters of hybrid seed varieties. Ideally, these subjects are randomly selected from a larger population of subjects. The default of the *t*-test using the *t.test* command in R is that no equal variance, independent samples, and two-sided tests. We generate the results using the following command in R.

```
> t.test(foodsec~adopters)

        Welch Two Sample t-test

data:  foodsec by adopters
t = -3.0289, df = 189.9, p-value = 0.002796
alternative hypothesis: true difference in means is not equal to 0
95 percent confidence interval:
 -9.605425 -2.028812
sample estimates:
mean in group 0 mean in group 1
       48.32967        54.14679
```

In this example, the t-statistics is -3.0289, with 189.9 degrees of freedom. The corresponding two-tailed *P*-value is 0.002796, which rejects the null hypothesis. So, we can conclude that the means are different for the two groups. It interprets that food security between these two groups is statistically different, and we conclude that hybrid maize adopters are better off in terms of food security.

Independent group t-test assuming equal variance

We compare the mean food security index between the groups of adopters and nonadopters of hybrid seed varieties with equal variance conditions. We conduct the *t*-test, assuming that variances for the two populations are the same. Since the default of the *t*-test in R assumes unequal variances, we need to add the argument *var.eq* = *T* to conduct the t-test under the same variance conditions. We generate the results using the following command in R.

```
> t.test(foodsec~adopters,var.eq=T)

        Two Sample t-test

data:  foodsec by adopters
t = -3.0358, df = 198, p-value = 0.002722
alternative hypothesis: true difference in means is not equal to 0
95 percent confidence interval:
 -9.595855 -2.038383
sample estimates:
mean in group 0 mean in group 1
       48.32967        54.14679
```

In this example, the t-statistics is -3.0358, with 198 degrees of freedom. The corresponding two-tailed *P*-value is 0.002722, which rejects the null hypothesis. So, we can conclude that the means are different for the two groups. It interprets that food security between these two groups is statistically different, and we conclude that hybrid maize adopters are better off in terms of food security.

Exercises

1. Based on the studies presented in this chapter, discuss the main findings related to effects of technology adoption on food consumption and nutrition. Discuss the policy implications of promoting technology adoption among rural households.
2. Compute the FOODSEC index with the following weights as variances of each individual component: NBR = 0.2, RUNDUM = 0.45, and LIVSTOCKSCALE = 0.35. Undertake an independent sample *t*-test to determine if there are significant differences among technology adopters and nonadopters in generating food security. Also, undertake a Levene's test to determine if the null hypothesis of equal variances across populations holds true.
3. The climatic conditions in southern parts of Malawi are not very favorable for maize production, while the northern regions are more productive in adopting hybrid maize technology. Write a syntax for the *t*-test that examines the impact of adopting hybrid maize on food insecurity and regional variations in the production of maize. In other words, undertake an independent sample *t*-test to determine if there are significant differences among technology adopters and nonadopters in generating food security and productivity in the northern and southern regions. *Hint:* Use the dummy for the northern (dnorth) and southern regions (dsouth) as the test variables in your analysis.
4. Check the food balance sheet for your country from the FAO website and prepare a commentary on the food availability for the past decade.

64 SECTION | I Food security policy analysis

5. Read the paper by Staatz et al. (1990) and describe how the food security index is constructed in their paper. What would be the impact of technology adoption in Malawi on food security using that index?
6. Do hybrid maize adopters have lower incidence of child malnutrition than nonadopters in Zambia? Explain.

STATA exercise

Consider the following data on food security for adopters (1) and nonadopters (0). Use the commands in STATA to check if the null hypothesis that the difference in the means of food security is zero for both these groups, assuming equal and unequal variances.

Observations	ADOPTERS	FOODSEC
1	1	34
2	1	39
3	0	63
4	1	44
5	0	47
6	1	47
7	0	57
8	1	39
9	0	48
10	1	47
11	0	34
12	0	37
13	1	47
14	0	47
15	0	39
16	0	47

The codes in STATA for this problem:

summarize
tabulate adopters, summarize(foodsec)
ttest foodsec, by(adopters)
ttest foodsec, by(adopters) unequal

The STATA output for each corresponding line is reproduced as follows:

. summarize

Variable	Obs	Mean	Std. Dev.	Min	Max
obs	16	8.5	4.760952	1	16
adopters	16	.4375	.5123475	0	1
foodsec	16	44.75	7.81025	34	63

```
. tabulate adopters, summarize(foodsec)
```

| | Summary of foodsec | | |
adopters	Mean	Std. Dev.	Freq.
0	46.555556	9.2751161	9
1	42.428571	5.159365	7
Total	44.75	7.8102497	16

The t-test assuming equal population variances is generated as follows:

Two-sample t test with equal variances

Group	Obs	Mean	Std. Err.	Std. Dev.	[95% Conf. Interval]
0	9	46.55556	3.091705	9.275116	39.42607 53.68504
1	7	42.42857	1.950057	5.159365	37.65695 47.20019
combined	16	44.75	1.952562	7.81025	40.58821 48.91179
diff		4.126984	3.921998		-4.284865 12.53883

```
        diff = mean(0) - mean(1)                                 t =   1.0523
Ho: diff = 0                                    degrees of freedom =       14

    Ha: diff < 0              Ha: diff != 0              Ha: diff > 0
Pr(T < t) = 0.8447       Pr(|T| > |t|) = 0.3105       Pr(T > t) = 0.1553
```

The t-test assuming unequal population variances is generated as follows:

Two-sample t test with unequal variances

Group	Obs	Mean	Std. Err.	Std. Dev.	[95% Conf. Interval]
0	9	46.55556	3.091705	9.275116	39.42607 53.68504
1	7	42.42857	1.950057	5.159365	37.65695 47.20019
combined	16	44.75	1.952562	7.81025	40.58821 48.91179
diff		4.126984	3.65532		-3.775604 12.02957

```
        diff = mean(0) - mean(1)                                 t =   1.1290
Ho: diff = 0                         Satterthwaite's degrees of freedom = 12.9076

    Ha: diff < 0              Ha: diff != 0              Ha: diff > 0
Pr(T < t) = 0.8603       Pr(|T| > |t|) = 0.2794       Pr(T > t) = 0.1397
```

Note once again, that STATA generates the t-values for equal population variances as follows:

$$t = \frac{\overline{x_1} - \overline{x_2}}{\left[\frac{(n_1-1)s_1^2 + (n_2-1)s_2^2}{n_1+n_2-2}\right]^{1/2}\left[\frac{1}{n_1}+\frac{1}{n_2}\right]^{1/2}}$$

$$= \frac{46.55 - 42.42}{\left[\frac{(9-1)9.27^2 + (7-1)5.15^2}{9+7-2}\right]^{1/2}\left[\frac{1}{9}+\frac{1}{7}\right]^{1/2}} = \frac{4.13}{3.91} = 1.05$$

And for unequal population variances, we have

$$t = \frac{\overline{x_1} - \overline{x_2}}{\left[\frac{s_1^2}{n_1}+\frac{s_2^2}{n_2}\right]^{1/2}} = \frac{46.55 - 42.42}{\left[\frac{9.27^2}{9}+\frac{5.15^2}{7}\right]^{1/2}} = \frac{4.13}{3.651} = 1.13$$

Check https://www.stata.com/manuals13/rttest.pdf (page 9) for the computation of the degrees of freedom in the latter case. Since the calculated $t < 1.96$, we cannot reject the null hypothesis of equality of means between adopters and nonadopters.

Chapter 3

Effects of commercialization of agriculture (shift from traditional crop to cash crop) on food consumption and nutrition—application of chi-square statistic

Chapter outline

Introduction	68
A few concepts	70
What is commercialization?	70
Recent cases of commercialization	71
Commercialization effects	73
Effects of commercialization of agriculture on food consumption	73
Effects of commercialization of agriculture on nutrition	74
Review of selected studies	74
Organic farms and commercialization in the United States	79
Organic farming in a global context	81
Empirical analysis	82
Data description and analysis	83
Descriptive analysis: cross-tabulation results	84
Chi-square tests	87
Chi-square tests using STATA	89
Conclusion and policy implications	93
Technical appendices	95
Pearson's chi-square (χ^2) test of independence	95
Student's t-test versus Pearson's chi-square (χ^2) test	96
Limitations of the chi-square procedure	96
Descriptive analysis: cross-tabulation results	97
Chi-square tests using R	99
Exercises	101
STATA exercise	102

By furthering the use of ethanol, farmers are presented with the opportunity to produce a cash crop by collecting their agricultural wastes.

—Richard Lugar

Introduction

Recent approaches to food security attempt to connect smallholder farmers to the markets though value chain development. Such approaches call for commercializing smallholder sector and further consider each faming unit as a business entity. Food system transformation involves smallholders moving from subsistence-oriented faming to commercial farming either in the choice of commodities they produce or in the approach to marketing their produces. Growing cash crops in lands where food is traditionally grown can have a profound impact on food security (Tshirley and Thieraut, 2013). Effects of this shift, known as commercialization of agriculture, on food consumption and nutrition vary; a number of studies have documented disastrous effects, while others found a positive or neutral effect. Critics of commercialization of agriculture contend that if the resources that are used to produce agricultural export crops were used instead to produce food for the local economy, the problem of malnutrition in many countries could be reduced. Proponents, on the other hand, argue that by exploiting comparative advantage, commercialization could raise farm incomes and improve nutrition (see, for example, von Braun et al., 1994; Bouis and Haddad, 1994; Babu and Rajasekaran, 1991b).

The emphasis on cereal production over the past three decades in many developing countries has resulted in low output prices and profitability and has dampened agricultural growth (the growth in cereal production was 2.3% during the period 1965—80 and declined to 1.9% per year during the period 1996—2000; see, for example, Barghouti et al., 2004).

Additionally, investment in the agricultural sector has also declined during the past three decades. To reverse this trend, agricultural commercialization has been identified as one of the strategies by the donor agencies (World Bank, 2002a; DFID, 2002). Commercialization is seen as a common and powerful means to increase rural household income and food access, as well as diversify production and reduce risks of income and food shortfalls (Ali and Farooq, 2003).

von Braun et al. (1994) emphasize that the process of commercialization raises income. Increased incomes further improve welfare, food security, and nutritional status—all of which could have been worse in the absence of commercialization of agriculture for rural populations. Recent studies (Govereh and Jayne, 2003) have also demonstrated that commercialization of agriculture can lead to improved productivity of other crops through household level synergies and regional spillover effects, thereby improving food security.

To conclude about the positive benefits of agricultural commercialization on food security and nutrition, it is important to answer a few specific questions: Is commercialization of smallholder agriculture a legitimate policy tool for improving food and nutrition security? As commercialization proceeds, which social and economic groups benefit from higher wages and incomes?

In other words, what does the distribution of average income look like—Is it skewed toward specific economic or social groups and, if so, toward which group? Which socioeconomic group adopts commercialization and what are the implications for resource ownership and sharing? Does higher household income from commercialization result in improved food consumption and better nutrition for all household members?

The issue is critically important since cash crops contribute to livelihood diversification and improve food and nutrition security by directly increasing the farm household's income earning potential, which, in turn, increases the household's spending potential. Second, since most cash crops tend to be labor-intensive, cash cropping entails a substantial expansion of the demand for hired labor. This employment effect for households that hire out labor may represent significant livelihood improvement (Masanjala, 2006). Third, the introduction of cash crops contributes to the development of rural financial markets, which partially relieves the cash constraints (Goetz, 1993). Finally, cash cropping opportunities are also accompanied by improved technology.

However, there are also reasons to suspect that the impact of cash crop liberalization on the welfare of households may be more limited than is generally acknowledged (Orr, 2000). First, a critical assumption made is that as farm households earn more income, the market will widen the scope for its welfare maximization since increases in income will assure household food security by increasing the farm household's access to food through the market. However, this impact chain is not at par with the empirical evidence from the food security literature, which suggests that cash cropping is associated with missed opportunities for improving household welfare. The income and employment benefits of commercialization are not spread equally between households due to imperfect or missing factor markets. Second, due to weak financial markets for expenditure and consumption smoothing, when cash cropping opportunities increase household incomes, allocating more income to food purchases is not automatic, as the literature seems to suggest (Paolisso et al., 2001). We also expand the issue of commercialization to the idea of organic farming in the United States and to other countries in Europe. We also look at the economics of organic farming in the United States with regard to the sector's viability.

In light of the aforementioned set of issues, this chapter examines the relationship between cash crop growing and household food security and nutrition situation using a chi-square statistical technique. The purpose of a chi-square test is to compare the observed frequencies with the expected frequencies derived under the hypothesis of independence. It is appropriate in examining the relationship between two bases of classification, since we are examining the relationship between cash crop growers and food security of the household and then determining the independence between cash crop growers and children's nutritional status. The chi-square procedure can be legitimately applied only if the categories in which the observations are sorted are independent of each other, i.e., only if the placement of each observation into a particular category does not depend on the placement of any of the other observations. In other words, the categories must be both exhaustive and mutually exclusive—each observation must fit into one or another of the categories and no observation fits into more than one. This procedure is appropriate to use since food security, cash crop production, and nutritional status variables can be classified into mutually exclusive and exhaustive cases. We also demonstrate the application in STATA with a few illustrative examples that link the food security and the characteristics of the farmers.

A discussion of the concepts of commercialization and its impact on food consumption and nutrition and an overview of some of the main studies serves as a motivation for the empirical analysis. We conclude with a few policy implications from the results of this chapter.

A few concepts

What is commercialization?

Commercialization of agriculture is mainly a process of production of cash crops. A cash crop is simply a crop produced for sale. Agricultural commercialization can be defined as the "proportion of agricultural production that is marketed" (Govereh et al., 1999). Commercialization of agriculture involves moving from subsistence-oriented patterns to increasingly market-oriented patterns. The underlying assumption behind this shift is that markets allow households to increase their incomes by producing those commodities that generate the highest returns and then use the cash to buy household consumption items (Timmer, 1997). Recent trends in globalization of smallholder agriculture including connecting smallholder farmers with domestic and international market chains are a clear indication that commercialization of smallholder agriculture will continue to increase in the developing world (von Braun, 2005).

Commercialization of smallholder agriculture by bringing the resource poor farmers to markets is not entirely a new phenomenon. Many agricultural processes have been implemented in developing countries for production of cash crops. Abbott (1994) provides a few examples in the context of commercialization in developing countries:

- Hanapi and Sons enterprise for rice production in Malaysia: This enterprise specializes in mill-polished rice, bran, very fine rice, broken rice, and husk from paddy.
- Corn Products Corporation International enterprise for maize production in Kenya and Pakistan: This enterprise converts maize to starch, sweeteners, oil-and-gluten feed, and flour for livestock.

- Siam Food Products enterprise for pineapple in Thailand and Rose's Lime Products enterprise for limes in Ghana: These enterprises specialize in canning of pineapples, processing of limes into lime juice, lime oil, etc.
- Dabaga Fruit and Vegetable Canning Company for fruits and vegetables in Tanzania: The main activities are processing products (fruits and vegetables) into jams, pickles, juices, and soups.
- Commercialization of agriculture can also include dairy farming enterprise such as the successful Anand Dairy Cooperative Union in India that processes and markets milk, butter, milk powder, ice cream, and other dairy products.

Recent cases of commercialization

1. Supermarket chains have opened in China, especially in the rural areas. The government of China is actively encouraging the development of rural retail networks, including the transformation of rural market fairs into modern supermarkets (USDA, 2005). The commercialization of Chinese agriculture has been accompanied by an improvement in calorie and protein intake, which is consistent with evidence from other developing countries. Thus, food insecurity is a relatively rare phenomenon in rural China.
2. With the opening of the Indian industrial sector during the early 1990s, the Punjab Agro-Industries Corporation in 1988 went into a joint venture with Pepsi (a US multinational corporation) with the third partner being Voltas, a domestic corporate firm (Kumar, 2006). The entry strategy of Pepsi was to procure and process certain fruits and vegetables grown in the state. By the early 1990s, these activities led to the emergence of contract farming for tomatoes. Also, another local entrepreneur named Nijjer entered the market with financial support from the Punjab Financial Corporation and set up a tomato processing plant and began processing and procuring tomatoes from the farmers under contract farming. Presently, Pepsi is procuring chilli and basmati rice and plans to enter the business of ginger and garlic paste and the processing of several fruits such as mango, guava, orange, and other citrus fruits. FritoLay, which is a subsidiary to Pepsi, is procuring potatoes from the farmers under contract farming, while Hindustan Lever Limited (HLL), a subsidiary of Unilever, is contracting farmers to procure basmati rice for exporting it abroad (Kumar, 2006).

Two types of contract farming can be distinguished: The first type consists of firms that have a direct contract relationship with the farmers, and they are responsible for providing inputs/extension services. These firms procured the end product directly from the farmers. Examples include Pepsi/FritoLay; Hindustan Lever Limited; Chambal Agritech Limited, and A.M. Todd.

The second type of firm was operating through the Punjab Agro Foodgrains Corporation (PAFC). PAFC tied up with three types of companies for providing inputs and extension services to the farmers and to buy back their contracted crops.[1] Thus, in the latter type, there was a tripartite agreement between the farmer, the PAFC, and the companies providing seed, extension services, and a buyback guarantee to the farmers on behalf of PAFC.

Out of these two types of contract farming, it was found that the former was operating efficiently. The extension services provided were up to the mark in the case of firms having direct contract with the farmers. Similarly, the buyback guarantee was honored at the predetermined procurement price in the case of direct contract firms. In the case of the PAFC model, although there was a written contract between the farmers and the service providers, there was a lack of commitment on the part of service providing companies. In most cases, no extension services were provided to the farmers, and as a result, farmers were forced to sell their commodities in the open market. Thus, the system of contract farming was heavily skewed toward medium and large farmers, and the smaller farmers were the losers.

Additionally, the productivity data clearly showed superiority of contract farmers over their counterpart noncontract farmers in all size classes, and this was clearly dominant in the case of crop productivity. One can conclude that the experiment of contract farming in Punjab has generally benefited the farming community. However, the system is heavily skewed toward the medium, and the large farmers and the small farmers were net losers in the process.

Commercialization as a process is not limited to cash crop production. It extends beyond cash crops and may also include traditional crops (grown for self-consumption) if one markets the produced surplus or adopts a purchased input technology. There are also instances where farmers have retained significant produce of a cash crop for home consumption, such as groundnuts in West Africa. Moreover, commercialization of agriculture does not necessarily mean commercialization of the rural economy. There are rural settings where there is negligible production of cash crops but a high level of nonfarm and nonagricultural employment. An example of such a setting could be a village that produces traditional crop for household consumption and derives its income mainly from nonagricultural commercial activities such as production of handicrafts or employment in textile mill for instance (von Braun et al., 1994).

---

1. The first set were seed companies, the second set were consultant companies providing consulting services on agronomic practices, while the third set belonged to the buyers/export companies who were tied up for a buyback guarantee to contract farmers and for providing forward linkages for their produce.

Commercialization effects

The effects of commercialization on income, consumption, food security, and nutrition are very complex in nature and mainly depend on household preferences and intrahousehold allocations (von Braun et al., 1994).

Some of the exogenous factors of commercialization are population change, adoption of new technologies, infrastructure and market-based development, and macroeconomic and trade policies. The endogenous factors of commercialization mainly depend on three characteristics of intrahousehold decision-making: First, the household determines the proportion of income spent on food and nonfood (mainly health and sanitation) expenditures. Second, the household determines the allocation of food expenditure among the various types and quantities of foods. Third, the household decision-maker also determines how the food and other consumption items are distributed among the household members. There are other endogenous consequences of commercialization, most important of which are the gender allocation of time, labor, and control of income.

Effects of commercialization of agriculture on food consumption

The decision of intrahousehold allocations, i.e., how to allocate household resources such as land, labor, time, and capital toward production activities and how to share household income among members, can have a significant impact on household food consumption (Haddad et al., 1997). Since household food availability determines food consumption, it is important to evaluate the impact that commercialization of agriculture has on household food availability. Such an impact can be either positive or negative on the individual members of the family depending on who controls the household income and how they are allocated to food consumption and nonfood items among the members.

On the positive side, commercialization can produce considerable real income gains, thus enhancing a household's capacity to acquire food. However, the income—food consumption relationship is not so direct. It is influenced by many factors such as who controls the income, the proportion of money spent on food and nonfood items, and whether the increased income results in higher intake of calories or intake of more expensive calories.

Furthermore, if the household shifts from a traditional crop production to cash crop production (particularly a nonfood cash crop with a long growing cycle), it allocates the majority of its land resources to such a commercialization process. In the absence of nonfarm income, the household's food supply may be affected negatively in the short and medium term. However, in the presence of nonfarm income, the household can compensate for such a loss by purchasing foods. In the worst-case scenario, nonfarm income is not available, and trading of cash crops takes place in the presence of unfavorable production conditions. This can adversely affect the household's food consumption and nutrition. The findings of studies on how commercialization affects rural food consumption in various developing countries are presented in later sections.

Effects of commercialization of agriculture on nutrition

The complexities involved in commercialization and food consumption relationship are also applicable to the commercialization and nutrition relationship (Peters and Herrera, 1994). The nutritional effect of commercialization can be attributed to the impact commercialization has on an individual's nutrient intake. There are important factors other than diet that need to be considered in analyzing the effect of agricultural commercialization on human nutrition, particularly on the nutritional status of women and children. The effects of commercialization on children's welfare are partly mediated through the income—consumption link, since increased incomes are found favorably to affect children's nutritional status. The effect of commercialization on nonfood expenditures can influence the health and sanitation environment positively, which, in turn, improves children's nutritional status. At the same time, the additional time allocated by women in pursuing productive activities can adversely affect the child's nutritional status, as less time is allocated to child care.

The general view is that the commercialization of agriculture increases household income, which, in turn, influences food consumption and nutritional adequacy. It is typically assumed that this income-mediated effect on nutritional improvement operates through two main pathways. First, increased income can be used for increased food expenditures that directly increase food consumption, which, in turn, may improve nutritional status by higher energy and nutrient intake. Second, increased income can result in increase in nonfood expenditures such as health and sanitation that indirectly have positive health effects (Kennedy and Haddad, 1994).

Review of selected studies

von Braun and Immink (1994) examined the impact of export-oriented vegetable production on household food security and nutritional status in Guatemala. Vegetable production appeared to be a promising option because of rising foreign demand during the 1980s, for two main reasons. First, for the smallholder sector, such production created opportunities for adoption of labor-absorbing production techniques. Second, the decline in market prices of traditional exports, an increasing external debt, and contracting foreign exchange reserves prompted the government to undertake policies toward production of vegetable crops to reduce the risk factors that are associated with declining income and food insecurity of farm households.

The cooperative was formally formed in 1979 with 177 farmers. Its membership rose to 1600 in 1989 comprising eight villages. In 1985, the cooperative farmers were cultivating almost 300 ha of export vegetables, mainly snow peas, cauliflower, broccoli, and parsley. The results of the study are based upon household-level surveys conducted in 1983 and 1985 in six villages of the Cuatro Pinos region in the western highlands of Guatemala. The cooperative was active in all these six villages, and the survey constituted of two rounds (1983 and 1985) in the same season with the same households.

In examining the impact of commercialization on income, the study found that export crop—producing households earn higher income than others and food expenditure as a share of total expenditure decreases with an increase in income levels. In addition, the calorie intake was increasing with the rise in incomes across all households. While the cooperative households, on average, increased their calorie intake by 2.8% with 10% more income compared to 4.4% by nonmembers, this difference in calorie intake was not significant.

Overall, although the favorable effect of an export producing commercialization scheme is not very significant in the short term, it had a significant positive impact in the long run. Increase in incomes decreased stunting and weight deficiency, but the effect was less pronounced at higher income levels. Women-controlled income affected children's nutrition significantly. Finally, higher export crop income share did not have adverse effects on children's nutrition.

Based on the aforementioned findings, the authors conclude that production of export crops had a positive effect on household income, food security, and nutrition of farmers in Guatemala. However, commercialization is not risk free and increases the small farmer's dependence on market conditions for both inputs and outputs. In the absence of appropriate financing, such risks may be hard to bear. The production of export crops has the potential for significant losses if market prices are unstable, the crop fails, marketing institutions break down, or export markets collapse. In addition, government's fiscal, monetary, and foreign trade policies can have a significant effect on the revenues from export crops in either direction (positive or negative). The study suggests a few policy implications of commercialization effects.

To attain a sustained income growth from export crop production, the policy structure should consist of the following:

1. Increasing the efficiency of domestic and international marketing channels through investment in infrastructure, transportation facilities, and development of effective marketing organizations.
2. Easier access to rural credits and innovation of credit schemes that address the specific needs of small and large farm households.
3. Macroeconomic and foreign trade policies that encourage production of export crops and a simplified export licensing structure.

4. Development of rural financial and social institutions that provide easier access to extension programs, savings schemes, and financing of community development projects.

The aforementioned policy structure accompanied by simultaneous and complementary social investments in health, environment, and sanitation not only can attain sustainable income growth for rural farmers but can also reinforce the positive effect of increased income on food security and nutrition.

Peters and Herrera (1994) conducted a study in the Zomba district of Southern Malawi to evaluate the impact of commercialization of tobacco on household food security and nutritional status. The sample consisted of households that were tobacco growers and nongrowers. Household consumption of maize (which is the main staple food) came mainly from own farm production.

The research site was located in the southern part of Zomba district, which was about 15 miles from the town of Zomba. The area was covered by one of the largest white-owned estates in the Shire Highlands—the Bruce Estates. A sample of 210 households was selected with 148 of them being nontobacco growing households, while the remaining were tobacco growing households.

The results of the study indicated that income was positively correlated to nutritional adequacy, with per capita income and per capita expenditures being the most important determinants of child nutrition. Interestingly, though, tobacco growers had both higher per capita incomes and expenditures, and their children were not significantly different in nutritional status from nongrowers. Overall, the study does not find a significant effect of cash cropping on children's nutrition.

Since Malawi is a very poor country, any agricultural initiative that guarantees an income increase for the rural population is encouraged. This is corroborated by the findings that income plays a significant role in influencing household food consumption and nutrition. As most of the tobacco growers in the sample, on average, still produce maize as the main crop, the cash crop component of income is not a significant part of production and income plans. Policy makers must address the important risks that small farmers face when recommending cash crop production at higher levels.

An effective policy plan should include development of efficient distribution and marketing of inputs (such as fertilizers, seeds, etc.) and outputs (processed tobacco), an easy access to credit, and training programs and policies that promote a sustainable and stable market for cash crops. In addition to market risks, farmers also face risks related to food security. Policy plans to encourage commercialization must consist of measures that reform the food supply inefficiencies.

Paolisso et al. (2001) address how the commercialization of the vegetable and fruit cash crop program (VFC) affects male and female time allocation and thereby child nutrition in Nepal. The overall goal of the VFC program is to increase the commercial value of vegetable and fruit production and raise household incomes of targeted farmers. The program also provides production inputs, training, and technical assistance to both men and women farmers.

The data came from three communities that represented different agroecological zones, ethnic groups, and cultural practices. A sample of 264 households was chosen using a random spot observation method by recording the activity of the households within the 6:30−18:30 time period. The activity of the households was recorded during the aforementioned time period by visiting them randomly 30 times during a 12-month period.

The study addressed three questions: First, what factors determined household participation in the VFC program; second, how male- and female-headed households allocated time among various activities depending on their VFC participation status; and third, how VFC participation affected male- and female-headed households' labor allocation to various activities after controlling for a number of individual and household characteristics. A probit analysis[2] was undertaken to estimate the likelihood that a household would have received VFC training.

The study used an instrumental variable estimation strategy to address the impact of VFC participation on time allocation by male- and female-headed households. The main result is that VFC participation increased the time allocated in the production of fruits and vegetables by both men and women; less time was allocated to cereals and livestock, and greater time was devoted to care of children less than 5 years by women. At the same time, men spent moderately less time caring for children. Although the study does not measure nutritional status of the children directly, it seems that participation in the VFC program by women gives them additional time to allocate for child care. This in turn possibly improves the nutritional status of preschoolers.

A study by Govereh and Jayne (2003) examined how agricultural commercialization affects food productivity through household level synergies and regional spillover effects for a northern district in Zimbabwe called Gokwe. Household level synergies occur when a household's participation in a commercialized crop scheme enables it to acquire resources that otherwise would not be available. For example, Strasberg (1997) points out that under credit and input market failures in northern Mozambique, cotton outgrower schemes were the primary method of acquiring cash for use in food production.

---

2. A probit analysis uses a transformation where each observed proportion is replaced by the value of the standard normal curve (z value) below which the observed proportion is found. Probit coefficients represent the difference a unit change in the predictor makes in the cumulative normal probability of the outcome, i.e., the effect of the predictor on the z value for the outcome. This probability depends on the levels of the predictors. For example, a unit change at the mean of the predictor has a different effect on the probability of the outcome than a unit change at the extreme value of the predictor.

Regional spillover effect occurs when commercialization schemes attract investments to a region that generate benefits to all farmers regardless of whether they engage in that commercialization scheme. The study examined the determinants of cotton commercialization at the household level and the contribution of commercialization to food crop yields and production. The data come from a survey of 480 rural households in 1996, using a stratified sampling approach. Gokwe district was selected since it was a major cotton-producing area.

The study used an instrumental variable estimation approach to address how commercialization affected food productivity. The results showed that the effect of households adopting commercialization had a positive and significant impact on food productivity. This can possibly be attributed to cotton producers having access to key inputs such as credit and training through the cotton schemes that are either not accessible to nonparticipating farmers or simply not available to them at all. Second, farm size had a negative and significant impact on yields, which suggests that smaller farms are more productive than larger farms. Third, regional spillovers had a positive and significant impact on food productivity, and an additional cotton input retailer in the area boosts maize grain output significantly. This is mainly because cotton retailers provide a range of services for farmers growing food crops, including inputs used in maize production. Thus, commercialization of cotton appears to be associated with higher grain productivity.

The main implication of the study is that cash crop production may not come at the expense of household food security. Participation in cash crop programs may allow farmers to overcome failures such as access to inputs and credit supply. The author thus suggests that synergies between cash crops and food crops research extension programs may have important implications for promoting smallholder food crop productivity growth and thus alleviate the problem of food insecurity significantly.

From the aforementioned studies, it is apparent that the impact of commercialization on food security and nutrition is income mediated. Research findings (Govereh et al., 1999) also point to the synergies between commercial cash crop production and food crop production, which can improve food crop productivity and thus have a positive impact on household food security. Commercialization at the household level can also contribute to farm capital formation, which is an additional source of productivity growth. However, for commercialization to be successful, investment in specific areas is necessary, such as irrigation and drainage, science and technology, rural infrastructure, and changing the policy and institutional environment. All these investments cannot be expected to come from the public sector. Thus, national governments have to create an enabling environment for the private sector to provide inputs and services so that farmers can have the incentives for commercialization.

Organic farms and commercialization in the United States

In the United States, consumer demand for organic food has risen quickly over the past decade (Greene et al., 2009). Organic goods have expanded into newer markets including big retail stores. Since the late 1990s, US organic production and demand have more than doubled. For example, Greene et al. (2009) point out that organic food sales have more than quintupled, increasing from $3.6 billion in 1997 to $18.9 billion in 2007.[3] For example, Genti et al. (2011) use US county level data and show that favorable natural amenities, water for irrigation, adjacency to metro areas, and government payments have a positive effect on **organic farming**. They also show that **organic farming** is more popular among young farmers.

However, the organic sector has had to contend with high price premiums for organic goods and bears the brunt of fluctuating economic conditions. Weakening economic conditions makes it difficult to attract new consumers into the market. The low organic adoption rate for grain crops continues to be a bottleneck for expansion of the US organic livestock sector, as organic livestock producers struggle to find reliable sources of affordable feed grains.[4]

One of the major challenges for the sector is the price premium that it has to contend with, as compared with nonorganic or conventional items. For example, at the retail level, organic produce and milk receive significant price premiums over conventionally grown products. Economic Research Service (ERS) analyzed organic prices for 18 fruits and 19 vegetables using 2005 data on produce purchases and found that the organic premium as a share of the corresponding conventional price was less than 30% for over two-thirds of the items. In contrast, organic price premiums for a half-gallon container of milk ranged from 60% for private-label organic milk above branded conventional milk in 2006 to 109% for branded organic milk above private-label conventional milk. While the organic sector enjoys a price premium at the retail level, recent empirical analysis by Uematsu and Mishra (2012) shows evidence that the organic farmers earn significantly higher household income than conventional farmers.

Finally, organic producers also face competition from new labels like the "locally grown" label. The USDA continues to fine-tune organic regulations, so as to maintain sustainability of the environment and the viability of the industry. Indeed, researchers such as Hanawa et al. (2012) demonstrate that organic grain producers had more than a single motivation and that younger organic farmers are more likely to be motivated by environmental and lifestyle

---

3. More than two-thirds of US consumers buy organic products at least occasionally, and 28% buy organic products weekly, according to the Organic Trade Association, also see Greene et al. (2009).
4. Only 0.2% of corn and soybean crops were grown under certified organic farming systems in 2005, according to ERS estimates.

goals than older farmers. Organic grain producers exhibited a diversity of motivations, including profit and stewardship. The USDA must consider these overarching goals on the part of the organic farmers, as it refines the incentives in the system.

Section highlights: is organic farming worth it?

In recent years, there has been an increase in the demand for organic products, particularly in the dairy sector. McBride and Greene (2009) point out organic milk production has been one of the fastest growing segments of organic agriculture in the United States. For example, their study conducted for the time period 2000–05 tells us that the number of cows on US farms increased by an annual average of 25%. However, their study also indicates that because of economic considerations, many organic dairies adapt their operations to be more like conventional dairies in terms of size, location, and the types of technologies used. In other words, organic milk producers usually begin as conventional dairy operators but are at a disadvantage of a challenging and costly transition. Producers have to qualify for organic certification based on the guidelines formed by National Organic Program (NOP). Producers must make changes in animal husbandry, land and crop management, input sourcing, and certification paperwork.

In addition to these challenges, organic milk producers must now contend with the impact of a weaker US economy on the demand for organic food products. McBride and Greene (2009) examine the challenges faced by this industry by developing a model to characterize the industry's cost structure. They then compare the comparative cost advantages of this sector with respect to conventional dairy farms. Their study provides a lot of interesting insights:

1. Organic dairies are smaller than conventional dairies (82 cows compared with 156 cows).
2. Organic dairies produce about 30% less milk per cow than conventional dairies (13,601 pounds per organic cow compared with 18,983 pounds per conventional cow).
3. Organic dairies use more pasture-based feeding, where more than 50% of dairy forage feed is from pasture than conventional dairies (63% compared with 18%).
4. Organic dairies paid $6.37 per cwt more than conventional dairies in operating and capital costs, including transition costs.
5. The average price premium for organic milk was $6.69 per cwt in 2005.
6. Total economic costs of organic dairies in 2005 were $7.65 per cwt higher than for conventional dairies, nearly $1 per cwt higher than the average price premium for organic milk.
7. Pasture-based organic dairies' total economic costs were about $4 per cwt higher than conventional pasture-based dairies, much lower than the average price premium for organic milk in 2005.

The researchers also found that the certification paperwork and compliance costs were reported by many producers as the most challenging aspect of organic milk production. Finding new organic input sources, such as dairy replacement

> **Section highlights: is organic farming worth it?—cont'd**
> and feed, higher costs of production, and maintaining animal health, is also an important concern. Interestingly, certification paperwork was a lesser concern for pasture-based dairies and more educated operators. The NOP, which develops, implements, and administers national production, handling, and labeling standards for organic agricultural products, is in the process of taking all these issues into consideration, and the future growth of the organic production sector continues to evolve.

Organic farming in a global context

The dilemma between sustainability and economic viability of organic farms continues to be a debating point between policy makers, farmers, and social activists in many parts of the world. Consequently, researchers constantly try to estimate the full economic and the environmental effects of organic farming practices, with a hope toward informing the general public and policy experts. In an interesting case study, Parra-Lopez et al. (2008) examine the performance of alternative olive-growing systems in Andalusia, Spain, on the basis of the assessments of different groups of experts. They test the hypothesis that the region exhibits greater sustainability of organic and integrated farming over conventional farming systems. Indeed, their results for Spain show a greater global performance of organic and integrated agriculture despite differences in the ideological tendencies of the experts, thus providing a scientific basis for endorsing institutional and social support for the promotion and implementation of these farming techniques.

The implications for incentives and policy can be very revealing as evidenced from an interesting study conducted for French organic farmers by Blondel et al. (2012). This research examined the impact of the Contrat Territorial d'Exploitation (CTE), land use contract, which was a significant experiment in France in the early 2000s. In this experiment, farmers were offered financial incentives in exchange for more organic farming practices, as part of a sustainable development (SD) approach. The research uncovers the true motivation for signing a CTE: Was it financial or ecological? In other words, were farmers simply maximizing their profits or was there a broader objective? Their study indicates the former to be true. This means that a policy may encourage SD, but farmers cannot be expected to adhere to it without a financial incentive.

Auerbach (2018) points out the advantages of organic farming in Africa, where high costs and limited supply of agricultural inputs make organic farming economically viable. Auerbach (2018) compares organic and conventional farming in cabbage, sweet potato, and cowpea crops. The relative advantages of organic farming are in less dependency in external inputs, lower soil acidity, carbon sequestration, water retention, improved microbiology, biodiversity, and a strengthening of the food chain. The difficulties with organic farming are mainly in pest and disease control. Similar to the recommendations from all other studies in this context, organic farming crucially depends on technical and training support for small farmers, market linkages, and assistance with quality maintenance.

In related studies for Wanzai county in China, Qiao et al. (2019, 2018) point out that, regardless of how profitability is measured, organic farming contributes to higher farm incomes for small-scale and medium-scale farmers. Overall, among organic farmers, those who belonged to cooperatives performed much better in economic terms than farmers not in any cooperatives. However, as noted across all studies, the local governments at every level at Wanzai provided institution, financial, technical support, and broadened market channels. Clearly, the model followed at Wanzai can serve as a template for all countries.

While examining the ecological aspect of organic farming in India and the United Kingdom, Toke and Raghavan (2010) find that the organic food industry not only helps the economy manage unsustainable growth but also reduces the negative environmental impact of traditional agriculture and at the same time improves the living standards of poor farmers. As noted by Azam and Shaheen (2019), lack of adequate economic resources, restrictive social and institutional factors, lack of marketing orientation, and government policy continue to thwart India's organic farming from becoming an economic alternative.

Empirical analysis

To generate policy recommendations, this chapter undertakes an empirical analysis based on the chi-square test to determine if cash crop adoption by rural households in Malawi has a positive impact on food security and nutrition of children.

Since policy makers are keenly interested in knowing whether commercialization creates multiplier effects by changing opportunities created by new production technology or market signals, it is important to understand whether this improved productivity leads to households being more food secure (especially the marginal farmers) after a commercialization scheme is launched.

Thus, the central questions addressed are as follows:

- Is it more likely for a cash crop—growing household than a traditional crop—growing household to be food secure?

- Is it more likely for a cash crop—growing household than a traditional crop—growing household to have children with adequate nutrition, i.e., absence of malnutrition?

To answer the aforementioned questions, one needs information on household characteristics such as incomes by family members, expenditure on food and nonfood items, demographic characteristics of the members, and food intake by family members. Additionally, one also needs suitable measures of the children's nutritional status. Pearson's chi-square test is used to determine whether the observed relationship between the nominal or categorical variable is statistically significant or is due to random variability. A brief description of the Pearson's chi-square test is presented in the technical appendix.

Data description and analysis

We use the household-level data for Malawi as before. The variables used in the analysis are as follows:

CASHCROP: Tobacco, groundnuts, cotton, and plantain are the major cash crops in Malawi. The variable is defined as CASHCROP = 1 if the household grows at least one of these four major cash crops and 0 otherwise. CASHCROP will thus represent commercialization of agriculture at the household level.

Two measures of household food security are computed:

1. The first measure of "food security" is a combination of dependency ratio and the number of meals that a household consumes as defined in Chapter 2.
2. The second measure relies on per adult equivalent calorie intake for households. Food security is defined as households being able to satisfy at least 80% of the requirement for calorie intake. This variable is coined CALREQ. If a household's daily per adult equivalent calorie intake (CALADEQ) is at least 80% of the requirement, 2200 kcal, then the household is qualified as "food secure."[5] The variable is defined as CALREQ = 1, if the household is food secure and CALREQ = 0, otherwise.

---

5. The percent intake of the average energy requirement is used as a means of examining calorie adequacy. Individual energy requirement is estimated according to the age—sex-specific recommendation of the joint FAO/WHO/UNO committee. The standard for an adult male equivalent is 2200 kcal.

CALREQ is computed as follows:

$$PCALMET = \left(\frac{CALADEQ}{2200}\right) \times 100 \qquad (3.1)$$

CALREQ is then defined by

$$CALREQ = \begin{cases} 1 & \text{if } PCALMET \geq 80 \; (\text{Indicating food } security) \\ 0 & otherwise (Indicating\ food\ insecurity) \end{cases} \qquad (3.2)$$

The following three nutrition measures are also used: ZHANEW, ZWANEW, and ZWHNEW: the Z-scores that identify malnutrition in children ZHA (height-for-age Z-score), ZWA (weight-for-age Z-score), and ZWH (weight-for-height Z-score). The details of computing the Z-scores are given in Appendix 7.

ZHANEW indicates presence or absence of stunting, ZWANEW indicates if the child has low weight for age, and ZWHNEW indicates the presence or absence of wasting. These three variables are constructed as follows:

1. All the Z-scores that are above 5 and below -5 are excluded. By rule of thumb, all the Z-scores above 5 and below -5 are considered outliers and are excluded from the analysis.
2. The three indicators of malnutrition are defined by creating two categories, one with Z-score less than -2 and the other with Z-score greater than or equal to -2, i.e.,

$$ZHANEW = \begin{cases} 1 & \text{if } ZHA \geq -2 (Normal\ Z - scores\ indicating\ absence\ of\ stunting) \\ 0 & otherwise\ (Low\ Z - scores\ indicating\ presence\ of\ stunting) \end{cases}$$

$$ZHANEW = \begin{cases} 1 & \text{if } ZWA \geq -2 (Normal\ Z - scores\ indicating\ absence\ of\ under\ weight) \\ 0 & otherwise (Low\ Z - scores\ indicating\ presence\ of\ under\ weight) \end{cases}$$

$$ZHANEW = \begin{cases} 1 & \text{if } ZWH \geq -2 (Normal\ Z - scores\ indicating\ absence\ of\ wasting) \\ 0 & otherwise (Low\ Z - scores\ indicating\ presence\ of\ wasting) \end{cases}$$

$$(3.3)$$

Descriptive analysis: cross-tabulation results

First, we investigate the relationship between CASHCROP, which is the independent variable, and the two food security measures. Later, we will investigate the relationship between CASHCROP production and child nutritional levels. It is important to note that all the aforementioned variables are nominal or categorical variables. The hypothesis is that there is no relationship

TABLE 3.1 Cross-tabulation results of cash crop growers and CALREQ.

Cashcrop			No	Yes	Total
CALREQ	No		225	126	351
			64.1%	35.9%	
	Yes		169	84	253
			66.8%	33.2%	
	Total		394	210	604 = n

between commercialization (CASHCROP) and food security (CALREQ and INSECURE). This is tested using cross-tabulation procedures (Table 3.1).

The cross-tabulation results indicate that 35.9% of the cash crop—growing households is food insecure (as measured by CALREQ) compared with 64.1% of the non—cash crop—growing households. It is likely that growing cash crop generates additional income for the household, who can sell these crops at local markets. This in turn allows them to purchase more food. Next, we investigate the relationship between CASHCROP and INSECURE (the second measure of food security) (Table 3.2).

From Table 3.2, it is evident that cash crop growers are relatively more food secure compared with non—cash crop growers. This is possibly because the household's participation in a commercialized crop scheme enables it to acquire resources that otherwise would not be available.

TABLE 3.2 Cross-tabulation results of cash crop growers and INSECURE.

Cashcrop			No	Yes	Total
INSECURE	Secure		49	26	75
			65.3%	34.7%	
	Moderately insecure		64	23	87
			73.6%	26.4%	
	Highly insecure		116	77	193
			60.1%	39.9%	
	Totally insecure		165	84	249
			66.3%	33.7%	
	Total		394	210	604 = n

Next, we want to investigate whether cash crop production results in achieving higher nutritional levels for the children as well as increased food security for the household members. We undertake cross-tabulation tests for CASHCROP and ZHANEW, ZWANEW, and ZWHNEW. All the aforementioned variables are dichotomous nominal variables. The hypothesis is that there is no relationship between commercialization (CASHCROP) and child nutrition as measured by the aforementioned indicators. Tables 3.3—3.5 show the relationship between them.

The cross-tabulation results indicate that 53.7% of preschoolers of the households not growing cash crops are stunted, while 46.3% of preschoolers for households growing cash crops are stunted. It is likely that the extra income generated through sale of cash crops achieves greater income, which helps in moderating food insecurity of the household. Household members can obtain higher energy intake as well as greater dietary diversity. The higher energy intake results in better child nutritional status.

TABLE 3.3 Cross-tabulation results of cash crop growers and height-for-age Z-scores for children under 5 Years.

Cashcrop		No	Yes	Total
ZHANEW	Low	66	57	123
		53.7%	46.3%	
	Normal	54	43	97
		55.7%	44.3%	
	Total	120	100	220 = n

TABLE 3.4 Cross-tabulation results of cash crop growers and weight-for-age Z-scores for children under 5 Years.

Cashcrop		No	Yes	Total
ZWANEW	Low	97	79	176
		55.1	44.9%	
	Normal	44	38	82
		53.7%	46.3%	
	Total	141	117	258 = n

TABLE 3.5 Cross-tabulation results of cash crop growers and weight-for-height Z-scores for children under 5 Years.

Cashcrop		No	Yes	Total
ZWHNEW	Low	121	91	212
		57.08%	42.92%	
	Normal	7	15	22
		31.82%	68.18%	
	Total	128	106	234 = n

The incidence of underweight preschoolers was 55.1% for households who did not grow cash crops and 44.9% for households who grew cash crops. The results indicate that underweight children are less likely to occur in cash crop−growing households relative to non−cash crop−growing households.

The cross-tabulation results of cash crop production and wasting are given in Table 3.5. It is evident that households who grew cash crops had a lesser incidence of wasting (42.9%) compared with households who did not grow the crops (57.1%). However, from the aforementioned results, we cannot conclude if the results were significant or only due to random variability. Thus, we undertake chi-square tests to determine this.

Chi-square tests

Similar to the cross-tabulation tests, we first investigate whether the relationship between CASHCROP and food security (Table 3.6) is significant and then determine whether the relationship between CASHCROP and nutritional indicators (such as ZHANEW, ZWANEW, and ZWHNEW) is significant or if it is just due to random variability. For the sake of brevity, we will just concentrate on one food security measure, namely per capita adult equivalent calorie intake falling below 2200 kcal per day.

TABLE 3.6 Chi-square tests between cashcrop and CALREQ.

	Value	P value
Test statistic	0.471	0.492
Number of valid cases	604	

The null hypothesis is given by H_0: No relationship exists between growing cash crops and food security, i.e., incidences of observed food insecurity among households are not statistically different between cash crop growers and non—cash crop growers. The value of Pearson chi-square is 0.471 with significance level (*P* value) of 0.492.[6] The *P* value is the smallest level of significance for which the observed data indicate that the null hypothesis should be rejected. Since the significance level is greater than 0.1, the null hypothesis cannot be rejected at the 10% level. It is important to report the *P* value, since it gives more information to the reader than stating whether the null hypothesis was rejected or not for some level of significance (for example, $\alpha = 0.05$) chosen by the researcher. The incidence of food insecurity is not statistically different between cash crop growers and non—cash crop growers. Although we find that cash crop growers have better food security using the cross-tabulation tests, we cannot infer that this relationship is statistically significant. Next, we undertake chi-square tests between cash crop growers and the various indicators of nutrition to determine whether the observed relationship is statistically significant or is simply due to random variability.

It is evident from Table 3.7 that the *P* value of the test statistic is quite high (0.766). Thus, we are unable to reject the null hypothesis and infer that there is no observed pattern of relationship between cash crop—growing and stunted preschoolers. Although, from the cross-tabulation results, we find that cash crop growing reduces stunting, this relationship is not significant and is only due to random variability.

It is apparent from Table 3.8 that the *P* value of the chi-square test statistic is 0.827 and is greater than 0.1. Thus, the null hypothesis that there is no observed pattern of relationship between cash crop—growing and underweight preschoolers cannot be rejected. Hence, the incidences of underweight preschoolers are not statistically different between these two groups.

TABLE 3.7 Chi-square tests between CASHCROP and height-for-age Z-scores for children under 5 Years.

	Value	*P* value
Test statistic	0.089	0.766
Number of valid cases	220	

---

6. The *P* value for the chi-square test is si4.gif, the probability of observing a value at least as extreme as the test-statistic for a chi-square distribution with $(r-1)(c-1)$ degrees of freedom. The *P* value was similarly computed for the *t*-distribution in the previous chapter.

TABLE 3.8 Chi-square tests between CASHCROP and weight-for-age Z-scores for children under 5 Years.

	Value	P value
Test statistic	0.048	0.827
Number of valid cases	258	

TABLE 3.9 Chi-square tests between CASHCROP and weight-for-height Z-scores for children under 5 Years.

	Value	P value
Test statistic	5.131	0.023
Number of valid cases	234	

We find that the P value of the chi-square test statistic is 0.023 (Table 3.9) and is significant. It can be reasonably concluded that cash crop growing reduces the incidence of wasting among preschoolers. This is probably because the WHZ is a short-term indicator of nutritional status and, at least in the short run, cash crop growing can benefit households by generating greater income and achieving food security. Improvement in food security status leads to greater distribution of food and other resources at the intrahousehold level, which, in turn, alleviates the problem of malnutrition for preschoolers.

Chi-square tests using STATA

In this section, we perform chi-square tests using STATA for a sample of 200 farmers, some of whom grow cash crops. We first break the data down using the summarize command in STATA (see Table 3.10):

The variable id simply tracks each farmer in the same with an identification number. We have 200 observations in our sample, where a unit of observation is a farmer. All the variables in this sample are exactly those defined in the previous section. We can perform the cross-tabulation and the chi-square test between cash crop and food security, using a simple line command in STATA:

. tabulate insecure cashcrop, column chi2

This produces the following output in STATA (see Table 3.11).

The STATA output provides the actual numbers with each category, with the percentages below these numbers. Thus, there are 13 farmers who do not grow a cash crop, and are fully secure, which is 14.29% of all the cash croppers. The total and the cumulative percentages are given in the last column on the right.

TABLE 3.10 STATA input and output: descriptive statistics.

Summarize					
Variable	Obs	Mean	Std. Dev.	Min	Max
id	200	100.5	57.87918	1	200
Cashcrop	200	0.545	0.4992205	0	1
Insecure	200	3.43	1.039472	1	4
Calreq	200	0.64	0.4812045	0	1
Zhanew	200	0.235	0.4250628	0	1
Zwanew	200	0.475	0.5006277	0	1
Zwhnew	200	0.29	0.4549007	0	1

TABLE 3.11 STATA output: cross-tabulation and chi-square tests between cashcrop and food security levels.

	Cashcrop		
Insecure	0	1	Total
1	13	11	24
	14.29	10.09	12.00
2	3	8	11
	3.30	7.34	5.50
3	7	13	20
	7.69	11.93	10.00
4	68	77	145
	74.73	70.64	72.50
Total	91	109	200
	100.00	100.00	100.00
Pearson chi2(3) = 3.2040 Pr = 0.361			

Recall that we have four different levels of food security, which we have coded as follows (1 = secure, 2 = moderately insecure, 3 = highly insecure, and 4 = totally insecure). In our sample data set here, we find that about 83% of cash croppers are insecure, and 82% of non—cash croppers are insecure.

The cross-tabulations also indicate that the numbers and percentages between cash and non—cash croppers are very similar for other levels of food security. Our arbitrary data set cannot therefore reject the null hypothesis that there is no difference in food security between these two types of croppers. Indeed, STATA returns a Pearson chi-square value of 3.2, which is less than the table value of 6.635. Similar to our concerns in the previous section, we want to investigate whether cash crop production results in achieving higher nutritional levels as well as increased food security in the data. We undertake cross-tabulation tests for CASHCROP and ZHANEW, ZWANEW, and ZWHNEW, using STATA. All the aforementioned variables are dichotomous nominal variables, as seen from the output from the summarize statement. The hypothesis is that there is no relationship between commercialization (CASHCROP) and child nutrition as measured by the aforementioned indicators. We continue with four similar commands in STATA for each of our variables:

.tabulate calreq cashcrop, column chi2
.tabulate zhanew cashcrop, column chi2
.tabulate zwanew cashcrop, column chi2
.tabulate zwhnew cashcrop, column chi2

The results in each case from STATA are reproduced in Tables 3.12—3.15.

In our sample data set here, we find that about 33% of cash croppers and 40% of non—cash croppers are food insecure, as their CALREQ scores are low. The cross-tabulations also indicate that the numbers and percentages between cash and noncash croppers are very similar when CALREQ is high.

TABLE 3.12 STATA output: cross-tabulation and chi-square tests between cashcrop and CALREQ.

	Cashcrop		
CALREQ	0	1	Total
0	36	36	72
	39.56	33.03	36.00
1	55	73	128
	60.44	66.97	64.00
Total	91	109	200
	100.00	100.00	100.00

Pearson chi2(1) = 0.9187 Pr = 0.338

TABLE 3.13 STATA output: Cross-tabulation and chi-square tests between cashcrop and ZHANEW.

ZHANEW	Cashcrop 0	1	Total
0	76	77	153
	83.52	70.64	76.50
1	15	32	47
	16.48	29.36	23.50
Total	91	109	200
	100.00	100.00	100.00

Pearson chi2(1) = 4.5725 $Pr = 0.032$

TABLE 3.14 STATA output: cross-tabulation and chi-square tests between cashcrop and ZWANEW.

ZWANEW	Cashcrop 0	1	Total
0	44	61	105
	48.35	55.96	52.50
1	47	48	95
	51.65	44.04	47.50
Total	91	109	200
	100.00	100.00	100.00

Pearson chi2(1) = 1.1522 $Pr = 0.283$

Our data set cannot therefore reject the null hypothesis that there is no difference in CALREQ between these two types of croppers. This is evidenced via the calculated Pearson chi-square value of 0.91 provided by STATA.

Likewise, in our data, we find that about 71% of cash croppers and 84% of non—cash croppers have a very low Z-score for ZHANEW. The cross-tabulations also indicate that the numbers and percentages between cash and non—cash croppers are very similar when ZHANEW = 1. Our data set cannot

TABLE 3.15 STATA output: cross-tabulation and chi-square tests between cashcrop and ZHANEW.

	Cashcrop		
ZWHNEW	0	1	Total
0	62	80	142
	68.13	73.39	71.00
1	29	29	58
	31.87	26.61	29.00
Total	91	109	200
	100.00	100.00	100.00

Pearson chi2(1) = 0.6671 Pr = 0.414

therefore reject the null hypothesis that there is no difference in ZHANEW between these two types of croppers. This is evidenced via the calculated Pearson chi-square value of 3.2 provided by STATA.

Similar to the previous results, we find from Tables 3.14 and 3.15 that we cannot reject the null hypothesis that there is no difference in ZWANEW and ZWHNEW between these two types of croppers. The advantage with STATA is that it provides us the cross-tabulation results and the chi-square tests in one statement.

Conclusion and policy implications

"Commercialization of agricultural systems is a universal and irreversible phenomenon that is triggered by economic growth" (Pingali and Rosegrant, 1998). The rate of commercialization differs across countries and its impact on food security and nutrition is also different. Agricultural commercialization has complex linkages with food security and nutrition. The relationship operates through its effect on household income, expenditures, and intra-household labor and resource allocations. Each of the aforementioned three factors can be broken down into further important categories.

Commercialization in agriculture has a significant impact on per capita income, income of men, and income of women. A positive effect on either of these variables has different effects on food consumption and nutrition. The income available per household member for consumption is a better measure than income. Men spend relatively more on nonfood expenditures that include nutrition-enhancing items such as health and sanitation in addition to alcoholic drinks. On the other hand, women on an average spend a higher share of their expenditures on food, which directly affects children's nutrition (see, for example, Peters and Herrera, 1994).

Similarly, household expenditures constitute food expenditures (more specifically, quantity and quality of calories), nonfood expenditures (especially health, sanitation, and education), and investments in agricultural inputs. Though higher income in many cases has resulted in higher food-related expenditures by rural populations, it is critical to understand the nature of these expenditures. Higher intake of calories by purchasing more staple food, say maize, may not have the similar effect on nutritional adequacy as by purchasing more expensive calories such as fish, fruit, meat, and so on. The effect of commercialization will obviously be more beneficial if the increased income is used to spend more on health and sanitation than on alcoholic drinks. Since a shift to cash crop from traditional crop requires investment in technology and various other inputs that are not easily affordable to small farmers, they look for access to the credit markets. Channeling a proportion of increased income to reduce debt and future investments in better technologies can also have significant effects on household food security and nutrition in the long run.

From the cross-tabulation and chi-square tests, we find that cash crop growing is usually beneficial to the household in achieving both greater food security and higher nutritional status for children. This is possibly mediated through higher incomes as well as the additional synergy of access to greater credit for the cash crop growers. However, undertaking the chi-square tests, we find that, except for ZWH, there is no observed pattern of relationship between cash crop growing and food security and between cash crop growing and nutritional status of children in general. This does not imply that cash crop growing has no benefits to the household at all. In the short run, cash crop growing can benefit households by generating greater income and achieving food security. Increase in food security leads to better distribution of food and other resources at the intrahousehold level, which can reduce the problem of malnutrition.

A few critical policy implications must be emphasized concerning the impact of commercialization on food security and nutrition. An efficient domestic and international marketing channel, easier and equal access to credit markets, macroeconomic and foreign trade policies that simplify export licensing and encourage export production, and promotion of a sustainable and stable domestic market for cash crops are vital for commercialization to work. Food security is a primary concern for small farmers in allocating land for cash crops. The policies advocating commercialization for increasing rural incomes and those reforming food supply inefficiencies must be juxtaposed. Most importantly, simultaneous and complementary social investments in health and environment not only can attain sustainable income growth for rural farmers but can also reinforce the positive effect of increased income on food security and nutrition.

Policy makers should also encourage women's participation in the commercialization process by providing incentives that hinder gender bias in decision-making and income sharing. Governments should offer easier information access to women about farming and agricultural production of cash crops. Policies that provide equal access to inputs and credit for women farmers should be brought into effect, since increasing women's income is associated with better child nutrition (Spring, 2000).

The nature of agricultural commercialization, socioeconomic environment, and agroclimatic conditions specify a certain policy structure. Thus, a generalized plan cannot be prescribed for all commercialization processes. The adverse impact of cash crop production on food consumption and nutrition is not necessarily the result of inherent fallacies in commercialization opportunity but can be due to bad policies or underinvestment in specific sectors such as rural infrastructure, irrigation, and drainage. Government actions, such as unnecessary trade constraints, inaccessible credit markets for rural poor, and absence of appropriate price supports for inputs and outputs, can be disadvantageous for the commercialization process to work effectively.

Technical appendices

Pearson's chi-square (χ^2) test of independence

The Pearson chi-square is the most common test for significance of the relationship between nominal variables. The purpose of the χ^2 test is to answer the question by comparing observed frequencies with the expected frequencies derived under the hypothesis of independence. The test statistic is given by $\chi^2_{statistic} = \sum (f_0 - f_t)^2 / f_t$, where f_0 is an observed frequency and f_t is the expected frequency. The expected frequencies are computed as follows:

$$f_t = \frac{(Row\ total)(Column\ total)}{Total\ sample\ \text{size}(n)}$$

For example, from Table 3.2, the expected frequencies for each of the four cells are computed as follows: For households who are cash crop growers and have food security, the expected frequency is given by (75) (210)/604 = 26.07. Similarly, for the households who are not cash crop growers but are still food secure, the expected frequency is given by (75) (394)/604 = 49. For the households who are cash crop growers but are not food secure, the expected frequency is (249) (210)/604 = 86.5. Finally, for the households who are not cash crop growers and who are food insecure, the expected frequency is given by (249) (394)/604 = 162.4.

The number of degrees of freedom must be determined to apply the aforementioned test. In this example with a 2 × 2 table, the degrees of freedom v = (2 − 1) (2 − 1) = 1. In general, for r rows and c columns, the number of degrees of freedom is $(r − 1) (c − 1)$. For example, the chi-square value between cash crop growing and food security was 0.471. The critical value at the 0.01 level of significance is $\chi^2_{0.01,1} = 6.635$. Since the actual value is less than the critical value, we are unable to reject the null hypothesis of no observed pattern of relationship. The chi-square test measures the discrepancy between the observed cell counts and what one would expect if the rows and columns were unrelated.

The main assumptions underlying the use of the chi-square test are that the sample is randomly selected and expected frequencies are not very small. The reason for the latter assumption is that the chi-square inherently estimates the underlying probabilities in each cell of the cross-tabulation—when the expected cell frequencies fall below 5, those probabilities cannot be estimated with sufficient precision. In our analysis, none of the cells in the cross-tabulations has a frequency below 5.

Student's *t*-test versus Pearson's chi-square (χ^2) test

When the primary concern is the dependence or independence rather than the absolute difference in mean between variables, the χ^2 test should be employed. This distinction becomes somewhat elusive sometimes. In fact, when the test for difference in means involves nominal (categorical) variables, both *t*-test and χ^2 tests can be applied. In general, however, the difference between the two is that a *t*-test is more applicable with the difference in means between variables, whereas the χ^2 test is designed to test the independence between variables (Lowry, 2003).

The χ^2 test should also be distinguished from various statistical procedures that measure the association between variables. While the former tests if there is dependence relationship among variables, the latter is intended to quantify such dependence, if it exists.

Limitations of the chi-square procedure

Although chi-square procedures are computationally simple, they rest on a complex logical structure and thus impose certain limitations. First, chi-square procedures can be applied when the *N* observations are independent of each other, i.e., putting one observation in one particular category does not depend in any way on placement of any of the other observations. The categories must be exhaustive and mutually exclusive. Second, validity of the chi-square tests is greater when the value of f_t, the expected frequencies within the cells, is fairly large. Although statisticians do not always agree where to draw the line between "large enough" and "too small," chi-square procedures can be applied if $f_t \geq 5$. For the special case of 2 × 2 contingency tables, this limitation can be avoided using the Fisher exact probability test (Lowry, 2003).

Descriptive analysis: cross-tabulation results

In this section, we perform chi-square tests using R for a sample of 200 farmers, which was randomly designated to conduct chi-square tests. (Variables: Cashcrop, INSECURE, CALREQ, ZHANEW, ZWANEW, ZWHNEW.) First, we investigate the relationship between CASHCROP, which is the independent variable, and the two food security measures (CALREQ and INSECURE) by using the cross-tabulation results. The variable is defined as CASHCROP = 1 if the household grows at least one of four major cash crops, which are tobacco, groundnuts, cotton, and plantain, and 0 otherwise. CALREQ is defined as households being able to satisfy at least 80% of the requirement for calorie intake that is 2200 kcal. If a household intakes more than 80% of calories, we can interpret the household is qualified as "food secure." The variable is defined as CALREQ = 1, if the household is food secure and CALREQ = 0, otherwise.

To conduct the cross-tabulation in R, we can use table command with the variables that we would like to examine. The Table(A, B) in R is for a two-way frequency table. A will be rows, and B will be columns. The result of table command is as follows:

```
> table(CALREQ,Cashcrop)
         Cashcrop
CALREQ    0   1
     0   63  38
     1   52  47
```

After conducting the table command, we save the results as table1 for further analysis to calculate shares.

```
> table1 <-table(CALREQ,Cashcrop)
```

Once we save the table1, we can examine probability tables on cell percentage (using *prop.table* (*table1*)), row percentage (using *prop.table* (*table1,1*)), and column percentage (using *prop.table* (*table1,2*)) arguments.

```
> prop.table(table1)
         Cashcrop
CALREQ      0     1
     0   0.315 0.190
     1   0.260 0.235
```

```
> prop.table(table1,1)
         Cashcrop
CALREQ        0         1
     0   0.6237624 0.3762376
     1   0.5252525 0.4747475
```

```
> prop.table(table1,2)
         Cashcrop
CALREQ        0         1
     0   0.5478261 0.4470588
     1   0.4521739 0.5529412
```

The cross-tabulation results show the probability of variable combinations on three different types of the denominator, which are the cell, row, and column from the left side. When we examine the results in the middle, the cross-tabulation results can interpret that 37.62% of the cash crop—growing households are food insecure. It is likely to interpret that growing cash crop generates additional income for the household, who can sell these crops at local markets. Additional income, in turn, allows them to purchase more food. Next, we investigate the relationship between CASHCROP and INSECURE (the second measure of food security). We employ the *table* command to look at the frequency of two variables (INSECURE, CASHCROP) and then save the results as table2. After, we calculated shares on rows to see portions of cash crop growing on each insecure level.

```
> table (INSECURE, Cashcrop)
         Cashcrop
INSECURE  0  1
       1 20  1
       2 18  5
       3 48 33
       4 29 46
```

```
> prop.table(table2,1)
         Cashcrop
INSECURE          0          1
       1 0.95238095 0.04761905
       2 0.78260870 0.21739130
       3 0.59259259 0.40740741
       4 0.38666667 0.61333333
```

table2: Results of table (INSECURE, CASHCROP)

INSECURE variable is the other variable to show food security. INSECURE = 4, if food security is secure, INSECURE = 3, if food secure is moderately insecure, INSECURE = 2, if food secure is highly insecure, INSECURE = 1, if food secure is totally insecure. From the results, it explains that cash crop growers are relatively more food secure compared with non—cash crop growers.

Next, we want to investigate whether cash crop production results in achieving higher nutritional levels for the children as well as increased food security for the household members. We undertake cross-tabulation texts for CASHCROP and ZHANEW (0 = low, 1 = normal), ZWANEW (0 = low, 1 = normal), and ZWHNEW (0 = low, 1 = normal). All the aforementioned variables are dichotomous nominal variables. The hypothesis is that there is no relationship between commercialization (CASHCROP) and child nutrition, as measured by the aforementioned indicators. The next parts show the relationship between them.

```
> table (ZHANEW, Cashcrop)
       Cashcrop
ZHANEW  0  1
     0 57 49
     1 58 36
```

```
> prop.table(table3,1)
       Cashcrop
ZHANEW         0         1
     0 0.5377358 0.4622642
     1 0.6170213 0.3829787
```

table3: Results of table (ZHANEW, CASHCROP)

The cross-tabulation results indicate that 53% of preschoolers of the households not growing cash crops are stunted, while 46.2% of preschoolers for households growing cash crops are stunted. It is likely that the extra income generated through sale of cash crops achieves greater income, which helps in moderating food insecurity of the household.

The incidence of underweight preschoolers was 59.4% for households who did not grow cash crops and 40.6% for households who grew cash crops. The results indicate that underweight children are less likely to occur in cash crop–growing households relative to non–cash crop–growing households.

```
> table (ZWANEW, Cashcrop)             > prop.table(table4,1)
       Cashcrop                               Cashcrop
ZWANEW  0  1                          ZWANEW          0         1
     0 57 39                               0  0.5937500 0.4062500
     1 58 46                               1  0.5576923 0.4423077
```

table4: Results of table (ZWANEW, CASHCROP)

The cross-tabulation results of cash crop production and wasting are as follows. The results present that households who grew cash crops had a lesser incidence of wasting (44.8%) compared with households who did not grow the crops (55.2%).

```
> table (ZWHNEW, Cashcrop)             > prop.table(table5,1)
       Cashcrop                               Cashcrop
ZWHNEW  0  1                          ZWHNEW          0         1
     0 53 43                               0  0.5520833 0.4479167
     1 62 42                               1  0.5961538 0.4038462
```

table5: Results of table (ZWHNEW, CASHCROP)

However, from the aforementioned results, we cannot conclude if the results were significant or only due to random variability. Thus, we undertake chi-square tests to determine this.

Chi-square tests using R

In this section, we perform chi-square tests on the data set with farmers, some of whom grow cash crops. We first examine the data using the *summary* command in R.

The null hypothesis is given by H_0: No relationship exists between growing cash crops and food security. In other words, the incidence of observed food insecurity among households is not statistically different between cash crop growers and non–cash crop growers.

100 SECTION | I Food security policy analysis

```
> summary(data_chapter3)
     Obs             Cashcrop            INSECURE            CALREQ
Min.   :  1.00   Min.   :0.000      Min.   :1.00      Min.   :0.000
1st Qu.: 50.75   1st Qu.:0.000      1st Qu.:3.00      1st Qu.:0.000
Median :100.50   Median :0.000      Median :3.00      Median :0.000
Mean   :100.50   Mean   :0.425      Mean   :3.05      Mean   :0.495
3rd Qu.:150.25   3rd Qu.:1.000      3rd Qu.:4.00      3rd Qu.:1.000
Max.   :200.00   Max.   :1.000      Max.   :4.00      Max.   :1.000
    ZHANEW              ZWANEW              ZWHNEW
Min.   :0.00     Min.   :0.00       Min.   :0.00
1st Qu.:0.00     1st Qu.:0.00       1st Qu.:0.00
Median :0.00     Median :1.00       Median :1.00
Mean   :0.47     Mean   :0.52       Mean   :0.52
3rd Qu.:1.00     3rd Qu.:1.00       3rd Qu.:1.00
Max.   :1.00     Max.   :1.00       Max.   :1.00
```

With saved tables 1 through 5 (table1—CALREQ and CASHCROP, table2—INSECURE and CASHCROP, table3—ZHANEW and CASHCROP, table4—ZWANEW and CASHCROP, table5—ZWHNEW and CASHCROP), we can conduct chi-square tests by using the *chisq.test* command in R. The results of table1 is are follows. The degree of freedom is 1, a Pearson chi-square value is 1.60, and the *P*-value is 0.2055. According to the *P*-value, our arbitrary data set cannot reject the null hypothesis that there is no relationship in food security between CASHCROP and CALREQ.

```
> chisq.test(table1)

        Pearson's Chi-squared test with Yates' continuity correction

data:  table1
X-squared = 1.6027, df = 1, p-value = 0.2055
```

The results of table2 (CASHCROP and INSECURE) are as follows. The degree of freedom is 3, a Pearson chi-Square value is 27.283, and the *P*-value is close to zero. According to the *P*-value, we can reject the null hypothesis and accept the alternative hypothesis that there is a significant relationship in food security and cash crop growing. We also want to investigate whether cash crop production results in achieving higher nutritional levels as well as increased food security in the data. We undertake chi-square tests for CASHCROP and other variables.

```
> chisq.test(table2)

        Pearson's Chi-squared test

data:  table2
X-squared = 27.283, df = 3, p-value = 5.135e-06
```

The chi-square results of table3 are as follows. The degree of freedom is 1, a Pearson chi-square value is 0.978, and the *P*-value is 0.32. Our data set cannot reject the null hypothesis that there is no relationship in food security between ZHANEW and two types of croppers.

```
> chisq.test(table3)

        Pearson's Chi-squared test with Yates' continuity correction

data:  table3
X-squared = 0.97764, df = 1, p-value = 0.3228
```

The chi-square results of table4 in R are as follows. A Pearson chi-square value is 0.139, and the *P*-value is 0.71. Our data set cannot reject the null hypothesis that there is no relationship in food security between ZWANEW and two types of croppers.

```
> chisq.test(table4)

        Pearson's Chi-squared test with Yates' continuity correction

data:  table4
X-squared = 0.13853, df = 1, p-value = 0.7097
```

Lastly, there are the results of the chi-square tests in the following. A Pearson chi-square value is 0.237, and the *P*-value is 0.63. Our data set cannot reject the null hypothesis that there is no relationship in food security between ZWHNEW and two types of croppers.

```
> chisq.test(table5)

        Pearson's Chi-squared test with Yates' continuity correction

data:  table5
X-squared = 0.2369, df = 1, p-value = 0.6265
```

Exercises

1. Based upon the literature listed in the analysis, discuss the main findings related to effects of agricultural commercialization on food consumption and nutrition. Discuss the policy implications of promoting commercialization among rural households.
2. Write a chi-square syntax to examine the relationship between CASHCROP and INSECURE. Is the test statistic significant? How many degrees of freedom does the test statistic have?
3. Recollect the food security index developed in Chapter 1. Can a chi-square test be applied between CASHCROP and this index? Discuss your answer in detail.
4. From the empirical results of this chapter, is it more likely for a cash crop—growing household than a traditional crop—growing household to be more food secure in Malawi? Discuss the findings critically.
5. Is it more likely for a cash crop—growing household than a traditional crop—growing household to have children with adequate nutrition? Discuss the results with reference to Malawi.

STATA exercise

Consider the first 25 observations on Food Security for Cash Croppers with the corresponding information on CALREQ, ZHANEW, ZWANEW, and ZWHNEW. The variable "obs" in the first column is just a variable that tracks each observation in the sample data, where each unit of observation is typical respondent in the survey. Use the commands in STATA to check for cross-tabulations and the Pearson chi-square tests between (1) CASHCROP and insecure (2) CASHCROP and ZHANEW (3) Cashcrop and ZWANEW and, finally, (4) Cashcrop and ZWHNEW. Which of the null hypothesis can be rejected at 95% confidence?

Obs	Cashcrop	INSECURE	CALREQ	ZHANEW	ZWANEW	ZWHNEW
1	1	1	1	1	0	1
2	1	1	1	0	1	0
3	0	1	1	1	0	1
4	1	1	1	1	0	0
5	0	1	1	1	0	0
6	1	1	1	1	0	1
7	0	1	1	0	1	0
8	1	1	1	1	0	1
9	0	1	1	0	1	0
10	1	1	1	0	1	0
11	0	1	1	0	1	1
12	0	1	1	0	1	0
13	1	1	1	0	1	0
16	0	1	1	1	0	1
17	1	1	1	0	1	0
18	0	1	1	0	1	1
19	1	1	1	1	0	0
21	0	1	1	0	1	1
22	0	1	1	0	1	0
23	1	2	1	1	0	1
24	0	2	1	0	1	0
25	1	2	1	0	1	1

The STATA code for this problem is as follows:

tabulate insecure cashcrop, column chi2
tabulate zhanew cashcrop, column chi2
tabulate zwanew cashcrop, column chi2
tabulate zwhnew cashcrop, column chi2

The output for each line along with a brief explanation of the results is given in the following:
    .tabulate insecure cashcrop, column chi2

INSECURE	Cashcrop 0	1	Total
1	10 90.91	9 81.82	19 86.36
2	1 9.09	2 18.18	3 13.64
Total	11 100.00	11 100.00	22 100.00

Pearson chi2(1) = 0.3860   Pr = 0.534

How does STATA compute the Pearson chi-square value? From the aforementioned table, the row totals and the column totals are first used to compute the *expected frequencies*:

	Cashcrop	
Insecure	0	1
1	(19*11)/22 = 9.5	(19*11)/22 = 9.5
2	(3*11)/22 = 1.5	(3*11)/22 = 1.5

$$\chi^2 = \frac{(10 - 9.5)^2}{9.5} + \frac{(1 - 1.5)^2}{1.5} + \frac{(9 - 9.5)^2}{9.5} + \frac{(2 - 1.5)^2}{1.5}$$

$$= 0.026 + 0.166 + 0.026 + 0.166 = 0.38$$

We cannot reject the null because the computed chi-square value is smaller than the table value = 3.841 (at 0.05 level of significance with degrees of freedom = 1)

.tabulate zhanew cashcrop, column chi2

ZHANEW	Cashcrop 0	1	Total
0	8 72.73	5 45.45	13 59.09
1	3 27.27	6 54.55	9 40.91
Total	11 100.00	11 100.00	22 100.00

Pearson chi2(1) = 1.6923   Pr = 0.193

Likewise, in this case, the chi-square value is computed as follows, by first computing *expected frequencies*:

	Cashcrop	
Insecure	0	1
1	(13*11)/22 = 6.5	(13*11)/22 = 6.5
2	(9*11)/22 = 4.5	(9*11)/22 = 4.5

$$\chi^2 = \frac{(8-6.5)^2}{6.5} + \frac{(3-4.5)^2}{4.5} + \frac{(5-6.5)^2}{6.5} + \frac{(6-4.5)^2}{4.5}$$

$$= 0.34 + 0.5 + 0.34 + 0.5 = 1.69$$

We cannot reject the null because the computed chi-square value is smaller than the table value = 3.841 (at 0.05 level of significance with degrees of freedom = 1)

```
.tabulate zwanew cashcrop, column chi2
                    Cashcrop
      ZWANEW        0           1        Total

           0        3           6            9
                27.27       54.55        40.91

           1        8           5           13
                72.73       45.45        59.09

       Total       11          11           22
               100.00      100.00       100.00

          Pearson chi2(1) =   1.6923   Pr = 0.193
```

The calculations are similar to the one shown for the previous command
```
.tabulate zwhnew cashcrop, column chi2
                    Cashcrop
      ZWHNEW        0           1        Total

           0        6           6           12
                54.55       54.55        54.55

           1        5           5           10
                45.45       45.45        45.45

       Total       11          11           22
               100.00      100.00       100.00

          Pearson chi2(1) =   0.0000   Pr = 1.000
```

Likewise, there are no differences between actual expected frequencies in this case, yielding a chi-square = 0. The null hypothesis cannot be rejected in this case either.

Chapter 4

Effects of technology adoption and gender of household head: the issue, its importance in food security—application of Cramer's V and phi coefficient

Chapter outline

Introduction	105
Review of selected studies	107
Female farm operators in Kenya and Ethiopia: recent evidence	109
Rights, norms, and institutions: beyond technology adoption	110
Female farm operators in the United States	112
Women in agriculture: the global scene	113
Uganda's coffee market: a case study	116
Empirical analysis	118
Data description and analysis	119
Descriptive analysis: cross-tabulation results	120
Cramer's V and phi tests	121
Conclusion and policy implications	123
Section highlights: Covid-19, women's burden, and the digital divide	123
Cramer's V in STATA	126
Technical appendices	128
Phi coefficient and Cramer's V	128
Phi coefficient (Φ)	128
Cramer's V	128
Applications in R	128
Exercises	131
STATA exercise	131

A message to young women: Be Brave, Be Bold, Be courageous, Take up your space unapologetically.

— Jemimah Njuki Director for Africa, IFPRI (2020).

Introduction

Gender mainstreaming and gender empowerment are not just buzzwords in development. They have profound impact on how the development process improves the livelihoods of women. Interventions to empower women must assess the context-specific needs of women and youth and their inclusion in the development process. In the context of achieving food security, it is increasingly recognized that women could effectively improve food access, availability, and utilization if they are empowered with capabilities and resources within the household. Women produce more than half the food grown in the developing countries (Stringer, 2000; Saenz, 2013). Women farmers in sub-Saharan Africa produce more than three-quarters of the region's basic food and manage about two-thirds of the marketing and at least one-half of the activities for storing food and raising animals (Saitio, 1994). In Asia, women account for more than two-thirds of food production, while they contribute to about 45% of production in Latin America and the Caribbean (FAO, 2000). Nonetheless, findings suggest that male-headed households are more likely to adopt a new technology or participate in a commercialization scheme than female-headed households (see, for example, von Braun, 1988; Kumar, 1994; David, 1998; Doss and Morris, 2001).

Most farmers in general and women in particular face limited choices and constraints at the household level. The gender differences in technology adoption rates can be attributed to two main reasons (Doss and Morris, 2001). Men and women may inherently have different preferences toward technology and women may be more risk averse (Babu et al., 1993).

The market orientation could be different between men and women, with women growing food for household consumption and men for commercial sale. First, even perceptions about tastes, appearance, and storage of crops can explain the difference in preferences between men and women. Second, even if men and women have the same preferences toward technology choices, their access to complementary inputs, such as land, labor, and credit and extension services, may be unequal. Understanding how gender affects farmers' access to land, labor, and other inputs and how this changing access affects the adoption of new technology and commercialization processes will help in designing gender-mediated policies and programs that improve food security (Quisumbing and Otsuka, 2001).

Additionally, the benefits of technological adoption are not always distributed equally among males and females. Such biased effects can have serious adverse implications for family welfare and nutrition of children. Men tend to spend relatively more on nonfood expenditures, which also include tobacco and alcoholic drinks. On the other hand, although women spend more on clothes, they also spend relatively more on food items that directly affect children's nutrition (see, for example, Kumar, 1994; Peters and Herrera, 1994; Lilja and Sanders, 1998).

Intrahousehold allocations and production structure in agriculture are the driving forces behind the gender-biased effects of agricultural innovations such as technological change (Kumar, 1994). The household characteristics that influence the allocation of resources are mainly women's share of resources (crop ownership) and household head's gender, education, and age.

These factors combined with other household characteristics, such as time spent in household activities by men and women, access to protected water, and health and sanitation conditions, are the key determinants of children's nutritional status (World Bank, 2007).

Since the effects of commercialization or technology adoption vary greatly by gender, it is critically important to understand the linkages between the two. The main issues addressed in this chapter are as follows:

1. Are male-headed households more likely to be technology adopters than the female-headed households?
2. Among technology adopters, are male-headed households more likely to be food secure than female-headed households?

Addressing the above issues can help in program and policy design since gender differences in technology adoption rates have significant implications for agricultural productivity and food security.

The next section presents some case studies that examine critically the role of gender in technology adoption on household resource allocation and thus on household food security and nutrition. We present interesting findings from the developing countries and from the United States. The third section presents the empirical analysis and provides the implementation via STATA. The final section draws some conclusions and policy implications.

Review of selected studies

In an extensive study, Kumar (1994) examined the gender effects of technological change in maize production in Zambia and the role of gender in food consumption and nutrition. Maize is a staple food crop in Zambia, and the traditional maize growers primarily rely on increasing the available land to improve productivity. Increasing the land area by itself is not sufficient for sustained growth in maize production. Thus, adoption of new technology is critical in improving yields. The technology adoption considered in this study is the production of high-yielding hybrid maize.

The Eastern Province of Zambia is one of the major agricultural regions and provides about 90% of household income through production or employment on farms. Additionally, the majority of the households derive food consumption from their own farm production. Maize is the most important crop grown in the province, with 83% of overall land devoted to its production. Local maize constitutes about 60% of land devoted to production, with the remaining land allocated to hybrid maize production.

The rate of hybrid maize adoption is 34% for male-headed households compared to 22% for female-headed households. While this is true for small farms with farm sizes between 1 and 3 ha, female-headed households had a higher adoption rate for farm sizes between 3 and 5 ha. Interestingly, a 100% adoption rate is observed for all female- and male-headed households with farm sizes above 5 ha. The study concluded that male-headed households have a higher likelihood of adoption of hybrid maize than female-headed households after controlling for other important observable factors. This was concluded after conducting logistic and two-stage regression models. The results implied that if there were two households with all observable characteristics and their decision to adopt new technology was equal and they differed only in respect to the gender of the household head, then the male-headed household is more likely to adopt hybrid maize production than their female counterpart.

The study found that the smaller farms are resource constrained on average. However, it is more likely that women would adopt new technology compared to men when the constraints are mitigated. The lower adoption rates for female-headed households can possibly be explained by women's tendency to be more risk averse about new technologies. It could also imply that access to critical inputs (such as land and labor) and other inputs such as training and credit is harder for women to obtain.

Agricultural innovations such as technological changes and commercialization have complex linkages with food security and nutrition and are determined mainly by their effect on household income, household expenditures, and intrahousehold labor and resource allocations. A gender perspective plays an important role in analyzing such effects. The fundamental intrahousehold processes of labor and income allocation are gender-specific, and they differ significantly between adopters and nonadopters (Katz, 1995).

Since the early 1970s, hedgerow intercropping (also known as alley cropping)[1] has been a major agro-forestry technology promoted by the various CGIAR (International Institute for Tropical Agriculture (IITA), International Livestock Center for Africa (ILCA), International Livestock Research Institute (ILRI), and the International Center for Research in Agroforestry (ICRAF)) centers. It is crucial that extension agents take into account the impact of adopting this technology on intrahousehold decision-making processes, given the complex nature of this technology due to its composite nature and multiple outputs.

Based on a study in southern Nigeria and western Kenya, David (1998) observes that the gender division of labor and decision-making determines not only whether the household adopts a new technology or shifts to a cash crop production but also the distribution of its benefits among the household

---

1. Hedgerow intercropping involves the planting of nitrogen fixing trees in a hedge environment with the purpose of using foliage as mulch on crops planted between the hedges. The main objectives of this technology are as follows: (i) continuous cropping on depleted soils by introducing trees that help in preventing soil erosion and (ii) add nutrients to the soil through nitrogen fixation, nitrogen recycling, and mulch. Unlike other technologies, it links three components of the farming system: crop production, land and soil management, and livestock husbandry.

members. Women's active role in agricultural commercialization (either as a household head or as a codecision-maker) tends to increase household food security. The adoption of hedgerow intercropping was more widespread and lasting in the south-western part of Nigeria relative to the south-eastern part. In the south-western part, male-headed households adopted the new technology. In contrast, in the south-eastern part, gender-related decision-making was clearly the prominent factor that constrained the diffusion of new technology. The reason behind the low diffusion rate was that women farmers were not happy about mixing fodder trees with food crops—especially cassava.

For western Kenya, on the other hand (an area characterized by subsistence level mixed crop farming system), most of the households are female-headed due to the high rate of male emigration. An estimated 30%—60% of households were female-headed. Since a high percentage of households are female-headed and women are responsible for most of the farm work, male emigration constitutes an important constraint in the adoption of hedgerows. The authors cite two main reasons behind this low rate of adoption:

1. Women were unable to find additional labor (labor availability constraint) for cutting hedges and small trees for firewood.
2. The task generally involves male participation, since coppicing and spreading the mulch involve physical strength.

Female farm operators in Kenya and Ethiopia: recent evidence

Low efficiency is a problem in most developing agriculture, and is one of the reasons for food insecurity. Nyariki (2011) provides information on smallholder production efficiency in Kenya. The estimates from this study indicate high levels of inefficiency between farm sizes, seasons, and adopters and nonadopters of "modern" farming technologies. Further, the results also show that the major factors influencing performance are the level of education, gender, market access, and off-farm capital. A comparison of various farming practices shows that use of modern inputs and livestock-based capital could significantly improve farmers' performance. Thus, policies aimed at improving education, rural infrastructure, as well as assuring farmers of income through improved livelihood opportunities, and therefore reduced perceived uncertainty, could improve farm-level efficiency. Most importantly, the findings also provide support for prioritizing issues of farm production associated with women in policymaking.

Kassie et al. (2009) use data from Ethiopia to investigate the factors influencing farmers' decisions to adopt sustainable agricultural production practices, with a particular focus on conservation tillage and compost. Their estimates indicate differences among factors that influence the choice to use either tillage or compost. In particular, their study indicates that poverty and access to information, among other factors, significantly impact the choice of farming practices. Interestingly, they also find evidence that the impact of gender on technology adoption is technology-specific, while the significance of plot characteristics indicated that the decision to adopt particular technologies is location-specific. Furthermore, their analysis supports the hypothesis that sustainable farming practices enhance productivity, relative to the use of chemical fertilizers. This study is very useful for policymakers because the findings imply that we have to investigate all factors that influence adoption practices and to use this knowledge to formulate policies that encourage adoption. This is particularly true for developing countries where farmers face severe resource constraints, and where sustainable farming can present desirable alternatives.

Rights, norms, and institutions: beyond technology adoption

Theis et al. (2018) examine why technology adoption is not sustained in many parts of Ethiopia, Ghana, and Tanzania. The researchers focus on small-scale irrigation technologies, and particularly on intrahousehold gender disparity, and bring issues of power, control, and property rights (*use* vs. *fructus*) to this important subject. The study observes several interesting features about technology diffusion and adoption, and challenges policymakers to question the underlying assumptions about societal norms and traditions that belie policy implementation.

Theis et al. (2018) note that women have fructus rights only over small horticulture plots, with labor-intensive techniques. Further, the said rights:

  i. are always in exchange for providing all manual work, with limited or no assistance from the male head, or hired help
 ii. have no real significance, since women have to trade off farmwork for familial obligations
iii. are superfluous, since women do not possess any control over the sale of their output
 iv. are constructed and established without any provisions for renegotiation.

Substantial differences in control over technology, and informational asymmetry in output prices, time allocation, mobility and economic security, between genders persist at the household level. Importantly, access to technology is never equal between the sexes, and often suffers from elite capture. Consequently, policies designed to promote technology adoption among women, although well intended, focus primarily on *use* rights, and ignore the larger context, wherein other essential complementary rights are always absent. Hence, the study makes it clear that technology adoption cannot be an end in and of itself.

Similar findings from recent research in Africa and India merit consideration:

- Shikuku (2019) shows that the information dissemination regarding modern technology and adoption among farming communities in Uganda is much faster when the information-disseminating farmer is a female.
- Agricultural extension programs in South Kivu, Congo, are more effective in terms of technology adoption, if the participating female farmers are from female-headed households, rather than from male-headed households (Lambrecht et al., 2016).
- Female farmers in Uganda are more likely to adopt labor-intensive techniques, and agricultural extension programs are not successful in reversing this trend (Lambrecht et al., 2016 and 2018).
- Arimi and Olajide (2016) also stress the importance of institutional, motivational, innovational, attitudinal, environmental, and the male—female decision-making rights with respect to the success in the adoption of improved rice production technology in Nigeria.
- Large landowners and wealthy farmers elicit high self-perception scores, while the poor, less-educated, and female farmers have low self-perception scores, with respect to adoption of submergence-tolerant rice, in Eastern India (Yamano et al., 2015)[2]
- Lessons from adoption of biogas digesters in rural Ethiopia reemphasize the point that the technology diffusion process should be carefully designed, and account for local conditions and contexts, rather than attempting to fit a standard adoption prototype for all settings (Kelebe et al., 2017).

Recent research by Bryan and Garner (2020) covering women empowerment and rights over technology also echo similar concerns about institutional norms and gender preferences that undermine technology intervention programs. Overall, women value irrigation, particularly for crops and plots over which they have direct control in management. Women do show a preference for mechanized technology and financial independence. In those instances, where women have higher bargaining power, the farms had a greater likelihood of using modern time-saving techniques. Some possible solutions to address empowerment, *fructus* rights, and control for women farmers are as follows:

- provide increased market access
- entry into formalized value chains
- improved financial services including digital alerts about prices and transactions

---

2. Also see Hidrobo et al. (2020) regarding the low willingness to pay for information and communication technologies among female farmers in Ghana.

- easier and transparent documentation of sales
- community engagement on gender roles
- support groups for women along with training to build up social capital
- support groups that disseminate information
- informing men about potential benefits of joint ownership and cooperative rights.

Female farm operators in the United States

In recent years, we witnessed a larger participation among women in the farming sector, in the United States. For example, according to the USDA the number of women as principal operators was around 209,000 in 1997, and rose to around 307,000 by 2007. For instance, Pennsylvania lost 2000 farms between 1997 and 2002, but gained 1000 farms operated by women (Economic Research Service, 2004). In an interesting study, Trauger et al. (2010) note that in some parts of the United States, the number of women principal operators has grown, while during this time period the number of male-operated farms has fallen. Further, the fact that the number of women operators has increased, during a time when the number of farms in the whole economy has decreased, raises some interesting issues for social scientists. Additionally, Trauger et al. (2010) also note that this rise in the number of women in farming has coincided with the rise in the number of organic and sustainable farming operations and farmers markets in the United States.

Several interesting insights can be gleaned from these trends. Research findings suggest that women-operated farms differ in many ways than the traditional units operated by men. Women tend to operate smaller farms, tend to be involved in livestock production, and are less likely to be the primary operator of farms that produce major commodities such as dairy, cotton, corn, soybeans, and hogs.

Most importantly, Trauger et al. (2010) note that most women farmers engage in a type of agriculture that is different from conventional and commodity farming. They use data from interviews in Pennsylvania and identify what is called "civic agriculture" that provides the necessary impetus for the role of women entrepreneurs in recent years. Civic agriculture refers to embedded social and economic strategies that provide economic benefits to farmers, while at the same time providing socioenvironmental benefits to the community.

This new movement in the food systems, or "civic agriculture," engages in building local markets through direct sales to consumers. These practices promote community social and economic development in ways that commodity agriculture cannot. For example, direct marketing, locality-based food processing and procurement, community gardens, farmers markets, community-supported agriculture, box schemes, and preordered and bulk meat purchases are among the activities that support this movement.

Further, these researchers have also found evidence that women in sustainable agriculture take on nontraditional managerial roles, with primary responsibilities for the work and decision-making related to business development and management, resource allocation, production of crops and livestock, marketing of products, and development of new value-added businesses. They also find that women farm operators are more likely to engage in sustainable agriculture because they were supported and affirmed in their identities as farmers in the sustainable community.

Civic agriculture changes the concept of traditional agriculture by making the farm itself become an integral part of the service and consumption, thereby making the farm a public space for civic work. In this process, women have redefined successful farming in terms of providing services to their community, as well as in terms of profit and productivity. Women often lead the way in innovation on small and medium-sized farms as business owners and community leaders, and this is connected to larger-scale movements fostering gender equity. The idea of women managing and making economic decisions in agriculture has attracted a lot of attention among development economists who wish to promote gender equity and poverty reduction.

Women in agriculture: the global scene

The FAO Report (2012) provides illuminating information about the role of women in agriculture, choice of techniques, and poverty. Women make essential contributions to agriculture in developing countries, but their roles differ significantly by region and are changing rapidly in some areas. Their contribution to agricultural work varies even more widely depending on the specific crop and activity.

Closing the gender gap in agriculture would generate significant gains for the agriculture sector and for society. If women had the same access to productive resources as men, they could increase yields on their farms by 20%–30%. This could raise total agricultural output in developing countries by 2.5%–4%, which could in turn reduce the number of hungry people in the world by 12%. The potential gains would vary by region depending on how many women are currently engaged in agriculture, how much production or land they control, and how wide a gender gap they face. Policy interventions can help close the gender gap in agriculture and rural labor markets. The FAO reports information about the global trends on women's role in agriculture:

1. Women comprise 43% of the agricultural labor force, on average, in developing countries; this figure ranges from around 20% in Latin America to 50% in parts of Africa and Asia, but it exceeds 60% in only a few countries (FAO, 2010).
2. One generalization about women in agriculture holds true across countries and contexts: compared with their male counterparts, female farmers in all regions control less land and livestock, make far less use of improved seed

TABLE 4.1 Percent use of fertilizer and mechanical equipment in female- and male-headed households across countries.[a]

Country	% Fertilizer Use F	M	% Mech Equip Use F	M
Bolivia	20	30	Na	Na
Ecuador	11	25	10	22
Guatemala	45	60	15	33
Nicaragua	15	30	11	32
Panama	9	20	9	20
Bangladesh	20	55	1	5
Nepal	50	65	2	4
Pakistan	12	40	3	9
Tajikistan	39	42	3	4
Viet Nam	71	81	35	49
Ghana	11	22	0	4
Madagascar	15	29	6	18
Malawi	60	65	3	7
Nigeria	15	38	2	10

The FAO Report (2012) pages 35, 36, Figs. 14 and 15.
[a]*Notes: F and M refer to the female- and male-headed households. The entries are percentages for each of the household type.*

varieties and purchased inputs such as fertilizers, are much less likely to use credit or insurance, have lower education levels, and are less likely to have access to extension services. See Table 4.1 for the marked differences in the subtle aspects of gender equity.
3. There is substantial discrimination in the rural labor markets. In most countries and in keeping with global figures, women in rural areas who work for wages are more likely than men to hold seasonal, part-time, and low-wage jobs, and women receive lower wages for the same work. Further, agriculture continues to be the most important source of employment for women in rural areas in most developing countries.
4. Putting more income in the hands of women yields beneficial results for child nutrition, health, and education. Other measures, such as improving education, that increase women's influence within the household are also associated with better outcomes for children. Empowering women is a well-proven strategy for improving children's well-being.

5. It appears that in sub-Saharan Africa women are less likely than men to suffer from chronic energy deficiency (CED), while in South America and Asia, women are more likely than men to suffer from CED. For example, in Asia and the Pacific a larger share of girls than boys are underweight, whereas the opposite is true in sub-Saharan Africa.

Section highlights: female farmers and choice of techniques

The FAO Report (2012) provides a comprehensive survey of the state of women in global agriculture. The study covers many countries and provides an in-depth look at many issues related to gender equity and agriculture. The study finds that gender gaps exist for a wide range of agricultural technologies, including machines and tools, improved plant varieties and animal breeds, fertilizers, pest control measures, and management techniques. The study identifies a number of constraints that lead to gender inequalities. The inequities in the system are linked to the access to and adoption of new technologies, as well as in the use of purchased inputs and existing technologies.

The use of purchased inputs depends on the availability of complementary assets such as land, credit, education, and labor, all of which tend to be more constrained for female-headed households than for male-headed households. Adoption of improved technologies and inputs may also be constrained by women's lower ability to absorb risk.

The study points to significant gender differences in the adoption of improved technologies and the use of purchased inputs. For example, male-headed households show much wider use of fertilizers than their female counterparts in all the countries covered by the study, and the difference is the largest in Southern Asia (Bangladesh and Pakistan) and in West Africa (Ghana and Nigeria). In Ghana, for example, only 39% of female farmers adopted improved crop varieties (compared with 59% of male farmers) because they had less access to land and extension services.

Several studies from Kenya show that female-headed households have much lower adoption rates for improved seeds and fertilizers. The share of farmers using mechanical equipment and tools is quite low in all countries, but it is significantly lower for farmers in female-headed households, sometimes by very wide margins. For example, in Kenya, the value of farm tools owned by women amounted to only 18% of the tools and equipment owned by male farmers. Furthermore, lack of access to transportation technology often limits the mobility of women and their capacity to transport crops to market centers. The study provides us several key messages for policy:

- Across diverse regions and contexts, women engaged in agriculture face gender-specific constraints that limit their access to productive inputs, assets, and services. Gender gaps are observed for land, livestock, farm labor, education, extension services, financial services, and technology.
- For those developing countries for which data are available, between 10% and 20% of all landholders are women, although this masks significant differences among countries even within the same region. The developing countries having both the lowest and highest shares of female landholders are in Africa.

Continued

> **Section highlights: female farmers and choice of techniques—cont'd**
> - Among smallholders, farms operated by female-headed households are smaller in almost all countries. The livestock holdings of female farmers are much smaller than those of men in all countries, and women earn less than men from their livestock holdings. Women are much less likely to own large animals, such as cattle and oxen that are useful as draught animals.
> - Farms run by female-headed households have less labor available for farm work because these households are typically smaller and have fewer working-age adult members and because women have heavy and unpaid household duties that take them away from more productive activities.
> - Education has seen improvements in gender parity at the national level, with females even exceeding male attainment levels in some countries, but in most regions women and girls still lag behind. The gender gap in education is particularly acute in rural areas, where female household heads sometimes have less than half the years of education of their male counterparts.
> - Smallholders everywhere face constraints in accessing credit and other financial services, but in most countries the share of female smallholders who can access credit is 5%–10% points lower than for male smallholders. Access to credit and insurance are important for accumulating and retaining other assets.
> - Women are much less likely to use purchased inputs such as fertilizers and improved seeds or to make use of mechanical tools and equipment. In many countries women are only half as likely as men to use fertilizers.

Uganda's coffee market: a case study

The FAO Report (2012) presents an interesting case study based on Uganda's coffee market and the divergences in the economic outcomes between men and women in this sector. Coffee is Uganda's largest export, providing employment to an estimated five million people. Only 23% of the sector is managed by female-headed households, as reported by Hill and Vigneri (2009). For small farmers, coffee is usually intercropped with staples such as banana, plantain, beans, sweet potatoes, and maize. Farming methods for coffee production are simple, with minimum use of purchased inputs such as fertilizer or pesticides. Female-headed households had less labor, land, and coffee trees than male-headed households; they also had lower levels of wealth and education. Women household heads tended to be older; many were wives who had taken over when their husband had died. As a result of these basic differences in scale, liquidity, and human capital, we may expect crop choice, production methods, and access to markets to be quite different for male- and female-headed households.[3]

---

3. For the importance of adequate food supply in Uganda, see WFP (2013a,b).

The share of labor allocated to coffee production and the proportion of trees harvested were comparable between male- and female-headed households, as was the yield per producing tree. However, because female-headed households farmed on a much smaller scale, women sold smaller amounts than men. Access to production centers and markets was a big advantage for male-headed farms. Several farmers transported their coffee to market, which allowed them to sell it at a higher price. Members of male-headed households were more likely than those of female-headed households to travel to market to sell their coffee. This may be because men were more likely to own a bicycle and could therefore travel to the market more easily than women. Farmers received a higher price for their coffee if they chose to mill it at the market before selling it.

The study concludes that gender differences in marketing are largely explained by the fact that women market smaller quantities of coffee and do not own bicycles. It also finds that a major constraint facing women is their relative difficulty in accessing marketing channels that allow added value. By engaging in marketing channels in which they add value, male-headed households received 7% more per kilogram of coffee.

In a comprehensive study for Ghana, Doss and Morris (2001) examined the following set of issues:

1. Does gender add any additional understanding of the technology adoption process?
2. To what extent are the observed differences in the rates at which men and women adopt technology attributable to gender-linked differences in access to complementary inputs, such as land, labor, and extension services?

The above questions have significant practical implications, since men and women could have fundamentally different preferences in technology adoption and access to complementary inputs, which prevent them from adopting technology at the same rate.

The case study is undertaken for maize production in Ghana, since it is the most important cereal crop both in terms of production and consumption. Data were collected on the adoption of modern varieties (MVs) and chemical fertilizer using a three-stage randomized procedure. A sample of 420 maize farmers located in 60 villages in the country was selected. A two-stage probit model was specified, where adopting MV and chemical fertilizer were related to each other with a vector of explanatory variables that explain technology adoption. The explanatory variables were gender of the farmer, farmer's age, farmer's education, amount of land owned by the farmer, level of infrastructure, and household size measuring labor availability.

Overall, the results indicated that technology adoption decisions depended critically on access to resources, rather than on gender per se. The above finding should be interpreted with caution, since it does not necessarily mean that MV and fertilizer adoption are gender-neutral technologies. This is because if adopting MV and fertilizer depends on the critical access to resources (such as availability of land, labor, and extension services) and men have better access to resources than women, then under these conditions technologies will not benefit men and women equally.

The use of gender of the household head as a proxy can be effective for two main reasons. First, the data are readily available from most surveys. Second, it is usually the case that the majority of decision-making is vested on the household head. Posel (2001) observed that for female-headed households, in particular, heads have a final say over the majority of the decisions even when they may not be earning most of the income in South Africa. In fact, the study argued that the highest income earner is usually a worse predictor of decision-making than the household head. The gender of the household head is thus very useful in distinguishing households on their access to economic resources and hence their welfare outcomes.

Given that the impact of technology adoption on food security and nutrition is essentially mixed and is highly dependent on variables such as the nature of the crop, the control of production and income, the allocation of household labor, the maintenance of subsistence production, land tenure policies, and prices of food and cash crops, it is important to understand the gender dimension of the technology adoption process on household food security and nutrition. It is the intervening factors and not crop choice that appear to be crucial in understanding the impact of technology adoption on nutrition. From a policy perspective, food and agricultural policies and programs that target the most vulnerable segments of the population are most likely to have a positive benefit on food security and nutrition.

In this chapter, we investigate whether commercialization and technology adoption varies by gender and how such a decision influences household food security and nutritional status through Cramer's V and Phi coefficient statistical technique.

Empirical analysis

To evaluate empirically the impact of agricultural innovations on the gender of the household head, one needs to address the following questions:

- Are male-headed households more likely to be technology adopters than the female-headed households?
- Among technology adopters, are male-headed households more likely to be food secure than female-headed households?
- Is cash crop commercialization more likely to be adopted by male-headed households than female-headed households?

The first question investigates if men are more likely to adopt new technology than women. The second question investigates if the effects of technology adoption are gender-biased, i.e., if they favor men more than

Effects of technology adoption and gender of household head Chapter | 4 119

women. The final question addresses whether male-headed households are more likely to undertake commercialization than producing for home consumption.

We again use the "cross-tabulation procedure" (similar to Chapter 3) in order to address the above set of issues. This method is appropriate since the relationship between two or more categorical variables[4] can be established with cross-tabulation. Cramer's V and Phi test statistic will be introduced in addition to cross-tabulation and chi-square test statistic in this chapter. This is because we are investigating if male- or female-headed households are more likely to be technology adopters, whether the different households (male or female) are more food secure and finally we want to determine if male- or female-headed households are more likely to commercialize crops and thereby receive greater income from the proceeds.

Data description and analysis

The main variables used in this analysis are as follows:

1. HYBRID: whether a household grows hybrid maize (HYBRID = 1) or not (HYBRID = 0) and represents technological adoption at the household level.
2. FEMHHH: whether the household head is male (FEMHHH = 0) or female (FEMHHH = 1) and hence represents gender of the household head.
3. CASHCROP: tobacco, groundnuts, cotton, and plantain are the major cash crops in Malawi. The variable is defined as CASHCROP = 1 if the household grows at least one of these four major cash crops and 0 otherwise. CASHCROP thus represents commercialization of agriculture at the household level.
4. CALREQ: the measure of food security used relies on per adult equivalent calorie intake for households[5]. It is defined as households being able to satisfy at least 80% of the requirement for calorie intake. This variable is coined CALREQ. If a household's daily per adult equivalent calorie intake (CALADEQ) is at least 80% of the requirement, 2200 kcal, then the household is qualified as "food secure"[6]. The variable is defined as CALREQ = 1 if the household is food secure and CALREQ = 0 otherwise.

---

4. Categorical variables are those where distinct categories exist, such as gender (female, male), ethnicity (Whites, Asian, Hispanic), and cash crop commercialization (commercialize, do not commercialize), etc.
5. Since the major focus of this chapter is to understand the gender dimension of new technology adoption and commercialization, we concentrate our analysis on only one measure of food security, namely per capita adult equivalent calorie intake falling short of 2200 kcal per day
6. The percent intake of the average energy requirement is used as a means of examining calorie adequacy. Individual energy requirement is estimated according to the age-sex specific recommendation of the joint FAO/WHO/UNO committee. The standard for an adult male equivalent is 2200 kcal.

Descriptive analysis: cross-tabulation results

We initially investigate the relationship between FEMHHH (the gender of the household head), which is the independent variable, and HYBRID (growing hybrid maize), which is a measure of technology adoption. Later, we will also examine the relationship between FEMHHH and food security and the relationship between FEMHHH and CASHCROP (a measure of agricultural commercialization). Examining the above relationships will help us in determining through what channels gender linkages affect technology adoption and commercialization process in agriculture. It is important to note that all the above variables are nominal or categorical variables.

Table 4.2 reports the cross-tabulation results of technology adopters by household head. The results indicate that almost 82% of the male-headed households are technology adopters compared to 18% of female-headed households. This result confirms to the existing literature that the male-headed households are more likely to adopt new technology relative to female-headed households. It is not possible to determine from the data why this is the case. One possibility is that female-headed households, on average, are generally risk averse in adopting new technology or they may not have critical access to the inputs required to adopt new technology such as land, labor, and credit. It is also likely that a combination of risk aversion and access to critical inputs are detrimental factors for women in adopting hybrid maize. Next, we examine the relationship between gender of the household head (FEMHHH) and food security measure (CALREQ) to understand which groups of households are relatively more food secure.

About 28% of the households grow hybrid maize in the sample. Among them, only 18% of the households are female headed. The cross-tabulation results in Table 4.3 indicate that, among hybrid maize growers, almost 86% of male-headed households are food insecure compared to 13.6% of the female-headed households. This interesting result indicates that, although a higher proportion of male-headed households adopt hybrid maize technology, a lower proportion of female-headed households among the adopters is food insecure. Next, we examine whether female-headed households are also less likely to adapt to cash crop commercialization just as the literature predicts.

TABLE 4.2 Cross-tabulation results of technology adopters and gender of household head.

		FEMHHH		
		Female	Male	Total
HYBRID	No	128	305	433
	Yes	31	140	171
	Total	159	445	604 = n

TABLE 4.3 Cross-tabulation results of food security and gender of household head.

		FEMHHH		
		Female	Male	Total
CALREQ	INSECURE	39	247	286
	SECURE	22	140	162
	Total	61	387	448 = n

Cramer's V and phi tests

The value of the chi-square itself is difficult to interpret, since it is a function of the sample size, the degree of independence between the variables, and the degrees of freedom (Hamburg and Young, 1994). To overcome this difficulty of interpretation, several statistics have been created that measure the "degree of association" between any two nominal variables. Two of these measures, namely phi and Cramer's V, are based on the chi-square itself, and the value of these statistics is the same for any sample size. Since the above measures are all based on the chi-square statistic, the significance level is also based on the significance of the chi-square statistic. In Tables 4.4–4.6, we want to determine if the relationship among the variables (the cross-tabulation results) is

TABLE 4.4 Tests between technology adopters (HYBRID) and gender of household head (FEMHHH).

	Value	P value
Phi	−0.117	0.004
Cramer's V	0.117	0.004
Number of valid cases	604	

TABLE 4.5 Tests between food security (CALREQ) and gender of household head (FEMHHH).

	Value	P value
Phi	−0.001	0.987
Cramer's V	0.001	0.987
Number of valid cases	448	

TABLE 4.6 Tests between cash crop commercialization (CASHCROP) and gender of household head (FEMHHH).

	Value	Significance
Phi	−0.097	0.017
Cramer's V	0.097	0.017
Number of valid cases	604	

significant or if it is just due to random variability. The null hypothesis is given by: H_0: no relationship between technology adoption and gender of the household head, i.e., incidences of hybrid maize adoption are not statistically different between the male- and female-headed households.

Since the significance level (*P* value) is less than 0.01, the null hypothesis is rejected at the 1% level of significance. Thus, we conclude that the incidences of hybrid maize adoption are statistically different between the male- and female-headed households. Although the value of the Phi coefficient is low (−0.117), it is statistically significant at the 1% level.

Next, we want to examine the null hypothesis of the relationship between food security and gender of the household head for hybrid maize growers. This is given by: H_0: no relationship between food security and gender of the household head for hybrid maize growers, i.e., both male- and female-headed households are not statistically different in regard to food security.

Since the significance level (for both Cramer's V and Phi statistic) is greater than 0.1, the null hypothesis cannot be rejected even at the 10% level. For both groups of households (male- and female-headed), the incidences of food security are not statistically different among hybrid maize growers. We find no pattern of relationship emerging between gender of the household head and food security for the technology adopters for this sample. One cannot conclude that technology adoption has a gender-biased impact on household food security.

Finally, we examine the null hypothesis of the relationship between cash crop growing and gender of the household head. This is given by: H_0: no relationship exists between cash crop growing and gender of the household head, i.e., incidences of cash crop commercialization or adoption are not statistically different between male- and female-headed households.

Since the significance level (both Cramer's V and Phi statistic) is 0.017, the null hypothesis cannot be rejected at the 5% level. The incidences of cash crop commercialization are statistically different for both the groups of households (male- and female-headed). We can conclude that the incidences of cash crop commercialization are statistically different between male- and female-headed households. Although the value of the Phi coefficient is low (−0.097), it is statistically significant at the 5% level.

Conclusion and policy implications

As pointed out by Doss (2001), understanding the linkages between female-headed households and technology adoption/commercialization is inherently complex. The African farm household is a multifaceted entity that undertakes various agricultural and nonagricultural activities and operates with different accessibility to various inputs, such as land, labor, credit, and extension services. It is clear, however, that gender matters in new technology adoption and commercialization patterns. The gender role and responsibilities may change in response to change in agricultural technology or urbanization, but gender remains an important analytical concept. Many interventions and projects that were designed to improve the conditions of rural women in Africa failed, since these projects did not consider the complex role women played in households and communities in adopting the new technology (Doss, 2001). Identifying the technologies that are beneficial to improve the economic conditions of women thus remains a challenge to researchers and policymakers in developing countries. This is because new technologies can have both positive and negative impacts as they may increase output but, at the same time, may increase women's labor input.

From the recent studies that examine the conditional relationship between gender of the household head and technology adoption decision, the evidence is generally mixed (World Bank, 2007). After controlling for other relevant characteristics, female-headed households in general are less likely to adopt new technology compared to male-headed households (Asfaw and Admassie, 2004; Chirwa, 2005). Very few studies find that female-headed households are more likely to adopt new technology than male-headed households (Bandiera and Rasul, 2006).

Section highlights: Covid-19, women's burden, and the digital divide

The income and livelihood loss among daily wage earners, due to Covid-19, has placed thousands of families at the risk of starvation and malnutrition. Using data from India, Acharya (2020) and Limaye et al. (2020) note that the major portion of this burden is borne by women.

Acharya (2020) notes that, compared to men, a greater portion of women report food shortage and reduced intake. Consequently, the pandemic has reinforced and widened the preexisting differences in intrahousehold gendered allocations of food. Pregnant women report consuming only three food groups on average, instead of the recommended five food groups. Pregnant women also consume less meat and dairy products, which are likely to produce anemia, micronutrient deficiency, and malnutrition among women and children.

Crucially, Acharya (2020) notes that there has been a decrease in the number of pregnant women who were provided with free meals under the preexisting government distribution scheme. There has been a big gap between the number of women who need, and those that received nutrition services.[7]

Alvi et al. (2020) indicate that impacts Covid-19 on women go beyond economic outcomes. Indeed, the very nature of data collection through phone surveys itself is beset with gendered issues, creating difficulties with response rates, response bias, and thereby skewing data quality, particularly if the respondents are women.

Trying to uncover the impact of Corona on women's livelihood through phone surveys exposed the researchers to the inherent systemic failures that create a digital gender divide. The primary difficulty was in unreliable cellphone networks, creating difficulties with reaching potential women respondents. Besides poor network services, a large portion of women do not have their own cellphones, or do not have an active service, or airtime, thereby affecting the survey response rates.

Indeed, as the researchers point out, with a reduction in income due to the pandemic, the first expense that is removed from the budget is the female member's cellphone. Family pressures from the spouse or the in-laws often force a major portion of the female respondents to keep the phone speakers on, thereby providing skewed responses. For example, with the speakers on, a major portion of women indicate that their husbands are the primary decision-makers on earnings.

From the empirical analysis in this chapter, it is evident that male-headed households in Malawi have a higher propensity to adopt new technology and sell cash crops to earn more cash revenues than female-headed households. The former (greater propensity to adopt new technology) is possibly due to men having greater access to key inputs such as land and labor as well as availability of complementary resources such as credit and more information from extension services. The latter (cash crop commercialization) is probably because men tend to specialize in cash crop production for earning greater cash revenues and providing income for the members of the household, while women specialize in subsistence production (maize and other food crop production) to improve the nutritional status of children. Additionally, we find no relationship between the gender of the household head and food security for the technology adopters. It is more likely that technology adoption has no gender-biased impact on household food security.

A few lessons and programmatic implications for the agricultural research community deserve special attention (Doss, 2001). First, there is no silver bullet that will resolve all agricultural productivity issues in Africa. Instead, new technologies will have a differential impact on men and women farmers depending on their access to the critical inputs. Second, researchers and

---

7. See Arrieta et al. (2020) for an impact of Covid-19 on women's mental health in the nonfarm sector, and the importance of SHG for women in this context.

extension agents must continually interact with local farmers to understand how adoption of a new technology may affect the dynamic within the household and community members. Third, it is also crucial to understand the patterns of labor and land allocation and how that affects individual members of the household before and after the introduction of a new technology. Fourth, technology that has a more direct impact on women's welfare should be taken into consideration in program design. This is because innovations that decrease women's labor hours in the farm can actually improve women's well-being. Finally, in addressing the gender dimension of new technologies, it is important to understand why women do not adopt certain technologies—is it because they have preferences in growing certain crops or is it the case that they face different constraints relative to men? The policy solution might be to address those constraints. It is evident that the simple dichotomies such as men's crops and women's crops, cash crops and food crops, male- and female-headed households will not provide sufficient insight by themselves.

The FAO Report (2012) also provides key policy proposals to alleviate these issues. Policy proposals cover (i) gender equality is good for agriculture, food security, and society and (ii) collaboration between governments, civil society, the private sector, and individuals. Basic principles for achieving gender equality and empowering women in agriculture included in the Report are as follows:

1. Eliminate discrimination against women under the law: Governments and civil society must work together to ensure that women are aware of their rights and have the support of their governments, communities, and families in claiming their rights.
2. Strengthen rural institutions and make them gender-aware: Efforts are required to ensure that women and men are equally served by rural institutions such as producers' organizations, labor unions, trade groups, and other membership-based organizations such as extension services, animal health services, and microfinance organizations, should consider the specific needs of men and women to ensure that their activities are gender-aware.
3. Free women for more rewarding and productive activities: Investments in basic infrastructure for essential public services can liberate women from the drudgery of difficult manual work, and free them for more rewarding and productive work. The building of the human capital of women and girls is fundamental, and the development of general education and the ongoing transfer of information and practical skills will broaden the range of choices women can make and give them more influence within their households and communities.
4. Bundle interventions: Some assets are complementary, and the constraints women face are often mutually reinforcing. It is impossible to separate women's economic activities from their household and community roles

and responsibilities. The gender-related constraints women face due to power relations within the family and community affect their ability to engage in economic activities and retain control over the assets they obtain.
5. Improve the collection and analysis of sex-disaggregated data: Understanding of many gender issues in agriculture—including crop, livestock, fisheries, and forestry sectors—is hindered by the lack of sex-disaggregated data, and inadequate analysis of the data that exist. Agricultural censuses should focus more attention on areas in which women are relatively more active and collect sex-disaggregated data on ownership of, access to and control over productive resources such as land, water, equipment, inputs, information, and credit.
6. Make gender-aware agricultural policy decisions: Virtually any agricultural policy related to natural resources, technology, infrastructure, or markets affects men and women differently because they play different roles and experience different constraints and opportunities in the sector. Making women's voices heard at all levels in decision-making is crucial in this regard.

Cramer's V in STATA

For illustrative purposes we use the following 17 observations on GENDER, CALREQ, and HYBRID, for maize production and estimate the Cramer's V coefficient in STATA:

Obs	GENDER	CALREQ	HYBRID
1	1	1	1
2	1	1	0
3	0	0	1
4	1	0	1
5	0	1	1
6	1	0	1
7	0	1	0
8	1	0	1
9	0	0	0
10	1	1	0
11	0	0	0
12	0	0	0
13	1	0	0
16	0	1	1
17	1	0	0
18	0	1	0
19	1	1	1

In the above table, all variables are categorical variables with male-headed households coded by setting GENDER = 1. The other variables are defined in the previous section. It is better to start by exploring the data and look at the descriptive statistics, which we can obtain, using the summarize command in STATA. The input and the output from STATA are given below for the 17 observations from the Table given above.

We can see from the output that there are 17 observations and that all the variables of interest are categorical variables. We can undertake the cross-tabulations and the Cramer's V using the following command STATA:

.tab gender calreq, column nokey chi2 V

As seen in the previous chapters, the tab command produces cross-tabulations in STATA. The column command produces the percentages for the row and column components. The nokey command suppresses the key that accompanies each table. The chi2 and the V command produce the chi-square test and the Cramer's V coefficient for the data. We produce the STATA input command line and the output as a screenshot for our test between GENDER and CALREQ below.

Although the Cramer's V coefficient indicates a strong negative relation between GENDER and CALREQ, the relation is not statistically significant because of the high p[8] value of 0.8^6. Thus we cannot reject the null hypothesis, which assumes that there is no relationship between technology adoption and gender of the household head, i.e., incidences of hybrid maize adoption are not statistically different between the male- and female-headed households.

In this example, the null hypothesis cannot be rejected at the 1% level of significance. Thus, we conclude that the incidences of hybrid maize adoption are not statistically different between the male- and female-headed households. Similarly, the chi-square test and the Cramer's V coefficient for the categorical variables GENDER and HYBRID are produced in the STATA input command line and the output in the screenshot below:

The results from STATA and the Cramer's V coefficient's estimate indicate that there is a strong positive correlation between HYBRID and GENDER, but once again, even in this case, we cannot reject the null hypothesis that there is no relationship between food security and gender of the household head for hybrid maize growers, i.e., both male- and female-headed households are not statistically different in regard to food security. Since the significance level of the Pearson chi statistic is greater than 0.1, the null hypothesis cannot be rejected even at the 10% level. For both groups of households (male- and female-headed), the incidences of food security are not statistically different among hybrid maize growers. We find no pattern of relationship emerging between gender of the household head and food security for the technology adopters for this sample. One cannot conclude that technology adoption has a gender-biased impact on household food security.

---

8. See Exercise 8.

Technical appendices

Phi coefficient and Cramer's V

While the chi-square test is useful for determining whether there is a relationship, it does not tell us the strength of the relationship. Symmetric measures such as Phi coefficient and Cramer's V attempt to quantify this relationship and are based on the chi-square statistic that controls for the sample size. They are designed for use with nominal data and with chi-square they jointly indicate the strength and the significance of a relationship. While these measures give some sense of the strength of the association, they do not, in general, have an intuitive interpretation (Lowry, 2003).

Phi coefficient (Φ)

The Phi coefficient is a measure of the degree of association between two binary variables. Phi is the ratio of the chi-square statistic to the total number of observations, i.e., $\phi = \sqrt{\chi^2/N}$. The range of phi varies between -1 and $+1$ for 2×2 tables. Conceptually, phi is the application of the Pearson r to dichotomous variables. If one ran Pearson r on these data, they would get exactly the same result. Since phi has a known sampling distribution, it is possible to compute its standard error and significance. SPSS and other major packages report the significance level of the computed phi value.

The general rule of thumb for Phi coefficient of correlation is:

- -1.0 to -0.7 strong negative association
- -0.7 to -0.3 weak negative association
- -0.3 to $+0.3$ little or no association
- $+0.3$ to $+0.7$ weak positive association
- $+0.7$ to $+1.0$ strong positive association.

It is the most "optimistic" of the symmetric measures and, unlike most association measures, does not have a theoretical upper bound when either of the variables has more than two categories.

Cramer's V

Cramer's V is usually appropriate for tables that are larger than 2×2. Cramer's V is a rescaling of phi so that it varies between 0 and 1. The formula for Cramer's V is $V = \sqrt{\chi^2/N(k-1)}$, where N is the total number of observations and k is the smaller of the number of rows and columns. For 2×2 tables, Cramer's V is equal to the absolute value of the Phi coefficient. This is because since $k = 2$, the $(k-1)$ term becomes 1.

Applications in R

We again use the cross-tabulation procedure, which is used in Chapter 3, to address the validity of different groups. In this chapter, Cramer's V and Phi test statistics will be introduced in addition to the cross-tabulation and chi-square test statistics. These two statistics have been created to measure the degree of association between any two nominal variables because Chi-square does not indicate how significant and important relation is. Cramer's V and phi coefficient are a posttest to give additional information. First, we examine three sets of frequencies between the two variables (HYBRID-FENHHH, CARLEQ-FEMHHH, and Cashcrop-FEMHHH) to examine if male- or female-headed households are more likely to commercialize crops and thereby receive higher income from the proceeds. We use the *table* and *summary* command for investigating descriptive statistics.

```
> table(HYBRID,FEMHHH)          > table(CALREQ,FEMHHH)
        FEMHHH                          FEMHHH
HYBRID   0  1                   CALREQ   0  1
     0  82 50                        0  63 38
     1  42 26                        1  61 38
> table(Cashcrop,FEMHHH)
          FEMHHH
Cashcrop   0  1
       0  69 46
       1  55 30
```

```
> summary(data_chapter4)
      Obs            CALREQ           FEMHHH          HYBRID          Cashcrop
 Min.   :  1.00   Min.   :0.000   Min.   :0.00   Min.   :0.00   Min.   :0.000
 1st Qu.: 50.75   1st Qu.:0.000   1st Qu.:0.00   1st Qu.:0.00   1st Qu.:0.000
 Median :100.50   Median :0.000   Median :0.00   Median :0.00   Median :0.000
 Mean   :100.50   Mean   :0.495   Mean   :0.38   Mean   :0.34   Mean   :0.425
 3rd Qu.:150.25   3rd Qu.:1.000   3rd Qu.:1.00   3rd Qu.:1.00   3rd Qu.:1.000
 Max.   :200.00   Max.   :1.000   Max.   :1.00   Max.   :1.00   Max.   :1.000
```

We can install the *lsr* packages to calculate the Cramer's V with the *CramersV* command. To install a package, we can use the *install.packages* command and load the downloaded package with the *library* command as below.

```
> install.packages("lsr")
```

```
> library(lsr)
```

Since the Cramer's V and phi tests are all based on the chi-square statistic, the significant level is also based on the significance of the chi-square statistic. Firstly, we can check the significance through chi-square analysis, and then we can use the *CramersV* command to calculate the Cramer's V to identify how much two variables are related. From the test, we can examine if the relationship between the variables (the cross-tabulation results) is significant or if it is just due to random variability.

The first example is the relationship between HYBRID and FENHHH variables (table1). The null hypothesis is given by: H_0: incidences of hybrid maize adoption are not statistically different between the male- and female-headed households. The significance level from the chi-squared test is 1, which means the null hypothesis is not rejected. Thus, we conclude that the incidences of hybrid maize adoption are not statistically different between the male- and female-headed households. Since there are no significant differences between the two variables, we cannot interpret the Cramer's V value.

```
> table1 <-table(HYBRID,FEMHHH)
> chisq.test(table1)

        Pearson's Chi-squared test with Yates' continuity correction

data:  table1
X-squared = 4.8846e-31, df = 1, p-value = 1

> cramersV(table1)
[1] 4.941957e-17
```

The second example is the relationship between CALREQ and FENHHH variables (table2). The null hypothesis is given by: H_0: no relationship between food security and gender of the household head. The significance level from the chi-squared test is also 1, so we cannot reject the null hypothesis. We find no pattern of relationship emerging between the gender of the household head and food security for this sample.

```
> table2 <- table(CALREQ,FEMHHH)
> chisq.test(table2)

        Pearson's Chi-squared test with Yates' continuity correction

data:  table2
X-squared = 0, df = 1, p-value = 1

> cramersV(table2)
[1] 0
```

The last example is the relationship between Cashcrop and FENHHH variables (table3). The null hypothesis is given by: H_0: no relationship exists between cash crop growing and the gender of the household head. The significance level from the chi-squared test is 0.6, so we cannot reject the null hypothesis. Since there are no significant differences between the two variables, we cannot interpret the Cramer's V value. However, if it is significant, we can interpret there is a 0.03 relationship within two variables.

```
> table3<- table(Cashcrop,FEMHHH)
> chisq.test(table3)

        Pearson's Chi-squared test with Yates' continuity correction

data:  table3
X-squared = 0.28137, df = 1, p-value = 0.5958

> cramersV(table3)
[1] 0.03750822
```

Exercises

1. Recollect the food security indicator INSECURE developed in Chapter 2. Undertake a cross-tabulation procedure and determine which households are more insecure. Is there any relationship emerging between gender of the household head and the INSECURE variable?
2. After studying the paper by Doss (2001), carefully examine how gender affects technology adoption among African farmers. What are the main challenges that women farmers face? Suggest policy measures that can benefit the adoption of new technology for women farmers.
3. From the empirical results of this chapter, discuss critically the role of gender in technology adoption process. Discuss whether male-headed households are more likely to be food secure than their female counterparts. What can you infer from the role of gender in cash crop commercialization from the results?
4. Refer to http://www.fao.org/docrep/013/i2050e/i2050e.pdf to get the FAO Report (2012) used in this chapter. Use Table A5 in this report and identify the countries that have the highest share of rural households that are female-headed.
5. Refer to the FAO Report and identify a few more characteristics of female poverty in agriculture. Which countries are the worst affected in this context?

STATA exercise

Use the STATA commands developed in this chapter and conduct a chi-square test for the following data given in the table below, with 10 observations on GENDER, CALREQ, and HYBRID, for maize production:

Obs	GENDER	CALREQ	HYBRID
1	1	1	1
2	1	1	0
3	0	0	1
4	1	0	1
5	0	0	0
6	1	1	0
7	0	0	0
8	0	0	0
9	1	0	0
10	1	1	0

The STATA code for the above problem is:
.summarize
.tab gender calreq, column nokey chi2 V

.tab gender hybrid, column nokey chi2 V
STATA output for each of the lines is:

```
. summarize

    Variable |        Obs        Mean    Std. Dev.       Min        Max
-------------+--------------------------------------------------------
         obs |         10         5.5     3.02765          1         10
      gender |         10          .6    .5163978          0          1
      calreq |         10          .4    .5163978          0          1
      hybrid |         10          .3    .4830459          0          1
```

The first set of results is the descriptive statistics. The Cramer's V is given below for a test between gender and Calreq:

```
. tab gender calreq, column nokey chi2 V
```

| | calreq | | |
gender	0	1	Total
0	4	0	4
	66.67	0.00	40.00
1	2	4	6
	33.33	100.00	60.00
Total	6	4	10
	100.00	100.00	100.00

```
            Pearson chi2(1) =   4.4444    Pr = 0.035
              Cramér's V =   0.6667
```

As explained in the last chapter, a table of expected frequencies is computed, and then the other results are obtained from this information:

From the output, the row totals and the column totals are first used to computed the *expected frequencies*:

Gender	Calreq 0	1
1	(4 × 6)/10 = 2.4	(4 × 4)/10 = 1.6
2	(6 × 6)/10 = 3.6	(4 × 6)/10 = 2.4

$$\chi^2 = \frac{(4-2.4)^2}{2.4} + \frac{(0-1.6)^2}{1.6} + \frac{(2-3.6)^2}{3.6} + \frac{(4-2.4)^2}{2.4}$$
$$= 1.06 + 1.6 + 0.77 + 2.67 = 4.44$$

Cramer's
$$V = \sqrt{\frac{4.44}{10 \times 2 - 1)}} = 0.44$$

We can reject the null because the computed chi-square value is larger than the table value = 3.841 (at 0.05 level of significance with degrees of freedom = 1). Since Cramer's V is between +0.3 and +0.7, we can conclude that there is a weak positive association between the two variables.

```
. tab gender hybrid, column nokey chi2 V

             |     hybrid
    gender   |    0         1    |   Total
   ----------+-------------------+--------
          0  |    3         1    |    4
             | 42.86     33.33   |  40.00
   ----------+-------------------+--------
          1  |    4         2    |    6
             | 57.14     66.67   |  60.00
   ----------+-------------------+--------
     Total   |    7         3    |   10
             |100.00    100.00   | 100.00

        Pearson chi2(1) =   0.0794   Pr = 0.778
          Cramér's V    =   0.0891
```

We can follow the same steps given above to generate the output results in this case as well. Since the Chi-square value is smaller than the table value, we cannot reject the null hypothesis. Further, the result is also reestablished with the Cramer's V = 0.08 which lies in the interval [−0.3, +0.3] indicating no association.

Chapter 5

Changes in food consumption patterns: its importance to food security—application of one-way ANOVA

Chapter outline

Introduction	135	One-way ANOVA in STATA	153
Determinants of food consumption patterns and its importance to food security and nutritional status	138	Conclusion and policy implications	158
		One-way ANOVA	159
		Underlying assumptions in the ANOVA procedure	160
Impact on food security	139	Decomposition of total variation	160
Review of selected studies	140	Number of degrees of freedom	161
Food consumption patterns for developing countries	141	F-test and distribution	161
Nutritional and economic outcomes	144	Relation of F to T-distribution	161
		STATA workout	162
Food consumption patterns in the United States	145	Fruit intakes per week for three income groups (F_1, F_2, and F_3)	162
Food consumption patterns in India and China	147	Compute mean square between and mean square within	164
Empirical analysis and main findings	150	Compute the calculated value of F	164
Data description	150	One-way ANOVA in R	165
Analysis method	150	Exercises	168
Results	150		

Consumption patterns continue at the expense of the environment and peaceful co-existence. The choice is ours.

—Wangari Maathai, Nobel Peace Laureate.

Introduction

Recent trends in globalization of food systems, increased agricultural trade, and income increases of various segments of the population have profound implications of what people consume and how they change their consumption patterns. For example, increase in household incomes could result in an increase in average daily food intake (Kumar and Joshi, 2013). The average global calorie intake had reached 2800 kcal (kcal/person/day) at the end of 2000 with developing country average expanding by more than 30% (Pingali and Stringer, 2003). The substantial increase in food consumption can be attributed to a combination of economic growth, increased use of irrigated land, long-term declines in food prices, and rapid growth of imports from developed economies. However, not all countries have shared the benefit of the increase in consumption. In sub-Saharan Africa, the intensity of food insecurity[1] remains a serious challenge during this century.[2] Recent food insecurity data are greatly worrisome, and estimates from FAO suggest that 43 countries will have average food consumption levels of less than 2500 kcal/day by 2015, with the number of actually undernourished increasing by 9 million (Pingali and Stringer, 2003). Most of this food insecurity is a result of food shortages caused by civil unrest, wars, and other natural calamities, and recent food price increases are likely to exacerbate the situation. As rising food prices affect the poor directly, the greatest concern is their impact on food consumption. The short-term impacts are alarming, with income falling by 25% and food consumption by almost 20% (Overseas Development Institute, 2008). In the light of chronic food shortages, changes in food consumption patterns can have important implications for achieving food security.

The study of consumption patterns is important for a number of reasons. First, since total consumption accounts for more than two-thirds of national income in many countries, it is the largest macroeconomic aggregate, having great significance for business conditions and the state of the economy as a whole (Clements and Selvanathan, 1994). Second, the pattern of consumption contains useful information regarding economic welfare and living standards and, thus, is an objective way of measuring and assessing economic performance. Finally, understanding the price responsiveness of consumption is critical for a number of microeconomic policy issues, which include the measurement of distortions, optimal taxation, and the treatment of externalities.

From a food security standpoint, understanding the trends in food consumption patterns is extremely crucial since policy makers may be interested

---

1. The FAO's primary indicator of food security is the number of people who consume sufficient number of calories for a healthy diet. This indicator is based on a combination of country-level estimates of average per person dietary energy supply (DES) from local food production, the number of calories required by different age and gender groups, and country-specific coefficients of income/expenditure distribution.
2. A recent study by IFPRI (2008) estimates that the increase in crop prices resulting from expanded biofuel production is accompanied by a net decrease in both food availability and access to food for all regions. For sub-Saharan Africa, calorie availability will decline by more than 8% if biofuels expand dramatically.

in knowing whether an increase in average per capita food availability is due to an increase in domestic production or imports. Food security can be thought of as improving when per capita food consumption increases due to an increase in domestic production or food imports. On the other hand, if per capita food consumption decreases due to a decline in domestic production or food imports, the situation may be unsustainable for long-run food security. Dependence on food imports can create a food insecurity problem at the national level if a country faces difficulty in paying for these imports, thereby losing valuable foreign exchange earnings.[3] From a policy perspective, it is crucial to understand how different socioeconomic groups (especially the poorest segments) in the population are affected by changes in food imports and changes in food prices in the international markets.

Along with the increase in per capita availability of food, a change in the composition of diet can affect food security. These changes occur due to demographic and epidemiological transitions (Popkin, 2003). While the demographic transition occurs due to a shift from a pattern of high fertility and mortality to one of low fertility and mortality, the epidemiological transition occurs due to a shift from a pattern of high prevalence of infectious disease associated with malnutrition, to one of high prevalence of chronic and degenerative disease associated with urban lifestyles (Olshansky and Ault, 1986). The consequence of these transitions is reflected in nutritional outcomes, such as changes in body composition and morbidity. In the developing world, changes in the diet pattern are occurring at a very rapid pace. While the share of cereals and roots and pulses declined, the share of vegetables, meat and poultry, and vegetable oils has increased remarkably during the past decade. Mittal (2008) estimates that for India, with a 9% GDP growth over the next two decades, the demand for edible oil is likely to increase almost threefold by 2026. The growing dietary diversity in many Asian and Latin American economies can be attributed to an increase in incomes. However, diet in sub-Saharan Africa has changed only marginally with cereals, starch roots, and pulses (the low-cost food) comprising 70% of the region's calorie consumption, while the share of meat and dairy products (higher cost food) continues to be very low. A recent research report by Smith et al. (2006) from a sample of 12 countries in sub-Saharan Africa found that the problems of diet quality are widespread in most of the countries. While low diet diversity appears to be relatively minor for West African countries, prevalence is higher in the East and Southern African countries, with the highest incidence found in Mozambique. The average per capita income in the region was around $500 per year during 2003, and the distribution of income is extremely skewed.

---

3. A recent study by Rosen and Shapouri (2008) of the USDA finds that with the sharp increase in food prices, prospects are not so bright for the 70 low-income, highly import-dependent countries. Any decline in import capacity arising from increase in food prices can have significant food security implications.

Inequality in purchasing power adversely affects the food security and nutritional status in the region. To raise food consumption and achieve a greater dietary diversity (especially for the poorest segments of the population), it is critical to promote policies that accelerate agricultural productivity and improve distribution in income.

Determining whether food security is sustainable or not requires an understanding of the trends of per capita food consumption (i.e., whether the increase in consumption is due to increased domestic production and falling imports). In the next section, we examine the determinants of food consumption and its linkages to food security and nutritional status. The third section provides a brief review of the selected studies (both regional-level and country-specific studies) that examine the linkages between food consumption on food security and nutritional status. The fourth section presents the empirical analysis using the one-way analysis of variance (ANOVA) approach. This approach is an extension of the two-sample t-test (introduced in Chapter 2). The last section makes some concluding remarks and draws policy implications of the results.

Determinants of food consumption patterns and its importance to food security and nutritional status

Food consumption refers to the quantity and quality of food intake by households or individual members. It is conceptually closer to "food intake" as measured by calories or broken down into different nutrients. Household-level food consumption is often proxied by calorie or nutrient availability.

As economic development proceeds over time, average per capita income and expenditure exhibit an increasing trend. This typically shifts the consumption patterns of the population. In other words, the food consumption basket changes from commodities with a low-quality dietary content to food commodities with a higher-quality dietary content. In the first phase, consumption of coarse cereals and starchy root crops increases, followed by a second phase of increase in staple food such as wheat- and rice-based products (Timmer et al., 1983). In this phase, the relative consumption level of starchy root crops exhibits a declining trend. In the final phase, consumption of cereals falls, and there is a shift toward higher dietary value (such as protein food, fruits, vegetables, etc.). Other factors, apart from income and expenditure, which may significantly affect food consumption, are rural—urban migration, changes in demographic structures and improvements in education, transport and communications, and marketing infrastructure. The recent surge in urbanization has increased the demand for animal-based products. Huang and Bouis (1996), using cross-sectional data for Taiwan (disaggregated by rural and urban area) over the period 1981—91, found that demand for food substantially changed not only by increase in incomes and price changes but also by the relative increase in urbanization.

They found that per person consumption of rice fell by 35.3 kg for rural areas. This is because income and urbanization worked in opposite directions. However, income and urbanization worked in a similar direction for meat. Total per person consumption rose by 24.2 kg, with 18.2 kg contributed from income effects and the remaining effect came from urbanization.

Dietary Diversity and Nepal's Success Story:

Nepal is a small farm economy with significant rural poverty, along with low-quality energy-dense staple diet among its population. About 54% of Nepalese faced chronic food insecurity. The country's women and children experience high levels of undernutrition, given that Nepal's agricultural products are mostly staple cereals such as rice, maize, wheat, and pulses. Given these constraints, the following nutritional outcomes are noteworthy:

- Nepal has the world record in generating the fastest reduction in stunting (from 56% to 41%) during 1995 and 2011.
- Increased Dietary Diversity and a significant transition in food consumption patterns during this time period: The share of food expenditures on staples has decreased, while that of fruits, vegetables, animal, and plant proteins has increased, and a consequent significant decline in the share of starchy staples in an average Nepalese diet (from 83% to 72%).

In an exhaustive study, Kumar et al. (2020) attributes these significant and positive accomplishments to changes in agricultural practices and households' socioeconomic characteristics: improved seeds, increasing crop diversity, farming wages, households hiring labor, parents' education, proportion of households with TVs and telephones, poverty status, male-headed household, family size and farming status with land ownership, and a quick access to paved roads.

The study suggests the following policies: setup programs for improvement in parental education, cash transfers, market access, agricultural centers, broader TV, and mobile device coverage, particularly for the vulnerable groups in remote rural areas.

Impact on food security

Trends in per capita food availability can have important consequences for food consumption patterns in developing countries. The FAO data indicate that food availability improved over the period 1990−99 for the least developed economies of sub-Saharan Africa. For 18 out of the 22 countries, the undernourished population declined during the aforementioned time period. (The countries where the proportion of undernourished actually increased were Botswana, Morocco, Senegal, and Uganda.) Per capita food consumption also depends on how income is distributed among the population. If a significant percentage of the population earns less income (more inequality in the distribution), then it is more likely that per capita food consumption may not reflect how food is distributed among the population. Thus, a large section of the population may remain food insecure (although with the yardstick of per capita food consumption, it may seem that households are not food insecure). Additionally, from a food security standpoint, it is crucial to understand how dietary pattern is changing over time. If a significant percentage of the population is deriving calories out of meat, dairy products, and fruits and vegetables instead of cereals and other root crops, then it can be inferred that the country is relatively more food secure. Thus, dietary diversity, which occurs as a result of income changes and rapid growth in urbanization, can be a suitable proxy for understanding food security over time. In this chapter, we investigate the studies that link dietary concerns to food security. We present the findings from recent studies in China and India. We also discuss the food security and food consumption patterns in the United States and present important policy findings.

Review of selected studies

This section divides the study of food consumption patterns first for developing countries and then for developed countries.

Nutritional and Economic Outcomes:

Findings from several clinical and experimental research studies confirm that diet is an important factor determining the start of chronic diseases, such as cardiovascular diseases, obesity, and diabetes. Hence, policy makers can reduce healthcare costs by exploring the link between nutrition and health. Furthermore, dietary habits and health are also tied to socioeconomic and demographic factors. Furthermore, improved nutritional intakes do support increased birth weights, improved hemoglobin concentration, reductions in the incidence of hypertension, and reduction in mortality rates.

The overarching findings from the spectrum of studies in this context provide some insights into the observed differences in nutritional outcomes.

Undernutrition among children and diet quality among infants are also important aspects. Using an extensive GIS-based community-level data set from 42 countries, Choudhury et al. (2019) show that the nutritional status of 6- to 24-month infants correlates strongly with maternal education, household wealth, and access to health and information. These factors are also correlated with consumption of meat, eggs, and vitamin A—rich fruits and vegetables. The study also makes an interesting link between a child's nutritional outcome and climate conditions, namely, children in the hottest temperature tercile are less likely to consume a fruit, vegetable, or legume. The study recommends that parents should be encouraged to include DGL vegetables and fish into their children's diets, and the success hinges on women's education.

Food consumption patterns for developing countries

One of the earlier studies of changes in food consumption patterns was by Delgado and Miller (1985). They studied the aggregate trends in production, consumption, and net imports of food grains and examined the determinants underlying consumption shifts for West African countries over the period 1960−80. The analysis on cereal trends showed that rice was gaining rapidly as a staple food with production lagging behind consumption growth. Thus, imports were required to close the gap between domestic supply and demand.

The major factors that explained substitution toward rice and wheat consumption were per capita GDP growth and high rates of urbanization. Additionally, relative prices for cereals were an important factor in the substitution process. The results demonstrated that governments in many West African countries kept the prices of rice and wheat artificially cheaper compared with coarse grains. The distortion in relative prices was a result of policies to subsidize urban consumption. These countries imported grains at overvalued exchange rates that made it cheaper relative to domestically produced coarse grains. The study identifies other policy factors that explained the consumption shift. First, advertising trends in West African countries made consumption of wheat and rice more "fashionable." Second, wheat and rice typically made up a large proportion of food aid, which introduced new irreversible dietary habits in the population.

The authors suggested some policy measures that could slow down the rate of consumption shifts and reduce the volume of imports from abroad. They were as follows:

1. income transfers,
2. consumer taxes,
3. tariffs and quotas, and
4. food aid policies.

Income transfers improve food security by taxing the wealthier segments of the population. Consumer taxes on food grains alter the relative price of coarse grains and superior cereals in favor of the former and thus affect the food consumption of poorer households. Tariffs and quotas on wheat and rice imports are an effective means of raising the domestic prices. Tariffs are analogous to direct taxes on food imports, revenues of which pass to the government. Quotas, on the other hand, can be an instrument to generate rents for government officials involved and may not be an effective instrument in improving the welfare of the poorer households. Appropriate food aid policies can also help in slowing down consumption shifts. In nonemergency situations, prices of superior cereals, such as rice or wheat, can be kept high and the revenues used to subsidize coarse grain production. The aforementioned policy instruments can be optimally designed to achieve a greater variety of cereals and a better distribution of per capita availability.

Ray (2007) examined the changes in the nature and quantity of food consumption in India during the reform decade of the 1990s and analyzed their implications for calorie intake and undernourishment. The analysis is motivated by the failure of expenditure and income-based poverty magnitudes that do not truly depict the food and nutrition security situation in a period of rapid changes in food consumption patterns. The study provides evidence at both the state and all-India levels on the magnitude and trends in food consumption. The prevalence of undernourishment in both the urban and rural areas is estimated using calorie intake.

The data sets used for the analysis were from the 43rd (July 1987−88), 55th (July 1999−2000), and 57th (July 2001−02) rounds of the National Sample Survey (NSS). The results of the study can be summarized as follows: First, cereals consumption was generally much higher in rural than in urban areas, mainly due to higher consumption of rice by the rural household. However, for vegetables and fruits, meat, and eggs, per capita consumption was higher in urban areas. Second, there has been a marked decline in the consumption of all cereal items over the period 1987−88 to 2001−02 in nearly all states—the reduction being particularly sharp in the case of insignificant cereals, such as barley, maize, and so on. Third, consumer preferences had shifted from cereal items to noncereal items such as meat and fish and fruits/vegetables, with the aforementioned trend holding true for the whole country. The aforementioned trend clearly demonstrates that the food share in total expenditure registered a sharp decline, especially in urban areas. The significant decline in calorie consumption can be attributed to the switch from calorie-intensive cereals to noncereals due to an increase in food prices during this period.

In examining the prevalence of undernutrition during the 1990s, the study found that the prevalence of undernutrition was especially acute in rural areas. At the all-India level, the prevalence of undernutrition increased from 57.6% in 1999−2000 to 66.9% in 2001−02. An important result of the study was that a significant number of households, even in the top expenditure decile, were unable to meet their daily calorie requirement.

Since a significant number of households, even in the top expenditure deciles, suffer from undernourishment, a reassessment of the strategies of directing the public distribution System (PDS) exclusively to households below the poverty line cannot be rationalized. By providing subsidized rice and wheat through the "fair price shops," there is more room for designing a more effective strategy for the PDS to target households above the poverty line.

Another implication of the result is that, despite a sharp decline in the expenditure of rice and wheat during the 1990s, both these cereals continue to provide the dominant share of calories, especially for the rural poor. Thus, it is important to go beyond the money metric measure of welfare to assess the changes in the living standards of households during a period of rapid structural change in the economy.

In a recent study, Chand (2007) estimates the future demand for basic food in India. The study also provides the demand projections for food grains toward the end of the Eleventh Five-Year Plan by 2020—21 and examines its impact on food security.

Food grains are important for household food and nutrition security for four main reasons:

1. As cereals and pulses are staple food, there is no possibility of substitution between staple food and other food.
2. Increased consumption of other food fills dietary deficiency due to the inadequate level of intake of almost all food by the poor.
3. Food grains are the cheapest source of energy and protein compared with other food, which has implications for food and nutrition security for the low-income classes.
4. As increased production and consumption of livestock products resulting from an increase in per capita incomes are usually used as feed for livestock, food grains still remain the main source of food security. Any decline in production in food grains results in price shocks, which consequently affects consumption adversely.

As the demand for cereals and pulses is expected to grow by almost 2% during the next two decades owing to increase in population growth and rise in indirect demand, it is imperative that growth in production keeps pace with it. The demand for food grains has outpaced supply, and this has created serious imbalances between domestic production and supply, with the consequence that it has cut down the stocks and exports of food grains. The implication is that if the growth rate in domestic production of food grains does not keep up with increased demand for food grains, it could eventually lead to an increased dependence on imports of essential food grains, such as wheat and rice.

Dien et al. (2004) examined the food consumption trends, prevalence of stunting (low height for age) for children below 5 years of age and women's nutritional status using a body mass index (BMI) criterion for the urban Vietnam region. The comparison was made based on the Vietnam Living Standards Survey (VLSS) during 1988 and 1998. From 1989 onward, Vietnam's transition to a market economy resulted in massive output gains in food production. For example, paddy production increased from around 23 million tons in 1993—94 to more than 30 million tons in 1999.

The main trend in consumption was a decrease in the consumption of rice and substitution toward consumption of other cereals, mainly wheat. While roots and tubers showed a declining trend, the consumption of animal products increased appreciably. Second, the study found that food consumption outside the home represents about 20% of food expenditure in urban areas compared with only 5% in the rural areas. The main reason behind this pattern of consumption was time savings, low price, and choice of variety. Third, in regard to the nutritional status, although stunting in children decreased substantially, there was an increase in obesity among adult urban women. However, this situation was not homogeneous across regions. Regions in the south had higher levels of income but still had a high rate of stunting. Thus, in spite of a remarkable growth in food supply, the nutritional status was still unsatisfactory due to health-related problems.

In a recent study, Asfaw (2008) examines the patterns and determinants of fruit and vegetable (F&V) availability for human consumption in Latin American and Caribbean (LAC) countries during the period 1991−2002. As inadequate intake of F&V is a leading cause of micronutrient deficiency, obesity, and chronic diseases, understanding the trend and determinants of F&V consumption is important to find effective solutions for the low-level F&V consumption and the related health problems.

The data came from FAOSTAT of FAO and the World Development Indicator of the World Bank. A panel data analysis was undertaken to understand the determinants of F&V availability for human consumption in LAC countries during the aforementioned time period.

The results of the study indicate that overall mean availability of F&V in LAC countries was 167.24 kg/capita/year, with the amount of F&V available large enough to meet the WHO's F&V intake level if no wastage was assumed. However, there was substantial variation in F&V available by countries, with some countries having significant shortages—notably countries such as Nicaragua, El Salvador, Panama, Peru, Honduras, and Uruguay—with others having surpluses. The policy implications of these results are that for countries where F&V availability is in serious shortage, public policy should focus on encouraging production and trade. In contrast, for countries where the apparent availability levels are higher than the recommended level, such as Belize, Costa Rica, Mexico, and Chile, nutrition education and distribution issues should be given higher priority.

Nutritional and economic outcomes

Findings from several clinical and experimental research studies confirm that diet is an important factor determining the start of chronic diseases, such as cardiovascular diseases, obesity, and diabetes. Hence, policy makers can reduce health care costs by exploring the link between nutrition and health. Furthermore, dietary habits and health are also tied to socioeconomic and demographic factors. Furthermore, improved nutritional intakes do support increased birth weights, improved hemoglobin concentration, reductions in the incidence of hypertension, and reduction in mortality rates.

The overarching findings from the spectrum of studies in this context provide some insights into the observed differences in nutritional outcomes.

Undernutrition among children and diet quality among infants are also important aspects. Using an extensive GIS-based community-level data set from 42 countries, Choudhury et al. (2019) show that the nutritional status of

6- to 24-month infants correlates strongly with maternal education, household wealth, and access to health and information. These factors are also correlated with consumption of meat, eggs, and vitamin A—rich fruits and vegetables. The study also makes an interesting link between a child's nutritional outcome and climate conditions, namely, children in the hottest temperature tercile are less likely to consume a fruit, vegetable, or legume. The study recommends that parents should be encouraged to include DGL vegetables and fish into their children's diets, and the success hinges on women's education.

Food consumption patterns in the United States

In an interesting study of poverty in the United States, Morton et al. (2008) reveal poor urban households are more likely to access food through the redistribution economy than poor rural households. Furthermore, they show that reciprocal nonmarket food exchanges occur more frequently in low-income rural households studied compared with low-income urban ones. The rural low-income purposeful sample was significantly more likely to give food to family, friends, and neighbors and obtain food such as fish, meat, and garden produce from friends and family compared with the urban low-income group. Furthermore, 58% of the low-income rural group had access to garden produce while only 23% of the low-income urban group reported access. In a rural random sample of the whole population in the two high-poverty counties, access to garden produce increased chances of attaining recommended vegetable and fruit servings controlling for income, education, and age. Access to a garden also significantly increased the variety of fruits and vegetables in diets.

Section highlights: are Americans eating unhealthy foods?

Palma and Knutson (2012) examine how we use our choices, when it comes to eating. For example, the USDA issues dietary guidelines for nutrition and general information. The goal of the dietary guidelines is to promote wellness and decrease the risk of dietary- and obesity-related diseases such as diabetes, some cancers, and heart disease. The recommendations promote the consumption of fruits, vegetables, and seafood products, together with the need for physical exercise. The guidelines encourage Americans to balance calories across food groups, increase the consumption of fruits and vegetables, make at least half of their grains whole, switch to fat-free or low-fat milk (1%), and reduce sodium and sugary drink consumption.

Palma and Knutson (2012) look at the alternative policy options regarding proper implementation of these guidelines. The economy's healthcare costs are estimated at $2.5 trillion in 2009 and those associated with obesity estimated at $147 billion. Would USDA's guidelines make any difference?

Duffy et al. (2012) examined the potential impact of the dietary guidelines in reducing the "obesogenic environment," which promotes consumers to over-consume high-energy food and underconsume physical activity. There are three

Continued

Section highlights: are Americans eating unhealthy foods?—cont'd

reasons for this: (1) Food expenditures as a percentage of income have been decreasing over time, (2) highly caloric food prices have been decreasing more rapidly, making it cheaper and, therefore, creating incentives for Americans to increase their consumption, and (3) unfortunately, at the same time, the relative price of other food categories beneficial to human health, such as fruits and vegetables, has increased.

Before exploring the various possible options, it is important to know that the USDA has several programs in place whose costs run into several billions of dollars. For example, the food assistance programs (FAPs) focusing on low-income households and other nutritionally susceptible groups costs roughly $92.8 billion annually. Included is the Supplemental Nutrition Assistance Program (SNAP), which itself costs roughly $75 billion. Another example included is the Women, Infants, and Children (WIC), which provides vouchers for certain types of food, and costs almost $7 billion. The school lunch program that benefits over 30 million children costs $10.8 billion. Finally, consumption of food away from home, especially related to working mothers, is also an important consideration. Guthrie et al. (2002) found that food away from home tends to be more energy dense, whose ill effects are more than equally borne by minority groups and females (Dunn, 2010). A few policy options are listed in the following:

1. Reducing/limiting advertising of "unhealthy food"

Although this is a good policy, this option may have some problems due to protection of free speech. One option for pursuing this strategy is to reduce TV advertising exposure during children-watching hours. However, according to Chou et al. (2008), this ban would reduce the number of overweight children ages 3–11 by 18%; and for adolescents 12–18, by 14%.

2. Tax unhealthy food

A tax on unhealthy food, such as soft drinks, solid fats, and added sugars, would potentially raise prices and create a substitution effect, decreasing the demand for these commodities. As a consequence of a tax, it is reasonable to expect a reduction in both caloric intake and consumption of the goods being taxed. This option also generates tax revenue, which can be used to fund some nutrition-related programs.

3. Create awareness, improve nutrition education, and promote physical activities

This option would include adding nutrition and health courses to the curriculum in schools. There should be more media education on nutrition and promotion of physical exercise to assist prevention and reduce treatment costs of chronic diseases related to obesity. In terms of consequences, Shiratori and Kinsey (2011) studied the media impacts of nutrition information on food choices. They showed a positive and significant effect of popular media on consumer food choices and suggested that popular media may be an effective communication approach to promote consumer's health. Dharmasena et al. (2011) also found positive effects of the guidelines in reducing caloric and nutrient intake of nonalcoholic beverages.

> **Section highlights: are Americans eating unhealthy foods?—cont'd**
>
> 4. Create incentives in FAPs
>
> The FAP incentives option would restrict the type of food that can be purchased using program funds, similar to the policies adopted by the WIC program. This would limit program participants to use their funds or vouchers to buy healthy food only. There are a few studies that have estimated the potential effects of financial incentives on healthy eating for SNAP participants.
>
> **Alternative option 1: maintaining the status quo**
>
> What are the implications of maintaining the status quo? The current obesity rate in the United States is over 40% for adults and 17% for children. Consequently, more aggressive measures may be necessary beyond the periodic updating of healthy eating guidelines.
>
> **Alternative option 2: free market option**
>
> Under this policy option, US government has no role in making nutrition and dietary recommendations and would stop releasing dietary guidelines and follow-ups. What are the consequences of this policy? Americans live in an environment that promotes the consumption of calorie-dense unhealthy food and low physical activity. People in poverty and minority groups are more likely to be pushed into such environments. We should not forget that food costs, as a percentage of income, have fallen, but energy-dense unhealthy food costs have been decreasing more rapidly in comparison with healthy food. Given this trend, it is reasonable to expect an increase in consumption of unhealthy food. The result would be an increase in obesity rates, chronic diseases, and medical costs.

Taylor et al. (2017) note that millions of households in the United States suffer from food insecurity. Obviously, issues surrounding food insecurity also relate to increased incidence of overweight, obesity, and diet-related chronic disease. The study links the food insecurity and poor health outcomes to higher prices of low-energy-dense, high-nutrient-dense foods. Furthermore, the study examines the food consumption of different food-insecure groups, to aid in developing targeted policy interventions. Not surprisingly, adults with very low, low, and marginal food security have lower intakes of fruits and vegetables, larger intakes of energy-dense grains, high fats, sweetened beverages, and dairy products. Overall, high costs of fruits and vegetables dictate the choice of such energy-dense high-fat foods. Consequently, US policy makers must examine appropriate nutrient intervention for targeted groups, and interventions must take into account the nutrient portfolio of different food-insecure groups.

Food consumption patterns in India and China

In recent years, India has experienced rapid economic growth and has become a major player in the global markets. Many economists and social scientists want to investigate whether the success of India's growth has trickled down equitably across all groups in the population. In a recent study by Maitra et al. (2013), we find evidence that despite economic success, India has not made progress toward meeting its Millennium Development Goal targets of reducing undernourishment, particularly among children.

The researchers use national data sets to link child nutritional outcomes and calorie intakes. They find that an improvement in the "height-for-age Z-scores," but a worsening in "weight-for-height Z-scores" for children aged 0—3 over the period 1998—2006. There is also evidence of a sharp decline in per adult equivalent calorie intake from the principal food items. Moreover, this decline was observed across all the expenditure quintiles. Consequently, they conclude that there is a comovement of declining nutritional intake for both adults and children and a lack of progress in improving nutritional outcomes of children.

Just as in China, Nepal, and in other economies, India also exhibits very poor dietary diversity, as the population sticks to cereals and vegetables, with very little inclusion of fruits, milk, and nonvegetarian products. Parappurathu et al. (2019) observes that households in eastern India who are better nourished are those with larger farm sizes, ownership of livestock, higher educational level of the head of the household, higher per capita incomes, and access to formal credit and related markets. While Santhosh (2018) shows that southern states are better than their northern counterparts in terms of dietary diversity, there are substantial disparities between different income classes, along rural—urban divide (see Hasan and Singh, 2017).

Zhen (2010) provides another account of food insecurity in rural China that links food consumption patterns to arable land available for farmers. Zhen (2010) identifies deficiencies of animal protein and fat intake, by linking land requirements relating to food consumption patterns. Zhen (2010) examines food consumption patterns and arable land requirements of Guyuan district, a remote rural area of Western China, where population growth and rapid economic development have increasingly reduced the land available for primary production, thereby creating potentially serious risks for China's food security. Population size, consumption patterns, land resources, and the level of farm intensification determine the amount of land required to produce food. Using data from household surveys, Zhen (2010) finds that food consumption involved only meeting basic requirements for sustenance, with grains, potatoes, vegetables, fruits, and plant oils being the most commonly consumed food. The study also finds that the per capita intake of calories measured via the daily intake of protein and fat was below the recommended standard. Furthermore, the study also finds that farmers are forced to sell their limited livestock to earn enough income to meet their daily consumption needs.

Generally, the results of the study demonstrate substantial scope for promoting the consumption of F&V in LAC countries through economic incentives, such as reducing the relative price of F&V and/or decreasing poverty and inequality.

Socioeconomic variables, particularly income, affect food choices in China. There has also been a concern about food consumption patterns and consequent health outcomes. Much of the heterogeneity in food consumption in China is attributed to changes in income distribution. Ren et al. (2018) note the high price and income elasticities for food, particularly among the household in the lowest-income group. In an exhaustive study covering 1989—2009, Streeter (2017) also shows how income growth and education lead to increases in the consumption of lower-calorie diets of animal foods, fruit, and dairy, with accompanying decreases in cereals. The Chinese economy also exhibits a rural—urban gap in caloric intake, with urban dwellers enjoying food with lesser carbohydrates. Streeter (2017) notes that the better educated, higher-income Chinese households consume more fat and animal foods, unlike their counterparts in the United States, where the tendency is to prefer less fat in high-quality meats.

China's food consumption has primarily focused on grains and vegetables. As in Streeter (2017), Yuan et al. (2019) demonstrate that this pattern has now shifted more toward animal products. Consumption of foods away from home and fast and processed foods has all gone up. While these patterns are ascribed to increase in incomes across the economy, this same economic force many also lead to obesity.

Can safety net programs help the lower-income groups in better food choices? China's Rural Minimum Living Standard Guarantee program (Dibao) is the largest social safety net program in the world.[4] Wang et al. (2019) examine the consumption patterns among households with per capita income at or below 1.5 times the national poverty line. The Dibao program seems to have shifted the consumption patterns of this sample, by prioritizing food and health over other expenses.

To generate similar policy implications, we examine the determinants of food consumption across various food groups and for the different expenditure brackets using a statistical technique called the one-way ANOVA approach. This approach is appropriate since we want to test whether the means of the share of calories across various food groups differ across expenditure brackets or if they are identical.

---

4. For a detailed evaluation of Diabao, see Kakwani et al. (2019).

Empirical analysis and main findings

The main question addressed is as follows:

Does the share of calories from the various food groups differ across households at different expenditure brackets?

The aforementioned question addresses whether the population means of any calorie source differs across households at different expenditure quartiles. This is important since we want to know if households at different expenditure levels are deriving calorie intake (a measure of dietary energy supply) from the same source or from different sources.

Data description

The variables used to address the aforementioned question are per capita expenditure (PXTOTAL), quartile of per capita expenditures (NPXTOTAL), and percent share of calories from various food groups. Per capita expenditure was derived by dividing total household expenditure by the number of members in the household. This is done to control for the number of household members, as membership varies widely in the sample. Expenditure quartiles were obtained by dividing the sample into four equal subsamples based on annual per capita expenditures and ranked into four socioeconomic quartiles, namely poor, lower middle, upper middle, and highest. The share of calories from major food groups was calculated by dividing total calories from each food group by the total calories consumed and then the ratio multiplied by 100. Based on this, the share of calories consumed from seven food groups was constructed: the seven groups were maize (PCALMZ), other cereals (PCALCR), roots/tubers (PCALRT), meat/fish/eggs (PCALMT), milk (PCALMLK), vegetables (PCALVEG), and pulses (PCALPUL).

Analysis method

Since we want to compare if households at different expenditure brackets derive the same calories from the various food groups, an ANOVA is appropriate. This is because we want to test whether the means of the share of calories across various food groups differ across expenditure brackets or if they are identical. The null hypothesis is that means of the share of calories derived from various food groups are identical across all expenditure groups, while the alternative hypothesis is that they are not all equal. The ANOVA procedure is also appropriate since we are examining the variability of the sample means to draw conclusions about the population means.

Results

Table 5.1 shows the mean and standard deviations of the food shares from the various groups. As a proportion of the total household calorie intake, the mean share of maize was almost 64%, which is not an unexpected finding. This is

TABLE 5.1 Food group shares (%) of household calorie intake for Malawi.

Food group	Mean share	Standard deviation
Maize	63.66	38.71
Other grains	9.76	21.91
Roots/tubers	16.54	27.01
Meat, fish, and eggs	4.91	16.43
Milk	1.56	7.73
Vegetables	1.98	9.92
Pulses	0.76	5.24

The mean share of calories does not sum to 100% since the calories from other food are not available.

because maize is the dominant staple food crop in Malawi. Next in importance are roots and tubers (cassava, plantains, and sweet potato), which constitute about 16.5% of total calorie intake, followed by other grains such as rice and sorghum.

Table 5.2 shows that most of the calories consumed by households (across all expenditure groups) originate from maize. Roots and tubers occupy the second place as a source of calorie followed by other grains (such as rice and sorghum). This is possibly because maize has been the major food crop in terms of the policy agenda and hectares of plants. For the hybrid maize variety, both poor and nonpoor households retain most of it for home consumption. Additionally, a greater percentage of the population retains local maize compared with hybrid maize. Thus, consumption preferences show that most of the produce is used for consumption by all quartiles of the population. From Table 5.2, we can also infer that, for the higher expenditure groups, the share of calories from maize, which is a staple diet, declines while the share of calories from animal products, vegetables, and dairy products increases. This is possibly because with increase in incomes, households substitute a better variety of food (in calorie and dietary content) compared with maize, which usually has lower dietary value.

Next, we investigate our central question whether the share of calories consumed from the various food groups differs across expenditure groups. This is done through a one-way ANOVA. The null hypothesis is H_0: Population means of the share of calories from various food groups are identical for all the expenditure groups.

The alternative hypothesis (H_1) is that population means of the share of calories are not identical for the income groups. The ANOVA procedure utilizes an F-test with $(k-1)$ and $(N-k)$ degrees of freedom, where k is the number of groups and N is the total number of observations. Whenever the observed F ratio is large enough, we would reject the null hypothesis that

TABLE 5.2 Mean share of calories from various food groups by expenditure brackets.

Per capita expenditure quartiles	Maize	Other grains	Roots/Tubers	Meat, fish, and eggs	Milk	Vegetables	Pulses
1 (lowest)	62.12	11.10	15.03	8.64	1.51	1.05	0.42
2 (lower middle)	72.17	6.75	13.52	3.64	0.64	1.73	0.36
3 (upper middle)	64.8	8.11	15.45	4.43	2.75	1.88	1.58
4 (highest)	57.3	12.3	21.97	1.97	1.27	3.40	0.72
Total	63.67	9.77	16.55	4.91	1.57	1.98	0.76

TABLE 5.3 One-way ANOVA for share of calories across expenditure quartiles.

	F	Sig.
Maize	3.565	0.014
Other grains	1.97	0.117
Roots/tubers	2.837	0.037
Meat/fish/eggs	5.09	0.002
Milk	1.82	0.142
Vegetables	1.59	0.191
Pulses	1.66	0.174

the share of calories is identical for all the four income groups in favor of the alternative hypothesis that the share of calories is not identical.

From Table 5.3, we can reject the null hypothesis that the shares of calories from maize for the four expenditure groups are the same at the 5% level. Additionally, we also reject the null hypothesis that calorie shares of roots and tubers and meat, fish, and eggs for households in all the expenditure groups are the same at the 5% and 1% levels, respectively. The groups for which the null hypothesis cannot be rejected are other grains, milk, vegetables, and pulses.

In other words, the share of calories for the aforementioned food groups is identical across all the expenditure groups. If we reject the null hypothesis, this implies that at least one of the means is different. However, the ANOVA method does not tell us where the difference lies.

One-way ANOVA in STATA

Once again, for illustrative purposes, we use the following 30 observations on food intakes (Converted to Retail Commodities Data) from the USDA and ERS estimates (www.ers.usda.gov), given in Table 5.4.

ERS tracks the supply of commodities available for consumption in the United States and examines consumer food preferences by age, income, region, race, place where food is eaten, and other characteristics. The USDA's Agricultural Research Service (ARS) and ERS developed the Food Intakes Converted to Retail Commodities Databases (FICRCD) to provide commodity content for food intake data as recorded in national dietary surveys. FICRCD reports food.

We have presented a sample from these data for different food categories (fruit, dairy, grains, meat, and vegetables). We have classified the food intakes across three income groups: low (income = 1), medium (income = 2), and high (income = 3). We wish to examine the food consumption patterns across income groups using the F-test from one-way ANOVA estimation in STATA.

TABLE 5.4 Consumption of different food categories across income groups.

Obs	Income	Fruit	Dairy	Grains	Meat	Vegetables
1	1	57.37	125.76	118.35	58.48	16.97
2	1	26.23	1.05	83.27	28.90	9.06
3	1	7.07	27.12	14.64	53.72	8.73
4	1	21.26	5.49	4.06	48.07	3.27
5	1	24.25	8.59	14.67	5.38	20.97
6	1	125.42	.	.	14.63	5.60
7	1	20.46	.	.	47.28	14.54
8	1	8.49	.	.	.	11.97
9	1	15.37	.	.	.	8.09
10	1	.	.	.	.	13.60
11	2	43.72	108.28	121.69	63.47	21.19
12	2	25.65	1.43	90.27	33.44	13.65
13	2	9.28	30.58	12.14	48.21	10.43
14	2	23.43	7.08	4.68	42.91	3.78
15	2	27.88	10.88	12.43	5.30	25.66
16	2	128.78	.	.	16.37	5.80

17	2	21.09	.	.	49.38	20.01
18	2	10.35	.	.	.	16.59
19	2	18.43	.	.	.	7.08
20	2	.	.	.	.	15.78
21	3	56.25	72.92	129.08	61.29	20.00
22	3	28.73	1.86	96.58	30.69	13.92
23	3	12.46	33.79	11.87	56.98	11.67
24	3	25.37	13.44	5.57	49.94	5.13
25	3	26.62	13.15	11.93	6.87	26.30
26	3	143.15	.	.	19.08	6.77
27	3	31.75	.	.	52.96	27.93
28	3	11.22	.	.	.	20.89
29	3	13.46	.	.	.	6.37
30	3	.	.	.	.	17.03

We look at the descriptive statistics, which we can obtain, using the summarize command in STATA. The input and the output from STATA are given in the following for the 30 observations from Table 5.4.

```
. summarize
```

Variable	Obs	Mean	Std. Dev.	Min	Max
obs	30	15.5	8.803408	1	30
income	30	2	.8304548	1	3
fruit	27	35.68667	37.16003	7.07	143.15
diary	15	30.76133	39.73893	1.05	125.76
grains	15	48.74867	50.23136	4.06	129.08
meat	21	37.77857	19.57051	5.3	63.47
vegetables	30	13.626	7.020305	3.27	27.93

The STATA output confirms the information in the table by noting that there are missing observations for several of the food categories. Similarly, we find that the standard deviations are large for across all food categories. We can undertake the cross-tabulations and one-way ANOVA estimation using the following command STATA, between income and fruit consumption:

. one-way fruit income, tabulate

The input line and the output from STATA are produced in the following screenshot:

```
. oneway fruit income, tabulate
```

Income	Summary of fruit Mean	Std. Dev.	Freq.
1	33.991111	37.295959	9
2	34.29	36.870322	9
3	38.778888	41.487667	9
Total	35.686666	37.160035	27

Analysis of Variance

Source	SS	df	MS	F	Prob > F
Between groups	129.486809	2	64.7434047	0.04	0.9576
Within groups	35773.0861	24	1490.54525		
Total	35902.5729	26	1380.86819		

Bartlett's test for equal variances: chi2(2) = 0.1312 Prob>chi2 = 0.936

Note that STATA ignores the three missing observations and estimates the cross-tabulations and the ANOVA for the 27 observations. We also find that the intake of fruits across income groups does not vary, as the means and the standard deviations are very closely aligned in the cross-tabulations. The same conclusion can be drawn when we see that the F value in the ANOVA results, which indicate that $F = 0.04$. Note that the table value of at 95% or $F(2, 24) = 3.40$. Consequently, we cannot reject the null hypothesis in this case, which states that the intake of fruit is identical for all the three income groups.

The alternative hypothesis in this problem states that the intake of fruit is not identical across these income groups. In other words, the intake of fruits is identical across all income groups. In one sense, this finding is very comforting that the lower-income groups have the same intake as the higher-income groups with respect to the calories from fruits. But the flip side is that if the amount of consumption does not produce sufficient calories, then why do the higher-income groups consume the same amount of fruits as the low-income groups? Either way, the ANOVA is helpful in directing our attention to this important issue concerning nutrient sufficiency.

The STATA output also provides us with the Bartlett's test of equal variances, which is approximated by chi-square test statistic. Bartlett's test is used to test if k samples have equal variances. Equal variances across samples are also known as homogeneity of variances. Some statistical tests, for example, the analysis of variance, assume that variances are equal across groups or samples. The Bartlett test can be used to verify that assumption.

Bartlett's test is sensitive to departures from normality. That is, if your samples come from nonnormal distributions, then Bartlett's test may simply be testing for nonnormality. In our STATA output, we find that the calculated chi-squared value is 0.13, which is smaller than 5.99 or the 95% chi-square table value with 2 degrees of freedom. Consequently, we cannot reject the null hypothesis of equal variances.

Similarly, we use the aforementioned data in Table 5.4 to perform the cross-tabulations and one-way ANOVA estimation using the following command STATA, between income and dairy consumption:

. one way dairy income, tabulate

The input line and the output from STATA are produced in the following screenshot:

```
. oneway fruit income, tabulate
```

income	Summary of fruit		
	Mean	Std. Dev.	Freq.
1	21.8	.44721317	5
2	20.4	.50990203	5
3	19	.54772238	5
Total	20.4	1.2716693	15

Source	Analysis of Variance				
	SS	df	MS	F	Prob > F
Between groups	19.6	2	9.8	38.68	0.0000
Within groups	3.03999802	12	.253333168		
Total	22.639998	14	1.61714272		

Bartlett's test for equal variances: chi2(2) = 0.1488 Prob>chi2 = 0.928

Just as we noted from the previous estimation, we find that the intake of dairy across income groups does not vary, as the means and the standard deviations are also very close, particularly for the low- and middle-income groups. To see if there are any group differences, we have to do an ANOVA, and we find that the same conclusion can be drawn from the $F = 0.04$ from the ANOVA portion of the STATA output. Note that the table value of F at 95% or $F(2, 12) = 3.88$. Consequently, we cannot reject the null hypothesis in this case, which states that the intake of dairy is identical for all the three income groups. Is it possible to find a food category which does not reject the null hypothesis? We leave that to you as a problem for further investigation.

Conclusion and policy implications

From a food security standpoint, understanding the trends in food consumption is extremely important since we want to know whether the increase in average per capita availability of food is due to an increase in domestic production or imports. Dependence on food imports, especially for developing economies, can create a food security problem at the national level if a country has difficulty in paying for these imports, thereby losing valuable foreign exchange earnings. Many countries, especially in sub-Saharan Africa, lack the purchasing power to meet their needs fully. Unless new actions are taken, the gap between demand and supply may widen, resulting in less per capita availability of food. These actions can be improving long-term agricultural productivity, sound and consistent trade, macroeconomic policies, property rights to land and other natural resources, improving rural infrastructure, and various other incentives for small farmers. It is also critical to improve women's access to land and mitigate labor supply constraints, since these factors affect agricultural productivity directly. From a policy perspective, it may be crucial to understand how different socioeconomic groups (especially the poorest segments) in the population are affected by changes in food imports and changes in food prices in the international markets.

At the same time, studying urban consumers' purchasing behavior and food consumption patterns may be very fruitful, since an increasing proportion of the developing world's population will be located in the urban areas in the next quarter of a century. Urbanization affects dietary and food demand patterns by increasing opportunity cost of women's time, changes in food preferences due to increased income and a more diversified diet. A consumption shift from basic cereals, such as sorghum and millet, to other cereals, such as rice and wheat, milk and livestock products, and fruits and vegetables, is a result of rapid urbanization. This dietary diversity is not only an indicator of enhanced food security but may also be an indicator of improved nutritional status of the population (especially women and children).

From our present analysis on the trends of consumption across households in different regions of Malawi, we find that most of the households consume maize as part of a staple diet to derive calories. From the ANOVA, we find that calorie shares of maize for households in different expenditure groups are not the same. This implies that poorer households derive most of the calories from cereals such as maize. As income of the households increases, there is greater substitution toward vegetables, milk and meat, fish, and eggs. Thus, there is a tendency toward greater dietary diversity as technical appendices income increases.

One-way ANOVA

The analysis of variance technique (ANOVA) uses sample information to determine whether three or more treatments[5] yield different results. Let us introduce some useful notations from the exercise conducted in this chapter. (This section is largely based on Lowry (2003), available at http://faculty.vassar.edu/lowry/overview.html.) Let X_{ij} be the share of calorie (say, for example, maize) of the i th household from the jth expenditure group, where $i = 1, \ldots, 600$ and $j = 1, 2, 3, 4$. Since there are four expenditure groups, the totals of the columns are denoted as ΣX_{i1}, ΣX_{i2}, ΣX_{i3}, and ΣX_{i4}, respectively. The subscript i indicates the total of each of the columns summed over the respective row. Let the means of the four columns be denoted by $\overline{X}_j$. Finally, we denote the grand mean $\overline{X}$ as the mean of all the observations. $\overline{X}$ is thus the sample mean of $\overline{X}_1, \overline{X}_2, \overline{X}_3,$ and $\overline{X}_4$, respectively. The null hypothesis is given by $H_0 : \mu_1 = \mu_2 = \mu_3 = \mu_4$. The alternative hypothesis is that the population means are not all equal. We want to determine if the differences among the calories consumed (maize) by the various expenditure groups can be attributed to chance error of samples from the population having the same

---

5. The term treatment is used in agricultural experimentation in which treatments may be different types of fertilizer applied to plots of land and feeding methods for animals and so on.

means. If the sample means differ significantly, we conclude that maize consumption differs across households from different expenditure groups.

Underlying assumptions in the ANOVA procedure

The main assumptions are as follows:

1. Normality: The samples are assumed to be drawn from normally distributed populations.
2. Independence: The samples are independent. This implies that the share of calories obtained from maize by one expenditure group does not affect the share of calories obtained by another expenditure group.
3. Homoscedasticity: The populations have equal variances.

Decomposition of total variation

The term "variation" in statistics is referred to as the sum of squared deviations or simply sum of squares. When a measure of variation is divided by the appropriate number of degrees of freedom, it is referred to as variance or mean square. There are two kinds of variation in an ANOVA framework.

Between treatment variation: This is calculated as follows:

$$SSB = \sum_j r\left(\overline{X}_j - \overline{X}\right)^2 \qquad (5.1)$$

where $r =$ sample size involved in the calculation of each column mean, which has been defined earlier. In our example, SSB for maize is 15,825.60.

Within treatment variation: This is a measure of the random errors of the individual observations around their column means. It is computed as follows:

$$SSW = \sum_j \sum_i \left(X_{ij} - \overline{X}_j\right)^2 \qquad (5.2)$$

where X_{ij} denotes the value of the observation in the i th row and the j th column. The double summation means that the squared deviations are first summed over all sample observations and then summed over all columns. In our example, for maize, the SSW is 881,963.9. The total variation (TSS) is computed by adding the squared deviations of all the individual observations from the grand mean $\overline{X}$. The formula is given as follows:

$$TSS = \sum_j \sum_i \left(X_{ij} - \overline{X}\right)^2 \qquad (5.3)$$

In our example, the TSS is 897,789.5.

Number of degrees of freedom

The next step in the procedure is to determine the number of degrees of freedom with each of the measures of variation. The number of degrees associated with SSB is $(k-1)$. This is since there are k groups; there are k sums of squares involved in measuring the variation of these column means around the grand mean. Since the sample grand mean is an estimate of the unknown population mean, we lose 1 degree of freedom. Thus, there are $(k-1)$ degrees of freedom. The total number of degrees of freedom associated with SSW is $(N-k)$, where N is the total number of observations. Since in the present illustration N is 600 and $k=4$, the degrees of freedom for SSW is 596. The total number of degrees of freedom for TSS is $(N-1)$, which is the sum of degrees of freedom for SSW and SSB.

F-test and distribution

The comparison of the SSB with the SSW is done by computing their ratio denoted as F. It is given by

$$F = \frac{SSB/(k-1)}{SSW(N-k)} \tag{5.4}$$

Under the null hypothesis, the population means are equal. If the computed F exceeds the critical F (at 1 or 5% level of significance), we will reject the null hypothesis that population means of the share of calories from various food groups are identical for all the expenditure groups. We can determine how large the test statistic F needs to be to reject the null hypothesis by looking at the probability distribution of the random variable F. In the case of maize, the computed $F = \frac{15825/3}{881963.9/596} = 3.565$. Since the critical value of F at 5% level of significance is 2.60, we can reject the null hypothesis that the population means of the share of calories from maize are identical for the various expenditure groups.

Relation of *F* to *T*-distribution

Since the F-test is just an extension of the t-test for more than two groups, the relation between them is as follows.

With two groups, $F = t2$, and this applies to both the critical and observed values. For example, consider the critical values for degrees of freedom (1, 15) with $\alpha = 0.05$, $Fcrit(1, 15) = tcrit(15)2$. Then, obtaining the values from the respective tables, we obtain $4.54 = (2.131)^2$. Since the t-test is a special case of the ANOVA and will almost always yield similar results, most researchers prefer ANOVA, since the technique is more powerful in complex experimental designs.

STATA workout

The following table provides the amount of fruits consumed per week, by five households in three income groups: F_1, F_2, and F_3. Is it possible that income really does not matter and that the difference in fruit consumption, *on average*, among F_1, F_2, and F_3 is purely random and is not something that is systematic? Verify this claim.

Fruit intakes per week for three income groups (F_1, F_2, and F_3)

(1) Obs	(2) F_1	(3) F_2	(4) F_3
1	21.6	20.4	18.8
2	22	20.8	19
3	22.4	20	19.6
4	21.8	21	19.4
5	21.2	19.8	18.2

The STATA code along with the output is as follows:

```
. oneway fruit income, tabulate

                    Summary of fruit
    income  |    Mean     Std. Dev.      Freq.
    --------+------------------------------------
         1  |    21.8     .44721317        5
         2  |    20.4     .50990203        5
         3  |    19       .54772238        5
    --------+------------------------------------
     Total  |    20.4     1.2716693       15

                   Analysis of Variance
    Source           SS         df     MS         F       Prob > F
    ------------------------------------------------------------------
    Between groups   19.6        2     9.8       38.68    0.0000
    Within groups    3.03999802  12    .253333168
    ------------------------------------------------------------------
    Total            22.639998   14    1.61714272

Bartlett's test for equal variances:  chi2(2) =   0.1488  Prob>chi2 = 0.928
```

How does STATA produce the results and how do we verify the claim? The following "sum of squares" are computed:

- Sum of squares **Within (SSW)**

    SSW represents total variance "within" each sample. Since we have three samples, we will have three variances "within each sample." When we add these variances, we get SSW.

- Sum of squares **Between (SSB)**

    SSB represents total variance "between" the groups.

- **Total** sum of squares (**SST**) where **SST = SSW + SSB**
- SST is the total variation in the sample: "within" + variation "between" the groups

Compute **SSW**: First, compute the mean of each sample. Here we have three samples. So we have to compute three sample means: $\overline{F}_1$, $\overline{F}_2$, and $\overline{F}_3$. Check the following table for this:

(1) i	(2) F_1	(3) F_2	(4) F_3	(5) For SS1	(6) For SS2	(7) For SS3
1	21.6	20.4	18.8	0.04	0	0.04
2	22	20.8	19	0.04	0.16	0
3	22.4	20	19.6	0.36	0.16	0.36
4	21.8	21	19.4	0	0.36	0.16
5	21.2	19.8	18.2	0.36	0.36	0.64
Sum	109	102	95	0.8	1.04	1.2
Mean	21.8	20.4	19			

Since there are three samples, there will be three sums of squares: SS1, SS2, and SS3. The results are presented in row titled **Sum**, in columns (5), (6) and (7):

$$SS1 = \sum_{i=1}^{5}\left(C_{1i} - \overline{C}_1\right)^2 = (21.6 - 21.8)^2 + (22 - 21.8)^2$$
$$+ (22.4 - 21.8)^2 + (21.8 - 21.8)^2 + (21.2 - 21.8)^2 = \mathbf{0.8}$$

$$SS2 = \sum_{i=1}^{5}\left(C_{2i} - \overline{C}_2\right)^2 = (20.4 - 20.4)^2 + (20.8 - 20.4)^2$$
$$+ (20 - 20.4)^2 + (21 - 20.4)^2 + (19.8 - 20.4)^2 = \mathbf{1.04}$$

$$SS3 = \sum_{i=1}^{5}\left(C_{3i} - \overline{C}_3\right)^2 = (18.8 - 19)^2 + (19 - 19)^2 + (19.6 - 19)^2$$
$$+ (19.4 - 19)^2 + (18.2 - 19)^2 = \mathbf{1.2}$$

Hence, **SS1** = 1.2, **SS2** = 1.04 and **SS3** = 0.8, and further:
SSW = SS1 + SS2 + SS3 = 1.2 + 1.04 + 0.8 = **3.04**.

Compute **SSB**: To compute SSB or the sum of squares "between" the samples: compute the overall mean of the entire data $\left(\overline{F}\right)$:

$$\overline{F} = \frac{\sum_{i=1}^{5}F_{1i} + \sum_{i=1}^{5}F_{2i} + \sum_{i=1}^{5}F_{3i}}{n_1 + n_2 + n_3} = \frac{95 + 102 + 109}{5 + 5 + 5} = \frac{306}{15} = 20.4$$

The next step is to compute SSB: The formula is

$$SSB = n_1\left(\overline{F}_1 - \overline{F}\right)^2 + n_2\left(\overline{F}_2 - \overline{F}\right)^2 + n_3\left(\overline{F}_3 - \overline{F}\right)^2$$

$$SSB = 5.(19 - 20.4)^2 + 5.(20.4 - 20.4)^2 + 5.(21.8 - 20.4)^2$$

SSB = 9.8 + 0 + 9.8 or **SSB = 19.6**.

Compute mean square between and mean square within

MSB and **MSW** are computed as follows:

$$MSB = \frac{SSB}{k-1} \quad \& \quad MSW = \frac{SSW}{n-k}$$

$$MSB = \frac{SSB}{k-1} = \frac{19.6}{3-1} = \mathbf{9.8} \quad \& \quad MSW = \frac{SSW}{n-k} = \frac{3.04}{15-3} = 0.253$$

Compute the calculated value of F

$$F = \frac{MSB}{MSW} = \frac{9.8}{0.253} = 38.73$$

Now compare this to the table value, let $\alpha = 0.5$. From the F-table, $F_{0.05,\,2,\,12} = 3.89$.

(Note: $v_1 = k - 1 = 3 - 1 = 2$; $v_2 = n - k = 15 - 3 = 12$).

We reject the null since the calculated F falls way inside the critical region. This implies there are systemic differences in the intakes of fruits across income groups.

In addition to the aforementioned results, STATA also produces the Bartlett's test for equal variances. The test is designed to verify whether the sample variances of fruit intake across the three income groups are the same:

$$H_0: \sigma_1^2 = \sigma_2^2 = \sigma_3^2$$

H_0 : At least one pair is not equal.

The test statistic in this case is

$$2.3026 \frac{(n-k)\log\left(s_p^2\right) - \sum_{i=1}^{n}(n_i - 1)\log\left(s_i^2\right)}{1 + \frac{1}{3(k-1)}\left[\sum_{i=1}^{n}\frac{1}{n_i - 1} - \frac{1}{n-k}\right]}$$

where $s_p^2 = \sum_{i=1}^{n} \frac{(n_i - 1)s_i^2}{n-k}$.

Or $s_p^2 = \dfrac{4(0.199 + 0.26 + 0.299)}{15 - 3} = 0.2526$, and the final test statistic equals

$$\dfrac{(15 - 3)\ln(0.2526) - 4 * (\ln 0.199 + \ln 0.26 + \ln 0.299)}{1 + \dfrac{1}{3(3-1)}\left[\left(\dfrac{1}{4} + \dfrac{1}{4} + \dfrac{1}{4}\right) - \dfrac{1}{15-12}\right]} = 0.14.$$

STATA reports that the computed chi-square for this test $= 0.1488$. The test statistic is distributed as a chi-square with $k - 1$ degrees of freedom. The table value for the two-tailed test or the "acceptance region" is in the interval [0.103, 5.991] with $\alpha = 0.1$. Since the computed values fall inside the interval, we cannot reject the null hypothesis, and this test hence verifies the equality of sample variances.

One-way ANOVA in R

For illustrative purposes, we use the following 30 observations on food intakes (Converted to Retail Commodities Data) from the USDA and ERS estimates (www.ers.usda.gov). We have presented a sample from these data for different food categories (fruit, dairy, grains, meat, and vegetables). We have classified the food intakes across three income groups: low (income $= 1$), medium (income $= 2$), and high (income $= 3$). We wish to examine the food consumption patterns across income groups using the F-test from one-way ANOVA estimation in R:

We look at the descriptive statistics, which we can obtain, using the *summa ry* command in R. The R output confirms the information in the table by noting that there are missing observations for several of the food categories. Similarly, we use the *apply* command in R to print the standard deviations for the variables. From the following results, we find that the standard deviations are large across all food categories. We can undertake the cross-tabulations:

```
> summary(data_chapter5)
      Obs              Income          Fruit            Dairy
 Min.   : 1.00    Min.   :1       Min.   :  7.07   Min.   :  1.050
 1st Qu.: 8.25    1st Qu.:1       1st Qu.: 14.41   1st Qu.:  6.285
 Median :15.50    Median :2       Median : 24.25   Median : 13.150
 Mean   :15.50    Mean   :2       Mean   : 35.69   Mean   : 30.761
 3rd Qu.:22.75    3rd Qu.:3       3rd Qu.: 30.24   3rd Qu.: 32.185
 Max.   :30.00    Max.   :3       Max.   :143.15   Max.   :125.760
                                  NA's   :3        NA's   :15
      Grains              Meat           Vegetables
 Min.   :  4.06    Min.   :  5.30    Min.   :  3.270
 1st Qu.: 11.90    1st Qu.: 19.08    1st Qu.:  7.332
 Median : 14.64    Median : 47.28    Median : 13.625
 Mean   : 48.75    Mean   : 37.78    Mean   : 13.626
 3rd Qu.: 93.42    3rd Qu.: 52.96    3rd Qu.: 19.258
 Max.   :129.08    Max.   : 63.47    Max.   : 27.930
 NA's   :15        NA's   :9


> apply(data_chapter5,2,sd,na.rm=T)
      Obs     Income      Fruit      Dairy     Grains       Meat Vegetables
 8.8034084  0.8304548 37.1600358 39.7389293 50.2313614 19.5705146  7.0203053
```

First, we undertake a one-way ANOVA estimation using the *oneway.test* command R, between income and fruit consumption. Note that the *oneway.test* command assumes to have a default of heteroscedasticity. Thus, to undertake one-way ANOVA, which assumes homoscedasticity, we need to add an argument: *var.equal = T*. This command also ignores the three missing observations and estimates the ANOVA for the 27 observations. We also find that the intake of fruits across income groups does not vary as we look at that the F value in the ANOVA results, which indicate that $F = 0.04$. Note that the table value of F at 95% or $F(2, 24) = 3.40$. Consequently, we cannot reject the null hypothesis in this case, which states that the intake of fruit is identical for all three income groups.

The alternative hypothesis in this problem states that the intake of fruit is not identical across these income groups. In other words, the intake of fruits is identical across all income groups. In one sense, this finding is very comforting that the lower-income groups have the same intake as the higher-income groups with respect to the calories from fruits. However, the flip side is that if the amount of consumption does not produce sufficient calories, then why do the higher-income groups consume the same amount of fruits as the low-income groups? Either way, the ANOVA helps direct our attention to this critical issue concerning nutrient sufficiency.

```
> oneway.test(Fruit~Income, var.equal=T)

        One-way analysis of means

data:   Fruit and Income
F = 0.043436, num df = 2, denom df = 24, p-value = 0.9576


> bartlett.test(Fruit~Income)

        Bartlett test of homogeneity of variances

data:   Fruit by Income
Bartlett's K-squared = 0.13123, df = 2, p-value = 0.9365
```

We also undertake Bartlett's test of equal variances, which is approximated by the chi-square test statistic. Bartlett's test is used to test if k samples have equal variances. Equal variances across samples are also known as homogeneity of variances. Some statistical tests, for example, the analysis of variance, assume that variances are equal across groups or samples. Bartlett's test can be used to verify that assumption.

Bartlett's test is sensitive to departures from normality. That is, if samples come from nonnormal distributions, then Bartlett's test may simply be testing for nonnormality. In our R output, we find that the calculated Bartlett's K-squared value is 0.13, which is smaller than 5.99 or 95% on the table value with 2 degrees of freedom. Consequently, we cannot reject the null hypothesis of equal variances.

Similarly, we use the aforementioned data to perform the one-way ANOVA estimation using the different command R (*aov*), but the same one-way ANOVA result between income and dairy consumption:

```
> aov(Dairy~ factor(Income))
Call:
   aov(formula = Dairy ~ factor(Income))

Terms:
                factor(Income)  Residuals
Sum of Squares         113.835  21994.720
Deg. of Freedom              2         12

Residual standard error: 42.8123
Estimated effects may be unbalanced
15 observations deleted due to missingness
```

We should use *factor* (*Income*) instead of a variable name (Income) to categorize the value of the variable. Without *factor* argument, the variable will be considered as a continuous variable. In addition, keep in mind that *aov* command assumes homogeneous variance met. To examine in detail, we can save the results of one-way ANOVA as ANOVA2 and get more information by using *summary* command as follows. Note that one can investigate details of an analysis by using the *summary* command as well as descriptive statistics.

```
> ANOVA2 = aov(Dairy ~ factor(Income))
> summary(ANOVA2)
                Df Sum Sq Mean Sq F value Pr(>F)
factor(Income)   2    114    56.9   0.031   0.97
Residuals       12  21995  1832.9
15 observations deleted due to missingness
```

```
> bartlett.test(Dairy~Income)

        Bartlett test of homogeneity of variances

data:  Dairy by Income
Bartlett's K-squared = 1.3301, df = 2, p-value = 0.5143
```

To see if there are any group differences, we find that the $F = 0.03$ from the ANOVA portion of the R output. Note that the table value of at 95% or $F(2, 12) = 3.88$. Also, the *P*-value is 0.97. Consequently, we cannot reject the null hypothesis in this case, which states that the intake of dairy is identical for all three income groups. We find that the calculated Bartlett's *K*-squared value is 1.33, and the *P*-value is 0.51; thus, we cannot reject the null hypothesis of equal variances.

Exercises

1. The following table provides the amount of fruits consumed per week, by 5 households in three income groups: F_1, F_2, and F_3. Do the observed differences in the sample means provide evidence against the null hypothesis? Test using the STATA code.

Obs	F_1	F_2	F_3
1	21.6	18.8	20.4
2	22	19	20.8
3	22.4	19.6	20
4	21.8	19.4	21
5	21.2	18.2	19.8

2. Recall the food security variables CALREQ and INSECURE developed in Chapter 2. Write an ANOVA syntax for CALREQ and per capita expenditure quartiles and INSECURE and per capita expenditure quartiles. Does food security differ across the different expenditure groups? Interpret the results.
3. Select member number $= 3$ as the unit of analysis, which denotes children. Write the syntax with member number $= 3$ as the unit of analysis. Define a new weight for height variable (ZWHNEW) such that if the weight for height variable falls below 2 standard deviations, the variable assumes a value of 1 or the child is wasted. Otherwise, the variable has a value of zero. Write a syntax relating ZWHNEW with per capita expenditure quartiles using ANOVA. Does wasting differ across different expenditure groups? Interpret your results in the light of whether food security translates into nutrition security.
4. From the study of Delgado and Miller (1985), identify the main factors behind the consumption shift in West African countries.
5. What are the determinants of food consumption patterns? What are the linkages between food consumption patterns and nutritional status?

Chapter 6

Impact of market access on food security—application of factor analysis

Chapter outline

Introduction	**170**	Yield/technology indicators	189
Assessing the linkages of market reforms on food security and productivity	**171**	Market access indicators	189
		Household level characteristics	190
		Factor analysis by principal components	**190**
Review of selected studies	173	Step 1: Computing the observed correlation matrix	190
Food deserts in the United States	175	Step 2: Estimating the factors	192
Access, information, and food security in Africa	178	Principal components analysis	193
The role of the informal sector and food security	180	**Examining eigenvalues**	**193**
		Scree plot	193
Food security issues in Middle East and North Africa	180	Step 3: Making the factors easier to interpret: rotation procedure	197
Food insecurity in South Asia: case studies of India and Afghanistan	181	Varimax orthogonal rotation	197
Food security, caloric intake, and wheat prices in Afghanistan	181	Step 4: Computing factor scores	198
		Parallel analysis and robustness check	201
Access and food security in India	182	**Principal components analysis in STATA**	**201**
Empirical analysis	**185**	**Conclusion and policy implications**	**204**
Technical concepts	**186**	**Technical appendices**	**206**
Eigenvalues and eigenvectors	186	Factor analysis decision process	206
Properties of eigenvalues	187	Principal components analysis in R	206
Data description and methodology	**188**	**Exercises**	**211**
Food indicators	189	STATA workout	212
Asset indicators	189		

Food Security, Poverty and Nutrition Policy Analysis. https://doi.org/10.1016/B978-0-12-820477-1.00006-1
Copyright © 2022 Elsevier Inc. All rights reserved.

Markets matter for food security. And the formal and informal rules of the game—the enabling environment—shape the makeup and performance of market systems.

Katie Garcia, in Agrilinks (2018).

Introduction

Market access determines how farmers can get better prices for their produces, which in turn determines their income. Improving market access for farmers can have a profound impact on their food security (Jordan and Jayne, 2013). During the past two decades, donors and international lending agencies have strongly advocated the reduction of direct state intervention in agricultural marketing and pricing in developing countries. The objectives of such policies were to redress the bias against producers and enhance economic development (World Bank, 1981). A wide body of empirical literature has demonstrated the impacts of food market reforms, which includes both the withdrawal of state agencies from pricing and marketing activities and the relaxation of regulatory restrictions on private trade (Kherallah et al., 2002; Babu and Sanyal, 2008). These reforms have led to increased entry of private traders into the food trade, increased producer prices, and improved market integration (Barrett, 1994; Jones, 1996). However, the question that still lingers on is whether the entry of these private traders has increased productivity and improved food security.

The rationale for entry of private traders to function efficiently is not only for direct intensification of agriculture by smallholder farmers in poor rural areas but also ways through which production, consumption, and investment linkages can be promoted. In this chapter, we examine the role of private traders (a measure of market access) in the agricultural food markets and how their participation affects food security.

The issue is critically important, since an efficient marketing system can play an important role in supplying yield enhancing modern inputs at reasonable prices and in assuring remunerative prices for farmers. Second, there is a broader consensus that agriculture in these countries is responsive to economic incentives, yet incentives fail when the risks are quite high. For example, a shift to high value crops entails higher risks, and failure of production could prove fatal to the livelihoods of small and poor farmers, consequently affecting their food security. Private traders could improve the efficiency of output and input markets by generating greater market information for small farmers, thereby reducing the price risks that they face.

In this chapter, we use a multivariate statistical technique called "factor analysis using principal components (PCs)" to study the impact of market access on food security. We also illustrate the workings of the econometric methodology using STATA. The main purpose of using the factor analysis approach is to reduce a large number of variables, which are the main determinants of food security, to a smaller set of underlying components or factors that summarize the essential information in the variables. The derived factors will thus be linear combinations of sets of original, highly correlated variables. Factor analysis is undertaken for two main reasons:

1. Since household food security is determined by a complex set of variables, such as food availability, assets of the household, technology indicators, and market access variables, it is important to condense the information contained in a large number of variables, which are bundled into a smaller set of factors representing the underlying dimensions.
2. Since the set of variables used in the analysis is linearly related, the Kaiser—Meyer—Olkin (KMO) measure (a measure of sampling adequacy that should be greater than 0.5) indicates that factor analysis is an appropriate technique to use. Both the summarization and data reduction are achieved for analyzing household food security with this analysis.

In what follows, we first introduce the basic concepts of market reforms and their linkages to food security. Next, we review a few case studies that investigate the role of food market reforms on food security during the past two decades. A discussion of the methods and the results of these studies serves as motivations for the analysis presented next. We relate the material to recent findings in the United States, Africa, Middle East and North Africa (MENA), and India.

Assessing the linkages of market reforms on food security and productivity

Agricultural market reforms occur within the context of domestic and external political economy interests, global economic trends, and macroeconomic adjustments (Kherallah et al., 2002). Understanding the process and impact of these reforms cannot be viewed in isolation from these important linkages. Reforms need to be examined in terms of actors, influences, and outcomes. During the process of a reform, national governments and other interest groups come in either cooperation or conflict with external actors, such as bilateral donors, lending agencies, and nongovernmental organizations. This, in turn, influences the pace and scope of reforms as well as the timing and sequence of the reform process. Fig. 6.1 depicts the mechanism through which market reforms affect market outcomes. Abandoning price controls by state agencies, allowing private sector entry into agricultural trading activities, and eliminating or easing trade restrictions are some of the reforms undertaken during the past two decades. The main issue is what the outcomes of this reform process were.

Kherallah et al. (2002) point to the outcome process at three main levels. First, the liberalization of agricultural markets affected the nature and efficiency of the market itself. The liberalization process affected the agricultural price levels, the extent of price transmission across markets, the stability of

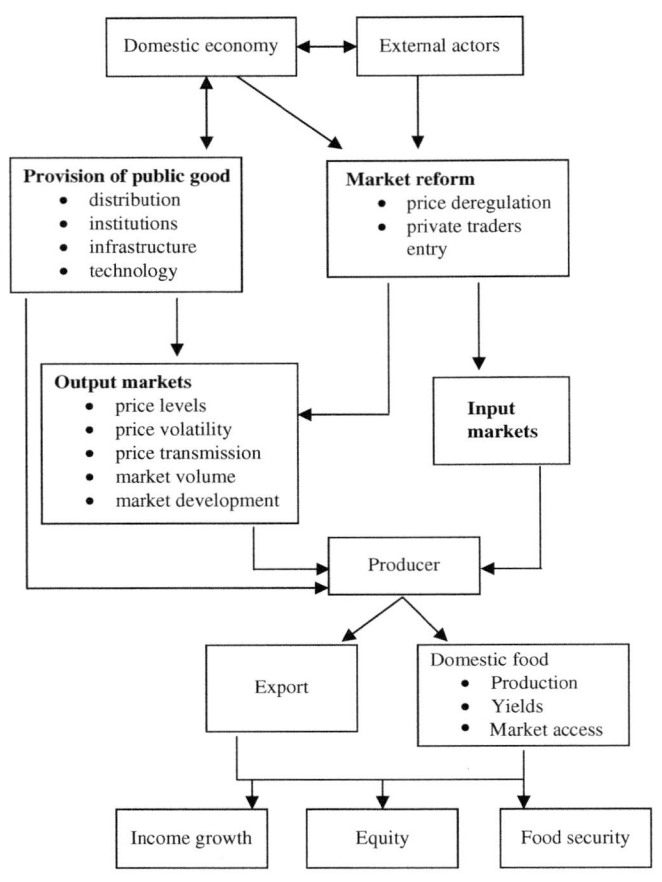

FIGURE 6.1 Agricultural market reform linkages to market outcomes. *Source: Adapted from Kherallah, Mylene; Delgado, Christopher L.; Gabre-Madhin, Eleni Zaude; Minot, Nicholas; and Johnson, Michael. 2002. Reforming agricultural markets in Africa. Baltimore, MD: Johns Hopkins University Press; Washington, DC: International Food Policy Research Institute (IFPRI). Reproduced with permission from the International Food Policy Research Institute (IFPRI) www.ifpri.org. The original figure is available online at https://ebrary.ifpri.org/digital/collection/p15738coll2/id/126303.*

market prices, and investments by private traders so as to improve the functioning of markets. Second, the impact of market reforms has to be evaluated in regard to changes in supply levels and agricultural productivity. Third, these reforms need to be assessed in terms of welfare changes. There are gains and losses associated with privatization to both producers and consumers, through price changes. Besides, the impact of reforms on food security can be assessed by looking at changes in the marketed surplus of food and access to food supplies (both domestic and imported). It is also important to determine the resulting increase or decrease in producer's income due to changes in producer prices after a liberalization program is initiated.

As Kherallah et al. (2002) point out, the outcome of market reforms needs to be examined in light of the changes in the provision of public goods. Withdrawal of the public sector can have significant implications on the positive functions that it carries out, namely information dissemination, law enforcement, product standardization, investments in infrastructure, and other social services. Reforms that might reduce the provision of public goods include dismantling the operation of state marketing boards, loss of quality control, and the increased transaction and price risks faced by producers due to abandonment of the public buffer stock schemes (see Kherallah et al., 2002, pp. 24—25 for further details). In light of the aforementioned, we investigate some selected case studies that examine the impact of agricultural market reforms on price stabilization, productivity, and food security. In particular, we look at one aspect of this reform process, namely the entry of private traders on the food security outcomes.

Review of selected studies

We first examine some of the cross-country studies that study the role of agricultural market reforms on the outcome variables, namely price stabilization, productivity, and food security, and later look at the country-specific studies for the sub-Saharan African region. The case studies provide the motivation for the subsequent empirical analysis.

Beynon et al. (1992) examined the impact of market reforms (removal of controls over food pricing and marketing) on food production and private sector development. The authors distinguished Eastern and Southern African (ESA) countries into two broad groups based on their policy stance. In the first group were countries whose policy stance has been generally favorable to food production, but marketing reforms have been incremental. The second group comprises of countries whose prereform marketing and pricing policies were biased against food production. For this group, liberalization of the private sector has been associated with the collapse of the public or state agencies (as prices have left them uncompetitive). Thus, the role of market reforms on private sector development is unclear. The first group of countries is Kenya, Malawi, Zimbabwe, and Zambia, while the second group consists of Somalia, Ethiopia, and Tanzania. The main policy implication that emerges is that price reforms should be the starting point rather than market liberalization for a viable public sector marketing institution to exist after liberalization.

Stifel et al. (2003) examined the impact of transaction costs on household consumption (a measure of poverty) and agricultural productivity for Madagascar. The data were collected during the period September to November 2001 for a sample of 5080 households. "Transaction cost" in this paper was defined by transportation costs (and proxied for isolation). It was captured using three measures:

1. Travel time to the nearest primary urban center.
2. Cost of transporting a 50 kg sack of rice to the nearest primary urban center.

3. A remoteness index, which was the result of a factor analysis on the various measures of access, or lack of it.

A major implication of the study was that, although little can be done with respect to distances to markets, policy interventions could substantially improve road quality (for example, through building new roads and maintaining existing ones). Such an investment in infrastructure would reduce marginal transaction costs and is likely to have a positive impact on market integration, productivity, and poverty.

In one of the major reports, IFPRI and CSR (2003) assessed the impact of ADMARC (the state-led marketing body of Malawi) marketing activities on household welfare in Malawi. ADMARC is a parastatal organization, which was created in 1971 to assist the development of the smallholder agricultural sector through marketing activities and investments in agro-industry enterprises. Besides, it was mandated to generate food security (particularly in the maize markets) through its role as a buyer and seller in remote areas. To achieve this marketing mandate, ADMARC developed an extensive network and infrastructure of markets across the country. The market infrastructure included regional offices, divisional offices, area offices, parent markets, unit markets, and seasonal markets.

Small farmers had traditionally depended on ADMARC for the purchase of inputs and marketing of crops. However, since the late 1980s, the dependence of smallholders on ADMARC started declining due to two main reasons:

1. ADMARC's maize trading activities started declining from the early 1990s.
2. ADMARC began experiencing financial problems in the early 1980s due to deterioration in the terms of trade, changes in government pricing policy, and the liberalization of agricultural marketing to private traders in 1987 (Scarborough, 1990).

ADMARC was restructured several times during the 1990s in response to the liberalized environment. However, the impact of these reforms (such as entry of private traders) is not clear on smallholder farms.

To address whether reforms improved household welfare, the study used a multivariate regression model, where the data were collected over a 12-month period during 1997/98 and 2002 for 6586 households. The dependent variable was the logarithm of per capita expenditure, while the explanatory variables were nonlabor assets, household labor characteristics, and several dummy variables representing ADMARC facilities. The hypothesis tested was that the presence of ADMARC facilities would have a positive impact on per capita household consumption.

The results of the study were highly unusual. First, improvement in household welfare between the two time periods was strongly and significantly affected by changes in the size of the household and changes in cultivated land area. Second, proximity to ADMARC had no impact on household expenditure. This result is striking as it suggests that proximity to ADMARC is less important in remote areas. Thus, differential access to ADMARC facilities did not have any impact on changes in household welfare, although simply looking at level regressions ADMARC does have a significant impact on household welfare.

From the aforementioned studies, one can reasonably conclude that the incidence of food insecurity increases with market isolation and decreases with improvement in market access. In addition, productivity of crops tends to decline as one gets farther from the markets, which, in turn, can have an adverse impact on food security. To understand the critical determinants of food security, it is important to construct a subset of underlying factors from the original set of factors (such as food indicators, asset indicators, technology indicators, market access indicators, and household level determinants). This can be accomplished with factor analysis, which tries to construct a subset of the underlying factors from the set of original factors to derive the determinants of food security.

Food deserts in the United States

In the United States, the USDA's Economic Research Service has identified about 6500 food desert tracts, based on the data on locations of supermarkets, supercenters, and large grocery stores. These food deserts are areas where people have limited access to a variety of healthy and affordable food. Dutko et al. (2012) point out the importance of increasing food access, and how it is important to understand the characteristics associated with these areas, such as income, vehicle availability, and access to public transportation. Dutko et al. (2012) provide a comprehensive analysis covering the socioeconomic and demographic characteristics of food deserts and distinguish these food desert tracts from other low-income census tracts. Dutko et al. (2012) provide many interesting findings:

- Areas in the United States that experience higher levels of poverty are more likely to be food deserts regardless of rural or urban designation.
- The aforementioned result is especially true in very dense urban areas where other population characteristics such as racial composition and unemployment rates are not predictors of food desert status, because they tend to be similar across tracts.
- Related to the aforementioned, in all but very dense urban areas, the higher the percentage of minority population, the more likely the area is to be a food desert.
- Residents in the Northeast are less likely to live far from a store than their counterparts in other regions of the country with similar income levels.

- For other factors, such as vehicle availability and use of public transportation, the association with food desert status varies across very dense urban areas, less dense urban areas, and rural areas.
- Rural areas experiencing population growth are less likely to be food deserts.

It is clear from Dutko et al. (2012) that in the United States, as community development and infrastructure investment are neglected, residents remain in impoverished conditions. Concentrated poverty in the United States is a significant predictor of low access, or a tract with high poverty rates at a given point is much more likely to be a food desert.

Dutko et al. (2012) also show that in less densely populated urban areas and in rural areas, economic and demographic heterogeneity from tract to tract obfuscates the most important causal factors of low access. For example, high vacant-housing rates affect the probability of a food desert differently in rural areas than in less dense urban areas. That is, abandoned property and movement of population is more detrimental to already dispersed rural populations.

Concentrated poverty and minority populations emerge from our study as the critical factors in determining low access. For the affected groups, this is also evidenced via poor access to healthcare and fitness facilities, limited access to healthy food, low income, low education levels, and high unemployment. These areas are unattractive markets for supermarkets and grocery stores.[1]

Dutko et al. (2012) propose an effective policy to encourage market participation that lower barriers to access, such as providing better public transportation to enable access to retailers in surrounding areas or addressing education and employment shortcomings. Policies must also encourage smaller stores in food deserts to carry healthier products. Enhancing community development and infrastructure investment in areas of concentrated poverty are the most effective options to remove barriers to food retail development and to create healthier living environments in these areas.

We must also keep in mind that solutions must be community-specific, because different infrastructure failures are at fault in urban areas than rural areas. Recognizing these locational and regional differences is crucial. Most importantly, Dutko et al. (2012) stress for policies that link location and diet.

---

1. For instance, food desert tracts have a *greater concentration* of all minorities, including Hispanics. In urban food deserts, this difference was nearly 53% in 2005—09. The proportion of minorities in rural food desert tracts is around 65% greater than in nonfood desert tracts in 2005—09. For racial and ethnic aspects of location and access, see *Access to Affordable and Nutritious Food: Measuring and Understanding Food Deserts and Their Consequences—*Report to Congress (USDA, 2009).

We need active policies that attract and encourage consumer participation with other outlets where fresh, healthy foods may be obtained, such as farmers' markets, mobile markets, or corner stores with expanded healthy options.

> **Section highlights: what do we really know about food security?**
>
> Although a lot of research has uncovered the issue of global food security and its relation to socioeconomic factors, there is still a lot of hard work that remains undone. For example, Carlo (2013) points to the multiplicity of opinions and estimated impacts that have followed the 2008 food crisis, and how the "embarrassment of riches" makes policy formulation very difficult. The main difficulty Carlo (2013) points to is the lack of adequate information and data. Since policies have to be informed by solid evidence based on firm empirical grounds, we need to concentrate on data quality and coverage. We must also pay close attention to coverage methods, standards, and tools for assessment. It is not possible to provide credible policies and coherent plans without a consensus on this basic methodological foundation.
>
> Carlo (2013) also provides us some insights as to how this can be achieved. We already have a lot of data which, if properly analyzed, will provide a lot of insights into the economic and social status of households. Furthermore, we can also glean a lot of information on household behavior, responses, adjustment strategies, and market access with respect to food consumption.
>
> It is important to identify and systematize the available data so as to be able to devise a comprehensive food security information system based on a key set of core indicators. This task, albeit difficult, will provide a better understanding of the determinants and impacts of food price volatility on food security. Carlo (2013) also shows how this data management program could adopt common standards in the collection, validation, and dissemination of data on agricultural prices, production, trade, and uses and on food consumption patterns. Here are the preliminary steps that have to be accomplished, before any strategy to confront food insecurity is entertained.

The difficulties with market access, particularly for the lower socioeconomic status (SES) groups, are often very subtle. In a study on Milwaukee's housing markets, Warsaw and Phaneuf (2019) show how the food security considerations seep into premiums in housing prices. For instance, the study uncovers that an additional grocery store within 0.5 miles of a home increases its sale price by 1.10%. Further, for similarly positioned homes, the grocery store premium jumps to 2.25%, as the proportion of African American and Latino American population increases in the area. Hence, market access to affordable quality diet, proximity, and affordability becomes a part of larger issues surrounding inequality and business decisions.

Would market access to affordable food alone contribute to nutritional improvement within the low SES groups? Allcott et al. (2019) perform the following counterfactual simulation: expose the low SES groups to the same products that are available to the high-income groups. Results show that the low SES groups experience only a marginal reduction in nutritional inequality, of about 10%. This implies that differences in demand, and not the locational constraints of grocery stores, account for a major portion of nutritional inequality in the United States.

Fitzpatrick et al. (2019) note that out of roughly 39 million citizens living in food deserts, about 5 million are elderly Americans. The elderly population is particularly affected by food deserts, given this group's lack of mobility in terms of transportation and other constraints related to fixed incomes, attachments to neighborhoods, and physical difficulties in shopping and food prep. Those elderly affected with chronic physical ailments are also most likely to be affected with limited access to grocery stores.

However, Fitzpatrick et al. (2019) show that food deserts are not a major driver of poor health among low-income elderly Americans. In particular, elderly residents within urban food deserts do not significantly suffer from poor health outcomes. Consequently, the researchers argue for carefully calibrated and targeted policies to combat the health and food security concerns of the elderly.

The results from Allcott et al. (2019) and Fitzpatrick et al. (2019) show that the impact of food deserts deserves a much deeper analysis. Indeed, De Master and Daniels (2019) argue that a totally different perspective has to be developed and applied to the concept of food deserts. The researchers reframe the approach using visually geographical positional mapping systems that link community assets, critical cartography, countermapping, and radical cartography. The authors demonstrate that the current boxed-in versions of research into "food deserts" suffer from problematic assumptions that restrict the space of communities to mere access to stores and transportation. The researchers call for a more nuanced and textured understanding of community network and food access.

Access, information, and food security in Africa

The importance of market access to economic development is brought out by a detailed study covering Tanzanias maize market by Shireen (2012). Shireen (2012) notes that prior to liberalization, procurement in African markets was mostly set up via producer cooperatives and public buying networks. However, after liberalization, these networks disappeared, and many producers found themselves geographically isolated and distant from markets.

Consequently, farm-gate buying became a prevalent arrangement, allowing private buyers to quickly set up buying networks in rural villages. In Tanzanias maize market, likewise, the cooperatives dissolved and farm-gate buying became the main marketing institution for most maize growers. Currently, only a small proportion of Tanzanias maize growers sell their output at a marketplace.

Shireen (2012) investigates the transaction costs of farm-gate buying. That is, Shireen (2012) calculates the extent to which distance to market affects farm-gate prices. In other words, the physical distance to markets can serve as an instrument for household remoteness or access to markets. The estimates indicate that greater distance to market depresses farm-gate prices but that it is a relatively modest effect. Furthermore, the effect of distance is procyclical, namely, it is stronger during the harvest season when prices are lowest.

An important aspect pointed out by Shireen (2012) is the role of search and information costs, which account for a large portion of the transactions cost. Policy measures have to focus on transaction costs related to sequential search patterns, so as to reduce the price dispersion in Tanzanias maize market.

Moreover, reducing search costs is important for reducing transaction costs that affect maize producers whose distances are remote. Policy measures must also help to establish market places in rural villages, alongside farm-gate buying. These markets would reduce search costs for many buyers, increase price information for producers, and create a reliable retail market for food in the village. Rural districts with a high proportion of food producers and villages that are distant from transportation networks will be the biggest beneficiaries of this measure.

Finally, Shireen (2012) is also careful to point out that there are many other threats from external agencies, such as local government taxes and market operations by traders to gain control of market shares. Therefore, what is needed is a combination of well-governed markets along with farm-gate buying, so as to increase the access of maize farmers to competitive markets and to lower their transaction costs.

In recent years, the rapid expansion of information technology has helped the agricultural sector. Access to such modernization is important for poorer households. Kiiza and Pederson (2012) examine this issue for small farm holders in Uganda.

Kiiza and Pederson (2012) note that access to information and communication technology (ICT)—based market information is crucial to the adoption of agricultural seed technologies for maize, beans, and groundnut and to improve smallholder farmer yields and income.

Furthermore, ICT-based market information is provided by FM radio stations and by market information centers through media such as mobile phones and Internet facilities. Factors that positively affect the probability of access to ICT-based market information include access to microfinance loans, membership in a farmer association, government awareness campaigns, and wealth.

Kiiza and Pederson (2012) find that the distance to the trading centers or district capital negatively affects this access. Importantly, the likelihood of access to ICT-based market information declines with female-headed households. Access to market information has a positive and significant impact on the intensity of adopting improved seed for all crops. Adoption of improved seed has a positive and significant effect on farm yields and gross farm returns.

In the case of Uganda, the research by Kiiza and Pederson (2012) shows that to generate meaningful improvements in food security and farm incomes, we need policies to promote ICT-based market information along with yield-augmenting agricultural seed technologies.

In an extremely interesting empirical analysis, Asfaw et al. (2012); Asfaw et al., (2012a, 2012b) find the determinants of output and input market participation on pigeonpea diversity in Kenya. The study uses cross-section data and links market participation to household welfare. Asfaw et al. (2012); Asfaw et al., (2012a, 2012b) find that input and output market participation decisions and their determinants are quite distinct. Output market participation is influenced by household demographics, farm size, and radio ownership, while input market participation is determined by farm size, bicycle ownership, and access to a salaried income. Asfaw et al. (2012); Asfaw et al., (2012a, 2012b) show that there is a positive and significant impact of output market participation on pigeonpea diversity, while input market participation had a negative and significant impact on diversity. The results indicate that output market participants have significantly higher food security status than nonparticipants.

The role of the informal sector and food security

A recent report from the IIED by Vorley (2013) notes that in less developed economies, the informal sector has grown in tandem with formal markets. The informal sector is central to rural and urban food security, livelihood generation, and job creation. Furthermore, this sector is also where much of the produce that reaches the formal sector originates. But the report notes that current development policy considers this sector to be a public "bad." The report highlights, however,

- how poor people adopt informality as a choice to secure their livelihoods and food security, despite the evolution of markets toward formality,
- how the institutions and governance of "traditional" and informal food markets can be improved, based on the situation and perspectives of farmers and low-income consumers, rather than the perspectives of agribusiness.

The report also points out that there is a need to understand the type of investments needed for access to wholesale markets. Informal markets are resilient against risks, losses, climate change, economic crisis, and energy costs. The report stresses for more research and evidence.

Food security issues in Middle East and North Africa

Sudden inflation in food prices can be a major issue for any economy. Larson et al. (2012) note that if commodity markets become highly volatile, then governments often try to protect their populations. A potential solution for food-deficit countries is to hold strategic reserves, which can be called on when international prices spike. But how large should strategic stockpiles be? Larson et al. (2012) derive a dynamic storage model for wheat in the MENA region, where imported wheat dominates the average diet.

Larson et al. (2012) analyze a strategy that allows for wheat stockpiles, which can be used when needed to keep domestic prices below a targeted price. Larson et al. (2012) demonstrate that if the target is set high and reserves are adequate, the strategy can be effective and robust. It is important to note that strategic storage policies are countercyclical and, when the importing region is sufficiently large, a regional policy can smooth global prices, which is the case for the MENA region.

In a related study, Ianchovichina et al. (2012) estimate the pass-through coefficients from international to domestic food prices by country in the MENA region. They show that despite the use of food price subsidies and other government interventions, a rise in global food prices is transmitted to a significant degree into domestic food prices in many countries within MENA. Domestic food prices are highly downwardly rigid.

Asymmetric price transmission indicates that not only international food price levels but also food price volatility matter. High food pass-through increases inflation pressures, where food consumption shares are high. Domestic factors, storage, logistics, and procurement, play a major role in explaining high food inflation in MENA.

Food insecurity in South Asia: case studies of India and Afghanistan

Food security, caloric intake, and wheat prices in Afghanistan

Researchers D'Souza and Jolliffe (2012) from the World Bank indicate that wheat is the staple food in Afghanistan, contributing approximately 54% of average daily caloric intake. Furthermore, Afghanistan has the highest prevalence of stunting in the world among children under 5 years old (D'Souza and Jolliffe, 2012). In 2008, due to increasing global food prices, regional export bans from key trading partners like Pakistan, and domestic drought, the domestic wheat grain and flour prices doubled. These sharp price increases constituted a serious economic shock to Afghan households, who spend the majority of their budgets on food. D'Souza and Jolliffe (2012) evaluate the impact of this price shock on household food security across Afghanistan.

D'Souza and Jolliffe (2012) note that households reduced the value of food consumption in response to wheat flour price increases and that this reduction in the value of consumption is the result of reducing the quality and quantity of food consumed in approximately equal proportions. Clearly, declines in consumption and nutrition indicators have long-term repercussions, particularly for vulnerable populations such as children, lactating women, and the elderly and those on the cusp of poverty and malnutrition (D'Souza and Jolliffe (2012)).

Most importantly, D'Souza and Jolliffe (2012) take into account that households at different points of the distribution employ different coping strategies to deal with shocks. Consequently, D'Souza and Jolliffe (2012) use an estimation methodology that disaggregates the estimated behavioral responses, by allowing the responses to vary across the distribution of the dependent variable after conditioning on the observed covariates. The quantile regression (QR) estimator that D'Souza and Jolliffe (2012) adopt does not assume constant marginal response, like the typical ordinary least squares estimator. The QR estimator is very useful for policy prescriptions concerning food security, because the estimator helps us to focus on a particular portion of the distribution, such as the lower tail, or a particular threshold, such as a poverty line or some fixed nutritional benchmark. The econometric methodology of D'Souza and Jolliffe (2012) provides some key insights regarding behavioral responses concerning quality and quantity adjustments:

- Households at the top of the food consumption and calorie distributions experience the largest declines in each of these measures, as might have been expected given that these households can afford to cut back.
- In contrast, households at the bottom of the calorie distribution cannot afford to make substantial cuts to caloric intake, since they are close to the minimum daily energy requirements; accordingly, we see a very small price effect for these households.
- Finally, very poor households at the bottom of the dietary diversity distribution experience large declines in dietary diversity as a result of the wheat flour price increases. Since households living at subsistence levels cannot make major cuts to caloric intake, they must adjust the compositions of their diet to maintain energy levels.

D'Souza and Jolliffe (2012) note the following implications from their results: First, policy makers should not focus exclusively on changes in caloric intake that result from price shocks. Poorer households do not cut back on calories very much, but they reduce dietary quality.

This implies that we need to provide micronutrient interventions, fortification of staples, and vitamin distributions during periods of high food prices.

Second, household survey consumption modules include questions on the quantity of food items consumed or the expenditure on food items, but not on the frequencies with which the food items are consumed. We have to augment household surveys, particularly for populations that are vulnerable to food insecurity, to include anthropometric data.

Access and food security in India

Hiroyuki and Nagarajan (2012) note that sometimes market participation may raise income for farmers, but it may also reduce on-farm varietal diversity. However, for underutilized crops like minor millets, Hiroyuki and Nagarajan (2012) show that market participation encourages growers to increase on-farm diversity through better access to new varieties exchanged at local markets and higher returns from varieties already grown.

Hiroyuki and Nagarajan (2012) collect data from two different agroecological niches, the Plains and the Hills in southern India. Their empirical results indicate that, in the less fertile plains, market participation improves on-farm varietal diversity of minor millets and increased net revenues. Market development had no effect on varietal diversity, in the fertile hill ecosystems. Hiroyuki and Nagarajan (2012) conclude that policies must be designed for on-farm conservation of underutilized crops in their own agro-ecosystems and call for active stakeholder participation.

Darshini (2012) also finds evidence of the link between food security and human development. Darshini (2012) shows that India's food security is related to people's access to the stock powers of food grains. It is clear from her analysis that lack of food is due to the lack of access. Closely resembling Sen's entitlement failures, we see that in India, food insecurity is due to slow growth of purchasing power rather than food availability. In particular, liberalization and recent neoliberal policies, together with counterproductive government action, have created India's food insecurity. Policies are needed to rejuvenate Indian agriculture.[2]

Mobile devices, market access andand nutrition

Sekabira and Qaim (2017) demonstrate how female mobile phone use in Uganda strongly and positively influences women empowerment, household income, food security, and dietary quality. The study also shows that women users of mobile phones gain substantially more than male users, in terms of market access and dietary quality. Access to mobile technologies allows women to influence their bargaining position within the household and hence improve gender equality and nutrition. Substantial economic success with mobile phones with respect to agricultural extension programs are evidences in Niger (Aker, 2010), and within fisheries in South India (Jensen, 2007).

The aforementioned studies along with related research on transportation economics have provided sufficient evidence that any infrastructure development not only has multiplier effects that work beyond just employment creation but also generates well-being in terms of nutrition and crop diversity. Atsushi and others (2018) show how development in transportation networks in rural Ethiopia improves market access to input and output markets, thereby improving agriculture and nonagriculture income and employment.

Does women's empowerment help with market access? This very question receives an affirmative response in Murugani and Thamaga-Chitja's (2019)

---

2. For the impact of intervention with fortified wheat on tribal children in India, see WFP (2011) and also WFP (2013); WFP, 2013a; WFP, 2013b, http://www.wfp.org/countries/india.

study in Limpopo Province, South Africa. The study shows a positive correlation between women's market access and food security. Market access also directly influences crop diversification and a widening nutrient portfolio. The authors suggest building community-based empowerment systems, advocacy for dietary diversity, and nutrition education.

COVID-19, frontline workers, farmers' organizations, informal markets and technology: ASEAN experience

India's food product market is mostly composed of small-scale informal traders and street vendors. Narayanan (2020) notes that these informal traders have managed the crisis successfully, relative to modern front stores and dominant online retailers.

Reliance on family labor, ease of entry into the market, and the adoption of mobile technology have all played a major role in aiding informal markets. Narayanan (2020), Manavi and Avinash (2020), and Limaye et al. (2020) note that local informal traders and farmers began delivery using mobile messaging and were aided by informal trades, consumer groups, and farmer producer organizations through Twitter and Facebook. Several small-scale private online delivery platforms also mushroomed along the way, to streamline orders and delivery.

Boss et al. (2020) note that another noteworthy feature in coping with the COVID-19 crisis is the role of farmers' organization (FOs) in ASEAN countries. The researchers note that FOs adapted to the crisis and minimized disruptions in the supply chain through the use of technology and in-house family labor.[3] The economies of scale enjoyed by FOs were particularly helpful toward in Vietnam, Malaysia, Thailand, and Laos, particularly with the supply of rice, pineapple, banana, coffee, cashew nuts, and fishery products. Some FOs also became entrepreneurs in food safety and value addition, through collaborations with the private sector.

Boss et al. (2020) also indicate, compared with India and other European countries, the supply chain disruptions in ASEAN have been much smaller and better managed, especially in the in the agribusiness extending to input suppliers, processors, agri-transporters, and extension officers.

Consequently, increasing opportunities to exploit the comparative advantage enjoyed by the FOs are a potential policy instrument. FOs can help with increasing awareness and help pivot farmers' requirements to appropriate programs and schemes.

For instance, Vietnam Farmers' Union and Asian Farmers Association (ASA), Pambansang Kilusan Ng Mga Samahang Magsasaka, and Kababaihang Dumagat ng Sierrra Madre, a women-led start-up in Philippines, are good

---

3. See Boss et al. (2020) for a list of FOs and their roles in ASEAN.

examples of such efforts. As in India, ASEAN countries continue to face substantial postharvest losses.

Hence, governments all across must invest in postharvest storage facilities including refrigeration, modified atmosphere packaging, fermentation, and canning. Taken together, the studies indicate that FOs have to be incentivized to strengthen their operations and capacities; forge new collaborations with private agencies, NGOs, and other countries; and develop new methods to maximize their potential.

Using Bangladesh's successful policy implementation during the pandemic as a case study, Avula et al. (2020) stress the importance of frontline workers' (FLWs') access to technology to support delivery. Governments across all countries must fund development of innovative and efficient delivery modalities through FLWs and create groups of FLWs to specifically target nutrition-specific interventions and other groups to deliver health services to pregnant women.

Naturally, the aforementioned steps increase pressure on FLWs, and governments have taken adequate steps to ensure the health and safety of FLWs. Several related issues remain in the forefront among researchers, as countries start relying a lot more on remote technologies to reach the local populations. For instance, it is not clear whether the informational videos reach the intended audiences and whether the viewers understand the content. Likewise, it is not obvious as to whether the suggestions made in the online programs are practical, and there has been no practical vehicle to track the participation rates. Finally, the uptake and reach of the programs are rendered difficult with unreliable electricity power supply and weak Internet service.

By and the large, the ASEAN experience has provided ample policy guidance for the way to move forward. There is critical need to develop and conduct implementation research alongside major programs for nutrition and health services, with detailed records or what works and what does not. Secondly, as technology takes the primary role in these circumstances, more research is needed to assess the relative merits of remote communication over FTF interaction. Constant updating of the knowledge-informational base through emerging data is also important for efficient design. Policy makers must also include and exploit local knowledge networks and resources to establish nutrition and health services.

Empirical analysis

The main objective of this chapter is to derive factor scores (which can be used for further hypothesis testing) from a subset of highly correlated variables from the original set of variables, while there is very little correlation between the factors. Each factor then comprises a group of variables that represents a single construct, responsible for the interrelationships between the variables. For example, if there are P variables, $X_1, X_2, \ldots, X_p$, factor analysis aims to find a group of variables $F_1, F_2, \ldots, F_k$, (which are the factors) where $k < P$ to describe the interrelationship between the original variables. Thus, the main aim of the analysis is as follows:

- Extract the factors.
- Compute the scores for each factor.
- Interpret the factors.

Thus, a good factor analysis must be simple and interpretable. A factor analysis involves the following steps. First, based on the correlation matrix for all variables, the appropriateness of the factor analysis is evaluated. Second, it is necessary to decide which factor model should be used and the number of factors to be extracted and to assess how well the model fits the data. The criteria used to extract the factors can be maximizing variance or minimizing residual correlations. The main techniques available in many statistical software packages, such as SPSS, are as follows:

1. *Principal components:* proportion of variance accounted for by the common factors.
2. *Principal axis factoring:* the communality of a variable is estimated from its correlation with other variables.
3. *Maximum likelihood:* produces estimates that are most likely to have produced the observed correlation matrix.

The technical concepts and analysis of this chapter largely follow from Hair et al. (1998).

Technical concepts

Eigenvalues and eigenvectors

Suppose that S is a symmetric matrix of order P. If v is also a column vector of order P and λ is a scalar such that $Sv = \lambda v$, then λ is called the eigenvalue (also called characteristic root) of S and v is the corresponding eigenvector (also called characteristic vector) of S. If $v'v = 1$, then v is said to be the standardized eigenvector of S. The eigenvalues can be arranged as diagonal entries in a diagonal matrix D (a square matrix where all the diagonal elements are zero) with the corresponding eigenvectors arranged as columns in V. Matrix V is orthogonal meaning:

$V'V = I$. Thus, the eigenstructure of S can be written in matrix form as

$$SV = VD \quad (6.1)$$

Additionally, it is also the case that

$$S = VDV' \quad (6.2)$$

Eq. (6.1) is also known as the eigenequation.

Properties of eigenvalues

1. If all the P eigenvalues of a symmetric matrix S are positive, then S is termed positive definite.
2. If any eigenvalue is zero, then S is singular.
3. The sum of diagonal elements of a symmetric matrix S is equal to the sum of its eigenvalues.

The most frequently used method for factor analysis is PC factor extraction, and it is used in the present analysis. The number of factors to be extracted is somewhat arbitrary but can be based on the following for PCs.

- Eigenvalue ($\lambda_j > 1$), i.e., factors with a variance less than 1 are no better than a single variable.
- *Scree test criterion:* This is done by plotting eigenvalues against the number of factors in their order of extraction. The number of extracted factors is determined by the point on the curve where the slope becomes horizontal. This point indicates the maximum number of factors to be extracted.

It is important to interpret the factors. This is done through a factor rotation procedure, which simplifies the factor structure giving more insight into each factor. The simplest case of rotation is the orthogonal procedure in which the axes are maintained at 90°. Varimax is one common rotational method under the orthogonal procedure and maximizes the sum of the variances of the required loadings of the factor matrix. Table 6.1 shows guidelines that are usually used for significant factor loadings based on sample size.

TABLE 6.1 Significance of factor loadings based on sample size.

Factor loading	Sample size needed for significance[a]
0.30	350
0.40	200
0.45	150
0.50	120
0.55	100
0.60	85
0.65	70
0.70	60
0.75	50

[a]*Significance based at 0.05 level.*
Source: Extracted from computations made with solo power analysis (Hair et al., 1998).

Finally, the standardized variables $Z_1, Z_2, \ldots, Z_p$ can be formed as linear combinations of the factors $F_1, F_2, \ldots, F_k$. The magnitude of each of these coefficients in the factor loading matrix is a weight measuring the importance of the jth/kth variable Z_j to the kth factor F_k. Scores on factors can be estimated once the factor loading matrix is available. These scores are standardized to have zero mean and unit standard deviation.

Data description and methodology

Fig. 6.2 helps in understanding how we can derive a food security indicator from the underlying variables that are highly correlated and derive the underlying factors, which explain the relationship among the original variables.

Fig. 6.2 shows a comprehensible mechanism through which food security is affected, which can include various food-related indicators, such as staple food left in storage, the number of meals that the household has on a daily basis, expenditure by the household on food, assets of the household—land area owned, livestock owned, other physical assets, yield or technology indicators—quantity harvested per household size, adoption of hybrid maize, use of modern chemical inputs such as fertilizer use, market access indicators, such as road quality, distance to an agricultural marketing facility for buying or selling, and household characteristics such as education of the household head, age of the members, whether the household is headed by a female. Some of these indicators may be highly correlated. The purpose of factor analysis is to compute a linear combination of the original variables (called factors) from which factor scores can be constructed. The following variables are used in undertaking the present analysis.

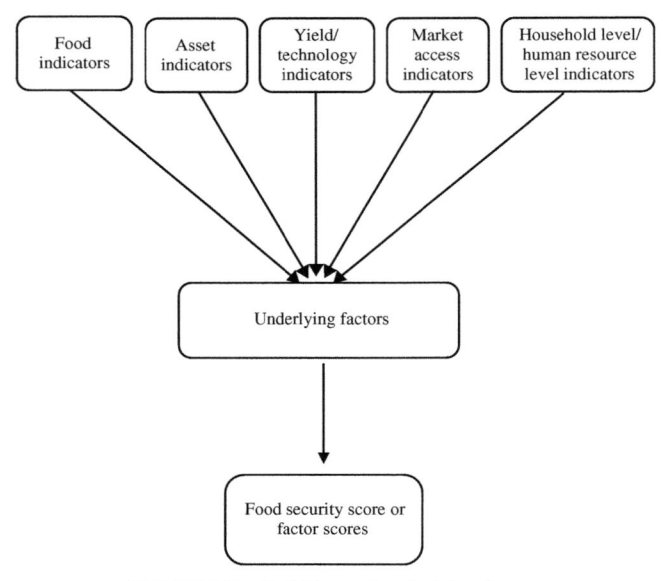

FIGURE 6.2 Variables and underlying factors.

Food indicators

The following indicators are used: per capita expenditure in food (PXFOOD), defined as the ratio of the total expenditure by the household on food to household size. Additionally, staple left over (STAPLEFT) is also another indicator of vulnerability. This is a dichotomous variable assuming two values 0 and 1; 1 indicates whether some staple is left over for the household for own consumption and 0 indicates no food left.

Asset indicators

Two asset indicators are used, namely size of land owned (LANDO)—a categorical variable, assuming values from 0 to 5. Higher values indicate that the household has more land. For example, food crops such as maize require a lot of land, as land is an essential input in the production system and is directly related to food security. Second, we also use LIVSTOCKSCALE as an additional variable for asset ownership by the household.

Yield/technology indicators

Three indicators are used to measure technology. Fertilizer use (FERTILIZ) is an essential input in the production of both food and cash crops. Thus, its use can be considered a pathway to modernizing production systems. FERTILIZ is a dichotomous variable assuming two values 0 and 1, with 1 denoting that the household uses chemical fertilizer as an input in the production process. Second, hybrid maize adoption (HYBRID) is a dichotomous variable and denotes whether the household adopts new technology. The ratio of total quantity harvested of local maize (LNPRODLMAIZ) to household size is considered as a measure of yield (since local maize is the main crop produced and consumed in Malawi) and the natural logarithm of this variable computed.

Market access indicators

Proximity to market is an important aspect of market access. Various indicators can be used to measure market access such as distance to the local market, distance of the household from ADMARC—the parastatal agency—and the distance of the household to the private traders' selling point. The closer the households are to roads, the more likely it is that transaction costs can be reduced, since food crops can be bought and sold. Since the main focus of the present analysis is to understand the role private traders play in enhancing food security, we consider the distance of the household to a private traders' selling point (SELPOINT) as a measure of market access. This is a categorical variable ranging from 0 to 5, with higher values indicating that the household is farther off from a private traders' selling place.

Household level characteristics

Finally, human resources of the household like age of the household head, gender, education/skills, and household size can play a critical role in food security. We consider two household level characteristics for the analysis: female-headed households (FEMHHH), which is a dichotomous variable with a value of unity indicating that the household head is a female and 0 otherwise. There is some evidence that, as females are usually discriminated against in the labor and land markets compared with males, their wages are lower relative to men, and thus, reduced income can affect food security adversely. The next characteristic that we consider is education of the spouse (EDUC-SPOUS). Since higher levels of education can translate into higher income, improving the levels of education can have implications on food security. This variable is also categorical in nature, ranging from 1 to 7, with 1 denoting no education.

Factor analysis by principal components

We now undertake factor analysis with each of the steps discussed earlier.

Step 1: Computing the observed correlation matrix

Computing the observed correlation matrix is the first step. Table 6.2 shows the observed correlation matrix for the eight selected variables in the present analysis. For lack of space, we do not report the correlations of female-headed households and per capita expenditure of households with the remaining variables.

Since the sample size is reasonably large in the present analysis, most of the coefficients are statistically different from zero. However, we are interested in groups of variables that are correlated to each other. The largest correlation coefficient is between selling point and staple left, -0.58 (third row, last column). This is possibly because approximately 40% of the households live at a distance of more than 15 km from the private traders' selling point and it is likely that they have less access to markets, which increases their vulnerability to food insecurity.

To understand if the set of variables is linearly related, the KMO measure is computed (Hair et al., 1998). This is an index that compares the size of the observed correlation coefficients to the sizes of the partial correlation coefficients. The KMO measure of sampling adequacy is defined as

$$\text{KMO} = \frac{\sum_{i \neq j} \sum r_{ij}^2}{\sum_{i \neq j} \sum r_{ij}^2 + \sum \sum a_{ij}^2} \quad (6.3)$$

The numerator is the sum of all the squared correlation coefficients (r_{ij}), while the denominator is the sum of all the squared correlation coefficients

TABLE 6.2 Matrix of observed correlation coefficients.

Variables	LNPRODLMAIZ	FERTILIZ	SELPOINT	LANDO	LIVSTOCKSCALE	HYBRID	EDUCSPOUS	STAPLEFT
LNPRODLMAIZ	1.00	0.238[a] (0.00)	−0.023 (0.254)	0.212[a] (0.00)	0.269[a] (0.00)	0.246[a] (0.00)	0.07[b] (0.02)	0.137[a] (0.00)
FERTILIZ		1.00	0.547[a] (0.00)	0.193[a] (0.00)	0.322[a] (0.00)	0.355[a] (0.00)	0.365[a] (0.00)	−0.47[a] (0.00)
SELPOINT			1.00	0.093[a] (0.004)	0.121[a] (0.00)	0.048 (0.089)	0.413[a] (0.00)	−0.58[a] (0.00)
LANDO				1.00	0.324[a] (0.00)	0.323[a] (0.00)	0.107[a] (0.001)	−0.039 (0.133)
LIVSTOCKSCALE					1.00	0.423[a] (0.00)	0.111[a] (0.001)	−0.112[a] (0.001)
HYBRID						1.00	0.171[a] (0.00)	−0.06[b] (0.036)
EDUCSPOUS							1.00	−0.428[a] (0.00)
STAPLEFT								1.00

[a] Denotes significant at the 1% level.
[b] Denotes significant at the 5% level.

TABLE 6.3 Kaiser–Meyer–Olkin (KMO)–Bartlett test.

KMO measure of sampling adequacy	
Bartlett test of sphericity	0.732
Approx chi-square	1669.045
Df	45
Significance of bartlett	0.00

plus the sum of all of the squared partial correlation coefficients (a_{ij}). If the ratio is close to 1, the partial correlation coefficients are small and indicate that the variables are linearly related. Small values of the KMO measure will thus show that the factor analysis of the variables may not be a good idea, since observed correlations between pairs of variables cannot be explained by other variables. The Bartlett test is done to test the null hypothesis that observed data come from a multivariate normal population in which all the correlation coefficients are zero. This test requires the assumption of multivariate normality and is sensitive to deviations from this assumption.

Table 6.3 gives the KMO value to be 0.732, which is closer to 1, and thus, we can undertake factor analysis.

Step 2: Estimating the factors

Once it is established that the variables are related to each other, we are ready to look at the factors that explain the observed correlations. One of the assumptions of factor analysis is that the observed correlations between the variables result from sharing of these factors. The general model in a factor analysis is of the following form:

$$X_i = A_{i1}F_1 + A_{i2}F_2 + \ldots + A_{ik}F_k + U_i \tag{6.4}$$

where X is the score of a typical household i, Fs are the common factors, U is the unique factor, and the As are the coefficients of loadings used to combine the k factors (Rencher, 2002). The unique factors are uncorrelated with each other and with the common factors. The factors in turn are inferred from the observed variables and are estimated from the linear combination of the variables. For example, a factor k can be expressed as

$$F_k = W_1 X_1 + W_2 X_2 + \ldots + W_p X_P \tag{6.5}$$

It is possible that all the variables contribute to factor k, but we expect only a subset of variables to have large coefficients. The Ws are known as *factor score coefficients* (Rencher, 2002).

Principal components analysis

There are different algorithms for extracting factors from a correlation matrix (see, for example, Tabachnick and Fidell, 2001). The simplest method is called PCs analysis. We will discuss this approach in some detail in the technical appendix at the end of the chapter. For additional technical details, the reader can consult Tabachnick and Fidell (2001). In this approach, linear combinations of the observed variables are formed. The first PC is the combination that accounts for the largest variance in the sample. The second component accounts for the next largest amount of variance and is uncorrelated with the first. Successive components thus explain progressively smaller portions of the sample variance and are uncorrelated with each other. In this chapter, we will use PCs to extract factors, with the output labeled as components, instead of factors. Factor analysis analyzes shared variance among the variables, while PCs just restructure all of the observed variance by forming linear combinations of the observed variables. However, for simplicity, the terms are used interchangeably.

Since one can calculate as many PCs as there are variables, the researcher does not gain any additional insight if all the variables are replaced by their PCs. Thus, one needs to determine how many factors are needed to represent the data, i.e., to reproduce the original correlations.

There are two main criteria for deciding how many factors to extract. They are as follows.

Examining eigenvalues

A criterion of eigenvalue greater than 1 suggests that only factors that account for variances greater than 1 should be included. Factors with a variance of less than 1 are not better than individual variables, since each variable has a variance of 1. From Table 6.4, we find that almost 57% of the total variance is explained by the first three factors and thus three factors can be retained for the analysis. We will name these factors later in the chapter.

Scree plot

The eigenvalues are plotted against the number of factors in their order of extraction. The number of extracted factors is determined by the point on the curve where the slope becomes horizontal. This point indicates the maximum number of factors to be extracted.

Fig. 6.3 depicts the scree plot of the eigenvalues against the factors. The graph is useful for determining how many factors to extract. It can be seen that the curve begins to flatten between *factors 3 and 4*. Thus, only the first three factors are retained.

TABLE 6.4 Explained common variance.

Component	Initial eigenvalues			Sum of squared factor loadings		
	Total	Percent of variance	Cumulative%	Total	Percent of variance	Cumulative%
1	2.814	28.144	28.144	2.814	28.144	28.144
2	1.828	18.275	46.419	1.828	18.275	46.419
3	1.015	10.149	56.568	1.015	10.149	56.568
4	0.985	9.846	66.414			
5	0.770	7.702	74.117			
6	0.690	6.895	81.012			
7	0.636	6.363	87.375			
8	0.530	5.303	92.678			
9	0.393	3.928	96.606			
10	0.339	3.394	100.00			

Extraction method: principal component analysis.

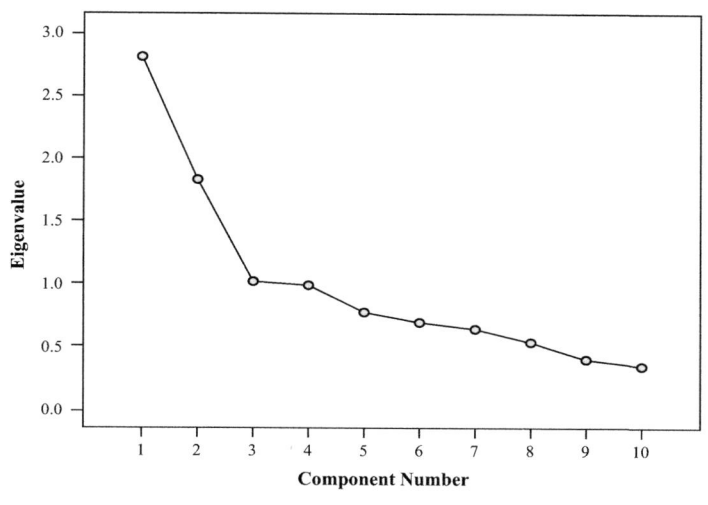

Impact of market access on food security—application **Chapter | 6** **195**

FIGURE 6.3 Scree plot of eigenvalues.

Now, to determine how the variables and factors are related (using the coefficients produced by the factor extraction method), one can express each variable as a linear function of the factors. The component matrix in Table 6.5 shows the relationship between the variables and the underlying factors.

TABLE 6.5 Component matrix.

	Component		
Variables	Factor 1	Factor 2	Factor 3
FERTILIZ	0.815		
SELPOINT	0.710	−0.440	
STAPLEFT	−0.660	0.472	−0.104
EDUCSPOUS	0.631	−0.151	0.174
LIVSTOCKSCALE	0.538	0.483	
LNPRODLMAIZ	0.251	0.603	0.241
HYBRID	0.478	0.559	−0.138
LANDO	0.370	0.496	−0.416
FEMHHH	0.304	−0.329	0.115
PXFOOD		0.349	0.840

Extraction method: principal component analysis. Three components extracted.

For example, we have

$$\text{LANDO} = 0.370(\text{factor } 1) + 0.496(\text{factor } 2) - 0.416(\text{factor } 3) + U_{LANDO}$$
$$\text{SELPOINT} = 0.710(\text{factor } 1) - 0.440(\text{factor } 2) + U_{SELPOINT}$$

The coefficients are called "factor loadings," as they determine how much weight is assigned to each factor for each variable. If the factors are orthogonal (uncorrelated with each other), the factor loading coefficients can be interpreted as the correlation coefficients between the factors and the variables. Thus, the variable LANDO has a correlation of 0.37 with *factor 1*, 0.496 with *factor 2*, and −0.416 with *factor 3*. Similarly, the variable SELPOINT (households' distance to a private traders' selling point) has a correlation of 0.71 with *factor 1* and of 0.44 with *factor 2* and is independent of *factor 3*.

Once a smaller number of factors are extracted, we can evaluate how well these three factors describe the original variables. In other words, we want to determine the proportion of variance of each variable that is explained by the three common factors. This is also known as *communalities*. Since the factors are uncorrelated, the total proportion of the variance explained for a variable is just the sum of the variance proportions explained by each factor. Table 6.6 shows the communalities of each variable.

As mentioned before, the proportion of variance explained by the common factors is known as *communality* of the variable. Communalities can range from 0 to 1, with 0 denoting that the common factors do not explain any of the variance, while 1 denotes that all of the variance is explained by the common

TABLE 6.6 Communalities table.

	Initial	Extraction
LNPRODLMAIZ	1.00	0.485
FERTILIZ	1.00	0.666
SELPOINT	1.00	0.698
LANDO	1.00	0.556
LIVSTOCKSCALE	1.00	0.527
PXFOOD	1.00	0.829
HYBRID	1.00	0.561
FEMHHH	1.00	0.214
EDUCSPOUS	1.00	0.451
STAPLEFT	1.00	0.670

Extraction method: principal component analysis.

factors. The variance that is not explained by the common factors is attributed to the unique factor for each variable. If there are variables that have very small communalities, it implies that they cannot be predicted by the common factors. Thus, it is reasonable to remove them from the analysis, as they are not linearly related to the other variables. Again, considering the variable LANDO as an example, we find that the communality can be calculated as

$$(0.37)2 + (0.496)2 + (0.416)2 = 0.556 \text{(rounded)}$$

Step 3: Making the factors easier to interpret: rotation procedure

From the component matrix in Table 6.5, we find that some of the variables are more correlated with a few factors than others. Ideally, we want to see the groups of variables with large coefficients for one factor and small coefficients for the others. This will enable us to assign meaning to the factors. The purpose of rotation is to identify the factors that have large loadings in absolute value for only some of the variables so that we can differentiate the factors from each other. If several factors have high loadings on the same variables, it is difficult to determine how the factors differ.

There are two kinds of rotation: orthogonal and oblique. If the axes are rotated and are perpendicular to each other, the rotation is called orthogonal. However, if the axes are not maintained at right angles, the rotation is called oblique. The consequence of an oblique rotation is that the factors are correlated and thus makes them difficult to interpret. The factor axes are rotated so that variables with the largest correlations are associated with a smaller number of factors (Kline, 1994).

Varimax orthogonal rotation

The VARIMAX method of rotation is the most frequently used rotation method (Hair et al., 1998). It minimizes the number of variables that have high loadings on a factor, so that the factors can be interpreted more easily. The relationship between the test points remains the same as before. However, the axes are altered to interpret the factors more easily. It is similar to describe people's position in the classroom in relation to the walls. The people retain the same relative position to each other, but the description of their position changes.

Table 6.7 presents the rotated component matrix. We report only the correlations above 0.5. To make it easier to identify the factors, we find that the use of inputs such as fertilizer, human capital inputs, staple left, and distance of the household to the private traders' selling point are highly correlated with *factor 1*. Similarly, ownership of livestock, land owned, and growing hybrid maize is clearly correlated with *factor 2*, and per capita expenditure on food is

TABLE 6.7 Rotated component matrix.

Variables	Component		
	VA	ASSETTECH	FOOD
FERTILIZ	0.691		
SELPOINT	0.820		
STAPLEFT	−0.815		
EDUCSPOUS	0.635		
LIVSTOCKSCALE		0.696	
LNPRODLMAIZ			
HYBRID		0.741	
LANDO		0.716	
FEMHHH			
PXFOOD			0.909

Extraction method: principal component analysis. Three components extracted.

correlated with *factor 3*. Since, for the first factor, input use, household characteristics, market access, and staple left over have higher loadings, it is difficult clearly to name this factor. However, since the absolute magnitude of private traders' selling point and staple left over are greater, we will call this factor *vulnerability and access* (*VA*). Similarly, we can call the second factor as *assets and technology adoption* (*ASSETTECH*). Finally, the third factor can be termed as the *food component* (*FOOD*). By comparing the component matrix with the rotated component matrix, we find that rotation improved the factor pattern, as the coefficients of the variables for *factors 2 and 3* increased substantially.

Another useful way to examine the success of the orthogonal rotation procedure is to plot the variables using the factor loadings as the coordinates. Since there are three factors, SPSS produces a three-dimensional plot of these factors. The coordinates in Fig. 6.4 are the factor loadings for the VARIMAX rotated solution. As can be seen from the figure, the first two factors have strong clusters of variables associated with them, while the variables that define *factor 3* do not cluster tightly.

Step 4: Computing factor scores

The final step is to compute the factor scores. The factor scores represent how much of each factor a case has. They are standardized measures with a mean of 0 and a standard deviation of 1.0, computed from the factor score coefficient

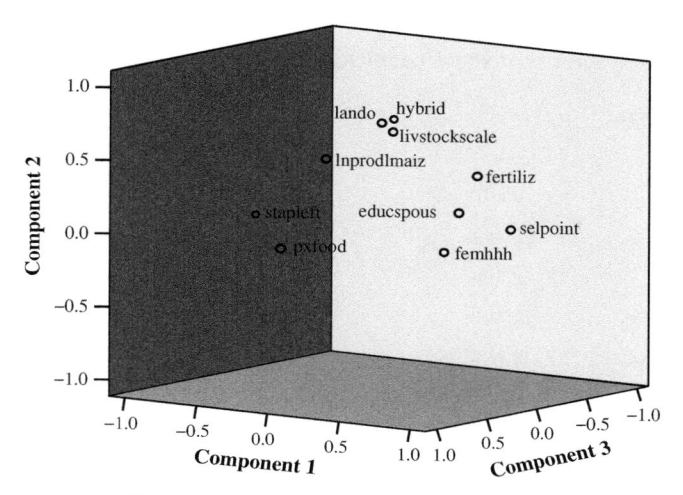

FIGURE 6.4 component plot in rotated space.

matrix. There are several methods for estimating factor score coefficients.[4] Since PC extraction is used in this analysis, all the methods will result in the same factor score. Thus, factor scores can quantify individual cases using a Z-score scale, which ranges from approximately −3.0 to +3.0. The factor score for household j for factor k can thus be derived as follows:

$$\widehat{F}_{jk} = \sum_{i=1}^{p} W_{ji} Z_{ik} \qquad (6.6)$$

where the Ws are the factor score coefficient and all the variables and factors are standardized. The component score coefficient matrix is given in Table 6.8.

For any household j, the first factor score vulnerability and access (VA) can be computed as follows:

$$\widehat{F}_{jVA} = 0.237(\text{fertiliz}) + 0.342(\text{selpoint}) + \ldots + 0.104(\text{pxfood}) \qquad (6.7)$$

where the subscript VA denotes *vulnerability and access*. Similarly, the factor scores can be computed for other households and factors.

In summary, we undertook the following steps in factor analysis. In the first step, we determined whether factor analysis was a suitable approach. This is undertaken using the KMO measure. We find that the KMO value was 0.732, which ensured that factor analysis approach is appropriate. Once established

---

4. SPSS contains three methods, namely regression, Andersen-Rubin, and Bartlett. The Anderson-Rubin method produces uncorrelated scores with a standard deviation of 1. The regression method factor scores can be correlated even when factors are orthogonal. Bartlett factor scores minimize the sum of squares of the unique factors over the variables.

TABLE 6.8 Component score coefficient matrix.

Variables	Component 1	Component 2	Component 3
FERTILIZ	0.237	0.159	−0.087
SELPOINT	0.342	−0.054	−0.081
STAPLEFT	−0.388	0.164	−0.028
EDUCSPOUS	0.340	−0.052	0.301
LIVSTOCKSCALE	−0.057	0.387	−0.208
LNPRODLMAIZ	−0.048	0.319	0.247
HYBRID	−0.025	0.369	0.006
LANDO	−0.085	0.347	−0.017
FEMHHH	0.002	0.055	−0.558
PXFOOD	0.104	−0.004	0.577

Extraction method: principal component analysis. Rotation method: VARIMAX with Kaiser normalization.

that the variables were related to each other, in the second step, we wanted to examine the factors that explained the observed correlation among the variables. We used a PC approach to extract the factors. We found that the first three components explained almost 57% of the total variation, and thus, we retained these components. We also wanted to determine how the variables and factors were related. This was done by expressing each variable as a linear function of the factors. The purpose of this exercise was to extract a small number of factors that described the original variables. We found that the following variables had higher factor loadings with the first component: FERTILIZ, SELPOINT, STAPLEFT. LNPRODLMAIZ, HYBRID, and LANDO had higher factor loadings with the second component. Finally, per capita expenditure on food (PXFOOD) had very high factor loadings with the third component. In the third step, we undertook a VARIMAX method of rotation to minimize the number of variables that have high loadings on a factor. The purpose was to interpret the factors more easily. The first factor was highly correlated with the following set of variables: use of fertilizer, human capital, staple left, and distance of the household to the private traders' selling point. However, since the absolute magnitude of private traders' selling point and staple left over are greater, we call this factor *vulnerability and access* (*VA*). Similarly, we call the second factor *assets and technology adoption* (*ASSETTECH*). The third factor is called the *food component* (*FOOD*). Finally, we derive the factor scores for each of these components.

Parallel analysis and robustness check

One of the important aspects of factor analysis is the number of factors to retain, to reduce the dimensionality of the data. The Kaiser rule and the scree plot are standard procedures to retain eigenvalues greater than unity for the PCs. The parallel analysis developed by Horn (1965) has also received a significant consensus among researchers as a valuable method to validate the number of retained PCs. Horn's (1965) method checks whether the eigenvalues generated via factor analysis could suffer from sample bias, due to finite sample size. Horn (1965) proceeds by generating a sufficiently large number of randomly generated data sets, with the same number of observations and variables, and performs a parallel factor analysis and averages the results. In the following section, we apply Dinno's (2009) implementation of Horn's parallel analysis using STATA.

Principal components analysis in STATA

There are many possible ways to implement the PCs procedure in STATA. As an illustration, we can start with the following command lines:

- factor $x1$ $x2$ $x3$ $x4$ $x5$ $x6$ $x7$ $x8$ $x9$, pcf
- greigen, yline (1)
- factor $x1$ $x2$ $x3$ $x4$ $x5$ $x6$ $x7$ $x8$ $x9$, pcf factor (2)
- rotate, varimax
- predict Bartlett

Assume we have 9 variables, listed as $x1$, $x2$, ..., $x9$, and wish to apply factor analysis to the data. The first command **factor** analyzes the 9 variables, $x1$ through $x9$, using iterated PCs analysis (**pcf**). The second command **greigen** produces the scree plot along with the eigenvalues by factor number. The option **yline(1)** produces a horizontal line parallel to the x-axis, where the eigenvalue = 1. Based on the results of the first two commands, we can make a decision, say in this case, to retain two common factors with the third command. The fourth command provides an orthogonally rotated solution, based on the most recent **factor** command. The last command produces factor scores for each of the three factors using Bartlett's algorithm. As we mentioned before, there are many variations and options within the procedure. We shall illustrate the aforementioned steps for the following example.

We motivate the implementation using a simple example with observations on food security and economic conditions from 9 states. Our illustrative data are given in Table 6.9.

We have nine observations, which capture information on Food Security (given by FSEC as an index), the percentage of the population receiving Food Stamps, given by SNAP, the percentage of population that is in rural areas, given by RURAL, monthly disposable per-household income as INC, and the percentage of female-headed households, given by FEM.

TABLE 6.9 Information on food security and economic conditions from 9 states.

State	FSEC	SNAP	RURAL	INC	FEM
A	5.42	23.52	57.9	2439	10
B	5.25	24.69	65.3	6050	10
C	5.52	24.78	62.5	6378	11
D	3.94	23.81	22.7	3398	12
E	1.314	27.28	35.6	7190	16
F	0.69	26.76	14.27	6000	17
G	1.19	25.94	28.65	2614	15
H	1.66	26.01	44.3	2475	18
I	1.2	22.04	59.22	1550	11

We begin by using the following command line to perform the test of sampling adequacy:

STATA's output indicates a *P*-value of 0.012, which indicates that we can reject the null hypothesis and proceed with factor analysis. Furthermore, the KMO value 0.541 is not an excellent recommendation, but it is not miserable either. Since the value is close to 0.6, we can assume that we have a mediocre recommendation for factor analysis and proceed to implement that.[5]

We implement the PC factor modeling in STATA using the following commands, so as to replicate the basic steps outlined before:

- factor fsec snap rural inc fem, pcf
- greigen, yline (1)
- fapara reps (10) factor fsec snap rural inc fem, pcf factor (2)
- rotate, varimax
- predict Bartlett

STATA produces an output for each command line, and we examine the results in the following. We begin with the **factor** command line.

We obtain the PCs using the command line, and the output shows us that only the first two components have eigenvalues greater than 1. These first two components explain 86% of the combined variance in the five variables.

---

5. KMO measures below 0.5 are unacceptable. For the appropriateness of the KMO measure, see http://www.utexas.edu/courses/schwab/sw388r7/Tutorials/PrincipalComponentsAnalysisintheLiterature_doc_html/027_Measures_of_Appropriateness_of_Factor_Analysis.html.

Impact of market access on food security—application Chapter | 6 203

The third, fourth, and the fifth PCs may be dropped from subsequent analysis. Note that FSEC, SNAP, RURAL, and FEM load heavily with the first factor, while INC loads heavily with the second factor. We can prune these results further. We look at the following scree plot produced by the following command:

- greigen, yline (1)

The horizontal line at eigenvalue $= 1$ produces the usual cutoff for retaining PCs. Once again, the scree plot tells us that we have to retain the first two components. The command line to produce the factor components with just two factors also gives the same results.

Now we conduct the parallel analysis using the input command line:

- fapara reps (10)

As noted earlier, parallel analysis is a method for determining the number of components or factors to retain from factor analysis. STATA creates a random data set that replicates the original data. A correlation matrix and eigenvalues of this matrix are computed for the randomly generated data. Several replications of the data generation are performed. In the aforementioned input line, the fapara command is performed with 10 replications. STATA produces the following graph, which compares the original eigenvalues with the replicated results:

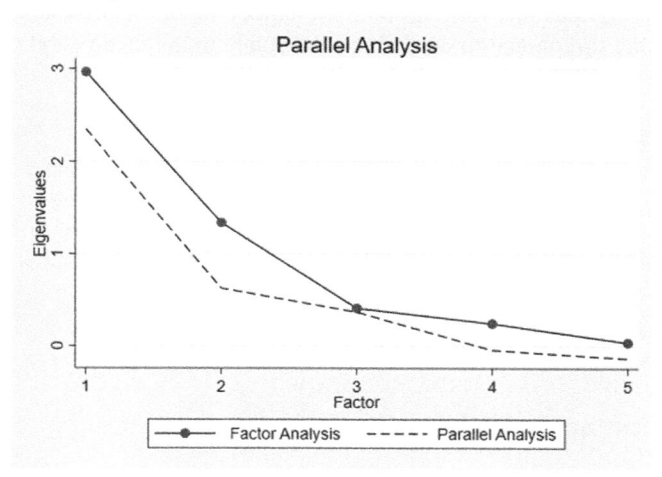

The dashed line for parallel analysis crosses the solid line from factor analysis, at the third component. Hence, the parallel analysis for the food security data set also indicates that two components should be retained.

FSEC, RURAL, and FEM load heavily on *factor 1*, while SNAP and INC load heavily on factor 2. Note that FSEC, RURAL, and FEM are highly sociological aspects, while SNAP and INC represent economic conditions. We are able to see an interesting way in which the data tell us to cluster the variables.

The last command produces the factor scores, which are the linear communalities, which are formed by standardizing each variable to zero mean and unit variance, and then weighting them with factor score coefficients and summing for each factor. The output for the last command line is given in the following.

Examining the factor scores, we can see that FSEC is 0.37 standard deviations above average on the *factor 1* dimension, and 0.23 standard deviations above *factor 2* dimension. Similarly, SNAP is 0.14 standard deviations below the average on *factor 1* dimension, and 0.41 standard deviations above average on the *factor 2* dimension.

We can also produce a clearer picture of how the data can be classified based on the components. To do this, we write the following commands in STATA, which is a slight variation to the previous commands:

- quietly pca fsec snap rural inc fem
- rotate varimax
- predict f1 f2, score
- graph two way scatter f1 f2, yline(0) xline(0) mlabel(state)

The first command line reproduces the same PCs and eigenvalues, but we suppress the output this time around with **quietly**. We also rotate the components using orthogonal rotation. We then compute the factors and call them **f1** and **f2**. The **predict** lines help us to store the factor scores for later use. In this case, we graph the PCs to identify the clusters in our data. The last command line produces the following graph.

Recall that the first factor loads heavily on the "socio" variables, while the second component loads heavily on the "income" or the "economic" variables. The inner states such as A, D, and I cluster together at the lower left. The states, F and E at the upper right, are completely opposite. This provides us a good idea as to how the data can be clustered and indicates some important aspects of our data. For instance, state I is low on both aspects of the socioeconomic front.

Conclusion and policy implications

Food marketing policies need to be understood as part of a broader development strategy that can affect food security and poverty alleviation. As pointed out byJayne and Jones (1997), food market liberalization in sub-Saharan Africa has generated more successes than originally recognized. Grain retailing and milling sectors are examples where consumers have gained from the lower milling margins of small-scale hammer mills. Additionally, there is a greater availability of food grains in many deficit areas due to the strengthened interrural private grain trade. However, success of these reforms can only be sustained if the private sector's expectation of payoffs and risks to future investment is incorporated in these reforms. This is because, historically, in most countries, small-scale cereal trading always coexisted with official marketing activities, while inputs were distributed only by the state. Although private traders have penetrated the fertilizer and seed markets in many countries, input marketing activities are still dominated by state-owned enterprises or multinational firms (Kherallah et al., 2002). In many instances, owing to various external shocks such as droughts and wars, countries reversed their reform policies and reimposed controls on the private sector. For example, Ethiopia and Zambia reintroduced fertilizer subsidies and allowed the state enterprises to distribute fertilizer.

Additionally, countries did not have a clear understanding of the appropriate timing and sequence of policies since reforms took place. Countries that simultaneously eliminated fertilizer subsidies and devalued their currencies witnessed a significant decline in the use of fertilizer (for example, in Malawi and Nigeria). Additionally, a devaluation of the currency without an appropriate liberalization strategy of the main export crop led to a shift of resources to other sectors of the economy. When reforms were implemented, the vested interests of the civil servants often led to reversal of reforms, which made it extremely difficult for the private sector to participate effectively in the reform process. Although in all the reforming countries market entry by private traders occurred in the food and cash crop markets, the wholesale and more capital-intensive marketing activities (such as motorized transport or external trade) were limited to groups that had strong social networks and state connections. Thus, all of the aforementioned factors did not lead to successful participation by the private sector in the reform process, since their expectation of the payoffs were low ex-ante.

To address the aforementioned constraints and policy reversals, government actions to reduce price instability can include improving the transport infrastructure, promotion of regional trade, market information systems that expand information on prices across borders, trade flows, and improving communication infrastructure.

In this chapter, we undertook a factor analysis exercise to determine the underlying factors that explain the correlation among a large number of variables, which influences food security. From our analysis, we found that vulnerability to food insecurity and market access, technology adoption, and assets owned by the household and per capita consumption expenditure on food are all critical components/factors in determining food insecurity. The analysis identified the dimensions through which households can become food insecure. Additionally, it is possible to undertake hypothesis tests in subsequent analysis using the factor scores. For example, one can test the null hypothesis that men and women have the same average values for each of the three factors using a t-test (as in Chapter 2). Besides, regression analysis can be undertaken to determine if these factor scores are significant determinants of food insecurity (as in Chapter 10).

Technical appendices
Factor analysis decision process

Factor analysis is an analytical technique that has been extensively used by various social scientists. Factor analysis using PCs is a method of data reduction in which many variables are chosen initially and they are explained by a few "factors" or "components." When conducting a factor analysis study, a number of issues should be considered. Hair et al. (1998) concentrate on five main issues: the choice of how many variables to include; the choice of a factor model to be used; the decision about the number of factors to retain; the methods of rotation; and the interpretation of the factor solution.

The number of factors to be chosen will be a subset of highly correlated variables from the original set of variables, while there is very little correlation between the factors. Each factor then comprises a group of variables that represent a single construct, responsible for the interrelationships between the variables. These factors should be easy to interpret and lack complex loadings. In the second step, the researcher must decide which factor model to use. There are two different approaches: common factor analysis and PC. The PC model assumes no unique or error variance in the data. In contrast, the common factor model assumes that the variance in a variable can be explained by common and unique components, with the unique variance being further divided between specific and random error variance (Hair et al., 1998).

In the next step, the number of factors to be retained prior to rotation affects the outcome of a factor analysis (Tabachnick and Fidell, 2001). The KMO criterion of retaining factors with eigenvalues greater than 1 is often considered as the most appropriate for PC analysis. The screen test can determine the number of factors to retain.

In the next step, the rotation of factors is undertaken to improve the meaningfulness, reliability, and reproducibility of factors (Hair et al., 1998). There are two methods of rotation: orthogonal and oblique. If the axes are rotated and are perpendicular to each other, the rotation is called orthogonal. However, if the axes are not maintained at right angles, the rotation is called oblique. The factor axes are rotated so that variables with the largest correlations are associated with a smaller number of factors. While orthogonal rotation is simple and conceptually clear, oblique rotation can sometimes portray the complexity of the variables, as factors in the real world are rarely uncorrelated.

Finally, interpreting the factors is important to provide meanings or labels to the factors. Large sample sizes are highly recommended for factor analysis studies. Recommendations vary from five observations per variable to a ratio of 10 observations per variable (Tabachnick and Fidell, 2001).

Principal components analysis in R

We motivate the implementation using a simple example with 32 observations and 8 variables, listed as $x1, x2, ..., x8$, to apply factor analysis to the data. We begin to perform the summary and correlation table to examine the characteristics of the data. The table shows the minimum, mean, max, and quantile values for each variable.

```
> summary(data_chapter6)
      x1                x2                x3                x4
 Min.   :0.000    Min.   :0.000    Min.   :0.0000    Min.   :1.000
 1st Qu.:0.000    1st Qu.:1.000    1st Qu.:0.0000    1st Qu.:1.750
 Median :1.000    Median :2.000    Median :1.0000    Median :3.000
 Mean   :0.625    Mean   :2.031    Mean   :0.6875    Mean   :2.469
 3rd Qu.:1.000    3rd Qu.:3.000    3rd Qu.:1.0000    3rd Qu.:3.000
 Max.   :1.000    Max.   :3.000    Max.   :1.0000    Max.   :4.000
      x5                x6                x7                x8
 Min.   :0.000    Min.   :0.000    Min.   :1.000    Min.   :1.000
 1st Qu.:2.000    1st Qu.:2.000    1st Qu.:1.000    1st Qu.:2.000
 Median :3.000    Median :3.000    Median :3.000    Median :3.000
 Mean   :2.781    Mean   :2.688    Mean   :2.219    Mean   :2.438
 3rd Qu.:4.000    3rd Qu.:3.250    3rd Qu.:3.000    3rd Qu.:3.000
 Max.   :4.000    Max.   :4.000    Max.   :3.000    Max.   :3.000
```

The *cor* command in R produces the correlation table. The correlation coefficient is shown on the table. We save the results of *cor* command as A to conduct KMO measure analysis for later.

```
> cor(data_chapter6)
          x1          x2          x3          x4          x5          x6          x7          x8
x1  1.0000000  0.74318544  0.8703883  0.29868835  0.4399434  0.5562181 -0.65316243 -0.58297525
x2  0.7431854  1.00000000  0.7095857 -0.07181198  0.5163844  0.4688958 -0.69724080 -0.60053256
x3  0.8703883  0.70958571  1.0000000  0.38363854  0.4867647  0.6202043 -0.56850492 -0.50741482
x4  0.2986883 -0.07181198  0.3836385  1.00000000  0.3548503  0.4860293 -0.07698225  0.01830173
x5  0.4399434  0.51638445  0.4867647  0.35485027  1.0000000  0.8827520 -0.60685372 -0.53855152
x6  0.5562181  0.46889585  0.6202043  0.48602928  0.8827520  1.0000000 -0.49683831 -0.41384708
x7 -0.6531624 -0.69724080 -0.5685049 -0.07698225 -0.6068537 -0.4968383  1.00000000  0.85463401
x8 -0.5829752 -0.60053256 -0.5074148  0.01830173 -0.5385515 -0.4138471  0.85463401  1.00000000
```

```
> A <- cor(data_chapter6)
```

First, we should check the test of sampling adequacy by conducting a Bartlett test of sphericity and KMO measure of factor adequacy. We employ packages to simplify the analysis rather than calculating line by line. The package we used here is called "*psych*." You can download the package by coding *install.packages(psych)* and register with *library(psych)* command as follows:

```
> install.packages("psych")
```

```
> library(psych)
```

The *cortest.bartlett* command can be used to undertake a Bartlett test of sphericity with "*psych*" packages. R's output indicates 214.39 of chi-squared value with the degree of freedom 28, and a *P*-value of smaller than 0.01, which means that we can reject the null hypothesis and proceed with factor analysis.

```
> cortest.bartlett(data_chapter6)
R was not square, finding R from data
$chisq
[1] 214.3929

$p.value
[1] 1.25532e-30

$df
[1] 28
```

Another package called "parameters" can be used as well. To employ the packages, install and load the downloaded packages by using the *install.packages* and library commands. The command to produce the Bartlett test of sphericity is *check_sphericity*. The results from two different commands are the same. Both Bartlett's test of sphericity indicates that there is sufficient significant correlation in the data for factor analysis.

```
> install.packages("parameters")
```

```
> library(parameters)
> check_sphericity(data_chapter6)
OK: Bartlett's test of sphericity suggests that there is sufficient significant correlation
 in the data for factor analaysis (Chisq(28) = 214.39, p < .001).
```

As mentioned earlier, the result of the correlation table of the data is saved as A to conduct KMO factor adequacy. The results present that the overall MSA (measure of sampling adequacy) is 0.73, which is far above 0.5.[6] From both tests of sampling adequacy, we met a prerequisite to proceed with the factor analysis.

```
> A <- cor(data_chapter6)
> KMO(A)
Kaiser-Meyer-Olkin factor adequacy
Call: KMO(r = A)
Overall MSA =  0.73
MSA for each item =
    x1   x2   x3   x4   x5   x6   x7   x8
  0.80 0.71 0.80 0.46 0.65 0.70 0.81 0.74
```

We implement the PCs analysis by using the *princomp* command in R. Before producing the PCs analysis, we register x as a combination of all

---

6. KMO measures below 0.5 are unacceptable. For more information of KMO measure, see https://www.rdocumentation.org/packages/psych/versions/1.8.12/topics/KMO.

variables from $x1$ to $x8$. We then use scores and cor arguments to conduct the PCs analysis. The results indicate that only the first two components have eigenvalues greater than 1. To examine closer, one can save the results of PCs analysis and then use the *summary* command to investigate the results in detail.

```
> X <- cbind(x1,x2,x3,x4,x5,x6,x7,x8)

> princomp(X,scores=TRUE,cor=TRUE)
Call:
princomp(x = X, cor = TRUE, scores = TRUE)

Standard deviations:
   Comp.1    Comp.2    Comp.3    Comp.4    Comp.5    Comp.6    Comp.7    Comp.8
2.1805750 1.1857954 0.9168483 0.7195014 0.4414731 0.3569723 0.3212545 0.2348545

 8 variables and  32 observations.
```

```
> pca1 <- princomp(X,scores=TRUE,cor=TRUE)

> summary(pca1)
Importance of components:
                          Comp.1    Comp.2    Comp.3    Comp.4    Comp.5    Comp.6    Comp.7     Comp.8
Standard deviation     2.1805750 1.1857954 0.9168483 0.71950144 0.44147310 0.35697233 0.32125449 0.234854527
Proportion of Variance 0.5943634 0.1757638 0.1050763 0.06471029 0.02436231 0.01592866 0.01290056 0.006894581
Cumulative Proportion  0.5943634 0.7701273 0.8752036 0.93991389 0.96427621 0.98020486 0.99310542 1.000000000
```

As a result, there are standard deviation, proportion of variance, and cumulative proportion. These first two components explain 77% of the combined variance in the eight variables. The rest PCs may be dropped from subsequent analysis.

To illustrate with graphic, the command the screeplot produces the scree plot along with the eigenvalues by factor number. The next command the abline produces a horizontal red line parallel to the x-axis, where the eigenvalue (variances) = 1. We can examine that there are only two components which have greater variance values than 1. Thus, we have to retain the first two components.

```
> screeplot(pca1,type="line",col='blue',main="Scree Plot")
> abline(h=1,lty=2,col="red")
```

Scree Plot

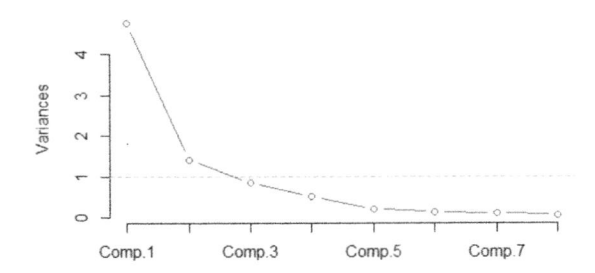

We can produce the factor analysis results by using the factanal command in R. In the bracket, X is a formula or a numeric matrix or an object that can be coerced to a numeric matrix. It is the variable matrix in this case. We identify how many factors should include aforementioned conducting factor analysis. The eigenvalue of components that are greater than 1 is two; thus, we include factors = 2 argument. We want to use the varimax, which is one common rotational method under the orthogonal procedure for rotation, so it is also included as an argument, rotation = "varimax." Note that ×1, ×2, ×3, ×7, and ×8 load heavily with the first factor, while the others load with the second factor.

```
> factanal(X, factors=2,rotation="varimax")

Call:
factanal(x = X, factors = 2, rotation = "varimax")

Uniquenesses:
   x1    x2    x3    x4    x5    x6    x7    x8
0.436 0.413 0.471 0.710 0.186 0.005 0.114 0.212

Loadings:
   Factor1 Factor2
x1  0.681   0.315
x2  0.741   0.196
x3  0.592   0.423
x4          0.538
x5  0.512   0.743
x6  0.391   0.918
x7 -0.930  -0.146
x8 -0.885

               Factor1 Factor2
SS loadings      3.426   2.027
Proportion Var   0.428   0.253
Cumulative Var   0.428   0.682

Test of the hypothesis that 2 factors are sufficient.
The chi square statistic is 53.5 on 13 degrees of freedom.
The p-value is 7.39e-07
```

The last argument *score* = *"regression"* produces the factor scores, which are the linear commonalities, and then weighting them with factor score coefficients and summing for each factor. The output for the last command line is given in the following. With the *head* command, we can print only six rows. However, there are indeed 32 observations, so 32 scores for both factors exist on the data. Examining the factor scores, we can see that $x1$ is -1.20 standard deviations below on the factor 1 dimension and 0.81 standard deviations above factor 2 dimension.

```
> factor <-factanal(X,factors=2,rotation="varimax",scores="regression")
> head(factor$scores)
         Factor1    Factor2
[1,]  -1.1955549   0.8158279
[2,]  -1.0606056   0.7672726
[3,]  -1.1955549   0.8158279
[4,]  -1.1955549   0.8158279
[5,]  -0.6411360  -2.4079120
[6,]  -0.7638105  -1.3642100
```

We produce a clearer picture of how the variables can be classified based on the components. To plot the figure with the loadings, we write the following commands in R.

```
> load <- factor$loadings[,2:1]
> plot(load,type="n")
> text(load,labels=names(data_chapter6),cex=1)
```

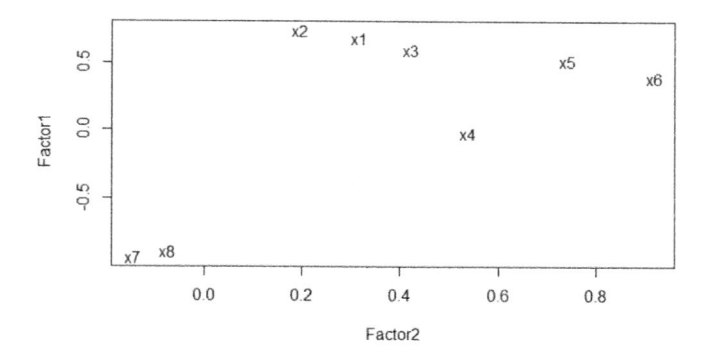

Exercises

1. Explain, in your own words, the various steps involved in a factor analysis. How are the factor coefficients extracted? How does one interpret the factors?
2. Based on the factor scores VA, ASSETTECH, and FOOD, undertake an independent sample t-test as in Chapter 2, to determine if there are significant differences between male- and female-headed households in the above scores.
3. Now undertake a multivariate regression analysis, with FOODSEC as the dependent variable and the factor scores VA, ASSETTECH, and FOOD as the independent variables. What is the adjusted R^2? Which variables are significant in your regression analysis? Interpret the results in light of

access to markets, technology adoption, and asset ownership as critical determinants of food security.
4. After studying the paper by Beynon et al. (1992), critically discuss the impact of the private sector on food market liberalization. What are the main policy implications that are suggested for private sector development?
5. Explain the methodology in Stifel et al. (2003) paper of how isolation affects rural poverty in Madagascar. What are the variables used in the study to measure transaction costs? What are the main findings of road quality improvement on rice production? Discuss critically.

STATA workout

Table 6.10 presents information on the standard-of-living conditions of 32 families. The variables are defined as follows:

1. ×1—Poverty Status Index
2. ×2—Food Security Index
3. ×3—Schooling Index
4. ×4—the number of children below age 5
5. ×5—the number of dependents
6. ×6—distance of drinking water

TABLE 6.10 Living conditions on 32 families.

ID	×1	×2	×3	×4	×5	×6	×7	×8
1	0	0	0	4	3	3	3	3
2	0	0	1	4	3	3	3	3
3	0	0	0	4	3	3	3	3
4	0	0	0	4	3	3	3	3
5	0	1	0	1	0	0	3	3
6	0	1	0	1	1	1	3	3
7	0	1	0	1	1	1	3	3
8	0	1	0	1	2	1	3	3
9	0	1	0	1	2	1	3	3
10	1	1	1	3	1	2	3	3
11	0	2	0	1	2	2	3	3

TABLE 6.10 Living conditions on 32 families.—cont'd

ID	x1	x2	x3	x4	x5	x6	x7	x8
12	0	2	0	1	3	2	3	3
13	1	2	1	1	4	4	1	1
14	1	2	1	3	1	2	3	3
15	1	2	1	3	2	2	3	3
16	1	2	1	3	2	3	3	3
17	1	2	1	3	3	3	3	3
18	0	3	1	2	3	3	3	3
19	1	3	1	2	3	3	1	2
20	1	3	1	2	3	3	2	2
21	1	3	1	2	4	4	1	1
22	1	3	1	2	4	4	1	1
23	1	3	1	3	3	2	1	1
24	1	3	1	3	3	2	1	1
25	1	3	1	3	3	3	1	2
26	1	3	1	3	3	3	1	2
27	1	3	1	3	4	3	1	2
28	1	3	1	3	4	4	2	3
29	1	3	1	3	4	4	2	3
30	1	3	1	3	4	4	3	2
31	1	3	1	3	4	4	1	2
32	1	3	1	3	4	4	1	2

7. x7—distance to a health clinic
8. x8—distance to the market
    a. Perform the KMO test of sampling adequacy. Does your KMO value lie above 0.6? What about the P-value from your chi-square test?
    b. Analyze the data using the PCs method using the STATA commands **factor** and **pca**.
    c. Use the **greigen** command and find the number of factors that have to be retained for the analysis.
    d. Use the **fapara** command and conduct a parallel analysis and compare the results with **pca**.

e. Compute the linear combinations or the communalities using **predict** and store the factors with the names **f1** and **f2**.
f. Cluster the families based on how they load on different factors. Identify and characterize the families that are particularly affected by poverty. The variables are defined as follows:

The STATA input codes for the above problems are as follows:

- global xlist x1-x8
- factortest $xlist
- factor x1-x8, pcf
- fapara, reps(100)
- quietly factor $xlist, pcf factor(2)
- rotate, varimax
- predict Bartlett
- graph two way scatter f1 f2, yline(0) xline(0) mlabel(id)

The screenshot from STATA output follows:

```
. factortest $xlist

Determinant of the correlation matrix
Det             =       0.000


Bartlett test of sphericity

Chi-square          =             214.393
Degrees of freedom  =                  28
p-value             =               0.000
H0: variables are not intercorrelated


Kaiser-Meyer-Olkin Measure of Sampling Adequacy
KMO             =       0.727
```

The determinant of the correlation matrix is zero, which indicates multi-collinearity in the data. Consequently, it informs us that we have to examine some of the variables and their correlations carefully. Hence, it is a good idea to reduce the dimension of the problem. The chi-square test and the KMO measure of sampling adequacy indicate that we can proceed with factor analysis.

The results of **factor x1-x8, pcf** are produced in the following. As in the previous example, the PCs model shows that the first two components have eigenvalues greater than 1. These two components explain 77% of the combined variance in the eight variables. Besides the first two, the rest of the components are dropped from the analysis. Variable $x4$ (number of children below the age of 5) loads heavily with the second factor, while rest of the variables load heavily with the first factor.

```
. factor x1-x8, pcf
(obs=32)
```

Factor analysis/correlation   Number of obs = 32
    Method: principal-component factors   Retained factors = 2
    Rotation: (unrotated)   Number of params = 15

Factor	Eigenvalue	Difference	Proportion	Cumulative
Factor1	4.75491	3.34880	0.5944	0.5944
Factor2	1.40611	0.56550	0.1758	0.7701
Factor3	0.84061	0.32293	0.1051	0.8752
Factor4	0.51768	0.32278	0.0647	0.9399
Factor5	0.19490	0.06747	0.0244	0.9643
Factor6	0.12743	0.02422	0.0159	0.9802
Factor7	0.10320	0.04805	0.0129	0.9931
Factor8	0.05516	.	0.0069	1.0000

LR test: independent vs. saturated: chi2(28) = 222.19 Prob>chi2 = 0.0000

Factor loadings (pattern matrix) and unique variances

Variable	Factor1	Factor2	Uniqueness
x1	0.8575	-0.0502	0.2622
x2	0.8092	-0.3703	0.2081
x3	0.8486	0.0937	0.2711
x4	0.3295	0.8460	0.1757
x5	0.7847	0.2736	0.3094
x6	0.7911	0.4454	0.1758
x7	-0.8407	0.3236	0.1886
x8	-0.7667	0.4050	0.2481

The **fapara, reps(100)** command is useful, since it produces the scree plot from the factor analysis and also from parallel analysis. The scree plot indicates that we can retain two factors from and is partially supported by parallel analysis also. The eigenvalues after 100 replications indicate that only the first factor is relevant. We continue with the analysis, assuming we can retain the first two factors.

```
. fapara, reps(100)

PA -- Parallel Analysis for Factor Analysis -- N = 32
PA Eigenvalues Averaged Over 100 Replications
            FA          PA          Dif
    1.   4.754908    1.067061    3.687846
    2.   1.406111     .714549     .6915616
    3.    .8406107    .4461759    .3944348
    4.    .5176823    .2250093    .292673
    5.    .1948985    .0352057    .1596928
    6.    .1274292   -.1199826    .2474118
    7.    .1032045   -.2528967    .3561012
    8.    .0551566   -.3567754    .4119321
```

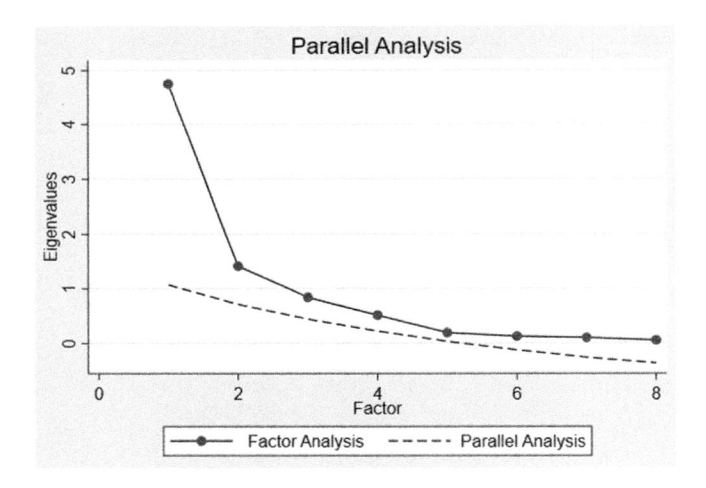

Parallel Analysis

Factor Analysis ----- Parallel Analysis

Factor rotation and factor loadings indicate that the following variables load heavily on factor 1: Poverty Status Index, Food Security Index, Schooling Index, distance to a health clinic, and distance to the market. The other variables, the number of children below age 5, the number of dependents, and distance of drinking water load heavily on factor 2. We can consider the first set of variables related to factor 1 as "economic" variables, while the second set that loads on factor 2 as "socio" variables.

```
. quietly factor $xlist, pcf factor(2)

. rotate, varimax

Factor analysis/correlation                          Number of obs    =    32
    Method: principal-component factors              Retained factors =     2
    Rotation: orthogonal varimax (Kaiser off)        Number of params =    15
```

Factor	Variance	Difference	Proportion	Cumulative
Factor1	3.91245	1.66388	0.4891	0.4891
Factor2	2.24857	.	0.2811	0.7701

```
LR test: independent vs. saturated:  chi2(28) =  222.19 Prob>chi2 = 0.0000
```

Rotated factor loadings (pattern matrix) and unique variances

Variable	Factor1	Factor2	Uniqueness
x1	0.7670	0.3867	0.2622
x2	0.8858	0.0855	0.2081
x3	0.6872	0.5067	0.2711
x4	-0.1393	0.8972	0.1757
x5	0.5416	0.6303	0.3094
x6	0.4610	0.7821	0.1758
x7	-0.8896	-0.1417	0.1886
x8	-0.8664	-0.0341	0.2481

Factor rotation matrix

	Factor1	Factor2
Factor1	0.8651	0.5016
Factor2	-0.5016	0.8651

Factor scores are formed by standardizing each variable to zero mean and unit variance and then weighting them with factor score coefficients and summing for each factor. The **predict Bartlett** results are produced in the following. For instance, the Poverty Status Index ($\times 1$) is 0.17 standard deviations above average on the *factor 1* dimension, and 0.05 standard deviations above average on the *factor 2* dimension.

```
. predict Bartlett
(regression scoring assumed)

Scoring coefficients (method = regression; based on varimax rotated factors)
```

Variable	Factor1	Factor2
x1	0.17391	0.05957
x2	0.27930	-0.14245
x3	0.12099	0.14714
x4	-0.24183	0.55527
x5	0.04517	0.25112
x6	-0.01493	0.35745
x7	-0.26839	0.11043
x8	-0.28397	0.16833

The list input line produces the following graph:

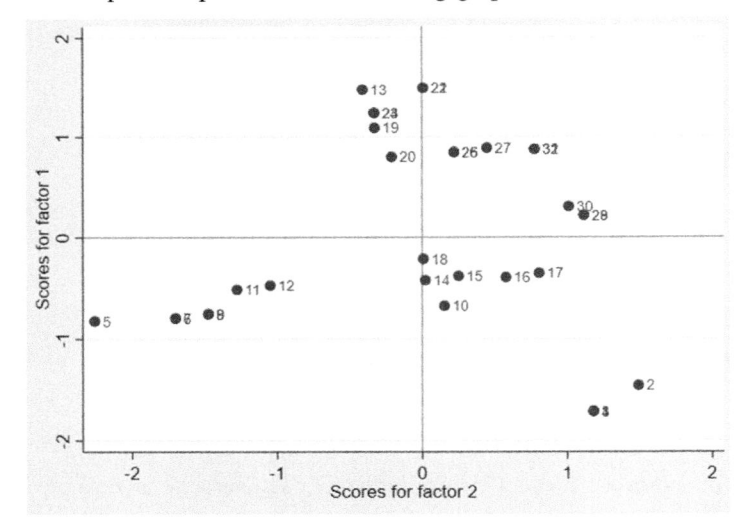

The first factor loads heavily on the "economic" variables, while the second loads heavily on the "socio" variables. Families (5, 6, 7, 8, 9, 11, and 12) are low in both dimensions. The aforementioned graph provides an interesting picture that illustrates the manner in which the data are clustered around these characteristics.

Section II

Nutrition policy analysis

Chapter 7

Impact of maternal education and care on preschoolers' nutrition—application of two-way ANOVA

Chapter outline

Introduction	222
Conceptual framework: linkages between maternal education, child care, and nutritional status of children	223
Possible linkages	223
Conceptual and measurement issues on child care	224
Measurement issues	225
Review of selected studies	227
Maternal education and nutrition status in the United States	229
Children's nutrition and maternal education in Africa	232
Kenya	232
Uganda and Nigeria	232
Substantive findings from Asia and Latin America	233
Empirical analysis	234
Educational level effect	234
Care effect	234
Interaction effect	234
Data description	234
Cross-tabulation of weight for height with mothers' educational levels	235
Two-way ANOVA results	237
Definition of main effect	237
Partitioning sum of squares	237
Interpreting the interaction effect and post hoc tests	241
Two-way ANOVA in STATA	242
Example 1	242
Example 2	246
Example 3	246
Conclusion	248
Technical appendices	249
Scoring system used to create the care index (Ruel et al., 1999) care index by age group	249
Two-way ANOVA model	249
Post hoc procedures	249
ANOVA in R	250
Exercises	253
STATA workout	254

Motherhood has been an exercise in guilt.

Felicity Huffman.

Introduction

Maternal education has the potential to improve the quality of care given to mothers and children resulting in better health of the family. It further can contribute to adoption of innovations and speeding up of social and behavioral change. Child malnutrition persists in many parts of the developing world despite various efforts to improve the nutritional status of mothers and children. The causes of malnutrition are multiple and include inadequate food, health, lack of sanitation facilities, high fertility rates, ignorance about child care practices, and lack of access to health services. WHO (2005) estimates show that poor water, sanitation, and hygiene account for 16% of deaths of children under 5 globally. Thus, understanding the causes and context of malnutrition is important in devising strategies that generate better child health and nutritional outcomes.

The importance of mothers' education for child health and survival through various pathways was first demonstrated by Caldwell (1979) in his seminal paper on Nigeria. This study suggested that education of women played an important role in determining child survival even after controlling for other socioeconomic characteristics. The various pathways suggested by the study of how maternal education can enhance child survival were implementation of health knowledge, an increased capability to interact in the modern world, and greater control over health choices for her children.

During the 1980s, the understanding of the association between maternal education and child health at the microlevel expanded greatly with the World Fertility Survey (WFS) program and from a UN study that used both survey and census data. Hobcraft et al. (1984) covered 28 WFS surveys, and Mensch et al. (1985) covered 15 countries. Both studies demonstrated that the association between maternal education and child survival was weaker in sub-Saharan Africa than in Asia and Latin America, where socioeconomic differences were generally larger. The aforementioned studies also suggested that there was no threshold level of maternal education that needed to be reached before advantages in child survival began to accrue. Recent research during the 1990s at both national and household levels has also demonstrated that increased maternal education has a positive impact on the nutritional status of young children, even after controlling for socioeconomic indicators and access to health services (see, for example, Christian et al., 1988; Ruel et al., 1992, 1999). This may be because children of educated mothers have a lower mortality risk, since educated women tend to marry and have their first child at a later age. Educated mothers are also likely to influence the decision-making within the family in favor of children's needs. Additionally, schooling makes women aware of the nearest health center, immunization of children against diseases, feeding children at the appropriate time and in right quantities, and taking early actions against infant diarrhea (Joshi, 1994).

From the policy and program intervention perspective, it is useful to know how various determinants of child malnutrition contribute independently and interact with each other in determining the final outcomes. Formal education of the mother may increase the care practices through knowledge. Yet educated mothers are also time constrained due to their participation in labor markets, and illiterate caregivers may have time but may have indigenous (or primitive) child survival practices. Understanding the interaction of such related variables has been a policy and programmatic challenge since such interactions are usually cultural and context specific (Cameron et al., 2001).

Understanding the issue is important since maternal education can affect child nutritional status through greater knowledge, greater provision of resources, and/ or change in a mother's status (Caldwell, 1979). Education provides knowledge which prevents illness and, at the same time, speeds recovery by consistent care of a sick child. Education also serves as a means to higher paying employment with which a household can command more resources through which it can improve the health of its members (Rosenzweig, 1995).

This chapter will examine the impact of maternal education and child care on children's nutritional status as measured by height-for-age (ZHA) and weight-for-height (ZWH) Z-scores, using a two-way ANOVA approach. The advantage of this approach is that one can simultaneously assess the effects of two (or more) independent variables on a single dependent variable and the possible combined effects of the independent variables on the dependent variable can be determined. This is known as the "interaction effect." Another way of stating the interaction effect is to say that the effect of one factor (for example, maternal education) depends on the level of the second factor (child care). In this chapter, we examine one such interaction effect: maternal education and child care on children's nutritional status. The two-way ANOVA approach will allow us simultaneously to determine the independent and combined impact of maternal education and child care on children's nutritional status. This chapter is organized as follows: In the next section, we provide a conceptual framework of the linkages between maternal education, child care practices, and nutritional status of children. We also examine the conceptual and measurement issues related to child care practices. We then review the few studies that examine the role of maternal education and child care practices on child health outcomes. We also examine the extent of the issue in the United States and Kenya. Next, we present the empirical analysis using a two-way ANOVA approach, followed by our conclusions. We provide examples and illustrations using STATA demonstrating the implementation of the two-way ANOVA methodology.

Conceptual framework: linkages between maternal education, child care, and nutritional status of children

Possible linkages

Fig. 7.1 demonstrates one possible pathway through which maternal schooling can affect the nutritional outcomes of children. Women with more education

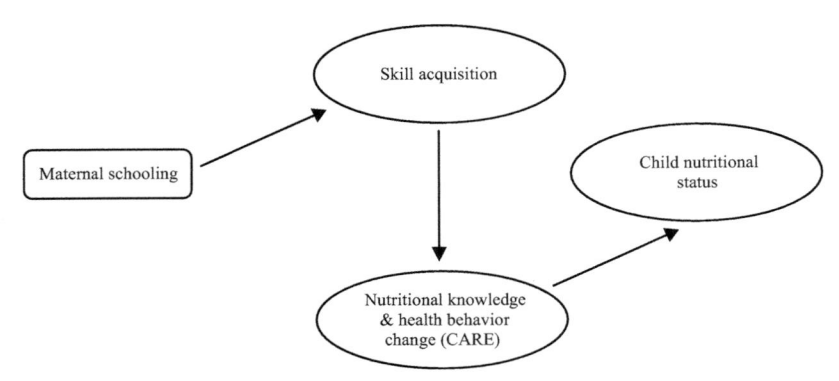

FIGURE 7.1 Linkages between maternal schooling, child care, and nutritional status. *Adapted from Levine, R.A., Dexter, E., Velasco, P., LeVine, S., Joshi, A.R., Stuebing, K.W., Tapia-Uribe, F.M., 1994. Maternal literacy and health care in three countries: a preliminary report. Health Trans. Rev. 4, 186–191.*

tend to be knowledgeable about healthcare and, if exposed to new information, can assimilate this improved knowledge into better care practices than women with lesser education. This additional skill level (especially for women) can make them aware of health services (such as health center facilities and availability of doctors) and generate additional nutritional knowledge (such as immunization of children against diseases, taking appropriate actions on incidence of infant diarrhea, feeding the child during sickness, and breastfeeding during early childhood). The aforementioned good care practices can, in turn, improve the nutritional status of children. Better healthcare practices are especially relevant for less educated mothers, for mothers with more dependents, and for children from households with limited resources, poor housing conditions, and lack of access to hygiene and sanitation services. One cannot assume, however, that mothers of malnourished children are necessarily ignorant of child care practices or that illiterate mothers, whether their children are healthy or malnourished, do not practice enough good care (Christian et al., 1988).

Conceptual and measurement issues on child care

Child survival, nutrition, and health depend on household food security, on a healthy environment and available health services, and care provided to women and children (UNICEF, 1990). The element of care provided to women is an important determinant in improving mothers' and children's nutritional status, and this element has received considerable attention since the 1990s.

The original conceptual model developed by the UNICEF defined care as "the provision in the household and the community of time, attention and support to meet the physical, mental, and social needs of the growing child and other household members." An "extended" conceptual model of care developed by Engle et al. (1999) provides a more detailed conceptual apparatus of both care practices and important household and community-level resources. They characterize behavior patterns into the following:

1. Care for pregnant and lactating women.
2. Breastfeeding and complementary feeding of young children.
3. Food preparation and food storage behaviors.
4. Hygiene behaviors.
5. Care for children during illness.

The resources for care were classified into six main categories:

1. Education, knowledge, and beliefs.
2. Health and nutritional status of the caregiver.
3. Mental health, lack of stress, and self-confidence of the caregiver.
4. Control of resources and intrahousehold allocation.
5. Workload and time constraints.
6. Social support from family members and the community.

Fig. 7.2 provides an extended UNICEF model of child care that includes not only an assessment of the caregiver's behavior but also the behavior of the child and the characteristics of the environmental context. All three factors play an important role in the nutritional status of the child.

Measurement issues

Two dimensions of caring behavior that have been identified are time spent (quantity of care) and the nature of the activities undertaken (quality of care). The "time spent on care" method is usually determined by assessing the time spent in specific activities with children (such as bathing, feeding, etc.) along with other activities of the household. Most of the studies do not find any significant association between child care time and nutritional status. Thus, time devoted to child care may not be a useful indicator of nutritional status. The "quality of care" approach tries to determine how specific practices lead to better nutritional outcomes for children. They are usually classified into caregiver and psychosocial care practices. Caregiver practices affect the child's nutrient intake through psychomotor capabilities (such as use of finger foods, spoon handling ability, and so on) and appetite (Engle et al., 1999). Additionally, the caregiver's ability to feed responsively may include encouraging the child to eat, offering additional foods, responding to poor appetite, and using a positive style of interaction with the child. Some studies in developing countries have also found a strong association between specific feeding behaviors (such as location of feeding, organization of feeding event) with mothers' educational status (Guldan et al., 1993).

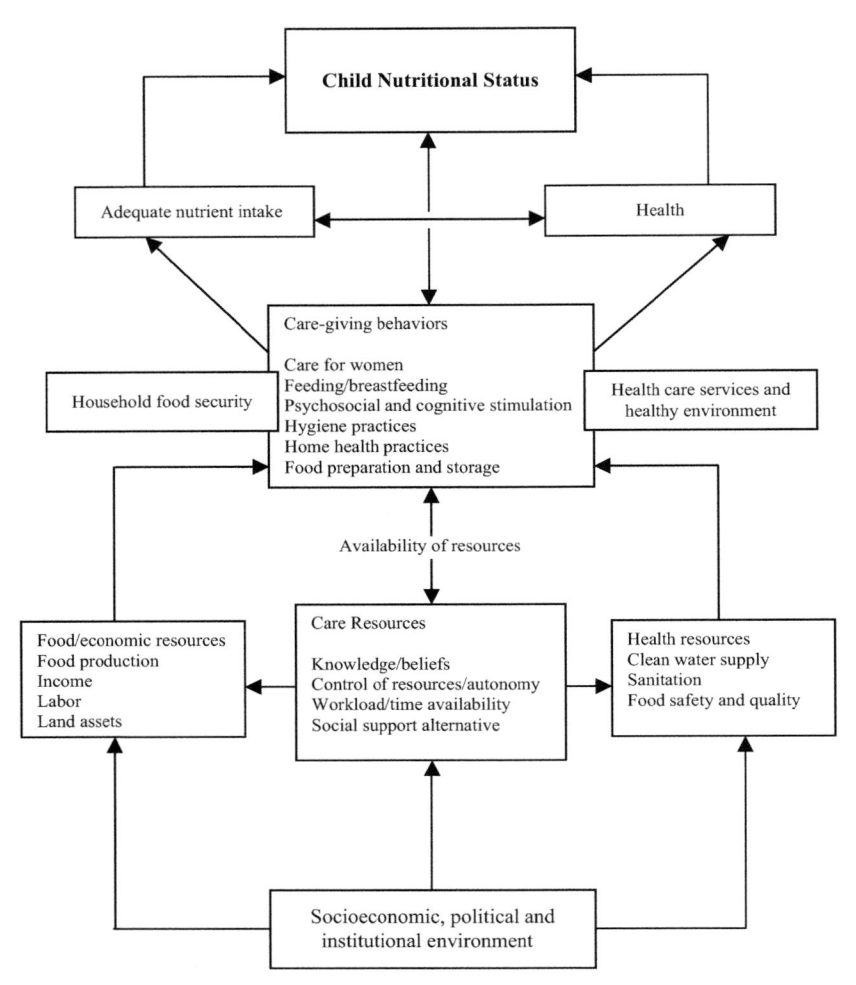

FIGURE 7.2 Extended model of child care. *Adapted from Engle, P.L., Menon, P., Haddad, L., 1999. Care and nutrition: concepts and measurement. World Develop. 27 (8), 1309—1337.*

Psychosocial care, on the other hand, refers to the provision of affection and warmth, responsiveness to the child and the encouragement of autonomy and exploration (Engle et al., 1999). Culture plays a central role in psychosocial care.

There are mainly three ways of measuring care practices: observation of specific practices, quantitative assessments of feeding behaviors, and behavioral ratings.

In the observation method, the examiner assesses the caregiver behaviors on a series of items. Bentley et al. (1991) developed a scoring system to measure child and caregiver behaviors for each food rather than for each eating event. They constructed a Guttman scale for child and caregiver behaviors. (The Guttman scale is based on the assumption that there is logical order among dichotomously coded items and the order is always unchanged.) For the child, the three-point scale was based on food refusal, food appetite, and food request. For the caregiver, the scale was no response, verbal encouragement, verbal pressure, and physical force.

The *quantitative assessment* method counts the number of instances of a behavior during a feeding episode. Sanders et al. (1993) rated the frequency of 14 parent and 17 children behaviors using the mealtime observation schedule in Australia. Behaviors that vary in duration can be coded as whether it is occurring after a fixed interval or not.

The *behavioral rating* method rates the overall quality of the child and caregiver interaction. The behavior rating scale (Bayley scale) for infant development has been used by Engle and Zeitlen (1996) for developing economies. In this method, a domain of behavior is defined by how the caregiver understands the child's need. For example, if a child protests against the way it is held, the parent can adjust the position. Next, we review specific studies that analyze maternal education and child care as determinants of child nutrition.

Review of selected studies

While the benefits from female education on child nutritional outcomes are well documented, the pathway through which it contributes to these outcomes has not been adequately analyzed. While many studies (Sahn and Alderman, 1997; Ruel et al., 1999) find strong positive linkages between maternal education and child nutrition, other studies controlling for different factors show little linkages between the two.

In Mali, Penders et al. (2000) found no significant effect of maternal education on ZHA, while Dargent-Molina et al. (1994) reported no beneficial impact on infant outcomes from improvement in maternal education. For the sake of brevity, we will explore only studies that find positive linkages from maternal education on child nutrition working through improved child care practices.

The study of Christian et al. (1988) was one of the earlier studies that examined the effects of mothers' literacy status and nutrition knowledge on the nutritional status of children in a two-way ANOVA framework. The study was carried out in rural villages of the Panchmahal district of Gujarat state in India. Education of women was categorized as literate or illiterate, while women's nutrition knowledge was evaluated on a scale of 0—9, with the maximum possible score being 9. Nutritional status of the child was categorized into five categories based on weight for age.

The study found that both female literacy and income were important intervening factors in the impact of nutrition knowledge on children's nutritional status. More importantly, the study found that nutritional knowledge exerted a stronger influence on children's nutritional status, which implies that, although all women do need formal education, nutrition education in the short run can significantly reduce the burden of child malnutrition.

In a comprehensive study undertaken in Accra, Ghana (in an urban setting), Ruel et al. (1999) investigated the impact of maternal education on nutritional status of children through the mediating role of child care practices. The main hypothesis tested was under what conditions are good care practices important for children's nutritional status.[1] The survey data were collected between January and March 1997 for a sample of 475 households with children 3 years or younger. The dependent variable was ZHA, and the independent variable of interest was care practices (the care index score). Two-way interaction terms between care practices and various other factors were used to test whether some groups of children benefited from good care practices.

The main finding of the study that good care practices had positive impact on children's nutritional status has important policy implications. Specific training in child feeding and use of preventive health services for mothers with no formal education could have a large impact on children living under poverty. Also, informal education could mitigate the negative effects of poverty and low maternal schooling on children's nutritional status. Thus, with better care practices, poor children from less educated mothers could improve their nutritional status.

Drawing on a large household survey consisting of 7200 households, using a multistage cluster sampling method in rural central Java, Indonesia, Webb and Block (2003) addressed three issues:

1. What is the impact of maternal schooling compared with maternal nutrition knowledge on child nutrition in the short and long run?
2. Are maternal and paternal schooling substitutes for good child nutrition?
3. How do maternal education and nutrition knowledge affect mothers' own nutritional status?

The motivation behind the aforementioned set of questions was that, while most mothers understood the importance of maintaining the protein energy intake to maintain nutritional status, a subset of mothers understood the importance of micronutrient-rich foods, and thus, their children were better protected from the crisis than others.

---

1. The authors hypothesize that good child care may be particularly important for children of less educated mothers; for time-constrained women; for households with more dependent children; and for children from households with limited resources, poor housing conditions, and lack of access to hygiene and sanitation.

The study estimated a reduced form nutritional production function, in which nutritional outcome was related to productive resources, maternal schooling, age, paternal schooling, and maternal nutrition knowledge. The model thus accommodated the possibility that nutrition knowledge could have differential impacts on nutritional status for different members of the household.

The main results can be summarized as follows:

1. ZWH was more related to care and feeding practices than household resources. In contrast, maternal schooling had no significant impact in the short term.
2. In the long run, nutritional knowledge contributed significantly to ZHA and nutritional status. In addition, maternal schooling and nutritional knowledge were neither substitutes nor complements in affecting short-term nutritional status. Thus, the study concluded that short-term and long-term child nutritional indicators were determined by different sets of factors. Short-term child nutritional status was much more responsive to maternal nutrition knowledge than maternal schooling, while formal schooling of mothers was more important in long-term child nutritional status.

The policy implications from the study are as follows. First, as nutrition knowledge clearly affects short-term child nutritional status more than household resources or maternal education, combining clear nutrition messages with other resources targeted to the poorest households in developing countries can have significant impact. Second, while maternal education is important for long-term child health status, the importance of the father's education should not be ignored, as investments in human capital are critical for development.

Maternal education and nutrition status in the United States

Olson (2004) investigates how the level of human resources and the diversion of financial resources away from food are related to the food security status of rural low-income households. Results show that the mothers who used a greater number of food and financial skills (managing bills, making a budget, stretching groceries, preparing meals) were more likely to have food-secure households, compared with the mothers who used fewer of these skills. Furthermore, maternal symptoms of depression and reported difficulty paying for medical expenses were related to increased risk of food insecurity.

Using a much larger data set with a broader scope, Carneiro et al. (2007) study the intergenerational effects of maternal education on children's cognitive achievement, behavioral problems, grade repetition, and obesity. They use data from the National Longitudinal Surveys and control for many variables including mothers' abilities and family backgrounds. Overall, they find substantial intergenerational returns to education. The study found income effects delayed childbearing. In addition, the study also demonstrated that maternal education led to substantial differences in maternal labor supply.

Section highlights: is maternal education related to children's obesity levels?

A natural causal analysis would suggest that supervising children's nutritional intakes is likely to reduce childhood obesity. However, this requires the supervisors, who are mostly women, to be aware of the importance of nutritional balance in diet and have the necessary wherewithal to implement these requirements on a daily basis. Consequently, researchers have tried to link the relationship between maternal education and childhood obesity levels. For instance, Fertig et al. (2009) investigate the channels through which maternal employment affects childhood obesity.

They use information from the Child Development Supplement of the Panel Study of Income Dynamics to conduct their analysis. Their data combine information on children's time allocation, children's BMI, and mother's labor force participation. Their results show that supervision and nutrition play significant but small roles in the relationship between maternal employment and childhood obesity.

However, the mechanisms through which child obesity relates to maternal education and employment are not as direct as it may seem. There are many interesting implications that the researchers bring to this issue. For instance, one is likely to conclude that maternal education is negatively associated with childhood obesity. However, after analyzing the data, the researchers find that educated mothers have children with lower BMI, only if they spend more time at home in child care. This implies that maternal education and child obesity are linked through the amount of time spent at home. For example, these researchers observe that the following:

- There is a pronounced correlation between skipping breakfast and overweight and obesity.
- Working married mothers spend a smaller share of their food budgets on vegetables, fruits, milk, and meat and beans than nonworking married mothers, suggesting that the content of the meals may be the important factor.
- Surprisingly, the effect of number of meals is stronger than the effect of TV watching or reading/talking/listening to music.
- More surprisingly, playing sports has no effect: An increase in mandated time for physical activity in school does not have a significant impact on children's BMI.

Their main findings are as follows:

1. The number of meals is significantly and negatively associated with percentile BMI, and mother's work hours are significantly and negatively associated with the number of meals. That is, more hours working increases children's BMI through the mechanism of fewer meals.
2. Reading/talking/listening to music is significantly and negatively associated with percentile BMI, and mother's work hours are significantly and negatively

> **Section highlights: is maternal education related to children's obesity levels?—cont'd**
>
> associated with the time spent reading/talking/listening to music. That is, more hours working increases children's BMI through the mechanism of reading/talking/listening to music.
> 3. For less educated mothers, the time in school is significantly and negatively associated with percentile BMI, and mother's work hours are significantly and positively associated with time spent in school. Among these mothers, more hours working is associated with their children having a lower BMI through the mechanism of school attendance.
> 4. TV watching is significantly and positively associated with high percentile BMI, and that mother's work hours are significantly and positively associated with more time spent watching TV. That is, more work hours are associated with a higher child's BMI through the mechanism of TV watching.
> 5. More work hours are significantly associated with more time spent in child care, but for the more educated mothers, more time spent in child care is associated with lower child BMI. Together, this means that more work hours is associated with lower child BMI through the mechanism of child care.
>
> Along with these interesting conclusions, the research also opens up complications that still have to be resolved. For example, for less educated mothers, the effect of mother's work hours on BMI is reduced when we control for the child's time spent in school. In contrast, for more educated mothers, the effect of mother's work hours is reduced if one considers the time spent in child care. This suggests that school absences are key factors, rather than attending before or after school programs. Thus, mothers who work more hours may ensure that their children do not miss school or child care and this reduces the children's BMI.
>
> However, it may be that parents who ensure that their children do not miss school or child care are also less likely to have overweight children. Or, alternatively, it may be that overweight children are more likely to miss school or child care for health reasons and this affects their mother's ability to work more hours consistently. The latter two interpretations would imply that a change in mothers' hours would not change their children's weight status. Future research can shed light on this endogeneity issue.

These results are of interest to policy makers and program managers who address food security issues in rural areas of the United States. Furthermore, female-headed households are a growing segment of the US population. It is important to note that education, health (nutrition) knowledge, income, and degree of urbanization are important predictors of the overall nutritional status of this population (see Ramezani and Roeder, 1995). Among many determinants, health knowledge is most significant, suggesting that improving such knowledge could lead to more informed decisions and an enhancement of diet quality.

Children's nutrition and maternal education in Africa

Kenya

The importance of maternal education on nutrition of children is also evidenced by Kabubo-Mariara et al. (2009) in Kenya. They examine the determinants of children's nutritional status based on demographic and health surveys, with a pooled data from 1998 to 2003. They estimate the impact of child, parental, household, and community characteristics on children's height and on the probability of stunting. They control sample design and heterogeneity arising from unobserved community characteristics correlated with children's nutritional status and its determinants. Interestingly, they find that boys suffer more malnutrition than girls, and children of multiple births are more likely to be malnourished than singletons. Furthermore, in the context of this chapter, their results indicate that maternal education is a more important determinant of children's nutritional status than paternal education. Household assets are also important determinants of children's nutritional status, but nutrition improves at a decreasing rate with assets. The use of public health services, such as modern contraceptives, is also found to be an important determinant of child nutritional status. Their econometric simulations suggest that there is big role for parental, household, and communities in reducing long-term malnutrition in Kenya. A correct policy mix may involve promoting strategies in the current high levels of malnutrition, and if Kenya is to achieve her strategic health objectives and millennium development target of reducing the prevalence of malnutrition, then Kenya must promote strategies for poverty alleviation, promotion of postsecondary education for women, and provision of basic preventive healthcare.

Uganda and Nigeria

Ssewanyana and Kasirye (2010) not that sub-Saharan Africa registers the widest inequalities in child nutrition, with a high incidence of stunting rates among the poorest children. Uganda particularly faces severe issues with estimates of about 2.4 million stunted children, who are less than 5 years old.

Ssewanyana and Kasirye (2010) use three different cross-sectional data for Uganda and note that maternal education attainment, along with household asset holdings, is a key driver of childhood nutrition and health.

Improvement in primary school completion rates for mothers, higher female education in secondary schooling, and universal free education programs for mothers are most effective in combating low child health status in Uganda.

Besides income inequality and poverty, Uganda also suffers from adverse political factors, ethnic fractionalization, and internal conflicts, which impede progress in the area of nutrition and health status, among children. Hoolda (2019) tracks the impact of the Lord's Resistance Army insurgency and shows how the conflict at the village level lowers Z-scores for both weight-for-age and weight-for-height measures. Besides low access to healthcare, poor maternal education are the main factors that inhibit childhood health, with lingering effects lasting for 5 years after the end of the conflict.

Fadare et al. (2019) also demonstrate how rural Nigeria experiences poor nutritional outcomes for Children, mainly due to lack of maternal education and to limited access thereof.

Substantive findings from Asia and Latin America

Overall, there is substantial evidence from various studies conducted across countries, which suggest the maternal education has a direct influence on childhood health status. Besides Uganda and Kenya, Favara (2018) has carefully calibrated similar findings for Peru, from a sample of a study from Young Lives. Favara (2018) notes that maternal education, along with maternal group participation, positively influences the child's height for age, particularly when children are 1 year old. Maternal education, along with the cooperation and association at the community level, helps mothers with low levels of education to foster better child health.

Interestingly, maternal education and community organization have substantial spillover effects on height for age, than just maternal education by itself. For instance, Favara (2018) notes that an 11-month-old child whose mother had no formal education, but was a member of a community organization, was at least 1.1 cm taller than a child whose mother had similar educational attainment, but who did not participate in any community organization.

Increased income and education may generate unanticipated "status bias" that might distort optimal allocation toward nutrition. For instance, Dasgupta et al. (2017) note that in Bangladesh, better maternal education and family status lead to an increase in the consumption of animal-source food, rather than fish, which is less expensive, and has substantial positive effects on child mortality and resistance childhood illnesses.[2] Likewise, Jansen et al. (2015) in an interesting study from Colombia show the early onset of menarcheal age attributed to maternal BMI and education.

Related to Favara (2018), Ervin and Bubak (2019) are careful to point out that maternal education alone cannot solve the problem of childhood malnutrition. With 13-year data from Paraguay, Ervin and Bubak (2019) that a multipronged strategy inclusive of healthcare improvements, access, family planning, and demographics alongside access to piped water, sanitation, and maternal education is much needed in eliminating the rural—urban inequality in the ZHA measure and in stunting in Paraguay.

---

2. Also see Hoddinott et al. (2018) for successful dietary interventions for preschoolers in Bangladesh, and Masters et al. (2018) for similar results in Ethiopia, Nigeria, and India.

Likewise, using data from Nepal, Miller et al. (2020) and Cunningham et al. (2017) show how dietary quality measured via dietary diversity and intake of animal-source food influence child development. In particular, the findings show that maternal education is a significant driver of home environmental quality and child development. Headey et al. (2016), Nguyen et al. (2018, 2016), Young et al. (2018), and related studies provide conclusive evidence in favor of maternal education alongside related services, to improve children's ZHA, across Bangladesh, Nepal, Pakistan, India, and Vietnam.

Empirical analysis

The main question addressed in the present analysis is how education level of the mother and child care affects child nutritional status. The two-way ANOVA approach can determine if there are overall differences in ZWH between different educational levels of the mother, between varying levels of child care, and whether there is an interaction effect of educational level and child care on improving child nutritional status (Karpinski, 2003). The interaction effect can be thought of as saying that the effect of one factor (e.g., educational level) depends on the level of the second factor (e.g., child care). For example, it may be the case that higher educated women provide better child care than lower educated women. Thus, we undertake the following hypothesis tests:

Educational level effect

H_0: mean ZWH does not differ by educational levels of the mother.
H_1: mean ZWH differs by educational levels of the mother.

Care effect

H_0: mean ZWH does not differ by care levels by the mother.
H_1: mean ZWH differs by care levels by the mother.

Interaction effect

H_0: there is no interaction between educational levels and care levels.
H_1: there is an interaction between educational levels and care levels.

Data description

The dependent variable in the analysis is ZWH, which is a measure of short-term child nutritional status. We convert this variable into a categorical variable as follows:

$$\text{ZWHNEW} = \begin{cases} 1 & \text{if ZWH} \geq -2 \text{ (normal } Z-\text{scores)} \\ 0 & \text{if ZWH} < -2 \text{ (low } Z-\text{scores)} \end{cases} \quad (7.1)$$

Thus, a value of 1 indicates absence of wasting, while a value of 0 indicates presence of wasting. The independent variables are as follows:

1. Education of the spouse (EDUCSPOUS): A categorical variable, the value of which ranges from 1 to 7. It measures the education level of the spouse (or mother) in number of years. (In the case of female-headed households in the two-way ANOVA, we separate out females who are heads of the household and thus are not the spouse of a male-headed household.) For example, the variable attains a value of 5 if the spouse completed secondary education. Thus, higher values indicate more number of years in schooling.
2. Child care index (CARE): A composite child care index was constructed on the lines of Ruel et al. (1999). The index was derived using variables related to child-feeding practices (such as breastfeeding and feeding the child during sickness) and preventive health-seeking behavior (whether the child was immunized). The index ranged on a continuous scale from -1 to $+1$, with -1 denoting poor child care practices and $+1$ denoting good care practices. For age groups where a particular practice (such as breastfeeding for children above 24 months of age and compulsory immunization to children below 9 months) is not likely to improve the growth of children, the component was assigned a value of 0 implying a neutral effect. The index was made age specific for each age group. The technical appendix provides details about the construction of this index. The composite care index was thus the sum of the individual components, namely the breastfeeding, child feeding during sickness, and compulsory immunization components. Terciles (NCARE, child care index) of this index were then created to distinguish among children whose mothers had poor caring practices from those with good care practices.

Cross-tabulation of weight for height with mothers' educational levels

First, we look at the prevalence of malnutrition (measured by ZWHNEW and ZHANEW); ZHANEW is constructed similar to Eq. (7.1) and differentiates between children who are stunted from those who are not by mothers' educational levels using cross-tabulation procedures.

Table 7.1 shows no marked differences in the prevalence of stunting between noneducated and educated women. For example, for mothers with no education, the prevalence of stunting is 51.3% relative to the presence of normal children of 48.1%. On the other hand, for some level of educational attainment of the mother (std 5–8) which in this sample is the highest educational attainment, the prevalence of stunting is 30% relative to nonprevalence (or normal children) of only 27.9%. The P-value is 0.457, and as the significance level is greater than 0.1, the null hypothesis cannot be

TABLE 7.1 Prevalence of stunting by mothers' educational level.

		ZHANEW		
		Low	Normal	Total
EDUCSPOUS	No education	41	50	91
	Adult literacy	1	6	7
	Std 1–4	14	19	33
	Std 5–8	24	29	53
	Total	80	104	184 = n

rejected at the 10% level. Thus, we can conclude that there is no significant difference in prevalence of stunting between educated and noneducated mothers.

Table 7.2 shows that there are significant differences in the prevalence of wasting between noneducated and educated women. For mothers with no education, the prevalence of wasting is almost 79% compared to nonprevalence (47.5%). Thus, it would appear that short-term nutritional status is significantly influenced by mothers' educational level. On the other hand, as educational level increases, there is much less prevalence of wasting. The P-value from the Pearson chi-square statistic is 0.006, and since it is less than 0.1, the null hypothesis that there is no difference between wasting among uneducated and educated mothers can be rejected. Thus, we conclude that

TABLE 7.2 Prevalence of wasting by mothers' educational level.

		ZHANEW		
		Low	Normal	Total
EDUCSPOUS	No education	15	84	99
		78.9%	47.5%	
	Adult literacy	2	5	7
		10.5%	2.8%	
	Std 1–4	0	35	35
		0.0%	19.8%	
	Std 5–8	2	53	55
		10.5%	29.9%	
	Total	19	177	196 = n

educational level matters for short-term nutritional status. In the next section, we investigate the role of mothers' education and child care on ZWH using a two-way ANOVA approach.

Two-way ANOVA results

The two-way ANOVA approach is suitable, since it allows us to determine the interaction effects of educational level of the mother and child care on ZWH and thus provides greater generalizability of results. In a two-way ANOVA, we will obtain three different statistical tests:

1. Main effect of educational level of the spouse.
2. Main effect of care levels by terciles of care.
3. Interaction effects between educational level of the spouse and care levels.

Definition of main effect

This is the effect of one independent variable on the dependent variable across the levels of the other independent variables. In our example, we want to determine if there is a difference in the mean ZWH by the educational levels of the mother averaging over the child care levels. In other words, ignoring the effect of child care levels, does ZWH differ between educated and non-educated mothers? Second, we want to determine if there is a difference in ZWH by child care terciles ignoring the educational levels of the mother. One way to understand the main effect is to examine the marginal means.

As evident from Table 7.3, we find the mean performance on ZWH is greater for mothers with some educational level relative to mothers with no education. Thus, we want to ask the following question: do the marginal means differ?

From Table 7.4, it is evident that the mean ZWH is highest for mothers in the medium care tercile (0.974) followed by the upper care tercile. The result possibly indicates that as care behavior improves (such as breastfeeding children below 2 years of age), nutritional status of children improves after controlling for educational level of the mother. Next, we determine whether the marginal means among the various child care terciles differ. This is done by calculating the sum of squares.

Partitioning sum of squares

For a two-way ANOVA, the model has additional components compared with a one-way ANOVA (as introduced in Chapter 5). Fig. 7.3 provides a useful way of understanding variance partitioning.

The breakdown of the total (corrected for the mean) sums of squares is summarized in Table 7.5.

TABLE 7.3 Effect of mothers' education on ZWHNEW.

EDUCSPOUS	Mean
No education	0.844
Adult literacy training	0.75
Std 1–4	1.00
Std 5–8	0.961

TABLE 7.4 Effect of child care on ZWHNEW.

NCARE	Mean
1	0.842
2	0.974
3	0.884

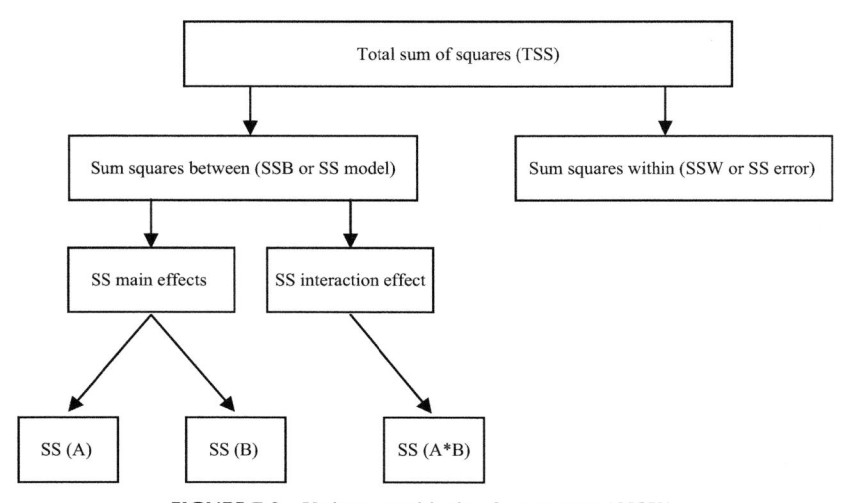

FIGURE 7.3 Variance partitioning for two-way ANOVA.

TABLE 7.5 ANOVA Table for an $a \times b$ Factorial Experiment.

Source	SS	df	MS
Factor A	SS(A)	$(a-1)$	MS(A) = SS(A)/$(a-1)$
Factor B	SS(B)	$(b-1)$	MS(B) = SS(B)/$(b-1)$
Interaction AB	SS(AB)	$(a-1)(b-1)$	MS(AB) = SS(AB)/$(a-1)(b-1)$
Error	SSW	$(N-ab)$	SSW/$(N-ab)$
Total (corrected)	TSS	$(N-1)$	

In our present analysis, the sum of squares between SSB is made up of three parts: the sum of squares of mothers' educational level SS(EDUCSPOUS), the sum of squares of terciles of child care SS(CARE), and the sum of squares of the interaction SS(interaction). Thus, the sum of squares of the interaction term is obtained as follows:

$$\text{SS(interaction)} = \text{SSB} - \text{SS(EDUCSPOUS)} - \text{SS(NCARE)} \quad (7.2)$$

The SSW or sum squares error is what is left over. Thus, we have

$$\text{SSW} = \text{TSS} - [\text{SS(EDUCSPOUS)} + \text{SS(NCARE)} + \text{SS(interaction)}] \quad (7.3)$$

To determine the mean squares, we have to find the respective sum of squares and divide it by the corresponding degrees of freedom. Thus, the degrees of freedom for TSS is $(N-1)$, where N is the total sample size. The degrees of freedom for SS(EDUCSPOUS) and SS(NCARE) are $(k-1)$, where there are four groups for mothers' education and three child care terciles. The degrees of freedom for the interaction are equal to the product of degrees of freedom of the SS(EDUCSPOUS) and SS(NCARE) and are six.

The next step is to calculate the three F-statistics for the tests. This is done by taking the mean squares for the two independent variables and the mean square of the interaction and dividing them by the mean squares within. To determine if a particular F-statistic is statistically significant, we compare the calculated F value with the critical value of F. If the obtained F value exceeds the critical value, we reject the null hypothesis and conclude that the independent variable has a significant impact.

Interpreting the output from a 2×2 factorial ANOVA involves examining the three F-values associated with the two main effects (EDUCSPOUS and NCARE) and the interaction effect (EDUCSPOUS*NCARE). For the main effect of mothers' education on ZWH, we obtain an F-value of 4.085 with the corresponding probability being 0.008 (Table 7.6). Since the probability is less than 0.05, we reject the null hypothesis that mean ZWH does not differ by educational levels of the mother. Thus, education of the mother has a

TABLE 7.6 Tests of between subject effects: dependent variable ZWHNEW.

Source	Type III sum of squares	df	Mean square	F	P-value
SSB	2.187	10	0.218	2.691	0.003
Intercept	64.330	1	64.330	798.914	0.000
EDUCSPOUS	0.987	3	0.329	4.085	0.008
NCARE	0.492	2	0.246	3.056	0.049
EDUCSPOUS*NCARE	0.708	6	0.118	1.468	0.202
SSW	14.971	185	0.081		
TSS	17.158	195			

$R^2 = 0.132$ (adjusted $R^2 = 0.085$).

Impact of maternal education and care on preschoolers' Chapter | 7 **241**

significant influence on ZWH. Taken together with the estimated marginal mean of Table 7.3, we can conclude that there is a significant difference in the mean ZWH between educated and noneducated mothers ignoring the impact of child care. Examining the main effect of child care terciles on ZWH, we find the F-value to be 3.056 with the associated significance level of 0.049. We thus reject the null hypothesis that the mean ZWH does not differ by care levels of the mother, and thus, child care levels have a significant impact on ZWH after ignoring the impact of mothers' education.

Interpreting the interaction effect and post hoc tests

Since there are more than two means, we want to determine through multiple post hoc comparisons (such as Tukey or LSD tests; these procedures are discussed in brief in the technical appendix at the end of this chapter) to determine which means (if any) are significantly different. Recall that there are four educational levels of the mother and three child care levels. Although, from Table 7.6, the overall interaction effect (EDUCSPOUS*NCARE) is not significant, there may be significant differences among the means of mothers' educational levels and child care. In other words, there may be differences among combinations of education levels and child care terciles. Post hoc tests are used when the researcher is exploring differences among group means; otherwise, the likelihood of type 1 errors increases. We will carry out an example of carrying out post hoc tests using educational level of the mother and child care terciles.

The group means for mothers with no education and standards 1–4 (some elementary schooling) are placed on the columns, while child care terciles (2 and 3) are placed in the rows (Table 7.7). It is apparent that mothers with more education relative to mothers with no education practice greater child care. The simple post hoc analysis compares a given pair of means. If they are significantly different ($P < .05$), different letters are placed next to these means to indicate that they are significantly different. Let us start with the cell in the upper left-hand corner with a mean of 0.929 and place the letter "**a**" next to it. Since the post hoc tests (both LSD and Tukey) indicate that the next highest mean for the combination of educational level of the mother being in standard

TABLE 7.7 Multiple comparison test.

		EDUCSPOUS	
		No education	Std 1–4
NCARE	2	0.929 **a**	1.00 **b**
	3	0.735 **c**	1.00 **d**

1—4 (some elementary schooling) and second child care tercile is significantly different, we place the letter "**b**" next to it. Then we compare the combination educational level of the mother (Std. 1—4) and child care (second tercile) with the next mean educational level of the mother (no education) and child (third tercile) and find the difference to be significant.

We place the letter "**c**" next to it. The general principle is that by comparing a given pair of means, if we find the difference to be significant ($P < .05$), we place a different letter next to it. On the other hand, if the difference is not significant ($P > .05$), then the same letter is placed next to these means to indicate that they are not significantly different. The interpretation of the interaction effect (EDUCSPOUS*NCARE) can be described as follows: For the same level of child care, higher education among mothers (elementary schooling) improves short-term nutritional status as measured by ZWH. On the other hand, for a given level of education (no education or some elementary schooling), mothers in the second child care tercile perform better than ones in the highest care tercile. The interpretation is that the impact of positive child care practices improves short-term nutritional status for households in the lower socioeconomic terciles compared with the upper socioeconomic terciles. This result is consistent with Ruel et al. (1999), where care was found to be more critical for children whose mothers had less than secondary schooling and for households in the lower two socioeconomic terciles for Ghana.

Two-way ANOVA in STATA

Example 1

For illustrative purposes, we use the following 36 observations given in Table 7.8 on ZWHNEW, EDUCSPOUS, and NCARE, where all these categorical variables are as defined in the previous section of this chapter.

TABLE 7.8 ZHNEW, EDUCSPOUS, and NCARE values from 36 families.

ZHNEW	EDUCSPOUS	NCARE
0	1	0
0	1	0
0	1	0
0	1	0
1	2	0
1	2	0
1	2	1

Continued

TABLE 7.8 ZHNEW, EDUCSPOUS, and NCARE values from 36 families.—cont'd

ZHNEW	EDUCSPOUS	NCARE
1	2	1
0	3	1
1	3	1
1	3	1
1	3	1
1	1	1
1	1	1
1	1	1
1	1	1
1	2	1
1	2	1
1	2	0
1	2	0
0	3	0
0	3	0
1	3	0
1	3	0
0	1	0
0	1	0
0	1	0
0	1	0
0	2	0
0	2	0
0	2	1
0	2	1
0	3	1
1	3	1
1	3	1
1	3	1

We look at the descriptive statistics, which we can obtain, using the summarize command in STATA. Together, we also obtain the two-way ANOVA table along with all the pairwise comparisons using the following commands in STATA:

1. summarize
2. anova zwhnew educspous ncare
3. pwmean zwhnew, over(educspous) mcompare(tukey) effects

The STATA summarize command produces, as we know all the important descriptive statistics. The anova command performs the ANOVA calculations and presents the results as a typical table, assuming ZWHNEW as the dependent variable. The pwmean command performs pairwise comparisons of means. In the aforementioned command, STATA computes all pairwise differences of the means of ZWHNEW over the combination of the levels of EDUCSPOUS. As mentioned in the previous section, the tests and confidence intervals for the pairwise comparisons assume equal variances across groups. The pwmean command in STATA also produces results from Tukey's method, which we discussed in the previous section.[3] The input lines and the output from STATA are given in the following for our two-way ANOVA for the 36 observations that we have in Table 7.8. We begin with the descriptive statistics produced by the summarize command:

. summarize

Variable	Obs	Mean	Std. Dev.	Min	Max
zwhnew	36	.5555556	.5039526	0	1
educspous	36	2	.8280787	1	3
ncare	36	.5	.5070926	0	1

Overall we find that there are large standard deviations for all the three categorical variables. We next look at the results from the ANOVA:

. anova zwhnew educspous ncare

```
                  Number of obs =      36     R-squared     =  0.2500
                  Root MSE      = .456435     Adj R-squared =  0.1797

   Source |  Partial SS    df       MS            F     Prob > F
----------+----------------------------------------------------------
    Model |  2.22222222     3   .740740741         3.56    0.0250
          |
 educspous|   .444444444    2   .222222222         1.07    0.3561
    ncare |  1.33333333     1   1.33333333         6.40    0.0165
          |
  Residual|  6.66666667    32   .208333333
----------+----------------------------------------------------------
    Total |  8.88888889    35   .253968254
```

---

3. Besides Tukey's method, STATA also allows for adjusting the confidence intervals and *P*-values to account for multiple comparisons using Bonferroni's method, Scheffe's method, and Dunnett's method.

we pointed out in the previous section, a particular F-statistic is statistically significant, if the calculated F value exceeds the critical value. In that case, we reject the null hypothesis and conclude that the independent variable has a significant impact. Interpreting the STATA output from a 2 × 2 factorial ANOVA involves examining the three F-values associated with the two main effects (EDUCSPOUS and NCARE) and the interaction effect (EDUC-SPOUS*NCARE), which are produced by STATA. For the main effect of mothers' education on ZWH, we obtain an F-value of 1.07 with the corresponding probability being 0.35. Since the probability is more than 0.05, we cannot reject the null hypothesis that mean ZWH does not differ by educational levels of the mother. Thus, education of the mother does not have a significant influence on ZWH. We can also compare the F-value of 1.07 with the table F-value, which at 99% significance (or $\alpha = 0.01$) with 2 degrees of freedom (numerator) and 32 degrees of freedom (denominator) equals 3.30. Since the calculated F-value is smaller than the table F-value, we cannot reject the null hypothesis.

Examining the main effect of child care ton ZWH, we find the F-value to be 6.4 with the associated significance level of 0.01. The table value of $F_{1,32}$ in this case is 4.15 at 99% significance, and since this is smaller than the calculated value, we can reject the null hypothesis. In other words, the mean ZWH does not differ by care levels of the mother, and thus, child care levels have a significant impact on ZWH after ignoring the impact of mothers' education.

Since there are more than two means, we want to determine a post hoc comparison, or a Tukey test, and determine which means are significantly different. Recall that there are three educational levels of the mother and two child care levels, and therefore, there may be significant differences among the means of mothers' educational levels and child care. In other words, there may be differences among combinations of education levels and child care levels. We will carry out an example using the pwmean command in STATA to generate the Tukey t-values:

```
. pwmean zwhnew, over(educspous) mcompare(tukey) effects

Pairwise comparisons of means with equal variances

over          : educspous
```

	Number of Comparisons
educspous	3

zwhnew	Contrast	Std. Err.	Tukey t	P>|t|	Tukey [95% Conf. Interval]
educspous					
2 vs 1	.3333333	.2010076	1.66	0.236	-.1598979 .8265646
3 vs 1	.3333333	.2010076	1.66	0.236	-.1598979 .8265646
3 vs 2	-5.55e-17	.2010076	-0.00	1.000	-.4932313 .4932313

From the *P*-values, we can conclude that for the same level of education, higher education among mothers does not improve nutritional status as measured by the *Z*-scores.

Example 2

We use the same data from the aforementioned and incorporate the interaction term EDUCSPOUS*NCARE. We undertake this in STATA using the following commands:

.gen x = educspous*ncare
.anova zwhnew educspous ncare x

The gen command in STATA is used whenever we want to create a new variable. In this case, we create a new variable called "x" defined as educspous*ncare, using the gen command. The next line is just a new anova command that incorporates the interaction term. The screenshot of the STATA results is given in the following:

```
. gen x = educspous*ncare

. anova zwhnew educspous ncare x
```

	Number of obs = 36	R-squared = 0.4188
	Root MSE = .414997	Adj R-squared = 0.3219

Source	Partial SS	df	MS	F	Prob > F
Model	3.72222222	5	.744444444	4.32	0.0044
educspous	1.66666667	2	.833333333	4.84	0.0151
ncare	.166666667	1	.166666667	0.97	0.3331
x	1.5	2	.75	4.35	0.0219
Residual	5.16666667	30	.172222222		
Total	8.88888889	35	.253968254		

The table value of $F_{2,30}$ at 99% significance is 3.32. Consequently, we can reject the null hypothesis for EDUCSPOUS and the interaction or the *x* variable terms, because their calculated *F* values are 4.84 and 4.35, respectively. The inclusion of the interaction term indicates that maternal education is important when it is taken in conjunction with NCARE. This indicates the subtle interaction between education and child care.

Example 3

Assume we have the following information on output of a farm that uses four different fertilizers and three different pesticides. It is usual to assume that each plot of land has the same probability of receiving each fertilizer–pesticide combination, to satisfy the completely randomized design portion of the experiment.

Our goal is to test the hypothesis that the population means for fertilizers are all identical and to test the hypothesis that the population means for pesticides are identical, using STATA. As a first step, it is important to type these data into STATA as columns. We can then use the anova command to derive the results. We accomplish this in the following example:

The data for the analysis in STATA must be entered in the following fashion[4]:

Pesticide	Fertilizer	Output
1	1	21
1	2	12
1	3	9
1	4	6
2	1	13
2	2	10
2	3	8
2	4	5
3	1	8
3	2	8
3	3	7
3	4	1

The input and the output information produced in STATA appear as follows:

```
. anova output fertilizer pesticide
```

Number of obs = 12   R-squared   = 0.8571
Root MSE   = 2.51661 Adj R-squared = 0.7381

Source	Partial SS	df	MS	F	Prob > F
Model	228	5	45.6	7.20	0.0161
fertilizer	156	3	52	8.21	0.0152
pesticide	72	2	36	5.68	0.0412
Residual	38	6	6.33333333		
Total	266	11	24.1818182		

The first hypothesis to be tested is $H_0: \mu_1 = \mu_2 = \mu_3 = \mu_4$ against the alternative H_1: $\mu_1, \mu_2, \mu_3, \mu_4$ that are not equal, where μ refers to the various population means of the four fertilizers. The table value of $F = 9.78$ at 99% significance, for degrees of freedom 3 (numerator) and 6 (denominator). We cannot reject H_0 because the calculated value is 8.21, which is smaller than the table value.

---

4. It is easier if you type this in EXCEL and save it as a csv file, and then read this file directly into the STATA online editor using the "insheet using" command. Type findit insheet using in STATA for more information. For more information, see Cameron and Trivedi (2010, page 37).

Likewise, for the second hypothesis we test, H_0: $\mu_1 = \mu_2 = \mu_3 = \mu_4$ against the alternative H_1: $\mu_1, \mu_2, \mu_3, \mu_4$ that are not equal, where µ refers to the various population means of the three pesticides. The table value of $F = 10.92$ at 99% significance, for degrees of freedom 2 (numerator) and 6 (denominator). We cannot reject H_0 because the calculated value is 5.68, which is smaller than the table value.

What is the implication of not rejecting the null hypothesis? We usually assume that the populations are normally distributed with equal variances. We can now view the four samples as coming from the *same* population.

Conclusion

The role of care practices in improving child nutritional status has increasingly been recognized during the 1990s with the UNICEF's nutrition conceptual framework. The framework suggested that not only were food security and healthcare services necessary for child survival, but care for women and children was equally important. Enhanced caregiving can balance the use of resources to promote good health and nutrition in women and children. Engle et al. (1999) improved on this conceptual framework by emphasizing the ways through which child care practices translate into improved child nutritional status. Six care practices (with subcategories) and three kinds of resources were identified as crucial determinants of child nutritional status. The modified framework emphasized the importance of complementary feeding practices, such as introduction of complementary foods and feeding frequency and caregivers' understanding of children's response to food intake.

The purpose of this chapter was to examine the roles of maternal education and child care as well as the interaction of maternal education and child care on short-term child nutritional status. For this purpose, we developed a composite index of child care by incorporating breastfeeding, feeding during sickness, and immunization of the child as components of this index. The index was made age specific for each age group of the children. Our results indicate that maternal education and child care have independent and significant influence on child nutritional status. After undertaking post hoc tests, we find that, for a given level of child care, education among mothers (elementary schooling) improves short-term nutritional status. On the other hand, for a given level of education, mothers in a lower child care tercile perform better than ones in the highest care tercile in improving child nutritional status.

However, specific issues need to be addressed regarding the determinants of caregiving and specific care practices for understanding child nutritional status. It is important to investigate the role of psychosocial care in promoting both physical growth and development in young children. Examining the linkages through which maternal schooling affects child nutrition, for example, through improved food selection or greater knowledge, would enhance the child care programs. Most importantly, at the policy level, programs that include care should make an effort to identify and support good care practices such as identifying the factors that motivate caregivers rather than simply advocating them.

Technical appendices

Scoring system used to create the care index (Ruel et al., 1999) care index by age group (see Table 7.10)

Two-way ANOVA model

Let us introduce the following notations for understanding the ≥ two-way ANOVA model:

a	the number of levels of the first factor (rows)
B	The number of levels of the second factor (columns)
N	The number of observations in each cell
X_{ijk}	The kth observation in the ith row and jth column
$\overline{X}_{ij}$	Mean of the n observations in cell (i, j)
$\overline{X}_i$	Mean of nb observations in row i
$\overline{X}_j$	Mean of na observations in column j
$\overline{X}$	Grand mean of the nab observations.

Then the total variation can be decomposed into

$$X_{ijk} = \overline{X} + \left(\overline{X}_{ij} - \overline{X}\right) + \left(\overline{X}_i - \overline{X}\right) + \left(\overline{X}_{ij} - \overline{X}_i - \overline{X}_j + \overline{X}\right) + \left(X_{ijk} - \overline{X}_{ij}\right) \quad (7.4)$$

Taking $\overline{X}$ to the left hand of Eq. (7.4), and squaring both sides of the equation and summing over all observations, we obtain

$$\sum_{i=1}^{n}\sum_{j=1}^{a}\sum_{k=1}^{b} \left(X_{ijk} - \overline{X}\right)^2 = na\sum_{j}\left(\overline{X}_j - \overline{X}\right)^2 + nb\sum_{i}\left(\overline{X}_i - \overline{X}\right)^2$$
$$+ n\sum_{i}\sum_{j}\left(\overline{X}_{ij} - \overline{X}_i - \overline{X}_j + \overline{X}\right)^2 \quad (7.5)$$
$$+ \sum_{i}\sum_{j}\sum_{k}\left(X_{ijk} - X_{ij}\right)^2$$

The left-hand side of Eq. (7.5) is the total sum of squares, while the right-hand side consists of SSB and SSW. The first three terms on the right hand are the sum of squares of factor A, factor B, and the interaction term, while the last expression denotes the sum of squares of the error term or the sum of squares within. After dividing the relevant expressions by the degrees of freedom, we obtain the mean square expressions. The corresponding F ratios are obtained by dividing the mean square of factor A, factor B, and the interaction term (An * B) by the mean square error.

Post hoc procedures

It is critical in the ANOVA procedure to reduce the probability of type 1 error to 0.05 or smaller. Post hoc procedures are used to assess which group means differ from each other after the overall F-test has demonstrated that at least one such difference exists. The group means refer to the means of the dependent variable for each of the k groups (a groups for factor A and b groups for factor B) formed by the categories of the independent variables. The possible number of comparisons is thus $k(k-1)/2$. The q-statistic[5] (also called the q range statistic) is commonly used for such comparisons. The q-statistic tests the probability that the largest and smallest mean among the k groups formed by categories of the independent variables were sampled from the same population. If the q-statistic computed for the two sample means is not greater than the critical q-value, then one cannot reject the null hypothesis that the groups do not differ at the given significance level ($P < .05$). The Tukey honestly significant difference test (HSD)[6] is undertaken when the sample sizes of the groups are highly unequal. It is a preferred method when the number of groups is large. However, many researchers prefer it for pairwise comparisons. When all pairwise comparisons are tested, this procedure is more powerful than other post hoc tests. For implementing the Tukey's HSD procedure, one needs to compare the actual t computed for any pairwise comparison to the critical value of the q-statistic. If $t_{computed} > (q_{critical})/\sqrt{2}$, we can reject the null hypothesis that the groups' means are the same. The least significant difference test (LSD), also called the Fisher's test, is based on the t-statistic and can be considered as a form of t-test. It compares all possible pairs of means after the F-test rejects the null hypothesis that groups do not differ. It can handle both pairwise and nonpairwise comparisons and does not require equal sample sizes. It is the most liberal of all the post hoc tests, and the results need to be interpreted with caution. Since it controls the type 1 error rate at a given level of significance, it is most likely to reject the null hypothesis in favor of finding groups that do differ. Toothacker (1993) recommends against use of LSD on the grounds that it has poor control over the level of significance when better alternatives exist.

ANOVA in R

For illustrative purposes, we used the following 36 observations given in Table 7.8 on ZWHNEW, EDUCSPOUS, and NCARE, where all these are categorical variables.

We look at the descriptive statistics, which we can gain, using the summary command in R. Together, we also obtain the two-way ANOVA table along with all the pairwise comparisons using the following command in R.

---

5. Tukey's insight was to determine a sampling distribution related to the t distribution, which is the pairwise maximum and is called the studentized q statistic. Mathematically, $(q = \sqrt{2t})$.
6. For a useful discussion of the Tukey HSD tests and Fisher's least significant difference test, the reader is encouraged to consult Karpinski (2003) Chapter 6 on planned contrasts and post hoc tests for one-way ANOVA.

The summary command produces vital descriptive statistics. We also undertake the apply command to get the information of standard deviation since the *summary* command in R does not show it automatically.

```
> summary(data_chapter7)
     ZWHNEW          EDUCSPOUS         NCARE
 Min.   :0.0000    Min.   :1       Min.   :0.0
 1st Qu.:0.0000    1st Qu.:1       1st Qu.:0.0
 Median :1.0000    Median :2       Median :0.5
 Mean   :0.5556    Mean   :2       Mean   :0.5
 3rd Qu.:1.0000    3rd Qu.:3       3rd Qu.:1.0
 Max.   :1.0000    Max.   :3       Max.   :1.0

> apply(data_chapter7,2,sd)
  ZWHNEW EDUCSPOUS      NCARE
0.5039526 0.8280787 0.5070926
```

The *aov* command that is the one used in the previous section for the one-way ANOVA is undertaken to perform the two-way ANOVA as well. Note that it should include the different formats of an outcome variable and independent variables. Before executing the *aov* command, we should manipulate numeric variables to factor ones because the numeric variable cannot be recognized as a categorical variable, which is EDUCSPOUS and NCARE. First, check the type of variables by using the *str* command. We can see that all variables are considered as numeric variables.

```
> str(data_chapter7)
Classes 'tbl_df', 'tbl' and 'data.frame':    36 obs. of 3 variables:
 $ ZWHNEW   : num  0 0 0 1 1 1 1 0 1 ...
 $ EDUCSPOUS: num  1 1 1 1 2 2 2 3 3 ...
 $ NCARE    : num  0 0 0 0 0 0 1 1 1 1 ...
```

We use the *as.factor* command to change the type of independent variables from numeric to factor. Then, recheck the variables by using the *str* command for two independent variables which are EDUCSPOUS and NCARE.

```
> EDUCSPOUS=as.factor(EDUCSPOUS)
> NCARE=as.factor(NCARE)

> str(EDUCSPOUS)
 Factor w/ 3 levels "1","2","3": 1 1 1 1 2 2 2 2 3 3 ...
> str(NCARE)
 Factor w/ 2 levels "0","1": 1 1 1 1 1 1 2 2 2 2 ...
```

Now we are ready to undertake the two-way ANOVA using the aov command. To perform the two-way ANOVA calculations, assuming ZWHNEW as the dependent variable and EDUCSPOUS and NCARE as the independent variables, type the command: $aov(ZWHNEW \sim EDUCSPOUS + NCARE)$. As we described in the earlier section, saving the results and using the *summary* command on the results presents more information.

```
> ANOVA2 <- aov(ZWHNEW~EDUCSPOUS+NCARE)
> summary(ANOVA2)
            Df Sum Sq Mean Sq F value Pr(>F)
EDUCSPOUS    2  0.889  0.4444   2.133 0.1350
NCARE        1  1.333  1.3333   6.400 0.0165 *
Residuals   32  6.667  0.2083
---
Signif. codes:  0 '***' 0.001 '**' 0.01 '*' 0.05 '.' 0.1 ' ' 1
```

The null hypothesis of the ANOVA is that the independent variable has no significant impact on the outcome variable. First, we perform the two-way ANOVA without an interaction variable. For the main effect of mothers' education on ZWH, the results show that F-value of 2.1 with the associated significance level of 0.14. We cannot reject the null hypothesis that mean ZWH does not differ by educational levels of the mother since the P-value is higher than 0.05. Therefore, the educational of the mother does not have a significant influence on ZWH.

The other main effect of child care on the dependent variable, we find the F-value to be 6.4, with the corresponding probability of 0.05. We can reject the null hypothesis at 95% significance, indicating there is a significant impact of child care levels on the ZWH. In other words, the mean ZWH does not differ by mothers' educational levels but differ by child care levels.

Since there are more than two means, we want to determine a post hoc comparison, or a Tukey test, to determine which means are significantly different among combinations of education levels and child care levels. We carry out an example by using the *TukeyHSD* command to generate the Tukey *t*-values:

```
> TukeyHSD(ANOVA2)
  Tukey multiple comparisons of means
    95% family-wise confidence level

Fit: aov(formula = ZWHNEW ~ EDUCSPOUS + NCARE)

$EDUCSPOUS
           diff         lwr       upr      p adj
2-1  3.333333e-01 -0.1245708 0.7912375 0.1894333
3-1  3.333333e-01 -0.1245708 0.7912375 0.1894333
3-2  4.440892e-16 -0.4579042 0.4579042 1.0000000

$NCARE
         diff        lwr       upr      p adj
1-0  0.3703704 0.06046083 0.6802799 0.0206784
```

From the results of Tukey's test, we can conclude that for the different levels of education, whether mothers received higher or lower education, there are no significant impacts on improving nutritional status.

We now use the same data from the aforementioned and incorporate the interaction term EDUCSPOUSE*NCARE. To include an interaction term, we can manipulate the independent variables by using *. The results of the ANOVA with the interaction terms are given as follows:

```
> ANOVA3 <- aov(ZWHNEW~EDUCSPOUS*NCARE)
> summary(ANOVA3)
                Df Sum Sq Mean Sq F value  Pr(>F)
EDUCSPOUS        2  0.889  0.4444   2.581 0.09244 .
NCARE            1  1.333  1.3333   7.742 0.00924 **
EDUCSPOUS:NCARE  2  1.500  0.7500   4.355 0.02185 *
Residuals       30  5.167  0.1722
---
Signif. codes:  0 '***' 0.001 '**' 0.01 '*' 0.05 '.' 0.1 ' ' 1
```

From the results, the P-values decrease in general so that we can reject the null hypothesis of all independent variables regarding P-value. For example, the mother's education level has an impact on weight for height at a significant level of 90%. Child care also has a significant effect on 99% significance. The inclusion of the interaction term indicates that maternal education is essential when it is taken in conjunction with NCARE.

Exercises

1. Discuss a methodology in constructing a CARE index after reviewing the existing literature.
2. Discuss how the total sum of squares is partitioned in a two-way ANOVA compared with a one-way ANOVA. What is meant by main effects and interaction effects?
3. Examine critically how Engle et al. (1999) improved on the UNICEF framework on the role of child care as an additional determinant in improving child nutritional status. What are the specific mechanisms discussed in the paper through which child nutritional status can be improved? Suggest a few dimensions of care that you will recommend to the Ministry of Health in your own country for improving child nutritional status.
4. Undertake a two-way ANOVA with each component of care (CARE1, CARE2, and CARE3) as independent variables separately with education of the spouse (EDUCSPOUS) as the other independent variable. Which component or components turns out significant? Carry out a post hoc analysis to determine which groups of CARE (CARE1, CARE2, and CARE3) terciles differ by educational status of the mother as analyzed in Table 7.8.
5. After undertaking a two-way ANOVA with per capita expenditure quartiles (NPXTOTAL) and EDUCSPOUS as the independent variables, we obtain the following results:

Source	df	SS	MS	F
NPXTOTAL		0.696		
EDUCSPOUS		0.915		
Interaction		0.850		
Error		16.925		
TSS	235	19.131		

- With the aforementioned information, calculate the degrees of freedom (df), SS(B), mean square errors, and the F-statistic.
- What are three main ways of measuring child care practices?
- Use the data given in Table 7.9 on fertilizers and pesticides for this problem. Enter the data in Excel and create a csv file. Use the insheet command and create a new file in STATA. Use the gen command and create an interaction term called x which equals fertilizer*pesticide. Use STATA and conduct an ANOVA using all the three factors: fertilizer, pesticide, and x. Are you able to reject H_0 in any of the cases? Use the pwmean command and generate the Tukey's test results.

The following table provides the overall caloric index of 36 children from their daily consumption habits. The children are selected from families that

TABLE 7.9 Output from different plots with four different fertilizers and three different pesticides.

	Fertilizer 1	Fertilizer 2	Fertilizer 3	Fertilizer 4
Pesticide 1	21	12	9	6
Pesticide 2	13	10	8	5
Pesticide 3	8	8	7	1

TABLE 7.10 Care index by age group.

	Scores Allocated by age group (in months)		
Practices included	0–8.9	9–23.9	≥24
Breastfeeding	No: −1 Yes: 1	No: −1 Yes: 1	No: 0 Yes: 0.5
Sickfeeding: cutoff point was 3	No: −1 Yes: 1	No: −1Yes: 1	No: −1 Yes: 1
Compulsory vaccination	No: 0 Yes: 0	No: −1Yes: 1	No: −1 Yes: 1

have different levels of income and maternal education. The information is provided as a 2 × 2 classification. A random sample of four children is selected for each of the three income levels (high, middle, low), and for three levels of maternal education (high, middle, and low). Each entry corresponds to the caloric index of the child based on the child's nutritional intake during a specific time period.

```
                        Maternal education levels
                High                Middle              Low
Income  High    0.36, 0.39, 0.39, 0.38  0.42, 0.40, 0.39, 0.42  0.32, 0.36, 0.35, 0.34
levels  Middle  0.38, 0.40, 0.41, 0.40  0.42, 0.45, 0.48, 0.47  0.37, 0.33, 0.33, 0.34
        Low     0.34, 0.32, 0.34, 0.35  0.34, 0.34, 0.30, 0.31  0.36, 0.35, 0.35, 0.33
```

Use the information in the aforementioned table and with the help of STATA's command, generate an ANOVA table and conduct F-tests for the presence of nonzero income effects, education effects, and interaction effects.

STATA workout

Suppose the following data on caloric intake from animal-source foods for 20 families is available. Furthermore, the families are classified into two different groups, or factors: based on maternal education (A_i) and income levels (B_j), which are labeled factor A and factor B.

Impact of maternal education and care on preschoolers' Chapter | 7 255

The first column in the table titled maternal education (A_i) classifies families into four groups, based on the level of education, alongside a corresponding categorical variable: $i = 1$ for no Schooling, *2* for Only preschool, *3* for only high school, and *4* for college degree.

The first row classifies the income distribution of the families (B_j): 1 for the lowest quintile, 2 for the second quintile etc.

Hence, the following table indicates how caloric intake from animal-source foods varies according to two groupings: across income and across maternal education levels.

Maternal education (A_i)		Income levels (B_j)				
		1	2	3	4	5
No schooling	1	78	82	88	72	80
Only preschool	2	70	76	78	66	70
Only high school	3	68	76	76	62	68
College degree	4	80	90	98	72	90

The basic questions the two-way ANOVA analysis examines are as follows:

1. Are the mean caloric intakes associated with various education levels (factor A) different?

2. Are the mean caloric intakes associated with various income levels (factor B) different?

The STATA input command and the STATA output for the two-way ANOVA are as follows:

```
. anova CAL ED Income

                  Number of obs =      20     R-squared     = 0.9352
                  Root MSE      = 2.94392     Adj R-squared = 0.8973

        Source |  Partial SS     df          MS          F      Prob>F
        -------+------------------------------------------------------
         Model |       1500      7    214.28571      24.73    0.0000
               |
            ED |        820      3    273.33333      31.54    0.0000
        Income |        680      4          170      19.62    0.0000
               |
      Residual |        104     12    8.6666667
        -------+------------------------------------------------------
         Total |       1604     19    84.421053
```

Here the input command line CAL represents the caloric intake, ED represents the categorical variable for maternal education (1, 2, 3, and 4), and finally, income represents the categorical variable representing income distribution (1, 2, 3, 4, and 5). The STATA output presents all the results including the sums of squares and the F-values. How does STATA generate these results?

The STATA results require the following calculations:

1. Compute the sum of squares for factors A and B (called SSA and SSB).
2. Compute the total sum of squares (written as SST).
3. Compute the unexplained sum of squares (SSE = SST − SSA − SSB).
4. Compute the mean squares of factors A and B (represented as MSA and MSB).
5. Compute the unexplained mean squares (or MSE).
6. Compute the two F-statistics.

Step 1: Here we compute the sum of Squares for factors A and B (called SSA and SSB). Both SSA and SSB are computed using the following formula:

$$\text{SSA} = (\#\text{columns}) \sum_{i=1}^{4} \left(\bar{a}_i - \overline{X}\right)^2 \text{ and SSB} = (\#\text{rows}) \sum_{i=1}^{5} \left(\bar{b}_i - \overline{X}\right)^2$$

where $\bar{a}_i$, $\bar{b}_i$, and $\overline{X}$ are the means of each row, column, and the overall mean of the whole data. These are calculated in the following table:

Income levels (B$_i$)

Maternal education (A$_i$)		1	2	3	4	5	Mean	For SSA: $\sum_{i=1}^{4}\left(\bar{a}_i - \overline{X}\right)^2$
No schooling	1	78	82	88	72	80	80	9
Only preschool	2	70	76	78	66	70	72	25
Only high school	3	68	76	76	62	68	70	49
College degree	4	80	90	98	72	90	86	81
Mean		74	81	85	68	77	77 = $\overline{X}$	164
		9	16	64	81	0	170	

For SSB: $\sum_{i=1}^{5}\left(\bar{b}_i - \overline{X}\right)^2$

Note first that mean of every row and column are computed under the title "mean" and the overall mean is computed, where $\overline{X} = 77$

$\text{SSA} = (\#\text{columns}) \sum_{i=1}^{4} \left(\bar{a}_i - \overline{X}\right)^2 = 5 \times 164 = \mathbf{820}$ &

$\text{SSB} = (\#\text{rows}) \sum_{i=1}^{5} \left(\bar{b}_i - \overline{X}\right)^2 = 4 \times 170 = \mathbf{680}$

These results are listed in the STATA output under "Partial SS"

Step 2: In this step, we compute the total sum of squares (written as SST). Each caloric intake entry is subtracted from the overall mean = 77, and

Impact of maternal education and care on preschoolers' Chapter | 7   **257**

squared, and summed over all entries. The following table presents these computations:

		\multicolumn{5}{c}{Income levels (B_j)}	
Maternal education (A_i)		1	2	3	4	5	Sum
No schooling	1	1	25	121	25	9	181
Only preschool	2	49	1	1	121	49	221
Only high school	3	81	1	1	225	81	389
College degree	4	9	169	441	25	169	813
Sum		140	196	564	396	308	**1604**

Each entry in the table is (caloric intake $- 77)^2$. The total across all rows and columns is **1604**. Hence, SST = 1604, which is the "total" under "Partial Sums" in the STATA output.

Step 3: Compute the unexplained sum of squares SSE = SST − SSA − SSB or SSE = 1604 − 820 − 680 = **104**, which is referred to as the, "Residual SS" in STATA output.

Steps 4, 5, and 6: The following table provides the details of the computations involving mean squares and the F-statistics:

	SS	Degrees of freedom	Mean square
Factor A (education)	SSA = 820	# columns − 1 = 5−1 = 4	$MSA = \frac{SSA}{df} = \frac{820}{4} = 273.33$
Factor B (income)	SSB = 680	# rows − 1 = 4 − 1 = 3	$MSB = \frac{SSB}{df} = \frac{680}{3} = 170$
Unexplained portion	SSE = 104	(# columns − 1) (# rows − 1) = 12	$MSE = \frac{SSE}{df} = \frac{104}{12} = 8.67$
Total	SST = 1604	N − 1 = 20 − 1 = 19	

$$F_A = \frac{MSA}{MSE} = \frac{273.33}{8.67} = 31.5 \ \& \ F_B = \frac{MSB}{MSE} = \frac{170}{8.67} = 19.6$$

Hence, all the results of the STATA output have been reproduced. Returning the two questions posed at the beginning of the section:

Are the mean caloric intakes associated with various education levels (Factor A) different?

The table value of $F_{0.05, 4, 12} = 3.26$. Since the computed value 19.61 > 3.26, we can reject the null hypothesis. In this case, the null states "mean caloric intakes associated with various education levels are NOT different." Thus, in the data, different education levels seem to have different caloric intakes.

Are the mean caloric intakes associated with various income levels (Factor B) different?

The table value of $F_{0.05,\ 3,\ 12} = 3.48$. Since the computed value $31.5 > 3.48$, we can reject the null hypothesis. In this case, the null states "mean caloric intakes associated with various income levels are NOT different." Thus, in the data, different income levels seem to have different caloric intakes.

Chapter 8

Indicators and causal factors of nutrition—application of correlation analysis

Chapter outline

Introduction	259	Data description and methodology	274
Review of selected studies	261	Concepts in correlation analysis	275
Food insecurity and nutrition in the United States	264	Inference about population parameters in correlation	276
Food insecurity in Brazil	269	Descriptive analysis	277
Global Monitoring Report on Nutrition and Millennium Development Goals	269	Main results	278
		Correlation analysis of the outcome variables	279
Impact of food price spike and domestic violence in rural Bangladesh	270	Estimating correlation using STATA	281
Malnutrition and chronic disease in India	271	Conclusion and policy implications	283
Malnutrition in Guatemala	272	Estimating correlation using R	286
Empirical analysis and main findings	273	Exercises	289
		STATA workout	289

The phrase 'correlation does not imply causation' goes back to 1880 …. However, use of the phrase took off in the 1990 and 2000s, and is becoming a quick way to short-circuit certain kinds of arguments.

Herbert West.

Introduction

Analysis of the correlates of indicators and causal factors of food security and nutrition is the subject of this chapter. Examining how various nutritional indicators are associated with each other and their determinants can be, among other things, useful for monitoring and evaluating food and nutrition interventions and their effect on a nutritionally vulnerable population. As mentioned in the earlier chapters, malnutrition remains a chronic and persistent problem in many parts of the developing world. From the most recent estimate of FAO, IFAD and WFP (2013), there are 854 million individuals who are undernourished, constituting about 12.6% of the world's population. Out of the 854 million people who do not get enough food, about 820 million are in developing countries (FAO, IFAD and WFP, 2013 and also http://www.wfp.org/hunger/stats). Additionally, in developing countries, about 160 million preschool children are underweight for their age. Furthermore, one in three children in developing countries suffers from stunting (de Onis et al., 2000).

An understanding of the determinants of malnutrition is important to improve human welfare and target intervention programs and policies. Malnutrition also causes a great deal of human suffering (both physical and emotional).[1] Apart from the human costs, chronic malnutrition has economic costs too. Deficiencies in vitamin A, protein, iron, and other micronutrients can cause prolonged impairment, thus reducing productivity of human capital.

Malnutrition is typically caused by a combination of factors, including inadequate food intake, mother's education and care, health status, and environmental factors. The immediate determinant of nutritional status is dietary intake (calories, protein, fat, micronutrients, carbohydrates, and vitamins). Dietary intake must be sufficient in quantity and quality, and nutrients must be consumed in appropriate combinations for the child to absorb them (see, for example, Smith and Haddad, 2000).

Inadequate food intake of a child is the result of households not having enough resources (such as own food production, income, or in-kind transfers of food) for gaining access to food. Although sustained income growth can improve the nutritional status of a child through the household's access to various resources, other factors such as women's education and nutritional knowledge play an equally important role. Women with at least secondary education tend to have fewer children and have better knowledge of feeding and caring practices. These knowledge and skills improve the caring practices and thereby positively influence the nutritional status of the child (Engle et al., 1999). A final underlying determinant of nutritional status is health environment and services, which constitute safe water, sanitation, healthcare, and environmental safety (Smith and Haddad, 2000).

A policy and program intervention challenge frequently faced by policy makers and program managers is which of the various factors has positive or negative association with child nutrition. It is important to understand the factors that are associated strongly with child nutrition, so that appropriate actions could be taken to improve those causal factors, thereby improving child nutrition.

---

1. Malnutrition has been associated with a 10%—45% increase in the incidence of diarrhea (de Onis et al., 2000). Similarly, vitamin A—deficient children are two to four times more susceptible to respiratory disease and twice as susceptible to diarrhea. Costs to the national health system due to poor nutritional status of mothers are substantial. The WHO estimates that 1.1 billion days of work time are lost worldwide as a result of various illnesses.

The purpose of this chapter is to examine how the various child nutritional indicators (weight-for-age [ZWA], height-for–age [ZWA], and weight-for-height [ZWH] Z-scores) are associated with each other using a correlation analysis. This kind of analysis can be useful for program monitoring and evaluation for the vulnerable segments of the population. For example, if a significant portion of children of a representative population group is found to be both underweight and stunted and program managers identify lack of child-care practices to be the primary cause, nutrition interventions in the form of health education to the caregiver may be appropriate in a given situation. We relate these indicators to the problems of nutrition and "hidden hunger" in the United States, India, Bangladesh, Guatemala, Brazil, and Senegal. We also provide the policy recommendations suggested by the World Bank that make the issue compatible with the Millennium Development Goals (MDGs).

In what follows, we first discuss the main studies that use correlation analysis as a tool for examining the causes of child nutritional status. The analysis is similar in logic to the application of multivariate regression. While in multivariate regression we are trying to predict the values of the Z-scores (height for age, weight for age, etc.) based on the different independent variables, in the correlation analysis we are looking at how the independent variables are associated with the outcome variables. We operationalize our statistical measures through related examples using STATA. Finally, we conclude with a few policy implications of the main results.

Review of selected studies

Explaining high levels of malnutrition in developing countries has been an area of interest to policy researchers and development practitioners. In the following, we discuss a few studies that use correlation analysis in addressing nutritional status.

Alderman and Garcia (1994), using a household production function approach, address how household food availability and health security affected the nutritional status of children in rural communities in Pakistan. Nutritional status is considered to be a process governed by two proximate factors, namely, nutrient availability and absence of infection.[2] The controls of the model were a family's nutrient availability, education of the parents, child's age, care, and household size. The data were collected over a 3-year survey for a sample consisting of 1200 households (from 52 villages chosen randomly), from the four poorest districts in Pakistan during the period 1986–87.

---

2. The five variables included for nutrient availability and absence of infection were the number of days the child was ill with diarrhea during the past 2 weeks, number of days the child had another illness during the past 2 weeks, whether the child was vaccinated, whether the child was breastfed exclusively, and whether the child was born in a hospital.

The study found that more than half of the children were stunted (or chronically malnourished) as defined by WHO reference standards, i.e., they fall below 2 standard deviations from the reference of each district. Additionally, 8.7% of the children in the sample were wasted (low weight for height). However, the study found low correlation between the two measures of malnutrition (-0.13), and this was not significant. Additionally, the correlation among the different determinants of nutrition was found to be very low.

The main result of the study was that child nutrition was highly responsive to health inputs rather than food availability at the household level. Since morbidity and poor nutritional status are interdependent, the study found diarrhea as the main culprit in reducing a child's height for weight, thereby curtailing long-run growth. In addition, the study found that if all mothers were educated up to the primary level, the level of child wasting would decrease by one-half of the current prevalence levels. Similarly, programs that decreased diarrhea occurrence by 1 day on an average over a 2-week period reduced the incidence of child wasting by 2.1% points.

A clear policy implication of the study is to improve community-level investments to develop the sanitary environment. Overall, the results of the study indicate that, in Pakistan, food security alone is not sufficient in improving nutritional status, particularly of children. Health and infection are equally or more important factors in determining child nutritional status.

Glewwe et al. (2003) examined the impact of income growth and improvement in health indicators on child nutritional status for Vietnam. Child nutrition is a key issue in Vietnam, since it is one of the poorest countries in the world with per capita GNP of about $370 during 1999. The study used the 1992–93 and 1997–98 Vietnam Living Standards Survey (http://www.worldbank.org/html/prdph/lsms/guide/select.html) with about 4300 households included in the sample. The sample was a large, nationally representative data set. ZHA (stunting) and ZWA (underweight) were considered as indicators of nutritional status of the population and were dependent on household per capita income, parents' schooling, local health environment, and child's innate healthiness as the main determinants.

One of the main questions addressed in the study was whether a rapid increase in household incomes and expenditures was the main cause of a decrease in stunting. Undertaking different estimation methods (ordinary least squares and instrumental variable), the study demonstrated that the impact of household expenditures on child nutritional status was not always significant. In regard to the other determinants of child stunting, the study found that child age had a strong relationship to stunting, implying the importance of genetic endowments on children's nutritional status. In examining the role of community characteristics (such as health services), the main finding was that the distance to the nearest pharmacies had a negative and significant impact on child health status. Additionally, the study found that providing health centers with good sanitation and ample supplies of oral rehydration salts had a substantial positive impact on child health.

Understanding correlations of water and sanitation conditions with child health variables can be useful for designing intervention programs and policies at the local level. A recent World Bank study (Shi, 2000) addressed how access to safe water and sanitation conditions affects child mortality using 1993 city-level data from the global urban indicators. The primary focus of this paper was to address the impact of the health environment on a health outcome variable: infant mortality. The data came from the global urban indicators for 237 cities in 110 countries on 46 key indicators. Infant mortality was modeled as a function of access to potable water or percentage of households having access to sewerage, a series of dummies that capture the type of organizations responsible for water or sewerage provision, and a set of control variables.

The results of the study demonstrated that child mortality was inversely correlated with access to potable water and sewerage connections by undertaking a correlation analysis. Thus, nutritional outcomes improved with progress in the health environment, as measured by water and sanitation conditions. In addition, the lack of health services facilities was also associated with higher child mortality. Cities with higher household income per capita had lower child mortality. Other findings showed that the type of organizations responsible for water service was also important for child mortality. Local government as a water service provider consistently lowered infant child mortality, while private service providers as a service provider were associated with a higher mortality rate.

Svedberg (2004) aimed at challenging the study by Behrman et al. (2004), which argued that infant and child nutrition could be significantly improved by the promotion of exclusive breastfeeding and supplementation of micronutrients such as iron, iodine, and vitamin A. The main hypothesis tested was that poverty (low income) was the crucial determinant of hunger and malnutrition.

The main focus of the study was to examine the relationship between the prevalence of stunting and underweight on the one hand and real per capita income on the other. The study provided the following rationales of income as a crucial determinant of child nutrition. First, with higher per capita income, households could exert a stronger effective demand for private consumption goods, including food with better nutritional contents. Second, higher gross national income can translate into higher government revenues and expenditures. If these expenditures are used to finance public investment and consumption in health- and nutrition-related services, it could have a positive impact on child nutritional status.

As the relationship between stunting and per capita income drifted downward during the 1990s, it is likely that nonincome factors might have played some role in reducing stunting. The author thus advocates that microlevel interventions and targeting methods must continue. However, in the absence of higher economic growth rates in the poorer countries, the hope of realizing the MDGs cannot be achieved.

Given the strong correlation between income growth and child nutritional outcomes at the cross-country level (Smith and Haddad, 2000; Svedberg, 2004) and the strong impact of women's education (operating through better child-care practices and improved maternal nutrition knowledge), it is important to understand which sort of interventions is important for improving child nutritional outcomes in the short and long term. Understanding the various factors that are associated with child nutritional outcomes will allow policy makers to undertake appropriate actions so that improvement in the causal factors enhances child nutritional outcomes. We examine a few case studies from developed and developing countries to illustrate the relationship between the various determinants of child nutrition (income, childcare, health environment) on nutritional outcomes (height for age, weight for height) using correlation analysis.

Food insecurity and nutrition in the United States

Food insecurity is defined as a lack of access to the kinds and amounts of food necessary for each member of a household to lead an active and a healthy lifestyle. Gundersen and Ziliak (2018) note that more than 41 million persons (about 12% of the population) were food insecure in 2016 and received around $ 66 billion in benefits. Food insecurity has been steady at around 17% for children. Annual food insecurity rates, Supplemental Nutrition Assistance Program (SNAP) participation rates, and total SNAP expenditures in the United States are given in Figs. 8.1 and 8.2.[3] Note the increase in food insecurity rates around 2008 coinciding with the Great Recession.

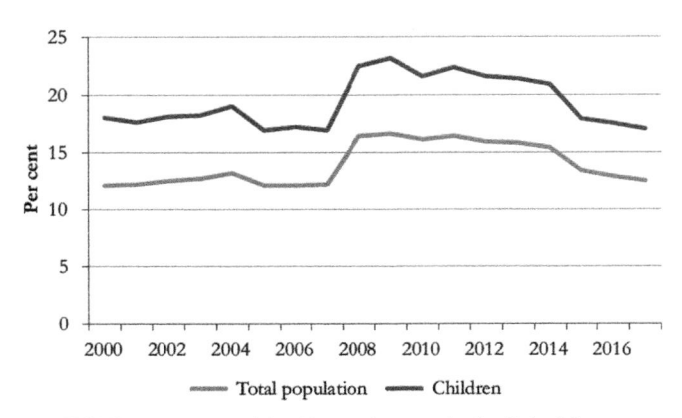

FIGURE 8.1 Annual food insecurity rates in the United States.

---

3. See Gunderson (2019a,b) for an exhaustive analysis of the SNAP program.

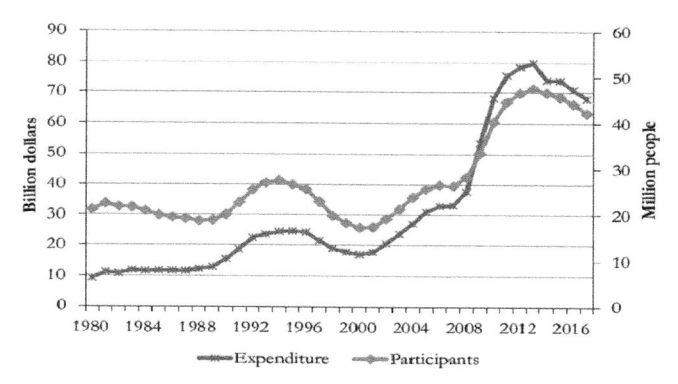

FIGURE 8.2 SNAP participants and total expenditures. *SNAP*, Supplemental Nutrition Assistance Program.

The general consensus from the extensive research undertaken within the United States suggests that food insecurity is associated poverty, low assets, low human capital, and low health outcomes across the age gradient. Gundersen (2019a) points out that increasing SNAP benefits by $ 41.62 per week for the recipients would lead to a reduction in food insecurity by 60%. Furthermore, that would be at an extra cost of $ 25 billion, although much of the cost will be overweighed by the reduction associated healthcare costs, which would prevail in the absence of the recommendation.

Gundersen et al. (2011) and Gundersen (2019a,b) have done extensive work on food insecurity in the United States and indicate the problems of adequate nutrition and health and access to food among many other determinants. They emphasize three important aspects of food insecurity in the United States. First, the SNAP reduces the prevalence of food insecurity. This should be kept in mind as reconstructions of SNAP are constantly being proposed. We must be careful that these proposals could compromise food security if more restricted food options discourage participation.

Second, the National School Lunch Program (NSLP) also reduces food insecurity, even though it is not explicitly designed for that purpose. Similar to the SNAP, policy makers should carefully weigh all anticipated benefits and costs, when they discuss modifications to the lunch program.

Third, alongside the direct benefits associated with reducing food, potential reductions in medical expenditures should be incorporated into relevant benefit—cost considerations of programs such as SNAP and NSLP.

Research into food insecurity in the United States has given policy makers many new insights into the causes and consequences of food insecurity. They have also developed a few approaches that are effective in alleviating food insecurity. Gundersen et al. (2011) and Gundersen (2019a,b) also prompt a lot of important questions that have to be answered for future policy considerations:

1. How is food insecurity distributed within a household?

Based on current evidence and research findings, we can conclude that there are differences in the distribution of food insecurity within households. Child-specific responses can lead to new insights into how families distribute food security status. There is no research on the effects of food prices.

2. How do food prices influence food insecurity?

There is a lot of variation in food prices across the United States. A report by Feeding America (2011) shows at least some correlation between food prices and food insecurity at a county level: 44 counties in the United States are in the top 10% of food prices and food insecurity rates. Beatty (2010) and Broda et al. (2009) show that food prices have an influence on the well-being of low-income consumers in developed countries.

3. How does food access influence food insecurity?

The effects of food deserts may be especially significant for three groups that may face mobility restrictions or live in remote areas: seniors, persons with disabilities, and American Indians living on reservations. We need more data and studies to be conducted on these groups to understand the importance of access. See the food desert map at http://www.ers.usda.gov/data-products/food-access-research-atlas/.

4. What types of coping mechanisms do low-income food-secure families utilize, and what are the effects of these mechanisms?

Seniors may be foregoing prescription drugs to feed themselves and other members of the household. Hence, food security combined with poverty should signal to policy makers that assistance may be needed and that food security does not indicate an absence of need.

5. Besides SNAP and NSLP, what are the effects of social safety net programs on food insecurity?

It is not clear how WIC and the School Breakfast Program affect food insecurity. There has also not been much research on the effects of other social safety net programs such as unemployment insurance and in-kind programs such as Medicaid and housing assistance.

6. How do the experiences of food insecurity in other countries differ from the United States?

Given the differences across countries in demographics, food prices, geography, and assistance programs, cross-country comparison may yield new insights akin to the new insights about this issue.

7. How do health limitations affect food insecurity?

More research is needed to get a full understanding on this aspect. Research using longitudinal data to connect mental health, unobserved characteristics, endogeneity to econometric work. The impact of disability among household members on food security status is also critical. Furthermore, the impact of the Affordable Care Act (ACA) with its individual mandate on food insecurity is still to be established.

8. What are the effects of private food assistance programs on food insecurity?

Feeding America is made up of 202 food banks, and thousands of agencies who receive food directly from major food companies, grocery stores, restaurants, commodity exchanges, and individual donors. They distribute food through emergency food pantries, emergency soup kitchens, and emergency shelters. The Feeding America system serves over 35 million people every year. Given the size of this program, research on the impact of these private food assistance programs on food insecurity would be of interest, especially to donors to these programs. Such research could further consider how these programs interact with public food assistance programs.

9. Further Concerns

Gundersen and Ziliak (2018) pose numerous questions that would add significant value to US policy on food insecurity:

- How efficient is the food distribution system in reducing food insecurity?
- Why are the food insecurity rates high among American Indians?
- How effective are the Charitable Food Assistance Programs?
- Why are older seniors more food secure than the younger seniors?
- What was the effect of the Great Recession on food insecurity?

Gregory and Smith (2019) point out how the extent of food insecurity may be mismeasured in the United States. The errors arise because of the "salience effect" inherent in the participants' responses to the survey in the USDA's Core Food Security Module (FSM).

Gregory and Smith (2019) apply prospect theory and show how survey responses are triggered by context, framing, and recent experiences. Respondents tend to downplay unpleasant experiences relative to recent pleasant experiences. Consequently, an index of food insecurity can become inherently biased, based on the timing of the survey.

Likewise, Deb and Gregory (2018) note the importance of heterogeneity in the response patterns of SNAP participants. While SNAP reduces food insecurity, by a significant 23% points for about 60% of the participants, it appears that the program does not significantly affect about 40% of the recipients. There could be two potential effects that are masked by data and the framing of questions. Particularly, the data for the latter group do not account for the short-term versus long-term effects of SNAP and also do not examine the usefulness of SNAP along the intensive margins. Similarly, Courtemanche et al. (2019) note the various ambiguities and misreporting that occurs among SNAP participants.[4] Finally, Gundersen (2019b) argues as to how SNAP needs to address food insecurity as a "basic rights" issue.

Section highlights: Is cigarette smoking in some way related to food insecurity?

Armour et al. (2007) examine this issue and show that the prevalence of smoking is higher among low-income families who are food insecure compared with low-income families who are food secure. To make informed decisions about policies and programs designed to reduce food insecurity, it is important to determine what other characteristics of low-income households are associated with food insecurity.

Armour et al. (2007) use data from the Panel Study of Income Dynamics (PSID) for 2001, to examine the association between food insecurity and smoking. Such a study will help state health departments and other organizations in formulating effective food security and tobacco-control policies. The researchers estimate multivariate logistic regression models, which we will study in Chapter 11, to estimate the risk of a family being food insecure. They included many exogenous variables such as age, race, education, marital status, number of children, whether the household head or spouse smoked cigarettes, weekly cigarette consumption, whether the household head or spouse drank alcohol on a daily basis, per-capita annual family income, etc.,

Their finding quantifies the relationship between food insecurity and smoking in the United States: cigarette consumption is associated with increased food insecurity. Furthermore, they find the following:

1. Low-income families with an adult smoker purchased approximately 10 packs of cigarettes per week. Assuming an average price of $3.37 per pack, these households spent approximately $33.70 per week on cigarettes.
2. The extent to which smokers in low-income families substitute cigarettes for food adversely affects household food security.
3. Smoking prevalence is 11.7% points higher among low-income families who are food insecure than low-income families who are food secure.

Health departments, medical professionals who serve low-income families, and state health plans such as Medicaid should increase their capacity to document tobacco-use status in medical charts. They should also develop tobacco-control strategies; provide free or low-cost access to smoking-cessation advice, counseling, and medication; and monitor quit attempts. They should also initiate community efforts, such as churches and youth groups, to target the needs and address these concerns.

---

4. The special issue *Southern Economic Journal* (2019), 86(1) is devoted to SNAP benefits and related topics.

> **Section highlights: Is cigarette smoking in some way related to food insecurity?—cont'd**
>
> It is of interest to note that the researchers found no statistically significant association between food insecurity and alcohol use. However, they found a negative and statistically significant association between food insecurity and the mean number of children residing in a household. This supports previous findings that children living in low-income households are not the same as hungry children. Once again, this finding raises a lot of interesting complications. For example, food-insecure households reporting hunger often have parents who go without food so the children can eat. Also, there may be an incentive for households with children to underreport hunger because of embarrassment or fear that child services might intervene and remove children from the household. Alternatively, perhaps this finding reflects that low-income families with children are eligible for other types of assistance that reduces food insecurity.
>
> Overall, for low-income families, with a household head, who smokes is associated with a 6% increase in the risk of the family experiencing food insecurity. Currently, US food assistance programs focus on alleviating food insecurity among low-income families, while tobacco-control programs focus on low-income individuals because they smoke at disproportionately higher rates than higher-income individuals.
>
> Given the findings that families near the federal poverty level spend a large share of their income on cigarettes, the researchers make a case for food assistance and tobacco-control programs to work together to help low-income people quit smoking and combat food insecurity.

Food insecurity in Brazil

Reis (2012) examines the evidence for the relation between income and health among children in Brazil. Reis (2012) demonstrates the lack of adequate nutrition among poor children and points to income levels as one of the possible explanations for this observation. Using data from the 2006 Brazilian Demographic and Health Survey, Reis (2012) shows that children living in households with **food insecurity** have worse **nutrition** and health indicators. In addition, the relationship between household income and many children's health and **nutrition** measures weakens but remains significant when controlling for **food insecurity**.

Global Monitoring Report on Nutrition and Millennium Development Goals

The World Bank's Global Monitoring Report (2012) on Food Prices and Nutrition provides an exhaustive analysis of the current global situation in food insecurity and nutrition. The Report traces the causes and the extent of the problem and provides comprehensive solutions for policy makers in different countries. Some of the consequences of higher food prices on undernutrition and the difficulties associated with meeting the MDGs noted in the Report are as follows:[5]

- As food prices increase, the purchasing power of the poor decreases, the composition of their diet worsens, and their food consumption may decrease. These changes directly affect all targets of MDG 1 on poverty, full and productive employment, and hunger.
- Malnutrition affects early childhood development and makes children more likely to drop out of school (MDG 2).
- An increase in food prices affects women and girls' consumption disproportionately (MDG 3).
- Undernutrition is linked directly to more than one-third of children's deaths each year (MDG 4).
- Pregnant women face heightened maternal mortality, through increased anemia, during a food price crisis (MDG 5).
- The adverse effects of a food crisis on the availability of health services and on health status bear on countries' and individuals' abilities to combat the HIV/AIDS epidemic (MDG 6).
- Undernutrition weakens the immune system and compounds the effect of diarrhea and waterborne diseases (MDG 7).
- Higher food prices have weakened intergovernmental coordination in food markets (MDG 8).

Impact of food price spike and domestic violence in rural Bangladesh

The Global Monitoring Report (2012) points out the example of Bangladesh, where the levels of child undernutrition are high (36% stunting, 16% wasting, and 46% underweight). Prices of key staples increased by as much as 50% from 2007 to 2008. On top of this, the country suffered floods in mid-2007 and a cyclone in November 2007, which reduced rice harvest. Export restrictions by India, one of the country's main rice providers, also raised rice prices.

An assessment of livelihood and nutrition security in Kurigram village (194 households) in 2005 and a follow-up assessment in November 2008 (250 households) show that the richest households *benefited* from the price hike, as rice producers. One-third to one-half of households had lower disposable income after the crisis, mainly because of the rice price hike and, to some minor extent, crop failure in one of the rice harvests.

---

5. See the Global Monitoring Report (2012, page 64).

The poorest quartile was no longer able to afford a diet that provided them with their energy and micronutrient needs. Children ate fewer meals, had less diverse diets, and received few nutrient-rich foods. Stunting was observed among children in the poorest households and was twice as high as in the richest households. A 7% point improvement in stunting rates was lost during the crisis.

Families responded to the price hike by sending children to work, taking children out of school, selling productive assets, and reducing their food intake. Even though the richest households benefited, agricultural labor wages did not increase enough to compensate poorer households for the price rise.[6]

In an interesting study, Lentz (2018) show how rural Bangladeshi women are forced to traverse and navigate complicated decision-making surrounding food allocation under "burdened agency" emanating from domestic violence. Hence, women suffer from undernutrition and hunger, because food is withheld from them in the form of violence and control.

Malnutrition and chronic disease in India

The Global Monitoring Report (2012) points out that about 42% of the 160 million children in India under the age of 5 are underweight. Prime Minister Manmohan Singh described the situation as a matter of "national shame" and undernutrition as "unacceptably high."

There are signs of progress—one in every five children has reached an acceptable healthy weight over the past 7 years in 100 focus districts, which were particularly worse off before. But the current figures point to the inadequacies and inefficiencies of government initiatives, such as the Integrated Child Development Scheme, the scale of the needs of India's child population, and the lack of awareness about nutrition.

The Report's observations are based on a longitudinal study of a cohort of births in South Delhi followed to age 32 found that those children who were thinner in infancy, with a body mass index (BMI) under 15, had an accelerated increase of BMI until adulthood. Although none was classified as obese by age 12, those with the greatest increase in BMI by this age had impaired glucose tolerance or diabetes by the age of 32.

The transition from a resource-poor environment to one that is less constrained may aggravate these risks. This report referring to the study of Ramachandran and Snehalatha (2010) indicates that India has the largest number of undernourished children in the world. It also has the largest

---

6. For a comprehensive analysis on Bangladesh, see the most recent household survey data put forth by IFPRI (2013a,b).

number of people with diabetes. As a warning, the Global Monitoring Report (2012) concludes that these numbers may be only the starting point of a much larger long-term problem.[7]

Aurino and Morrow (2018) show in an interesting study based on child-focused evidence that children living in food insecure households are aware of their situation. The children can describe their challenges, the consequences of their poor diet on health and education, the sacrifices mothers make to feed their families, the impact of inflation on food security, the limitations, and the benefits of social safety nets such as MGNREGS. The study recommends complete plugging of holes in the current distribution system through greater monitoring and accountability in the supply chain, and regular inspections of food safety and quality.

Malnutrition in Guatemala

Hoddinott et al. (2011) find evidence of growth failure in early life in rural Guatemala, as measured by low height for age (stunting) at 36 months. Hoddinott et al. (2011) examine how this failure in health affects a wide range of adult outcomes: education, choice of marriage partners, fertility, health, wages and income, and poverty and consumption. The findings were based on data collected through interviews between 2002 and 2004 of participants in a nutrition supplementation trial between 1969 and 1977. Hoddinott et al. (2011) provide us with very interesting insights from their intervention study:

1. Participants who had received nutritional supplementation (a high-protein energy drink with multiple micronutrients) and free preventive and curative medical care (including the services of community health workers and trained midwives, as well as immunization and deworming) were less likely to become stunted.
2. Participants who were stunted at 36 months of age had left school earlier and had significantly worse results on tests of reading and vocabulary and on nonverbal cognitive ability some 35 years later. They also married people with lower schooling attainment. Women had 1.86 more pregnancies and were more likely to experience stillbirths and miscarriages.
3. Individuals who were not stunted earned higher wages and were more likely to hold higher-paying skilled jobs or white-collar jobs. They were 34% points less likely to live in a poor household.
4. A one standard deviation increase in height for age lifted men's hourly wage by 20%, increased women's likelihood of operating their own business by 10% points, and raised the per capita consumption of households where the participants lived by nearly 20%.

---

7. The Report's findings are based on research by Ramachandran and Snehalatha (2010), and Alderman (2011) and the references contained there.

Leroy et al. (2019) examine the effects of POCOMIDA, a program to improve mother—child health and nutrition through a food assistance in Guatemala. The intervention was useful in preventing food insecurity and improved health outcomes through better nutrition in food-insecure households. However, the program produced unhealthy weight gains in food-secure populations, reestablishing the importance of adequate monitoring and targeting, as suggested by Aurino and Morrow (2018) for India.

Empirical analysis and main findings

The central question addressed in this chapter is to what extent the different indicators of nutritional status are associated with each other. For this purpose, a Pearson's correlation analysis is undertaken. Correlation analysis can be useful for the following reasons:

1. Formulating nutrition targets (such as targets within a development plan)
2. Planning social development programs/projects for the vulnerable sections of the population (for example, government and nongovernment organizations use nutrition indicators in implementing, monitoring, and evaluating social developmental programs)
3. Using it as baseline and benchmark data (nutrition indicators often reflect the current nutrition situation with which future data can be compared at the start of an intervention project)

Cross-sectional surveys are the preferred approach to collect data on health and nutritional variables, as large representative samples and the information on a range of topics can be obtained in a short time period. These surveys are also cost-effective compared with long-term longitudinal studies.

One important limitation of cross-sectional data in the context of health and nutrition surveys is that, unlike longitudinal surveys, they do not support assessment of the direct effect of a particular episode of illness on nutritional status of the child. For example, the assessment of the impact of illness on growth attainment requires knowledge of individual growth trends, which cannot be determined from a single measurement. For this reason, cross-sectional measurements are unlikely to reflect a consistent relationship of nutritional status with reports of illness, whereas a series of measurements obtained at different points in time are very likely to demonstrate a direct causal relationship between episodes of illness, especially diarrhea.

On the other hand, it is reasonable to use cross-sectional data to analyze the correlation of socioeconomic, demographic, or environmental factors with nutritional status (McMurray, 1996). Since the association is less direct, population patterns are likely to reflect an association with these factors. However, the method of analysis can affect the results. In particular, the use of a cutoff point to classify cases can blur the association because the causes of child nutritional status in any two groups are not uniform. Thus, given the nature of cross-sectional data and their limitations at the individual level, analysis based on cutoff points should be used in conjunction with other analytical techniques, which provides a better picture of nutritional patterns across the whole sample (Pelletier et al., 1994).

A better approach to the analysis of cross-sectional data is to treat the anthropometric indicators as continuous variables as done in the present chapter and to focus on patterns of covariation rather than on the odds of being in one discrete category rather than another.

We motivate the empirical analysis by first discussing the methodology of correlation analysis, then examining some descriptive analysis undertaken in various case studies, and then undertaking the correlation analysis.

Data description and methodology

The different indicators of nutrition may be classified as follows:

1. Outcome variables
2. Socioeconomic indicators that affect child nutrition
3. CARE indicators
4. Community characteristics

The latter three indicators are determinants of the child nutritional status and are also associated with the child nutritional status.

The *outcome* variables of child nutritional status are the anthropometric ZHA, ZWA, and ZWH.

We consider the following *socioeconomic* indicators that affect child nutritional status significantly. These are as follows:

1. Per capita expenditure on food (PXFD): Expenditure on food is a critical variable in models of child health and nutrition outcomes and is used as a proxy for income.
2. Education of the spouse (EDUCSPOUS): This is a categorical variable that has a value ranging from 1 to 7 and measures the education level of the mother in number of years. Higher values of this variable indicate greater levels of education.

The *CARE* indicators included in the analysis are as follows:

1. Clinic feeding (CLINFEED): This is a dichotomous variable denoting whether the child is fed in a clinic or not.
2. Breastfeeding (BFEEDNEW): This is also a dichotomous variable denoting whether the child is breastfed or not during his or her infanthood.

The *community characteristics* consist of the following set of variables:

1. Drinking distance (DRINKDST): This is a categorical variable assuming values from 1 to 5. Higher values of this variable denote that the distance to

a protected drinking source for the household is higher. For example, the variable attains a value of 4 if distance to a protected drinking source exceeds 3 km. We can expect that the greater the distance to a protected water source the more is the likelihood that children will suffer from malnutrition. This is because by reducing the risk of bacterial infections and diarrheal diseases, sanitation and clean water can indirectly contribute in improving a child's nutritional status.
2. Provision of sanitation (LATERINE): This is a dichotomous variable assuming two values 0 and 1, with 0 indicating absence of latrine from the household. Sanitation appears to be more important in nutritional outcomes than presence of protected drinking source, since it is directly related in preventing diarrhea, thereby improving children's nutritional status.
3. Diarrhea: This variable indicates whether the child has diarrhea and is a dichotomous variable assuming two values 0 and 1. One indicates that the child had diarrhea during the survey. Infections such as diarrhea can reduce the nutrients in the body and thus increase the likelihood of malnutrition further.
4. Distance to a health facility (HEALTDST): This is a categorical variable denoting the distance of the household to a health clinic and assumes 4 values. Higher values indicate that the household is located farther from the nearest health center. For example, a value of 4 indicates that the distance to the nearest health clinic for the household is more than 10 km.

Concepts in correlation analysis

Before proceeding to the analysis, it is important to explain a few concepts associated with the *correlation analysis* (Lowry, 2003).

Suppose we have two random variables X and Y with means $\overline{X}$ and $\overline{Y}$ and standard deviations S_X and S_Y, respectively. Then, the correlation coefficient can be computed as follows:

$$r = \frac{\sum_{i=1}^{n}\left(X_i - \overline{X}\right)\left(Y_i - \overline{Y}\right)}{(n-1)S_X S_Y} \tag{8.1}$$

Eq. (8.1) can be interpreted as follows: suppose that an X value was above average and that the associated Y value was also above average. Then the product $\left(X_i - \overline{X}\right)\left(Y_i - \overline{Y}\right)$ would be the product of two positive numbers, which is positive. If the X value and the Y value were both below average, then the product would also be positive. Thus, a positive correlation indicates that large values of X are associated with large values of Y, while small values of X are associated with small values of Y. The correlation coefficient measures the strength of a linear relationship between any two variables and is always between −1 and +1. The closer the correlation is to +1 or −1, the closer it is to a perfect relationship. One can also express r in terms of the regression coefficients in Chapter 9 as follows:

$$r = \beta_i \frac{S_X}{S_Y}$$

$$S_X = \sqrt{\sum \left(X_i - \overline{X}\right)^2} \quad S_Y = \sqrt{\sum \left(Y_i - \overline{Y}\right)^2} \quad (8.2)$$

The interpretation of r^2 (the square of r) can also be made in terms of variation in the dependent variable Y that is explained by the regression line. Suppose that the total deviation of an actual Y from its mean $\overline{Y}$ can be expressed as the sum of two nonoverlapping components $(\hat{Y} - \overline{Y})$ and $(Y - \hat{Y})$. The first component represents that part of the total difference explained or accounted by the relationship of Y with X; the other component represents that part of the total difference remaining after accounting for the relationship of Y with X. Thus, we get

$$\sum (Y - \overline{Y})^2 = \sum \left(\hat{Y} - \overline{Y}\right)^2 + \sum (Y - \overline{Y})^2 \quad (8.3)$$

Eq. (8.3) can be interpreted as the sum of total variation to be equal to the sum of explained and unexplained variation. The ratio $\sum \left(Y - \hat{Y}\right)^2 / \sum (Y - \overline{Y})^2$ is the proportion of total variation that remains unexplained by the regression equation. On the other hand, $1 - \sum \left(Y - \hat{Y}\right)^2 / \sum (Y - \overline{Y})^2$ represents the proportion of the total variation in Y that can be explained by the regression equation. The aforementioned ideas can be summarized as follows:

$$r^2 = 1 - \frac{\sum \left(Y - \hat{Y}\right)^2}{\sum \left(Y - \hat{Y}\right)^2} = 1 - \frac{Unexplained\ variation}{Total\ variation} = \frac{Explained\ variation}{Total\ variation}$$

(8.4)

Inference about population parameters in correlation

Let us assume a situation in which we have a random sample of n units from a population with paired observations of X and Y for each unit. We want to test the null hypothesis that the population correlation coefficient $\rho = 0$ against the alternative that $\rho \neq 0$. If the computed r values in successive samples from the population were distributed normally, we would have the standard error to perform the usual t-test involving the normal distribution. Thus, we have the following statistic:

$$t = \frac{r - \rho}{S_r} = \frac{r}{\sqrt{(1 - r^2)/(n - 2)}} \sim t_{n-2} \quad (8.5)$$

The standard error of r is given by $\sqrt{(1 - r^2)/(n - 2)}$. Note that the hypothesis testing procedure is in terms of r instead of r^2.

A few comments may be made concerning the hypothesis testing procedure. First, this technique is only valid for a hypothesized population value of $\rho = 0$. Other procedures such as Fisher's Z transformation[8] can be used when the population correlation coefficient is assumed to be different from zero. Second, even though the sample r is significant according to this test, in some instances, the amount of correlation may not be considered important. For example, in a large sample, a low value of r may be found to be different from zero. However, since little correlation was found between the two variables, it may not be appropriate to infer any association between X and Y for decision-making purposes. Third, the distributions of the t-values computed from Eq. (8.5) approach the normal distribution, as the sample size increases. This is because for large sample sizes, the t-value is approximately equal to z in the standard normal distribution. Thus, the critical values applicable to the normal distribution can be used instead. Finally, in the correlation analysis, both X and Y are assumed to be normally distributed random variables. Both variables should be random for hypothesis testing about the value of ρ. In the regression analysis, however, the independent variable X is not a random variable.

Descriptive analysis

We first compute the frequency distribution of the different determinants of nutrition to understand general nutritional situation of households in Malawi.

Table 8.1 displays the frequency distribution of the various indicators that might affect child nutrition. The main areas of concern are education level of the mother (all the mothers in the sample do not have education beyond the primary level), the CARE indicators (such as feeding the child in a clinic and breastfeeding and other complementary feeding), and availability of sanitation facilities. Since better educated mothers usually have better nutrition knowledge, they can make informed decisions regarding good child-care practices, which is one of the pathways through which maternal education might improve child nutrition. Thus, at least in the short and medium term, maternal education seems to be of great priority. Additionally, for most households

---

8. Fisher's Z transformation can be used as follows: $z = \frac{1}{2} \log_e \left(\frac{1+r}{1-r} \right)$. This statistic is approximately normally distributed with mean $u_z = \frac{1}{2} \log_e \left(\frac{1+\rho}{1-\rho} \right)$, and standard deviation $\sigma_z = \frac{1}{\sqrt{n-3}}$.

TABLE 8.1 Frequency distribution of nutritional indicators.

Indicator	Cases	Percent
EDUCSPOUS	No education	52.0
	Adult literacy training	2.8
	Primary education	45.2
CLINFEED	No	80.5
	Yes	19.5
BFEEDNEW	No	57.2
	Yes	42.8
LATERINE	No	61.1
	Yes	38.9
DIARRHEA	No	83.7
	Yes	16.3
DRINKDST	<2 km	71.1
	≥2 km	28.9
HEALTDDST	<2 km	19.8
	≥2 km	80.2

(about 80%), the distance to the nearest health clinic is very far (more than 2 km). Such a long distance to a health facility can be an important barrier in the decision to seek modern healthcare when needed. Thus, a comprehensive nutritional program that addresses health access problems (such as children attending clinics and setting up new health facilities) and improving the level of education of the mother can be very crucial in solving the problem of child malnutrition in Malawi.

Main results

Figs. 8.3 and 8.4 provide scatterplots (a geometric representation) of observations on two variables: namely incidence of wasting and distance to a protected water source and underweight with distance to a protected water source. The scatterplot is one of the most useful statistical graphs since it represents data on a two-dimensional plane. The bivariate scatterplot is a natural display of the relationship between any two quantitative variables. As evident from the diagrams, both the incidences of wasting and underweight increase, as the distance to the protected water source increases. This is

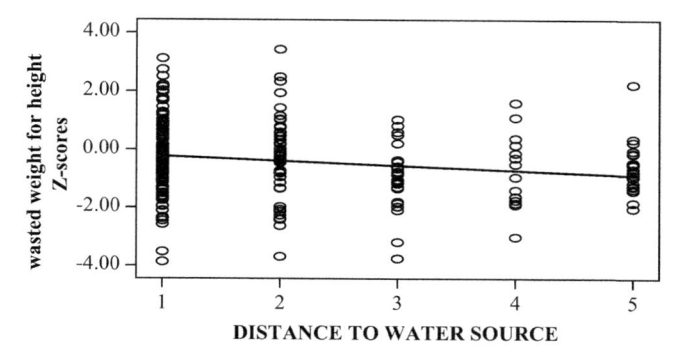

FIGURE 8.3 Scatterplot of wasting with distance to a drinking water source.

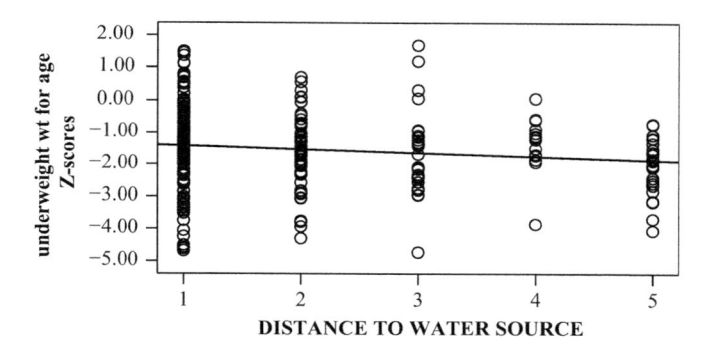

FIGURE 8.4 Scatterplot of underweight with distance to a drinking water source.

possibly because the risk of bacterial infections and diarrheal diseases increases as the household is located farther from a protected water source which, in turn, affects child nutrition adversely.

Correlation analysis of the outcome variables

We next undertake the correlation analysis among the various anthropometric measures or outcome variables (ZWH, ZWA, and ZHA) to understand their degree of association. There are three distinct possibilities in a correlation analysis. First, two sets of variables may show positive correlation (such as dieting and exercise). In other words, as the value of one variable increases, the value of the other variable also increases.

Second, two variables may exhibit negative correlation. In other words, a higher value of one is associated with lower values of the other. Finally, there may be no correlation, i.e., two or more variables may not have any relationship with each other. Correlation can be weak or strong depending on its relative magnitude.

As a rule of thumb, in absolute value terms, the following guidelines can be used:

- 0–0.2—weak to negligible correlation,
- 0.2–0.4—weak correlation (not very significant),
- 0.4–0.7—moderate correlation,
- 0.7–0.9—strong or high correlation, and
- 0.9–1—very strong correlation.

From Table 8.2, it is evident that stunting (defined as ZHA below -2) and wasting (defined as ZWA below -2) are very weakly correlated at the 1% level. Since stunting and wasting are usually considered as long-term and short-term indicators of the nutritional status of the child, the low correlation is possibly showing that the determinants of short-term and long-term factors of nutrition are different. On the other hand, the moderate correlation between

TABLE 8.2 Pearsons correlation coefficient among the various Z-scores.

		Stunted height-for-age Z-scores	Wasted weight-for-height Z-scores	Underweight weight-for-age Z-scores
Stunted height-for-age Z-scores	Pearson correlation	1		
	Sig. (two-tailed)	.		
	N	235		
Wasted weight-for-height Z-scores	Pearson correlation	−0.160[a]	1	
	Sig. (two-tailed)	0.014	.	
	N	235		
Underweight weight-for-age Z-scores	Pearson correlation	0.640[b]	0.565[b]	1
	Sig. (two-tailed)	0.000	0.000	.
	N	235	250	276

[a]Notes: Significant at the 0.05 level (two-tailed).
[b]Significant at the 0.01 level (two-tailed).

weight for age (underweight)[9] and stunting and weight for age and wasting possibly indicates that significant monitoring and evaluation are required, since there is a higher likelihood that a significant proportion of children in Malawi (especially in the rural areas) may suffer from long-term malnutrition problems.

From the descriptive and correlation analysis, it is evident that the main areas that the government of Malawi needs to intervene for a successful nutrition strategy to be effective are (Table 8.3) as follows:

1. Improving the educational level (especially that of females)
2. Communicating messages of better care practices (such as timely introduction of breastfeeding and other complementary feeding)
3. Improving sanitation facilities in communities where they are lacking, so as to prevent diseases such as diarrhea and other vector-borne diseases

Estimating correlation using STATA

In this section, we illustrate the usefulness of the following two commands in STATA: correlate and pwcorr.

We use a data set from the ERS concerning food security that we used in the last chapter to illustrate the usefulness of both commands. For illustrative purposes, we use a selected list of variables from that data set. The data we use have the following variables, and for completeness, we present their definitions:

Recall that ERS also provides these data for six types of countries (Table 8.4).

In this chapter, we have introduced an additional variable x6, which represents the percent of paved roads in each country, which was not included in Chapter 6. Before we implement the correlate and pwcorr procedures, we produce the descriptive statistics using the summarize command:

TABLE 8.3 List of variables with their codes in STATA.

No	Code	Definition
1	x3	Share of dietary energy supply derived from cereals roots and tubers
2	x4	Average protein supply
3	x5	Average supply of protein of animal origin
4	x6	Percent of paved roads over total roads

---

9. Weight for age is influenced by both the height of the child (height for age) and weight for height. Due to its composite nature, its interpretation is not easy.

TABLE 8.4 List of countries used for the STATA data set.

No	Code	Country definition
1	c1	Least developed countries
2	c2	Landlocked developing countries
3	c3	Small island developing states
4	c4	Low-income economies
5	c5	Lower middle income economies
6	c6	Low-income food-deficit countries

The first thing we note is that the data are not balanced—that is, we do not have the same number of observations for all the variables. Suppose we use the correlate and pwcorr commands for the variables x3, x4, x5, and x6, we can obtain the correlation table for these variables. The commands are

.correlate x3, x4, x5, x6
.pwcorr x3 x4 x5 x6, obs, sig star(.01)

The first command produces the correlation table for the four variables of interest. The second command, pwcorr, produces the pairwise correlation along with the number of observations and the significance level and places a star by those estimates, which are significant at 0.01 level or 99% confidence. The input and the output lines from STATA are given in the following.

In the first case the correlate command produces the correlation for all the variables, with only 41 observations. Note the first line of the output in parentheses, which states that only 41 observations were used. Although there are 109 observations for x3, x4, and x5, variable x6 has only 43 observations of which 41 are used for calculating the correlation for all the variables.

The correlate procedure retains only the full information that is available for all the variables of interest and drops the observations for which the data are not complete. However, since there are 109 observations for x3, x4, and x5, we might like to produce correlation estimates with all the observations. To this end, the pwcorr is useful because it produces a pairwise correlation for all the variable sets. In the output, we get the observations for each pair. As you can see from this portion, the estimates are different from the ones produced from the correlate command. In particular, the pairwise correlation between x4 and x5 is 0.69 and significant if we use all the 109 observations. On the other hand, the correlation between these two variables is only 0.34 from the correlate procedure. So researchers must be careful as to which procedure is useful and pay attention to the details concerning these procedural differences.

Note that we have produced the correlation estimates for the entire data set, which has all the countries in one group. Suppose we are interested in finding the correlation between these variables for each country, or say for one specific country. We can use STATA to produce those estimates by sorting according to the variable country, which is coded as "c" in our data set. After sorting the information, we can produce estimates using the "by" statement. The input lines that capture these are as follows:

.sort c
.by c: summarize x6

The first statement sorts the data set according to the countries in the data set. The next statement produces the descriptive statistics for each country, and in this case only for the variable x6. The output from these lines is reproduced in the following screenshots.

From the output in STATA, we can see that for the lower middle-income economies, or for $c = 5$, there are 17 observations for variable x6 in the data set. Therefore, it will be interesting to see the correlation between these variables $c = 5$.

We can generate the correlation for this country (or for $c = 5$) using the following command in STATA:

.pwcorr x3 x4 x5 x6 if c==5, obs sig star(.01)

The output from STATA for this statement is also provided in the following.

As you can see, the correlation between x6 and other variables for this country set is opposite in signs when compared with the same estimates for the overall sample.

Thus, this basic data exploratory analysis indicates that there are inter-country differences in these variables and that the researcher must take into account the heterogeneity across countries.

Conclusion and policy implications

Although much progress has been made during the past two decades in tackling the problem of malnutrition (especially in reducing micronutrient deficiencies), a significant proportion of South Asian and African children continue to be malnourished, and the numbers are expected to grow in the coming decades. More than 160 million preschool children are underweight and stunted, and many are anemic and have vitamin A deficiency. Since malnutrition is not directly observable, it is often overlooked until the problem becomes severe. There are three main reasons for the slow progress in tackling malnutrition. First, it is a complex issue as it encompasses biological and socioeconomic causes. Second, its severity and impact may be ignored even in countries with national nutrition plans. This is because policy makers may fail to understand the urgency and seriousness of the problem. Third, lack of effective organizational skills and design may result in inappropriate programs and strategy such as unaffordable food subsidies.

Thus, understanding the underlying causes of malnutrition may be extremely crucial in developing an effective nutritional strategy. From the existing studies, it is evident that sustained income growth is necessary for reducing malnutrition. However, income growth by itself is unlikely to reduce malnutrition. Along with income growth, complementary factors such as availability of healthcare services, maternal education, effective child-care, access to clean water and sanitation, women's relative status, and household food security are equally important in tackling the severe problem of malnutrition in developing and developed countries.

From the analysis in this chapter, it is evident that the main areas of concern in improving the nutritional status of children in Malawi are enhancing the levels of maternal education so that mothers and other caregivers have better health knowledge to undertake better child-care practices. Second, improved sanitation conditions within the household and the community can be an effective way of preventing diseases such as diarrhea, which, in turn, can have positive impact on child nutritional status. Also, setting up additional health clinics (both government and private) can be an important tool for modern healthcare. Plotting the incidence of wasting and underweight with respect to distance to a protected water source, we find that the incidence of wasting and underweight increases, as distance to the protected water source increases. From a policy perspective, it seems pertinent to develop a comprehensive nutrition strategy for Malawi that emphasizes not only steady income growth but additionally highlights improving the levels of mothers' education and setting up additional health facilities to tackle the problem of malnutrition. Based on a comprehensive research program, the Global Monitoring Report (2012) provides key recommendations:

1. Improve the information about nutrition status, practices, and interventions

A basic problem in designing interventions to mitigate the effects of food price hikes is the lack of quality data on basic nutrition indicators and the effects of both the price rise and some of the interventions to mitigate them. Unfortunately, only a few national surveys collect full food consumption data at the household and individual levels with the needed periodicity. Disaggregated data on costs and impacts also remain scarce. Although the MDG indicator is child underweight, recent findings confirm that stunting is the most appropriate measure for undernutrition. The first recommendation is to produce a multipurpose, nationally representative household survey with information on food consumption, nutritional status, including micronutrient information, and market exposure to monitor nutritional status and design appropriate targeted interventions.

2. Investing in nutrition offers high returns

The global costs of scaling up nutrition may seem high initially, but the costs of inaction are also high, and importantly the estimated returns are very high. Yet funding remains low. Basic nutrition capacity is also scarce. However, renewed interest is appearing from multilateral donors such as the World Bank; bilateral donors such as Canada, Denmark, France, Japan, Norway, and the United Kingdom; and NGOs such as Save the Children.[10]

3. Target the period from conception to 2 years of life

Many interventions have indirect effects on nutrition, but specific interventions for young children and their caregivers and for pregnant and lactating women are crucial, given the importance of that window as a foundation of human capital. There is sufficient evidence about the intensity of physical and sociocognitive development and the negative short-, medium-, and long-term impacts of undernutrition in utero and in the first 2 years of life. Most interventions during this time period have very high rates of return.

4. Tailor the intervention package to country implementation capacity and issues

While acute undernutrition triggers funding and relief interventions, countries also need to tackle chronic undernutrition. "Hidden hunger," or micronutrient deficiencies, requires a different set of interventions. The main micronutrient deficiencies that affect high shares of populations include iron, vitamin A, zinc, and iodine. The package of measures, recommended in SUN, includes supplementation to vulnerable groups in high prevalence areas (vitamin A and iron for pregnant women and children, zinc tablets for children with diarrhea), and fortification including iodized salt and fortified flour and sugar. Deworming is also important in settings where women and children have high worm burdens and develop anemia. Community-based programs are frequently the platform for behavior change interventions and nutrition surveillance.

5. Incorporate nutrition-sensitive approaches in multisectoral interventions

In developing a twin-track approach to nutrition and food security, countries need to weigh the benefits and costs of short-term relief and longer-term investments to raise productivity, especially for smallholder farms, and to work across sectors, especially to link nutrition to health, agriculture, and social protection. A variety of approaches can make interventions in health, agriculture, and social protection, including food aid. National markets also matter and need improved functioning and more involvement of the private sector. Locally, successful implementation will require an alliance of governments

---

10. See the Global Monitoring Report (2012) for the important intervention via the SUN program.

with the private sector, NGOs, and communities, especially because an increase in food prices will have disparate impacts depending on markets and production potential.

6. Senegal's Success and Lessons for the Rest of the World

Senegal in subSaharan Africa has made significant strides in the fight against undernutrition, where nutrition has broken out of the low-priority cycle. Senegal now has a Multisectoral Forum for the Fight against Malnutrition under the Prime Minister's Office; a national nutritional policy and a national executive office that ensures the day-to-day management, coordination, and monitoring of the policy; periodically updated, strategic plans for nutrition; multiple programs with multiple stakeholders from all sectors; a budget line currently projected to grow; donor contributions; national program coverage; and, importantly, a reduction in chronic undernutrition that is 16 times above the average reduction in Africa as a whole.

Estimating correlation using R

In this section, we illustrate correlation analysis in R. We use a data set from the exercise part of this chapter, consisting of 30 observations and 7 variables. Before we implement the correlation analysis, we produce descriptive statistics using the *summary* command for four variables: X3, X4, X5, and X6. To attain descriptive statistics for specific variables, we can add *[,4:7]* to print the results only from the fourth column to the seventh column. The result of the command is as follows:

```
> summary(data_chapter8[,4:7])
      X3              X4              X5              X6
 Min.   :44.00   Min.   :54.00   Min.   :10.00   Min.   :35.70
 1st Qu.:45.00   1st Qu.:55.00   1st Qu.:11.00   1st Qu.:40.85
 Median :62.00   Median :56.00   Median :14.50   Median :48.20
 Mean   :57.87   Mean   :56.97   Mean   :16.13   Mean   :45.17
 3rd Qu.:66.00   3rd Qu.:59.00   3rd Qu.:23.00   3rd Qu.:49.30
 Max.   :68.00   Max.   :61.00   Max.   :24.00   Max.   :51.60
```

Now we can obtain the correlation table for the variables of interest by using *cor* command in R. We add *[,4:7]* to present correlation of the variables of interest. Note that the default of *cor* command will show the Pearson correlation coefficient. If one wants to run other types of the correlation coefficient, add an option: *method = "spearman" or "kendall."*

```
> cor(data_chapter8[,4:7])
           X3          X4          X5          X6
X3  1.0000000 -0.2713070 -0.9790343 -0.7975773
X4 -0.2713070  1.0000000  0.4095501  0.5059641
X5 -0.9790343  0.4095501  1.0000000  0.7765873
X6 -0.7975773  0.5059641  0.7765873  1.0000000
```

We simplify the results table with smaller digits of the correlation coefficient by using the *round* command. We designate to print two digits on the results table as follows:

```
> simple_cor=round(cor(data_chapter8[,4:7]),2)
> simple_cor
      X3    X4    X5    X6
X3  1.00 -0.27 -0.98 -0.80
X4 -0.27  1.00  0.41  0.51
X5 -0.98  0.41  1.00  0.78
X6 -0.80  0.51  0.78  1.00
```

To show the results table with significance, we need to install a package named "*Hmisc*" and then attach it to R.

```
> install.packages("Hmisc")
```

```
> library(Hmisc)
```

The *rcorr* command allows us to examine the results according to the *P*-value. The variables of interest are the same as the previous analyses. With the *rcorr* command, we can investigate the correlation coefficient matrix, the number of observations, and the *P*-value of the correlation. Note that the default of this command also calculates the correlation with the Pearson coefficient.

```
> rcorr(as.matrix(data_chapter8[,4:7]))
      X3    X4    X5    X6
X3  1.00 -0.27 -0.98 -0.80
X4 -0.27  1.00  0.41  0.51
X5 -0.98  0.41  1.00  0.78
X6 -0.80  0.51  0.78  1.00

n= 30

P
      X3     X4     X5     X6
X3           0.1470 0.0000 0.0000
X4 0.1470           0.0246 0.0043
X5 0.0000 0.0246           0.0000
X6 0.0000 0.0043 0.0000
```

We have produced the correlation estimates for the entire data set, which has all the countries in one group. (In the data set, C variable means all countries according to their economic levels.) Suppose we are interested in finding the correlation between these variables for one specific country. We produce descriptive statistics on specific variable X6 for each country. It is performed by using the *by* command in R.

```
> by(data = X6, INDICES = C, FUN = summary)
C: 1
   Min. 1st Qu.  Median    Mean 3rd Qu.    Max.
  37.90   39.80   41.60   42.30   41.83   51.60
-------------------------------------------------
C: 2
   Min. 1st Qu.  Median    Mean 3rd Qu.    Max.
  35.70   38.35   44.40   43.42   48.80   49.40
-------------------------------------------------
C: 3
   Min. 1st Qu.  Median    Mean 3rd Qu.    Max.
  49.10   49.62   50.20   50.12   50.62   51.00
-------------------------------------------------
C: 5
   Min. 1st Qu.  Median    Mean 3rd Qu.    Max.
  48.00   48.40   48.75   48.75   49.25   49.30
-------------------------------------------------
C: 6
   Min. 1st Qu.  Median    Mean 3rd Qu.    Max.
  38.90   40.52   41.15   41.27   42.60   43.00
```

We then are interested to see the correlation between the variables C = 5. We can make a subset when C equals to 5 and then conduct the *cor* command to produce the correlation coefficient in R. The command used for the results is as follows:

```
> C5 <-subset(data_chapter8,C==5)
> cor=round(cor(C5[,4:7]),2)
> cor
      X3    X4    X5    X6
X3  1.00 -0.92 -0.89 -0.16
X4 -0.92  1.00  0.80 -0.15
X5 -0.89  0.80  1.00  0.36
X6 -0.16 -0.15  0.36  1.00
```

As you can see, the correlation between x3 and other variables for this country setting is opposite in signs when compared with the same estimates for the overall sample. Thus, this basic data exploratory analysis indicates that there are intercountry differences in these variables and that the researcher must take into account the heterogeneity across countries.

Exercises

1. Write a correlation syntax in STATA for the socioeconomic indicators (PXFD and EDUCSPOUS), CARE variables (BFEEDNEW and CLINFEED), and community characteristics (DRINKDST, LATERINE, DIARRHEA, and HEALTDST) with ZWA, ZWH, and ZHA. In other words, find the Pearson correlation coefficient among the aforementioned indicators with the outcome variables. Which indicators are significantly associated with underweight, wasting, and stunting? Based on your results, what nutrition interventions will you recommend to the Ministry of Health? How do your results compare with the correlation found with the anthropometric measures?
2. How is correlation analysis helpful in planning for social development programs and formulating nutritional targets?

STATA workout

The following table has 30 observations from 5 (numbered c = 1, 2, 3, 5, and 6) countries for the variables defined in the previous section. Use the STATA commands to produce the following estimates:

- descriptive statistics for the entire sample
- descriptive statistics for each country
- correlation estimates for the whole sample
- correlation estimates for each country

Are the correlation estimates different across countries and the whole sample? Which country's correlation between these variables comes close to the estimates from the whole sample with 30 observations?

obs	Year	C	X3	X4	X5	X6
1	1993	1	68	54	10	37.9
2	1994	1	67	54	10	39.2
3	1995	1	67	54	10	41.6
4	1996	1	66	54	10	41.9
5	1997	1	66	55	11	41.6
6	2003	1	46	61	24	51.6
7	1992	2	65	59	15	35.7
8	1992	2	44	57	24	48.5
9	1993	2	44	56	23	48.9
10	1994	2	44	56	23	49.4
11	1995	2	66	58	14	37.7
12	1996	2	66	58	14	40.3
13	1995	3	45	55	23	49.1
14	1996	3	45	56	23	49.5
15	1997	3	45	56	23	50.0
16	1998	3	45	57	23	50.4
17	2001	3	45	60	24	50.7

18	2002	3	45	60	24	51.0
19	2003	5	62	58	14	49.3
20	2004	5	62	59	14	48.0
21	2006	5	61	59	15	49.1
22	2007	5	61	60	15	48.4
23	2008	5	60	61	15	48.4
24	2009	5	60	61	16	49.3
25	1998	6	66	55	11	41.7
26	1999	6	66	55	11	42.9
27	2000	6	65	55	11	43.0
28	2001	6	65	55	11	38.9
29	2002	6	65	55	11	40.5
30	2006	6	64	56	12	40.6

The **summarize** and the **pwcorr** commands produces the following outputs:

```
. summarize
```

Variable	Obs	Mean	Std. Dev.	Min	Max
obs	30	15.5	8.803408	1	30
year	30	1999.067	5.023622	1992	2009
c	30	3.4	1.886431	1	6
x3	30	57.86667	9.615839	44	68
x4	30	56.96667	2.341284	54	61
x5	30	16.13333	5.506944	10	24
x6	30	45.17	4.947667	35.7	51.6

```
. pwcorr x3 x4 x5 x6, sig star(.01)
```

	x3	x4	x5	x6
x3	1.0000			
x4	-0.2713	1.0000		
	0.1470			
x5	-0.9790*	0.4096	1.0000	
	0.0000	0.0246		
x6	-0.7976*	0.5060*	0.7766*	1.0000
	0.0000	0.0043	0.0000	

Indicators and causal factors Chapter | 8 **291**

The pairwise correlation for the entire data indicates that there is a significant negative correlation between (x5, x3). Furthermore, x6 is also significantly correlated with all the other variables. Next the data are sorted by country using the **sort** command, and subsequently the **summarize** and the **pwcorr** commands are repeated.

```
. sort c

. by c: summarize x3 x4 x4 x6
```

-> c = 1

Variable	Obs	Mean	Std. Dev.	Min	Max
x3	6	63.33333	8.524475	46	68
x4	6	55.33333	2.804758	54	61
x4	6	55.33333	2.804758	54	61
x6	6	42.3	4.829078	37.9	51.6

-> c = 2

Variable	Obs	Mean	Std. Dev.	Min	Max
x3	6	54.83333	11.87294	44	66
x4	6	57.33333	1.21106	56	59
x6	6	43.41667	6.223317	35.7	49.4

-> c = 3

Variable	Obs	Mean	Std. Dev.	Min	Max
x3	6	45	0	45	45
x4	6	57.33333	2.160247	55	60
x6	6	50.11667	.7250294	49.1	51

```
-> c = 5
```

Variable	Obs	Mean	Std. Dev.	Min	Max
x3	6	61	.8944272	60	62
x4	6	59.66667	1.21106	58	61
x6	6	48.75	.5540749	48	49.3

```
-> c = 6
```

Variable	Obs	Mean	Std. Dev.	Min	Max
x3	6	65.16667	.7527727	64	66
x4	6	55.16667	.4082483	55	56
x6	6	41.26667	1.580717	38.9	43

Finally, the pairwise correlation for each country indicates that the variable x6 is significantly correlated with all the variables, only for $c = 1$ and $c = 2$. Furthermore, there are intercountry differences across select correlations. For instance, when $c = 5$, there is significant negative correlation between (x3, x4) that was not present for the entire data.

```
. by c: pwcorr x3 x4 x5 x6, sig star(.01)
```

```
-> c = 1
```

	x3	x4	x5	x6
x3	1.0000			
x4	-0.9926* 0.0001	1.0000		
x5	-0.9970* 0.0000	0.9974* 0.0000	1.0000	
x6	-0.9649* 0.0018	0.9509* 0.0036	0.9496* 0.0037	1.0000

-> c = 2

	x3	x4	x5	x6
x3	1.0000			
x4	0.8948	1.0000		
	0.0160			
x5	-0.9964*	-0.8552	1.0000	
	0.0000	0.0299		
x6	-0.9646*	-0.9482*	0.9487*	1.0000
	0.0019	0.0040	0.0039	

-> c = 3

	x3	x4	x5	x6
x3	.			
x4	.	1.0000		
	.			
x5	.	0.9562*	1.0000	
	.	0.0028		
x6	.	0.9151	0.7835	1.0000
	.	0.0105	0.0653	

```
-> c = 5
```

	x3	x4	x5	x6
x3	1.0000			
x4	-0.9232* 0.0086	1.0000		
x5	-0.8911 0.0171	0.8044 0.0536	1.0000	
x6	-0.1614 0.7600	-0.1490 0.7781	0.3596 0.4838	1.0000

Chapter 9

Effects of individual, household, and community indicators on child's nutritional status—application of simple linear regression

Chapter outline

Introduction	296
Conceptual framework and indicators of nutritional status	297
Household utility maximization framework	297
Core indicators of nutritional status	299
Anthropometric indicators	299
Health and demographic indicators	300
Health status indicators	300
Health service indicators	301
Water and sanitation indicators	302
Review of studies on the determinants of child nutritional status	303
Child's nutritional status in the United States	307
The economics of double burden	308
AIDS and double burden in Africa	309
Malnutrition and mortality in Pakistan and India	310
Social participation as social capital, women empowerment, and nutrition in Peru	312
Social capital and policy during the pandemic	313
Empirical analysis and main findings	316
Data description	316
Incidence of stunting and wasting	318
Normality tests and transformation of variables	319
Regression results	321
Simple regression in STATA	325
Example 1	325
Example 2	327
Conclusion	327
Simple regression in R	328
Exercises	330
STATA workout	331

A radical inner transformation and rise to a new level of consciousness might be the only real hope we have in the current global crisis brought on by the dominance of the Western mechanistic paradigm.

Stanislav Grof.

Introduction

This chapter is concerned with the study of statistical relationships between the nutritional outcomes and their determinants. Simple linear regression method is introduced in this chapter to develop an understanding of the role of causal factors and their relative impact on child nutritional status. The reduced form equations of the determinants of nutrition are derived from a household utility maximization framework, and the empirical analysis shows how the different individual, household, and community factors affect child nutrition. The concepts of independent and dependent variables are introduced along with normality tests and transformation of variables. Interpretation of the regression results along with policy implications is also discussed.

Chronic malnutrition in preschool children remains a substantial challenge for developing countries—178 million children under the age of 5 suffer from stunting (low height for age (H/A)) as a result of chronic undernutrition (Black et al., 2008). Stunting is associated with higher rates of illness and death, reduced cognitive ability accompanied with lower school performance in children, and lower productivity as adults (Cohen et al., 2008). Another 55 million preschool children in developing countries are wasted—lower than expected weight for height (W/H)—due to acute malnutrition (Black et al., 2008). In addition, each year over 19 million preschool children are born with low birthweights—less than 2.5 kg—accounting for 16% of the developing world's annual births (Cohen et al., 2008). These children face a significantly higher risk of neonatal death than normal birthweight children and, if they survive, have much higher rates of illness and stunting in both childhood and as adults (Black et al., 2008).

Policy makers study the causes of child malnutrition to mitigate their effects. Understanding the causes is important, since child malnutrition is one of the severe forms of material deprivations that have intergenerational implications on poverty. Various causes of child malnutrition include poor socioeconomic conditions, inadequate care, maternal malnutrition, large number of dependents, lack of nutritional knowledge, repeated infections, and lack of access to health services. It has been estimated that more than 80% of infant deaths are due to mild or moderate malnutrition (Pelletier et al., 1993). Thus, understanding the causal factors and their relative impact on malnutrition is pertinent to devise intervention strategies that can generate better child nutrition and health outcomes.

In addition, it is important to understand the causes of malnutrition, since there is little agreement in the literature on the relative importance of the factors affecting nutritional status. Based on empirical results, some studies stress the importance of parental education and/or nutritional knowledge, while others recommend the need to focus on improving the poverty/income status of households in poor countries (SC UK, 2003; Christiaensen and Alderman, 2004). For example, SC UK (2003) challenges the nutrition component of World Bank funded projects (in Bangladesh, Ethiopia, and Uganda), which incorporate growth monitoring as a key strategy to educate mothers as a means of reducing malnutrition in young children. The argument is that these projects are based on the questionable assumption that lack of knowledge, confidence, and capacity to solve problems are major causes of malnutrition and that provision of counseling and encouraging women to care for their children will significantly improve nutrition even when families remain trapped in poverty and health and sanitation services are very weak. Thus, while caring practices can contribute to improving child malnutrition, the naïve view that investments in growth monitoring to promote change in caregivers' behavior will necessarily have a significant impact on nutritional status is not well founded. It is therefore critical to understand the individual characteristics, socioeconomic determinants, and the community influences on child nutrition to develop a more comprehensive policy approach.

In light of the aforementioned, this chapter seeks to determine how the various factors (individual, household, and community characteristics) affect children's nutritional status as measured by the weight-for-height Z-score (ZWH). This will be undertaken using a simple linear regression framework. This approach is appropriate since we want to determine how the average score of ZWH changes when an independent variable (causal factor) increases by 1 standard deviation unit. Thus, we want to determine the unique contribution of the changes in individual, household, and community variables on child malnutrition. This chapter is organized as follows. In the next section, we discuss the conceptual framework and indicators of nutritional status. We then review a few studies that examine the role of individual characteristics (such as age of the child), household characteristics (such as maternal education, nutrition knowledge), and community characteristics (such as water and sanitation facilities) on child nutritional status.

This is followed by an examination of the role of individual, household, and community characteristics on child nutritional status using data from Malawi.

Conceptual framework and indicators of nutritional status

Household utility maximization framework

From an economic perspective, optimum nutritional status can be understood from the perspective of a household that seeks to maximize utility subject to its time and income constraints. Let the behavioral function of a typical household be as follows[1]:

$$U = u(N, H, F, Z, L) \tag{9.1}$$

where N, H, F, Z, and L denote nutritional status, health, food, other commodity consumption, and leisure time, respectively.

---

1. We omit superscripts and subscripts to make the analysis simple. The framework is an extended version of Smith and Haddad (2000).

The nutritional status of an individual (N) within the household (especially the child) depends on food intake, other commodity consumption (Z), care (C) for children, and the health environment vector (Ω). In other words:

$$N = n(F, Z, C, \Omega) \qquad (9.2)$$

N refers to nutritional intake as distinct from nutritional outcome such as weight or height. C denotes care received by a typical child, which may consist of exclusive breastfeeding, timely introduction of complementary foods. Ω denotes the health environment, consisting of availability of safe water, sanitation, and health services in the household's community.

Health (H) is produced by nutritional intake, food and other commodity consumption, income of the household (I), educational level of the household head (EM), time devoted to health care (TC), and the health environment (Ω). Thus, the health production function is as follows:

$$H = h(N, F, Z, I, EM, TC, \Omega) \qquad (9.3)$$

Household production of food F and other commodities (Z) will be dependent on the time needed to prepare it, relevant environmental variables, and possibly on the nutritional status of the household head. Additionally, we must have two resource constraints that limit household production and consumption possibilities. The first is the income constraint, which is that household income from all sources must be spent on expenditure on foods and other goods, expenditure on health, and caregiving. The second constraint is time, which states that household's time endowment is allocated between labor on the one hand and household production of N, H, F, and Z and leisure on the other. Maximization of Eq. (9.1) subject to the production constraints of health, nutritional status (as in Eqs. 9.2 and 9.3), food and other commodities, and the household members' time and income constraint determines the nutritional status of a child in any given year as follows:

$$N^* = n^*(\Omega, P, C, I, EM) \qquad (9.4)$$

In other words, optimum nutritional status depends on the environmental factors such as access to safe water, sanitation, prices (P) of food and other nonfood items, the household's total income (I), care given by the household to the child, and mother's educational level (EM). As explained before, a mother's educational level can have an important influence on feeding and caring practices. These knowledge and skills improve the caring practices and thereby positively influence the nutritional status of the child. Having conceptualized in general terms the important determinants of a child's nutritional status, it is useful to look at some of the critical indicators of nutrition. At the outset, it is important to distinguish between the nutritional outcomes (such as low H/A, low W/H, body mass index among adults) from the determinants of nutritional status (such as age and sex composition of the household, educational attainment of the household, income, and health environment).

Core indicators of nutritional status

A variety of methods, such as anthropometric, health, and demographic approaches, are used to assess the nutritional status of populations (Food and Nutrition Assessment Technical Assistance, 2003). While the anthropometric approach is often used as a proxy to assess the extent and severity of malnutrition, i.e., outcomes from lack of nutrition, the latter two approaches (health and demographic) help in understanding the causal factors and linkages through which malnutrition occurs. We discuss them each in turn.

Anthropometric indicators

Anthropometric indicators are useful at both the individual and population levels. At the individual level, they can be valuable for screening children for any intervention required. At the population level, these indicators can be used to assess the nutrition status within a community, country, or region and to study the determinants and consequences of malnutrition. Anthropometric survey data often contain measures of weight, height, and age of children. It is possible to use these physical measurements to assess the adequacy of diet and growth especially for infants and children. Comparing an individual with a "healthy" reference group and identifying extreme departures from this distribution could overcome the severity of malnutrition. The three most commonly used anthropometric indicators for children are W/H, W/A, and H/A.

Weight for height

"W/H" measures body weight relative to the height of the child. It is normally used as an indicator of current nutritional status and is useful for measuring short-term changes in nutritional status. Low (W/H) relative to a child in the healthy reference group is referred to as "thinness." An extreme value of low (W/H) is referred to as "wasting." Wasting is often associated with acute starvation as in the case of famine situations or severe disease and may also be the result of a sudden shock on children with chronic malnutrition.

Weight for age

"W/A" measures body mass relative to the age of the child. Low W/A is influenced by the height of the child (H/A) and weight for height. Due to its composite nature, its interpretation is not easy. Low W/A relative to a healthy child in the reference population is referred to as "lightness." This measure is commonly used for monitoring growth of the child and to understand the severity of malnutrition over time.

Height for age

"H/A" indicates chronic malnutrition or illness and is thus a cumulative indicator of physical growth. Low H/A relative to a healthy child of the reference population is referred to as "shortness." An extreme case of low H/A is referred to as "stunting." At the population level, high levels of stunting are associated with increased risk of illness and/or poor socioeconomic conditions over a prolonged period of time.

All the aforementioned three measures are usually expressed in the form of Z-scores, which compare the aforementioned indicators to a similar child from a healthy reference group. In other words, the Z-score of a child i is the difference between the height of the child (H_i) and the median height of a group of healthy children of the same age and sex from the reference population (H_r), divided by the standard deviation of the height of the same group of children from the reference population σ_r. Thus, the Z-score of height for age (H/A) is defined as

$$ZHA = \frac{H_i - H_r}{\sigma_r}$$

Similarly,

$$ZWA = \frac{W_i - W_r}{\sigma_r} \quad \text{and} \quad ZWH = \frac{(W/H)_i - (W/H)_r}{\sigma_r} \qquad (9.5)$$

Stunted children are commonly defined as more than 2 standard deviations below the median. The two preferred anthropometric indices are stunting and wasting, since they distinguish between long-run and short-run physiological changes. The "wasting index" (low W/H) has the advantage that it can be calculated without knowing the child's age. It is particularly useful in the short run in analyzing the current health status of a population and in evaluating the benefits of intervention programs. A disadvantage of this index is that it classifies children with poor growth in height as normal. Stunting (low H/A) on the other hand measures long-run nutritional status of a population, since it uses past nutritional status. WHO (1995) recommends stunting as a reliable measure of overall social deprivation.

Health and demographic indicators

Certain demographic and health indicators can be very useful in understanding various aspects of malnutrition. It is important to emphasize that these indicators should be conceived as determinants or factors that ultimately affect the nutritional status of a population to decline rather than outcomes of lack of nutrition. We discuss three sets of health and demographic indicators: health status indicators, health service indicators, and water and sanitation indicators.

Health status indicators

The most commonly used health status measures are discussed in the following (Skolnik, 2007).

Infant mortality rate

Infant mortality rate can be defined as the number of children less than 1 year old who die in any given year per 1000 live births. This indicator provides information regarding nutritional conditions such as weaning and reflects other socioeconomic conditions in which the infant grows.

Life expectancy at birth

Life expectancy at birth can be defined as the average number of additional years a person could expect to live if current mortality trends continued for the rest of the person's life. It is a summary indicator of overall health and physical well-being in a country.

Child health conditions (diarrhea and immunization coverage)

Occurrence of diarrhea is one of the main causes for stunted child growth and infant mortality and has a negative impact on child nutritional status. It is not an input of nutrition but rather the outcome of investments in other aspects of health that influence the productivity of inputs into nutrition, or the investment itself. Diarrhea is defined as more than three loose stools passed in a 24-hour period. For children below 5 years, "diarrheal incidence" may be defined as the ratio of the number of children below 5 years with diarrhea during the last 2 weeks to the total number of children below 5 years.

Similarly, immunization indicators are used to assist operational planning for full immunization coverage and to prevent further diseases. "Immunization coverage" for children below 1 year is calculated from the number of children fully immunized (defined as the first visit where all the vaccinations are completed) divided by the total number of children below 1 year. The primary course of immunization includes Bacille Calmette−Guérin (BCG), three doses of diphtheria−tetanus−pertussis (DTP) and polio, measles vaccine, and hepatitis B.

Health service indicators

The commonly used measures to understand availability of health services are as follows (Mcguire, 2006).

Utilization rate

This is defined as the number of visits per person per year to the nearest health facility. This is applicable only in areas and regions where a health facility is available at a reasonable distance.

Health personnel

Along with the data on the number and types of health facilities, data on health personnel are also used as an indicator of physical access to health service.

One way of measuring it is to compute the number of health workers as a percentage of total population. However, aggregate measures like this indicator (at the country or regional level) may hide all regional and subregional differences.

Per capita health expenditure

This measure denotes the amount spent on health per person per year. This is also an aggregate measure and does not distinguish between the proportions of population covered by medical schemes from those who are not.

Water and sanitation indicators

Water and sanitation improvements can have a positive and significant impact on nutritional status of a population by reducing a variety of diseases such as diarrhea, guinea worm, and skin diseases. Improvement in nutritional status can occur through a variety of mechanisms. Increasing the quantity of water allows better hygiene practices, while improving the quality reduces the ingestion of pathogens. With less disease, infants can eat and absorb more food, which improves their nutritional status. Additionally, a healthier adult population can be more productive, which, in turn, raises income and the capacity to acquire more food.

Improvements in sanitation can improve the health of the population by lowering the incidence of diarrhea, reducing parasitic infections, and ultimately leading to a reduction in morbidity and mortality. Thus, efforts to improve sanitation are worth undertaking, as they have community and individual level effects. Improvements in water and sanitation facilities, however, do not automatically translate into improvements in health and nutritional status. Hygiene education is a prerequisite to have health effects translate into greater nutritional status. Hygiene education consists of teaching the importance of hand washing, disposal of feces, and protection of drinking water (Billig et al., 1999). The most commonly used water and sanitation indicators are as follows.

Quantity of water used per capita per day

This indicator includes all water collected by or delivered to the household and used for drinking, cooking, personal and household hygiene, and sanitation needs by the members of the household. All adults and children in the household are counted. It is assumed that the amount collected is the amount used. It is defined as the ratio of volume of water (in liters) collected for domestic use per day by all households in the sample to the total number of persons in the sample. The aforementioned indicator will be measured more precisely if calculated for individual households first and then averaged for the total number of households sampled. This step accounts for the large variations in the number of individuals per household. There are some problems involved in the collection of data for water usage when water is piped directly into the house or compound. Such systems typically are not metered either at the source or at the household, and thus, total water used in the community cannot be calculated. Additionally, piped water may have leaks, or people outside the service area may take water from the household. Under such circumstances, distance to the water source may be an alternative indicator for water use (Billig et al., 1999).

Access to improved water source

Access to an improved water source implies that the home or compound is directly connected to a piped system or a public fountain, well, or any other water source that is located within a reasonable distance from the home. It is defined as the ratio of the number of households with access to an improved water source to the total number of households in the sample. The usage can be for drinking, cooking, cleaning, and other personal reasons. Unimproved surface water sources, such as rivers, lakes, and streams, are not calculated in this procedure.

Percentage of households with access to sanitation

A *sanitation facility* is defined as an excreta disposal facility, typically a toilet or latrine. Access implies that the household has a private facility or shares the facility with other people. For urban areas, access to sanitation is defined as being served by connections to public sewers or household systems such as flush latrines, septic tanks, communal toilets, and the like. Rural access consists of pour flush latrines, pit privies, etc. It is calculated as the ratio of the number of households with access to a sanitation facility to the total number of households in the sample (Billig et al., 1999).

Review of studies on the determinants of child nutritional status

In the following, we review a few case studies that examine the determinants of child nutritional status using the framework developed in the aforementioned section. The determinants can be classified under child-specific, household characteristics and community characteristics. The child-specific determinants of nutritional status are child's age and gender.

Sahn and Alderman (1997) examined the impact of age specificity (for children below 24 months and those between 25 and 72 months) on child nutritional status. The study used data from Maputo, Mozambique, using a randomly selected cluster method consisting of 1816 households. The study defined malnutrition as 2 standard deviations below the median for ZHA and found that 32.3% of males and 26.8% of females suffer from chronic malnutrition. The study estimated a reduced form nutrition production function using per capita calorie consumption and birthweight of the child in the aggregate sample (no age differences). Additionally, the production function was estimated with age-specific effects. A number of covariates were also included in the model, such as education dummy variables and information about sanitary facilities.

The main result of the study confirmed that nutrition responds to increases in income and the impact of income is only significant for children 2 years and older. For younger children, mother's education was significant. The aforementioned results imply that education and programs that aim at improving child-care practices are appropriate for mothers of younger children.

Guldan et al. (1993) studied how the interaction of maternal education and feeding practices improves child nutritional status. The study was conducted in the Manikganj district in Bangladesh—a rural lowland area situated about 60 miles west of Dhaka. A sample of 185 children aged 4—27 months was selected for a 6-month period, and the information was gathered through home visits. The main question addressed in the study was whether expansion of education for women promoted child survival through its association with improved infant- and child-feeding practices.

The results of the study consistently demonstrated that educated mothers fed their children in cleaner locations with fewer distractions, where they have more control over the child's meal. Maternal education was also associated with more caretaker initiations of feelings and more attentiveness to the child. Thus, all these behavioral changes would possibly improve child nutritional status. Additionally, the study found that both maternal and paternal education were significant predictors of child-feeding practices and care behavior in rural Bangladesh after controlling for household wealth.

The study provided some policy recommendations for improving child nutritional status. First, the formal education system should be strengthened with an increased emphasis on health-related components for all family members. Second, since education had a strong impact on feeding practices, it is pertinent to introduce large-scale nonformal education for adults and older children. This will ensure better behavior toward personal and household hygiene, proper weaning practices, feeding children appropriately during and after diarrheal episodes, etc.

Ruel and Menon (2002), using demographic and health survey data sets from five countries in Latin America (Bolivia, Colombia, Nicaragua, Guatemala, and Peru) between 1994 and 1999, examined the impact of child-feeding practices on child nutritional status. The main objectives of the study were threefold:

1. To assess the feasibility of an age-specific child-feeding index.
2. To estimate the strength of association between child-feeding practices and child nutritional status.
3. To evaluate whether better feeding practices were more important for some subgroups of children than others.

Effects of individual, household, and community indicators Chapter | 9 **305**

The main contribution of the study was to identify how caregiving (through better feeding practices) can be a crucial determinant of child nutritional status. The methodology of this study can be used to identify the vulnerable sections of the population that are more likely to benefit from interventions to promote improved child-feeding practices.

Gragnolati (1999), using data from the Guatemalan survey of family health for a survey of 2872 women between the ages of 18 and 35 and a set of community surveys for 60 rural communities for the period May to October 1995, investigates how individual, household, and community characteristics affect child nutritional status.

Overall, the results of the study indicated that ZHA among children varied widely among the communities, with the prevalence of stunting ranging from 20% to 88%. From the analysis of the community level variables, two main findings emerge:

1. More than 90% of the community-level variation can be explained by household and community variables.
2. Altitude accounts for the largest proportion of the overall variation.

Valdivia (2004) explored the impact of public investments in health infrastructure during the 1990s on ZHA. There are several ways through which expansion of public investments in health facilities affects the growth of children. First, by introducing new healthcare facilities, travel time barriers for mothers are significantly reduced (or the time they have to wait to consult a doctor). Second, building a health facility in a locality that did not have one before also helps in the organization of the delivery of social services.

The main finding of the study was that mothers' education had a positive and significant impact on child growth consistent with earlier studies. There was also a marginal positive impact of health infrastructure on child H/A. Undertaking the analysis separately for urban and rural areas, the study found that the factors that influence child nutritional outcomes differed significantly.

Section highlights: Were some World War I recruits better than others?

Haines and Steckel (2000) examine the role childhood mortality and nutritional status as indicators of standard of living from World War I recruits in the United States. They find large health differentials in the United States early in the 20th century, especially when using height as an indicator. Stature differences were in the range of two inches between Texas and Rhode Island, which the researchers consider to be large, based on modern standards. Their estimates on childhood mortality index also showed a very large variation across states, ranging from 0.57 in Vermont to 1.9 in New Mexico in 1910.

Haines and Steckel (2000) examine the relationship between anthropometric indicators of health and mortality. They link average heights with other social

Continued

Section highlights: Were some World War I recruits better than others?—cont'd

indicators such as per capita income and mortality rates. They find that overall a 1% reduction in childhood mortality index around 1895 would have resulted in a 0.27 inch increase in stature. For data around 1904, the increase in height would have been about 0.55 inch. They also compute this for data at the county level for the year 1910 and find that the height increase would have been around 0.29 inch. According to Haines and Steckel (2000), these are substantial effects of mortality on physical stature. They also find substantial regional clustering for both height and BMI. For instance, the shortest recruits came from the Northeast and eastern Midwest and the tallest from the western South, part of the Mountain region, and the upper Midwest. In contrast, the least robust recruits, with the lowest BMI, originated in the South and the most robust in the Northeast, upper Midwest, and far West.

What could account for these regional health differences? According to the researchers, socioeconomic factors, adult illiteracy, and urbanization provide interesting clues. According to the researchers, H/A and H/W are reasonable social indicators and can shed light on assessing the state of health and the biological standard of living within the United States in the early 20th century. This is a particularly important period, because the mortality transition was fully underway. Even during the critical period 1890–1920, however, urbanization continued to exhibit strong effects on both childhood mortality and stature in the expected directions. In other words, the "urban penalty" had not yet been eliminated.

Most interestingly, their results confirm that the urban mortality penalty, which had only begun to diminish substantially by the late 19th century, continued to influence the health conditions in the early 20th century. The shortest recruits came from the most urbanized and industrialized states and counties in the northeastern United States. The most robust recruits came from the western upper Midwest and western states of the United States, while the leanest originated in the South. Consequently, the researchers conclude that these physical attributes were related to the epidemiological, health, dietary, and general living conditions in those regions.

Their results for BMI were similar to that for height. This suggests that regional clustering of anthropometric measures of health was significant in the early 20th century but that childhood mortality was spatially diffused. They conclude that public health improvements affect mortality more quickly than nutrition and generalized standard of living. The latter measures are more effective for physical stature.

Socioeconomic factors do provide clues to the sources of these regional health differences. Their results based on R-squared values from different models provide these insights. It is also a good exercise for us to examine these, in the context of this chapter (see Problem 1 in the Problems section).

The age effect of the child was more pronounced in the rural areas relative to the urban areas, indicating that some unobserved environmental factors were affecting child growth. The health infrastructure variable was found to have a positive and significant impact on child growth in the urban areas but not in rural areas. There were two plausible reasons for the impact of health infrastructure on the urban areas only:

1. Reduction in distance barriers was not enough to benefit rural children.
2. Cultural factors (such as language barriers) could explain why rural families did not benefit much from augmenting the health infrastructure.

From the analysis of interaction effects, the study found that the only interaction that was significant was education of the mother interacted with health infrastructure, implying that higher benefits occurred to children of less educated mothers. The result can be interpreted as a propoor bias in the expansion of health infrastructure. The implication of the aforementioned findings suggests that reducing distance and waiting time barriers may be crucial in improving the conditions of the rural poor, but policies that included the indigenous groups must also be taken into account. From the aforementioned studies, it is evident that individual characteristics, such as age (younger children are more prone to malnutrition compared with older children; Sahn and Alderman, 1997), sex of the child, household characteristics, such as income, maternal education, childcare practices, and community characteristics such as health infrastructure, water, and sanitation facilities are all critical determinants of child nutritional outcomes. Since many variables (such as income and maternal education) are associated equally with child nutritional outcomes, it is important to use ordinary least squares regression to explore the bivariate association of one cause at a time. This will allow us to determine the potential explanatory variables that were associated with any of the dependent variable (child nutritional outcomes) for the multivariate regression analysis developed in Chapter 10.

Child's nutritional status in the United States

The impact of poverty and welfare has attracted a lot of researchers to examine the status of health and nutrition of children in the United States. Gundersen et al. (2011) in particular examine the effect of various government programs that affect children's health, including school lunch programs. Besharov (2003) in particular examines issues of food security among children and examines the role of US government's policies on poverty and hunger among US children. Consequently, US policy makers treat the role of economic conditions on children's nutritional status very seriously.

Evidence continues to mount in favor of SNAP's effects in reducing food insecurity. Kreider et al. (2012), Gundersen et al. (2012), Ziliak and Gundersen (2016), and Gundersen et al. (2017) have all shown the positive impact of SNAP on food security and health outcomes among children.

In an interesting study for the United States, Aizer et al. (2016) show how the initial cash transfer programs during 1911–1935 improved the longevity, schooling years, lifetime earnings, and overall health of male children of the participants from poor families.

The economics of double burden

Using comparable data across Ethiopia, Peru, India, and Vietnam, Schott et al. (2019) note how groups with higher income and maternal education are protected against the probability of high stunting. However, groups with higher income and urban residencies predict high probabilities of being overweight. Avula (2020) provides substantial insights into the causes of double burden, generating evidence from Bangladesh, India, Vietnam, Mexico, Ethiopia, and other countries.

The issue of double burden is of concern in India as well, where the levels of overweight are around 40% in some districts. Swaminathan and Menon (2020) report how India, while relatively successful in implementing programs to tackle undernutrition, has many challenges that are yet unmet. First is the wide variation in maternal and child undernutrition across states, compelling a look toward adequate and responsible nutrition for children under 5 years of age.

Second, about 10% of children under 19 years are prediabetic. This calls for a concerted program that targets food diversity, better nutrition, nutrition literacy, access to healthcare for children, pregnant women, and restricting the marketing of unhealthy food and drinks.

The aforementioned challenges will be even more difficult to address, given the silo-mentality of different ministries and the lack of information sharing and a lack of appreciation for all sectors to come together and establish accountability. Constantinides et al. (2019) provides an excellent case study demonstrating this difficulty from Tamilnadu, where there is very little awareness of double burden among policy makers, and where there is no clear strategy or a shared narrative that can serve for a cohesive and an efficient program.

Recently, many researchers including Ruel and Hawkes (2019) and Hawkes et al. (2010) note that the problem of double burden of malnutrition as an inadvertent consequence of unfocused nutrition intervention programs. For instance, Ruel and Hawkes (2019) demonstrate how cash and food transfer programs in Latin America and Egypt, although well intended, unfortunately led to excess intakes of high-energy-fat-sugar-dense foods with low content of micronutrients, leading to higher risks of obesity and noncommunicable diseases.

Hence, in recent years, there is a push toward double-duty policies to address the double-burden phenomenon. Hawkes et al. (2020) provide a comprehensive strategy that identifies 10 double-duty actions to prevent undernutrition, obesity, and diet-related noncommunicable diseases. The actions call for evaluating current strategies and their limitations, redesign existing strategies using the double-duty approach, develop steps to tackle malnutrition at all stages of the life cycle starting with women during pregnancy and lactation, and follow through with infants, preschoolers, school-age children, and adolescents.

AIDS and double burden in Africa

The impact of AIDS on economic development has been studied extensively by economists and social scientists. Zivin et al. (2009) examine the impact of medications and intervention on children's health in Kenya. The provision of antiretroviral medications is a central component of the response to HIV/AIDS, and it is important to test whether medications and interventions lead to better economic outcomes. Using longitudinal survey data from Kenya, the researchers examine the relationship between the provision of treatment to adults and the schooling and nutrition outcomes of children in their households. They find that school attendance increased by over 20% within 6 months after treatment was initiated for adult patients. They also find partial evidence that indicates improvement in young children's short-term nutritional status. Their results link the health improvements in adults to more efficient intrahousehold allocations of time and resources.

Custodio (2010) has observed an interesting phenomenon of a "double burden of over- and undernutrition" that occur at the same time, in Equatorial Guinea. Equatorial Guinea is a good example to examine this subject because it is a country that is in a socioeconomic transition. Custodio (2010) examines the trends in children's nutritional status, which included information on diet, health, and anthropometric measurements from which Z-scores for height-for-age (ZHA), weight-for-age (ZWA), and weight-for-height (ZWH) were calculated. Custodio (2010) finds that in the time period 1997−2004, the prevalence of child overweight for all children increased from 21.8% to 31.7%, especially in urban areas. The prevalence of stunting was still very high, particularly in the rural areas, even though the overall trend shows a decline. Estafina's results suggest that the country is undergoing a nutrition transition and acquiring the concomitant double burden of under- and overnutrition.

In a recent paper, Philip (2012) investigates the effect of child undernutrition on the risk of mortality in Burundi. Using anthropometric data from a longitudinal survey (1998−2007), Philip (2012) finds that undernourished children, measured by the ZHA in 1998, had a higher probability of death during subsequent years. Interestingly, Philip (2012) uses data that capture the risk of mortality via the child's length of exposure to civil war prior to 1998 as a source of exogenous variation in a child's nutritional status. Importantly, Philip (2012) finds that children exposed to civil war in their area of residence have worse nutritional status. The results indicate that 1 year of exposure translates into a 0.15 decrease in the ZHA, resulting in a 10% increase in the probability of death. For boys, there is a 0.34 decrease in ZHA per year of exposure, resulting in 25% increase in the probability of death.

Malnutrition and mortality in Pakistan and India

Poverty and living conditions should be considered as primary factors that affect a child's nutritional status. Khan and Azid (2011) examine the plight of malnutrition of primary school-age children in urban and slum areas in Pakistan. They construct a composite index of anthropometric failure and apply this to a large data set of 882 children. Their logit model suggests that children's nutritional status improves with age, birth order, female sex, and activity of the child other than schooling. The researchers also find that parents' education, specifically mothers' education, plays an important role for child's nutritional status. Finally, malnutrition is positively related with the number of members in the household, while provision of basic amenities such as electricity, safe drinking water, and underground drainage affects children's malnutrition negatively.

Toshiaki (2019) notes that among 10 Asian countries, Pakistan fares worst outcomes, with respect to child malnutrition. Arif et al. (2014) and Toshiaki (2019) note the importance of household asset size, and socioeconomic status as key drivers of child nutritional and health outcomes.

Importantly, female empowerment continues to drive children's nutritional outcomes in Pakistan. Haroon (2018) explores how female empowerment—measured via educational attainment, labor force participation, decision-making rights in the household, asset ownership and freedom of movement, and perceptions on domestic violence—leads to improvements in nutritional and health outcomes measured via incidence of stunting, wasting, and weight measures.

Hence, issues surrounding gender discrimination, cash transfer programs, microfinance and inclusion in agriculture, and livestock extension programs are crucial pathways that can relieve Pakistan's crisis with childhood malnutrition and health.

Sahu (2019) notes the terrible plight of undernutrition and nutritional anemia not just among tribal children but also among pregnant and nursing women in the tribal districts in Maharashtra. Boo and Canon (2012) use a very interesting data set from Andhra Pradesh, India, to link nutritional status to cognitive development. They use a value-added model of cognition—nutrition for children who are observed from 6 months to 8 years. First, an interesting aspect of this study is the way in which they calibrated the data, which were from three rounds of the Indian survey of the Young Lives (YL) project. In Round 1, starting in 2002, they surveyed 2000 children aged 6—18 months from the "Younger Cohort." They followed this with Round 2 and tracked the same children and surveyed them in 2006 at age 5. Finally, in Round 3, the researchers surveyed the children in 2010 at age 8. The data contained information on important socioeconomic aspects including caste. Consequently, with these data, the researchers were able to look at the nutrition—cognition link and at the relationship between caste and test scores. Their results indicate the following:

1. A 1 standard deviation increase in ZHA at the age of 5 leads to cognitive test scores that are about 16% of an SD higher at age 8.
2. The differences in income levels between castes found in adulthood arise early in childhood.
3. Upper caste children show a substantial advantage in vocabulary tests, but most importantly, they show a more pronounced gender inequality than their lower caste counterparts.
4. Upper caste families discriminate more against girls.
5. There are substantial family-fixed effects.

In a recent study, Imai et al. (2012) investigate whether mother's empowerment or relative bargaining power affects children's nutritional status. They use the NFHS and NCAER data for the years 1992—2006, covering rural India. Interestingly, they find that the relative bargaining index (defined as the share of mother's schooling years over father's schooling years) positively and significantly influences "ZWA" and "ZWH"—short-term measures of nutritional status of children. Their regression results suggest, however, that the bargaining power will improve a chronic measure of nutritional status, or "H/A" at the low end of conditional distribution of Z-score or those stunted. Not surprisingly, they also find that access to health scheme or health insurance and health-related facility, infrastructure, and environment are important factors in reducing child malnutrition.

Based on an empirical study in West Bengal, De and Sarker (2011) examine whether women's involvement in the microcredit program through self-help groups or SHGs makes any positive change on women's empowerment. They use various criteria such as power, autonomy, self-reliance, entitlement, participation, awareness, and capacity building as indicators of empowerment. De and Sarker (2011) find that women participating in the microcredit program through SHGs for 8 years or more have significant improvements in the level of women's empowerment. Most importantly, in the context of this chapter and the last, De and Sarker (2011) find that women's earnings from savings and credit have positive and significant effects on nutritional status of the children of women members of SHGs and on the protein intake for their household compared with that of among control groups. The results obtained from this study highlight the fact that the empowerment level of women participating in the microcredit program has positive and highly significant effect on the nutritional status, along with protein intake for their children. A major policy conclusion is that there is a need to establishing women's microcredit program through SHGs supported by various agencies throughout West Bengal and that these programs have to be sustained for long time periods.

Social participation as social capital, women empowerment, and nutrition in Peru

According to Favara (2012, 2018), social participation is particularly effective in improving child nutritional status when the mother's education and economic resources are limited, which is particularly true in poor communities. Similar to the study by Florencia and Maria (2012), Marta (2012), and Favara (2012, 2018) uses the Peruvian sample of the *YL project*, which is a longitudinal data set following children at two points in time, at age 1 and 5, and gathering information on child nutritional status and maternal social capital. In this context, Peru is a very good example to study the association between social capital and child health, given the high rate of both child malnutrition and social participation in this region.

Favara (2012, 2018) finds that maternal social capital is positively associated with 1-year-old children's ZHA for those children whose mothers have no education. Favara's (2012, 2018) study is interesting because it tackles the endogeneity issue that exists between maternal education and poverty: a mother's unobservable characteristics might drive both the decision to participate and the child nutritional status.

In other words, poor child health might prevent mothers from participating actively in community life or conversely, might encourage mothers to participate to have access to health services and advices. That is, the decision to participate might be endogenous given that becoming a member of any organization is a voluntary decision. To disentangle these effects, Favara (2012, 2018) uses a special data set that has information on maternal participation before child's nutritional status.

The results show that social participation is an effective instrument against children's undernourishment in the first year of life for those children born to uneducated mothers. According to the findings, a 11-month-old child whose mother has no formal education, but is a member of at least one community organization, is 1.1 cm taller than a child whose mother has the same educational level but does not participate in any community group. This effect is quite sizable considering that it is equivalent to the effect of maternal education, which has been proved to be one of the main indicators of child development. The descriptive statistics from Favara (2012, 2018) shows that social participation affects child ZHA, by providing access to resources to mothers network.

As mentioned in the previous chapter, the distribution of bargaining rights at the household is a good indicator of nutrition allocation. Novella (2019) notes that in Peru, nutritional outcomes are better for girls than for boys, in those homes where maternal decision-making power is higher. Aurino et al. (2019) link the same maternal power and rights to adolescent cognitive skills, cognitive attainment, and nutrition. Aurino et al. (2019) show that age-1 ZHA is positively related to cognitive scores and show evidence of no discrimination between boys and girls in math and reading scores. However, between ages 1 and 15, there are unpredicted improvements in ZHA and in higher cognitive scores for boys than for girls.[2]

"All these interesting studies from Peru indicate that community initiatives are a good way out of poverty among the poorest." Even in a very deprived context characterized by scarce human and physical capital, there is space for effective policy interventions in a relatively short term. Low educated mothers might be helped in raising healthy children, facilitating the cooperation and association at the community level.

Social capital and policy during the pandemic

As noted in the previous chapters, the corona outbreak has produced drastic reductions in global economic performance. Several countries resorted to lockdowns, resulting in massive supply—demand shocks. For example, India had imposed severe lockdowns during March to May 2020. As a consequence, Saroj et al. (2020) and Narayanan (2020) note that the Indian agricultural food supply chains from "farm-to-fork" had to face severe disruptions.[3]

Arndt et al. (2020) and Resnick (2020) cannot stress enough that social transfers during "black swan"—type shocks are critical and demonstrate how, in Africa, income transfers have been extremely successful in protecting the poorer and less-educated households from slipping into food insecurity and starvation during the pandemic.

Ray and Subramanian (2020) also indicate the importance of modest cash transfers in assisting vulnerable groups such as migrant workers and daily-wage earners to deal with the sudden income loss. Likewise, Dorosh (2020) prescribes a combination of enhancing PDS, alongside cash transfers to facilitate open market sales.

---

2. In addition to Peru, Aurino et al. (2019) establish this for India, Ethiopia, and Vietnam.
3. The supply chain disruptions, the accompanying fall in producer prices, and an increase in consumer prices for pulses, edible oils, and processed foods like flour are attributed to shortages of available inputs such as labor and machinery; closure of markets that regularly purchase the produce; lack of fresh and processed food for the retail sector; lack of demand from hotels, tourism sector, hospitality industry, catering; and resultant higher transportation costs. Lowe and Roth (2020) provide an excellent discussion on India's supply chain disruptions following the lockdown.

Section highlights: Pandemic, food security, and Nepal's success story

Nepal's adoption of a large-scale multisectoral nutrition program during the corona outbreak has been fairly successful and has been praised by many international agencies for the program's strategies, implementation, and outcomes.

Nepal's *Suaahara* "Good Nutrition" program combined the efforts of several groups: USAID, local and international partners led by Helen Keller International, and the government and other private agencies. Rana and Kshetri (2020) indicate *Suaahara's* key features that were instrumental in its success. The program directly helped mothers and children and created an integrated package (from birth of a child to its first 1000 days) to help with nutrition, hygiene, and health services. There were provisions of technology platforms and outreach, counseling, home visits, and community events. Finally, specific government staff were assigned to work closely with specific groups, districts, and municipalities at local levels.

Additionally, Rana and Kshetri (2020) note that the entire program was transferred to mobile devices and technology during the lockdown without any disturbance. Mobile technology established helplines and intervention to prevent domestic abuse/violence and assisted frontline workers (FLWs) with effective social messaging.

The program's success is evidenced by the facts that, even during the pandemic, the public distribution systems remained operational at the municipality level, and agricultural work continued without any disruptions.

Importantly, *Suaahara* recruited clinicians and lab technicians to deal with COVID-19 in terms of testing and the control of spread. *Suaahara* recognized early on the pandemic's effect on food insecurity and was able to partner with the government and other agencies and successfully track the most susceptible areas and groups. The government and Suaahara II continue to work together to
- ensure open, two-way communication between all partners in society along with experts, covering specific subgroups within the population;
- update information on essential and emergency services, particularly about food and health safety-net access and support for the most vulnerable sections;
- promote functional food systems, livelihood opportunities, robust supply chain linkages;
- provide national vitamin A and deworming distribution programs; and
- acquire and distribute protective equipment for female FLWs, and encourage uptake of the services at local community levels.

Nepal's *Suaahara* is a good example of an operational multisectoral program that simultaneously addresses food insecurity, mother—child nutrition, role of women in households, public safety-net programs, and the effective use of communication technologies. Nepal's program serves as a good template for other nations to replicate.

Indian government's policies toward strengthening social capital through relief programs are noted by Acharya (2020), Ray and Subramanian (2020).[4] Government's nutrition assistance programs provide holistic nutrition to vulnerable segments of population, including women, girls, and children. However, the lockdown prevented key residents from accessing the benefits, affecting pregnant and lactating women.

In response to the pandemic, both central and State governments have announced measures to ensure food security of the poor, such as direct cash transfers through existing schemes and additional grain allotment for 3 months.

The central government has increased fund allocation to the Midday Meal Scheme, and different State governments devised ways to ensure either home delivery of one-time meals (Kerala) or ingredients (e.g., in West Bengal) or cash in lieu of meals (e.g., in Bihar).

Available evidence indicates that the lockdown resulted in a reduction in the quantity and quality of food consumed and could serve to widen intra-household gender disparities in this regard. Support from existing government nutrition schemes such as take-home rations for pregnant women and young children and school meals for school-going children needs to be prioritized as well as strengthened.

In addition to a multilevel response to the issue, continued management of severely malnourished children, effective nutrition messaging that promotes balanced diets, adequate resource allocation for meeting increased demand for nutrition support, and effective delivery of supplementary nutrition are essential steps to preserve gains from years of nutrition programs.

There is now a huge opportunity to learn from the pandemic and build effective and resilient systems of nutrition program delivery. Among policy measures emanating from recent research to enhance social capital and food systems are

- revising farmer loan-repayment schedules;
- expansion of public distribution systems;
- expansion of employment guarantee schemes;
- widening and increasing door-to-door delivery of packaged food at affordable prices;
- buildup of nutrition schemes, food support systems for children and pregnant women investing in robust mobile technologies;
- deregulating markets to manage postharvest agricultural produce;
- investment in storage infrastructure to sustain postharvest output;
- large-scale procurement of crops and grains through assistance programs;
- loosening up restrictions on transport of commodities; and
- scale-up indirect transfers.

---

4. *Pradhan Mantri Garib Kalyan Yojana* (PM-GKY), *PM Kissan Samman Nidhi* (PM-KISSAN), *PM Jan Dhan Yojana* (PM-JDY), and *PM Ujjwala Yojana* (PM-UY) are some of India's direct transfer programs for different sections of the population.

Empirical analysis and main findings

Several studies such as Choudhury et al. (2019), Bird et al. (2019), and Toshiaki (2019) have gathered sufficient evidence across South Asian countries, all demonstrating stresses in child malnutrition. Overall, the consensus is that poor nutritional outcomes are the result of households' socioeconomic characteristics, community-level indicators of climate, infrastructure, and dietary diversity. The research findings make a strong call for public policy that encourages infant feeding and women education and develops healthier food environments.

Importantly, public policy action must also combat food systems that are affected by climate change. Bird et al. (2019) suggest actions similar to those taken by the Leveraging Agriculture for Nutrition in South Asia (LANSA), which link women empowerment, children's health, and nutritional outcomes to agriculture extension programs.

For instance, anemia is a serious public health problem among school-aged boys and girls in rural Ghana. Azupogo et al. (2019) note that the main factors responsible for this crisis are farm diversity score, availability of maize stock crop in the household, household asset, and the agroecological conditions. The authors call for a national campaign toward deworming, water hygiene, and sanitation programs to alleviate this health condition.

The critical question examined in the present analysis is how the various factors (individual, household, and community characteristics) affect child nutritional status. For this purpose, a bivariate simple regression analysis is undertaken to understand the relationship between the dependent variable (child's ZWH) and the different independent variables.

Regression techniques are particularly suitable in the present analysis even when there is correlation among the independent variates (such as in multiple regression framework). For example, even if income and education of the household head are highly correlated, the independent effect of each of these variables on child nutritional status can be properly ascertained in a simple regression analysis framework. In other words, we can determine the independent impact of each of these variables on child nutritional status by looking at the parameter estimates.

Another advantage of regression analysis is that it can be used for either continuous or dichotomous variables. A continuous variable, for example, can be converted into a dichotomous variable using specific threshold values of the continuous variate.

Data description

The data collected in Mzuzu, Salima, and Ngabu Agricultural Development District (ADD) during 1991–92 represent our sample. The data consist of 604 households, out of which 197 households had information on 304 children (their height, age, weight) below the age of 5.

The dependent variable in our analysis is ZWH. Since the data were collected on a monthly basis, it is more likely that short-term nutritional status will indicate vulnerability of children over a short period of time. ZWH is defined as the child's weight in relation to the median height of a reference population of that age. In other words:

ZWH = (Observed weight −median weight)/Standard deviation where both the median weight and the standard deviation (how the different values are distributed about the mean) are taken from the normalized growth curves derived from the NCHS/WHO reference values for the given height.

The individual characteristic of the child is measured by age of the child in months (AGEMNTH). The *household* (socioeconomic) characteristics are as follows:

1. *Education of the household head (EDUCHEAD)*: This is a categorical variable that has a value of 1−7 and measures the education level of the household head in number of years. Higher values of this variable indicate a greater number of years in schooling.
2. *Education of the spouse (EDUCSPOUS)*: This variable is similarly defined as the education of the household head and denotes the education level of the spouse in number of years. If father's and mother's schooling are complementary factors in improving child's nutritional status, we can expect the coefficients of these variables to be positive. However, if father's schooling substitutes mother's schooling (possibly due to time allocation constraints), we can expect one coefficient to be negative and the other positive.
3. *Total expenditure (LNXTOTAL)*: This is the total expenditure of the household which is converted into natural logarithm to ensure normality.
4. *Total food expenditure (LNXFD)*: This is the total expenditure devoted by the household on food and related items and is also converted into natural logarithm to ensure normality.
5. *Household size (HHLDQTY)*: It indicates the number of members within a household (including children and other relatives) and is a continuous variable.
6. *Food security (CALREQ)*: This variable was defined in Chapter 2 and denotes whether the household is able to satisfy at least 80% of the requirement for calorie intake. It is a dichotomous variable assuming two values 0 and 1, respectively, with a value of 1 denoting that the household is food secure.

The *CARE* variables included in the analysis are as follows:

1. *Clinic feeding (CLINFEED)*: This is a dichotomous variable denoting whether the child is fed in a clinic or not.
2. *Breastfeeding (BFEEDNEW)*: This is also a dichotomous variable denoting whether the child is breastfed or not during his or her infanthood.

3. *Number of times child fed during sickness (SICKFEED)*: This is a continuous variable denoting how many times a child is fed during sickness. It is not always the case that feeding a child more during periods of sickness improves nutritional status. Optimal feeding (frequency of feeding) accompanied with introducing complementary feeding may be desirable.

Community characteristics included in the analysis are water and sanitation indicators. We did not include the health source (such as hospital) or distance to a health center as independent variables since the coefficients were not significant and adjusted R^2 was negative. They are as follows:

1. *Water source (WATER)*: This is a dichotomous variable assuming two values—0 and 1, indicating unprotected and protected water sources, respectively.
2. *Drinking distance (DRINKDST)*: This is a categorical variable assuming values from 1 to 4. Higher values denote that the distance to a protected drinking source for the household is higher. For example, the variable attains a value of 4 if the distance to a protected drinking source exceeds 3 km.
3. *Diarrhea:* This is a dichotomous variable denoting whether the child has diarrhea or not.

Incidence of stunting and wasting

We first investigate the prevalence of stunting and wasting for Malawi children. Stunted children are commonly defined as ZHA more than 2 standard deviations below the median. Wasting, on the other hand, is defined as weight for height more than 2 standard deviations below the median. It usually results from acute food shortages and/or disease (UNICEF). These are the two most preferred measures of child nutritional status, since they distinguish between long-run and short-run physiological changes. The "wasting indicator" has the advantage that it can be calculated without knowing the child's age. It is particularly useful in the short run in analyzing the current health status of a population and in evaluating the benefits of intervention programs. A disadvantage of the index is that it classifies children with poor growth in height as normal. Stunting (low H/A) on the other hand measures long-run nutritional status of a population, and it takes into account past nutritional status.

It is quite evident that a significant proportion of children (44.1%)[5] suffer from long-term chronic malnourishment as measured by the ZHA falling below 2 standard deviations below the median (Table 9.1). However, only

---

5. Garrett and Ruel (2003) estimate the percent of children stunted to be around 55% using 1992 as the base year.

Effects of individual, household, and community indicators Chapter | 9 **319**

TABLE 9.1 Percentage of stunted and wasted children.

Percent of stunted children (ZHA ≤ -2)	Percent of wasted children (ZWH ≤ -2)
44.1	9.4

ZHA, height-for-age Z-score; ZWH, weight-for-height Z-score.

9.4% of the children suffer from short-term malnutrition as measured by the ZWH scores. They can also be verified by looking at the mean ZHA and ZWA. The mean ZHA is -1.76, while the mean of ZWH is -0.38. Thus, nutrition programs and interventions that can reduce the incidence of chronic malnourishment in the long run are extremely critical for children in Malawi (especially in the rural areas).

Normality tests and transformation of variables

Linear regression often requires that the outcome (dependent) and the independent variables be normally distributed for the analysis to be valid. The residuals (the difference between the predicted dependent variable and the observed independent variables) must be normally distributed for the *t*-tests to be valid. A common cause of nonnormally distributed residuals is nonnormally distributed dependent and/or predictor variables. Let us start by imposing a normal option to superimpose a normal curve on the dependent variable ZWH.

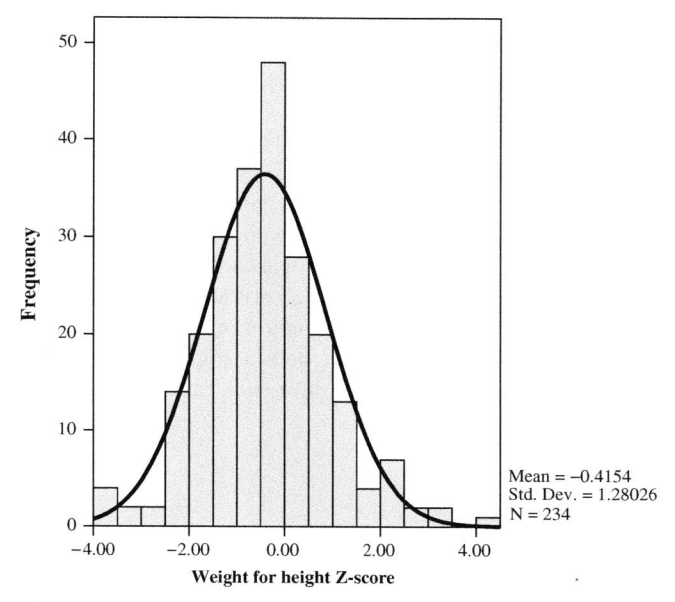

FIGURE 9.1 Testing normality of ZWH. *ZWH*, weight-for-height Z-score.

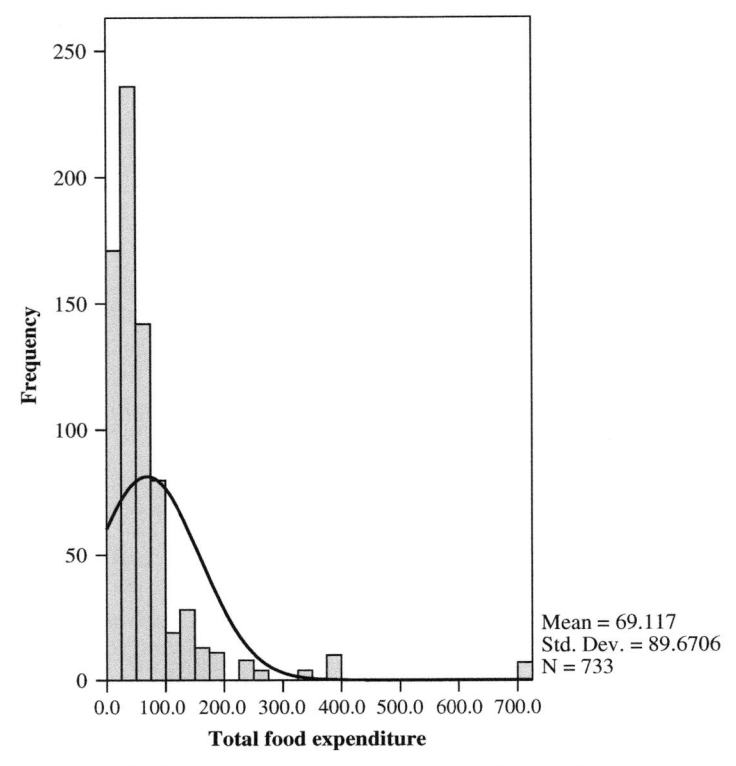

FIGURE 9.2 Testing normality of expenditure on food.

From Fig. 9.1, it is evident that ZWH is approximately normal. Now, let us undertake some normality test for total food expenditure (XFD) to determine if this variable is also normally distributed.

Fig. 9.2 tests for normality of the total food expenditure. As evident from the figure, total food expenditure by the households is skewed to the right and is not normally distributed. Hence, undertaking a regression analysis of ZWH as the outcome variable and total food expenditure as the independent variable will be inappropriate. Thus, we undertake a natural logarithm transformation of XFD and call it LNXFD to determine if this variable is normally distributed or skewed.

As evident from Fig. 9.3, LNXFD is much more normally distributed, and its skewness is near zero. Thus, taking the natural log of total food expenditure seems to have successfully produced a normally distributed variable and is suitable for regression analysis.

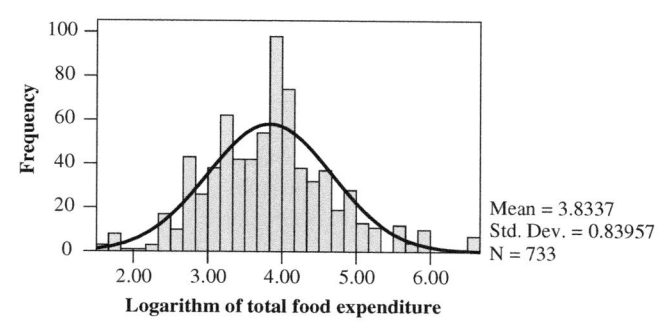

Effects of individual, household, and community indicators **Chapter | 9 321**

FIGURE 9.3 Testing normality of logarithm of expenditure on food.

Regression results

First, it may be worthwhile to introduce a few concepts before discussing the results. In the special case of a single independent variable, a regression equation takes the form (Hamburg and Young, 1994):

$$Y = \alpha + \beta X + \varepsilon \tag{9.6}$$

where Y is the dependent variable, β is the regression coefficient corresponding to the independent term, α is the constant or intercept, and ε is the residual term. Residual is the difference between the observed values and those predicted by the regression equation. The regression coefficient β is the slope of the regression line, and larger the coefficient, the more the dependent variable changes for each unit change of the independent variable. The coefficient β thus reflects the unique contribution of each independent variable on the dependent variable. The intercept α is the estimated value of Y when X has a value of 0.

t-tests are used to assess the significance of β to test the null hypothesis that $\beta = 0$. A common rule of thumb is to drop the variable or variables not significant at the 0.05 level or higher. **R^2**, also called the coefficient of determination, is the percent of the variance in the dependent variable that is explained uniquely by the independent variables. It can also be interpreted as the proportionate reduction in error in estimating the dependent variable after knowing the independent variables. Mathematically, $R^2 = 1 - (RSS/TSS)$, where RSS is the residual sum of squares and TSS denotes total sum of squares.

F-test is used to test the significance of R^2, which is the same as testing the significance of the regression model as a whole. If $prob(F < 0.05)$, the model is considered significantly better than would be expected by chance. One can then reject the null hypothesis of no linear relationship between Y and the independent variables. F is a function of R^2, the number of independent variables, and the number of cases. It is computed with k and $(N - k - 1)$ degrees of freedom, where k is the number of terms in Eq. (11.6) and N denotes the total number of observations. Mathematically:

$$F = (R^2/k)/(1-R^2)/(N-k-1) \qquad (9.7)$$

We next look at the regression estimates (along with the significance level) of the individual and household characteristics on ZWH. For the sake of brevity, we do not report the estimates of constants.

We undertake the regression analysis with ZWH score as the dependent variable and examine the individual impact of the individual child and household characteristics on ZWH. We select only the child-level observations (MBRREL = 3) in our analysis. The main results are reported in Table 9.2. First, we find that age of the child has a positive and significant impact on child nutritional status. This may be because younger children (below 2 years of age) are more likely to be susceptible to diseases than older children. This result is consistent with Alderman and Garcia (1994) that age-specific effects can have differential impact on child nutritional status.

Deprivations during pregnancy and selective mortality (where the less healthy among the poor die early) are possible explanations behind the age-specific effect on child nutritional status. Education variables (education of

TABLE 9.2 Impact of individual and household characteristics on weight-for-height Z-scores.

Variables	Coefficient	t-Statistic	R^2	Prob (F)
AGEMNTH	0.014	2.423	0.025	0.016
EDUCHEAD	−0.085	−1.26	0.007	0.208
EDUCSPOUS	0.100	1.596	0.012	0.112
EDUCHEAD	−0.139	−1.84	0.027	0.053
EDUCSPOUS	0.143	2.14		
LNXTOTAL	0.004	0.04	0.00	0.965
LNXFD	0.116	1.142	0.006	0.255
LNXFD (EDUCHEAD ≥ 4)	0.388	2.62	0.054	0.01
HHLDQTY	0.047	1.38	0.008	0.168
CALREQ	0.375	1.48	0.009	0.14
CLINFEED	0.324	1.55	0.01	0.123
BFEEDNEW	0.546	3.37	0.04	0.001
SICKFEED	−0.221	−2.959	0.037	0.003

the household head and the spouse) can have significant influence on child health. Parental education can increase total family resources, can affect the mother's cost of time, and may also affect the preferences for child health and family size. However, research on the impact of education on child nutritional status is generally ambiguous. Some studies find father's education to be complementary to mother's education, while other studies demonstrate substitution effect. Additionally, the education variable reveals the bargaining strengths of the household head and the spouse, and thus the signs of the coefficients for male and female education can be opposite (Mackinnon, 1995). From the regression analysis, we find that education of the household head (usually the male) is negative and insignificant, while the education of the spouse (usually the female) is positive and insignificant. Thus, we test the null hypothesis whether these two variables are jointly significant. In other words, our null hypothesis is $(\beta_{EDUCHEAD} + \beta_{EDUCSPOUS}) = 0$. This is done using the F-test. We reject the null hypothesis and find that, although the coefficients have opposite signs, the education of the spouse is positive and significant. The result possibly reflects that bargaining strengths of the household members (the head and the spouse) are different and that spousal education has stronger influence on child nutritional status.

Another major household resource is income. Since data on income were not available, the analysis relies on expenditure as a measure of permanent income. We use two measures of expenditure to reflect household resources:

1. Total expenditure
2. Total expenditure on food

Both coefficients were found to be insignificant. Given the overwhelming evidence that income has an independent impact on improving child health, this result is surprising. Thus, we estimate the impact of total food expenditure on ZWH for the cases where household head had some education level above elementary schooling. We find that one standard deviation change in total food expenditure of these households improves ZWH by almost 0.39 points, suggesting that the income effect (through greater spending on food) is coming through education of the household head in increasing the total amount of resources available to the household. The increase in available resources goes in rearing of children and improving their nutritional status.

We find household size and food security (as measured by per capita calorie adult equivalencies) to have no effect on child nutritional status. Thus, food security may not translate necessarily into greater nutritional security. An important point to note from Table 9.2 is that the coefficient of determination (R^2) is very low. This may be due to two main reasons:

1. The nature of the data is cross-sectional (one time period with households from various regions are chosen). In most cross-sectional studies on the

determinants of nutritional status, the coefficient of determination is usually low.

2. In the aforementioned regression framework, the effect of one independent variable on the dependent variable is investigated. In a multivariate setting (with more independent variables or controls), it is likely that the coefficient of determination will improve.

In regard to the CARE variables, we find that breastfeeding during infancy has a positive and significant impact on child nutritional status consistent with the recent empirical literature on the positive role of care practices on child nutritional status (Ruel and Menon, 2002). However, we find that the feeding during sickness (SICKFEED) has a negative and significant impact on ZWH. This is possibly either due to measurement error (this variable was originally constructed on a continuous scale without reference to either the age of the child or the kind of food served during sickness) or the lack of nutrition knowledge of the caregiver in the household.

Next, we examine the impact of the community-level determinants on child nutritional status.

Water and sanitation improvements can have a positive and significant impact on nutritional status of a population by reducing a variety of diseases such as diarrhea, guinea worm, and skin diseases (Table 9.3). With less disease, infants can eat and absorb more food, which improves their nutritional status. Improvements in water and sanitation facilities, however, do not automatically translate into improvements in health and nutritional status. Hygiene education is a prerequisite for health effects translating into greater nutritional status. From the regression analysis, we find that only distance to a protected drinking water source has a significant impact on child nutritional status. This is possibly because as drinking distance to a protected water source increases, the household has less access to cleaner water to feed the children. Lower access to protected water sources increases the likelihood of diseases and affects the child nutritional status adversely.

TABLE 9.3 Impact of community characteristics on weight-for-height Z-scores.

Variables	Coefficient	t-Statistic	R^2	Prob (F)
WATER	0.21	1.256	0.007	0.21
DRINKDST	−0.142	−2.30	0.02	0.02
DIARRHEA	−0.246	−1.04	0.005	0.296

Simple regression in STATA

Example 1

Assume we have the following five values or observations for farm output (we call it Y) and the price of fertilizers which is an input (called as X):

Variables/Obs	1	2	3	4	5
Corn (Y)	5	3	2	1	0.2
Fertilizer price (X)	1	2	3	4	5

Suppose we want to find out the impact of using fertilizer prices on corn output. Does inflation in the fertilizer market decrease corn output? By how much does an additional unit of price increase contribute to output decline? Are these statistically significant values?

To answer the aforementioned question, we have to estimate the coefficients of the following equation:

$$Y = a - bX$$

We first provide the steps involved in calculating the coefficients "a" and "b" in the aforementioned equations. There are many steps and calculations involved in linear regression, and we illustrate these calculations in the following table:

(1)	(2)	(3)	(4)	(5)	(6)
Y	X	$X - \bar{X}$	$Y - \bar{Y}$	$(X - \bar{X})^2$	$(X - \bar{X})(Y - \bar{X})$
5	1	−2	2.76	4	−5.52
3	2	−1	0.76	1	−0.76
2	3	0	−0.24	0	0
1	4	1	−1.24	1	−1.24
0.2	5	2	−2.04	4	−4.08
11.2	15	0	0	10	−11.6

We use the following steps to compute the regression coefficients:

- Compute $\bar{X}$ and $\bar{Y}$ (the means of X and Y).
- Compute $(X - \bar{X})$ and $(Y - \bar{Y})$.
- Compute $(X - \bar{X})^2$ and $(X - \bar{X})(Y - \bar{Y})$ and their sums.

We perform all these steps in the following table:

- Compute $\overline{X}$ and $\overline{Y}$. We begin with entries in columns (1) and (2). We compute the sums of X and Y in the last row. Based on this information, we get $\overline{X} = 2.24$ and $\overline{Y} = 3$.
- Compute $(X - \overline{X})$ and $(Y - \overline{Y})$. For each entry in column (2), we compute the deviation term $(X - \overline{X})$, and place it in column (3). We perform the same operation for each entry in column (1) and compute $(Y - \overline{X})$, and these entries are in column (4).

Compute $\sum (X - \overline{X})^2$. Each entry in column (3) is squared and is reported in column (5). We note that $\sum (X - \overline{X})^2 = 10$. Column (6) = (3) × (4). The total of this column or $\sum (X - \overline{X})(Y - \overline{Y})$ gives us -11.6. We now apply the formula:

$$b = \frac{\text{cov}(X, Y)}{\text{Var}(X)} = \frac{\sum (X - \overline{X}) \sum (Y - \overline{Y})}{\sum (X - \overline{X})^2} = \frac{-11.6}{10} = -1.16$$

Finally, the intercept: $a = \overline{Y} - b\overline{X} = 2.24 - (-1.16)(10) = 5.72$
Consequently, the estimated equation is
Corn output $= 5.72 - 1.16$ fertilizer price
We can derive these coefficients "a" and "b" in STATA using the following input line commands:
.summarize
.regress corn fertilizer
As usual, we begin with the summarize command to get a general idea about the data, and the regress command produces all the relevant coefficients and the regression estimates. The input and the output screenshots from this step are as follows:
We can see from the last two lines of the regression portion of the STATA output that the estimated equation for our data:

Corn output $= 5.72 - 1.16$ *fertilizer price*

Thus, a one-unit increase in the price of fertilizers will decrease output by 1.16 units. Furthermore, the t-value of the coefficient "b" is -9.02 with the P-value equal to 0.003. This implies that the estimate is significant at the 1% level, and hence we can reject the null hypothesis that there is no significant influence of price on output.

Example 2

In the following, we present a table with information on corn output together with two input levels: fertilizers and pesticides. Can we use this information in STATA and the regress command to answer the following questions?

- Will corn output increase if we increase the level of fertilizer usage?
- Will corn output increase if we increase the level of pesticide usage?

Obs	Corn	Fertilizer	Pesticide
1	40	6	4
2	44	10	4
3	46	12	5
4	48	14	7
5	52	16	9
6	58	18	12
7	60	22	14
8	68	24	20
9	74	26	21
10	80	32	24

We use the regress command in STATA to generate the following output: The estimated equation in this case is *Corn output* $= 27.12 + 1.65$ *fertilizer*.

In other words, as we increase the use of fertilizers, corn output increases. Furthermore, for a one-unit increase in fertilizer usage, corn output grows by 1.65 units. Finally, we note that all the t-values are significant at 99% confidence and that 97% of the variability in corn output is due to the variability in fertilizer usage (note the value of R-squared $= 0.97$). We proceed to investigate the impact of pesticide use on corn output, through the same regress command:

In this case, the estimated regression is *Corn output* $= 35.57 + 1.78$ *pesticide*.

We note that pesticide usage and corn output are positively related. A one-unit increase in the use of pesticides increases corn output by 1.78 units. All coefficients are significant at the 1% level, as shown by the t-values and the P-values. The R-squared value of 0.98 indicates that a large portion, or 98% of the variation in output, is accounted for by the variation in pesticide usage.

Conclusion

Improved food security does not necessarily translate to improved nutritional status of children (Haddad et al., 1996). Nutritional status of an individual can be effectively influenced not only by food factors but also through good care practices, improved healthcare, and clean water and sanitation. There is little doubt that household income and other resources play a crucial role in determining both child health and nutrition. Equally important are care factors, such as breastfeeding and complementary feeding practices in improving child nutritional status. Countries that have low rates of antenatal care and low feeding practices can expect substantial returns if nutrition knowledge to women is imparted.

Various authors have also emphasized the role of mothers' education (relative to fathers' education) on child nutrition. However, it is not clear from the existing evidence the mechanisms through which parental education leads to higher knowledge about health and nutrition. It is more likely that the impact of education comes from raising income and these additional resources are invested in improving child nutrition.

Another important determinant of child health and nutrition is health infrastructure and the availability of clean water and sanitation. By preventing infections and diarrhea, a cleaner water and sanitation environment leads to better nutritional outcome.

From our simple bivariate regression framework, it is evident that individual, household (socioeconomic), and community characteristics all significantly matter in influencing child nutritional outcomes. Age-specific effects matter in influencing child nutritional outcome, but so does education of the mother. Increase in food expenditures improves child nutritional outcomes, but only for education level of the household head above a certain threshold level. Thus, income effect is possibly coming through the education level of the household head in generating additional resources. We find care activities to significantly influence child nutritional outcomes, suggesting that it may be as important as income or education. Additionally, we also find community characteristics, such as distance to a protected drinking water source, to have significant impact on child nutritional status.

However, a multivariate regression model is more appropriate in examining how the various determinants affect child nutritional outcome or infant mortality, and this will be done in a later chapter.

Simple regression in R

We employ a simple data set to understand how regression operates in *R*. Assume we have the following five observations for farm output (Y) and the price of fertilizers, which is an input called X. To begin with, we draw a scatter graph by using the *plot* command to identify the relationship between the farm output and the price of fertilizers. From the graph, we can identify that X and Y have a negative relationship.

Effects of individual, household, and community indicators Chapter | 9 **329**

```
> summary(data_chapter9)
      Obs              Y               X
 Min.   :1      Min.   :0.20     Min.   :1
 1st Qu.:2      1st Qu.:1.00     1st Qu.:2
 Median :3      Median :2.00     Median :3
 Mean   :3      Mean   :2.24     Mean   :3
 3rd Qu.:4      3rd Qu.:3.00     3rd Qu.:4
 Max.   :5      Max.   :5.00     Max.   :5
```

```
> plot(X,Y,main="Scatterplot")
```

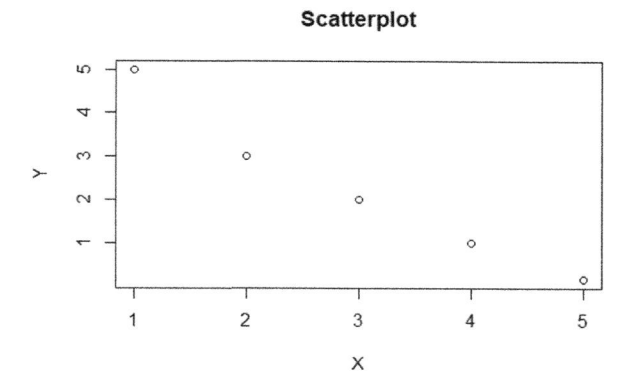

Scatterplot

Now we will derive these coefficients in *R* using the following command. Note that the dependent variable should place first. The *lm* command operates a regression in *R*.

```
> lm(Y~X)

Call:
lm(formula = Y ~ X)

Coefficients:
(Intercept)            X
       5.72        -1.16
```

To look at in detail about the regression, we can save the function and import in using *summary* as follows. We can examine general and specific information such as *F*-statistics, *R*-squared, adjusted *R* squared, and *P*-value. We can see that the estimated equation for our data:

$$Corn\ output = 5.72 - 1.16\ fertilizer\ price$$

```
> regression <-lm(Y~X)
> summary(regression)

Call:
lm(formula = Y ~ X)

Residuals:
    1     2     3     4     5
 0.44 -0.40 -0.24 -0.08  0.28

Coefficients:
            Estimate Std. Error t value Pr(>|t|)
(Intercept)   5.7200     0.4265  13.413 0.000896 ***
X            -1.1600     0.1286  -9.021 0.002876 **
---
Signif. codes:  0 '***' 0.001 '**' 0.01 '*' 0.05 '.' 0.1 ' ' 1

Residual standard error: 0.4066 on 3 degrees of freedom
Multiple R-squared:  0.9644,    Adjusted R-squared:  0.9526

F-statistic: 81.39 on 1 and 3 DF,  p-value: 0.002876
```

Thus, a one-unit increase in the price of fertilizers will decrease output by 1.16 units. Furthermore, the *t*-value of the coefficient is −9.02, with the *P*-value equal to 0.003. This implies that the estimate is significant at the 1% level.

Exercises

1. Based on the existing literature discussed in this chapter, discuss whether parental education and nutrition knowledge acts as substitute or complementary factors in improving child nutritional status. Justify your answer with appropriate reasoning.
2. What are the causes of child malnutrition? How can the causes of child malnutrition be addressed?
3. How does maternal education correlate with child-care practices in affecting child nutritional status? Discuss critically from the existing studies in this chapter.
4. Carefully examine the role of age specificity in determining child nutrition after studying the paper by Sahn and Alderman.
5. a. According to Haines and Steckel (2000), socioeconomic factors provide important clues to the sources of these regional health differences in the United States during the early part of the 20th century. To test this, they estimate many regression equations with the mortality index as the dependent variable. In an equation with only the socioeconomic variables, they find an *R*-squared value to be 0.50. When they included only

Effects of individual, household, and community indicators Chapter | 9 **331**

the regional dummy variables, the R-squared value was 0.39. When they included all the variables, the R-squared value was 0.505. What can we conclude from these three results?

b. They repeated these regressions for 1910 data and found that the regression with only the socioeconomic variables produced an R-squared equal to 0.4 and was 0.17 in the equation with only the regional dummy variables. When all the variables were included, the R-squared value was 0.8. Do these results from the 1910 data support the same conclusions from the pre-1910 data in part a?

c. The results for regressions that use height as a dependent variable were quite different. Socioeconomic variables alone explained 0.659 of the variation in heights for states in 1900, while the regional dummies accounted for 0.69. Both sets of variables together gave an R-squared of 0.806. What does this suggest about that regional clustering of anthropometric measures of health, such as height and BMI, in the early part of the 20th century? What about childhood mortality?

STATA workout

The following data provide us 12 observations on ZWH, EDUCHEAD, CALREQ, and SICKFEED, where these variables are defined as in the previous section in this chapter.

Obs	ZWH	EDUCHEAD	CALREQ	SICKFEED
1	37.5	1.00	0.833	0.936
2	37.5	1.00	0.875	0.944
3	40	0.92	0.917	0.952
4	45	0.75	1.000	0.944
5	37.5	1.00	0.792	0.952
6	30	1.08	0.625	0.944
7	42.5	0.75	0.833	0.960
8	42.5	0.83	0.958	0.976
9	42.5	0.83	0.792	0.984
10	40	0.92	0.750	0.984
11	45	0.75	0.917	0.992
12	50	0.67	0.958	1.000

Use these data and the STATA command regress to answer the following questions:

Is ZWH positively related to EDUCHEAD and is it significant?
What is the impact of CALREQ on ZWH?

Is SICKFEED a significant variable to explain the variation in ZWH? The input lines in STATA are as follows:

- regress ZWH EDUCHEAD
- estimates store m1, title(Model 1)
- regress ZWH CALREQ
- estimates store m2, title(Model 2)
- regress ZWH SICKFEED
- estimates store m3, title(Model 3)
- estout m1 m2 m3, cells(b(star fmt(3)) se(par fmt(2))) legend label varlabels (_cons Constant) stats(r2, fmt(2) label(R-sq))

We estimate three simple linear regressions: Model 1, Model 2, and Model 3. The estimates are stored in three different output data sets: m1, m2 and m3. The last line uses the **estout** command to recall the output data sets, with the options:

- stars are placed for statistical significance;
- parentheses are placed around the standard errors, reported at 2 decimal places;
- the label, "_cons" is replaced with the term "Constant"; and
- a line is added in the bottom to report the R-square for each model.

The relevant input lines and the corresponding output from STATA are as follows:

. regress ZWH EDUCHEAD

Source	SS	df	MS			
Model	250.568188	1	250.568188	Number of obs	=	12
Residual	28.5984784	10	2.85984784	F(1, 10)	=	87.62
				Prob > F	=	0.0000
				R-squared	=	0.8976
				Adj R-squared	=	0.8873
Total	279.166667	11	25.3787879	Root MSE	=	1.6911

ZWH	Coef.	Std. Err.	t	P>|t|	[95% Conf. Interval]
EDUCHEAD	-36.71329	3.922218	-9.36	0.000	-45.45253 -27.97404
_cons	72.95746	3.466488	21.05	0.000	65.23364 80.68127

. estimates store m1, title(Model 1)
.
. regress ZWH CALREQ

Source	SS	df	MS		
Model	170.389032	1	170.389032	Number of obs =	12
Residual	108.777634	10	10.8777634	F(1, 10) =	15.66
				Prob > F =	0.0027
				R-squared =	0.6103
				Adj R-squared =	0.5714
Total	279.166667	11	25.3787879	Root MSE =	3.2981

ZWH	Coef.	Std. Err.	t	P>|t|	[95% Conf. Interval]
CALREQ	37.19608	9.398231	3.96	0.003	16.25551	58.13664
_cons	9.061684	8.083919	1.12	0.289	-8.950409	27.07378

. estimates store m2, title(Model 2)

. regress ZWH SICKFEED

Source	SS	df	MS		
Model	126.581258	1	126.581258	Number of obs =	12
Residual	152.585409	10	15.2585409	F(1, 10) =	8.30
				Prob > F =	0.0164
				R-squared =	0.4534
				Adj R-squared =	0.3988
Total	279.166667	11	25.3787879	Root MSE =	3.9062

ZWH	Coef.	Std. Err.	t	P>|t|	[95% Conf. Interval]
SICKFEED	154.3674	53.59539	2.88	0.016	34.94945	273.7854
_cons	-107.9769	51.67826	-2.09	0.063	-223.1232	7.169481

```
.  estimates store m3, title(Model 3)

.  estout m1 m2 m3, cells(b(star fmt(3)) se(par fmt(2))) legend label varlabels(_cons Constant) stats
>  (r2, fmt(2) label(R-sq))
```

	Model 1 b/se	Model 2 b/se	Model 3 b/se
EDUCHEAD	-36.713***		
	(3.92)		
CALREQ		37.196**	
		(9.40)	
SICKFEED			154.367*
			(53.60)
Constant	72.957***	9.062	-107.977
	(3.47)	(8.08)	(51.68)
R-sq	0.90	0.61	0.45

* $p<0.05$, ** $p<0.01$, *** $p<0.001$

The last table summarizes the results of the three regressions. We see that ZWH is negatively related to EDUCHEAD and positively related to CALREQ and SICKFEED. All the three variables are significant at 99% level. Furthermore, Model 1 has the highest R-square value of 0.90, followed by Model 2. Hence, we have a clear unambiguous result that EDUCHEAD and CALREQ are significant drivers of ZWH.

Chapter 10

Maternal education and community characteristics as indicators of nutritional status of children—application of multivariate regression

Chapter outline

Introduction	336	Step 3: hypotheses testing	357
Selected studies on the role of maternal education and community characteristics on child nutritional status	337	Tests about the equation	357
		Tests about individual coefficients	358
Community characteristics and Children's nutrition in the United States	342	Part and partial correlation coefficients	361
		Step 4: Checking for violations of regression assumptions	362
Community characteristics and child nutrition in Kenya	345	Detecting influential observations	362
Financial crisis and child nutrition in East Asia	345	Checking normality of the errors	363
Double burden within mother–child pairs: Asian case	346	Checking for homogeneity of variance of the residuals	364
Empirical analysis	347	Multiple regression in STATA	364
Data description and methodology	350	STATA Output	368
Descriptive summary of independent variables	352	Example 1	368
		Example 2	369
Main results	353	**Conclusions**	**371**
Step 1: Estimating the coefficients of the model	353	Multiple regression in R	372
Step 2: Examining how good the model predicts	355	**Exercises**	**376**

We are not concerned with the very poor. They are unthinkable, and only to be approached by the statistician or the poet.

—E.M. Forster.

Introduction

The main purpose of this chapter is to demonstrate how to predict the nutritional outcomes when two or more independent variables are at work. This chapter discusses various topics including the derivation of the least squares estimates, estimation of coefficients of the multiple regression models, examination of power of the model prediction, hypothesis testing of individual coefficient, checking for the violation of regression assumptions, and interpretation of regression results. The literature on the welfare of children in developing countries emphasizes the importance of having literate parents and, in particular, of having a literate mother. The nutritional status of children is enhanced by having better educated parents, especially the mother (Borooah, 2002; Pongou et al., 2006). There are both immediate causes of poor health and nutrition, such as lack of access to food, low utilization of health facilities or the poor quality of these facilities, and the underlying factors that cause it, such as family income and educational status. The mechanisms through which maternal education affects child health are quite complex (whether through higher wages, better health knowledge, greater child care, or improving the health environment in reducing mortality rates) and thus requires a thorough analysis.

Schultz (1984) argues that mothers' education can influence child health through the following pathways:

1. Better educated mothers can be more effective in producing child health for a given amount and mix of health goods (for example, through improved child care for a given amount of health).
2. Schooling can reduce parents' preferences for fewer but healthier children.
3. Additional schooling raises family incomes either via increased wages or higher productivity, which, in turn, can improve child's health status.
4. Moreover, education raises the opportunity costs of time that tends to decrease the time for child care.

The purpose of this chapter is to examine the pathways through which maternal education improves child health and to understand the impact of community characteristics (such as presence of hospitals and better water and sanitation conditions) on weight for age and height for age using a multivariate regression model that controls for other individual characteristics such as child's age. As Schultz (1984) points out, community characteristics can play three major roles in affecting child health:

1. They can reduce the price of health inputs, directly through subsidization of the goods or services, or by increasing access to them indirectly, thereby reducing the travel costs.
2. They may provide information of how to produce health more efficiently, such as including information on new inputs, or on efficient practices with traditional inputs such as breastfeeding, bringing the child to a clinic, and so on.
3. They can alter the health environment, such as control of malaria and eradication of smallpox.

The issue is important since it addresses policy questions such as follows: Which individual characteristics—including occupation, income, and education—play bigger roles? Why do the community factors act differently for different socioeconomic groups? The answers to the aforementioned questions are important in setting up effective health interventions to improve child health. For example, living in a well-developed community can benefit the disadvantaged household (less educated, lower income) more and function as a substitute to those unfavorable individual characteristics.

This chapter is organized as follows. The next section provides a brief review of the main studies that examine the role of maternal education and community characteristics on child nutritional outcomes. We examine the importance of community, social networks, family status, and other indicators on nutrition adequacy in the United States, Kenya, and East Asia and Asia. The third section presents the empirical analysis of how individual/household and community characteristics affect child weight for age using a multivariate regression framework and provide illustrative examples and implementation via STATA, while the final section concludes.

Selected studies on the role of maternal education and community characteristics on child nutritional status

One of the earlier studies by Barrera (1990) addressed why and how mothers' schooling affected child health and whether this impact varied across child age groups. Additionally, the study also examined the pattern of interaction of maternal education with various public health programs in improving child nutrition.

The data came from the Philippines multipurpose survey of 1978 with a supplementary survey during 1981. The sample consisted of 3821 children below the age of 15 from 1383 households. Child health was measured by the height-for-age Z-scores. The education variable was measured by years of schooling. Water, sanitation, and healthcare variables were also included in the model as proxies for community infrastructure. Household income was measured by the sources of income of the family except mother's cash and noncash earnings. Mother's height was used as a proxy for her genetic traits and health endowment. A reduced form equation was estimated where the health of the child was modeled as a function of maternal education, community characteristics, and other controls.

Children of more educated mothers had better height-for-age Z-scores. The impact of maternal schooling, however, varied across the age groups, with preschoolers showing the greatest sensitivity. This is a robust finding as evident from the relatively large coefficients of maternal education for the youngest children. Children of less educated mothers derived greater health benefits from a cleaner environment and water connections, compared with children of more educated mothers. On the other hand, access to healthcare facilities and toilet connections benefited children of more educated mothers than for mothers with less schooling. These patterns of interactions between maternal education and public health programs suggest that the likely channels through which maternal education might affect child health are by affecting the productivity of health inputs and by lowering the cost of information.

Benefo and Schultz (1996) examined the impact of individual, household, and community characteristics that affect fertility in two West African countries—Cote d'Ivoire and Ghana—and analyzed the relationship between child mortality and fertility. The relationship between child mortality and fertility is difficult to conceptualize, since both variables may affect each other, may be modified by other factors, and are measured with errors. Thus, both least squares and instrumental variable estimation were used.

The data were obtained from the LSMS surveys conducted by the national statistical agencies in the World Bank during the periods 1985—87 for Cote d'Ivoire and two rounds of surveys for Ghana during the period 1987—88 and 1988—89. The sample consisted of 1943 women in Cote d'Ivoire and 2237 women in Ghana.

The main results of the study can be summarized as follows. First, in examining the determinants of mortality for Ghana, economic resources of households, maternal education, and access to markets were all associated with child mortality. Sanitation infrastructure increased child survival but only for children of less educated mothers. There were substitution effects between education of mother and water supplies, implying that more educated mothers were more efficient in reducing health risks. In Cote d'Ivoire, households living at a larger distance from the health clinics experienced higher mortality rates. Since children benefit more from the public health system in Cote d'Ivoire than in Ghana, household assets were not a significant predictor of child mortality. Child survival increased with greater maternal education in both countries. Second, in examining the determinants of fertility, the study found that women's education beyond the primary school level was strongly associated with declines in fertility rates in both countries. However, wealth and socioeconomic status had an opposite impact in both countries. In Cote d'Ivoire, assets and maternal health were positively related to fertility, whereas in Ghana, these variables were negatively related to fertility.[1]

---

1. Also see Malapit and Quisumbing (2015) for the positive effect of women's empowerment on children's nutrition in Ghana.

An indirect policy implication of the study was that a more egalitarian distribution of social services would hasten the decline in child mortality and fertility, if women's education increased more rapidly in the rural areas and rural sanitation and health problems were effectively addressed.

Adams et al. (2018) indicate the impact of unintended spillover effects of a specific nutrition-based intervention targeted to mothers and infants in Ghana. The main effects of the intervention were positive, in terms of expenditures on food, nutrition, nonfood and services, and labor income among fathers. Interestingly, there were positive spillover effects among the nontargeted children who registered higher height-for-age Z-scores, particularly within the recipient households, were the mothers were relatively taller. This study along with others detailed in Chapter 9 indicate that well-designed interventions may have unintended behavioral implications, which must be taken into account, to evaluate the full impact of the said interventions. Consequently, adequate ex ante data from the targeted and the nontargeted populations must be collected for adequate analytical understanding of the total effects of the said interventions.

Devi and Geervani (1994) examined the role of child characteristics, parental characteristics, and socioeconomic and environmental factors on child health. The study was conducted in four villages in the Medak district of Andhra Pradesh in India, using a 24 h recall method. One hundred and ninety-seven children below 4 years of age were selected from low-income households. Low-income households were selected based on whether the household had at least one preschool child.

From the study, differences in nutritional status of preschool children could be attributed to two main causes. First, children who were vulnerable to more infections during early infancy had higher morbidity and became malnourished. Second, children who did not eat adequate food (possibly because of lack of appetite) were highly vulnerable to diarrheal diseases and consequently became malnourished. Thus, the effects of chronic calorie deficiency and infection were the basic causes of poor nutritional status.

China, South Africa, and Central America: Zhao and Bishai (2004) explored how community factors interacted with various individual characteristics that affected child health. In other words, the study examined which community factors were substitutes or complements to maternal education in improving child nutritional status. The novelty of the study lies in comprehensively specifying community factors such as measures of infrastructure, market activities, water and sanitation conditions, and healthcare services that can affect child health. Additionally, some political and policy features related to child health were also considered as determinants of child health. Finally, the study used analysis from China, where such a study had not been undertaken before.

The data set used in the analysis came from the China Health and Nutrition Survey (CHNS). The survey covered nine provinces that varied substantially in geography, economic development, and health indicators. A cross-sectional analysis using data of children below 10 years of age for the 1993 survey was used in the analysis. A sample of 1848 children was chosen based on anthropometric information (children's height for age), which was then merged with the household and community data.

The main results of the study were as follows:

1. Increased access to local infrastructure and basic health services served as substitutes to the household educational levels mainly for the disadvantaged households in reducing child stunting.
2. Local services that required a certain level of knowledge and skill were complementary to the household level variables in affecting child stunting.
3. Family planning policy was an extremely important community component in China, and the results showed that there was a strong association between stunting and family planning policy.

David et al. (2004) examined the private and public determinants of child health in the context of two Central American countries—Nicaragua and Honduras. Both these countries executed major social programs at the beginning of the 1990s. Honduras executed two major programs, namely, the Honduran Social Investment Fund (FHIS) to provide poor communities with basic social infrastructure, such as schools, health centers, water supply and excreta disposal systems, and the PRAF (Family Entitlement Program) that focused more directly on helping poor families, such as providing cash transfers. Nicaragua also initiated similar programs executed by the FISE (Social and Economic Investment Fund) and MIFAMILIA (Ministry of Family). The rationale for these programs was that social investment is currently low in these countries, and thus, the social returns must be high. However, there was no indication in general of the relative importance of community factors (such as social infrastructure creation) and cash transfer programs on child health, which was the motivation behind this study.

Individual and household level factors (and not the community level factors) were the main determinants of child nutritional status. Maternal endowment, as measured by maternal height, was the key variable in the Honduras sample. Household income came second in relative importance. While educational levels were not significantly related to any of the nutritional status indicators in Nicaragua, in Honduras the primary education of the woman of the household had an impact on height-for-age and weight-for-age indicators.

The impact of interaction effects of household income and educational levels was also investigated. In Nicaragua, education of the household head had an independent impact, but only for the higher income bracket. In general, women's education was not significant. In the Honduran case, the educational variable acquired different levels of significance for different quintiles, without any systematic pattern of increasing or decreasing influence across the range of expenditures.

Only in the Honduras sample were few of the community-level variables individually significant. The proportion of households with access to tap water and the level of agricultural wages only were significant. In both countries, direct health/nutrition intervention did not have any impact on children's nutritional status. This could be because a food donation to mothers around US $4 monthly in Nicaragua could be too low to generate a significant change in the feeding patterns.

What is the overall evidence of maternal education and community characteristics on child health outcomes? The answer is mixed. In the case of the Philippines, Barrera found that less educated mothers obtained greater health benefits from water supply and a cleaner environment suggesting complementarity, while Benefo and Schultz (1996) found that better educated mothers obtained greater benefits from child health through better water supply suggesting substitution effects for Ghana. While Barrera found that more educated mothers could take advantage of sanitation facilities thus improving child health in the Philippines, suggesting substitution effect, Benefo and Schultz found that improvement in sanitation facilities improved child survival only for less educated mothers in Ghana, suggesting complementarities. Yet other recent studies, such as David et al. (2004), demonstrated that community factors were neither complementary nor substitutes with maternal education in affecting child nutritional status.

In light of the aforementioned, we investigate the role of maternal education and community characteristics (such as water and sanitation facilities and health infrastructure) on child nutritional status for the case of Malawi. The purpose of this chapter is to understand if maternal education acts as substitutes or complements to the community infrastructure variables in a multivariate regression framework.

Is catch-up possible? Handa and Peterman (2016) note that over one-third of all children under the age of 5 in developing countries suffer from nutritional deficiency, characterized by underweight (27%), stunting (31%), and wasting 910%). Do corrective public policies help in restoring children to get back on track and reach their full potential growth? The possibility of children getting back on track with their targeted growth curve, through a phase of rapid growth, is referred to as catch-up. If catch-up is possible, then there is an opportunity to establish corrective public policies to encourage the process. Using three different data sets from China, South Africa, and Nicaragua, Handa and Peterman (2016) show that catch-up growth in height among children is possible, especially when the child is less than 2 years old.[2]

---

2. See Choudhury et al. (2019), Bird et al. (2019), and Toshiaki (2019) for the importance of community and socioeconomic characteristics that affect malnutrition across South Asia.

In this context, Young et al. (2018) examine the importance of preconception maternal nutritional status (PMNS) on child growth for the first 1000 days, using an intervention through preconception micronutrient supplementation in Vietnam. If women's preconception height and weight are major drivers of offspring growth, then public policies must actively incorporate this preconception period to inform policy makers about appropriate interventions. Indeed, in Vietnam, Young et al. (2018) demonstrate that women whose preconception height is less than 150 cm or a preconception weight less than 43 kg were at increased risk of having a stunted child at 2 years. Hence, interventions must focus on early maternal nutrition of girls during the first 1000 days, and in adulthood, pre- and postconception, to influence childhood growth and development, and ultimately establish their catch-up.

Community characteristics and Children's nutrition in the United States

In the United States, there is an enigma surrounding the Supplemental Nutrition Assistance Program (SNAP), or the Food Stamp Program. The cause for concern is the positive associations between SNAP recipients and various undesirable health outcomes such as food insecurity. In other words, the choice of participation may depend upon the level of food security and health. However, economists have been unable to disentangle these effects because of endogenous selection into participation and extensive systematic underreporting of participation status.

Using data from the National Health and Nutrition Examination Survey (NHANES), Kreider et al. (2012) address these two issues. With interesting econometric techniques, they uncover the selection and classification error problems and find that SNAP actually helps improve child health. Furthermore, they demonstrate that the common perceptions about SNAP and poor health are misleading.

The impact of SNAP on immigrant health is also equally of concern in the United States. For example, Curtis (2012) notes that immigrants' children live in the fastest-growing segment of the under-18 population in the United States. Curtis (2012) also notes that immigrant family children are much more likely to experience economic deprivation than native family children. Furthermore, immigrant families eligible for federal and state income, work supports, and SNAP access them at significantly lower rates than do native families.

Section highlights: Latino immigrants, rural america, and food insecurity

Are all Latino families equally capable of meeting their food requirements? The answer given by Sano et al. (2011) is clearly no. These researchers point out that, firstly, the rate of food insecurity among Latino households is nearly double the national rate. Secondly, food insecurity among this group is correlated with poor quality housing, along with cultural and language barriers. Thirdly, Latino immigrant families tend to live in spatial isolation from non-Latinos, and hence, their access to information and local support services is limited.

Sano et al. (2011) examine why some Latino immigrant families have been successful in meeting their food needs, while others have not. The researchers use ecological theory of human development and identify four important factors that determine a family's capacity to meet its requirements: (1) microsystem concerning individuals and families, (2) mesosystem that includes social networks, (3) exosystem or the community, and (4) macrosystem or the larger cultural context.

The case study methodology by the researchers is a very interesting approach that demonstrates the interaction between ecological variables and food insecurity. The research findings support previous results that relate food security issues to families' human capital, employment opportunity, local economy, acceptance of public assistance, and immigration policies. However, the ecological model also looks at the complex web of relationships and highlights the difficulty in policy formulation. For instance, the ecological model indicates that even those families in the consistently food secure group are only one step away from food insecurity.

The study by Sano et al. (2011) indicates that Latino immigrants are a vital part of rural America and that to prevent food insecurity, we must consider multifaceted policies covering early childhood development, healthcare, and immigration reform. Policies must aim to strengthen the immigrants' basic literacy and life skills including knowledge of basic nutrition, preventative health practices, family planning, and language skills.

The researchers also point out that the Latino immigrant families must be integrated with the local business and government leaders, education, health, and social service family professionals, and to other Latino immigrant families. In other words, the ecological model stresses access to community resources as a key driver for the success of this group.

Proactive efforts to reach out to Latino immigrant families are necessary, as their immigration status significantly influences program participation. Children of Latino immigrant parents, especially those whose parents are undocumented, face a higher risk of not having basic needs met. Food security, housing, and access to healthcare are closely linked to positive child development. In the future, children of immigrant parents will make up a large portion of the future US workforce. Hence, it is critical to assist immigrant families to ensure the healthy development of their children. Finally, policy makers must develop a reasonable immigration policy for those immigrants who lack documentation.

Alsan and Yang (2018) demonstrate how the silent fear factor of deportation that spreads among the Hispanic population has substantially reduced the take-up of safety net programs, causing massive food insecurity within this network.

Curtis (2012) demonstrates that broad-based state legislation restricting immigrant rights reduces program participation among immigrant families. Policy recommendations include stronger outreach efforts by state program administrators to promote SNAP among immigrant groups. Policy calls for the administrators to make it easier for working parents to enroll in the program, so as to encourage greater participation in this important social safety net program.

Community characteristics often play an invisible role in health status. For instance, is it possible that the SNAP program in the United States can have different effects, based on the nature of the neighborhood or community the respondent is located in? This is an important question, because, as noted in previous chapters, the number of individuals who are obese or overweight in the low-income US population is disproportionately large and has raised interest in the influence of neighborhood conditions and public assistance programs on weight and health.

Vartanian and Houser (2012), for instance, combine neighborhood effects and program participation effects in a single study using the 1968–2005 Panel Study of Income Dynamics (PSID) data. Vartanian and Houser (2012) examines the long-term effects of SNAP neighborhood conditions and the interaction of these two, on adult body mass index (BMI).

Vartanian and Houser's (2012) findings provide key insights about SNAP and community characteristics. Firstly, relative to children in other low-income families, children in SNAP-recipient households have higher average adult BMI values. Secondly, the effects of childhood SNAP usage are sensitive to both residential neighborhood and age at receipt. For example, for those children growing up in advantaged neighborhoods, projected adult BMI is higher among those in SNAP-recipient households, than for children in low-income, nonrecipient households. Interestingly, for those growing up in less-advantaged areas, adult BMI differences between children in SNAP-recipient and those in low-income, nonrecipient households are small.

Schenck-Fontaine et al. (2017) note an interesting aspect of community characteristic in the United States, namely the prevalence of informal safety nets among SNAP recipients in Durham, North Carolina. SNAP recipients have to adjust to the economic instability that occurs outside the SNAP benefit cycle. Families buffer their insecurity by aligning formal with informal resources. Informal lending practices, food, and material support among similarly situated neighbors, friends, and community organizations are the main mechanisms frequently used by SNAP recipients to cope against food insecurity. The study shows that there is a need for policy that effectively reaches out to these communities that suffer from economic segregation.

Community characteristics and child nutrition in Kenya

Kabubo-Mariara et al. (2009) use a pooled sample of the 1998 and 2003 Demographic and Health Survey from Kenya to analyze the determinants of children's nutritional status. The study evaluates the impact of child, parental, household, and community characteristics on children's height and on the probability of stunting. The study also provides us with important findings:

1. Boys suffer more malnutrition than girls.
2. Children of multiple births are more likely to be malnourished than singletons.
3. Maternal education is a more important determinant of children's nutritional status than paternal education.
4. Household assets are also important determinants of children's nutritional status, but nutrition improves at a decreasing rate with assets.
5. Public health services and modern contraceptives are also important determinants of child nutritional status.

Kabubo-Mariara et al. (2009) develop many policy simulations to check the robustness of their results. Policy proposals in Kenya should stress the role of parental, household, and community characteristics in reducing long-term malnutrition in Kenya. The researchers stress that, if Kenya is to achieve her strategic health objectives and millennium development target of reducing the prevalence of malnutrition and poverty, then the promotion of postsecondary education for women and provision of basic preventive healthcare are critical concerns that have to be dealt with.

Financial crisis and child nutrition in East Asia

Is it possible that a financial crisis can worsen the status of children's nutritional levels in an economy? To answer this interesting and important question, Bhutta et al. (2009) look at the 1997 financial crisis in East Asia and link the crisis to potential risks to food, fuel, economic, child health, and nutrition in the region.

Using information available on the 1997 crisis, the researchers evaluate the effects of the crisis on nutrition status, reportable diseases, immunization status, and child mortality. Specifically, they find that the 1997 financial crisis has increased maternal anemia rates by 10%−20%, prevalence of low birth weight by 5%−10%, childhood stunting by 3%−7%, wasting by 8%−16%, and under-5 child mortality in severely affected countries from 3% to 11%. A major policy recommendation is to undertake a range of low-cost and high-impact interventions, including the establishment and improvement of primary care settings and access to nutrition.

Double burden within mother–child pairs: Asian case

In the previous chapter, we cited a very interesting study by Estefania (2010) who investigated the "double burden" of over- and undernutrition. In recent years, with economic development and urbanization, there is a greater incidence of under- and overnutrition within the same household. Jehn and Brewis (2009), for instance, use Demographic and Health Survey data sets from 18 lower- and middle-income countries to demonstrate this paradox. Specifically, they look for maternal overweight with child underweight or stunting in mother–child pairs.

Jehn and Brewis (2009) find several factors that are significantly associated with discordant mother–child pairs, including working in subsistence agriculture, low levels of maternal education, more siblings in the household, and relative household poverty. Incidentally, many of these factors are also highly correlated with poor nutritional status in mother–child pairs. Based on their analyses, they conclude that the paradoxical weight status between mothers and children can be best understood as a consequence of rapid secular increases in maternal weight, rather than a distinct nutritional condition with a discrete etiology.

> **Section highlights: Successful Nutritional Policy in Tamil Nadu, India**
>
> The largest number of stunted children in the world is in India, where the incidence of stunting is worse than sub-Saharan Africa, and this malady is paradoxical, given India's significant economic growth and advantages. An important aspect of child malnutrition in India is the large variation across states.
>
> Cavatorta et al. (2015) examine the heterogeneity across Indian states in nutritional outcomes. The researchers identify one particular state, namely Tamil Nadu, which has produced the best outcomes in child nutrition through appropriate nutrition-based interventions.
>
> It is reasonable to expect better nutritional outcomes to be associated with household assets, income, education, sanitation, and while Tamil Nadu and other states such as Kerala and Goa score high on these fronts, the relative strengths of these endowments on nutrition can vary across states. Is Tamil Nadu's superior performance in children's nutritional outcome purely the result of these endowments?
>
> Cavatorta et al. (2015) examine the variation in height-for-age Z-score (ZHA) distribution across states and note the states that perform poorly: Bihar, Uttar Pradesh, Madhya Pradesh, Odisha, and Gujarat, where the incidence of stunting ranges from 45% to 57% for the under-5 age group. On the contrary, Tamil Nadu, with a different political manifesto, has managed to provide a voice and mobilize the vulnerable groups, to take up nutrition-based policy interventions such as the Nutritious Noon Meal Scheme and the Tamil Nadu Integrated Nutrition Project (TINP).
>
> The differences in outcomes across India reflect the inherent heterogeneity in culture and politics, given that nutrition, agriculture, and health policies are all

> **Section highlights: Successful Nutritional Policy in Tamil Nadu, India—cont'd**
>
> decided at the state level. The "coefficient effects" from the regressions in Cavatorta et al. (2015) demonstrate that Tamil Nadu performed consistently better than other states with respect to health, food, and nutrition policies, because the state
> - established the TINP in the 1980s and integrated food supplementation, nutrition education, growth monitoring, and primary healthcare across the region;
> - managed to reach the under-3s much better than the central government programs (ICDS);
> - maintained a "two-worker" program to service under-3s and older children, while other states maintained a "one-worker" program in every ICDS center;
> - continued to provide universal coverage of the public distribution system (PDS), while other states used a targeted system based on a poverty measure, that reduced total take-up of food subsidy significantly in these regions;
> - plugged the leakages in the system by requiring food supplements to be consumed on premises; and
> - has a relatively larger portion of agricultural land ownership by the household, and this is an important aspect, since land ownership is strongly correlated with HAZ distribution.
>
> The positive outcomes obtained in Tamil Nadu can be successfully replicated and scaled up in other states in India and in other countries, given that the sample of states in Cavatorta et al. (2015) covers a sixth of the world's stunted group under the age of 5.

Empirical analysis

It is important to point out at the outset that, while multivariate regression is usually an appropriate framework for investigating the role of individual/household and community characteristics on child nutritional status, a potential endogeneity problem may emerge with a socioeconomic status variable such as income and wealth with nutritional status. In such a situation, it is appropriate to instrument out income using appropriate proxies in the first stage, and then the instrumented variable can be used with other variables in a two-stage least squares framework.

The main purpose of this chapter is to predict the value of the dependent variable (the outcome variable) from the values of two or more independent variables (predictor variables). In our case, we would like to predict the values of weight-for-age Z-score (ZWA) and height-for-age Z-scores (ZWH) from the values of individual/household and community characteristics. We introduce some technical details and notations before proceeding to the main analysis. The general form of a regression model is as follows (Tabachnick and Fidell, 2001):

$$Y_i = \beta_0 + \beta_1 X_{i1} + \beta_2 X_{i2} + \ldots + \beta_k X_{ik} + \varepsilon_i \quad (10.1)$$

where $i = 1, 2, \ldots, k$ and ε_i follows a normal distribution with mean 0 and variance $\sigma2$. X_{ij} denotes the jth independent variable for the ith observation. When the aforementioned equation is applied to the data, it yields a set of predicted values $\widehat{Y}_i$ for which the sum of the $(Y_i - \widehat{Y}_i)$ values over all the k cases is a minimum. β_0 is the mean response when all the other independent variables are equal to zero. β_i is the expected change in Y_i per unit increase in Xi. One can obtain the least squares estimates[3] for the regression coefficients in Eq. (10.1), but this will not be undertaken in the present analysis. The standard errors of the parameter estimates (β_i) can be calculated using the following formula:

$$SE(\beta_i) = \left(\frac{s_y}{s_i}\right) \sqrt{\frac{1}{(1-R_i^2)}} \sqrt{\frac{(1-R_y^2)}{n-k-1}} \quad (10.2)$$

where s_y is the standard deviation of the dependent variable, s_i is the standard deviation of X_i, R_i^2 is the multiple correlation coefficient between the dependent variable and all the independent variables, and R_i^2 is the multiple correlation between X_i and the other independent variables. One can also compare an estimated regression coefficient to a specific value and perform a hypothesis test of a point estimate as follows:

$$H_0: \beta_i = \beta_{\text{null}}$$
$$H_1: \beta_i \neq \beta_{\text{null}} \quad (10.3)$$
$$t = \frac{\beta_i - \beta_{\text{null}}}{SE(\beta_i)}$$

where β_i is the estimate of β_i, the standard error of β_i is given by Eq. (10.2), and the degrees of freedom are $(n - k - 1)$. The aforementioned test is undertaken to determine whether a given predictor variable can independently account for a significant amount of the variability of the dependent variable.

Applying the regression equation to the independent variables yields a set of estimated $\widehat{Y}_i$ values. The simple correlation between Y and $\widehat{Y}$ is the multiple correlation coefficient and is given by

$$R_y^2 = \sum_{i=1}^{k} R_{yi} S_i \quad (10.4)$$

The summation is over all the independent variables, and β_i is the standardized regression coefficient for X_i. The semipartial (or part) and partial

---

3. In matrix algebra, the parameter estimates can be obtained as $\widehat{\beta} = (X'X)^{-1} X'Y$.

correlation between each independent variables and the dependent variable is given by the following formulas:

$$Sr_i = S_i\sqrt{(1-R_i)} \tag{10.5}$$

$$pr_i = \frac{\beta_i\sqrt{(1-R_i)}}{\sqrt{1-R^2_{y\cdot(i)}}} \tag{10.6}$$

As defined earlier, R_i is the multiple correlation between X_i and all the other independent variables, and $R^2_{y\cdot(i)}$ is the multiple correlation between the dependent variable and all the independent variables except X_i. As explained in Chapter 7, the total sums of squares (TSS), i.e., the total amount of variability in the dependent variable, can be partitioned into the sum squares between (SSB) and sum of squares within (SSW) or the sum of squared errors. The coefficient of multiple determination (R^2_y) can also be computed as follows:

$$R^2_y = \frac{SSB}{TSS} = \frac{\sum(\widehat{Y}_i - \overline{Y})^2}{\sum(Y_i - \overline{Y})^2} = \sum_{i=1}^{k} R_{yi}\beta_i \tag{10.7}$$

Having obtained $R2$, the question arises with what degree of confidence can we assert the linear relationship between the set of k independent variables and Y being not zero in the population? In other words, we want to determine if the collection of the independent variables accounts for a significant portion of the variance in the dependent variable. Thus, the following hypothesis needs to be tested:

$$H_0: \beta_1 = \beta_2 \ldots = \beta_k = 0$$
$$H_1: \text{at least one } \beta \neq 0$$

To conduct this test, the following statistic needs to be computed:

$$F = \frac{MSB}{MSW} = \frac{\dfrac{\sum(\widehat{Y}_i - \overline{Y})^2}{k}}{\dfrac{\sum(Y_i - \widehat{Y}_i)^2}{(n-k-1)}} = \frac{R^2(n-k-1)}{(1-R^2)k} \tag{10.8}$$

The aforementioned distribution follows an F distribution with k as the numerator and $(n-k-1)$ as the denominator degrees of freedom. One needs to look at the tests of individual coefficients to determine if the overall F is significant.

Data description and methodology

Following Strauss and Thomas (1995), a reduced form demand function for child nutritional status (N^*) can be specified in terms of a vector of individual characteristics of the child, such as age (XIC), household-level variables such as income and educational level of the mother (XH), and community-level factors (XC), such as water and sanitation conditions, health services, and so on. Thus, we can write the reduced form demand function for child's nutritional status as follows:

$$N^* = n(X_{IC}, X_H, X_C, \varepsilon) \qquad (10.9)$$

ε is a random error term that reflects heterogeneity in individual taste, health endowment, and other unobserved factors.

The outcome variables for our analysis are ZWA and ZHA. These are computed in terms of Z-scores using the NCHS/WHO reference values. The Z-score is defined as the difference between the value for an individual (weight or height) and the median value of the reference population for the same age or height, and divided by the standard deviation of the reference population. A low height for age (often referred to as stunting) results from long-term poor health and nutrition and is often regarded as a stable indicator of child nutritional status, since it is not subject to short-term fluctuations. Stunting in infancy and early childhood often has adverse consequences for later life, such as reduced productivity and increased reproductive and maternal health risks (WHO, 1995).

On the other hand, low weight for age is influenced by both the height of the child (height for age) and weight (weight for height). Its composite nature makes the interpretation complex. One of the drawbacks of this measure is that it fails to distinguish between short children of adequate body weight and tall thin children. This indicator can also reflect the long-term nutritional experience of the population. The XIC vector consists of age of the child in months (AGEMNTH) and the squared term of child age (AGESQ). We include this quadratic term since there was a curvilinear relationship (with bivariate plots) between age of the child and nutritional status as measured by both ZHA and ZWA. The rationale for such a relationship stems from the fact that younger children are more vulnerable to various diseases such as diarrhea, malaria, and so on and thus are more prone to malnutrition than older children.

The XH vector consists of the following variables:

1. Per capita expenditure on food (PXFD): This variable is used as a proxy for income since per capita food expenditure varies between 75% and 90% of total per capita expenditure for most of the households. Income is a central variable in models of child health and nutrition outcomes. More resources available to the household can translate into higher expenditures on food and nonfood items (such as health). However, as suggested by

Chamarbagwala et al. (2004), including income can create two potential problems: First, households can smooth consumption over a certain period of time, and thus, expenditure is the preferred measure and, second, since income is endogenously determined with child nutrition, including income as a right-hand side variable in a multiple regression framework can result in biased estimates. Thus, we use per capita food expenditure as a proxy for income, since food constitutes an important component of household expenditure. The variable is computed by taking the ratio of total food expenditure to the household size.

2. Education of the spouse (EDUCSPOUS): A categorical variable with values ranging from 1 to 7. This variable measures the education level of the mother in number of years. Higher values of this variable indicate greater levels of education.
3. Food security (CALREQ): This variable is based on whether the household can satisfy at least 80% of the requirement for calorie intake. This variable is dichotomous assuming two values 0 and 1, respectively, with a value of 1 denoting that the household is food secure. We also use the INSECURE variable as another measure of food insecurity in the determination of stunting, since a large number of dependents accompanied with less number of meals can cause children to have less number of meals as well.
4. MBRSEXNEW: This variable is dichotomous in nature assuming two values 0 and 1, with 1 denoting that the member is male.

The CARE variables included in the analysis are as follows:

1. Clinic feeding (CLINFEED): This is a dichotomous variable denoting whether the child is fed in a clinic or not.
2. Breastfeeding (BFEEDNEW): This is also a dichotomous variable denoting whether the child is breastfed or not during his or her infanthood.
3. Child attending clinic (ATTCLINI): A continuous variable denoting how many times the mother took the child to a clinic during his or her sickness. One can expect that the greater the availability of public health services (such as general and specialized health facilities, number of maternity clinics, health clinics, and doctors) in a region, the lower the risk of the child suffering from malnutrition. For example, the WHO guidelines for the Integrated Management of Childhood Illness (IMCI) recognize that many children attending a health clinic and not having previous weight record should undergo monitoring of their growth chart for weight for age. This is important as severe malnutrition can be identified and the necessary steps can be taken for prevention.[4]

---

4. A WHO-supported study in Brazil demonstrated that IMCI nutrition counseling provided by health workers during routine health visits could prevent weight growth faltering for infants between 6 and 12 months. Additionally, it provided a basis for promoting weight growth in infants 1 year and older (Bryce et al., 2006).

The community characteristics vector (*XC*) consists of the following set of variables:

1. Drinking distance (DRINKDST): This is a categorical variable assuming values from 1 to 5. Higher values of this variable denote that the distance to a protected drinking source for the household is higher. For example, the variable attains a value of 4 if distance to a protected drinking source exceeds 3 km. We can expect that the greater the distance to a protected water source, the more the likelihood that children will suffer from malnutrition.
2. Provision of sanitation (LATERINE): This is a dichotomous variable assuming two values 0 and 1, with 0 indicating absence of latrine from the household. Sanitation appears more important in nutritional outcomes than the presence of protected drinking source, since it is directly related in preventing diarrhea, thereby improving children's nutritional status.
3. DIARRHEA: This is a dichotomous variable indicating whether the child has diarrhea. This variable assumes two values 0 and 1. One indicates that the child had diarrhea during the survey.
4. Distance to a health facility (HEALTDST): This is a categorical variable denoting the distance of the household to a health clinic and assumes four values. Higher values indicate that the household is located farther from the nearest health center. For example, a value of 4 indicates that the distance to the nearest health clinic for the household is more than 10 km.

Descriptive summary of independent variables

We first look at a few descriptive summaries of the independent variables, such as mean and standard deviation, to understand the individual, household, and community characteristics of the sample. Table 10.1 provides the summary.

The average age of the child in the sample is about 27 months. In regard to the household level (or socioeconomic) variables, we find that the mean per capita expenditure on food (which proxies income) is about 9 kwacha, with the range being 190.5. There is a significant variation in income among the households, with the standard deviation being 9.85. The average year of education for the mother is 2.23 years, which implies that most mothers do not have education beyond grades 1–4. The mean value of calorie requirements met (CALREQ) was 0.32, indicating that most households on an average were not food secure. Although the main pathways through which the nutritional status of children can be improved (such as income and maternal educational status) are not large enough, the mean of the CARE indicators (such as bringing the child to clinic during periods of sickness) is quite high. It is quite likely that good care practices can mitigate the negative effects of low income and low maternal schooling on child nutritional status.

TABLE 10.1 Means and standard deviations of variables: Malawi sample.

Variables	Mean	Standard deviation
BFEEDNEW	1.57	0.49
DIARRHEA	0.16	0.37
CALREQ	0.32	0.47
ATTCLINI	1.32	0.59
AGEMNTH	27.29	16.28
CLINFEED	2.29	4.57
DRINKDST	1.98	1.39
EDUCSPOUS	2.23	1.30
LATERINE	0.40	0.49
PXFD	9.00	9.85

In regard to the community-level variables, the mean distance to a protected water source is about 1.98, which indicates that, on an average, households have to travel anywhere between 1 and 2 km to reach a protected water source. Additionally, the mean value of LATERINE for this sample is about 0.4, indicating that, on an average, most of the households do not have adequate access to sanitation facilities.

Main results

As explained before, model formulation involves selecting a mathematical model that best fits the data. We first fit the model with weight for age as the dependent variable and then undertake the same exercise with height for age as the dependent variable. The following steps are required in a multivariate regression framework.

Step 1: Estimating the coefficients of the model

Tables 10.2 and 10.3 show a reduced form demand function for child nutritional status based on weight for age and height for age as the dependent variables, respectively. The coefficients are estimated using least squares method, which results in the smallest sum of squares differences between the observed and the predicted values of the dependent variable. We estimate two models for weight for age. In the first model, we include per capita expenditure on food as one of the regressors to examine the impact of household income on underweight. In the second model, we exclude the impact of income and

TABLE 10.2 Determinants of weight-for-age Z-scores.

Variables	Model 1 Coefficients	Std. Error	Model 2 Coefficients	Std. Error
Constant	−3.03	0.504	−2.93	0.492
EDUCSPOUS	0.173	0.058	0.156	0.057
ATTCLINI	0.439	0.144	0.439	0.142
DRINKDST	−0.143	0.051	−0.147	0.05
LATERINE	0.134	0.152	0.152	0.148
PXFD	0.009	0.008		
AGEMNTH	−0.084	0.021	−0.088	0.021
AGESQ	0.001	0.0003	0.001	0.0003
CLINFEED	0.717	0.181	0.699	0.179
DIARRHEA	−0.608	0.201	−0.620	0.198
BFEEDNEW	0.664	0.226	0.625	0.224
CALREQ			0.566	0.212
HEALTDST	−0.094	0.071	−0.06	0.072

examine the impact of per capita calorie requirements on underweight. This is done to determine if household food security independently affects child nutritional status excluding the impact of income.

From the aforementioned coefficients, the equations to predict the ZWA for Model 1, for example, can be written as

$$ZWA^{pred} = -3.03 + 0.173 \text{ EDUCSPOUS} + 0.439 \text{ ATTCLINI}$$
$$- 0.143 \text{ DRINKDST} + 0.134 \text{ LATERINE}$$
$$+ 0.009 \text{ PXFD} - 0.084 \text{ AGEMNTH} + 0.001 \text{ AGESQ}$$
$$+ 0.717 \text{ CLINFEED} - 0.09 \text{ HEALDST}$$

(10.10)

The regression coefficient for an independent variable tells us how much the standard deviation units of the dependent variable change for a one-unit change in the independent variable, when all the other independent variables are held constant. For example, ZWA *pred* changes by 0.173 units for a unit change in EDUCSPOUS when the other independent variables are held constant.

TABLE 10.3 Determinants of height-for-age Z-scores.

Variables	Model 1 Coefficients	Std. Error	Model 2 Coefficients	Std. Error
Constant	−2.159	0.677	−2.564	0.685
EDUCSPOUS	0.048	0.072	0.106	0.074
ATTCLINI	0.514	0.180	0.402	0.177
DRINKDST	−0.024	0.064	−0.018	0.062
LATERINE	0.072	0.192	0.114	0.185
PXFD	0.014	0.01		
AGEMNTH	−0.091	0.034	−0.094	0.034
AGESQ	0.001	0.0004	0.0009	0.0004
CLINFEED	0.686	0.227	0.761	0.224
DIARRHEA	−0.829	0.261	−0.932	0.260
BFEEDNEW	0.149	0.305	0.174	0.299
INSECURE			0.252	0.085
HEALTDST	−0.106	0.092	−0.127	0.091

Table 10.3 provides the estimates of the determinants of ZHA. As before, we estimate two models for ZHA. In the first model, the impact of income on stunting is examined while, in the second model, the income effect is separated out and the impact of food insecurity as measured by INSECURE is examined. This variable is included as the regressor, since it fits the data better as a determinant of long-term nutritional status.

Table 10.3 shows that the impact of maternal education on stunting significantly declines compared with the determinants of underweight. On the other hand, health environmental conditions, such as the prevalence of diarrhea, have a larger impact on stunting after controlling for the impact of the other independent variables.

Step 2: Examining how good the model predicts

A reasonable way to determine if the model fits the data is to compare the observed and predicted values of the dependent variable. Tables 10.4 and 10.5 provide some measures based on the correlation between the observed and predicted values for weight-for-age and height-for-age regressions.

TABLE 10.4 Summary of the model for determinants of weight-for-age Z-scores[a].

Models	R	R^2	Adjusted R^2	Std. Error of the estimate
1	0.488	0.238	0.203	1.09
2	0.506	0.256	0.222	1.079

[a] $\widetilde{R}^2 = 1 - (1-R^2)\frac{(n-1)}{(n-k-1)}$.

TABLE 10.5 Summary of the model for determinants of height-for-age Z-scores.

Models	R	R^2	Adjusted R^2	Std. Error of the estimate
1	0.418	0.174	0.13	1.285
2	0.448	0.201	0.157	1.265

Multiple R is the correlation coefficient between the observed and predicted values and ranges in value from 0 to 1. The coefficient of multiple determination (R^2) (Eq. 10.7) is given in the third column of Tables 10.4 and 10.5, respectively. It shows the proportion of variability in the dependent variable explained by the regression equation. It is the square of the coefficient of multiple determination and is an important statistic in regression analysis.

Although this coefficient is small for both ZWA and ZHA regressions, it is consistent with other studies.[5] Adjusted[1] R^2 ($\widetilde{R}^2$) corrects for R^2 being an optimistic estimate of how well the model fits the population. If the number of independent variables is greater, sample R^2 is likely to be larger than the population R^2. Thus, one naturally prefers an estimate of the population R^2 that is more accurate than the positively biased sample R^2. As can be seen from column (4) of the tables, this estimate is always smaller in magnitude than the sample R^2. The magnitude of this decline will be greater for small values of R^2 than for larger values, other things being equal. Additionally, the magnitude of the decline will also be larger as the ratio of the number of independent variables to the number of observations in the sample increases. Finally, the standard error of the estimate is the standard deviation of the residuals (standard deviation of $Y_i - \widehat{Y}_i$). For a successful regression model, the

---

5. Coefficient of multiple determination is of a similar magnitude in regressing child nutritional status on its determinants as in Glewwe (1999).

standard error of the estimate should be considerably smaller than the standard deviation of the dependent variable. In other words, we want the observations to lie near the regression line.

Step 3: hypotheses testing

If one needs to draw conclusions about the population from the sample results, the error term ε_i in Eq. (10.1) must satisfy the following properties:

1. The errors have a normal distribution.
2. The errors in the model are all independent.
3. The same amount of error is found at each level of X. The implication is that the average difference between the regression line and the observed values is constant across all values of the independent variables. Having equal variance at each level of X is known as "homoscedasticity," while having unequal variance is called "heteroskedasticity."

Tests about the equation

There are several hypotheses tests in a multiple regression output, but all of them try to determine whether the underlying model parameters are actually zero. The first question that may be asked is: "Is the multiple regression model any good at all?" To address the aforementioned question, we want to test the null hypothesis in Eq. (10.1) as follows:

$$H_0: \beta_1 = \beta_2 = \ldots = \beta_k = 0 \tag{10.11}$$

The alternative hypothesis is that the slope coefficients are not all equal to zero. The null hypothesis can be tested using the F-statistic as given by Eq. (10.8). The following analysis of variance (ANOVA) tables (Tables 10.6 and 10.7) divides the observed variability in the dependent variables (namely ZWA and ZHA) into two parts: the regression sum of squares (SSB) and the residual sum of squares (SSW). The total sum of squares is the sum of these two numbers. We can derive the coefficient of multiple determination (R^2) by dividing the regression sum of squares by the total sum of squares.

The observed significance level for the F-statistic tells us if the null hypothesis can be rejected as in Eq. (10.11).

As is evident from Tables 10.6 and 10.7, the coefficient of multiple determination (R^2) of Tables 10.4 and 10.5 can be obtained as the ratio of the regression sum of squares to the total sum of squares. For example, for Model 1 of determinants of ZWA, the ratio of regression sum of squares to the total sum of squares is (89.139/374.075) = 0.238. The computation of the F-values from Eq. (10.8) can be demonstrated as follows for Model 1 of the weight-for-age regressions:

$$F = \text{MSB}/\text{MSW} = 8.1035/1.1922 = 6.797$$

TABLE 10.6 Analysis of variance table for weight-for-age Z-scores.

	Models	Sum of squares	df	Mean square	F	Sig.
I	Regression	89.139	11	8.104	6.797	0.00
	Residual	284.936	239	1.192		
	Total	374.075	250			
II	Regression	95.757	11	8.705	7.475	0.00
	Residual	278.318	239	1.165		
	Total	374.075	250			

TABLE 10.7 Analysis of variance table for height-for-age Z-scores.

	Models	Sum of squares	df	Mean square	F	Sig.
I	Regression	70.948	11	6.45	3.901	0.00
	Residual	335.649	203	1.653		
	Total	406.597	214			
II	Regression	81.525	11	7.411	4.628	0.00
	Residual	325.072	203	1.601		
	Total	406.597	214			

The critical value of F for 11 and 239 degrees of freedom at the significance level $\alpha = 0.05$ is 1.84. Since the obtained F value exceeds the critical value, we can infer that the slope coefficients are not all zero and conclude that the multiple regression model is better than just using the mean.

Tests about individual coefficients

Even if the null hypothesis is rejected, it does not mean that all of the variables in the equation have regression coefficients that are significantly different from zero. To test whether a particular coefficient is zero, we first look at the column labeled t in Tables 10.8 and 10.9, which are derived from Tables 10.2 and 10.3, respectively. This column is computed by dividing each coefficient by its standard error and using Eq. (10.3). The degrees of freedom are given by $(n - k - 1)$. The purpose is to determine the independent impact of individual/household and community characteristics on child nutritional status as measured by ZWA and ZHA.

TABLE 10.8 Tests of individual coefficients for determinants of weight-for-age Z-scores.

Variables	Model 1 t-Stat.	Model 1 P value	Model 2 t-Stat.	Model 2 P value
Constant	−6.016	0.00	−5.956	0.00
EDUCSPOUS	2.987*	0.003	2.70*	0.007
ATTCLINI	3.047*	0.003	3.102*	0.002
DRINKDST	−2.817*	0.005	−2.938*	0.004
LATERINE	0.882	0.378	1.03	0.304
PXFD	1.185	0.237		
AGEMNTH	−4.041*	0.000	−4.282*	0.00
AGESQ	3.096*	0.002	3.386*	0.001
CLINFEED	3.966*	0.00	3.911*	0.00
DIARRHEA	−3.018*	0.003	−3.126*	0.002
BFEEDNEW	2.935*	0.004	2.791*	0.006
CALREQ			2.668*	0.008
HEALTDST	−1.32	0.188	−0.851	0.396

Denotes at 1% level of significance.

Comparing Tables 10.8 and 10.9, we find that while maternal education has a positive and significant impact on child nutritional status as measured by ZWA, it does not have a significant effect on ZHA.

The aforementioned result is puzzling given that many studies in the literature find a positive and significant impact of maternal education on stunting, which is a long-run measure of child nutritional status. We provide two reasons behind this result:

1. Maternal education may be instrumented out using suitable proxies of health knowledge (as in Glewwe, 1999), and then, two-stage least squares model may be applied.
2. It is possible that the community variables and the care factors included in the model may have taken away some of the associations between maternal education and stunting.

It is also important to emphasize that after controlling for the dummy variable for Mzuzu and excluding some community-level variables (such as distance to a health facility, distance to a protected water source, and sanitation

TABLE 10.9 Tests of individual coefficients for determinants of height-for-age Z-scores.

Variables	Model 1 t-Stat.	Model 1 P value	Model 2 t-Stat.	Model 2 P value
Constant	−3.187	0.002	−3.745	0.000
EDUCSPOUS	0.661	0.509	1.437	0.152
ATTCLINI	2.863*	0.005	2.273**	0.024
DRINKDST	−0.37	0.712	−0.286	0.775
LATERINE	0.375	0.708	0.615	0.539
PXFD	1.479	0.141		
AGEMNTH	−2.642*	0.009	−2.78*	0.006
AGESQ	1.949**	0.053	2.022**	0.044
CLINFEED	3.018*	0.003	3.394*	0.001
DIARRHEA	−3.171*	0.002	−3.584*	0.00
BFEEDNEW	0.489	0.625	0.580	0.563
INSECURE			2.977*	0.003
HEALTDST	−1.15	0.252	−1.395	0.164

* Denotes at 1% level of significance.
** Denotes at 5% level of significance.

facilities) as regressors from the model, both education and income turn out marginally significant (at the 10% level). Our results also indicate that per capita food expenditure (a proxy for household income), although positive, was not significant for both ZWA and ZHA. This may be because of two main reasons:

1. After controlling for mothers' education, income does not significantly affect child nutritional status.
2. Income and nutrition may be jointly determined, and income may be instrumented out using suitable asset proxies in the first stage, and then the instrumented value may be used in a two-stage least squares framework.

Turning now to the CARE variables (such as taking the child to a clinic, breastfeeding, and feeding the child in a clinic), we find that their individual impact is positive and significant on ZWA and ZHA. However, we did not find breastfeeding to be significant in the ZHA regressions. This is consistent with the existing literature (Ruel et al., 1999), which suggests that proper care of the child can be extremely crucial in improving child's nutritional status in the medium and long run. However, food security (as measured by per capita calorie requirements met or as measured by INSECURE) turns out to be positive and significant suggesting that, after controlling for the household and community level variables, food security at the household level has a positive and significant impact on child's nutritional status.

Examining the impact of age-specific effects on child nutritional status, we find that younger children are more prone to suffer from malnutrition compared with older children. This is evident from the coefficient of the age term (AGEMNTH) being negative and significant, while the coefficient of the squared term to be positive and significant in both the regressions. This is also consistent with the existing literature (see, for example Sahn and Alderman, 1997 and Glewwe, 1999) that younger children are more prone to malnutrition than older children. In regard to the community infrastructure variables, we find that, in general, most of the community-level variables are not significant, except for drinking distance of the household to a protected water source in the weight-for-age regressions. This variable, however, is not significant in the height-for-age regressions. However, since the impact of diarrhea prevalence on child malnutrition is negative and significant in both the regressions, it is likely that this variable is capturing the impact of poor water and sanitation conditions.

Part and partial correlation coefficients

Two other statistics that are usually used to assess the contribution of an independent variable to an existing model are part (or semipartial) and partial correlation coefficients, the formulas of which are given in Eqs. (10.5) and (10.6), respectively. Both of these measures can range in value from -1 to $+1$. Intuitively, the part correlation coefficient provides unique information about the dependent variable that is not available from the other independent variables in the equation. In other words, it is the correlation coefficient between the dependent variable and the residuals of an equation in which the other independent variables in the model are used to predict the dependent variable. The partial correlation coefficient, on the other hand, is the correlation between the independent variables and the dependent variable when the linear effects of the other independent variables have been removed from both the dependent and the independent variables (Lowry, 2003). We will only report the part and partial correlation coefficients when the income effect is considered (that is Model 1), for the weight-for-age and height-for-age regressions (Table 10.10).

It is important to note here that even if variables are related to each other, this does not necessarily imply that one causes the other.

TABLE 10.10 Part and partial correlation coefficients for weight-for-age and height-for-age.

	Weight for age		Height for age	
Variables	Part correlation	Partial correlation	Part correlation	Partial correlation
EDUCSPOUS	0.169	0.190	0.042	0.046
ATTCLINI	0.172	0.193	0.183	0.197
DRINKDST	−0.159	−0.179	−0.024	−0.026
LATERINE	0.05	0.057	0.024	0.026
PXFD	0.067	0.076	0.094	0.103
AGEMNTH	−0.228	−0.253	−0.168	−0.182
AGESQ	0.175	0.196	0.124	0.136
CLINFEED	0.224	0.248	0.192	0.207
DIARRHEA	−0.17	−0.192	−0.202	−0.217
BFEEDNEW	0.166	0.187	0.031	0.034
HEALTDST	−0.075	−0.085	−0.073	−0.08

Step 4: Checking for violations of regression assumptions

Detecting influential observations

Even before undertaking the regression analysis, it is important to diagnose outliers in the model. The presence of outliers (influential observations) can influence the slope and/or intercept of the regression model (Tabachnick and Fidell, 2001). Including these observations will thus result in the estimates being biased. The vertical distance of a point from a regression line is called the "residual or deviation," while the horizontal distance of a point from the mean is called "leverage." "Influence" is computed as the product of the residual and the leverage. The influence of a point is the amount by which the slope of the regression line changes when that point is removed from the data set, and thus, a new slope is computed. The amount of the change was developed by Cook and is called Cook's influence or Cook's distance. Intuitively, Cook's distance measures the change in the sum of squared differences for every observation, except when the relevant point is removed. In regression diagnostics, values greater than 1 are of concern and should be carefully investigated. When the regressions are run, the Cook's influence values will appear in the data set as a new variable labeled coo_1, coo_2. We delete the observations for which this distance was greater than 1 and undertake the regression analysis from the reduced sample.

Checking normality of the errors

The histogram of the residuals in addition to normal P-P plot of the standardized residuals can be plotted to check whether the errors are normal or not (Figs. 10.1 and 10.2). The latter plot must be approximately linear for the distribution to be normal. The normal probability plot compares the percentiles of the values from the distribution to the percentiles of the standard normal distribution (Spanos, 1998). For example, in a normal distribution, approximately 3% of the values are more than 3 standard deviations below the mean, 5% of the values are more than 2 standard deviations below the mean, and 33% of the values are more than 1 standard deviation below the mean. The plot will thus have percentiles of the standard normal distribution on the Y-axis and the percentiles of the variable on the X-axis. If the variable has an approximate normal distribution, then the normal probability plot will look like a straight line. We examine the normal probability plot from the weight for age (as an example) for Model 1, where per capita expenditure of food is one of the regressors.

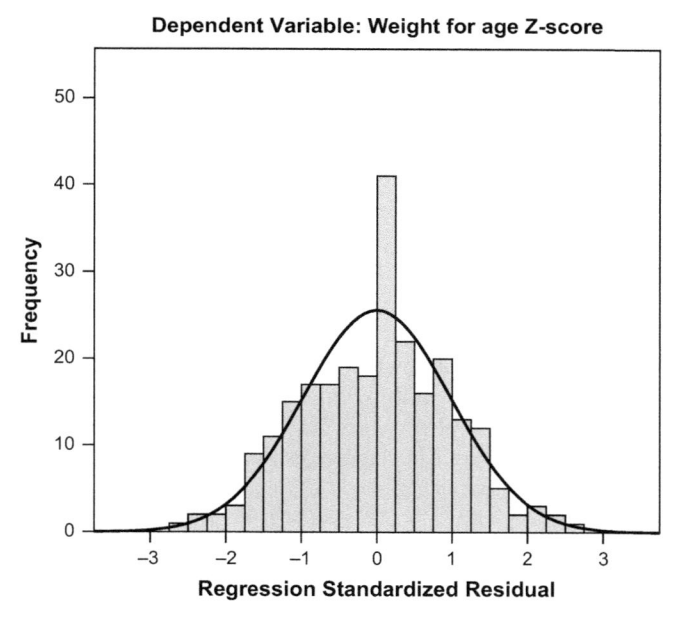

FIGURE 10.1 Histogram of standardized residuals of weight for age.

Dependent Variable: Weight for age Z-score

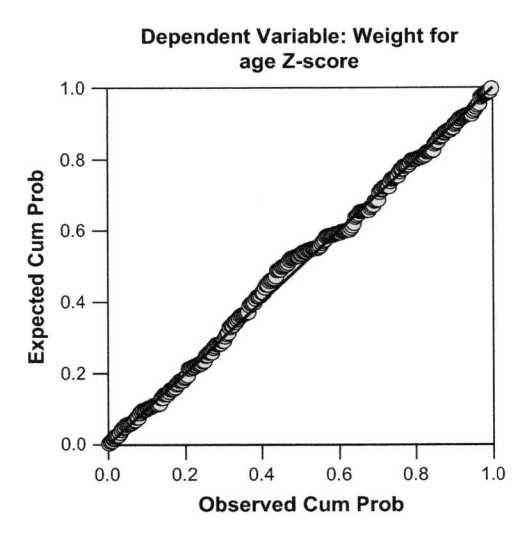

FIGURE 10.2 Normal P-P plot of regression standardized residuals.

As evident from the graphs, the residuals of weight for age are approximately normal. Regression is relatively robust to slight violations from normality as long as the sample size is reasonably large.

Checking for homogeneity of variance of the residuals

The easiest way to check for the homogeneity of variance (homoskedasticity) across different levels of the independent variables is to examine a scatterplot of the residuals against the predicted values. If the spread of the residuals appears to be constant across the different levels of the dependent variables, there is not much likelihood of a problem. It is important to mention that regression is also robust to violations of homoskedasticity. However, mild violations of this assumption are not worrisome (Tabachnick and Fidell, 2001).

From Figs. 10.3 and 10.4, it is evident that, while the assumption of homoskedasticity is not violated for the weight-for-age regressions, it is violated for the case of height-for-age regressions, since the spread thickens around the center.

Multiple regression in STATA

Multiple regression technique is very useful as discussed so far. But the calculations can be very lengthy and cumbersome. To show you the extent of calculations involved, we take a simple example with 10 observations on corn output that we examined in the last chapter.

Dependent Variable: Weight for age Z-score

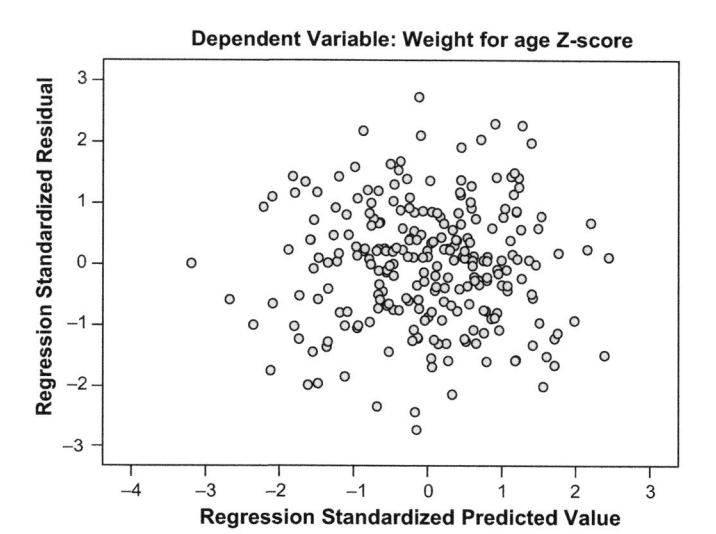

FIGURE 10.3 Residuals plotted against predicted values for weight for age.

Dependent Variable: Height for age Z-score

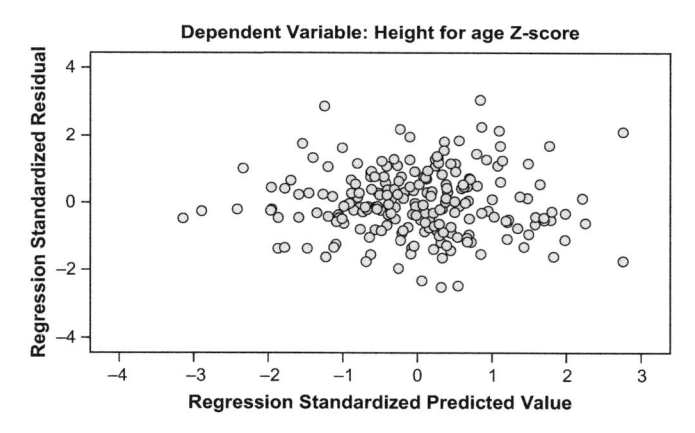

FIGURE 10.4 Residuals plotted against predicted values for height for age.

Recall in that context we had only 1 independent variable, namely Fertilizer. Now we add another variable, Pesticide, to these data. We are interested in estimating a regression equation that captures the following relationship:

$$\text{Corn} = a + b \text{ Fertilizer} + c \text{ Pesticide}$$

Note that in this case we have three parameters: a, b, and c. For simplicity, we represent the variables as follows: Corn = Y or the dependent variable,

TABLE 10.11 Multiple regression: Example with corn, fertilizer, and pesticide.

Obs	Corn (Y)	Fertilizer (X_1)	Pesticide (X_2)
1	40	6	4
2	44	10	4
3	46	12	5
4	48	14	7
5	52	16	9
6	58	18	12
7	60	22	14
8	68	24	20
9	74	26	21
10	80	32	24

Fertilizer = X_1 and Pesticide = X_2 are the two independent variables. The data for this purpose are given in Table 10.11.

Similar to the previous chapter, we will identify the steps involved in computing these parameters and show the calculations in detail. To estimate the parameters, we use these steps:

Step 1: Compute $\overline{X}_1$, $\overline{X}_2$, and $\overline{Y}$ (the means of X_1, X_2, and Y).

Step 2: Compute $x_1 = (X_1 - \overline{X}_1)$, $x_2 = (X_2 - \overline{X}_2)$, and $y = (Y - \overline{Y})$.

Step 3: Compute $x_1^2 = (X_1 - \overline{X}_1)^2$, $x_2^2 = (X_2 - \overline{X}_2)^2$, and their sums.

Step 4: Compute $x_1 y = (X_1 - \overline{X}_1)(Y - \overline{Y})$, $x_2 y = (X_2 - \overline{X}_2)(Y - \overline{Y})$, and $x_1 x_2 = (X_1 - \overline{X}_1)(X_2 - \overline{X}_2)$ and their sums.

We perform all these steps in the following table:

Step 1: Compute $\overline{X}_1$, $\overline{X}_2$, and $\overline{Y}$. We begin with entries in columns (2), (3), and (4). We present these calculations in Table 10.12 below. We compute the sums of X and Y in the last row. Based on this information, we get $\overline{X}_1 = 18$; $\overline{X}_2 = 12$, and $\overline{Y} = 57$.

TABLE 10.12 Detailed calculations of the parameters in a multiple regression.

Obs	Y	X_1	X_2	y	x_1	x_2	x_1^2	x_2^2	$x_1 y$	$x_2 y$	$x_1 x_2$
1	40	6	4	−17	−12	−8	144	64	204	136	96
2	44	10	4	−13	−8	−8	64	64	104	104	64
3	46	12	5	−11	−6	−7	36	49	66	77	42
4	48	14	7	−9	−4	−5	16	25	36	45	20
5	52	16	9	−5	−2	−3	4	9	10	15	6
6	58	18	12	1	0	0	0	0	0	0	0
7	60	22	14	3	4	2	16	4	12	6	8
8	68	24	20	11	6	8	36	64	66	88	48
9	74	26	21	17	8	9	64	81	136	153	72
10	80	32	24	23	14	12	196	144	322	276	168
Sum	570	180	120	0	0	0	576	504	956	900	524
Mean	57	18	12								

Step 2: Compute $x_1 = (X_1 - \overline{X}_1)$, $x_2 = (X_2 - \overline{X}_2)$, and $y = (Y - \overline{Y})$. For each entry in columns (2), (3), and (4), we compute the deviation terms and they are in columns (5), (6), and (7).

Step 3: Compute $x_1^2 = (X_1 - \overline{X}_1)^2$, $x_2^2 = (X_2 - \overline{X}_2)^2$ and their sums. Each entry in columns (6) and (7) is squared and is reported in (8) and (9) of Table 10.12 (their sums are 576 and 504).

Step 4: Compute $x_1 y = (X_1 - \overline{X}_1)(Y - \overline{Y})$, $x_2 y = (X_2 - \overline{X}_2)(Y - \overline{Y})$, and $x_1 x_2 = (X_1 - \overline{X}_1)(X_2 - \overline{X}_2)$ and their sums. These are done in the last three columns of Table 10.12. You should also check their values.

We now apply the formula that will give us the coefficients:

$$b = \frac{\sum x_1 y \sum x_2^2 - \sum x_2 y \sum x_1 x_2}{\sum x_1^2 \sum x_2^2 - \left(\sum x_1 x_2\right)^2} = \frac{(956)(504) - (900)(524)}{(576)(504) - 524^2} = 0.65$$

$$c = \frac{\sum x_2 y \sum x_1^2 - \sum x_1 y \sum x_1 x_2}{\sum x_1^2 \sum x_2^2 - \left(\sum x_1 x_2\right)^2} = \frac{(900)(576) - (956)(524)}{(576)(504) - 524^2} = 1.11$$

Finally, the intercept:

$$a = \overline{Y} - b\overline{X}_1 - c\overline{X}_2 = 57 - (0.65)(18) - (1.11)(12) = 31.98$$

Consequently, the estimated equation is

$$\text{Corn Output} = 32 + 0.65 \text{ Fertilizer} + 1.1 \text{ Pesticide}$$

As you can see that the calculations can become tedious if we have more observations and variables. However, we can derive these coefficients "*a*" and "*b*" in STATA. We illustrate this application in the following.

STATA Output

Example 1

We expand the regress command in STATA to include additional explanatory variables. We use the table from the previous section about Corn output together with two input levels: Fertilizers and Pesticides.

We can use this information in STATA and the regress command to answer the following question: how does corn output fluctuate when we include both Fertilizer and Pesticide levels in the same regression?

We use the following regress commands in STATA in this context:

- regress corn fertilizer pesticide
- regress corn fertilizer pesticide, vce(robust)

The first input statement simply expands the regress command that we used in the previous chapter to include more than one independent variable. In this case, we just list the dependent variable first (which is "corn") and the list of independent variables that we wish to include in the regression model.

The second statement includes the option vce(robust), which produces standard errors and significance levels that are valid even if the data are heteroskedastic. We look at the output from both the following statements:

The estimated equation in this case is

$$\text{Corn Output} = 32 + 0.65 \text{ Fertilizer} + 1.1 \text{ Pesticide}$$

In other words, Corn output is positively related to the use of fertilizers and pesticides. A one-unit increase in fertilizer usage increases corn output by 0.65 units, while a one-unit increase in pesticide increases corn output by 1.1 units.

Finally, we note that all the t-values are significant at 99% confidence and that 99% of the variability in corn output is due to the variability in fertilizer and pesticide usage (note the value of adjusted R-squared $= 0.99$).

We proceed to investigate the impact of pesticide and fertilizer use on corn output, through the same regress command, in the second line, which uses the vce(robust) command. The estimated equation in this case is still the same as before. However, the standard errors and t-values have changed, because STATA controls for heteroskedasticity in the data. In these data, however, we do not find evidence of this problem and both variables continue to be significant at 99% level, and the R-squared value is also very high.

Example 2

We now look at another example, which uses 12 observations on ZWH. We wish to examine the determinants of ZWH and so we include four exogenous variables: EDUCSPOUS (an index of education), CALREQ, SICKFEED where these variables are as defined in the main section of the chapter.

Finally, we also include a variable called GENDER, which is a dichotomous variable taking the value 1 if the observation is a male and 0 if female. The data are given in Table 10.13.

We first use the regress command and conduct the multivariate regression with three variables: EDUCSPOUS, CALREQ, and SICKFEED.

TABLE 10.13 Data for example 2 on ZWH, EDUCSPOUS, CALREQ, and GENDER for multiple regression.

Obs	ZWH	EDUCSPOUS	CALREQ	SICKFEED	GENDER
1	37.5	1.00	0.833	0.936	1
2	37.5	1.00	0.875	0.944	1
3	40	0.92	0.917	0.952	0
4	45	0.75	1.000	0.944	0
5	37.5	1.00	0.792	0.952	0
6	30	1.08	0.625	0.944	1
7	42.5	0.75	0.833	0.960	1
8	42.5	0.83	0.958	0.976	0
9	42.5	0.83	0.792	0.984	0
10	40	0.92	0.750	0.984	0
11	45	0.75	0.917	0.992	1
12	50	0.67	0.958	1.000	1

The output with vce(robust) estimates is given in the following as a screenshot. In the first output, we see that EDUCSPOUS is negatively related to ZWH. However, both CALREQ and SICKFEED are positively related to ZWH.

The $F(3, 8)$ is 61.25, which represents a significant model. All the variables are significant at the 5% level, and thus, we can conclude that ZWH is modeled appropriately with the chosen variables.

In the second portion of the output, we include GENDER and we find that this variable is negative, but insignificant, because the corresponding t-value is 0.32, which is much smaller than the table value of t (with 4 degrees of freedom at 99%) which is 4.6.

Consequently, we cannot reject the null hypothesis, thus controlling for other determinants of weight for height, gender is not significant.

Finally, we produce the plot of the residuals with the fitted values using the command rvfplot:

```
. regress zwh educspous calreq sickfeed, vce(robust)
```

Linear regression   Number of obs =   12
                                           F(3,   8) =   61.25
                                           Prob > F   = 0.0000
                                           R-squared   = 0.9592
                                           Root MSE   = 1.1934

| | | Robust | | | | |
zwh	Coef.	Std. Err.	t	P>|t|	[95% Conf. Interval]
educspous	-21.00372	3.717855	-5.65	0.000	-29.57711	-12.43033
calreq	17.16802	5.525261	3.11	0.015	4.426743	29.90929
sickfeed	55.82736	19.93295	2.80	0.023	9.861894	101.7928
_cons	-9.270335	24.76436	-0.37	0.718	-66.37704	47.83637

```
. regress zwh educspous calreq sickfeed gender,vce(robust)
```

Linear regression   Number of obs =   12
                                           F(4,   7) =   45.01
                                           Prob > F   = 0.0000
                                           R-squared   = 0.9597
                                           Root MSE   = 1.2683

| | | Robust | | | | |
zwh	Coef.	Std. Err.	t	P>|t|	[95% Conf. Interval]
educspous	-21.38463	3.60295	-5.94	0.001	-29.90426	-12.86501
calreq	16.75975	5.599069	2.99	0.020	3.52006	29.99945
sickfeed	54.43349	20.21948	2.69	0.031	6.622018	102.245
gender	-.2189027	.686133	-0.32	0.759	-1.841349	1.403544
_cons	-7.135169	24.72373	-0.29	0.781	-65.5975	51.32717

Note that all the residuals are within $(-2, 2)$ range, and hence, we can conclude that this is a good model.

Conclusions

The results presented in this chapter have key implications for health and nutrition monitoring in Malawi, i.e., intervention opportunities that might have a positive impact on child nutrition over the long run. Maternal education, as measured by the number of years in school, has a significant influence on child nutrition. This is in line with the past development research on this subject, which has shown that maternal education compared with fathers' education can have a significant influence on child nutritional outcomes. Although we do not find income to influence significantly child nutritional outcomes after controlling for the impact of maternal education, income might still be important. However, food security significantly influences child nutritional outcomes, and thus, improving food security at the household level seems crucial in the long run. One way to improve household food security is to raise agricultural productivity, especially for the small farmers.

An additional finding to note is that child care practices had a significant influence on child health and nutritional status. Thus, complementary interventions that improve women's income and wealth earning opportunities and provide better information on care and hygienic practices can be important in improving child nutritional status.

We also found the incidence of diarrhea to have a negative and significant influence on child nutritional status. This might be indirectly capturing poor water and sanitation conditions, which can affect child nutritional status adversely. Increasing education levels of mothers and appropriate hygiene and sanitation practices seems crucial. For example, a higher level of maternal knowledge about handwashing can have a positive association with child nutrition by preventing diarrhea. The effects of such water and sanitation interventions can be very beneficial for certain subgroups of the population, such as the low-income and the low-educated households. We also found some evidence of increase in distance to a protected water source adversely affecting weight for age but not height for age. Thus, reducing travel time to collect safe water can relax the time constraint of the household. The time saved can be reallocated to leisure, health production, and other agricultural activities, which, consequently, can improve household productivity and child nutritional status over a longer time horizon.

In conclusion, the results from this chapter can provide general guidelines to program managers, government officials, as well as researchers, as to what interventions are most likely to improve child health and nutrition outcomes. The results presented should help them fine-tune those interventions, both programmatically and for the targeted groups.

Multiple regression in *R*

We take a simple example with 10 observations on corn output that we examined in the last chapter. We now have two independent variables to perform a multiple regression in *R*. We are interested in estimating a regression equation that captures the following relationship:

$$Y = a + bX_1 + cX_2$$

We have three parameters to capture: a, b, and c. Note that we simplify the notations for variables: Corn = Y or the outcome variable, and Fertilizer = X_1 and Pesticide = X_2 are the two independent variables. We expand the *lm* command in *R* to include additional explanatory variables.

To perform multiple regression in *R*, the dependent variable should list first and list independent variables by using + sign between variables. As undertaken earlier, we save the results of multiple regression as regression_chapter10 and print it in detail in using the *summary* command as follows:

```
> regression_chapter10<-lm(Y~X1+X2)
> summary(regression_chapter10)

Call:
lm(formula = Y ~ X1 + X2)

Residuals:
    Min      1Q  Median      3Q     Max
-1.8199 -0.7304  0.1302  0.9173  1.8108

Coefficients:
            Estimate Std. Error t value Pr(>|t|)
(Intercept)  31.9807     1.6318  19.598 2.25e-07 ***
X1            0.6501     0.2502   2.599  0.03550 *
X2            1.1099     0.2674   4.150  0.00429 **
---
Signif. codes:  0 '***' 0.001 '**' 0.01 '*' 0.05 '.' 0.1 ' ' 1

Residual standard error: 1.397 on 7 degrees of freedom
Multiple R-squared:  0.9916,    Adjusted R-squared:  0.9892
F-statistic: 414.8 on 2 and 7 DF,  p-value: 5.356e-08
```

From the table, we substitute coefficients to the equation. Corn output is positively related to the use of fertilizers and pesticides. A one-unit increase in fertilizer and pesticide usages increases corn output by 0.65 and 1.1 units, respectively. We can see information about residuals, *t*-value of coefficient, and *P*-value. According to the value of adjusted *R*-squared, the change in corn output is due to the variability in fertilizer and pesticide usage at 99% confidence.

$$\text{Corn Ouput} = 32 + 0.65 \text{ Fertilizer} + 1.1 \text{ Pesticide}$$

Then, we proceed to investigate the impact of pesticide and fertilizer use on corn output by computing robust covariance matrix estimators. We use the variance estimator in a regression model by installing *lmtest* and *sandwich* package. With *coeftest* command from the "*lmtest*" package, one will get the same results as the *vce(robust)* option in STATA. With *vcovHC* from the sandwich package, we can employ a heteroskedasticity-consistent estimation of the covariance matrix to produce a robust regression. The command and results of controlling for heteroskedasticity in the data are as follows. We use the variance estimator of the previous regression named regression_chapter10 to adjust values.

```
> coeftest(regression_chapter10,vcov=vcovHC(regression_chapter10,type="HC1"))

t test of coefficients:

             Estimate Std. Error t value  Pr(>|t|)
(Intercept) 31.98067    1.12476 28.4332 1.711e-08 ***
X1           0.65005    0.24070  2.7007  0.030610 *
X2           1.10987    0.28814  3.8518  0.006278 **
---
Signif. codes:  0 '***' 0.001 '**' 0.01 '*' 0.05 '.' 0.1 ' ' 1
```

The estimated equation is still the same as before. However, the standard errors and t-values have changed because of the command controls for heteroskedasticity in the data.

We now look at another example with more independent variables. We will check the hypothesis test and violations of regression assumptions at the end of this chapter. First, we perform a multivariate regression on ZWH concerning four independent variables, which are EDUSCPOUS, CALREQ, SICKFEED, and GENDER. As we performed earlier, we use *lm* and *coeftest* command to produce robust regression results as follows:

```
> regression<-lm(ZWH~EDUCSPOUS+CALREQ+SICKFEED+GENDER)
> summary(regression)

Call:
lm(formula = ZWH ~ EDUCSPOUS + CALREQ + SICKFEED + GENDER)

Residuals:
    Min      1Q  Median      3Q     Max
-1.7985 -0.5288  0.1087  0.7388  1.3281

Coefficients:
             Estimate Std. Error t value Pr(>|t|)
(Intercept)   -7.1352    31.8950  -0.224  0.82937
EDUCSPOUS    -21.3846     5.9140  -3.616  0.00856 **
CALREQ        16.7598     5.6568   2.963  0.02102 *
SICKFEED      54.4335    25.9439   2.098  0.07408 .
GENDER        -0.2189     0.7590  -0.288  0.78138
---
Signif. codes:  0 '***' 0.001 '**' 0.01 '*' 0.05 '.' 0.1 ' ' 1

Residual standard error: 1.268 on 7 degrees of freedom
Multiple R-squared:  0.9597,    Adjusted R-squared:  0.9366
F-statistic: 41.64 on 4 and 7 DF,  p-value: 5.745e-05
```

```
> coeftest(regression,vcov=vcovHC(regression,type="HC1"))

t test of coefficients:

             Estimate Std. Error t value  Pr(>|t|)
(Intercept)  -7.13518   24.72375 -0.2886 0.7812494
EDUCSPOUS   -21.38463    3.60295 -5.9353 0.0005785 ***
CALREQ       16.75975    5.59907  2.9933 0.0201318 *
SICKFEED     54.43350   20.21949  2.6921 0.0309900 *
GENDER       -0.21890    0.68613 -0.3190 0.7590053
---
Signif. codes:  0 '***' 0.001 '**' 0.01 '*' 0.05 '.' 0.1 ' ' 1
```

As a result, we can get an equation. CALREQ and SICKFEED have a positive and significant impact on ZWH, 16.76, and 54.43 changes per unit, at the 5% level, respectively. EDUCSPOUS and GENDER have negative impacts on ZWH. However, GENDER variable is not statistically significant.

$$ZWH = -7.13 - 21.38 \text{ EDUCSPOUS} + 16.76 \text{ CALREQ} + 54.43 \text{ SICKFEED} - 0.22 \text{ GENDER}$$

We should check the normality of the errors to validate the regression. The *plot(regression model)* command is used to check the validity with a graph. *R* has a set of built-in regression diagnostic plot command with regression results allows us to produce a residual plot. X-axis is the predicted or fitted Y values. On the Y-axis, is the residual or errors. If the normality assumption is met, the line should be flat, and points should seem a cloud. If the variation is constant, there might be no pattern. However, in the following figure, there is a clear pattern, not making a relatively flat line. It is because there are not enough observations in the data set that we used. We cannot find any patterns of points, so we can say that we need more observations to ensure the validity of the regression.

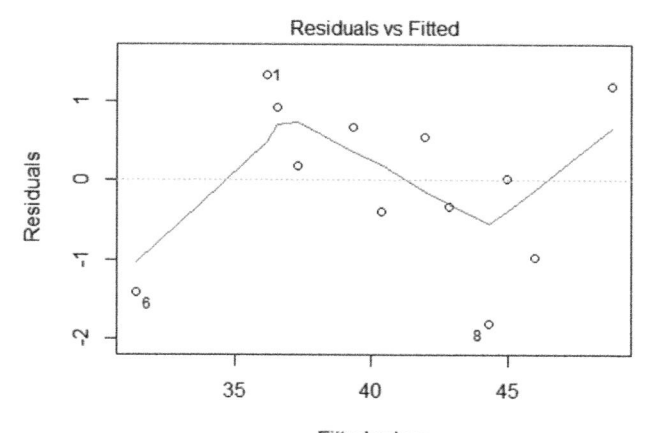

Residuals vs Fitted

Fitted values
lm(ZWH ~ EDUCSPOUS + CALREQ + SICKFEED + GENDER)

We can also examine the Q–Q plot or quantile–quantile plot by using the *qqline (regression model)* command. Y-axis is the standardized residuals, and X-axis is the ordered theoretical residuals. The points should follow a diagonal line if the errors or the residuals are normally distributed.

Normal Q-Q

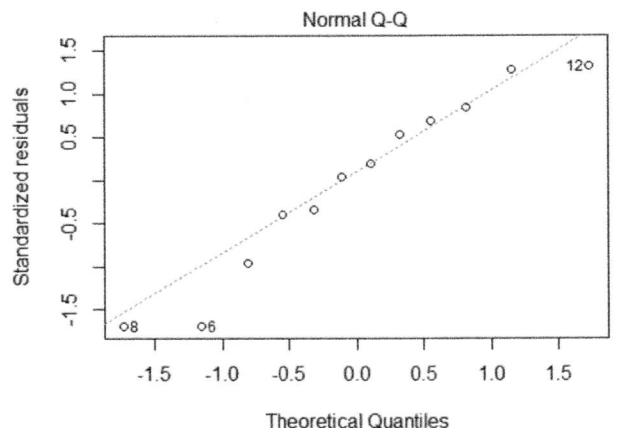

lm(ZWH ~ EDUCSPOUS + CALREQ + SICKFEED + GENDER)

Exercises

1. Undertake a multivariate regression analysis with ZWH (wasting) as the dependent variable and the same set of explanatory variables in this chapter. Which variables significantly influence wasting? What is the value of adjusted R^2 in your model? Plot the normal probability plot and the histogram of the weight-for-height residuals and examine if these residuals violate the assumption of normality.

2. Create interaction variables of maternal education (EDUCSPOUS) and the child care variables (ATTCLINI, CLINFEED, and BFEEDNEW) as the product of education of the mother and the child care variables (for example, EDUCSPOUS*ATTCLINI). Conduct a multiple regression analysis with EDUCSPOUS, ATTCLINI, and EDUCSPOUS*ATTCLINI as the independent variables and ZWH, ZHA, and ZWA as the dependent variables, respectively (there should be 9 regressions). Repeat the process with EDUCSPOUS*BFEEDNEW and EDUCSPOUS*CLINFEED as the independent variables. Is there a statistically significant interaction between education and the child care variables? Explain your results.

3. Suppose you are given the following regression equation: $\widehat{Y} = 15 + 1.5X_1 + 0.75X_2$. Can you tell the relative importance of X_1 and X_2 from the information given?

4. Study the papers by Strauss and Thomas (1995), Benefo and Schultz (1996), and Zhao and Bishai (2004). In light of the available evidence, explain whether maternal education and community characteristics are substitutes or complements in the determination of child nutritional status. (Your answer should not be more than two pages.)

5. What are the main determinants of fertility and mortality? (Study the paper by Benefo and Schultz (1996) to elucidate your answer.) Through what mechanisms does maternal education affect infant mortality and fertility? Discuss the community characteristics that are complementary and substitutes to maternal education.
6. Discuss the various steps for undertaking a multivariate regression analysis. Why is it important to check for violations of the regression assumptions?
7. How does improved water and sanitation condition translate into better child health outcomes? Critically discuss after studying the literature as presented in this chapter.
8. The following table presents an index of ZWA for 30 children. Along with this information, you are also provided the educational status of the head of the household (EDUC = 0 if illiterate, and EDUC = 1 if not illiterate), gender of the child (GENDER = 1 if male and GENDER = 0 if female), and finally, the age of the head of the household (AGE in years).

Obs	ZWA	EDUC	GENDER	AGE
1	11.19	1	0	29
2	10.00	1	0	33
3	5.77	0	1	30
4	1.56	0	1	32
5	14.96	1	0	31
6	8.66	1	1	26
7	7.79	0	1	31
8	17.79	0	0	33
9	11.06	0	0	29
10	12.08	0	0	30
11	14.42	1	0	30
12	7.93	1	1	30
13	13.74	1	1	34
14	7.21	0	1	26
15	23.56	1	0	29
16	9.16	1	0	31
17	11.69	0	1	34
18	6.73	0	0	32
19	12.50	0	0	30
20	8.65	1	1	28
21	11.78	1	0	26
22	3.99	0	1	31
23	19.20	0	0	34
24	10.97	0	1	29
25	6.01	1	0	31
26	1.89	0	0	31
27	7.21	0	1	34
28	15.67	0	0	30
29	5.10	1	1	25
30	12.98	1	0	28

Use the regress command and estimate the following models with robust standard errors:

Model 1: ZWA $= a + b$ AGE.
Model 2: ZWA $= a + b$ AGE $+ c$ EDUC.
Model 3: ZWA $= a + b$ AGE $+ c$ EDUC $+ d$ GENDER.

Which of these models is a useful indicator of ZWA? Does GENDER affect the variability in ZWA significantly? How about EDUC? Perform the necessary tests, taking the results from your STATA output.

Section III

Special topics on poverty, nutrition, and food policy analysis

Chapter 11

Predicting child nutritional status using related socioeconomic variables—application of discriminant function analysis

Chapter outline

Introduction	382
Conceptual framework: linkages between women's status and child nutrition	384
Linkages between women's status and child nutrition	384
Examples of linkages between women's status and children's nutrition	385
Review of selected studies	385
Direct linkage studies between women's status and children's nutritional status	385
Indirect linkages between women's status and child's nutritional status	388
USDA nutrition assistance programs: a case study from the United States	390
Case studies of women's status and child nutritional status from Africa, Asia, and Latin America	393
Food security and welfare in Africa: social customs, technology, and climate change	393
Childhood undernutrition and climate change in Asia	396
Adaptive strategies and sustainability lessons from Latin America	397
Can garden plots save Russia?	398
Empirical analysis and main findings	398
Data description and analysis	399
Descriptive statistics	401
Testing the assumptions underlying discriminant analysis model	403
Box's M Test	403
Tests of equality of group means	404
Summary of main findings	405
Relative impact of the predictor variables on ZWHNEW	407
Correlation between the predictor variables and discriminant function	408
Classification statistics	409
Classification function based on equal and unequal prior probabilities	410
Canonical discriminant analysis using STATA	412
Conclusions	416
Technical appendix: discriminant analysis	417

Food Security, Poverty and Nutrition Policy Analysis. https://doi.org/10.1016/B978-0-12-820477-1.00001-2
Copyright © 2022 Elsevier Inc. All rights reserved.

Discriminant analysis decision process		Stage 4: estimating the discriminant function and	
	417	assessing overall fit	421
Stage 1: objectives	418	Stage 5: interpretation of	
Stage 2: research design	418	discriminant functions	423
Discriminant analysis for several groups: a technical		Stage 6: validation of results	424
		Canonical discriminant analysis	
digression	419	**using R**	**424**
Stage 3: checking assumptions	420	**Exercises**	**426**
		STATA workout	**426**

You can tell the condition of a nation by looking at the status of its women.
Jawaharlal Nehru, the first prime minister of India.

Introduction

This chapter introduces discriminant analysis (DA) as a tool to identify the most important determinant of food security and nutritional status that can predict best among a large pool of such independent variables. Such variables need to be identified for context- and community-specific food and nutrition interventions.

Studies explaining child nutrition with related socioeconomic variables have gained momentum in the recent development literature for several reasons (Hobcraft et al., 1984). First, given the high cost of data collection on the nutritional status of children, which requires special equipment, there is a need for indicators that predict child nutrition in a cost-effective manner. Second, socioeconomic and demographic surveys that are conducted on a regular basis, such as national expenditure surveys and living standard measurement surveys (LSMS), do not often include nutritional status. Third, there has been increasing interest among nutrition economists to identify a set of robust variables that can explain child nutritional status in a community. Finally, for cross-country comparisons using existing data sets, variables that predict child nutrition consistently could be used in explaining global and regional trends in welfare indicators such as child nutrition.

In this chapter, we examine a variety of cross-country evidence from the United States, Africa, Asia, and Russia. We list the research findings from a variety of field research and link socioeconomic variables to nutritional status in the households. In this chapter, we introduce DA as a tool to identify the most important variables that can predict child nutrition among a large pool of variables.

The DA method is useful for predicting membership in naturally occurring groups and thus addresses whether a combination of variables (such as women's status, household income, and other community variables) can be used to predict group membership (such as child nutritional status). Once group means are found to be statistically significant, classification of variables is undertaken. The method automatically determines some optimal combination of variables so that the first function provides the most overall discrimination between groups, the second provides the second most, and so on. An important advantage of this method is that the functions will be independent or orthogonal, so that their contributions to the discrimination between groups will not overlap. The subjects will be classified in the groups where they had the highest classification scores. The maximum number of discriminant functions will be equal to the degrees of freedom or the number of variables in the analysis, whichever is smaller (Rencher, 2002).

Women's status in the society has been one of the socioeconomic variables that has gained recognition as a good predictor of child nutritional status (Haddad, 1999). Although much has been emphasized about the various causes of child malnutrition and the ways to eliminate it, little attention has been paid until recently to understand the role of women's status relative to men and its impact on child nutrition in developing countries.

Both anecdotal and empirical evidence suggest that income earned by women is more likely to be spent on food and other basic household needs than income earned by men, and thus, it can have a greater positive impact on children's nutritional status (Kumar and Quisumbing, 2012). In general, women in many parts of the world traditionally had major responsibility for feeding and caring for children. Furthermore, women in developing countries generally join the workforce out of economic necessity, so that their incomes must be devoted to survival needs. From the aforementioned facts, it is clear that an increase in income earned by women is likely to improve the nutritional status of children. Another reason to focus on women's income is that economically active women are more likely to act on many ideas given in nutrition and health projects than women whose activities are confined to the household sphere (Rogers and Youssef, 1988). The opportunity to earn an income will thus provide an initial incentive for women to change traditional patterns of behavior, which will bring perceptible benefits of improved home management and child care practices.

In what follows, we first provide a conceptual framework for briefly studying whether women's status is important for predicting child nutritional status. We then review a few studies that attempt to identify good predictors of child nutrition status. A discussion of the issues, data sets, analytical methods, and results of these studies serves as a motivation for understanding the use and application of the DA that follows. We provide a detailed implementation of the procedure in STATA. The final section provides concluding remarks and some implications of the results.

Conceptual framework: linkages between women's status and child nutrition

Linkages between women's status and child nutrition

Fig. 11.1 provides a conceptual framework for understanding the various causes of child malnutrition. The immediate causes are inadequate dietary intake and disease. Children can become malnourished due to insufficient food. Diseases, on the other hand, inhibit the absorption of nutrients and affect children's growth. At the household level, lack of food security, inadequate child care practices, and poor water and sanitation facilities are the underlying causes of malnutrition. Food security is defined as the access to sufficient food for a healthy life for all household members. Care is the provision in households and communities of time and attention for physical and other needs of the growing child. Feeding behavior (such as breastfeeding) can be critically important in the growth of the child. Also, a proper health environment, such as safe drinking water and sanitary facilities, can help in reducing diseases and thereby improve children's growth. Finally, the basic (distant) causes of child malnutrition can be attributed to the resources available in society and women's status (Smith et al., 2003).

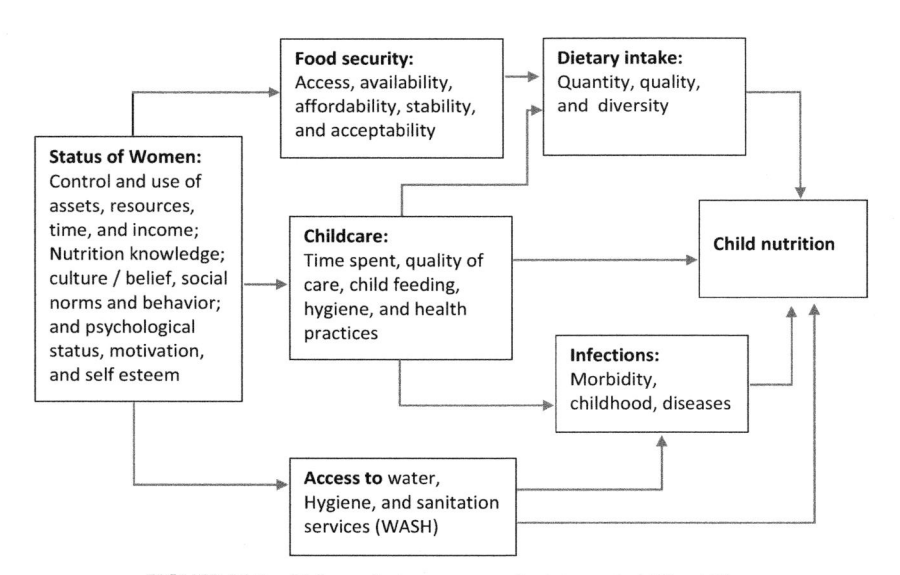

FIGURE 11.1 Linkages between women's status and child nutrition.

Examples of linkages between women's status and children's nutrition

In many societies, women are the primary caregivers for young children. At the same time, in parts of the developing world, women have relatively lower status compared with men. This has consequences for women's control over household time and income, knowledge and beliefs, and self-confidence. First, consider women's control over household resources. The greater a woman's control over her income, the more the likelihood that she will invest greater time in child care.

Studies (Haddad et al., 1997) have shown that income or assets accruing to women have a positive impact on child nutrition through greater expenditures on food, clothing, and healthcare for the child. However, under certain situations when resources are pooled jointly, a household acts as a unitary decision-maker. Under the aforementioned circumstances, women's status has no positive impact on the nutritional status of the child. On the other hand, if a woman is time constrained (which is possibly a consequence of low status), she can devote less time to quality care for the child. Less caregiving activities can have adverse consequences for child welfare.

Women with low status have less knowledge and more beliefs (Engle et al., 1999; Kishor, 2000), as they have less mobility and are less likely to go outside the home and engage in social interactions. They are more likely to be exposed to less health and nutrition knowledge, which can have unfavorable impact on the children's nutritional status.

Women's confidence and self-esteem in society can also affect the nutritional status of a child. The more a woman is dependent on her husband, the greater is the likelihood of disrespect. These repeated interactions with the spouse can lead to poor mental health and low self-esteem. Under such situations, it is less likely that the woman will make timely and independent decisions regarding healthcare treatment of a child (see, for example, Engle et al., 1999).

Indirect linkages, such as lower health status (malnourished or sick) of a woman, can have negative consequences on child nutrition. A woman who is malnourished may be less capable of breastfeeding successfully, with the consequence that they may give birth to underweight children. Moreover, women who are more time constrained are less likely to devote time for child care, which can affect their nutritional status adversely.

Review of selected studies

Direct linkage studies between women's status and children's nutritional status

Direct linkages should be interpreted as those affecting women's status directly, such as her income, work conditions, time constraints, and self-confidence, while indirect linkages can occur through lower healthcare for the woman, fertility regulation, and prenatal and birth care practices and is usually caused by women's lower status relative to men.

Kennedy and Haddad (1994) examined the impact of income level by gender of the household on child nutritional status for Kenya and Ghana. Six hundred and seventeen households were chosen for Kenya over a 3-year period in South Nyanza district (a rural area characterized by low population density, highly productive agricultural land, and a maize/sugarcane production mix).

For Ghana, the living standards survey (LSS) data were from a nationally representative survey of 3136 households. Households were disaggregated according to the self-reported household head's gender and months spent away from the residence by the self-reported male head. Thus, de facto female-headed households were classified as those where the self-declared male head was absent for at least 50% of the time, whereas de jure female-headed households were those in which a woman was generally considered as the legal and customary head of the household.[1] Female-headed households were 17% and 30% of all the households in the sample from Kenya and Ghana, respectively.

The main conclusion from the study was that the complex interactions of income, gender of the household head, and gender of the preschooler had an important influence on child nutritional outcomes. An interesting finding was that some poorer female-headed households were able to ensure better nutritional status for their children than wealthier male- and other female-headed households. An important implication of the study is that targeting interventions by gender of the household head may not be the most effective way to reach the poorest household, while interventions that promote the returns to appropriate nurturing behaviors (such as providing provision for public health services) can be extremely effective in short-term gains to child health and nutrition.

Barros et al. (1997) analyzed the characteristics and behavior of female-headed households in urban Brazil and identified the consequences of growth of these households for child welfare. The rationale for choosing Brazil was that female-headed households had a much higher probability of being poor compared with male-headed households. The main question addressed in the study is as follows: do female-headed households have a lower number of

---

1. De facto FHHs are households where the male member is absent for a significant part of the time. On the other hand, de jure FHHs are those where the self-reported female head does not have any legal or common male partner. These households are usually headed by widows, unmarried women, or those who are divorced or separated.

earners with less income-earning power or more mouths to feed? The main criteria for defining a household head were twofold. First, the person who had the highest income in the household was considered as the head. Second, the household head was the one who devoted the maximum effort (measured by the number of hours worked) on behalf of the household. The data used in the study were from the 1984 Brazilian household sample survey. Since there were high and variable rates of inflation during the 1980s in Brazil, the task of conducting intertemporal and interregional comparisons of income becomes vastly complicated. The study was conducted in three representative metropolitan areas of Brazil, with poverty defined according to relative income in each area.

One of the main conclusions from the study was that female-headed households were a heterogeneous and diverse group with substantial variations by region. Female-headed households were not, on average, a vulnerable segment of the population, and the extent of poverty among the female-headed households varied greatly among regions. In the northeast region, female-headed households were generally poorer, especially in the area of Recife. Female-headed households (FHHs) with children did display a greater income gap compared with other households. Second, female-headed households had lower income, as the head of the household earned less. Thus, the analysis showed that the best interventions for eliminating poverty for the targeted group would be to focus explicitly on ending wage discrimination and ending occupational segregation for women. Third, the results suggested that special intervention was required for children in female-headed households, since such children usually stayed out of school. Even after controlling for household income, the results demonstrated that children in female-headed households had poorer school attendance records than other children. This was especially true for older children. The aforementioned finding possibly indicates that female-headed households need to balance the trade-off between earning more income and the time constraint that they face in child care practices.

In a comprehensive study conducted between 1990 and 1998 for three major regions (South Asia, sub-Saharan Africa [SSA], and Latin America and the Caribbean [LAC] and included 36 countries), Smith et al. (2003) examined the role of women's status on child's nutritional status. The sample included 117,242 children less than 3 years of age and 105,567 women who were usually mothers. The sample children lived in two-parent households, while the sample women included those individuals who were married and had at least one child less than 3 years old. The study used both least squares and logistic regression methods. There were 25 dependent variables examined, with six of them being measures of child nutritional status. Various socioeconomic measures and household characteristics were controlled for as independent variables.

The major inference from this highly influential benchmark study demonstrated that women's status was critically important in improving children's nutritional status. However, the strength of the impact varied across regions. Women's relative decision-making power had a strong influence on child nutritional status in South Asia and moderate and positive impact in SSA, and affected only the short-term nutritional status in the LAC region. Gender equality appeared to have a weaker impact on child nutritional status, with the positive impact occurring only in South Asia. Finally, the study found that economic status and women's status have strong linkages in affecting child nutritional status among the poorer households.

Indirect linkages between women's status and child's nutritional status

Lower status of women translates into greater likelihood of them being malnourished and sick, which, in turn, reduces their energy level and responsiveness to needs of the child. A woman who is malnourished may be incapable of breastfeeding successfully. Moreover, if she is often pregnant, the nutrient reserves of the body decline, which, in turn, could lead to poor growth of the child. A woman with lower status usually does not have decision-making capability in fertility matters. Micronutrient deficiency of the woman can be passed on to the child during the pregnancy period and affect the nutritional status of the child adversely (such as low birthweight).

Ukwuani and Suchindran (2003) examined the relationship between women's work conditions and child nutritional status (stunting and wasting) for a sample of Nigerian children aged 0—59 months, using the 1990 Nigerian Demographic and Health Survey data set. Women's work conditions can serve as an important component of women's status, as it can influence the nutritional status of the child through the pathways of income and child care. The novelty of the study was in going beyond simple categorization of women's work, as it examined whether women's earned cash from work and taking their children to work had a significant impact on child nutritional status.[2]

---

2. The impact of women's work on child health is generally ambiguous. This is because women's work can have a positive impact on child health by providing greater income and child care. At the same time, more work can have a negative impact on child health through time constraints (such as time available for breastfeeding).

The main hypotheses examined in the study were as follows:

1. Women's work had a negative impact on child nutrition during infancy. This negative effect could be reduced when women combined work with child care.
2. Women's work had a positive impact during childhood when women could earn cash from work.

The dependent variable was child nutritional status, as measured by stunting and wasting. The primary independent variable was women's work. Other controls were socioeconomic factors, household characteristics, child care, and water and sanitation conditions.

The authors recommend provision of child care services for working mothers, as there is evidence of the negative impact of mother's work on child nutritional status during infancy. Additionally, since mother's work had a negative impact on stunting during childhood, better employment opportunities and empowering women remain a crucial agenda in Nigeria. Since the incidence of diarrhea caused wasting and lack of immunization had a strong effect on stunting and wasting, policy makers would be better off undertaking vigorous health interventions for maternal and child health care. Overall, the study concluded that empowering women through better work conditions could alleviate the problem of malnutrition to a significant extent.

In another study, Li (2003) examined the impact of fertility reduction (one-child policy) on the well-being of Chinese children. The major issues addressed in this study were as follows:

1. What was the impact of one-child policy on the health status of Chinese children? In other words, the author examined if there was a trade-off between quantity and quality of children.
2. Did the decrease in family size help in improving girls' well-being?

The data were collected from the China Health and Nutrition survey conducted during 1989, 1991, and 1993. About 190 communities from eight provinces were chosen with differing levels of economic development, public resources, and health indicators. A total of 3800 households were chosen consisting of 16,000 individuals.

A reduced-form specification was estimated using least squares, where the nutrient intake (measured alternatively by average calorie intake, protein intake, and fat intake) and health status (measured by the height for age Z-scores) were regressed on individual characteristics of the child such as age, sex, and birth order; household characteristics such as mother's education, household income; and community characteristics such as one-child policy, water and sanitary conditions, and unobserved individual, household, and village-specific fixed effects.

The paper evaluated indirectly the impact of women's status (fertility reduction) on child nutrition with a focus on gender differences. The findings indicate the following facts: First, parent's trade-off quality and quantity of children in China. By reducing family size, the one-child policy (a reduction in fertility rate) significantly improved children's well-being. Second, sex selection in China did not eliminate the discriminating treatment toward girls compared with boys.

Kim et al. (2020), Menon et al. (2020), Nguyen et al. (2020), and Frongillo et al. (2019), in a sequence of related studies, provide details of a successful nutrition intervention program called Maternal, Neonatal, and Child Health (MNCH) in Bangladesh, which focuses particularly on antenatal care. The program was designed for pregnant women and lactating mothers, to assist their health, and also aid infant development. The nutrition-focused MNCH program is multilayered and has several detailed components: antenatal care visits, interpersonal communication, community mobilization, monitoring of weight gain, maternal diet quality, micronutrient intake, breastfeeding practices, health of the fetus, medical complications, and building support structure, emergency preparedness, and information diffusion among social networks, The program's success is evidenced by Frongillo et al. (2019) by reductions in food insecurity among the participants: by 20% compared to the baseline, and by 22% compared to the standard MNCH groups.[3]

USDA nutrition assistance programs: a case study from the United States

The US Department of Agriculture (USDA) administers 15 domestic nutrition assistance programs, which form a nutritional safety net for millions of children and low-income adults, a role that is especially important when the economy falters and many Americans lose jobs and income. Hanson and Oliveira (2012) investigate the relationship between economic conditions and participation at the national level across USDA's five largest nutrition assistance programs. The goal of this research is to examine how changes in program policies and other factors such as demographics affect participation.

To accomplish this, the researchers used national-level administrative data on program participation collected by USDA's Food and Nutrition Service (FNS) and data on unemployment rates published by the Bureau of Labor Statistics. The study covered the period 1976–2010, which includes four

---

3. Menon et al. (2020) and Kim et al. (2020) obtain similar results for Ethiopia and Vietnam.

complete business cycles, each consisting of economic growth characterized by a falling unemployment rate and a period of economic decline characterized by a rising unemployment rate.

The authors corroborate the business cycle data with the fluctuation in program participation and the unemployment rate. The researchers then integrate various publications and regulations to determine how program policy and demographics influence the relationship between program participation and the unemployment rate.

For instance, Hanson and Oliveira (2012) suggest that, to varying degrees, economic conditions, as measured by the unemployment rate, influence participation in all the major nutrition assistance programs. As mentioned in Chapter 8, Gundersen and Ziliak (2018) and Gundersen (2019a,b) note that more than 41 million persons (about 12% of the population) were food insecure in 2016 and received around $ 66 billion in benefits. Food insecurity has been steady at around 17% for children. Note the increase in food insecurity rates around 2008 coinciding with the Great Recession.

1. The general consensus from extensive research suggests that food insecurity is associated with poverty, low assets, low human capital, and low health outcomes across the age gradient.[4] Other findings include the following: The increase in SNAP participation during the 2008–10 period of economic decline was consistent with the increase during previous periods of economic decline, at 2–3 million participants per 1% point increase in the unemployment rate.
2. Policy changes pertaining to rules of eligibility, benefit levels, outreach, and the application certification process tend to augment the increase in SNAP participation due to economic conditions in each period of economic decline.
3. Before being fully funded in the late 1990s, WIC participation was rationed by the program's budgetary limits and expanded as the budget grew. The introduction of infant formula rebates in the late 1980s lowered the cost of the WIC food package, enabling more people to participate within the program's budget and fueling an increase in the annual average growth in participation.
4. After reaching full funding in the late 1990s, WIC caseloads became more sensitive to economic conditions, increasing by nearly 2.5% or about 200,000 participants per 1% point increase in the unemployment rate.
5. The number of births also had a strong influence on the number of participants; for instance, during the recession of 2008–09, the low number of births tended to counter the growth in participation prompted by economic conditions.

---

4. Also see *Southern Economic Journal*, 2016, 82(4) for a symposium on food security among children in the United States.

6. The percentage of participants receiving free and reduced-price meals in the child nutrition programs is related to economic conditions, rising with the unemployment rate during periods of economic decline.
7. Total participation in the child nutrition programs has steadily increased during periods of both economic growth and decline. These programs serve both low- and high-income children.
8. The National School Lunch Program (NSLP) participation is linked to school enrollment, while availability of the program in schools has been a key to the growth of the participation in School Breakfast Program (SBP).
9. Increasing SNAP benefits by $ 41.62 per week for the recipients would lead to a reduction in food insecurity by 60%.
10. The Free Lunch Program reduces food insecurity by at least 6%, poor health by at least 33%, and obesity by at least 21%.

Section highlights: What about home-cooked meals?

One of the key determinants of child's nutritional status is the quality of food. Given the importance of the educational status of the parents in this context, a natural question arises as to whether good quality food can be regularly provided if both parents work full time. In other words, parents with low income should be able to prepare home-cooked meals because their opportunity cost of time should be low. This feature is an important economic aspect of the problem because deciding whether or not to cook a meal at home depends on not only the cost of the ingredients used to prepare the meal but also the time it takes to prepare the meal, serve it, as well as clean up afterward. Since time is scarce and costly, it is important to account for the time cost of food preparation in addition to the direct cost of ingredients in the context of nutritional assistance program policy.

How does this aspect fit in with the issue of child's nutritional status? To put this in perspective, Raschke (2012) points out that during 2010, over 40 million people in the United States received benefits from the Supplemental Nutrition Assistance Program, or Food Stamps each month.

Food stamps allow the recipient to purchase food at grocery stores, produce markets, and other stores using a debit card system that electronically deducts money spent on groceries from the recipient's monthly allotment. However, food stamps cannot be used to purchase prepared meals in the market or nonfood items also available at grocery stores.

Therefore, when using food stamps, the recipient must necessarily incur a time cost of preparing the food at home. A one-dollar benefit can be used to buy a one-dollar cheeseburger at a fast food restaurant, or a dollar's worth of ingredients can be purchased at a grocery store, to prepare the food at home. In the latter case, the recipient's total benefit is one dollar minus the time cost of transforming the ingredients into a meal and cleaning up afterward.

Now it is easy to criticize SNAP because it is possible that the cooking time required to prepare all meals from scratch is excessive. This opportunity cost can

> **Section highlights: What about home-cooked meals?—cont'd**
>
> be high and can cut into the labor force participation of low-income individuals. Whether the total opportunity cost of home-cooked meals is high enough to be a disincentive to work is ultimately an empirical question, and Raschke (2012) investigates the issue.
>
> Raschke (2012) estimates the time cost and total cost of food-at-home for food stamp recipients as well as nonrecipients using a structural model of individuals' time allocation decisions. Raschke (2012) combines very interesting and different data sets to estimate a time allocation model that includes various components of cost and benefits for SNAP recipients and nonrecipients. Raschke (2012) provides us the following insights regarding the microeconomic problem of optimal time allocation:
>
> - The time cost of preparing food at home is a significant proportion of the total cost incurred by an individual preparing food at home.
> - The average household not participating in SNAP incurs about $28.27 in time cost per day by preparing meals at home. In terms of individuals, the average time cost is $11.51 per person per day across households.
> - Food stamp recipients incur a significantly smaller time cost of $24.56 per day per household or $9.40 per day per individual.
> - Time costs account for 64% of the total cost of food at home for SNAP nonparticipants, whereas time costs account for about 61% of the total cost of food at home for SNAP participants.
>
> What can we conclude from this analysis? Is the total cost of food preparation higher for food stamp recipients compared with nonrecipients due to the increased amount of time that is spent preparing food? Raschké's results show that food stamp recipients have an estimated shadow wage of home production that is significantly *lower* than nonrecipients. Furthermore, the shadow wage of home production of food stamp recipients is low enough so as to offset the increased amount of time spent in the kitchen. Consequently, the total cost of food preparation at home is still significantly lower for food stamp recipients compared with nonrecipients. Needless to say, these aspects have direct relevance for child's nutritional status, quality of life, and economic outcomes.

Case studies of women's status and child nutritional status from Africa, Asia, and Latin America

Food security and welfare in Africa: social customs, technology, and climate change

Hoddinott et al. (2012) examine Ethiopia's Food Security Program, which provides income transfers through public works in its Productive Safety Net

Program (PSNP) as well as targeted services provided through the Other Food Security Program (OFSP) and the Household Asset Building Program (HABP). The income transfers through these programs are designed to improve agricultural productivity. Hoddinott et al. (2012) point out that there is a trade-off between these two types of transfers. The trade-off is between short-term improvements in food security and longer-term food security achieved through increased agricultural productivity.

The researchers quantify the trade-off by calculating the relative impact of PSNP transfers alone and joint transfers from the PSNP and OFSP/HABP on agricultural output, yields, fertilizer use, and agricultural investment for farmers growing cereals in Ethiopia from 2006 to 2010.

The results for Ethiopia indicate that access to the OFSP/HABP program along with high levels of payments from the PSNP led to considerable improvements in the use of fertilizer and enhanced investments in agriculture. Hence, this strategy improves agricultural productivity among households that receive both programs. However, the researchers find mixed effects of participation in both programs in terms of impacts on yields. Furthermore, high levels of participation in the PSNP program alone have no effect on agricultural input use or productivity and limited impact on agricultural investments. Consequently, the study indicates that the combined transfers are more effective at increasing yields.

Social customs in Ethiopia also have consequences on economic welfare. In an interesting study, Kumar and Quisumbing (2012) examine the role of men's and women's asset inheritance in poverty and well-being in rural Ethiopia. They collect data from the Ethiopian Rural Household Survey (1997, 2004, and 2009) and investigate the long-term impact of gender differentials in inheritance on household consumption, poverty, and food security. The study also identifies significant differences in poverty and well-being between male- and female-headed households. Most interestingly, amounts of inheritance received, and not whether women inherit at all, have the most profound impacts. The area of land inherited is particularly important for women's long-term well-being. Obviously, the results of Kumar and Quisumbing´s (2012) underscore the importance of women's rights to inherit equally with men.

Asfaw et al. (2012a,b) examine the impact of adoption of improved legume technologies on rural household welfare measured by consumption expenditure in rural Ethiopia and Tanzania. In particular, Asfaw et al. (2012a,b) estimate the true welfare effect of technology adoption by controlling for the role of selection problem on production and adoption decisions. The researchers show that adoption of improved agricultural technologies has positive impact consumption expenditure in rural Ethiopia and Tanzania. The study confirms the potential role of technology adoption in improving rural household welfare as higher consumption expenditure from improved technologies translate into lower poverty, higher food security, and greater ability to withstand risk. Inadequate local supply of seed, access to information, and perception about the new cultivars are the major constraints that discourage technology adoption.

For SSA, Simatele et al. (2012) note that many urban dwellers increasingly resort to a wide range of informal sector activities to ameliorate food insecurity and generate household income. Among these activities is urban agriculture, which serves as a source of basic foodstuff and income generation. Despite its significance and contribution to the urban household food basket, urban agriculture in many SSA cities has not been integrated into urban development and planning policy. In addition to the absence of a supportive local government policy, over the past two decades, this activity has come under increasing pressure from extreme weather-related events such as droughts and flooding. Simatele et al. (2012) undertake field-based research in Lusaka and demonstrate the plight of urban agriculture given extreme weather conditions, and its negative impact on the livelihood options of urban poor.

Zereyesus et al. (2017) note the importance of women empowerment and well-being on the health status of their children. Using a Women's Empowerment in Agriculture Index, the researchers examine the impact of a Multiple Indicators Multiple Causes (MIMIC) model on children's health. The MIMIC model indicates that the associations between children's health status, particularly height-for-age and control variables such as mother's education, child's age, household's hunger scale, and residence locale, are statistically significant.

Evidence from several studies continue to support the connection underlying this theme. Narayanan et al. (2019), Gupta et al. (2019), and Prakash and Jain (2016) note the slow decline in the proportion of malnourished children in India, with a steady increase in economic inequalities among this group. The studies uniformly find that mothers' illiteracy and women empowerment in agriculture and household decisions are key drivers of their own lower BMI, anemia, iron deficiency, and a consequent malnourishment of their children. Inequalities within the group are often related to low diet diversity score, share of wheat and rice in the diet, and the lack of iron intakes. Anwar et al. (2015) and Haroon (2018) reiterate similar findings for Bangladesh, Nepal, and Pakistan. Overall, substantial evidence across countries indicate the following:

- Women empowerment (via education, labor force participation, asset ownership, conditional cash transfers, microfinance, ownership of rights in household, agriculture and livestock decisions, and freedom of movement and from violence) is strongly related to reductions in children's stunting, wasting, and related health outcomes
- Maternal BMI is significantly and positively related to child nutritional status.
- Mother's education, working status, health, availability of safe drinking water, family size, and vaccination have significant effect on child health.

Childhood undernutrition and climate change in Asia

Menon (2012) points out that South Asia carries the world's highest burden of child nutrition. In her study of South Asia, Menon (2012) examines the problem by combining insights from nutrition, biological, and social sciences. Menon (2012) demonstrates that child nutrition is an outcome of immediate factors such as food and nutrient intake and illness, which in turn are influenced by underlying factors such as household food security and poverty, women's status and access to health, nutrition, and social services. Furthermore, basic societal factors such as institutions, governance, politics, and culture are also strong determinants of childhood undernutrition.

Similar to the climate change problems in SSA, climatic impact on agricultural production is a serious concern in Asia also. Barnwal and Kotani (2013) examine the issue of climate change and link it to food security and poverty, by quantifying the impact on crop yield distributions. The researchers examine the case of rice yield in Andhra Pradesh, India. They select Andhra Pradesh because it is an important state, which produces rice as a main crop, and hence, is highly vulnerable to climate change.

Sam et al. (2019) note how drought severely affects food security of rural household in the state of Odisha, how states across South Asia are still underprepared in the wake of severe drought risk. Venakatasubramanian and Ramnarain (2018) demonstrate how climate change impacts women, and particularly, women in pastoral communities across the state of Gujarat. Hence, women in these communities carry the burden of coping with climate change, alongside unequal gendered norms. Fan (2016) calls for an abandonment of the silo approach in agriculture policies in Pakistan and emphasizes the need for exploiting intersectoral linkages among food, water, and energy to reduce costs associated with climate change and foster nutrition security.

Barnwal and Kotani (2013) collect an interesting data set to link climate conditions and agriculture output. Using the interesting data set that relates climate change to agricultural output, the authors find three key results:

1. There is a substantial heterogeneity in the impacts of climatic variables across yield distribution.
2. The direction of the climatic impact on rice yield highly depends on agroclimatic zones.
3. Seasonal climatic impacts on rice yield are significant. More specifically, a monsoon-dependent crop is more sensitive to temperature and precipitation, whereas a winter crop remains largely resilient to changes in the levels of climate variables. These findings clarify the idiosyncratic climatic impacts on agriculture in India, and the researchers stress the need for location- and season-specific adaptation policies in India.

Adaptive strategies and sustainability lessons from Latin America

How do low-income households that are vulnerable to food insecurity make adjustments in their lives to cope with their plight? This question has always interested researchers, and as an example, Floro and Bali (2013) contend that household members can undertake adaptive strategy using occupational choices to help ensure their access to food. The authors collected data on self-employed women and men in 14 predominantly slum communities in Bolivia, Ecuador, Philippines, and Thailand. The study demonstrates that the household vulnerability to food insecurity significantly affects the choice of business. Women in vulnerable households are more likely to engage in food enterprises.

The findings of Floro and Bali (2013) suggest that urban low-income households in Latin America mitigate the risk of food shortage, through the selection of an enterprise activity that earns money income, and is a direct source of food for consumption.

Another study using the case study of Brazil examines to what extent the country has reduced hunger and food and nutrition insecurity. Rocha et al. (2012) also indicate that, in the past few years, Brazil has made significant progress in reducing hunger and food and nutrition insecurity. By the end of 2009, it had met the first United Nations Millennium Development Goal of reducing poverty and malnutrition by half—6 years ahead of the 2015 deadline.

Much of the aforementioned progress in Brazil has been achieved through innovative policies and initiatives championed by civil society organizations for over the two decades. From Rocha et al. (2012), we find that civil society organizations have high levels of participation in Brazil due to the institutionalization of these organizations in local- and national-level councils. Through their participation in food and nutrition security councils, civil society organizations promote a more integrated perspective on public policy planning and healthy food practices. As a result of these efforts, Rocha et al. (2012) point out that Brazil was able to add the "Right to Food" to the country's constitution in 2010, making it the duty of the state to ensure that all Brazilians enjoy adequate food.

Focusing on conditions for small family farmers, Rocha et al. (2012) describe the main elements of these government programs as well as relevant civil society initiatives.

The following are the key insights of the study:

- Supporting small farmers in developing countries not only improves food security but also creates employment opportunities and sustainable rural development.
- Organized civil society should be involved in planning, implementing, and monitoring policies.

- Civil society organizations need information, training, and education to be effective participants in policy development.
- Governments need to set up processes whereby civil society's participation in policy development, implementation, and monitoring is institutionalized.

Can garden plots save Russia?

Takeda (2012) investigates whether food production on garden plots, a traditional activity in rural Russia, functions as a safety net for rural households in the event of an income shock. The study also uses interesting microdata from the Russian Household Budget Survey of 2004 and 2009 for the empirical study. The main finding of this study is that poor rural households are more active in carrying out the food security function of their garden plots in the event of an income shock. This shows that production on garden plots could help rural households recover from an income shock and help poor rural households in particular escape poverty. This finding has many applications for other economies also.

Overall, what can we conclude from the existing studies of the relationship between women's status and child nutrition? As pointed out by Buvinic and Gupta (1997), and Fuwa (2000), the answer is at best ambiguous. On the one hand, if FHHs are indeed poorer than non-FHHs in terms of both consumption and leisure, then children's welfare in FHHs tends to be lower through lower consumption (including food consumption, which could have long-term effects), lower educational expenditures, etc. On the other hand, children within the FHH households could be better off than their non-FHH counterpart due to systematic differences in the patterns of household resource allocation as a result of differential preferences between women and men. Thus, there are mixed results from the existing empirical studies on the relationship between female headship and the welfare of children. This chapter uses DA as an analytical tool to determine how child nutritional status can be predicted from a set of group variables, with one of the grouping variables being women's status. In this analysis, if the means for a variable are significantly different in various groups, then this variable discriminates between the two groups, allowing the use of that particular variable to predict group membership.

Empirical analysis and main findings

We undertake DA to understand which series of variables best predicts the nutritional status of the child.[5] A DA approach is suitable since the dependent variable nutritional status (weight-for-height Z-scores [ZWH]) can be

---

5. A very useful web resource for multiple discriminant analysis can be found at David Garson's webpage at http://www2.chass.ncsu.edu/garson/pa765/mda.htm.

categorized into discrete groups, and one can predict group membership from a set of predictor variables (both discrete and continuous). In a DA approach, we want to determine a function that maximizes the distance between the groups. In other words, we want to come up with a function that has strong *discriminatory power* among the groups.

It is essentially identical to multiple regression, but, in reality, the two techniques are quite different. In linear regression, we estimate the parameters to minimize the residual sum of squares, whereas in the DA procedure, we estimate the parameters that minimize the within-group sum of squares. A DA model looks as follows:

$$Z = a + W_1 X_1 + W_2 X_2 + \ldots + W_k X_k \tag{11.1}$$

where Z is the discriminant score, i.e., a number that can be used to predict group membership, a is a discriminant constant, W_k denotes the discriminant weight or coefficient, which is a measure to the extent that X_k discriminates among the groups of Z. X_k is a predictor variable, which can be either discrete or continuous. The various steps involved in the DA are as follows:

- Specify the dependent and the predictor variables.
- Determine the method of selection and criteria for entering the predictor variables in the model.
- Test the model's assumptions a priori.[6]
- Estimate the parameters of the model.
- Determine the significance of the predictors.
- Validate the results.

Data description and analysis

The ZWH is considered as the group variable and is divided into three categories as follows: ZWHNEW = 0 if ZWH $\geq$ 0. This category represents good nutritional status. ZWHNEW = 1 if $-2 <$ ZWH < 0. The aforementioned category represents moderate to low wasting. ZWHNEW = 2 if ZWH < -2. This category will be called severe wasting. The reference group for comparison is ZWHNEW = 0 (good nutritional status). (We categorized ZWHNEW initially into five different categories by differentiating between good nutritional status, low wasting, moderate wasting, high wasting, and severe wasting with different threshold values of ZWH. However, the discriminating function was unable clearly to differentiate the cases between the different groups. Additionally, some of the covariance matrices of the

---

6. The main assumptions in the discriminant analysis are (1) predictor variables are multivariate normal; (2) predictor variables are noncollinear; (3) absence of outliers; (4) sample is large enough (at least 30 cases) for each predictor variable; and (5) variance covariance matrices of the predictor variables across the various groups are homogeneous.

groups were singular, possibly indicating some linear dependency among the variables. Thus, the aforementioned classification was not undertaken in subsequent analysis.) The predictor variables are classified into individual characteristics (X_{IC}), household or socioeconomic variables (X_H), and community-level factors (X_C), similar to the multivariate regression model. Thus, the discriminant model looks as follows:

$$\text{ZWHNEW} = a + W_1 X_{IC} + W_2 X_H + W_3 X_C \qquad (11.2)$$

The *individual* characteristics or the individual predictor variable (X_{IC}) is the age of the child in months (AGEMNTH). The *household or socioeconomic* vector (X_H) consists of the following set of variables:

1. Per capita expenditure on food (PXFD): Income is a central variable in models of child health and nutrition outcomes. As more resources become available to households, they can incur higher expenditures on food and nonfood items (such as health). We use per capita food expenditure as a proxy for income, since food constitutes an important component of household expenditure in Malawi. This variable is computed by taking the ratio of total food expenditure to the household size.
2. Women's status (S): This is the key predictor variable in our model and is measured by the difference in educational status between the household head and the spouse. It is a continuous measure with values ranging from -3 to $+3$. Since education equips individuals with skills that allow them to understand better and operate in a modern environment to undertake quality child care, it is important to understand how women are educated relative to men. Additionally, this variable may also reflect intrahousehold bargaining power, which translates into differences in relative power. This measure is defined to have a threshold level such that a woman has as much or more education than her husband. Child care variables are also critical in improving nutritional status of children. The following variable is used as a proxy for child care:
3. BFEEDNEW: This is a dichotomous variable indicating whether the child is breastfed or not.

The *community characteristics* variables (X_C), such as water and sanitation indicators, usually affect the health production function and, consequently, influence the nutritional status of the child positively. The different community characteristics considered in the present analysis are as follows:

1. Drinking distance (DRINKDST): This is a categorical variable assuming values from 1 to 5. Higher values of this variable denote that the distance to a protected drinking source for the household is higher. We can expect that the greater the distance to a protected water source, the more the likelihood that children will suffer from malnutrition.

2. Provision of sanitation (LATERINE): This is a dichotomous variable assuming two values 0 and 1, with 0 indicating absence of latrine from the household.
3. DIARRHEA: This variable indicates whether the child has diarrhea and is a dichotomous variable assuming two values 0 and 1. 1 indicates that the child had diarrhea during the survey.
4. In addition to the aforementioned measures, we also consider the provision of health and infrastructural facilities, which indirectly contribute to improved child nutritional status. They are as follows.
5. HOSPITAL: This is a dichotomous variable assuming two values 0 and 1, with 1 indicating presence of modern medical care. Since all the households in the sample have some form of medical care (either modern facilities or traditional doctors in the community), there is no variation among households with this measure. Thus, we expect that this variable may not be able to discriminate clearly among the groups.
6. SADMDIST: This is a categorical variable assuming values from 1 to 5. Higher values indicate that the distance to the nearest ADMARC market for private traders to sell their products is higher.

Descriptive statistics

We first look at a few descriptive statistics relating the various characteristics of the households (households who have information pertaining to both food security and nutritional status of children, market access variables and the community characteristics variables) with the gender of the household head.

Table 11.1 shows that female-headed households are relatively much less educated compared with the male-headed counterparts, and this difference is significant using the chi-square tests. Additionally, 70% of the female-headed households have no latrine in the house compared with about 60.5% of the male-headed counterparts, while 62.5% of the female-headed households have unprotected drinking water sources. Thus, female-headed households are relatively at a greater disadvantage than their male counterparts. Having looked at some of the characteristics of the household with the gender dimension, let us investigate whether children from female-headed households have better nutritional status.

Table 11.2 shows some consistent patterns. Children from the female-headed households are less prone to being underweight and wasted than male-headed households. In contrast, children from male-headed households are less prone to be stunted relative to the female counterparts (although this difference is not significant after undertaking chi-square tests). This may be because wasting and stunting measure different dimensions of child nutritional status and the underlying determinants can differ. One plausible reason why female-headed households may have a greater percentage of stunted children

TABLE 11.1 Gender of the head and household characteristics.

Characteristics of the household	Categories	Male headed	Female headed
Educational levels	No education	19.75	50.0
	Adult literacy training	0.64	2.5
	Std 1–4	28.02	17.5
	Std 5–8	45.22	30.0
	Secondary education	6.37	0.00
Latrine in house	No	60.51	70.0
	Yes	39.49	30.0
Drinking water source	Unprotected	48.41	62.5
	Protected	51.59	37.5

Note: The numbers in the table denote percentage.

TABLE 11.2 Gender of the head and nutritional status of children aged 0–5.

Nutritional status measures	Categories	Male headed	Female headed
ZWHNEW	Good	32.42	38.89
	Low to moderate wasting	58.24	52.78
	Severe wasting	9.34	8.33
ZHANEW	Good	26.04	23.53
	Low to moderate stunting	29.59	29.41
	Severe stunting	44.37	47.06
ZWANEW	Good	8.91	15.79
	Low to moderate underweight	55.94	63.16
	Severe underweight	35.15	21.05

Note: The numbers in the table denote percentage.

is that both income and educational level of these households are lower compared with the male-headed households. The aforementioned differences translate into lower nutritional status of children in the long run, as female-headed households have less resources. At the same time, these households are not aware (possibly less exposed to the media sources) of how to improve children's nutritional status (for example, through better care practices), since they are less educated on an average than male-headed households.

Testing the assumptions underlying discriminant analysis model

Box's M Test

This procedure tests the assumption of homogeneity of the variance—covariance matrices. For g number of groups, the null hypothesis is of the form:

$$H_0: \sum_1 = \ldots = \sum_g \tag{11.3}$$

Box (1949) derived a test statistic based on the likelihood ratio test, and it is known as Box's M. The details are given in the technical appendix of this chapter.

For moderate to small sample sizes, an F approximation is used to compute its significance.

Table 11.3 shows that the test is significant, and approximately $F = 1.71$, $P = .00$. Thus, we conclude that the three ZWHNEW groups differ in their covariance matrices, which violates an assumption of DA. However, DA analysis is robust even when the homogeneity of variance assumption is not met since we have eliminated the outliers from the data set. Also, since the number of observations is reasonably large, small deviations from homogeneity will be found significant.

TABLE 11.3 Box's M Test results.

	Box's M		139.94
F	Approx.		1.714
	df1		72
	df2		7355.08
	Sig.		0.00

Tests null hypothesis of equal population covariance matrices.

Tests of equality of group means

The rules for variable selection in the stepwise method are as follows:

- Eligible variables with higher inclusion levels are entered before eligible variables with lower inclusion levels.
- The order of entry of the eligible variables with the same even inclusion level is determined by their order on the ANALYSIS specification.
- The order of entry of the eligible variables with the same odd level of inclusion is determined by their value on the entry criterion and the variable with the "best" value for the criterion statistic is entered first.
- Prior to including any eligible variables, all the entered variables, which have level one inclusion numbers, are examined for removal. A variable can be removed if its F-value to remove is less than the F-value for variable removal. Otherwise, if the probability criterion is used, the significance of the variable F to remove exceeds the specified probability level.

If more than one variable is eligible for removal, the particular variable is removed, which leaves the "best" value for the criterion statistic for the remaining variables. This process continues unless no other variables are eligible for removal.

On the other hand, a variable with an odd inclusion number is considered ineligible for inclusion if the following conditions are satisfied:

- its F to enter is less than the F-value for a variable to enter value; or
- the significance level associated with its F to enter exceeds the probability to enter.

One of the most commonly used metrics is to test the null hypothesis that the group mean vectors are equal. In other words, we want to test the null hypothesis:

$$H_0: \mu_1 = \mu_2 = \ldots = \mu_g \qquad (11.4)$$

where μ_i is the p-length population mean for group i. The nonrejection of the null suggests that the distribution of the variables does not differ significantly across the groups. In other words, the values of the variables are not useful in explaining or predicting groups.

Table 11.4 tests for the equality of group means of the predictors. The smaller the Wilks' lambda (λ), the more important the effect of the independent variable to the discriminant function. The null hypothesis can be rejected when the value of the test statistic is greater than the critical value of F for the desired level of significance. In the present case, λ is significant by the F-test for women's status as measured by the educational difference between the household head and the spouse. Thus, women's status is a significant predictor of the discriminant function. Additionally, λ is also significant for the CARE variable as measured by the breastfeeding of the child. Thus, both women's status and

TABLE 11.4 Tests of equality of group means.

	Wilks' lambda	F	df1	df2	Sig.
PXFD	0.992	0.828	2	199	0.438
EDUCDIFF	0.938	6.531	2	199	0.002
BFEEDNEW	0.975	2.513	2	199	0.084
AGEMNTH	0.969	3.157	2	199	0.045
DIARRHEA	0.996	0.397	2	199	0.673
DRINKDST	0.933	7.173	2	199	0.001
HOSPITAL	0.995	0.515	2	199	0.598
SADMDIST	0.960	4.197	2	199	0.016
LATERINE	0.996	0.356	2	199	0.701

child care are significant predictors of the child nutritional status. Moreover, the presence of protected drinking water source and the availability of infrastructural facilities in the household and the community appear to be significant predictors for the various groups. Interestingly, however, we do not find sanitation facilities to be a significant predictor of the discriminant function.

Summary of main findings

One discriminant function will be computed from the lesser of $(g-1)$ (number of dependent groups minus 1) or k (the number of independent variables). Since the dependent ZWHNEW has three groups, the number of discriminant functions computed is two. Now to ascertain which of these discriminant functions is significant, we undertake the eigenvalue (Δ) and the corresponding Wilks' lambda tests. In matrix algebra, an eigenvalue is a constant, which when subtracted from the diagonal elements of a matrix results in a new matrix whose determinant is zero.

The Wilks' lambda test is given as follows: Suppose that there are random samples of sizes for each of the g-groups, and let be the sample mean vectors for each group, then Wilks' lambda is defined as

$$\Delta = \frac{W}{T} = \frac{\sum_{i=1}^{g}\sum_{n=1}^{n_i}\left(X_{in} - \overline{X}_i\right)\left(X_{in} - \overline{X}_i\right)'}{\sum_{i=1}^{g}\sum_{n=1}^{n_i}\left(X_{in} - \overline{X}\right)\left(X_{in} - \overline{X}\right)'} \quad (11.5)$$

where the numerator denotes the within-group sum of squares matrix, while the denominator is the total deviation sum of squares matrix. To test the significance of the discriminating functions after the first *case*, the test statistic is given as

$$\chi^2 = -(n-(q+g)/2-1)\ln\Delta_k \qquad (11.6)$$

where n is the total sample size and q is the number of variables selected at each step. The test statistic is distributed as χ^2 with $(q-m+1)(g-m)$ degrees of freedom at the mth step. The exact computational procedure will be discussed in the technical appendix.

In the present case, since the group variable (ZWHNEW) consists of three groups, two eigenvalues can be extracted. The eigenvalue is the ratio of between the group sum of squares to the within group sum of squares. The larger the eigenvalue, the greater the discriminatory power of the model.

For the first and second functions, the eigenvalues are, respectively, 0.135 and 0.092 (Table 11.5). The variances explained by the two functions are 59.5% and 40.5%, respectively. Now, to know which eigenvalues are significant, two statistical indicators can be derived, namely canonical correlation (η) and Wilks' lambda (λ). The canonical correlation (η) can be computed as

$$\eta = \sqrt{\frac{\lambda}{(1+\lambda)}} = \sqrt{\frac{SSB}{TSS}}$$

where *SSB* denotes the sum of squares between the groups, while *TSS* denotes the total sum of squares. For our example,

$$\eta = \sqrt{\frac{0.135}{1.135}} = 0.344.$$

TABLE 11.5 Eigenvalues and Wilks' lambda.

Function	Eigenvalue	Percentage of variance	Cumulative (%)	Canonical correlation
1	0.135	59.5	59.5	0.344
2	0.092	40.5	100.0	0.290

First two canonical discriminant functions were used in the analysis.

Test of Function(s)	Wilks' lambda	Chi-square	Df	Sig.
1 through 2	0.807	41.829	16	0.000
2	0.916	17.150	7	0.016

Wilks' lambda is used to test the null hypothesis that the populations have identical means on the discriminant function. It is given by $\lambda = WSS/TSS = 1 - \eta^2$ and is distributed as (X^2_{k-1}), where k is the number of parameters estimated. WSS denotes the within-group sum of squares among the various groups. For the first function, $\chi^2_{16} = 41.829$, with the associated $P < .0001$. Thus, the null hypothesis that in the population the $SSB = 0$, $\eta = 0$, is rejected. For the second function, $\chi^2_7 = 17.15$, with the associated $P = .016$. Thus, for both the functions, the impact of the predictor variables on the various groups of child nutritional status as measured by ZWH will be assessed.

Relative impact of the predictor variables on ZWHNEW

Since the predictor variables per capita food expenditure are in monetary units, while age of the child is in months and feeding practice, availability of health facilities, infrastructural facilities, incidence of diarrhea, and distance to protected water source are dichotomous or discrete variables and are measured in different units, the relative difference among the discriminant coefficients cannot be compared. We need a standardized discriminant function, which is of the following form:

$$ZWHNEW_{STD} = W_4 X_{IC} + W_5 X_H + W_6 X_C \quad (11.7)$$

The coefficients can then be interpreted as the beta weights in a multiple regression model. The $ZWHNEW_{STD}$ has been transformed using a weighted unstandardized discriminant function coefficient and will be discussed in greater detail in the technical appendix. Notice that there is no constant term in Eq. (11.7). This is because the mean of a standardized variable equals zero, while the variance is unity.

From the Wilks' lambda test, we use the associated predictors that were significant in the test for equality of group mean in our analysis (Table 11.6). The variables that were not significant in the test for equality of group means will not be included in the standardized discriminant function analysis.

TABLE 11.6 Standardized canonical discriminant function coefficients.

Predictors	Function 1	Function 2
EDUCDIFF	0.664	0.337
BFEEDNEW	0.315	−0.042
AGEMNTH	0.275	0.335
DRINKDST	−0.582	0.463
SADMDIST	−0.071	0.508

The two standardized discriminant functions in this case are given as follows:

$$\text{ZWHNEW}_{STD1} = 0.664 \text{ EDUCDIFF} + 0.315 \text{ BFEEDNEW}$$
$$+ 0.275 \text{ AGEMNTH} - 0.582 \text{ DRINKDST}$$
$$- 0.071 \text{ SADMDIST}$$

$$\text{ZWHNEW}_{STD2} = 0.337 \text{ EDUCDIFF} - 0.042 \text{ BFEEDNEW}$$
$$+ 0.335 \text{ AGEMNTH} + 0.463 \text{ DRINKDST} \quad (11.8)$$
$$+ 0.508 \text{ SADMDIST}$$

From Eq. (11.8), it is evident that, for the first standardized discriminant function, the maximum impact of the child's nutritional status comes through educational difference between the household head and the spouse. A unit increase in educational difference between the household head and the spouse raises the standardized ZWH by 0.664 units. The aforementioned result signifies that the greater the educational differences, the lower the possibility that women will be aware of child care practices. At the same time, it is also possible that women may have less bargaining power in the intrahousehold decision process. Both these effects reinforce the negative impact on child's nutritional status.

For the second standardized discriminant function, we find the maximum effect occurring from the community-level variables such as distance to a protected water source and availability of infrastructural facilities. The independent impact of DRINKDST and SADMDIST is, respectively, 0.463 and 0.528. This indicates that, as the distance to a protected water source increases for the household, the health environment deteriorates, which has adverse consequences on the child nutritional status. We can thus name the first discriminant function as women's status and the second discriminant function as community infrastructure.

Correlation between the predictor variables and discriminant function

Table 11.7 shows the structure matrix, which is the correlation between each predictor with the discriminant function. The structure coefficients indicate the simple correlation between each variable and the first discriminant function. The higher the absolute value of the coefficient, the greater the discriminatory impact of the independent variable on the dependent variable. In the determination of the two discriminant functions, the first function has the highest correlation with educational differences, breastfeeding practices, and age of the child, while the second function has the highest correlation with distance to protected water source and selling point distance of the private traders to

TABLE 11.7 Structure matrix.

	Function	
Predictors	1	2
PXFD	0.032	0.299[a]
EDUCDIFF	0.601[a]	0.431
BFEEDNEW	0.412[a]	0.162
AGEMNTH	0.448[a]	0.226
DIARRHEA	−0.025	0.206[a]
DRINKDST	−0.487	0.662[a]
HOSPITAL	0.196[a]	−0.018
SADMDIST	−0.191	0.638[a]
LATERINE[b]	0.045	0.056

Pooled within-groups correlations between discriminating variables and standardized canonical discriminant functions. Variables ordered by absolute size of correlation within function.
[a]Largest absolute correlation between each variable and any discriminant function.
[b]This variable not used in the analysis.

ADMARC, which reinforces our earlier finding from the standardized discriminant function analysis that women's status and community infrastructure are significantly correlated with the discriminant functions.

Classification statistics

It is also informative to examine the mean values of each of the functions for each group. These means are called group centroids. Group centroids provide an indication whether the discriminant function will contribute significantly to the separation of groups. The main objective of this analysis is to maximize the amount of variance between each group and grand mean for all the groups, while simultaneously minimizing the amount of variance that exists within each group.

Table 11.8 shows that the first discriminant function discriminates well between the healthy child group from those who are severely wasted. For this function, the mean value of group 0 (the healthy group) is 0.428, while the mean value of group 2 (the severely wasted group) is −0.837. Thus, most of the discriminatory power of this function comes from differentiating between the children with good nutritional status with that of the severely wasted group. On the other hand, the second discriminant function does not differentiate significantly between the healthy children and the children who are severely wasted. However, it does a good job in differentiating the groups of children who are low to moderately wasted (ZWHNEW = 1) with that of

TABLE 11.8 Functions at group centroids.

Weight for height category	Function 1	Function 2
0.00	0.428	−0.230
1.00	−0.121	0.239
2.00	−0.837	−0.668

Note: Unstandardized canonical discriminant functions evaluated at group means.

children in the severely wasted category (ZWHNEW = 2). Thus, most of the discriminatory power of this function comes between differentiating the children who are severely wasted to the children who are low to moderately wasted.

Classification function based on equal and unequal prior probabilities

To assess the performance and in interpreting the results of a DA model, an internal classificatory analysis can be used to assess the probabilities of correct classification (hit rates). The most common hit rate is based on the actual sample-based conditional hit rate, which represents the hit rate for a given sample-based classification rule. (For a more comprehensive treatment of the different kinds of hit rates, see, for example, Huberty et al., 1987). SPSS uses the leave-one-out (L-O-O) method, which involves a two-step process. First, each observation is deleted in turn from a sample of size n, and the classification functions are determined on the remaining $(n - 1)$ observations. Then, in the second stage, the classification functions are used to classify the deleted observations into one of the g groups.

Table 11.9 classifies the cases when all the three groups are assumed to be equally likely in the population, while Table 11.10 classifies cases when the groups are assumed to have different prior probabilities based on their actual frequency. In the first case, we find that only 49.5% of the original cases are correctly specified. For children who have good nutritional status (ZWHNEW > 0) and for children who are severely wasted, the correct classification rates (hit ratio) are 60.3% and 66.7%, respectively. However, it fails to classify children in the low and the moderately wasted group and predicts wrong group membership (about 40.5%).

For the case where groups are assumed to have different prior probabilities, we find that 60.4% of the original cases are correctly predicted to belong to their respective groups. This method of assigning unequal prior probabilities based on the actual frequency now increases the correct classification rate for

TABLE 11.9 Classification results based on S.[a]

		Weight for height category	Predicted group membership 0.00	1.00	2.00	Total
Original	Count	0.00	41	7	20	68
		1.00	42	47	27	116
		2.00	4	2	12	18
	%	0.00	60.3	10.3	29.4	100.0
		1.00	36.2	40.5	23.3	100.0
		2.00	22.2	11.1	66.7	100.0

[a]49.5% of original grouped cases correctly classified.

TABLE 11.10 Classification results based on unequal prior probabilities.[a]

		Weight for height category	Predicted group membership 0.00	1.00	2.00	Total
Original	Count	0.00	23	45	0	68
		1.00	19	96	1	116
		2.00	2	13	3	18
	%	0.00	33.8	66.2	0.0	100.0
		1.00	16.3	82.8	0.9	100.0
		2.00	11.1	72.2	16.7	100.0

[a]49.5% of original grouped cases correctly classified.

the children belonging to the moderate and low wasting groups (almost 83%). However, correct classification of the healthy children and children who are severely wasted suffers from correct classification. Since around 57% of the children in the sample are low to moderately wasted, that information has overwhelmed the information in the discriminant scores, leading to classification in the largest group. When sample proportions are markedly unequal, their use as prior probabilities always increases the rate of correct classification.

From the discriminant function analysis, we find that improving women's status in Malawi (such as providing them with greater educational opportunities) can improve child nutritional status, possibly through the linkages of greater decision-making within the household. Since improvement in educational opportunities can improve both nutritional knowledge and better care practices, it can indirectly improve child nutritional status even in the short term. Additionally, we found that better health infrastructure (such as lower distance to a protected water source) can improve child health by reducing the exposure to infections from waterborne pathogens and releasing the labor of the caregiver for various productive activities.

Canonical discriminant analysis using STATA

In this section, we outline the steps in identifying the discriminant functions using STATA for a sample of 200 farmers, some of whom grow cash crop. We have already investigated these data in Chapter 3 and also in later chapters for our ANOVA and t-tests. We first recap the data using the summarize command in STATA (see Table 11.11):

The variable id simply tracks each farmer in the same with an identification number. We have 200 observations in our sample, where a unit of observation is a farmer. All the variables in this example have been defined before. We start by using the manova command in STATA and the output is:

manova cashcrop calreq zhanew zwanew = insecure

The overall multivariate test is significant because we can see that the F-values are high and the P-values are also significant, which means that differences between the levels of the variable **insecure** are significantly different from 0.

W represents the Wilks' lambda coefficient, which shows the proportion of the variance in the outcomes that is not explained by an effect. P is the value of Pillai's trace, which is also another statistic that captures the outcomes not explained by the group effect. L and R are also similar statistics produced by STATA.

TABLE 11.11 STATA input and output: descriptive statistics.

Summarize						
Variable	Obs	Mean	Std. Dev.	Min	Max
id	200	100.5	57.87918	1	200
cashcrop	200	0.545	0.4992205	0	1
insecure	200	3.43	1.039472	1	4
calreq	200	0.64	0.4812045	0	1
zhanew	200	0.235	0.4250628	0	1
zwanew	200	0.475	0.5006277	0	1
zwhnew	200	0.29	0.4549007	0	1

We can also perform post hoc tests to see the source of the differences seen in the W and P statistics. As an example, we compare multivariate test of group 1 versus the average of groups 2, 3, and 4. To do this, we must first apply the **manova, showorder** command in STATA to determine the order of the elements in the design matrix. We have to use this order and create another matrix for the post hoc tests. So we type the next command in STATA and get:

manovatest, showorder

From the aforementioned output, we can try to compare the treatment group (insecure = 1) to an average of the control groups (insecure = 2, 3, and 4). Our goal is to test the hypothesis that the mean across these groups is equal.

From the output, we see that the fifth element in the matrix is the constant, so in the following **matrix** command, this particular element will set it to 0. In the next step, we create a matrix and call it **c1**. We then apply the **manovatest** command to test **c1**.

The results indicate that control group 1 is not statistically significantly different from control group 2, or we cannot reject the null hypothesis of equal means. We can conduct various MANOVA tests between different control groups and treatment groups to identify where the differences lie. We provide some hints in the Problems section and leave these as exploratory exercise to the reader using a simple data set. We proceed toward determining the discriminant functions for our data.

The input lines in STATA and the corresponding output are given in the following:

```
. matrix c1 = (2,-1,-1,-1,0)

. manovatest, test(c1)

Test constraint
 (1)    2*1.insecure - 2.insecure - 3.insecure - 4.insecure = 0

                    W = Wilks' lambda       L = Lawley-Hotelling trace
                    P = Pillai's trace      R = Roy's largest root

         Source | Statistic    df   F(df1,    df2) =     F   Prob>F
      ----------+-----------------------------------------------------
      manovatest| W   0.9753    1    4.0     193.0    1.22  0.3020 e
                | P   0.0247         4.0     193.0    1.22  0.3020 e
                | L   0.0254         4.0     193.0    1.22  0.3020 e
                | R   0.0254         4.0     193.0    1.22  0.3020 e
      ----------+-----------------------------------------------------
        Residual|              196
      ----------+-----------------------------------------------------
                    e = exact, a = approximate, u = upper bound on F

. candisc calreq zhanew zwanew, group(insecure)
```

In STATA, we use the **Candisc** procedure for canonical linear DA. We have to list the variables that are useful for the analysis, and use the **group()** command to apply the analysis to the categorical variable of interest. Following the input command line, STATA produces a series of calculations. We look at each output portion sequentially.

The first output STATA produces is the canonical DA. In the first column, we have Fcn, which represents the first, second, or the third canonical linear discriminant function. The number of functions is equal to 1 less than the number of levels in the group variable or the number of discriminating variables, if there are more groups than variables. In our example, **insecure** has four levels and we used four discriminating variables, and hence, three functions are calculated. Each function acts as projections of the data onto a dimension that best separates or discriminates between the groups.

The second column produces the canonical correlations of the functions under the title Canon. Corr. Basically, our canonical correlation tells us how the sets of variables relate to each other using pairs of linear combinations of the variables from each set. In canonical correlation, each pair of linear combination is generated so as to produce the maximum correlation. So from the input command, the STATA output will include the canonical correlations we see in this portion of the output.

For our example, the canonical correlations are 0.419, 0.119, and 0.032.

It is also a usual practice to see which of the discriminating variables can be grouped and be part of DA. These correlations are closely associated with the eigenvalues of the functions and tell us how much discriminating power the functions possess.

In the next column, we have the proportion of discriminating power of the three continuous variables found in a given function, Prop. In this example, the first function accounts for 93% of the discriminating power of the discriminating variables, and the second and the third functions together account for the remaining 0.7%. The item Cumul. in the next column is the cumulative proportion, produced from the Prop. column.

The next column produces the likelihood ratio of a given function. It can be used as a test statistic to evaluate the hypothesis that the current canonical correlation and all smaller ones are zero in the population. We can also use the next column or the F statistic. This is also a test that the canonical correlation of the given function is equal to zero. In other words, the null hypothesis is that the function and all functions that follow have no discriminating power. This hypothesis is tested using the F statistic, which is generated from the likelihood ratio, along with the P-values in the last column. The null hypothesis that a given function's canonical correlation and all smaller canonical correlations are equal to zero is evaluated with regard to this P-value. If the P-value is less than the specified alpha (say 0.05), the null hypothesis is rejected. If not, then we fail to reject the null hypothesis. In our example, we cannot reject the null hypotheses that the canonical correlations of functions 1 and 2, and of 1 and 3 are zero. Thus, only function 1 is helpful in discriminating between the groups found in **insecure** based on the discriminant variables in the model.

The next output produced by STATA is:

Standardized canonical discriminant function coefficients

	function1	function2	function3
calreq	.7653343	.6381721	−.1856197
zhanew	.7187148	−.6898374	.6150299
zwanew	.275989	.2546189	1.120883

The output table titled standardized canonical discriminant function can be used to calculate the discriminant score for a given record. For our example, the standard discriminant function and the function scores would be calculated using the following equations:

Insecure$_{STD1}$ = 0.765***calreq** + 0.718***zhanew** + 0.275***zwanew**
Insecure$_{STD2}$ = 0.638***calreq** − 0.689***zhanew** + 0.254***zwanew**
Insecure$_{STD3}$ = −0.185***calreq** + 0.615***zhanew** + 1.12***zwanew**

As pointed out in the main section of the chapter, the distribution of the scores from each function is standardized to have a mean of zero and standard deviation of one. The magnitudes of these coefficients indicate how strongly the discriminating variables effect the score. For example, we can see that the standardized coefficient for **calreq** in the first function is greater in magnitude than the coefficients for the other two variables. Thus, **calreq** will have the greatest impact of the three on the first discriminant score.

STATA also produces two other results:

Canonical structure

	function1	function2	function3
calreq	.7835261	.5358081	−.3146372
zhanew	.642594	−.7660441	.0157916
zwanew	−.2228392	.50906	.8313848

Group means on canonical variables

insecure	function1	function2	function3
1	.7021663	.1433975	.0603624
2	.5076106	.3033512	−.0988252
3	.8943882	−.2556709	−.0241359
4	−.2780929	−.0114827	.0008352

The portion of the output called the canonical structure gives us canonical loading or discriminant loadings of the discriminant functions. It represents the correlations between the observed variables (the three continuous discriminating variables) and the dimensions created with the unobserved discriminant functions (dimensions). This is the correlation between the predictor variables and the discriminant functions. Once again, we find function 1 with high correlation with groups 1 and 2. The variable ZHANEW is a good predictor for our analysis.

STATA also produces the group means on canonical variables. These are the means of the discriminant function scores by group for each function calculated. That is, group 1 has a mean of 0.702, and group 2 has a mean of 0.507.

Finally, STATA also produces a summary of the classification.

The first column produces the frequencies of groups found in the data: 24 in group 1, 11 in group 2, 20 in group 3, and 145 in group 4. Across each row, we see how many in the group are classified by our analysis into each of the different groups. For example, of the 11 records that are in group 2, 7 are classified correctly by the analysis as belonging to group 2 and 8 are classified incorrectly as not belonging to group 2.

The predicted frequencies are given inside each cell under the classified. For example, of the 62 records that were predicted to be in group 4, 59 were correctly predicted, and 3 were incorrectly predicted. Finally, all the prior proportions assumed for the distribution of records into the groups are by default equally distributed among the categories.

Conclusions

Smith et al. (2003) have convincingly demonstrated that women's status can have a positive and long-term impact on children's nutritional status across major developing regions. This is possibly because women with greater status have more control over resources in their households, are less time constrained, have better mental health and self-confidence, and live in areas with greater availability of health services, which cater to their needs. Improvement in women's status relative to men can have a significant impact on children's nutritional status through a wide range of caring practices, such as complementary feeding of children in a timely fashion, treatment of illness of children, and timely initiation of breastfeeding practices.

Education can equip individuals with the necessary skills that allow them better to understand, interpret, and operate in a modern environment. Differentials between male and female education rates can cause women to be substantially disadvantaged not only in terms of future employment prospects but also with fewer resources (especially for female-headed households).[7] Additionally, an uneducated female is more likely to lack nutritional knowledge and good child care practices.

We find that educational difference, a measure of women's status, turns out to be an important discriminating factor in affecting children's nutritional status. Since there is a substantial gap in the number of years of education between the household head and the spouse, it is likely that, in Malawi, a significant portion of females within male-headed households as well as female-headed households are illiterate. Thus, they lack both nutritional knowledge and good child care practices that are conducive to child growth. We find breastfeeding to be an important discriminating factor for child nutritional status, which might be indirectly capturing the lack of nutritional knowledge arising out of lack of education of females. Additionally, community-level variables are significant discriminatory factors in improving children's nutritional status. The aforementioned variables possibly capture the improvement of the health environment for the community as a whole, which affects the health production function and subsequently influences children's nutritional status.

An important implication that emerges from the present analysis is that improving women's educational status (both formal and informal) remains a critical agenda for Malawi, since a significant percentage of women either are illiterate or do not have basic primary educational levels. Improving educational status of women may have long-run implications on maternal nutritional knowledge, which can translate into better care practices. Additionally, access to gainful employment can help women in bargaining situations, while rightful ownership of land and ownership of productive assets can improve the likelihood of women receiving a larger share of resources.

Technical appendix: discriminant analysis

Discriminant analysis decision process

This section follows from Rencher (2002) as we have permission to reproduce from Wiley Publishers. The application of DA can be understood from the following six-stage model building perspective: objectives, research design, assumptions, estimation, interpretation of discriminant functions, and validation of discriminant results. The basic idea in a DA is how to determine the

---

7. UNICEF (1998) has identified 4 years of primary education as the threshold point, since women with less than 4 years of education are more likely to be illiterate than those who complete at least primary education. This is because literacy is a skill that can only be realized after the minimum number of years of education.

variables that discriminate between two or more naturally occurring groups. For example, a financial researcher may be interested in determining what characteristics best discriminate between good and bad credit customers, a medical practitioner may be interested in knowing which variables best predict whether a cancer patient will recover completely, partially, or not at all. DA is useful in the following types of situations:

1. Incomplete knowledge of future situations
2. Unavailable or expensive information

DA is the appropriate statistical technique for testing the hypothesis that the group means of a set of independent variables for two or more groups are equal. For this purpose, the analysis multiplies each independent variable by its corresponding weight and adds these products together. The result is a single composite discriminant Z-score for each entity. By averaging the discriminant scores for all the entities within a particular group, one arrives at the group mean. The group mean is referred to as a *centroid*. The centroids determine the typical location of an entity from a particular group, and a comparison of the group centroids determines how far apart the groups are along the dimension being tested. The statistical significance of the discriminant function is a generalized measure of the distance between the group centroids. Now let us explain the various steps of a DA.

Stage 1: objectives

DA, in general, has the following research objectives:

1. To determine if statistically significant differences exist between average scores on a set of variables for two (or more) a priori defined groups
2. To determine which independent variables account for most of the differences in the average score profiles of two or more groups
3. To establish the number and composition of the dimensions of discrimination between groups formed from the set of independent variables

This method is most appropriate when there is a single categorical dependent variable and several categorical or continuous independent variables. For understanding group differences, this technique provides insight into the role of individual variables as well as defining the combinations of variables that represent dimensions of discrimination between groups.

Stage 2: research design

To successfully apply this technique, the researcher must specify which variables are independent and which variable is dependent. The number of dependent variable groups can be two or more, but these groups must be mutually exclusive and exhaustive. The independent variables can be selected in two main ways. The first approach involves identifying variables that are justified based on a theoretical model. In the second approach, the identifying variables may consist of selecting them on an intuitive basis, but that logically might be related to predicting the groups for the dependent variable. DA is also quite sensitive to the ratio of sample size to the number of predictor variables. Many studies suggest a ratio of 20 observations for each predictor variable. While this ratio is difficult to maintain in practice, the results become unstable as the sample size decreases relative to the number of independent variables. In addition to the overall sample size, the sample size of each group must be taken into account. In practice, the smallest group size must exceed the number of independent variables. Before discussing the assumptions of the DA, it may be useful to understand some technical aspects to it. The following section closely follows Rencher (2002) and is reproduced based on permission from Wiley Interscience.

Discriminant analysis for several groups: a technical digression

For several groups, we want to find the linear combination of the independent variables that best separate the g groups. Then a subset of the original variables is constructed that best separates the groups. The variables are then ranked in terms of their relative contribution to group separation, and the new dimensions represented by the discriminant function are interpreted. Without loss of generality, let there be g groups with n_i observations in the ith group. Let $Z_{ij} = w'X_{ij}$, $i = 1,2,\ldots,g$; $j = 1,2,\ldots,n_i$, and the means $\overline{Z}_l = w'\overline{X}_l$, where $\overline{X}_l = \sum_{i=1}^{n_i} X_{ij}/n_i$. The vector w maximally separates the g groups. To express separation among the groups, let the squared distance function between any two groups be expressed as

$$\frac{\left(\overline{Z}_1 - \overline{Z}_2\right)^2}{S_Z^2} = \frac{w'\left(\overline{X}_1 - \overline{X}_2\right)\left(\overline{X}_1 - \overline{X}_2\right)'w}{W'S_P W} \tag{11.9}$$

where S_P denotes the pooled common population variance—covariance matrix. Let us assume that g independent random samples of size n are obtained from k-variate normal populations with equal covariance matrices and let matrix H be the between sum of squares (SSB) on the diagonal for each of the k variables. Off-diagonal elements are sums of products for each pair of variables. On the other hand, let matrix E be the within sum of squares (WSS) for each variable on the diagonal, with analogous sums of products off-diagonal. (For additional details on how these matrices are constructed, refer to Rencher, pp. 160–161.) Extending Eq. (11.9) to g groups, we have

$$W'(HW - \lambda EW) = 0 \qquad (11.10)$$

where H is substituted for $(\overline{X}_1 - \overline{X}_2)(\overline{X}_1 - \overline{X}_2)'$ and E is substituted for S_p.

$$(E^{-1}H - \lambda I)W = 0 \qquad (11.11)$$

Eq. (11.11) is derived by premultiplying Eq. (11.10) by E^{-1}.

The solutions of Eq. (11.11) are the eigenvalues $\lambda_1, \lambda_2, \ldots \lambda_s$ associated with the eigenvectors $W_1, W_2, \ldots W_s$. From the s eigenvectors $W_1, W_2, \ldots W_s$ of $E^{-1}H$ corresponding to $\lambda_1, \lambda_2, \ldots \lambda_s$, one obtains s discriminant functions. These discriminant functions are uncorrelated but are not orthogonal. The relative importance of each discriminant function Z_i can be assessed by considering its eigenvalue as a proportion of the total. By the aforementioned criterion, often two or three discriminant functions can describe group differences.

Stage 3: checking assumptions

In DA, the key assumptions to be met are multivariate normality of the independent variables and equal covariance structure for the groups as defined by the dependent variable. If the sample sizes are small and the covariance matrices are unequal, then the significance of the estimation process is adversely affected. This effect can be minimized by increasing the sample size and by using group-specific covariance matrices for classification purposes. Formally, this is given as follows: for g multivariate populations, the null hypothesis of equality of covariance matrices is given by Eq. (11.3). For independent samples, one can calculate

$$M = \frac{|S_1|^{v_1/2}|S_2|^{v_2/2}\ldots|S_g|^{v_g/2}}{|S_p|^{\sum_i v_i/2}} \qquad (11.12)$$

where S_i is the covariance matrix of the ith sample, S_p is the pooled sample covariance matrix. The statistic M is a modification of the likelihood ratio and varies between 0 and 1. Values near 1 favor H_0 as in Eq. (11.3), while values near 0 lead to rejection of H_0. Box (1950) provided both χ^2 and F approximations for the distribution of M, and these approximate tests are called *Box's M-test*. (The details of deriving the approximate F-statistic are beyond the scope of the current chapter. The interested reader can consult Rencher, pp. 257–258.)

The χ^2 approximation is given by

$$\ln M = \frac{1}{2}\sum_{i=1}^{g} v_i \ln|S_i| - \frac{1}{2}\left(\sum_{i=1}^{g} v_i\right)\ln|S_p| \qquad (11.13)$$

Stage 4: estimating the discriminant function and assessing overall fit

To derive the discriminant function, the researcher must decide on the method of estimation and then determine the number of functions that can be retained. The overall fit of the model can be assessed by either looking at the discriminant Z-scores or by comparing the group means on the Z-scores to measure discrimination between groups.

Two *computational* methods are available in SPSS in deriving a discriminant function, namely the simultaneous and stepwise methods. In the simultaneous estimation, the discriminant function is computed with all the independent variables considered simultaneously. This method is appropriate when the researcher wants to include all the independent variables in the analysis and is not interested in looking at intermediate variables based on the most discriminating factors.

Stepwise estimation, on the other hand, involves entering the independent variables into the discriminant function one at a time on the basis of their discriminating power. This approach begins by choosing the single best discriminating variable. The best discriminating variable is then paired with each independent variable one at a time, and the variables that are best able to improve the discriminatory power of the function in combination with the best discriminating variable are retained. This process is repeated when all the independent variables contribute significantly to the discriminating power. This method is useful when the researcher sequentially selects the best discriminating variable at each step and the variables that are not useful in discriminating between the groups are eliminated.

Standardized discriminant functions: The relative contribution of the X_s to the separation of several groups is informative only if the variables are measured on the same scale with comparable variances. For the case of two groups, the discriminant function in terms of the standardized variables can be expressed as

$$Z_{1i} = a_1^* \frac{X_{1i1} - \overline{X}_{11}}{s_1} + a_2^* \frac{X_{1i2} - \overline{X}_{12}}{s_2} + \ldots + a_k^* \frac{X_{1ik} - \overline{X}_{1k}}{s_k}$$

$$Z_{2i} = a_1^* \frac{X_{2i1} - \overline{X}_{21}}{s_1} + a_2^* \frac{X_{2i2} - \overline{X}_{22}}{s_2} + \ldots + a_k^* \frac{X_{2ik} - \overline{X}_{2k}}{s_k} \qquad (11.14)$$

where $\overline{X}_1' = \left(\overline{X}_{11}, \overline{X}_{12}, \ldots, \overline{X}_{1k}\right)$ and $\overline{X}_2' = \left(\overline{X}_{21}, \overline{X}_{22}, \ldots, \overline{X}_{2k}\right)$ are the mean vectors for the two groups and sr is the within sample standard deviation of the rth variable, obtained as the square root of the rth diagonal element of S_p. In vector form, the coefficient a can be expressed as

$$a^* = (diagS_p)^{1/2} a \qquad (11.15)$$

Evaluating group differences: from Eqs. (11.10) and (11.11), it is maximized by λ_1; the largest eigenvalue of $E^{-1}H$ and the remaining eigenvalues $\lambda_2, \ldots \lambda_s$ correspond to other discriminant dimensions. These eigenvalues are the same as that of the Wilks' test for significant differences among the mean vectors. The test is given as follows: Since, Δ_1 is small if one or more $\lambda_i s$ are large, Wilks' Δ tests for significance of the eigenvalues and thus for the discriminant functions. The s eigenvalues denote s dimensions of the separation of the mean vectors. The χ^2 approximation for Δ_1 is given by

$$V_1 = -\left[n - 1 - \frac{1}{2}(q+g)\right] \ln \prod_{i=1}^{s} \frac{1}{(1+\lambda_i)}$$

$$= \left[n - 1 - \frac{1}{2}(q+g)\right] \sum_{i=1}^{s} \ln(1+\lambda_i) \qquad (11.16)$$

which is approximately χ^2 with $q(g-1)$ degrees of freedom. q denotes the number of variables selected at each step. If the aforementioned test leads to rejection of the null hypothesis, then one can conclude that at least one of the λs is significantly different from zero. Thus, there is at least one dimension of separation of the mean vectors. To test the significance of $\lambda_2, \ldots \lambda_s$, the procedure is to delete λ_1 from Wilks' Δ and the associated χ^2 approximation for Δ_2 is given by

$$V_2 = -\left[n - 1 - \frac{1}{2}(q+g)\right] \ln \prod_{i=1}^{s} \frac{1}{(1+\lambda_i)}$$

$$= \left[n - 1 - \frac{1}{2}(q+g)\right] \sum_{i=2}^{s} \ln(1+\lambda_i) \qquad (11.17)$$

which is approximately χ^2 with $(q-1)(g-2)$ degrees of freedom. If the test leads to rejection of the null hypothesis, one can conclude that λ_2 is significant along with the associated discriminant function. The test statistic at the mth step is given by

$$V_m = -\left[n - 1 - \frac{1}{2}(q+g)\right] \ln \prod_{i=1}^{s} \frac{1}{(1+\lambda_i)}$$

$$= \left[n - 1 - \frac{1}{2}(q+g)\right] \sum_{i=m}^{s} \ln(1+\lambda_i) \qquad (11.18)$$

which is approximately distributed as χ^2 with $(q - m + 1)(g - m)$ degrees of freedom.

Stage 5: interpretation of discriminant functions

If the discriminant function is statistically significant, one needs to interpret the findings. In other words, it is important to examine the relative importance of each independent variable in discriminating between the groups. Three methods are usually proposed in the literature:

1. Standardized discriminant weights
2. Structure correlations
3. Partial F-values

Discriminant weights: The sign and magnitude of the discriminant coefficient (coefficients in Table 11.6) is called the discriminant weight. Each weight shows the relative contribution of its associated variable to that function. Variables with relatively larger weights contribute more to the discriminatory power of the function than do variables with smaller weights. The sign only denotes whether a variable makes a positive or negative contribution. The interpretation of the discriminant weights is analogous to the interpretation of the beta weights in a regression analysis and is subject to the same criticisms. It is important to interpret these weights with caution while interpreting the results of a DA.

Structure correlations: This measure (often called discriminant loadings) is the simple linear correlation between each independent variable and the discriminant function. These correlations reproduce the t- or F-statistic for each variable and thus show how each variable separates the groups by ignoring the presence of the other variables. However, these correlations do not provide information about how the variables contribute jointly to the separation of the groups and can become misleading in interpreting the discriminant functions.

Partial F-values: For any variable X_r, one can calculate a partial F-test showing the significance of this variable after adjusting for the other variables. For the several groups case, the partial Δ for X_r adjusted for the other $(k - 1)$ variables is given by

$$\Delta(X_r | X_1, \ldots X_{r-1}, X_{r+1}, \ldots, X_k) = \frac{\Delta_k}{\Delta_{k-1}} \tag{11.19}$$

where Δ_k is Wilks' Δ for all the k variables and Δ_{k-1} involves all the variables except X_r. The corresponding partial F is given as follows:

$$F = \frac{1 - \Delta}{\Delta} \frac{(n - g - k + 1)}{(g - 1)} \tag{11.20}$$

The partial F is distributed as $F_{g-1, n-g-k+1}$. The partial F-values are not associated with a single dimension of group separation as the standardized discriminant function coefficients. However, the partial *F*-values will often rank the variables in the same order as the standardized coefficients for the first discriminant function if it is very large such that the first function accounts for most of the separation among the groups.

Stage 6: validation of results

The final stage of the DA involves validating the discriminant results so that the results have both internal and external validity (see Tables 11.9 and 11.10). Cross-validation is an essential step in achieving validity. The most widely used procedure in validating the discriminant function is to divide the groups randomly into analysis and holdout samples. This procedure involves developing a discriminant function with the sample in the analysis and then applying it to the holdout sample. The rationale for dividing the sample into groups is that an upward bias will occur if the prediction accuracy of the discriminant function used in developing the classification matrix is the same as those used in computing the function. In other words, the classification accuracy will be higher than is valid for the discriminant function if it was used to classify a separate sample. SPSS uses the "leave-one-out" principle in which the discriminant function is fitted to repeatedly drawn samples of the original sample. The most prevalent method is to estimate $(n - 1)$ samples, eliminating one observation at a time from a sample of n observations. This approach can only be used when the smallest group size is at least three times the number of predictor variables.

Canonical discriminant analysis using R

In this section, we outline the steps in identifying the discriminant functions using R for a sample of 200 farmers, which was used in Chapter 3. We first recapitulate the data using the *summary* command in R.

```
> summary(data_chapter3)
      Obs              Cashcrop           INSECURE
 Min.   :  1.00    Min.   :0.000     Min.   :1.00
 1st Qu.: 50.75    1st Qu.:0.000     1st Qu.:3.00
 Median :100.50    Median :0.000     Median :3.00
 Mean   :100.50    Mean   :0.425     Mean   :3.05
 3rd Qu.:150.25    3rd Qu.:1.000     3rd Qu.:4.00
 Max.   :200.00    Max.   :1.000     Max.   :4.00
     CALREQ             ZHANEW             ZWANEW              ZWHNEW
 Min.   :0.000     Min.   :0.00      Min.   :0.00      Min.   :0.00
 1st Qu.:0.000     1st Qu.:0.00      1st Qu.:0.00      1st Qu.:0.00
 Median :0.000     Median :0.00      Median :1.00      Median :1.00
 Mean   :0.495     Mean   :0.47      Mean   :0.52      Mean   :0.52
 3rd Qu.:1.000     3rd Qu.:1.00      3rd Qu.:1.00      3rd Qu.:1.00
 Max.   :1.000     Max.   :1.00      Max.   :1.00      Max.   :1.00
```

Then, we start by using the *manova* and *summary* command in R that shows a class for the multivariate analysis of variance. The output is

```
> manova<-manova(cbind(Cashcrop,CALREQ,ZHANEW,ZWANEW)~INSECURE)
> summary(manova)
            Df  Pillai approx F num Df den Df   Pr(>F)
INSECURE     1 0.14453   8.2361      4    195 3.705e-06 ***
Residuals  198
---
Signif. codes:  0 '***' 0.001 '**' 0.01 '*' 0.05 '.' 0.1 ' ' 1
```

We can see that the default of the *manova* command in R is the value of Pillai's trace. By using other arguments, we undertake several other multivariate tests. We can add test = "Wilks"/"Hotelling-Lawley"/"Roy" to produce other statistics that capture the outcomes.

```
> summary(manova,test="Wilks")
            Df   Wilks approx F num Df den Df   Pr(>F)
INSECURE     1 0.85547   8.2361      4    195 3.705e-06 ***
Residuals  198
---
Signif. codes:  0 '***' 0.001 '**' 0.01 '*' 0.05 '.' 0.1 ' ' 1
> summary(manova,test="Hotelling-Lawley")
            Df Hotelling-Lawley approx F num Df den Df   Pr(>F)
INSECURE     1          0.16895   8.2361      4    195 3.705e-06
Residuals  198

INSECURE   ***
Residuals
---
Signif. codes:  0 '***' 0.001 '**' 0.01 '*' 0.05 '.' 0.1 ' ' 1
> summary(manova,test="Roy")
            Df     Roy approx F num Df den Df   Pr(>F)
INSECURE     1 0.16895   8.2361      4    195 3.705e-06 ***
Residuals  198
---
Signif. codes:  0 '***' 0.001 '**' 0.01 '*' 0.05 '.' 0.1 ' ' 1
```

In R, we use the *candisc* command to undertake canonical linear DA.

```
> candisc(manova)

Canonical Discriminant Analysis for INSECURE:

   CanRsq Eigenvalue Difference Percent Cumulative
1 0.14453    0.16895                100        100

Test of H0: The canonical correlations in the
current row and all that follow are zero

  LR test stat approx F numDF denDF   Pr(> F)
1      0.85547   8.2361     4   195 3.705e-06 ***
---
Signif. codes:  0 '***' 0.001 '**' 0.01 '*' 0.05 '.' 0.1 ' ' 1
```

Exercises

1. Carry out a DA by categorizing height for age (ZHANEW) and weight for age (ZWANEW) into three or four different categories similar to the exercise undertaken in this chapter. What are the discriminatory factors in each of the cases? Interpret the classification statistics based on equal prior probabilities and unequal prior probabilities for each group.
2. Define women's status as the product of per capita total expenditure (PXTOTAL) and the female-headed household dummy (FEMHHH). Call this variable PEXPGEND. Undertake a DA with PEXPGEND as the variable measuring women's status along with the other predictors. Is this variable a significant predictor of child nutritional status as measured by ZWHNEW? How many discriminant functions can be derived? Interpret the classification statistics based on equal prior probabilities and unequal prior probabilities for each of the groups.
3. Now, define women's status as the product of land owned (OWNED) and the female-headed household dummy variable (FEMHHH). This is an interaction variable that shows the amount of land owned by female-headed households relative to male-headed households. Call this variable LANDGEND. Is this variable a significant predictor of child nutritional status as measured by ZWHNEW? How many discriminant functions can be derived? Interpret the classification statistics based on equal prior probabilities and unequal prior probabilities for each of the groups.
4. Additionally, undertake a similar analysis with ZHANEW and ZWANEW by forming similar groups like ZWHNEW, with LANDGEND as one of the predictors along with the other predictors of the present chapter. Do you find this variable to be a significant predictor of child nutritional status as measured by ZHANEW and ZWANEW? How many discriminant functions can be derived for each case? Interpret the classification statistics for both ZHANEW and ZWANEW based on equal prior probabilities and unequal prior probabilities for each of the groups.
5. Define women's status in your own words. How does improvement in women's status translate into better child nutritional outcomes? What are the different policy interventions needed to improve women's status?
6. Describe the various steps in undertaking a DA? How does Wilks' lambda test help us in determining which discriminant functions are significant?

STATA workout

1. Consider the first 25 observations on Food Security for Cash Croppers with the corresponding information on CALREQ, ZHANEW, ZWANEW, and ZWHNEW. The variable "obs" in the first column is just a variable that tracks each observation in the sample data, where each unit of observation

is typical respondent in the survey. We have four groups for INSECURE. Use the following commands in STATA to perform a DA, and derive the discriminant functions.

manova cashcrop calreq zhanew zwanew = insecure
manovatest, showorder
matrix c1 = (2,-1,-1,-1,0)
manovatest, test(c1)
candisc calreq zhanew zwanew, group(insecure)

2. Include zwhnew in the analysis. What do you find? Can you adjust your program and check for the new discriminant function?

id	cashcrop	insecure	calreq	zhanew	zwanew	zwhnew
1	1	1	1	1	0	0
2	1	1	1	0	1	0
3	0	1	1	1	0	0
4	0	1	1	0	0	1
5	0	1	1	0	0	1
6	0	1	1	0	0	1
7	1	2	1	1	0	0
8	0	2	1	0	1	0
9	1	2	1	0	1	0
10	1	2	1	0	0	1
11	1	2	1	0	0	1
12	1	2	1	0	0	1
13	1	3	1	1	0	0
14	1	3	1	1	0	0
15	0	3	1	1	0	0
16	1	3	1	0	0	1
17	0	3	1	0	0	1
18	1	3	1	0	0	1
19	0	4	1	0	1	0
20	1	4	1	0	1	0
21	0	4	1	0	1	0
22	0	4	0	0	0	1
23	0	4	0	0	0	1
24	1	4	0	0	0	1
25	0	4	0	0	0	1

The STATA input lines along with the screen shot of the STATA output are as follows:

```
. manova cashcrop calreq zhanew zwanew = insecure

                 Number of obs =       25

                 W = Wilks' lambda      L = Lawley-Hotelling trace
                 P = Pillai's trace     R = Roy's largest root

      Source  | Statistic        df    F(df1,     df2) =     F    Prob>F
     ---------+---------------------------------------------------------
     insecure |W   0.2353         3    12.0      47.9      2.91  0.0043 a
              |P   0.9311              12.0      60.0      2.25  0.0199 a
              |L   2.5726              12.0      50.0      3.57  0.0007 a
              |R   2.2957               4.0      20.0     11.48  0.0001 u
     ---------+---------------------------------------------------------
     Residual |                        21
     ---------+---------------------------------------------------------
        Total |                        24

          e = exact, a = approximate, u = upper bound on F
```

The F-values indicate significance, which implies that the differences between the levels of the variable **insecure** are significantly different from 0. The Wilk's lambda coefficient indicates that 25% of the variance in **insecure** is still unexplained. On the other hand, the Pillai's trace indicates that 93% of the outcomes remain unexplained by group effects.

L represents Lawly-Hotelling trace, and R represents Roy's largest root.

```
. manovatest, showorder

   Order of columns in the design matrix
        1: (insecure==1)
        2: (insecure==2)
        3: (insecure==3)
        4: (insecure==4)
        5: _cons

. matrix c1 = (2,-1,-1,-1,0)
```

The **manovatest** is to test the linear combination of the underlying design matrix of the MANOVA model. The null hypothesis states that the mean vector for the control group is not significantly different from the mean vectors of the other three groups. The *P*-values indicate that we cannot reject the null hypothesis. Matrix c1 can be redefined to check the differences between the mean vectors of different subgroups.

```
. manovatest, test(c1)

Test constraint
 (1)   2*1.insecure - 2.insecure - 3.insecure - 4.insecure = 0

                     W = Wilks' lambda      L = Lawley-Hotelling trace
                     P = Pillai's trace     R = Roy's largest root

         Source | Statistic        df    F(df1,    df2) =     F   Prob>F
     manovatest |W     0.7640       1     4.0      18.0     1.39 0.2771 e
                |P     0.2360             4.0      18.0     1.39 0.2771 e
                |L     0.3088             4.0      18.0     1.39 0.2771 e
                |R     0.3088             4.0      18.0     1.39 0.2771 e
       Residual |                         21

         e = exact, a = approximate, u = upper bound on F
```

Next, STATA produces the canonical DA with the command line:
.candisc calreq zhanew zwanew, group(insecure)

The number of canonical discriminant functions is one less than the number of levels in **insecure**. The canonical correlations are 0.83 and 028, which indicates how the sets of variables relate to each other using pairs of linear combinations of the variables from each set. There is also multicollinearity between function 1 and function 3. The column titled Cumul indicates that 96% of the discriminating power of the discriminating variables is accounted for by the first function, while the second function accounts for the remaining 4%.

The calculated F values are also provided in the output. The null hypothesis states that a given function's canonical correlation and all smaller canonical correlations are equal to zero. In this example, we cannot reject the null hypothesis that the canonical correlations of functions 1 and 2 and of 1 and 3 are zero. Thus, only function 1 is helpful in discriminating between the groups found in **insecure**.

Canonical linear discriminant analysis

Fcn	Canon. Corr.	Eigen-value	Variance Prop.	Cumul.	Likelihood Ratio	F	df1	df2	Prob>F
1	0.8332	2.26973	0.9618	0.9618	0.2805	3.5355	9	46.39	0.0021 a
2	0.2877	.090265	0.0382	1.0000	0.9172	.44158	4	40	0.7778 e
3	.	-1.7e-16	-0.0000	1.0000	1.0000	-9.3e-15	1	21	1.0000 e

Ho: this and smaller canon. corr. are zero; e = exact F, a = approximate F

Standardized canonical discriminant function coefficients

	function1	function2	function3
calreq	-1.056346	.4743409	.0638877
zhanew	-.144782	-.680932	.7474825
zwanew	.8388031	.3055066	.7690144

The standardized canonical discriminant function coefficients yield the following equations:

Insecure$_{STD1}$ = −1.05 calreq − 0.144 zhanew + 0.83 zwanew
Insecure$_{STD2}$ = 0.474 calreq − 0.68 zhanew + 0.30 zwanew
Insecure$_{STD3}$ = 0.63 calreq + 0.74 zhanew + 0.76 zwanew

Since the standardized coefficient of **calreq** is the largest, this variable has the highest impact on the first discriminant score than the other two variables.

STATA also provides the correlation between the predictors and the discriminant functions. All the three functions have high correlations with the groups. The variable **calreq** is a good predictor.

Canonical structure

	function1	function2	function3
calreq	-.6382329	.6266249	.4472136
zhanew	-.2930215	-.7349234	.6115766
zwanew	.3378397	.6622877	.6687595

Group means on canonical variables

insecure	function1	function2	function3
1	-.845028	.0329947	-2.22e-16
2	-.4620809	.4230358	-8.88e-16
3	-1.227975	-.3570463	4.44e-16
4	2.172929	-.0848436	0

Resubstitution classification summary

STATA also provides the group means on canonical variables. The predicted means are provided in the last table under each cell. The first column produces the frequencies of groups found in the data; 6 in the first 3 groups and 7 in the last group. The rows indicate how many in the group are classified by the analysis into each of the different groups. Of the 6 records in group 2, 2 are classified correctly as belonging to group 2, and 4 are classified incorrectly, as not belonging to group 2.

Key
Number
Percent

True insecure	Classified 1	2	3	4	Total
1	3	1	2	0	6
	50.00	16.67	33.33	0.00	100.00
2	3	2	1	0	6
	50.00	33.33	16.67	0.00	100.00
3	3	0	3	0	6
	50.00	0.00	50.00	0.00	100.00
4	0	3	0	4	7
	0.00	42.86	0.00	57.14	100.00
Total	9	6	6	4	25
	36.00	24.00	24.00	16.00	100.00
Priors	0.2500	0.2500	0.2500	0.2500	

In the next command, we include the variable **zwhnew**:

```
. manova cashcrop calreq zhanew zwanew zwhnew = insecure

                    Number of obs =        25

                    W = Wilks' lambda      L = Lawley-Hotelling trace
                    P = Pillai's trace     R = Roy's largest root

         Source | Statistic      df    F(df1,     df2) =    F    Prob>F
        --------+-------------------------------------------------------
        insecure| W  0.2279        3    15.0      47.3    2.24  0.0181 a
                | P  0.9533             15.0      57.0    1.77  0.0626 a
                | L  2.6341             15.0      47.0    2.75  0.0041 a
                | R  2.3360              5.0      19.0    8.88  0.0002 u
        --------+-------------------------------------------------------
        Residual|                       21
        --------+-------------------------------------------------------
           Total|                       24

             e = exact, a = approximate, u = upper bound on F
```

The F-values indicate significance, which implies that the differences between the levels of the variable **insecure** are significantly different from 0. The Wilk's lambda coefficient indicates that 22% of the variance in **insecure** is still unexplained. The standardized canonical discriminant function portion of the following output indicates that **zhanew** is a very good predictor, followed by **calreq**.

Standardized canonical discriminant function coefficients

	function1	function2	function3
calreq	-1.024211	.2745669	.4717165
zhanew	.1111355	-1.379374	1.021039
zwanew	1.027367	-.275413	1.050451
zwhnew	.3368112	-1.026336	1.906512

Chapter 12

Measurement and determinants of poverty—application of logistic regression models

Chapter outline

Introduction	434
Dimensions and rationale for measuring poverty	435
Defining and measuring poverty	435
Monetary approach	435
Basic needs approach	436
Capability approach	436
Participatory poverty approach	437
Combined approaches	438
Rationale for measuring poverty	439
Targeting interventions	439
Designing programs and policies	439
Indicators in measuring poverty	439
Income measure	440
Consumption expenditure	440
Construction of poverty lines using food energy intake and cost of basic needs approaches	442
Poverty lines in theory	442
Absolute and relative poverty	443
Referencing and identification problems	443
Deriving a poverty line	444
Food energy intake method	444
Cost of basic needs method	446
Identifying regions	447
Constructing the food poverty lines	447
Constructing the nonfood poverty lines	448
New measures of poverty based on the engel curve	449
Measures of poverty	449
Headcount measure	449
Poverty gap index	450
Squared poverty gap index	451
Computing poverty measures for Malawi	452
Characteristics of poor households	453
Selected review of studies on determinants of poverty	454
Poverty and welfare in the United States	458
Agriculture and poverty in Laos and Cambodia	463
Financial crisis and poverty in the Russian Federation	465
Poverty in Europe	465
Poverty in developing countries: China and India	467
Determinants of poverty—binary logistic regression analysis	468
Dichotomous logistic regression model	469
An example with the Malawi dataset	469

Food Security, Poverty and Nutrition Policy Analysis. https://doi.org/10.1016/B978-0-12-820477-1.00031-0
Copyright © 2022 Elsevier Inc. All rights reserved.

Expected determinants of household welfare	470	Estimating logistic regression models in STATA	478
Empirical results	472	Example 1	478
Measuring model fit	472	Example 2	482
Log-likelihood ratio	472	Conclusions and implications	485
Hosmer–Lemeshow goodness-of-fit test	473	Technical appendices	486
		Technical notes on logistic regression model	486
Generalized coefficient of determination	474	Estimating logistic regression models in R	488
Classification table	475		
Interpreting the logistic coefficients and discussion of results	476	Exercises	489
		STATA workout	490

Through education, learning, and skill formation, people can become much more productive over time and this contributes greatly to the process of economic expansion.

Amartya Sen, Nobel Laureate in Economics (1999).

Introduction

The analysis of measurement and determinants of poverty and its relationship to food security and nutritional outcomes have become a major area of investigation by household welfare analysts. This chapter introduces various approaches to identifying the poor, measurement of poverty, construction of poverty lines, and deriving various measures of poverty including headcount index, poverty gap index, and squared poverty gap index. Poverty lines are calculated using food energy intake (FEI) and cost of basic needs (CBN) approaches. Analysis of determinants of poverty using binary logistic regression is also demonstrated.

Poverty reduction is considered to be one of the most important goals of development and of development policy. According to the World Bank (2000), poverty is defined as "pronounced deprivation in well-being," where well-being can be measured by individual or households" possession of income, health, education, assets, and certain rights in a society, such as freedom of speech. For simplicity, poverty is usually referred to as "whether individuals or households have enough resources or abilities to meet their needs" (Asian Development Bank, 2001). Although the aforementioned two definitions are clearly related, in the latter definition, comparison of individuals' income, consumption, education, or other attributes with some threshold level is considered and, if individuals fall below this threshold level with a certain attribute, they are considered poor. Additionally, poverty can also be conceived of as lack of opportunities, powerlessness, and vulnerability. Thus, poverty is a multidimensional phenomenon requiring multidimensional interventions to improve the well-being of individuals (Hulme and Shepherd, 2003).

In this chapter, we introduce concepts and methods for measuring and understanding poverty and its causes. The next section addresses the various dimensions of poverty and the causes of poverty and justifies the rationale for measuring poverty. In the following section, we explain the steps in measuring poverty and the various indicators that are widely used to measure poverty. We then discuss how poverty lines are constructed after a suitable indicator is chosen. The derivation of the poverty line based on the CBN approach is also described. This is followed by discussion on the various measures of poverty such as the headcount ratio, poverty gap, and squared poverty gap, and the derivation of these measures is illustrated using Malawi data. This section also looks at the various socioeconomic characteristics of the poor.

We also review the conditions of poverty in the United States, Asia, and Europe. In recent years, economists have started linking poverty to global financial crisis and family stressors. We examine some of this for Russia. The literature on the determinants of poverty using a logistic regression framework is briefly reviewed. We then use a logistic regression model to examine the determinants of poverty using the Malawi data set and finally conclude with some implications for future poverty research. We also provide examples and implementation via STATA.

Dimensions and rationale for measuring poverty

Defining and measuring poverty

Poverty can be understood from two broad approaches: the income and basic needs approach and the capability approach. The income and basic needs approach (often coined as the means indicators) is mainly characterized by quantitative indicators, while the human capability approach (often coined the ends indicators) is characterized by both quantitative and qualitative indicators. The capability approach usually incorporates more qualitative indicators that supplement the income and basic needs approach. The income indicators are used in the monetary approach and the basic needs approach to measure poverty, while in the capability approach, welfare is understood as an expansion of human capabilities. The latter approach assesses well-being and policy objectives in terms of freedom of individuals to live lives that are valued in realizing their true potential. This approach can be further divided into the capability approach and the participatory approach. We discuss these four approaches in turn.

Monetary approach

This is the most commonly used method in identifying and measuring poverty. In this approach, poverty is identified as a shortfall of income or consumption from some poverty line (Ravallion, 1998). Valuation of the different components of income or consumption is done at market prices that require imputing the monetary values for the items. For goods that cannot be valued at market prices (such as subsistence production and public goods), imputing monetary values is crucial for measuring poverty. The appeal of this approach to economists lies in being compatible with utility maximizing behavior of households with expenditure reflecting the marginal value that individuals place on commodities. Welfare is then measured as the total consumption enjoyed either by individuals or households, and poverty is defined as the shortfall below some minimum level of resources—the poverty line.

Basic needs approach

The basic needs approach is a natural extension of the monetary approach to evaluate global poverty. "Basic needs" are defined to include not only food, water, shelter, and clothing but also access to assets such as education, health, participation in political process, security, and dignity of individual (Streeten et al., 1981).

The Human Poverty Index (HPI) created by the United Nations Development Programme (UNDP) aimed to reduce the problem of income measures by using indicators that show the depth of deprivation across countries (UNDP, 1997). HPI is a composite index that uses three indicators in measuring poverty—a short life, lack of basic education, and lack of access to public and private resources. The first component relates to survival and vulnerability to death at an early age. In developing countries, the index represents the percentage of people expected to die before the age of 40, while in developed economies, the index represents the percentage of people expected to die before the age of 60. The second component relates to knowledge acquisition and is measured by the percentage of adults in the country who are illiterate. The final component of the index relates to overall standard of living and is a combination of three variables: the percentage of people with access to health services and safe water and the percentage of malnourished children below the age of 5.

Capability approach

The "capability approach," developed by Sen (1985, 1999), is a natural extension of the HPI approach. In this approach, monetary income as a measure of well-being is rejected, and indicators of the freedom to live a valued life are emphasized. In this framework, poverty is defined as deprivation in the space of capabilities or failure to achieve certain minimal or basic capabilities. "Basic capability," according to Sen, "is the ability to satisfy certain important functioning up to certain minimally adequate levels."

In this approach, well-being is seen as the freedom of individuals to live lives that are valued in terms of realization of the human potential and is thus an ends-based approach. Monetary resources are considered only as means to

```
                    Private monetary income                    Social income
                                        ↘   Utility   ↙
                                         ↗          ↖
                           Commodities              Public goods
                                    ↘              ↙
                              Characteristics of commodities
                                          ↓
                                  Feasible utilizations
                                          ↓
                                    Capability set
                                          ↓
                        Individual choice within the capability set
                                          ↓
                                     Functionings
```

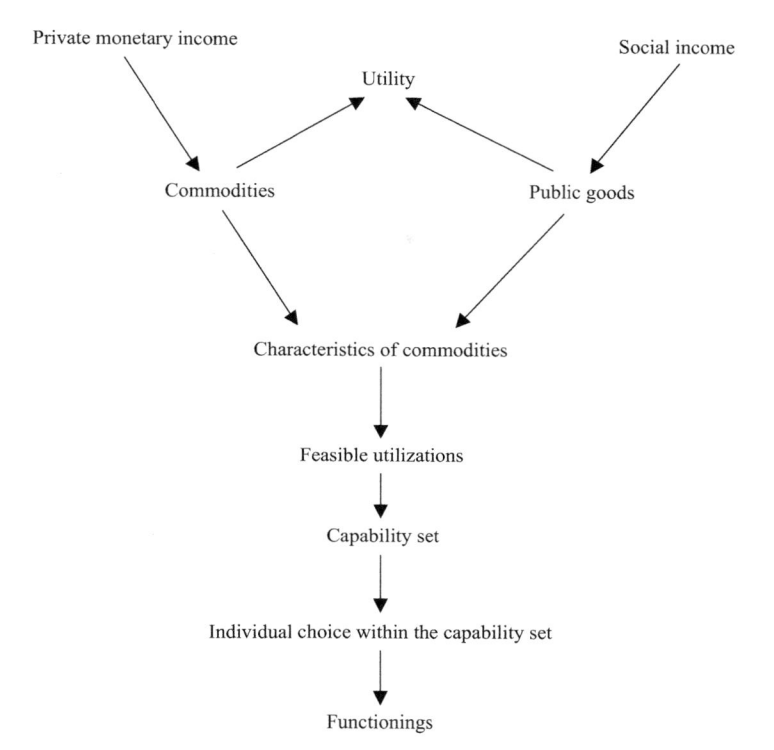

FIGURE 12.1 Capability approach. *Adapted from Laderchi, C.R., Saith, R., Stewart, F., 2003. Does It Matter that We Don't Agree on the Definition of Poverty? A Comparison of Four Approaches. Queen Elizabeth House Working Paper No. 107. University of Oxford, Oxford.*

enhance well-being, rather than the actual outcome of interest. These resources may not be considered as reliable indicators of capabilities since achievements (or functionings) can differ based on individual characteristics or contexts in which individuals live in. For example, able-bodied and handicapped individuals need different amounts of resources to obtain the same outcome. Contexts in which individuals live can also differ, such as areas where basic public services, are provided versus areas where such services are absent. The conceptual framework of the capability approach is illustrated in Fig. 12.1. Both monetary income and public goods along with individual's own personal characteristics (such as gender, age, and physical capacities) determine the capability set of the individual.

Participatory poverty approach

Conventional measures of poverty (such as the income approach) rely mainly on statistical information contained in household surveys, combined with an arbitrary cutoff point (the poverty line) in separating the poor from the nonpoor.

A recent empirical approach adopted by the World Bank asks people what to them constitutes poverty.

Participatory surveys are designed to study how individuals learn from various social groups to assess their own poverty, how various survival strategies work, and what kind of poverty reduction strategies people prefer and are prepared to support. These surveys draw their methodology from analytical instruments used by the World Bank such as beneficiary assessments and participatory rural appraisals (World Bank, 2000).

Combined approaches

The ideas of basic needs and Sen's capabilities approach have spawned an immense amount of research and furthering of developments with respect to understanding of poverty. For example, Rao and Min (2018) combine both these approaches and the complementary notion of Decent Living Standards and incorporate living conditions, social participation, minimum wage and reference budgets, and environmental changes to evaluate and address poverty.

Lutafali et al. (2016) extend the basic needs model to combat poverty in India, Pakistan, and Bangladesh, by extending microfinance to unbanked entrepreneurs, through an interesting organogram model. Meyer and Keyser (2016) expand on yet another received Lived Poverty Index using the basic needs methodology and identify many newer components that adequately describe poverty in South Africa.

Likewise, the capabilities approach has received a lot of attention among researchers that has led to several multidimensional measures of inequality. Islam and Al-Amin (2019) show how capability deprivation drives Bangladeshi tea garden workers to poverty and social vulnerability. Van Phan and O'Brien (2019) use the capability approach to expand single-dimensional analysis of well-being and inequality and develop a polychoric principal component analysis for Vietnam.

In recent years, the capability approach has motivated community-driven programs to combat poverty. Pham (2018) uses the capability approach to evaluate the effectiveness of community-driven programs. Community-driven programs have the potential to be effective because of the underlying principles of a wider informational base for normative policies and a greater sensitivity to gender differences.

For example, in an interesting study by Vansteenkiste and Schuller (2018), note the differences in frameworks adopted by two women's organizations, in comparison with two mixed-gender organization, in Haiti. The two women's organizations adopted practical reasoning and affiliations to create space that values opportunities and functions for women, which are usually unavailable, thus enhancing gendered capabilities. The advantages of community-based programs are also evidenced by Biggeri et al. (2018) for India. Von Jacobi (2018) demonstrates how the capabilities approach fits in-between macrolevel policies and local-state level decisions. We will examine the construction of a multidimensional index of poverty, later in the book, which incorporates all of these ideas, and which received a lot of attention in recent years.

Rationale for measuring poverty

There are two main reasons for studying poverty:

- Targeting interventions
- Designing programs and policies to reduce poverty (Ravallion, 1998)

Targeting interventions

Having data on household poverty status allows a researcher to evaluate the impact of programs on the poor and determine whether these programs achieve the goals with respect to targeting a certain group of households. Targeting can be an important step in effectively reaching the disadvantaged groups and backward areas. The most important use is to reduce aggregate poverty through regional targeting. An example of targeting is employment targeting. Employment targeting can help the vast majority of households in a certain country to escape poverty by their earnings through employment.

Designing programs and policies

The main purpose of this process is to inform policy makers on poverty measurement and diagnostics so as to reduce poverty over a sufficiently long-run period. For example, the World Bank introduced the concept of Poverty Reduction Strategies Paper (PRSP) for the highly indebted poor countries (HIPCs) in 1999, setting out a strategy for fighting poverty. The main principle of the PRSP was that fighting poverty should be country-driven, result-oriented, and comprehensive. It must also be based on the participation of civil societies along with partnerships from donors. The emphasis of the PRSP process is to have a long-term development plan to reduce poverty. The main step is to understand the characteristics and causes of poverty once it is determined how many poor there are. Once the poor are identified, the next step in the PRSP process is to choose public actions and programs that have the greatest impact on poverty.

Indicators in measuring poverty

For quantitative analysis of poverty, household income and consumption expenditure provide direct measures of household welfare. To assess poverty based on household expenditure, the use of an expenditure function can be a convenient tool (Deaton, 1997). An expenditure function shows the minimum cost of achieving a given level of utility u, which is derived from a vector of goods q and prices p. It is obtained by minimizing an objective function (in this case expenditure), subject to given level of utility, assuming prices to be fixed. Typically, in household surveys, the actual level of expenditure is used for information on consumption.

Income measure

While household income is a direct welfare measure, its definition remains a problem. The most accepted measure of income formulated by Haig and Simons and documented by Rosen (2002) is

$$\text{Income} = \text{consumption} + \text{change in ne tworth}$$

The main problem with the aforementioned definition, however, is choosing an appropriate time period. Should it be defined over a 1-year, 5-year period, or over a lifetime? An additional problem concerns measurement. While it may be easy to measure wages and salaries, interest, dividends, and income from self-employment, it is difficult to measure the value of housing services or capital gains (such as increase in the value of house, increase in the value of stocks, etc.). For these reasons, income is largely understated.

For countries with significantly large agricultural populations, income is seriously understated, because of the following:

- Individuals may be reluctant to disclose the full extent of their income because of the tax-collector or other external problems.
- People forget, especially in a single interview, about the items they may have sold or money they have received up to a year before (Ravallion, 1998).

Consumption expenditure

Consumption includes both goods and services that are purchased and those provided from one's own production. Consumption can be a better indicator of lifetime welfare than income, since income fluctuates from year to year (it typically rises and then falls in the course of an individual's lifetime, whereas consumption remains fairly stable). However, consumption can also be systematically understated, since households may underdeclare what they spend on luxuries and other illicit items (Deaton, 1997).

Durable goods can be another challenge in the computation of total consumption expenditure by households. This is because goods such as refrigerators, televisions, and cars can be bought at a point in time and can be used over a period of several years. Consumption should thus include the amount of durable goods that are used up during the year, which can be measured by the change in the value of asset during the year plus the cost of locking up the money in the asset. For example, if a bicycle was worth $75 a year ago and is worth $60 now, then $15 worth was used. Suppose the individual used the money and earned 5% in interest during the year from a savings account. Then the true cost of the bicycle was $18.75.

The value of durable goods as part of consumption expenditure is used to achieve comparability across households. If this value were not included, one might have the impression that a household that spends $100 on food and $15 on renting a bicycle is better off than a household that spends $100 on food and owns a bicycle (that it could rent out for $15), when in fact both households were equally well off.

In measuring the value of housing services, one needs to ask how much they would have paid in rent for the house if they had not owned it. The standard procedure to impute the value of rent is to estimate rent as a function of housing characteristics such as the size of the house, the year in which it was built, the type of roof, number of bedrooms, and so on. For households owning their house, the imputed rental along with the costs of maintenance and other repairs will represent the annual consumption of housing services (Deaton and Zaidi, 2002). For households that pay interest on a mortgage, the imputed rental, costs of maintenance, and other repairs will represent consumption, but the mortgage interest payments should not be included in the computation (since this would represent double counting).

Household composition can be an important problem in measuring poverty. To avoid this problem, a direct approach is to convert household consumption to individual consumption by dividing total expenditure by the number of people in the household (Deaton and Zaidi, 2002). Then, expenditure per capita is the measure of welfare assigned to each member of the household. This approach, however, is not satisfactory, since individuals have different needs. For example, a young child does not have the same food need as an adult. Also, there are economies of scale in consumption (especially for nonfood items). It costs less to house a couple than to house two single individuals.

To address this problem, a system of weights called equivalence scales may be used. For a household of a given size and demographic composition, an equivalence scale measures the number of adult males which that household is deemed to be equivalent to. Thus, each member of the household is counted as some fraction of an adult male. Hence, household size is the sum of these fractions and is measured in the numbers of adult equivalents and not in numbers of persons. One common way of measuring this is the OECD scale of adult equivalents (AE), which can be written as follows:

$$AE = 1 + 0.7 * (N_{adults} - 1) + 0.5 * N_{children} \qquad (12.1)$$

Thus, a one-adult household would have an adult equivalent of one, and a two-adult household would have an adult equivalent of 1.7. 0.7 thus represents economies of scale. 0.5 weight is given to children, since presumably they have lower needs (such as less food, space, and housing).

Construction of poverty lines using food energy intake and cost of basic needs approaches

Poverty lines in theory

This section closely follows and utilizes Ravallion (1998) and Ravallion and Bidani (1994). A poverty line can be defined as the minimum cost to the household for attaining the poverty level of utility at prevailing prices and for given household characteristics. Formally, let a household's utility function be represented by $u\,(q, x)$, where q denotes a bundle of goods in quantities and x represents the household's characteristics. The function $u(.)$ assigns a single number to each possible q, given x. Let the household's expenditure function be denoted by $e\,(p, x, u)$, which is the minimum cost to the household with characteristics x to attain a given level of utility u when facing a price vector p. When evaluated at the actual utility level, $e\,(p, x, u)$ is the total expenditure on consumption, $y = pq$, for a utility maximizing household. Let u_z be the reference utility level required for escaping poverty. Then the poverty line is given by

$$Z = e(p, x, u_z) \qquad (12.2)$$

The aforementioned definition shows how to arrive at poverty in terms of money to poverty in terms of utility. However, the poverty level of utility (u_z) is still not defined. To measure poverty, one needs to combine the poverty line with the distribution of consumption expenditures. The two ways of combining the poverty line with the distribution of consumption expenditure are as follows:

- *Welfare ratio approach*: In this approach, the money income or total expenditure (pq) can be deflated by the poverty line z so that the ratio (y/z) is the welfare ratio. Alternatively, one can compute the true cost of living index as $e\,(p, x, u_z)/e\,(p^r, x^r, u_z)$, where p^r and x^r denote the base prices and base characteristics at a given point in time and location and $e\,(p^r, x^r, u_z)$ is the base poverty line. Then, the cost of living index is the ratio of each household's poverty line to the base poverty line. All money incomes can then be normalized to comparable monetary units, and a single poverty line can be applied with respect to the base.
- *Equivalent expenditure method*: In this approach, welfare is computed using the expenditure function given by

$$y^e = e[(p^r, x^r, v(\cdot)p, x, y)] \qquad (12.3)$$

where $v(\cdot)$ is the indirect utility function[1] obtained as a function of prices and expenditure. Since both p^r and x^r are fixed, y^e is a monotonically increasing function of utility. The welfare ratio can then be calculated as the ratio of y^e and the base poverty line to obtain the equivalent welfare ratio.

Absolute and relative poverty

Absolute poverty is defined in terms of the standard of living in a location at a point in time, while the relative poverty increases with average expenditure. An absolute poverty line is used for the antipoverty measures so that valid comparisons can be made between one country and another or one region to another. However, the absolute poverty line being invariant to welfare measure does not mean that it is invariant to average expenditure as well. If welfare depends on the expenditure relative to the mean of a reference group, then the real value of the poverty line will also vary with the mean. Formally, this is demonstrated as follows:

$$u = f\left(y, \frac{y}{\bar{y}}\right) \quad (12.4)$$

where $\frac{y}{\bar{y}}$ denotes an individual's or household's relative expenditure and $\bar{y}$ is the mean expenditure of the reference group. If poverty line is absolute in the utility space, then

$$u_z = f\left(z, \frac{z}{\bar{y}}\right) \quad (12.5)$$

Then assuming invertibility, the implicit function relating poverty line to the mean expenditure is given by

$$z = f^{-1}(\bar{y}, u_z) \quad (12.6)$$

In other words, for the poverty line to be absolute in the space of welfare, the commodity-based poverty line (z) has to increase as $\bar{y}$ increases.

Referencing and identification problems

From the aforementioned analysis, we can see that poverty lines cannot be derived without first determining how "utility" needs to be ascertained. The *referencing problem* asks the question: what is the appropriate value of u_z so that an individual can escape poverty? While it can be said that the choice is

---

1. The indirect utility function is obtained by maximizing the utility function subject to a budget constraint and then substituting the commodity bundle as a function of income and prices back into the utility function. This function gives the maximum utility achievable for a given level of prices and income.

essentially arbitrary, Ravallion (1998) points out that, in the practice of poverty measurement, the choice of reference is far from arbitrary but is crucial for poverty measure. The choice of this variable can allow for qualitative comparisons of poverty in different regions and can make priorities for policy makers for geographic targeting.

The second problem *(identification)* relates to the correct value of z, i.e., the commodity value of the poverty line. The problem arises as households vary in size and demographic composition, which influence welfare.

Deriving a poverty line

Given the aforementioned problems, how does one determine a poverty line? A solution to the aforementioned question is through constructing an objective poverty line.[2] The key idea is that poverty lines should be set at a level that enables households to achieve certain capabilities, such as healthy and active lives and full participation in a society. There are two main approaches in constructing an objective poverty line: (1) the FEI method and (2) the CBN method. We demonstrate how to compute the poverty lines using the aforementioned approaches for Malawi and Mozambique, respectively.

Food energy intake method

In this method, the poverty line is set by finding the consumption expenditure or income level at which FEI is just sufficient to meet predetermined food energy requirements. Determining the food energy requirements can be difficult since requirements vary across individuals and over time for an individual. The basic idea is illustrated in Fig. 12.2, which shows a calorie income function.

In Fig. 12.2, the vertical axis is FEI, which is plotted against total income or expenditure on the horizontal axis. The function shows that as income (or expenditure) increases, FEI also rises, but more slowly. Thus, if k denotes FEI, with $k = f(y)$, then, for a given minimum adequate level of calorie intake k_{min} (such as 2100 kcal per day), the poverty line is given by the following equation:

$$z = f^{-1}(k_{min}) \tag{12.7}$$

---

2. In subjective poverty line construction, the respondent is asked questions such as "What income level do you personally consider to be absolutely minimal?" We will not discuss this issue in the current chapter, since we are interested in objective poverty lines. Second, in most developing country studies, objective poverty lines have been constructed, while a few attempts have been made for OECD countries for the construction of subjective poverty lines. For an extensive discussion of how poverty lines are constructed using this approach, see Ravallion (1998).

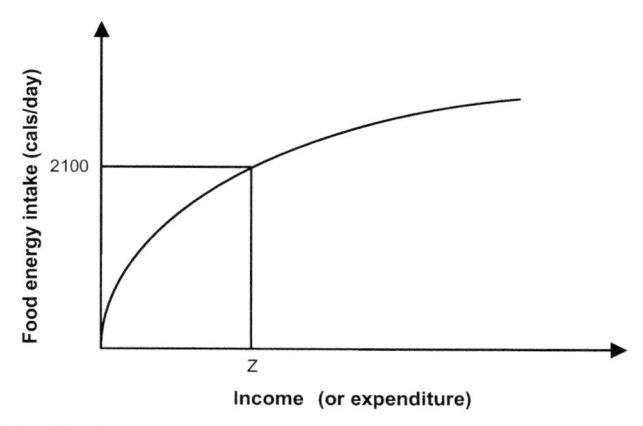

Measurement and determinants of poverty—application **Chapter | 12 445**

FIGURE 12.2 Calorie income function. *From Ravallion, M., 1998. Poverty Lines in Theory and Practice. LSMS Working Paper No. 133. World Bank, Washington, DC.*

This approach is parsimonious since it does not require any information about the prices of goods consumed. Let us now apply this method using the Malawi dataset. Let k denote the calories per adult equivalent for the household and let x be the total household expenditure. Then the cost of calories function can be represented by:

$$\ln x = a + bk \tag{12.8}$$

Let the minimum calorie requirements be set at 2250 kcal per day. Then the poverty line is given by:

$$z = \exp(a + bL) \tag{12.9}$$

where z denotes the cost of buying the minimum calorie intake L (which is assumed to be 2250 kcal per adult equivalent). Estimating Eq. (12.8) using ordinary or weighted least squares, one can determine the relationship between x and k.

We estimate Eq. (12.8) using the Malawi dataset, by making a logarithm transformation of the calorie per adult equivalent. (We weight the observations by the household size to take into account economies of scale within the household). Our estimated coefficients are $\hat{a} = 2.158$ and $\hat{b} = 0.328$. Then the poverty line based on FEI method is given as:

$$Z = \exp(2.158 + .0328 * 7.718) = 107.93 \tag{12.10}$$

Thus, using this method, the poverty line is determined as a total expenditure of 107.93 kwacha per person per month.

Ravallion (1998) points out the following problems with this method. First, the relationship between FEI and income will shift according to the differences in tastes, relative prices, publicly provided goods, or other determinants of affluence besides consumption expenditure. For example, relative prices can differ across locations, which alter the demand behavior at a given real expenditure level. The prices of certain non-food items are usually lower relative to food items in urban areas than rural areas. This may imply that the demand for food and FEI will be lower in urban areas relative to rural areas for a given level of real income. However, this does not mean that urban households are poorer at a given expenditure level. The real income at which an urban resident typically attains a given level of caloric requirement is higher than in rural areas even if the cost of basic consumption needs is not different between rural and urban areas. The problem with the FEI method is that the poverty lines may not be defined in terms of the command over commodities. Access to publicly provided goods with the FEI method can also result in inconsistency of poverty estimates. For example, in urban areas, access to better health care and schooling can mean that individuals tend to consume a diet that is nutritionally better balanced with fewer calories and more micronutrients. In this case, using the FEI method, more people will be considered as poor in urban areas with a higher poverty line.

Secondly, mobility from rural to urban areas (such as migration) can result in poverty lines being higher in urban than rural areas using this method. For example, suppose that an individual who is above the FEI poverty line in the rural sector moves to the urban sector and obtains gainful employment. Though the person is better off (in the sense of greater purchasing power over commodities), the aggregate measure of poverty across sectors will show an increase since the migrant is deemed as poor. This is because the FEI poverty line has higher purchasing power in terms of basic needs in urban than in rural areas. Thus, using this method one can arrive at a measured increase in poverty in the urban sector when, in fact, none of the poor are worse off and at least some are better off. To address the above problems, the CBN approach to poverty is given below. Ravallion (1994, 1998) and Ravallion and Bidani (1994), have demonstrated that the CBN approach does not suffer from the problem of inconsistent poverty comparisons when the FEI method is used to set poverty lines.

Cost of basic needs method

This method stipulates a consumption bundle that is considered to be adequate for basic consumption needs, and then the cost to each of the subgroups is considered for generating a poverty profile. This method can be interpreted in two distinct ways. First, it can be interpreted as the "cost of utility" under special assumptions about preferences. The second interpretation is a socially determined normative minimum for avoiding poverty. Among the infinite number of consumption vectors that yield any given set of basic needs, a vector is chosen that is consistent with the choices actually made by some reference group (say the bottom 60% of the population). Poverty is then measured by comparing actual expenditures to the CBN.

The derivation of the total poverty line at the national level can be organized in three main parts:

- Identifying regions for the definition of poverty lines.
- Steps in the construction of the food poverty lines.
- Construction of the nonfood poverty lines.

Identifying regions

Defining a single poverty line in nominal terms for a country where standards of living vary greatly by regions is not a useful strategy. This is because markets are not often spatially integrated and there are substantial variations in regional prices. Since consumption patterns vary widely due to differences in relative prices across regions, differences in consumption patterns should be considered in assessing cost of living differentials. Thus, an appropriate level of spatial disaggregation is critical for constructing the poverty lines.

In constructing region-specific poverty lines, two considerations are important. First, a rural—urban distinction is necessary since consumption expenditure varies greatly between the rural and urban areas. Second, provinces or regions need to be grouped together which are relatively homogeneous. (See Simler et al. (2004) for an illustration of deriving regional-specific poverty lines. This section closely follows procedures given in this publication).

Constructing the food poverty lines

In the CBN approach, the food poverty line is constructed by determining the calorie intake requirements of the poor, the calorie content of a typical diet of the poor in the region, and the average cost of a calorie when consuming that diet. The food poverty line is computed as the product of the average daily per capita caloric requirement and the average cost per composite calorie. In other words, the food poverty line is the cost of meeting the "minimum caloric requirements" when consuming the average food bundle for the poor in a particular region.

The immediate consideration that arises is the minimum per capita calorie requirements in each region. For example, a region with a large proportion of children will typically require fewer calories per capita than a region with a higher proportion of middle-aged adults, since children usually have lower caloric requirements. In principle, while computing caloric requirements, one needs to take into account the age and sex composition of the population and the physical activity level of women (i.e., whether they are pregnant or are in the first 6 months of breastfeeding). For the sake of simplicity, the average per capita calorie requirement could be set at 2100 kcal per person per day. In deriving the *average price or cost per calorie*, it is critical to define the poorer households' consumption bundle. Poorer households are generally defined as the bottom 60% of the population in terms of per capita expenditure, often known as the "reference group." In this method, detailed consumption data including the total food expenditure levels and the quantities of the food items actually consumed are necessary.

For operationalizing this method, in the first step, the resulting consumption data need to be aggregated at the household level, with the price per calorie determined. In the second step, the household level data set can be further aggregated to the regional level, which will give the mean price per calorie for each region. The food poverty line will then be the product of the mean cost per calorie and the minimum calorie requirements.

Constructing the nonfood poverty lines

The nonfood poverty line is derived by looking at the nonfood consumption among those households whose per capita total expenditure was equal to the food poverty line. This is also known as the lower bound of the nonfood poverty line. An upper bound can be similarly constructed by looking at households whose per capita food expenditure was equal to the food poverty line. The rationale for using this method is that if a household's total consumption was only sufficient to purchase the minimum amount of calories for a typical food bundle consumed by the poor, then any expenditure on nonfoods is possibly displacing food expenditure or forcing the household to buy a food bundle that is inferior to the bundle usually consumed by the poor. In either case, the nonfood consumption of such a household displaces typical food consumption and thus can be considered as "essential" (Simler et al., 2004).

The neighborhood where per capita total consumption is equal to the food poverty line is defined as 80%—120% of the food poverty line. The cost of the minimum food bundle is then estimated as the weighted average of nonfood expenditure. Using a kernel estimation procedure with triangular weights (Datt et al., 2001), observations that were closer to the food poverty line are given a higher weight. (For an extensive discussion on kernel estimation, the following website is useful: http://www.quantlet.com/mdstat/scripts/anr/html/anrhtml node11.html).

For example, households which consumption is within 2% of the food poverty line are given a relative weight of 10, whereas households which consumption is between 18% and 20% of the food poverty line are given a weight of 1. Observations that are more or less than 20% of the food poverty line are given a weight of 0, since these households are presumed to be further away from the food poverty line. The weighted average of the nonfood consumption per capita for each of the regions is determined by weighting the household level observations by the expansion factor, which is the product of the sampling weight and the triangular weights. Having information on per capita consumption and a poverty line, one could proceed to determine an appropriate aggregate measure of poverty.

New measures of poverty based on the engel curve

While the calorie intake method and the CBN methods are extremely popular in measuring poverty, a recent approach by Kumar et al. (2008) tries to measure poverty based on the Engel curve of a community.

The approach considers an individual as a member of a community and the situation of the individual within the community that determines the norm that individuals set for themselves. While any point on an Engel curve depicts the average consumption for a given level of income for that community, a *saturating Engel curve* depicts the average consumption of an essential commodity if the individual is not constrained by low income. The cumulative shortfall of actual consumption of an essential commodity from this norm thus constitutes consumption deprivation. The consumption deprivation of an entire community with respect to all essential commodities constitutes the poverty of the commodity.

While this poverty line is subjectively chosen by the researcher, the deprivation index is determined by the socioeconomic setting where the household is situated. The index is objectively derived from the observed Engel curve and not on nutritional or other normative considerations. Secondly, while the traditional measures of poverty are related to consumption deprivation through an indirect link between income and consumption deprivation, the index is such that if there are more households with greater consumption deprivation, greater is the contribution of the group to poverty. The index thus measures severity of deprivation.

Although this approach has only been applied to Indian households, it can be used for other countries in different socioeconomic settings.

Measures of poverty

There are a number of aggregate poverty measures that can be computed based on the assumption that the survey was a random sample drawn from the population (Foster et al., 1984). They are as follows.

Headcount measure

The headcount index is the proportion of the population that is counted as poor and is often denoted by P_0. Formally, it is given as follows:

$$P_0 = \frac{1}{N} \sum_{i=1}^{N} I(y_i \leq z) = \frac{N_p}{N} \quad (12.11)$$

where N is the total population and $I(.)$ is an indicator function that takes on a value of 1 if expenditure (y_i) is less than the poverty line (z) and is 0 otherwise. Np denotes the total number of poor. The main advantage of this measure is that it is simple to construct and easy to comprehend. However, this measure does not take into account the intensity of poverty. For example, consider the following two income distributions of two hypothetical regions within a country, where the poverty line is 130.

		Expenditures for individuals				P_0
Region expenditure	A's	90	110	200	200	50%
Region expenditure	B's	115	125	200	200	50%

From the aforementioned example, it is evident that there is greater poverty in region A. However, the headcount index does not capture this phenomenon. Second, the index violates the transfer principle. In other words, if a somewhat poor household transferred some income to another poor household, the index would remain unaltered, although one can presume that overall poverty has been reduced. Finally, the index does not indicate how the poor are and thus does not change if people below the poverty line become poorer.

Poverty gap index

This is a popular measure of poverty and shows the extent to which individuals fall below the poverty line and computes that as percentage of the poverty line (Foster et al., 1984). Formally, let the poverty gap (G_n) be defined as follows:

$$G_n = (z - y_i)\, I(y_i \leq z) \qquad (12.12)$$

Then the poverty gap index can be written as follows:

$$P_1 = \frac{1}{N} \sum_{i=1}^{N} \frac{G_n}{z} \qquad (12.13)$$

The aforementioned measure is the mean proportionate gap in the population (where the nonpoor have zero poverty gap). Some researchers consider this measure as the cost of eliminating poverty, since it shows how much money needs to be transferred to the poor so that their income (or expenditure) can be brought up to the poverty line. However, the minimum cost of eliminating poverty using targeted transfers is the sum of all poverty gaps within a population. However, policy makers do not always have enough information to make the transfers to the targeted segments of the population.

The advantage of this measure is that there is no discontinuity in the poverty line. However, an important drawback of this measure is that it may not capture the differences in the severity of poverty among the poor. For example, consider two regional income distributions of a country as follows: region A = (1, 2, 3, 4) and region B = (2, 2, 2, 4) and let the poverty line $z = 3$. Then the poverty gap index for both the regions is 0.25. However, the poorest person in region A has only half the consumption of the poorest in region B. Thus, one can consider B being generated from A by a transfer from the least poor of the poor persons (individual with an expenditure of 3 in A) to the poorest. Thus, the main drawback is that it ignores inequality among the poor.

In summary, the poverty gap index is the average of all the individuals of the gaps between poor people's standard of living and the poverty line, expressed as a ratio to the poverty line. The smaller the poverty gap index, the greater is the potential to allocate budget in identifying the characteristics of the poor so as to target benefits and programs.

Squared poverty gap index

To address the problem of inequality among the poor, the squared poverty gap index is often used (Foster et al., 1984). This is a weighted sum of poverty gaps (as a proportion of the poverty line) where the weights are directly proportional to the poverty gaps themselves. For example, a poverty gap of 20% of the poverty line is given a weight of 20%, whereas a poverty gap of 60% of the poverty line is given a weight of 60%. By squaring the poverty gap index, the measure imposes more weight on observations that fall well below the poverty line. Formally, this is defined as follows:

$$P_2 = \frac{1}{N} \sum_{i=1}^{N} \left(\frac{G_n}{z}\right)^2 \quad (12.14)$$

The advantage of this measure is that it reflects the severity of poverty and is sensitive to the distribution among the poor. However, the measure lacks intuitive appeal and is harder to interpret.

All the aforementioned measures of poverty can be thought of as being generated by a family of measures proposed by Foster et al. (1984), also called the FGT measure. Formally, it can be defined as follows:

$$P_\alpha = \frac{1}{N} \sum_{i=1}^{N} \left(\frac{G_n}{z}\right)^\alpha \text{ with } \alpha \geq 0 \quad (12.15)$$

where the parameter α is a measure of sensitivity of the index to poverty. The poverty gap for individual j is given by $G_j = z - y_j$ (with $G_j = 0$ if $y_j > z$). When the parameter $\alpha = 0$, the measure in Eq. (12.15) boils down to the headcount index. When $\alpha = 1$, we have the poverty gap index P_1, and when $\alpha = 2$, we have Eq. (12.15) boiling down to the poverty severity index.

Properties of $P\alpha$:

- For all $\alpha > 0$, $P\alpha$ is strictly decreasing in the living standard of the poor.
- For $\alpha > 1$, increase in measured poverty due to a fall in the standard of living will be greater the poorer one is. In other words, it is said to be "strictly convex" in incomes.

The measures of depth and severity of poverty give us complementary information on the incidence of poverty. Thus, some income groups might have higher poverty incidence but low poverty gap, while other income groups might have lower poverty incidence but higher poverty gap. Thus, interventions needed to help these two groups will have to be different in nature.

Computing poverty measures for Malawi

We now demonstrate how the various measures of poverty are computed using Eqs. (12.11), (12.13), and (12.14) respectively. Table 12.1 gives the poverty headcount, the poverty gap index, and the poverty severity index for the full sample of households of all the regions combined together and for Mzuzu, Salima, and Ngabu regions, respectively.

The aforementioned estimates show that 73.2% of Malawi's population lived in poverty during 1991−92. Although regional comparisons are more difficult to understand, given the confounding effect of the presence of both rural and urban households within a given region, the incidence and severity of poverty was highest in the Salima add-code[3] (which is located in the central

TABLE 12.1 Individual poverty measures by regions using the 604 household data set.

Regions	Poverty headcount (in percent)	Poverty gap index	Poverty severity index
Full sample	73.18	0.41	0.28
	(0.01)	(0.01)	(0.01)
Muzzy	62.21	0.28	0.15
	(0.03)	(0.02)	(0.01)
Salima	83.04	0.54	0.40
	(0.02)	(0.02)	(0.02)
Ngabu	73.89	0.40	0.27
	(0.035)	(0.027)	(0.02)

Standard errors are corrected for sample design and are given in parentheses.

---

3. This result is also consistent with Babu and Chapasuka (1997), since the measure of poverty is based on insufficient food expenditure to meet the daily nutritional requirement of 2250 kcal. They found the estimates of headcount to be 38.5%, 67.8%, and 62.9%, respectively, for the Mzuzu, Salima, and Ngabu regions based on their sample of households. In other words, the relative ranking of the poverty headcount has not changed in our sample.

part of Malawi) followed by Ngabu (which is located in the southern part of Malawi).[4] Relative to these two regions, Mzuzu had a lower proportion (62.2%) of poor people. Additionally, both the poverty gap and severity index were lower in Mzuzu relative to the rest of the regions. A possible reason for this difference in intensity and severity across regions can be attributable to the densely populated areas in the southern and central regions of the country, which have the lowest mean incomes. Having investigated the measures of poverty across regions, let us turn our attention to some of the characteristics of poor households.

Characteristics of poor households

The analysis presented in the following shows some of the economic and demographic characteristics of the poor households compared with the nonpoor households using the cross-tabulation procedure.

Table 12.2 shows that the Malawian smallholders own very few productive assets and have only small plots of land. Compared with only 40% by the households who are not poor and owning less than 1 ha of land, the poorer households constitute about 59%. The difference was found to be significant using the chi-square tests. The computed chi-square value was 19.44, while the critical value at 1% level of significance is 15.08 (for 5 degrees of freedom). Since the computed value is greater than the critical value, one can conclude that there is significant difference in landholdings by poor and nonpoor households. One of the possible conclusions can be that access to more inputs,

TABLE 12.2 Cross-tabulation results of land owned by the number of poor people.

	No. of people	Not poor	Poor	Total
Size of land owned	<1 ha	65	259	324
		40.12%	58.6%	
	1–2 ha	86	158	244
		53.09%	35.75%	
	>2 ha	11	25	36
		6.79%	5.66%	
Total		162	442	$604 = n$

---

4. This result is consistent with the poverty profile report of the World Bank, which indicated that poverty was deep and severe in the central region.

TABLE 12.3 Cross-tabulation results of dependency ratio by the number of poor people.

		No. of people	Not poor	Poor	Total
Dependency ratio	<0.5		79	189	268
			48.77%	42.76%	
	≥0.5		83	253	336
			51.23%	57.24%	
Total			162	442	604 = n

such as the size of land owned, could raise farm productivity significantly (possibly through adoption of improved agricultural technology) and can be used as a policy measure for lowering poverty.

Dependency ratio is defined as the ratio of the number of children and individuals above 60 years of age to the total household size. From Table 12.3, it is evident that poorer households have a higher percentage (about 57%) of dependents compared with the nonpoor households (about 51%). This difference was also found to be significant using the chi-square tests. The computed chi-square value was 83.93, while for 50 degrees of freedom, the area to the right of a chi-square value at 0.01 level of significance is 76.19. One can conclude that there is a significant difference in the dependency ratio between the poor and nonpoor households, since the computed value is greater than the critical value. Thus, a higher dependency ratio may be positively correlated with the level of household poverty.

Selected review of studies on determinants of poverty

In this section, we review a few studies that examine the determinants of poverty using a logistic regression framework.

Rodriguez and Smith (1994) examined the factors affecting the levels of poverty among urban, rural, rural farm, and rural nonfarm families using several logit models. The choice of Costa Rica (although it was a middle income country during 1993) was primarily motivated by the fact that it was adversely affected by an economic recession during the 1980s, which resulted in reversals in gains in poverty reduction achieved during the previous decade.

The analysis was undertaken using a logistic regression framework. The dependent variable was the poverty status of the family, which assumed two values: one being poor and zero being nonpoor. Two models were estimated. The first examined whether employment of the head was a significant determinant of poverty, while the second investigated whether participation of the household head in the labor market affected poverty.

The results from the study point out that poverty is a multidimensional phenomenon and several approaches are needed to address it effectively. On the one hand, trends toward lower child dependency ratio should reduce poverty levels. On the other hand, education and employment policies cannot be considered as separate efforts in reducing poverty but should be mutually reinforcing. While education and employment-related variables had relatively more influence in reducing urban and rural nonfarm poverty, it does not imply that efforts should be less in rural and farm settings. On the contrary, in the presence of large rural to urban migration, farm residents could be better off if they had some education. Thus, in combination with greater regional development, rural and farm families could have a better chance of obtaining gainful employment if they were educated.

Grootaert (1997) assessed the role of household endowments in determining the poverty status of households after controlling for the macroeconomic factors and the household's socioeconomic status. The incidence of poverty greatly increased in Cote d'Ivoire during the 1980s. In the second half of the decade, household consumption declined by more than 30%, and poverty rose sharply. Headcount increased from 30% during 1985 to 46% during 1988. By the end of the decade, half of the population was estimated to live in poverty. The data came from the Cote d'Ivoire Living Standards Survey (CILSS), which was conducted annually from 1985 to 1988. Each year, the survey covered a representative sample of 1600 households, and detailed information on employment, income, expenditure, assets, and other socioeconomic characteristics was collected.

A probit model was used to examine the determinants of poverty to address how the parameters differed across different segments of the distribution. Additionally, the study undertook sensitivity analysis by varying the poverty line and then reestimating the probit model to determine whether an increase in a given explanatory variable reduced the probability of being poor regardless of the poverty line. The study was undertaken separately for urban and rural regions for two time periods, namely 1985 and 1988.

The main results were that in both years and for both regions, urban and rural, education reduced the probability to be poor. In urban areas, the coefficients were largest at the lowest poverty line while, in rural areas, the effect tended to increase with a higher poverty line. Second, for urban areas in particular, a strong location effect persisted, implying that an otherwise similarly endowed household will be poorer in other cities compared to Abidjan. Finally, the results indicated that the probability for a non-Ivorian household to be poor in rural areas increased between 1985 and 1988.

Justino and Litchfield (2002) examined the impact of trade-related reforms on poverty, controlling for household- and community-level characteristics.

The two important trade-related reforms considered in the study were as follows:

- Liberalization of agricultural markets for outputs and inputs (including removal of price controls on rice and other crops and fertilizers.
- Liberalization of export markets (through removal of export quotas and tariffs).

The study used data from the Vietnam Living Standards Measurement Survey (VLSS), conducted during 1992–93 and 1997–98. Given that data for two time periods were available, it was possible to examine households that remained poor, had moved out of poverty, or fell into poverty and determine which household characteristics were associated with these movements. The analysis developed in the paper was based on a panel of 3494 rural households, and the standard of living indicator considered was household expenditure per capita, using the food and nonfood poverty lines.

A multinomial logit model[5] was postulated for analyzing household poverty dynamics. Poverty dynamics between the two periods was divided into four mutually exclusive outcomes:

- Being poor in both periods.
- Being nonpoor in the first period and poor in the second period.
- Being poor in the first period and nonpoor in the second period.
- Being nonpoor in both periods.

This model was used to compute the odds ratio of the household escaping and falling into poverty.

The log-odds ratio can be computed as $\ln\left(P_{ij}/P_{i0}\right) = X'_j(\beta_j - \beta_k)$

The study empirically examined the impact of trade-related reforms on household poverty dynamics using a multinomial logit regression model. In particular, two important reforms were highlighted:

- Liberalization of agricultural markets and prices of agricultural crops.
- Liberalization of export markets followed by the removal of export quotas and tariffs, which influenced employment patterns of households. The main findings can be summarized as follows:
- Employment effects were positive and had significant poverty reduction effects for households employed in the main export sectors—seafood, food processing, textiles and garments, and footwear. These effects benefited not

---

5. These regression models are used to model processes that involve a single outcome among several alternatives, which cannot be ordered such as occupational choices, modes of traveling, and so on. The multinomial model determines the probability that household i experiences one of the j outcomes. This probability is given by $P(Y_i = j) = \exp(\beta_j X_i) / -\sum_{k=1}^{j} \exp(\beta_k X_i)$ for $j = 0, 1, 2, ..., j$, where Y_i is the outcome experienced by the i th household, and X_i includes household characteristics as well as choices.

only households that in 1992–93 were involved in the export industries, but also households that increased the number of members employed in 1997–98.
- Trade reforms that affected the agricultural sector through an increase in the price of agricultural products had a strong impact on the poorest households. In particular, increases in agricultural diversification between 1992–93 and 1997–98 benefited the rural households significantly. The decrease in poverty was noticeable among households that diversified away from producing rice.

As policy measures, the study recommends further diversification of rural incomes by encouraging the establishment of small-scale farm and nonfarm enterprises for reducing poverty further.

Bigsten and Shimeles (2004) addressed factors that are related to the dynamics of income poverty using a unique household panel data for both urban and rural areas of Ethiopia over the period 1994–97. The novelty of the study was to understand vulnerability to poverty, as an integral component in the analysis of poverty. Vulnerability, as defined by Pritchett et al. (2000), is the probability of being below the poverty line at any given year.

Additionally, poverty was decomposed into a chronic and a transient component, where each was defined over a stream of income for an individual over the entire period. The measures of vulnerability were then compared with chronic poverty to understand the persistence of poverty. The data set consisted of 3000 households from both the urban and rural areas and divided equally between them.

The method used to derive the poverty spell was to compute the probabilities of falling into poverty, given certain states and other characteristics of households. Thus, a duration analysis was undertaken to estimate the entry and exit probabilities for being in poverty.

Formally, let X be a random variable indicating the duration of poverty or the length of time an individual had been in poverty, and let the distribution and density functions be

$$F(x) = Pr(x < X), \quad f(x) = dF(x)/dx$$

The conditional probability is given by

$$\theta(x) = Pr(x \leq X / X \geq x) = \frac{f(x)}{1 - F(x)}$$

Assuming θ to follow a logistic structure, with $x_{idt} = \alpha_{id} + \beta_{it} Z_{it}$, where the length of the probability depends on the duration effects, and other variables Z that vary across people and time. i, t, and d denote individuals, time, and the number of years in poverty, respectively. Then the probability of exiting poverty will be given by the following hazard function:

$$\theta_{idt} = \frac{\exp(\alpha_{id} + \beta_{it} Z_{it})}{1 + \exp(\alpha_{id} + \beta_{it} Z_{it})}$$

The aforementioned exit probabilities are estimated by maximizing the relevant log-likelihood function for all observations, which are quite similar to the logit estimates.

To complement the analysis of transitory poverty, the study then investigated the determinants of chronic poverty. The main determinant of chronic poverty was the education of the household head. Additionally, chronic poverty was reduced by variables such as crop sales, ownership of assets, and market access. An interesting finding was that off-farm activity was associated with higher chronic poverty, suggesting that off-farm activity was a survival strategy and not an indication of a household moving up the income scale.

One of the main results was that poverty was more persistent in urban areas compared with rural areas in Ethiopia. The aforementioned result clearly demonstrated that different approaches were required to fight poverty in urban and rural areas. Security issues were more important in rural areas, while expanding opportunities were more appropriate in the urban areas.

Thus, for policy purposes, depending on the nature of poverty (chronic or transitory), different measures need to be undertaken. If the nature of poverty is chronic, one needs to invest in long-run projects and structural reforms. This can include building up of human capital through education and health services apart from investment in physical and financial assets. On the other hand, if poverty was found to be transitory, temporary interventions to support households during bad periods are required. The measures could include different forms of safety nets, credit, and insurance schemes.

A recent paper by Rhoe et al. (2008) analyzes poverty in Kazakhstan using LSMS data. Poverty measures based on a food poverty line and total poverty line are compared, and the determinants of poverty are analyzed. They find that although there are some variations among the determinants of poverty under the two poverty lines, the explanatory power of the common determinants reduces when nonfood expenditures are included in deriving the poverty line.

Poverty and welfare in the United States

Ben-Shalom et al. (2011) provide a comprehensive assessment of the different programs in the United States that try to address the issue of poverty. Ben-Shalom et al. (2011) provide a list of the major programs in the United States.

There are many governmental benefit programs in the United States that attempt to reduce poverty. The two main types of programs are called the means-tested programs and the social insurance programs. Means-tested programs provide benefits to those with low income or assets with the objective of helping those in most need. Social insurance programs provide benefits to the population as a whole and are intended to insure individuals against the risk of unemployment, disability, and old age and inability to work.[6]

From Ben-Shalom et al. (2011), we note that a good example of the means-tested program is the Temporary Assistance for Needy Families (TANF) program, which provides cash benefits to families with low income and assets who have children in the household. Generally, recipients are single-mother families, where the father of the child is not present. Most funds for the program come from a federal block grant, but states also supplement these funds out of their own revenues. The program has work requirements, and the recipients who do not comply with the requirements are faced with benefit reduction penalties.

The Supplemental Nutrition Assistance Program (SNAP), formerly known as the Food Stamp Program, is another example of the means-tested program. It provides food assistance to individuals and families with low income and assets. Unlike the TANF program, the SNAP is an "in-kind" program, which provides assistance to all individuals and families, regardless of marital status or the presence of children.

The SNAP is entirely a federal program and is funded out of federal revenues, although states contribute a small amount for administrative costs. The federal government sets eligibility requirements and the benefit formula. We note from the last chapter that there are also other important food-based assistance programs in the United States. The School Lunch Program, the School Breakfast Program, and the Supplemental Nutrition Program for Women, Infants, and Children (WIC) are some examples of the SNAP.

Besides the TNAF, there is also the Supplemental Security Income (SSI) program, which also provides cash benefits to low-income, low-asset individuals who are above age 65, or who are blind or disabled adults or children. SSI recipients are generally also automatically eligible for the Medicaid program that provides subsidized medical care. The largest recipient group is low-income mothers and children. For the most part, recipients receive the full set of medical services with a zero copayment as long as their income and

---

6. While not directly aimed at helping those in most need, the social insurance programs in the United States have a major impact on poverty because of their large scale. The leading means-tested programs are Temporary Assistance for Needy Families, the Supplemental Nutrition Assistance Program, Supplemental Security Income, Medicaid, the Earned Income Tax Credit, and programs for assistance with housing, job training, and child care. The leading social insurance programs are the Social Security Retirement Program, Social Security Disability Insurance Program, Unemployment Insurance, Workers' Compensation, and Medicare.

assets make them eligible, and lose benefits entirely if their income and assets rise above the eligibility point.[7]

Another example of the means-tested program is the Earned Income Tax Credit (EITC), which provides benefits to individuals and families who have earnings below a threshold. The credit is proportional to earnings up to a cutoff point and later declines with higher earnings, eventually reaching zero.

Besides the aforementioned programs, there are other important means-tested programs in the United States. The government provides housing vouchers for housing assistance. The Head Start program provides preschool children from poor families with school readiness programs, nutritional assistance, and health screening.

As mentioned earlier, besides the means-tested programs, the government also has social insurance programs that are provided universally to all those who meet relatively minor employment thresholds. The programs insure against risks of unemployment, disability, and old age. As noted by Ben-Shalom et al. (2011), these programs have a strong impact on poverty, purely because of the size of the benefits and the number of recipients.

Old-Age Social Security (OASI) is a very huge program, which provides monthly cash payments to individuals who have made sufficient contributions to the system through their earnings over their lifetime. More than 95% of all workers in the United States are part of the system. Most importantly, the program provides not only for retirement benefits for the insured individual but also for his or her spouse, children under 18, and survivors, regardless of their earnings histories.

Part of the Social Security program is Medicare, which provides medical assistance to those above 65 and to Social Security Disability Insurance recipients below 65. The benefits provide are payments for hospital expenses, prescription drugs, and physician charges (coverage for the last of these is voluntary and requires premium payments).

There are also many other social insurance programs in the United States. For example, the Social Security Disability Insurance (DI) program provides cash assistance to workers who have experienced a mental or physical disability, which prevents them from working. The Unemployment Insurance (UI) program is another example, and this program provides cash payments to the involuntarily unemployed. The program is financed by a state tax on employers.

Ben-Shalom et al. (2011) note that the United States has no single comprehensive cash transfer program that covers all poor families and individuals. Instead, there are only cash programs that support specific groups,

---

7. A closely related program enacted more recently is the State Children's Health Insurance Program (SCHIP), under which the federal government pays a share of state costs for programs that provide medical care to low-income children who are not eligible for Medicaid (see Ben-Shalom et al. (2011)).

such as single mothers, the disabled, and so on. Also, the in-kind programs subsidize certain types of household expenditure such as food, medical care, and housing. Ben-Shalom et al. (2011) indicate that this is a kind of "patchwork system" and that there is clearly a danger of missing people who may not be covered.

Ben-Shalom et al. (2011) examine the expenditure trends of various programs and assess the effectiveness of the means-tested and social insurance programs provided by the government of the United States. From their study conducted with 2004 data, we observe that the combination of the means-tested and social insurance transfers in the system has a major impact on reducing poverty. Furthermore, this impact is only negligibly affected by work incentives.

The OASI program has the largest impact. OASI reduces poverty dramatically and reduces deep poverty and the poverty gap among the elderly almost to zero, which is very remarkable given that the elderly had pretransfer poverty rates of 55% in 2004. The DI program along with The SSI, TANF, Food Stamp, EITC, and housing assistance programs all have significant impacts, though the size gets smaller based on the size of the targeted population.

Along with these highlights, Ben-Shalom et al. (2011) also find that the nonelderly, nondisabled families with no continuously employed members are highly underserved. While such families are eligible for some means transfers such as TANF, Food Stamps, and housing, they are generally ineligible for other benefits. Their poverty rates are over 80% before transfers and 67% after transfers.

However, the findings of Ben-Shalom et al. (2011) also suggest that there are major shifts in the distribution of transfers within and across demographic groups. Within single-parent and two-parent families, as well as those with nonemployed members, they find a notable shift in transfers *away from* those in deep poverty *toward those* at higher income levels, so that the posttransfer deep poverty rates for these groups have actually risen.

Furthermore, they find that the system favors groups with special needs, such as the disabled and the elderly. These groups are "perceived as especially deserving" and hence receive disproportionate transfers, and those transfers have been increasing over time. Along with this trend, they find that the system favors workers over nonworkers, and in-kind transfers for food, medical care, and housing over pure cash transfers.

The result of these preferences is that the system differs dramatically from the "universalist ideal" envisioned by promoters of the negative income tax and others proposing similar systems. Under a universalist ideal, one would provide cash benefits purely based on income and not on the basis of any other characteristic. This would serve all poor families equally. Instead, the current system appears to be "paternalistic," preferring to impose the preferences of the general public on the poor, and appears to be heavily influenced by perceptions of "deservingness."

Finally, the system also prefers to subsidize the low-income employed, and to disfavor providing subsidies to nonemployed men or to women to remain at home with their children. The latter preference may arise from the increasing labor force attachment of middle-class women, and consequent changes in expectations about whether women should work.[8]

Section highlights: the antebellum paradox

As mentioned in the previous chapter, height is an important indicator of the standard of living, and social scientists often use this indicator to examine economic well-being. Since height is a biological standard, most social scientists are comfortable using this standard because it presents an objective view of the economic conditions. Historians and economists often use height of different populations to track intergenerational inequality across groups.

In this context, the term "antebellum puzzle" or "the antebellum paradox" has gained a lot of attention among economic historians. As an example, Marco (2011) examines the data of female passport applicants in the 19th century in the United States. Marco (2011) tracks the height of all the applicants during this entire time period and finds that the heights in this group increased substantially, and proportionately more than the rest of the population. Indeed, Marco (2011) substantiates his results by showing that the height of the rest of the population was getting shorter. It is possible that diseases during this time period may have affected the height and the physical stature of everyone in the society, but the heights of elite women did not decline, and in some cases even increased. This suggests that wealthy families were able to shield their women from diseases, particularly during the time of food inflation.

Carson (2011) uses historical prison records from the 19th century and compares the heights of African American females with those of white females. Even though this was a time period of relatively rapid economic development, Carson (2011) finds that white females were consistently taller than black females by about 1.5 cm.

Carson (2011) also observes regional differences with the whites from the Great Lakes and Plains, and the Southwestern black females as the tallest. The recovery in the physical stature was also first for females than for males. Carson (2011) concludes that the physical stature of lower-class men and women did not move in the same way during a period of economic growth.

Similarly, Haines et al. (2011) examine 8592 adult black recruits in the United States Colored Troops during the Civil War. The important aspect of these recruits is that they were mostly ex-slaves. The researchers link the characteristics of these recruits to the information from the 1840 and 1850 Censuses. The results indicate that these recruits showed evidence of a decline in heights from the birth cohorts of the 1820s onward, much unlike their slave counterparts from the coastal areas.

---

8. For more on this and for the distributive implications of the social insurance programs, see Ben-Shalom et al. (2011).

> **Section highlights: the antebellum paradox—cont'd**
> Further, unlike the native-white recruits, the characteristics of their counties of birth had relatively less power in explaining differences in heights.[9] The authors conclude from this that there is a support for the mortality hypothesis, but the nutrition hypothesis needs to be reinterpreted, because slave owners generally have a strong incentive in monitoring and controlling the diet of their slaves.

Agriculture and poverty in Laos and Cambodia

In a detailed study for Laos and Cambodia, Gaiha et al. (2012) detail the performance of agriculture and related issues in poverty and development. It is important to note both these economies have been transitioning to a market-oriented system. Furthermore, both are agrarian economies with agriculture contributing about one-third of the gross domestic product (GDP). Gaiha et al. (2012) assess the prospects of achieving Millennium Development Goal 1 (MDG1) in these economies. To achieve the aforementioned objective, the researchers examine the potential of agricultural growth in attaining the MDGs.

The researchers examine both these economies in detail and obtain insights into determinants of poverty, such as access to markets, returns to crops, education, land size, nonfarm activities, ethnic affiliation, and rural infrastructure. In their models, Gaiha et al. (2012) base poverty on consumption expenditure, which includes several explanatory variables such as returns on crops harvested, village price of crops sold, age of the head of household, a Gini coefficient, education of household head, and so on. Along with these, the study also includes demographic characteristics, access to credit, assets, and so on.

Their findings indicate that higher returns on all crops reduce poverty. There are also village effects, such as average village level producer prices of all crops and of glutinous rice, and presence of a credit bank in the village. Presence of a credit bank in a village consistently reduced poverty regardless of the measure used. Similarly, the level of education of the head of the household, land holdings, and the number of adults employed in nonagricultural activities all have a negative effect on poverty.

In their analysis of Cambodia, the researchers include age of household head, human capital endowments (educational attainments of household head),

---

9. Also see Craig and Weiss (1997) who demonstrate an interesting relationship between agricultural surpluses, nutritional status, and height, for white and black recruits in the Antebellum United States. This is an interesting application of a multivariate analysis, because the results show that for the white sample, a protein surplus of one standard deviation above the mean yielded an additional 0.10 in. in adult height, and a similar deviation in surplus calorie production yielded an additional 0.20 in. For blacks, however, the effect was probably negligible. Also see Zehetmayer (2011) for newer evidence concerning the Antebellum paradox.

land owned, whether the sources of income are diversified, security of land title, location in terms of distance from an all-weather road, rural infrastructure (e.g., proportion of households using electricity), and vulnerability to disasters.

Their findings suggest that while demographic characteristics matter, land holding size, a land title certificate, educational attainments of household heads, and ethnic affiliation matter considerably more. Some village characteristics also matter a great deal, especially access to all-weather roads. From the perspective of vulnerability, two results are significant: (1) diversified households face lower risks of poverty and (2) in disaster-prone villages, the risk of being pushed into poverty, other things being given, is moderately high.

Gaiha et al. (2012) note important differences between Laos and Cambodia. Cambodia is more exposed to external risks and is already a member of the World Trade Organization (WTO), while Laos is landlocked and still in preparation for joining WTO. Both are poor, agrarian economies, in which smallholders dominate the sector. Furthermore, farmers are easily susceptible to catastrophic and increasingly market risks, lack of human and financial capital, and rural infrastructure. Their integration into high value chains is rendered difficult by high transaction costs. Ethnic divisions are deep and impede collective action. Institutional quality is low given high levels of corruption and weak accountability mechanisms.

The most important governmental policy should be to provide public goods such as infrastructure, food safety standards, and a favorable environment for enforcing contracts. The policy should help farmers facilitate collaboration with private markets in providing inputs and transferring technology to smallholders. Cooperatives and farmers associations would also help improve quality and marketing of produce.

Government policy should focus on reducing the barriers between large nd smallholders by minimizing transaction costs. An accelerated transition to a more market-oriented policy regime may promote not just a more efficient agriculture but also a more equitable outcome.

Both Laos and Cambodia have high incidence of "energy poverty," wherein households do not have access or are unable to afford energy. Phoumin and Fukunari (2019) and Khanna et al. (2019) note that in the ASEAN region, energy poverty is most severe in Cambodia. Oum (2019) further shows that energy poverty is also associated with low-quality fuel, which impacts health, education, and earnings, particularly for rural poor in Laos. Jain and Koch (2020) identify local entrepreneurial ventures that can engage communities and restore cleaner energy and employment opportunities.

Distress Health Financing is another facet of economic hardship in both countries. Ir et al. (2019) note social health protection and services is often nonexistent, and hence, puts households in a financial difficulty, as they meet their outpatient and inpatient care expenses. More than 55% of consumers of healthcare are forced to borrow for underground markets that fleece them with usurious interest rates. Nguyen et al. (2020) demonstrate how households smooth out these idiosyncratic health shock expenditures through natural resource extraction. Chantarat et al. (2019) demonstrate an absolute need in these countries for developing and implementing safety nets and development interventions, for large segments of the population, particularly in the face of natural disasters.

Financial crisis and poverty in the Russian Federation

In the previous chapter, we examined how the East Asian Financial Crisis affected children's nutritional status. Similarly, Mills and Mykerezi (2009) examine the impact of the Russian Federation's financial crisis from 1990 to 1994 on poverty in Russia. The severity of poverty in the Russian Federation is found to stem largely from transient, rather than chronic, spells of economic hardship.

Mills and Mykerezi (2009) find that a household's exposure to transient poverty is strongly influenced by household levels of workforce participation, educational and physical capital, geographical isolation, and local economic conditions. Workforce participation and physical capital also mitigate exposure to chronic poverty. The importance of these determinants of transient and chronic poverty appears to change in the precrisis and postcrisis periods.

Mills and Mykerezi (2009) also demonstrate that postsecondary education below a university degree and involvement in own enterprises are more effective buffers against transient poverty in the recovery period than they were in the period leading up to the financial crisis.

Transient and chronic poverty appear to lie closer together on the continuum of exposure to poverty in Russia than in the United States and many Western European nations. Importantly, Mills and Mykerezi (2009) show that employment-related shocks do not affect transient or chronic poverty, suggesting that social protection strategies should focus on developing household educational and physical capital assets, rather than focus solely on adjusting to labor market shocks.

Fortunately, the findings of Mills and Mykerezi (2009) indicate that the severity of chronic poverty does not increase during the period of analysis. This suggests that the combined movement toward a market economy and the financial crisis did not generate a new underclass of chronically poor households.

Poverty in Europe

D'Ambrosio et al. (2011) examine the extent of poverty within five European countries, Belgium, France, Germany, Italy, and Spain. They use three different multidimensional approaches, with a variety of explanatory variables and logit regression analysis to investigate the nature of poor households in the region. They also undertake a Shapley decomposition procedure to determine the exact marginal impact of the selected explanatory variables. This is an important line of enquiry, because there are many alternative approaches to measuring poverty, and it is important to compare these approaches. For example, we need to check to what extent each measure identifies the same households as poor. D'Ambrosio et al. (2011) perform this using many explanatory variables, such as size of the household, age of the head of the household, her gender, marital status, and status at work.

Interestingly, the study finds that out of the three multidimensional approaches adopted, 80% of the households defined as poor by two approaches are identical. The study also finds a U-shaped relationship between poverty and the size of the household as well as between poverty and the age of the individual. Unemployed individuals have a much higher probability of being poor while the probability of being poor seems to be lower among self-employed than among salaried workers. Finally, ceteris paribus, married individuals, whatever their gender, have a lower probability of being poor than singles.

Coromaldi and Zoli (2012) conduct a similar analysis for Italy and derive indicators of multiple deprivations, by applying a particular multivariate statistical technique, the nonlinear principal component analysis. Their analysis identifies the poor in Italy by analyzing deprivation both as a distinct phenomenon in different life domains and as a single multidimensional concept. Their findings along with Algieri and Aquino (2011) indicate that there are regional differences of Italian poverty.

For example, between 1992 and 2008, an increase by 1% in the unemployment rate and a lower level of education raise the risk of being trapped in poverty by 0.42% and 1.7%. Conversely, social protection benefits and higher levels of education attainments significantly reduce this risk.

Algieri and Aquino (2011) find that the low rate of employment in "footloose activities" is one of the main culprits of the poor economic conditions in Southern Italy. The researchers suggest fiscal measures that aim at promoting automatic incentives to employment in manufacturing activities. These findings corroborate those of Claudio et al. (2011), who also demonstrate that the degree of diversity of deficiencies is higher in Italy's "poorer" southern provinces and lower in "richer" northern ones. Hence, in Italy's case, there is a need to try different approaches to poverty measurement so as to identify those areas, which, more than others, need structural interventions.

Regional differences in poverty through alternative measurements are also observed by Losa and Soldini (2011), who investigate poverty across seven regions in Switzerland. Stark et al. (2009) relate aggregate relative poverty to migration using Polish regional data. They find that the Gini coefficient and migration are positively correlated. Likewise, Caglayan and Dayioglu (2011), using Turkish Household Budget Survey prepared by the Turkish Statistical Institute for the year 2008, find that the most important determinants of poverty are the working status and occupation of household head, income, and ratio of worker in household and region.

In recent years, many economists have started to integrate newer econometric techniques to study poverty in Europe. In a detailed study for the European Union that uses panel data, Dafermos and Papatheodorou (2013) relate the macroeconomic and institutional determinants of inequality and poverty. In particular, the study looks at the effects of macroeconomic environment, social protection, and labor market institutions on poverty. The empirical analysis shows that the social transfers in cash, and principally the transfers that do not include pensions, exert a prominent impact on inequality and poverty. Also significant is the effect of the GDP per capita. The impact of employment or union density on inequality and poverty is not significant. Most importantly, the results support the view that the social protection system acts as a catalyst in determining the effectiveness of social spending and the distributive role of economic growth and employment.

Poverty in developing countries: China and India

In a richly detailed study, Dao (2009) estimates the determinants of rural and national poverty, of income distribution, and of agricultural growth in developing countries and finds that the percentage of the rural population living below the national rural poverty line in a developing country is dependent upon per capita purchasing power parity gross national income and the region in which it is located, per capita agriculture value added, the share of women in the agricultural labor force, the amount of permanent arable cropland, and the share of agricultural employment in total employment.

Researchers, such as Glauben et al. (2012), point out the problem of persistent poverty in rural China. For example, they use rural household panel data from three Chinese provinces and look for the determinants of long-term poverty. They also test the duration dependence on the probability to leave poverty. Interestingly, they find that the majority of population is only temporary poor. However, the probability of leaving poverty is dependent upon regional and provincial differences, ranging from no duration dependence in Zhejiang to highly significant duration dependence in Yunnan.

The number of nonworking family members, education, and several village characteristics are the key determinants of rural poverty in China. There are many other interesting links in this chain as well. For example, Demurger and Fournier (2011) relate consumption of firewood to poverty status in rural China. The researchers find strong support for the poverty—environment hypothesis since household economic wealth is a significant and negative determinant of firewood consumption. Firewood can therefore be considered as an inferior good for the whole population rural areas.

Besides economic wealth, the analysis also shows that the own price effect is important in explaining firewood consumption behavior, the price effect gaining importance with rising incomes. Finally, the study also shows that increasing education is a key factor in energy consumption behavior, especially if policies emphasize energy-switching behavior.

There is also a high degree of vulnerability to poverty in rural India. Jha et al. (2012) employ a unique panel data for rural India for the periods 1999−2006, to study vulnerability to poverty. Jha et al. (2012) quantify household vulnerability and investigate the determinants of ex post poverty as well as ex ante vulnerability. Importantly, the study examines how the effects of the determinants of vulnerability vary at different points across the vulnerability distribution. The researchers find that over time economic growth reduces the incidence of poverty. Although chronic poverty is relatively small, there is a high incidence of transient poverty. The study identifies a number of factors that affect household vulnerability and lists policy variables that can be targeted to reduce the incidence of vulnerability in rural India.

Determinants of poverty—binary logistic regression analysis

In this section, we analyze the determinants of poverty using a logistic regression model with a binary-dependent variable with mutually exclusive and exhaustive outcomes (Tabachnick and Fidell, 2001). The dependent variable is the poverty status of household i, which is 1 if the household is poor and zero otherwise. Let us consider the following levels regression of the form:

$$y_i = \beta x_i + \varepsilon_i \tag{12.16}$$

where y_i is household expenditure per capita, β denotes the vector of parameters, x_i is the vector of household characteristics, and ε_i is the error term.

The aforementioned equation can be estimated by least squares assuming normally distributed error term. The aforementioned specification can, however, be extended in the analysis of household welfare relative to some predetermined poverty line as follows:

$$\begin{aligned} S_i &= 1 \text{ if } Y_i \leq z \\ S_i &= 0 \text{ otherwise} \end{aligned} \tag{12.17}$$

where s_i is the categorical poverty indicator for household i and z is the poverty line. The binary specification can then be written as follows:

$$\Pi_i = P(y_i = 1) = F(z - \beta x_i) \tag{12.18}$$

where PI is the probability that the household is poor and F is the cumulative probability function. The aforementioned model can then be estimated by probit or logit, assuming logistic distribution of the error term. Before proceeding on to the main analysis, it may be useful to substantiate on the dichotomous logistic regression model.

Dichotomous logistic regression model

The logistic regression model can be written in terms of the log of the odds (odds are simply defined as the probability of a "success" outcome divided by the probability of a "failure" outcome), called the logit, as follows:

$$\log\left(\frac{\Pi_i}{1-\Pi_i}\right) = \beta_0 + \beta_1 X_1 + \beta_2 X_2 + \cdots + \beta_k X_k \quad (12.19)$$

with the aforementioned model, the logit is just the natural logarithm of the odds, and the range of values of the left-hand side of Eq. (12.19) is between $-\infty$ and $+\infty$. An alternative way of writing the aforementioned model in terms of the odds is as follows:

$$\frac{Pr(y=1)}{Pr(y=0)} = \frac{\Pi_i}{1-\Pi_i} = \exp(\beta_0 + \beta_1 X_1 + \beta_2 X_2 + \cdots \beta_k X_k) \quad (12.20)$$

The range of values is between 0 and ∞ that the right-hand side of Eq. (12.20) can assume.

Rearranging Eq. (12.20), the underlying probability of a success outcome is given by

$$\Pi_i = \frac{\exp(\beta_0 + \beta_1 X_1 + \beta_2 X_2 + \cdots \beta_k X_k)}{1 + \exp(\beta_0 + \beta_1 X_1 + \beta_2 X_2 + \cdots \beta_k X_k)} \quad (12.21)$$

Eqs. (12.19)–(12.21) are identical in interpretation. However, for practical purposes, Eqs. (12.19) and (12.21) are usually computed, since it provides not only the logit estimates but also the probability of success.

An example with the Malawi dataset

Suppose we ran a regression similar to Eq. (12.19) with the probability that the household is poor as the dependent variable and the size of the land owned as the only independent variable. Then the regression equation becomes

$$\log\left(\frac{\Pi_i}{1-\Pi_i}\right) = 1.83 - 0.299 \text{ LANDO} \quad (12.22)$$

The probability that an individual is poor using Eq. (12.21) is given by

$$P(y=1) = \frac{\exp(1.83 - 0.299 \text{ LANDO})}{1 + \exp(1.83 - 0.299 \text{ LANDO})} \quad (12.23)$$

The probability that a household is poor with a size of land owned = 3 (i.e., owning land between 1 and less than 1.5 ha) using Eq. (12.23) is 0.717, while the probability that a household is poor with a size of land owned = 5 (i.e., owning land greater than 2 ha) is 0.582. The probability of being poor declines as the household owns more land. Fig. 12.3 shows the plot of the size of land owned with the probability that the household is poor.

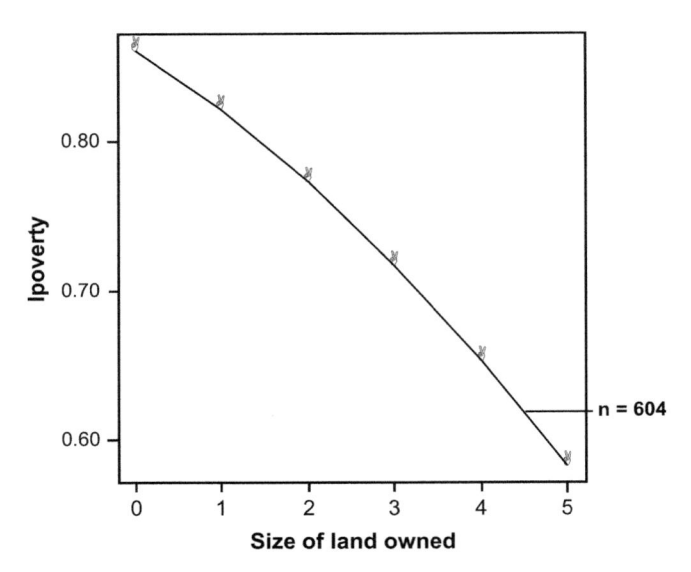

FIGURE 12.3 Logistic regression plot of the probability of being poor with size of land owned.

As evident from Fig. 12.3, as the size of land owned by the household increases, the probability of being poor decreases monotonically at a decreasing rate. This is due to the functional specification as in Eq. (12.21), as multiplying successive values of the size of land owned decreases the probability of poverty exponentially.

Expected determinants of household welfare

This section will summarize the expected determinants of household welfare using a logistic specification. The model is as follows:

$$P(y=1) = \frac{\exp(\beta_0 + \beta_1 X_1 + \beta_2 X_2 + \cdots \beta_k X_k)}{1 + \exp(\beta_0 + \beta_1 X_1 + \beta_2 X_2 + \cdots \beta_k X_k)} \quad (12.24)$$

where $y=1$ if the family is poor and zero if the family is nonpoor. A household was found to be poor if its total expenditure was 107.93 kwacha per person per month using the FEI method described before. $X_1, X_2, \ldots X_k$ are the set of explanatory variables, and β_k is the parameter associated with X_k. The main determinants of household welfare will relate to household and demographic characteristics, area owned (a measure of assets of the household), technology (as measured by maize productivity for the maize growing households), household food security, and community variables.

The *household level characteristics and demographic variables* are age of the household head, household size, and education of the spouse, gender of the household head and marital status of the household head. The age of the household head can have a significant impact on welfare, since older heads of households are likely to have more experience and respect in the community.

The effect of household size on household welfare can be either positive or negative and depends in part on the degree of rivalry in consumption among household members. If all consumption is public, every marginal increase in consumption benefits all household members. An example of such consumption could be increased security within the community or provision of a tap providing clean drinking water.

In contrast, where all consumption is private (with only one person benefiting from any consumption activity), only one member's welfare increases and not the entire household. An example might be nutrition. In such a case, household welfare decreases with household size.

Moreover, there may be synergies from larger household size, in both production and consumption activities. Working in groups can be more productive through pooling tools and experience, or through higher motivation. Returns to scale can have an impact on household welfare via household size for a given degree of rivalry in production and consumption. Thus, this variable will be included in determining whether rivalry or scale effect dominates in affecting household welfare.

Household education is likely to have a positive effect on household welfare. We include the education of the mother since it generally has a larger positive effect on household food consumption than the male head. The gender of the household head can be an important determinant of household welfare, since female-headed households suffer from the twin constraints of labor supply (time constraint) and caring activities for the children. The twin constraints can reduce their effective time at work and thus affect their wages and other earnings, consequently reducing household welfare.

Land owned (as a measure of asset) can be directly linked to household welfare through the quality and characteristics of the cultivated land and through the total area farmed per household. These characteristics affect household agricultural production, credit opportunities, and, indirectly, household labor availability and thus on welfare. The general expectation is that an increase in the area owned by the household will raise welfare.

Increase in *agricultural productivity* is an important determinant of household welfare in the long run by releasing household resource and other constraints. Since the main food crop produced in Malawi is maize, we consider yields from local maize (for only those households that grow maize) as a measure of productivity. This is defined as the total quantity harvested of local maize to household size (PRODLMAIZ). We also include production of composite maize (COMPOSIT) (which is a dummy variable) as an additional measure of productivity.

Intrinsically related to agricultural productivity, household food security can be an important determinant of household welfare. If a household is food insecure, it is more likely that the members of the household are undernourished, which can affect labor productivity adversely. This, in turn, can reduce household welfare. The measure of household food security considered is based on per adult equivalent calorie intake for households. It is defined as households being able to satisfy at least 80% of the requirement for calorie intake.

Finally, *community-level* variables are expected to be key determinants of household consumption. Both the absence of protected drinking water source and health infrastructure can affect household welfare adversely, and the variables captured for these effects are distance to the nearest health clinic (HEALTDST) and distance to a protected drinking water source (DRINKDST). In addition, since private traders can improve market access (through selling staple crops) and thereby ensure regular supply of low cost food to the households, the presence of private trader (PBTRADER) is also included as a community level variable. Finally, we also include a local market dummy variable (which assumes a value of one if the household is located more than 5 km from the nearest market place and zero otherwise) to understand how households' market access affects their welfare.

Empirical results

Two issues that seem critically important in logistic regression estimation are as follows:

- How good the model fits the data.
- How to interpret the fitted model.

Measuring model fit

The following four statistics are available for assessing overall model fit with two or more independent variables.

Log-likelihood ratio

The logistic regression model uses maximum likelihood method[10] to maximize the equation's log-likelihood (LL) function. Software packages differ in reporting either the LL value or minus twice the value (−2LL), which has distributional properties enabling us to apply the chi-square distributions. The LL function is always used in comparison with an alternative equation specification.

---

10. The maximum likelihood (ML) estimation maximizes a log-likelihood function, whose core expression is $-\sum_{i=1}^{N}(Y_i - \mu)^2$, where Y is the dependent variable and μ is a measure of central tendency of the parameter distribution. The second derivative of the aforementioned expression identifies the location of the parameter's maximum value. For additional details on the ML estimation procedure, see Freund and Walpole (1986).

A pair of multivariate equations is called *"nested"* if all the parameters included in the first equation also appear in the second equation. In other words, the first equation is "nested inside" the second. The difference in $-2LLs$ for a pair of nested equation tests whether the additional parameters specified in the second equation improve the fit to the data over the first equation's fit. In other words, we want to test the following null hypothesis:

$$H_0: (-2LL_1) - (-2LL_2) = 0$$
$$H_0: (-2LL_1) - (-2LL_2) > 0$$

The log-likelihood ratio for comparing the two nested equations is given by

$$G^2 = (-2\ln L_1) - (-2\ln L_2) \qquad (12.25)$$

The G^2 test statistic is distributed as chi-square with degrees of freedom equal to the difference in the two equations. In our example, the initial $-2LL$ value where the "constant only" equation is estimated is 163.37. The $-2LL$ value at the first iteration step is 135.73. Thus, the difference between these two $-2LL$ values is $163.37 - 135.73 = 27.64$, which is the output in the omnibus tests for model coefficients. The model chi-square has 11 degrees of freedom, since the first model has 1 degree of freedom (for the constant) and the second model has 11 degrees of freedom (the constant plus the 11 predictors). If we set $\alpha = 0.005$, the critical value of $x_{11}^2 = 26.75$. Since the computed value is greater than the critical value, we reject the null hypothesis and conclude that including the additional parameters specified in the second equation improves the fit to the data over the first equation's fit (with only the constant term).

Hosmer−Lemeshow goodness-of-fit test

A commonly used test of the overall fit of a model to the observed data is the *Hosmer and Lemeshow test*. The idea is to form groups of cases and construct a "goodness-of-fit" statistic by comparing the observed and predicted number of events in each group. The cases are first put in order by their estimated probability on the outcome variable. Then the cases are divided into 10 groups according to their estimated probability, i.e., those with estimate probability below 0.1 (in the lowest decile) form one group and so on, up to those with estimated probability 0.9 or higher (in the highest decile).

The next step is to divide the cases into two groups on the outcome variables (namely not poor and poor) to form a 2 × 10 matrix of observed frequencies. Expected frequencies for each of the 20 cells are obtained from the model. If the logistic regression model fits well, then most of the cases with outcome $= 1$ or the poor are in the higher deciles of risk and most with outcome $= 0$ or the nonpoor are in the lower deciles of risk.

TABLE 12.4 Contingency table for Hosmer and Lemeshow test.

		Poor1 = 0.00		Poor1 = 1.00		
		Observed	Expected	Observed	Expected	Total
Step 1	1	11	11.062	2	1.938	13
	2	9	8.217	4	4.783	13
	3	8	6.501	5	6.499	13
	4	3	4.652	10	8.348	13
	5	3	3.563	10	9.437	13
	6	3	2.548	10	10.452	13
	7	1	1.609	12	11.391	13
	8	2	0.979	11	12.021	13
	9	0	0.560	13	12.440	13
	10	0	0.308	17	16.692	17

The goodness-of-fit statistic is then calculated as follows:

$$\text{GFIT} = \sum_{\text{cells}} \frac{(f_0 - f_E)^2}{f_E}. \qquad (12.26)$$

where f_0 and f_E are the observed and the expected numbers in the cell as in Table 12.4. The idea is that closer the expected numbers to the observed, the smaller the value of this statistic and smaller values will indicate that the model is a good fit. The value of this test statistic with all the explanatory variables (both categorical and continuous) is 4.34, which is compared with the critical value from the chi-square distribution with 8 degrees of freedom (number of groups −2).

The *P* value is 0.825, and thus, we do not reject the null hypothesis that there is no difference between the observed and predicted values. Hence, we conclude that the model appears to fit the data reasonably well.

Generalized coefficient of determination

In linear regression models, one measure of usefulness of the model was the statistic R^2, which gives the proportion of variation in the outcome variable explained by the model. Several statistics have been proposed in the logistic regression framework that is almost equivalent in interpretation to the R^2 in the linear regression model. Cox and Snell (1989) proposed a generalization of the least squares coefficient of determination as follows:

$$R^2 = 1 - \left(\frac{L_0}{L_1}\right)^{2/N} \qquad (12.27)$$

where L_0 is the log-likelihood function for the "constant only" equation, L_1 is the log-likelihood for the equation with one or more predictors, and N is the sample size. Nagelkerke (1991) adjusted the aforementioned coefficient by rescaling it according to the largest value R^2 can achieve, thus enabling the measure to reach a maximum of 1. This is given as follows:

$$\overline{R}^2 = \frac{R^2}{R^2_{\max}} = \frac{R^2}{1 - (L_0)^{2/N}} \qquad (12.28)$$

Both coefficients should be viewed as purely descriptive statistics that provide a rough approximation for judging a model's predictive efficacy. No statistical test is available in the logistic regression model to test the null hypothesis that $R^2 = 0$ in the population as in the ordinary least squares (OLS).

From Table 12.5, both the measures of the coefficient of determination improve significantly over stage 1. The interpretation is that the model explains about 44% variation in the data. However, as explained before, there is no formal test that can tell us whether 44% is sufficient or not.

Classification table

Another way of assessing how well the model fits the data is to produce a classification table. This is a simple tool that indicates how good the model is at predicting the outcome variable (namely poor and nonpoor). OLS regression equations are usually used to predict the score of every case, which can then be compared with the observed value to see how accurate the prediction is. The logistic regression procedure, on the other hand, uses the estimated equation to decide if the expected probability is < 0.5, then the predicted score is 0. On the

TABLE 12.5 Coefficient of determination in successive steps.

Coefficient of determination	Step 1[a]	Step 2[b]
R^2	0.186	0.31
$\overline{R}^2$	0.265	0.44

[a]Based on comparing the log-likelihood for the constant model and the model with only continuous variables.
[b]Based on comparing the log-likelihood for the model with only continuous variable to the model with both continuous and categorical variables.

TABLE 12.6 Classification table.

		Predicted		
		Poor1		
Observed		0.00	1.00	Percentage correct
Step 1	poor1			
	0.00	25	15	62.5
	1.00	8	86	91.5
Overall percentage				82.8

other hand, if the expected probability is ≥ 0.5, then the predicted value is 1. The percentages of correctly predicted cases are then calculated and displayed in a classification table. If the equation "completely explains" the variation of the dependent variable, all cases would fall on the main diagonal and the overall percentage correct would be 100%. In other words, cases predicted to be equal to 0 would be observed 0s and predicted 1s would be the observed 1s.

Table 12.6 seems to indicate an impressively high level of correct predictions (82.8% overall). It predicted the poor correctly by 91.5% and the nonpoor by 62.5%. Thus, from all the aforementioned tests, the model fits the observed data very well.

Interpreting the logistic coefficients and discussion of results

Recall from the discussion of the first part of this section that the logistic regression model can be written on three different scales, namely logit, odds, or probability. We will report the results on the logit scale. The null hypothesis that the population value of a given parameter is zero is given by

$$H_0: \beta_k = 0$$
$$H_1: \beta_k \neq 0$$

In the logistic regression model, a *Wald* statistic is computed, to decide on whether to reject the null hypothesis. This statistic is distributed as a chi-square variable with 1 degree of freedom and is given by the following expression:

$$\text{Wald} = \left(\frac{b_j}{S_b}\right) \tag{12.29}$$

The aforementioned formula is just the square of the usual t-test (when the tested population parameter is 0). One can also construct the confidence intervals for the βs based on the estimated coefficients and the standard errors. We proceed to the determinants of poverty using Eq. (12.24). The results are reported in Table 12.7.

Measurement and determinants of poverty—application Chapter | 12 **477**

TABLE 12.7 Determinants of poverty: logistic regression results.

Variables	Coefficient	Wald-statistic	Significance level (P value)
Intercept	10.64[b](4.429)	5.775	0.01
mbrage	0.05(0.20)	0.063	0.80
mbragesq	−0.001(0.002)	0.184	0.67
educspous	−0.112(0.197)	0.325	0.568
hholdhed	−1.74[d](1.05)	2.739	0.098
mstatus	0.086(0.547)	0.025	0.875
hhldqty	−0.27[d](0.148)	3.347	0.067
prodlmaiz	−0.008[b](0.003)	7.29	0.007
composit	−1.27(0.813)	2.457	0.117
owned	−0.471[d](0.273)	2.97	0.085
calreq	−1.30[c](0.61)	4.55	0.03
healtdst	−0.369(0.237)	2.42	0.12
drinkdst	0.08(0.187)	0.207	0.649
pbtrader	−0.65(1.12)	0.336	0.562
mktgt	−2.86[a](0.776)	13.61	0.00

Values in parentheses are standard errors of the estimated coefficients.
[a]$P < .001$.
[b]$P < .01$.
[c]$P < .05$.
[d]$P < .10$.

Using Eq. (12.19), the logit model can be written as follows:

$$\log\left(\frac{\Pi_i}{1-\Pi_i}\right) = 10.64 + 0.05(\text{MBRAGE}) - 0.001(\text{MBRAGESQ})$$

$$- 0.112(\text{EDUCSPOUS}) - 1.74(\text{HHOLDHED})$$

$$+ 0.086(\text{MSTATUS}) - 0.27(\text{HHLDQTY})$$

$$- 0.008(\text{PRODLMAIZ}) - 1.27(\text{COMPOSIT})$$

$$- 0.471(\text{OWNED}) - 1.30(\text{CALREQ})$$

$$- 0.369(\text{HEALTDST}) + 0.08(\text{DRINKDST})$$

$$- 0.65(\text{PBTRADER}) - 2.86(\text{MKTGT})$$

As evident from Table 12.7, household characteristics such as age of the household head, marital status, and education of the mother have no significant effect on poverty. This result is possible, as education of the spouse is very low. The household may not find adequate income earning opportunities to improve on its welfare. However, the household head being a male has a significant effect on household welfare. In other words, there is a lower likelihood of being poor if the household head is a male compared with being a female. Having a greater household size reduces the likelihood of being poor, which possibly indicates that the scale effect is dominating over the rivalry effect.

From the estimates of the technology level variables, we find that agricultural productivity has a significant impact on household welfare suggesting that the growth in productivity in the long run remains a key mechanism in alleviating poverty. However, producing a composite crop had no significant impact on the likelihood of moving out of poverty.

It was expected that owning better and higher quality land would improve household welfare. We find evidence that increase in land owned can reduce the likelihood that the household is poor. This is the asset effect on household welfare. Improvement in food security can also be crucial in the long run in alleviating poverty, since better nourished household members can be more productive and have higher income earning capacity. Food security had a positive impact on the likelihood that the household escapes poverty.

Finally, we find no evidence of the community-level variables to affect household welfare with the exception of the local market dummy variable. This result possibly indicates that if the household is closer to market, then its food security situation can improve. This indirect linkage can result in the likelihood that the household will not be poor.

Estimating logistic regression models in STATA

In this section, we implement the logistic model estimation using STATA.

Example 1

In Table 12.8, we have 39 observations on three variables that we discussed in the text, in the context of the determinants of household welfare. The three variables are as follows:

1. poverty, which is a dichotomous variable that receives the value 1 if the family is poor and zero if the family is not poor.
2. drinkdst, an index that measures the distance to a protected drinking water source.
3. healthdst, an index that measures the distance to the nearest health clinic.

Measurement and determinants of poverty—application Chapter | 12 **479**

TABLE 12.8 Poverty, water and health clinic data for STATA.

Obs	poverty	drinkdst	healtdst
1	1	3.7	0.825
2	1	1.25	2.5
3	1	0.8	3.2
4	1	0.75	3.75
5	1	3.2	1.6
6	1	1.6	1.78
7	1	1.8	1.5
8	1	1.9	0.95
9	1	2.7	0.75
10	1	1.2	2
11	1	1.3	1.625
12	1	3.5	1.09
13	1	0.75	1.5
14	1	0.7	3.5
15	1	1.4	2.33
16	1	2.3	1.64
17	1	0.85	1.415
18	1	1.8	1.8
19	1	1.1	2.2
20	1	0.8	3.33
21	0	0.6	0.75
22	0	0.9	0.75
23	0	0.8	0.57
24	0	0.6	3
25	0	1.7	1.06
26	0	0.4	2
27	0	1.35	1.35
28	0	1.1	1.83
29	0	0.95	1.9
30	0	1.1	1.7

Continued

TABLE 12.8 Poverty, water and health clinic data for STATA.—cont'd

Obs	poverty	drinkdst	healtdst
31	0	0.9	0.45
32	0	0.55	2.75
33	0	0.95	1.36
34	0	1.5	1.36
35	0	0.6	1.5
36	0	0.95	1.9
37	0	1.6	0.4
38	0	2.35	0.03
39	0	0.75	1.9

Our goal is to relate the two independent variables (drinkdst and healthdst) to the poverty status of the household, similar to our exercise using the Malawi data. The logit command in STATA provides the estimates and all the key statistical information.

The command line in STATA is

.logit poverty drinkdst healtdst

The input and the corresponding output from STATA are given below. Using the results from STATA, we can write the logit model as follows:

Poverty = − 9.52 + 3.88 drinkdst + 2.64 healtdst

Similar to the Malawi data, we see that as the distance to drinking water and health facilities increases, the likelihood that a family is poor increases. We can include more variables in the data to check for the determinants of poverty more fully. We leave this as an exercise at the end of the chapter.

We can also use the results from STATA to check the reliability of the model. The STATA output produces the standard errors of the coefficients and the z-values alongside the *P*-values. From the output, we can see that *P*-values are all less than 0.05, and hence, we can reject the null hypotheses that both independent variables are individually not significant.

An overall χ^2 test evaluates the null hypothesis that all coefficients in the model, except the constant, equal zero, where

$$\chi^2 = -2(\ln\mathscr{L}_i - \ln\mathscr{L}_f)$$

where $\ln\mathscr{L}_i$ is the log likelihood at the initial iteration 0, and $\ln\mathscr{L}_f$ is final iteration's log likelihood value, which is given beside LRchi2(2) value in the STATA output:

$$\chi^2 = -2(\ln\mathscr{L}_i - \ln\mathscr{L}_f) = -2(-27.019918 - (-14.886152)) = 24.27$$

The probability of a greater chi-square value, with 2 degrees of freedom, tells us that the overall model is a good fit for the poverty analysis. For the overall fit of the model, we can also perform the goodness-of-fit test using the estat gof command in STATA:

```
. logit poverty drinkdst healtdst

Iteration 0:    log likelihood = -27.019918
Iteration 1:    log likelihood = -15.303968
Iteration 2:    log likelihood = -14.894701
Iteration 3:    log likelihood = -14.886165
Iteration 4:    log likelihood = -14.886152
Iteration 5:    log likelihood = -14.886152

Logistic regression                       Number of obs   =         39
                                          LR chi2(2)      =      24.27
                                          Prob > chi2     =     0.0000
Log likelihood = -14.886152               Pseudo R2       =     0.4491
```

poverty	Coef.	Std. Err.	z	P>|z|	[95% Conf. Interval]	
drinkdst	3.88215	1.428613	2.72	0.007	1.08212	6.682181
healtdst	2.649118	.91422	2.90	0.004	.8572799	4.440957
_cons	-9.529586	3.233201	-2.95	0.003	-15.86654	-3.192629

```
. estat gof
```

Logistic model for poverty, goodness-of-fit test

```
          number of observations =       39
    number of covariate patterns =       38
              Pearson chi2(35)   =    39.01
                   Prob > chi2   =   0.2941
```

When the goodness-of-fit test is done for the whole data without dividing the information into any groupings, we see that the P-value is 0.3, and hence, we are unable to reject the null hypothesis. This implies that the given specification is true. We can also create different groups and perform the same chi-squared test, or the Hosmer and Lemeshow test. We use the next example to illustrate this aspect and a few other commands in STATA for the logit analysis.

Example 2

We have discussed the WTP example for electricity previously, in Chapter 6. This example is from Gajanan et al. (2013). We provide a short review of the variables used for analysis (see Table 12.9).

TABLE 12.9 Variables and definitions for WTP analysis.

Variable	Variable description
Y	Dependent variable that tracks the willingness to pay for a reliable package of electricity (1 = "yes"; 0 = "no")
EDUC	Educational status of the respondent (head of the household): is a dummy variable = 1 if the head of the household is a literate; 0, if not
Household	Household type (1 = single, 2 = nuclear, 3 = joint family)
Income	The net annual household income in rupees
Debt_level	Debt (1 = the household has debt; 0 if not)
Irrigation	The source of irrigation before getting an electric pump set connection (1 = tank irrigation; 2 = canal irrigation; 3 = nonelectrical methods)
Farm Size	Farm size (1 = marginal; 2 = small; 3 = medium; 4 = large)
Pumps	The number of electric pump set connections (1, 2; 3 = more than 2)
Pump Use_1	Use of electric pumps everyday in the last month (yes = 1; no = 0)
Pump Use_2	The number of times the pump set was switched off yesterday
Pump Age	Age of the first electric pump set (in years)
Crop Pattern	Factors influencing the cropping pattern (1 = availability of water; 2 = power tariff; 3 = market conditions; 4 = others)
Well Quality	Quality of the well water at present (1 = excellent; 2 = good; 3 = satisfactory; 4 = bad; 5 = very bad)
Water Level	Sufficient water available for irrigation (1 = yes; 0 = no)
Power cut	Was there power cut yesterday (1 = yes; 0 = no)

Measurement and determinants of poverty—application Chapter | 12 483

TABLE 12.9 Variables and definitions for WTP analysis.—cont'd

Variable	Variable description
Burnout_1	Aware that voltage fluctuations cause motor burnouts (1 = yes; 0 = no)
Burnout_2	Motor burnout during the last year due to voltage fluctuation (1 = yes; 0 = no)
Reliability	Rate the reliability of power supply to agriculture (1 = very good; 2 = good; 3 = satisfactory; 4 = bad; 5 = very bad)
Power Need	Quality of power is indispensable to cheap irrigation (1 = yes; 0 = no)

We wish to examine the determinants of Y, or the willingness to pay for a reliable package of electricity. The data was collected from 450 farmers from two villages in Tamil Nadu, India. We show the various modeling procedures and tests in STATA under logit.

The first few lines are meant to define a vector of variables that can be used later:

global xlist1 educ household income debt_level
global xlist2 irrigation farm_size pumps pumpuse_1 pumpuse_2 pumpage
global xlist3 crop_pattern well_quality water_level
global xlist4 power_cut burnout_1 burnout_2 reliability powerneed
global is a macro in STATA that allows us to set the contents of a vector of variables using a shortcut. The first line global xlist1 sets into the contents, these variables: educ household income debt_level. We can now refer to these variable using $xlist1

The next three lines define three new lists of variables using xlist2, xlist3, xlist4. The logit command and the goodness-of-fit tests in STATA are generated using the command lines shown before in Example 1:

logit y $xlist1 $xlist2 $xlist3 $xlist4, vce(robust)
estat gof, table group(10)
test $xlist1

The aforementioned output indicates all the estimation results from the logit model for the entire set of variables in our data. The chi-square value of 88.1 indicates that the model is good fit. Several variables are significant, and most importantly, the coefficient of farm size is negative, indicating that rich farmers have very little incentive to pay for electricity.

The Hosmer–Lemeshow test of specification can be used for samples with different groups. The group(#) option specifies the number of quintiles to be used to group the data, with 10 being the default. The outcome is given by the *P*-value, which is 0.97 indicating that the null hypothesis that there is misspecification can be rejected.

The test $xlist1 command tests the null hypothesis that the overall model is insignificant. The results indicate the *P*-value < .05, and hence we can reject the null hypothesis. The output for the Wald test of overall significance is given in the following:

```
. test $xlist1

 ( 1)  [y]educ = 0
 ( 2)  [y]household = 0
 ( 3)  [y]income = 0
 ( 4)  [y]debt_level = 0

        chi2( 4) =    9.61
      Prob > chi2 =   0.0476
```

In the next few lines, we illustrate the steps involved in producing a likelihood ratio test that allows us to test between equations.

In the first line, we estimate an unrestricted model with all the variables in the analysis. The command quietly suppresses the output but stores all the information from the estimation. We can store all the estimates using the estimates store command. We can refer to the stored values in a box called full.

In the next two lines, we estimate a restricted model after dropping the variables in $xlist4. We store the variables from this estimation process in A.

The likelihood ratio test between the unrestricted and the restricted model is given by the command line lrtest A full

- quietly logit y $xlist1 $xlist2 $xlist3 $xlist4
- estimates store full
- logit y $xlist1 $xlist2 $xlist3
- estimates store A
- lrtest A full

The output from the restricted model is given in the following. We note that the z-value is significant for many of the variables including education, type of household, irrigation, and the characteristics of the pumps.

The results of the likelihood ratio are given following the logit results. Since the restricted model is nested in the unrestricted specification, we can conduct an LR test for our data. Since the estimated chi-square value is 56, we can conclude that the variables in $xlist4 are useful predictors and can be retained for future analysis.

Conclusions and implications

The purpose of this chapter was to understand the different dimensions of poverty, the various approaches to measure poverty, the characteristics of the poor, and the underlying causes of poverty. While we used a monetary-based approach in measuring poverty, in the future, capability-based approaches and participatory surveys can be supplemented with monetary-based approaches to understand the multidimensional aspects of poverty. This will allow researchers to evaluate the impact of programs on the poor and determine whether these programs achieve the goals with respect to targeting a certain group of households.

Looking at the headcount measure of poverty (along with other measures, such as the poverty gap and squared poverty gap), we found that poverty in Malawi can be classified as deep and pervasive. About 72% of the consumption level of the country's population was deemed insufficient to meet their basic needs. Additionally, by looking at the characteristics of the poor, we found that poorer households owned very few productive assets and had a higher dependency ratio. Thus, given limited resources, it is more desirable to reduce the consumption shortfall of a larger proportion of the poor than to eliminate the shortfall of a smaller proportion.

We next investigated the causes of poverty in Malawi using a logistic regression framework. Our results indicate that poverty reduction in Malawi will not occur unless there is rapid agricultural productivity growth accompanied with better quality and distribution of land for the smallholders. At the same time, it is desirable to improve the economic conditions (such as more equitable distribution of land, better schooling, etc.) of the female-headed households, since they are the vulnerable segments of the population. Thus, asset ownership and human capital formation are extremely relevant if a poverty reduction strategy is to work.

While eradicating global poverty is a noble goal, there are serious difficulties in measuring poverty. Indeed, Samman (2013) has produced a very interesting study from ODI indicating the practical difficulties encountered at the implementation level. The general agreement among the researchers

Samman (2013) points out is that there is a need to derive a globally defined poverty measure that identifies people who cannot fulfill their basic needs—i.e., absolute deprivation. People should be able to live not only free from starvation and disease but also in accordance with social norms. Another conclusion from the study is that a plurality of measures would also deflect criticism that current poverty lines exclude a large number of people who, if not actually destitute, might be hovering precariously around that line. Multiple dimensions of poverty will allow us to focus on the lack of adequate housing, improved sanitation, education, and the likelihood of survival. Samman (2013) also indicates the limitations associated with the PPP measure, which makes international poverty comparisons difficult.

Technical appendices

Technical notes on logistic regression model

A standard multiple linear regression model is inappropriate to use when the dependent variable is binary (Tabachnick and Fidell, 2001). This is because, first, the model's predicted probabilities could fall outside the range 0—1. Second, the dependent variable is not normally distributed and, in fact, a binomial distribution would be more appropriate. Third, the normal distribution cannot be considered as an approximation to the binomial model, since the variance of the dependent variable is not constant.

Now, let us consider y_i as a realization of the random variable Y_i that can take the values 0 and 1 with probabilities Π_i and $(1 - \Pi_i)$. Y_i is called a *Bernoulli distribution* with parameters Π_i and can be written in compact form as follows:

$$Pr(Y_i = y_i) = \Pi_i^y (1 - \Pi_i)^{1-y} \quad (12.30)$$

for $y_i = 0, 1$.

If $y_i = 1$, we obtain the probability to be equal to Π_i and, if $y_i = 0$, we obtain the probability to be equal to $(1 - \Pi_i)$. Now, let the logit of the underlying probability be defined as logit $(\Pi_i) = \log$

$$\text{logit}(\Pi_i) = \log\left(\frac{\Pi_i}{1 - \Pi_i}\right)$$

Thus, as the probability goes to zero, the logit function approaches $-\infty$. On the other extreme, as the probability approaches 1, the logit function also approaches $+\infty$. (Check that negative logits represent probabilities below one half, while positive logits correspond to probabilities above one half).

Let us further suppose that the logit of the underlying probability be given by

$$\text{logit}(\Pi_i) = X'_1\beta \tag{12.31}$$

where X_i is a vector of covariates and β is the vector of regression coefficients. Exponentiating Eq. (12.31), we obtain the odds ratio as

$$\frac{\Pi_i}{1-\Pi_i} = \exp(X'_i\beta) \tag{12.32}$$

The aforementioned expression implies that if one changes the jth predictor by one unit holding all the other variables constant, one needs to multiply the odds by $\exp(\beta_j)$. Transforming Eq. (12.32) in terms of the probability scale, one obtains:

$$\Pi_i = \frac{\exp(X'_i\beta)}{1+\exp(X'_i\beta)} \tag{12.33}$$

There is no simple way to express the right-hand side, since it is a nonlinear function of the predictors. An approximate effect of the predictor on the probability can only be given for continuous predictors and using the quotient rule for the jth predictor, we obtain

$$\frac{d\Pi_i}{dx_{ij}} = \beta_j \Pi_i (1-\Pi_i) \tag{12.34}$$

The effect of the jth predictor on the probability thus depends on the coefficient β_j and the value of the probability. The result thus approximates the effect of the covariate near the mean of the response.

A brief description of the maximum likelihood estimation procedure in the logistic regression framework is also warranted. Suppose in Eq. (12.30) there are n independent Bernoulli observations for the random variable Y_i.

Then the likelihood function is the product of the densities given by Eq. (12.30). Taking logs the likelihood function can be expressed as follows:

$$\log L(\beta) = \sum_{j=1}^{n}(y_i \log(\Pi_i) + (n-y_i)\log(1-\Pi_i)) \tag{12.35}$$

Substituting Eq. (12.33) for the value of Π_i, the logarithm of the likelihood function can be expressed alternatively as follows:

$$\log L(\beta) = \sum_i \beta_i \sum_j y_j x_{ij} - \sum_j \log\{1+\exp(X'_i\beta)\} \tag{12.36}$$

One can then find the maximum likelihood estimates by differentiating the left-hand side of Eq. (12.36) with respect to β_i and setting the partial derivative to be equal to zero and solving for the nonlinear equations for β_i.

Estimating logistic regression models in R

In this section, we implement the logistic model estimation in R. We have a data set with 39 observations on three variables in the context of the determinant of household welfare: poverty, drinkdst, and healthdst. Our objective of this part is to relate the two independent variables (drniksdt and healthdst) to the poverty status of the household. As mentioned in the previous chapters, we perform the *summary* command first to identify the descriptive statistics.

```
> summary(data_chapter12)
      Obs              Poverty          Drinkdst         Healtdst
 Min.   : 1.0    Min.   :0.0000    Min.   :0.40    Min.   :0.030
 1st Qu.:10.5    1st Qu.:0.0000    1st Qu.:0.80    1st Qu.:1.075
 Median :20.0    Median :1.0000    Median :1.10    Median :1.625
 Mean   :20.0    Mean   :0.5128    Mean   :1.36    Mean   :1.688
 3rd Qu.:29.5    3rd Qu.:1.0000    3rd Qu.:1.65    3rd Qu.:2.000
 Max.   :39.0    Max.   :1.0000    Max.   :3.70    Max.   :3.750
```

The *glm* command in R provides the estimates and statistical information. The input and the corresponding output from R are given in the following.

```
> logit<-glm(Poverty~Drinkdst+Healtdst,family='binomial')
> summary(logit)

Call:
glm(formula = Poverty ~ Drinkdst + Healtdst, family = "binomial")

Deviance Residuals:
     Min        1Q    Median        3Q       Max
 -1.50657  -0.73464   0.03997   0.48854   2.32935

Coefficients:
            Estimate Std. Error z value Pr(>|z|)
(Intercept)  -9.5296     3.2332  -2.947  0.00320 **
Drinkdst      3.8822     1.4286   2.717  0.00658 **
Healtdst      2.6491     0.9142   2.898  0.00376 **
---
Signif. codes:  0 '***' 0.001 '**' 0.01 '*' 0.05 '.' 0.1 ' ' 1

(Dispersion parameter for binomial family taken to be 1)

    Null deviance: 54.040  on 38  degrees of freedom
Residual deviance: 29.772  on 36  degrees of freedom
AIC: 35.772

Number of Fisher Scoring iterations: 6
```

We can get an equation with the results and interpret that the distance to drinking water and health facility increases, the likelihood that a facility is poor increases. The output produces the standard errors of the coefficients and the z-values alongside the *P*-values. From the output, we can see that *P*-values are all less than 0.05.

$$\text{Poverty} = -9.52 + 3.88 \text{ drinkdst} + 2.64 \text{ healthdst}$$

The Hosmer–Lemeshow test is a statistical test for goodness of fit for logistic regression models in R. We need to install the *"ResourceSelection"* package to produce the test.

```
> hoslem.test(Poverty, fitted(logit), g=10)

        Hosmer and Lemeshow goodness of fit (GOF) test

data:  Poverty, fitted(logit)
X-squared = 17.812, df = 8, p-value = 0.02268
```

Exercises

- What are the various approaches to measure poverty? What are the key differences in the concepts and indicators used in the income-based and capability-based approaches?
- Describe in your own words the characteristics and causes of poverty. Identify the main causes of poverty in your own country.
- What is the rationale for measuring poverty?
- Define poverty line. What is meant by the referencing and identification problems in deriving a poverty line?
- Define headcount measure, poverty gap index, and squared poverty gap index. Discuss the advantages and drawbacks of each measure. Convert the following probabilities into logits:

$$\Pi_i = 0.1, \Pi_i = 0.3, \Pi_i = 0.5, \Pi_i = 0.65$$

- Suppose you have the following logit equation specifying household poverty to educational attainment of the spouse:

$$\log\left(\frac{\Pi_i}{1 - \Pi_i}\right) = -5.3 + 0.3(\text{EDUCSPOUS})$$

- Calculate the predicted probability that the household is poor when EDUCSPOUS $= 2$ and when EDUCSPOUS $= 7$. Interpret your results.
- Suppose from the logistic regression equation you wanted to predict the probability of a household being poor using the following equation:

$$\log\left(\frac{\Pi_i}{1 - \Pi_i}\right) = 10.64 - 0.27(\text{HHLDQTY}) - 0.008$$

$$-0.471\,(\text{OWNED}) - 1.30\,(\text{CALREQ})$$

First, assume the following values for the explanatory variables: HHLDQTY $= 3$, PRODLMAIZ $= 40$, OWNED $= 2$, and CALREQ $= 0$. What is the expected probability that the household is poor? Next assume the following values for the explanatory variables HHLDQTY = 5, PRODLMAIZ = 180, OWNED = 5, and CALREQ = 1. What is the expected probability given that the household is poor? Compare both scenarios and interpret your results in the light of the discussion on determinants of poverty.

STATA workout

The following below has 27 observations on variables that have been defined in the main section of this chapter. We have information on all the important variables that determine the likelihood of poverty. Use STATA:

a. Estimate an unrestricted model and store the estimates.
b. Estimate a restricted model without the variables HEALTDST and DRINKDST, and store the estimates.
c. Use the likelihood ratio tests and test the null hypothesis.
d. Interpret the findings with respect to the relation between poverty and, the "access" variables.

Obs	Poverty	mbrage	prodlmaiz	calreq	healtdst	drinkdst
1	0	0.8	0.84	0.99	0.6	0.75
2	0	1	0.93	0.99	0.9	0.75
3	0	1	0.81	0.99	0.8	0.57
4	0	0.95	0.91	1.01	0.6	3
5	0	1	0.67	1	1.7	1.06
6	0	0.95	0.58	1.02	0.4	2
7	0	0.85	0.57	0.99	1.35	1.35
8	0	0.7	0.73	0.99	1.1	1.83
9	0	0.8	0.61	1	0.95	1.9
10	0	0.2	0.28	0.99	1.1	1.7
11	0	1	0.95	0.99	0.9	0.45
12	0	0.65	0.52	1.01	0.55	2.75
13	0	1	0.87	1	0.95	1.36
14	0	0.5	0.47	0.99	1.5	1.36
15	0	1	0.57	1	0.6	1.5
16	0	0.9	0.92	1.01	0.95	1.9
17	0	0.95	0.55	0.99	1.6	0.4
18	0	1	0.85	0.99	2.35	0.03
19	1	0.8	0.81	1	0.75	1.9
20	1	0.9	0.57	1	3.7	0.825
21	1	0.9	0.82	0.99	1.25	2.5
22	1	0.95	0.96	1	0.8	3.2
23	1	1	0.92	1.01	0.75	3.75
24	1	1	0.79	0.99	3.2	1.6
25	1	1	0.76	1	1.6	1.78
26	1	1	0.77	0.99	1.8	1.5
27	1	1	0.83	0.99	1.9	0.95

The input data code is as follows:

. global xlist1 mbrage prodlmaiz calreq healtdst drinkdst
. global xlist1 mbrage prodlmaiz calreq
. logit poverty $xlist1
. estimates store A
. logit poverty $xlist2
. estimates store B
. lrtest A B

The output portions for the corresponding input lines from STATA are given as follows:

```
. logit poverty $xlist1

Iteration 0:    log likelihood = -17.185883
Iteration 1:    log likelihood = -6.4863824
Iteration 2:    log likelihood =  -5.933922
Iteration 3:    log likelihood = -5.8981714
Iteration 4:    log likelihood = -5.8981008
Iteration 5:    log likelihood = -5.8981008

Logistic regression                             Number of obs   =         27
                                                LR chi2(5)      =      22.58
                                                Prob > chi2     =     0.0004
Log likelihood = -5.8981008                     Pseudo R2       =     0.6568
```

poverty	Coef.	Std. Err.	z	P>|z|	[95% Conf. Interval]
mbrage	4.417727	9.493874	0.47	0.642	-14.18992 23.02538
prodlmaiz	6.235905	8.976621	0.69	0.487	-11.35795 23.82976
calreq	-212.8624	175.0268	-1.22	0.224	-555.9087 130.1839
healtdst	3.957459	2.140029	1.85	0.064	-.236921 8.151839
drinkdst	4.039075	1.851807	2.18	0.029	.4095995 7.668551
_cons	190.8118	168.6648	1.13	0.258	-139.7652 521.3888

. estimates store A

Note that that P > |z| values indicate that both the access variables, healtdst and drinkdst, are significant. The estimates of the restricted model are given as follows:

```
. logit poverty $xlist2

Iteration 0:   log likelihood = -17.185883
Iteration 1:   log likelihood = -15.707733
Iteration 2:   log likelihood = -15.578981
Iteration 3:   log likelihood = -15.577849
Iteration 4:   log likelihood = -15.577849

Logistic regression                               Number of obs   =         27
                                                  LR chi2(3)      =       3.22
                                                  Prob > chi2     =     0.3595
Log likelihood = -15.577849                       Pseudo R2       =     0.0936
```

poverty	Coef.	Std. Err.	z	P>|z|	[95% Conf. Interval]
mbrage	4.039447	4.756515	0.85	0.396	-5.283151 13.36204
prodlmaiz	2.521413	3.406925	0.74	0.459	-4.156037 9.198864
calreq	-11.62457	51.71348	-0.22	0.822	-112.9811 89.73199
_cons	5.283209	51.80856	0.10	0.919	-96.25971 106.8261

```
. estimates store B
```

The restricted model indicates a much lower Pseduo R2 than the unrestricted model. The likelihood ratio test is conducted as follows, and the chi-square value and the P-value indicate that the null hypothesis that the two models are not different can be rejected. Hence, the access variables (access to health and drinking water) are significant drivers of the likelihood of being poor.

```
. lrtest A B

Likelihood-ratio test                  LR chi2(2) =      19.36
(Assumption: B nested in A)            Prob > chi2 =     0.0001
```

Chapter 13

Classifying households on food security and poverty dimensions—application of K-Means cluster analysis

Chapter outline

Introduction	493	Empirical analysis: *K*-Means clustering	506
Food hardships and economic status in the United States	495	Data description	507
		Initial partitions and optimum number of clusters	507
Food security, economic crisis, and poverty in India	499	Descriptive characteristics of the cluster of households	508
Cluster analysis: various approaches	501		
Hierarchical clustering method	501	Cluster centers	510
Single linkage (nearest neighbor method)	502	Cluster analysis in STATA	513
		Conclusion and implications	516
Complete linkage (farthest neighbor method)	502	Cluster analysis in R	517
		Exercises	519
Average linkage method	503	STATA workout—1	519
K-means method	503	STATA workout—2	523
Review of selected studies using cluster analysis	503		

Typology's usefulness in providing actionable insights can only be assessed when applied and confronted with other (typology) approaches or, more importantly, with an informed reality check on the ground.

Marivoet et al., (2019).

Introduction

This chapter introduces cluster analysis with a goal to find an optimal grouping of observations on food security and nutritional status that are similar in a cluster but are dissimilar to observations in other clusters. We motivate the chapter by discussing the extent of under reporting of food hardships in the

United States. It is shown how targeted food and nutrition interventions could be designed for groups of households on the basis of certain socioeconomic characteristics or when certain program interventions are desirable to improve the welfare of specific groups of the population. Initial partitioning of observations, optimum number of clusters, describing characteristics of the clusters, and cluster centers are some of the topics covered under this chapter. Several food and nutrition related case studies that use cluster analysis are also reviewed.

Use of typology of households based on their food security and poverty characteristics is a useful way for designing decentralized interventions (Marivoet et al., 2019). Yet, studying the relationship between food insecurity and poverty is complex. As pointed out by Smith et al. (2000), an important development goal is not only to improve food security but also to improve it in a manner that can be sustainable for the maximum number of people. From an analysis of a sample of developing countries during the 1990s, they found that poverty was the most binding constraint in improving food security. In addition to poverty, many countries faced problems of national food availability (especially the sub-Saharan African countries), while other countries faced nutrition insecurity problems linked to health and care (particularly South Asian countries). Policy objectives should be devised in such a way that synergies can be obtained using multiple interventions where there are multiple causes of problems (related to food insecurity and nutrition insecurity). An example where multiple interventions can have a synergistic effect is combining income-generating activities with nutrition education (von Braun et al., 1993).

The issue of addressing multiple policy objectives is critically important, since there may be trade-offs in reaching competing policy goals and an optimal balance is necessary for achieving such goals. A focus on food availability issues (ensuring food security) may lead to the direction of investment in new technology to increase food production in high-potential areas. On the other hand, a focus on poverty reduction can lead to investments that raise the incomes of poor people who often live in resource poor and low potential areas, where large amounts of food may not exist. Thus, policy combinations need to be optimally designed so that improvement in food and nutrition security can enhance long-running sustainable poverty reduction.

In the next section, we examine the various issues of food insecurity in the United States and in India. In both these countries, the problem of food insecurity forces us to focus attention on more than just increasing productivity. The issue of multiple complementary policies becomes particularly important when we examine the issues in the United States and in India, as good case studies.

In this chapter, a cluster analysis method is undertaken to group households on the basis of food security and poverty dimensions. Clustering is a process of grouping data into classes so that objects within a class have high similarity in comparison with one another but are very dissimilar with respect to objects in other clusters.

Cluster analysis will thus allow us to identify households that are vulnerable in the dimensions of food insecurity alone (such as lack of access to food), households that are vulnerable in the dimensions of poverty (such as access to productive assets such as land owned), and households that might possibly be vulnerable on both dimensions.

The patterns of income, expenditure, assets, and other related socioeconomic variables offer an insight into the dynamic and interrelated nature of poverty and food security that is lost in traditional multivariate analysis. Thus, with a more integrated analysis, interventions can be designed more effectively, so as to strengthen and complement households' own efforts to manage diversity. Cluster analysis can help policy makers better understand the complexity of individual and households' lives and can guide them to the design of poverty alleviation strategies that take this complexity into account. This sort of exercise could also be useful in providing the basis for long-term monitoring of panels of households and the impact of interventions on their welfare. We also provide the implementation of cluster analysis using STATA.

The chapter is organized as follows: We present the problems of food crisis and distribution in the United States and in India in the next section. Following that, we critically examine the different theoretical approaches to cluster analysis. We then review a few studies that highlight the linkages between food insecurity and poverty using cluster analysis. We also point out some of the studies that have used this method to analyze dietary patterns and determinants of child health. We next undertake the empirical analysis with our household dataset using a K-means cluster analysis and demonstrate the implementation of cluster linkage in STATA. The final section makes some concluding remarks and draws some implications from the current research.

Food hardships and economic status in the United States

In the United States, food insecurity and food insufficiency, according to Gundersen and Ribar (2011), appear to be real phenomena with serious consequences. Importantly, food insecurity and insufficiency are associated, with incomes, expenditures, and needs.

Gundersen and Ribar (2011) along with Gundersen et al. (2011) point out that most of the food hardships are underreported and that this underreporting takes place at the lower end of the expenditure distribution.

Specifically, Gundersen and Ribar (2011) point out that this data masking, "should be disquieting to researchers and policy makers." The implications of the underreporting among the poorer households are serious, because it prevents optimal policy interventions from taking place.

Following Amartya Sen's contributions, Gundersen and Ribar (2011) include many determinants to capture economic well-being. These measures include income and wealth, reports of material, financial, and food hardships.

However, although researchers frequently use these measures, Gundersen and Ribar (2011) find many anomalies that appear in the literature. For example, they question the finding that a high proportion of reported food hardships occur among households with moderate and high levels of income.

Similarly, another anomaly Gundersen and Ribar (2011) report is the finding that the average intakes for food insufficient households exceed 100% of the recommended daily allowances for most nutrients, or that the children in poor, food insufficient households had nearly the same Healthy Eating Index values as children in more affluent, food sufficient households.

According to Gundersen and Ribar (2011), these findings have to be reexamined closely, and to that end, they derive newer measures of validity and check for internal consistency, with the available data for the United States.

To check for reporting errors and patterns, Gundersen and Ribar (2011) use the food insufficiency measure and the food insecurity scale. The two measures are closely related, with each addressing households' food problems. Along with a question, "Do you obtain enough to eat," there are prompts to motivate the respondents to respond to adequacy and variety.

Which of these statements best describes the food eaten in your household?

1. Enough of the kinds of food we want to eat.
2. Enough but not always the *kinds* of food we want to eat.
3. Sometimes not enough to eat.
4. Often not enough to eat.

The researchers also develop subsequent analysis and food security scale to capture information regarding food intakes, anxiety, and financial stresses. They develop a food insecurity scale and, through a series of tests, factor analysis, and item response theory, derive external validity, and check for reliability and robustness of the scale. Gundersen and Ribar (2011) look more extensively at the strength of the relationships between food hardships and other measures and by formalizing one test for "reasonability."

Consistent with previous research, the researchers find that in the United States, food insecurity and insufficiency are related to economic conditions. However, they also find that the association between self-reported food hardships and objectively scaled food expenditures is weak and that the prevalence of hardships among households with low levels of income and objectively scaled expenditures is low. The highest incidence of food insecurity is among households with lower incomes but is never much above 50%. Furthermore, there is a much higher incidence of food hardships among households with low food expenditures.

What does this tell us about the self-reporting studies? Gundersen and Ribar (2011) conclude that reports of food hardships are internally consistent. Nevertheless, there is low level of reporting among households at the bottom of the expenditure distribution. Most likely, these reporting patterns are due to, "social-desirability bias," which occurs when survey subjects are uncomfortable reporting potentially embarrassing information.

What are some lessons that we glean from Gundersen and Ribar (2011)? First, researchers have to check whether some of the questions correspond to information on food expenditure. Second, researchers must maintain follow-up checks on those who report low levels of food expenditures and food security and those who report high levels of food expenditures and food insecurity. Third, researchers must include many other indicators of well-being and relate these to income and expenditure data.

A similar issue with information masked by data also appears in the way in which we have come to understand poverty and food insecurity in the United States. For example, Mills and Mykerezi (2009) point out that information on poverty is derived from annual data.

However, consider households that have average annual incomes or are able to sustain average annual consumption that is above the annual poverty line. These households will not be classified as poor if annual data are used. But it is possible that the same households are not able to smooth consumption to guarantee above-poverty consumption levels throughout the year. Such households would consume subpoverty amounts for at least part of the year. The incidence, severity, causes, and consequences of intraannual poverty are then a serious issue and Mills and Mykerezi (2009) estimate this using data from the Food Stamp Program (FSP).

Mills and Mykerezi (2009) link the FSP to consumption variation. Their strategy is as follows: If the FSP is primarily used as a short-term expenditure smoothing mechanism, then the FSP will reduce intraannual poverty. However, if the FSP is mainly used to support longer-term expenditure levels, it will reduce annual poverty.

We must note that intraannual poverty accounts for a large share of the economic hardship that US families face during any given year. Mills and Mykerezi (2009) find that the incidence of poverty for at least one quarter is thus double that of annual poverty alone. Furthermore, the intraannual poverty component accounts for over one-third of the total severity of poverty.

Section highlights: Stress, Comfort Foods, and Obesity

The rise in obesity in the United States has attracted a lot of attention in the media and among researchers across disciplines. However, as in every aspect of social sciences, there is not a one unique cause for the increase in obesity. Research points to technological change in the workplace, female labor force participation, availability of convenience foods, lower relative prices of energy-dense foods, and urban sprawl as some of the reasons for this social outcome.

Smith (2011) argues that the upward trend in economic insecurity faced by households in the United States since 1980 is the main cause for the increase in obesity and for the greater consumer demand for "fattening" foods. Smith (2011) links economic insecurity to mental stress that forces workers and consumers to consume fattening foods. Smith (2011) also provides references to various forms of stress that motivate consumers to seek out "fattening" foods and links these stressors to current obesity rates.

Gundersen et al. (2008) note that roughly 17.1% of US children between the ages of 2 and 19 years are obese, and another 16.5% are overweight, and furthermore, this rate has increased threefold for children since 1970. In terms of food insecurity and its links to obesity, Gundersen et al. (2008) point out that 1 in 5 children and 4 of 10 low-income children in the United States live in a food insecure household. Interestingly, their research also points to the idea that children in low-income families experience many psychosocial conditions that result in high levels of stress. Unfortunately, low-income families have difficulty dealing with all the stress and make the appropriate adjustments. Gundersen et al. (2008) evaluate the impact of maternal mental, physical, financial, and family structure stressors, to weight and childhood obesity.

Most interestingly, the empirical findings of Gundersen et al. (2008) show that food insecurity and maternal stressors are jointly related and that children in food secure households suffering from maternal stressors were more likely to be overweight or obese than children in food insecure households.

The aforementioned finding is important because the majority of low-income children in the United States live in food secure households. Thus, maternal stressors may be an important factor for children in the United States that are overweight or obese, particularly for those between the ages of 3 and 10.

A natural question arises as to why the findings are contrary to our expectations. Gundersen et al. (2008) provide many reasons for this finding. For example, even though children in both household types may have wanted to eat in response to maternal stressors, only children in food secure households may be able to do so. Furthermore, children in food-secure households may have a greater ability to consume more "comfort foods," which are often unhealthy, in response to the stressors they experience. It is also possible that food-insecure children experiencing undernourishment face the catabolic effects of cortisol in muscle, thus resulting in a lower BMI.

Gundersen et al. (2008) also show that when it comes to overweight and obesity, low-income children are indeed influenced by stress emanating from their mother. They find that the cumulative stress experienced by the child's mother is an important determinant of child overweight. Thus, policy measures are required for providing women with relevant medical care and counseling to combat childhood obesity in the United States.

The primary predictors of intraannual poverty are similar to those for annual dimensions of economic deprivation: low human capital, minority status, and involuntary unemployment of the household head. Interestingly, reductions in family size are associated with increased intraannual poverty, implying that such changes generate negative but short-term shocks to family well-being. Nonincome lump-sum receipts also appear to have only a temporary effect on household well-being, as they are effective only for intraannual poverty.

Many poor and near-poor families use the FSP as a short-term expenditure stabilization tool rather than for long-term expenditure support. Mills and Mykerezi (2009) conclude that the fact that the FSP's main impact is on intraannual poverty is somewhat problematic. That is, intraannual poverty and FSP are not usually discussed, and the connection is not obvious. Mills and Mykerezi (2009) suggest fast-track FSP that monitors eligibility and certification guidelines, so as to enhance short-term program use and improve the program's effectiveness for intraannually poor families.

Food security, economic crisis, and poverty in India

Based on Prof. Amartya Sen's work, it is clear that poverty encapsulates many dimensions of deprivation that relate to human capabilities. These include consumption and food security, health, education, rights, voice, security, dignity, and decent work. Basic measures of well-being must include an individual's physical, cognitive, and emotional development. Finally, we must not forget that all productive functions in humans require access to food of adequate quantity and quality.

Lone and Rather (2012) note that poverty is the sum total of a number of factors that include not just income but also access to land and credit, nutrition, health, literacy, education, safe drinking water, sanitation, and other infrastructural facilities. Lone and Rather (2012) also note that in India, poverty is conventionally defined in terms of income poverty that is related to hunger and measured in different ways. As of 2009–10, there were 32% of Indians below the poverty line.

Lone and Rather (2012) point out the mounting food grain stocks that exist with the Food Corporation of India, alongside recurring starvation deaths in different parts of India. This observation shows that the food security policies in India are not bearing fruit. That is, India has achieved food security at the national level, but not at the household level.

Lone and Rather (2012) use a set of internationally accepted measures to describe the food security situation in India. According to their results, the food security situation in India has worsened since 1995, and the country's substantial levels of undernourishment stand at 21% in 2007. Using the Global Hunger Index, the researchers note that India has moved from a situation of extremely alarming hunger (31.7%) to one of alarming hunger (24.1%). It is important to add that India is among the four countries in the world with the highest prevalence of underweight in children below 5 years of age (43.5%).

In terms of poverty reduction, researchers in India have noted that every additional kilometer of roads would lift 1.57 poor people out of poverty in irrigated areas but would lift 3.5 and 9.51 people out of poverty in high and low potential rainfed areas. Finally, the problem of food security in India is clearly not merely a problem of production, but one of access and entitlements.

The food security situation in India has also worsened since the economic crisis in 2008. Currently, there are regional inequalities, and the benefits of liberalization have not trickled down at the rate everyone expected. Sheereen (2012) notes that the increase in food insecurity in India since 2006 has been due to three key reasons:

1. Neglect of agriculture sector relevant to poor by the governments
2. Significant increase in food prices in past several years
3. The global financial crisis

The global financial crises affected returns in international agricultural markets because of the global flow of capital. It is important to note that in India, the food and fuel crisis of 2006 preceded the financial crisis. Thus, international trade flows and financial services fell for many countries including India. There was also a reduction in exports. Sheereen (2012) notes that the impact of the crisis was a loss of 8 million jobs.

Obviously, the result was not very good in terms of food security either. India's poor performance on the GHI is due to its high levels of child malnutrition and undernourishment resulting from calorie-deficient diets.

While India has been able to achieve success in combating transient food insecurity caused by droughts or floods, Sheereen (2012) notes that India has failed to make a dent on chronic food insecurity as reflected by high incidences of hunger and malnutrition. The improvement in nutritional status has been slow. Besides, it is interesting to note that the proportion of expenditure spent on food has declined in households with chronic undernourishment. The financial and commercial integration of the developing economies exposes the poor to risk of market uncertainties as evidenced during the economic crisis.

Public policy and programs should target the poor who have inadequate access. Adequate availability of food is required by ensuring production of cereals, pulses, and vegetables to meet nutritional needs and then making them available through the year. Nutrient deficiencies must be addressed by providing coarse grains, pulses, and iodized double-fortified salt.

Along with direct provision, optimal agricultural policies should also be put in place, which generate farm production and employment. Efforts must also include better utilization of agricultural inputs, proper marketing infrastructure and support, stepping up investment in agriculture with due emphasis on environmental concerns, and efficient food management.

Chandrasekar (2013) notes that given the current food situation in India, it is imperative for the government to step in with immediate corrective action. But paradoxically, there is a big rush to keep the government out of procurement and distribution of food and leave the entire channel to the market. Chandrasekar (2013) notes that "the evidence from across the world shows that this would not work. It would be a disaster in a country that is ranked 66 among 105 countries in the Global Hunger Index."

Cluster analysis: various approaches

The goal of a cluster analysis is to find an optimal grouping for which the observations or objects within each cluster are similar but are dissimilar to the objects in other clusters. Cluster analysis has been widely applied to data in many disciplines, such as market research, biology, medicine, economics, and engineering. For example, in biology, one can categorize genes with similar functions and gain insight into structures inherent in populations. On the other hand, for business, this analytical method can help a market researcher in identifying distinct groups of customers and characterize them based on their purchasing trends. There are two common approaches to cluster analysis: hierarchical and partitioning. Both approaches deal with the fundamental assumption of within-group and between-group similarity. Correlation, distance, and association measures are the main ways to measure similarity and will be discussed in this section. In the *hierarchical* procedure, one starts with a situation where all the cases form their own cluster. The procedure then identifies the two most similar cases, which form the first cluster. In the next stage, the two closest cases are grouped into another cluster. The procedure continues until all the cases are members of one group consistent with all cases in the analysis. In the *partitioning (K-means)* procedure, the cases are randomly assigned to the number of clusters that the analyst wants depending on the definition of the cluster characteristics. For each case, the distance between cluster centers is computed, and each case is assigned to the nearest cluster such that the error sum of squares does not change.

Hierarchical clustering method

Hierarchical clustering allows one to find "good" clusters in a data set. The main idea behind this technique is to start with each cluster comprising of exactly one object and then combine the two nearest clusters until there is just one cluster left consisting of all the objects. In a large data set, searching for all possible clusters is a tedious task. Let $X(m, h)$ be the number of ways one can separate out m items to form h clusters (the derivation of this equation can be found in Jensen (1969) and Seber (1984). Then,

$$X(m,h) = \frac{1}{h!}\sum_{i=1}^{h}\binom{h}{i}(-1)^{h-i}i^n \qquad (13.1)$$

The expression provides an approximation for Eq. (13.1). Thus, with the hierarchical method, one can arrive at a meaningful solution without searching for all possible solutions.

There are two kinds of hierarchical clustering algorithms: (1) agglomerative and (2) divisive. The divisive procedure is not discussed in additional detail, since SPSS uses the agglomerative hierarchical algorithm in the computation of clusters. The reader can refer to Rencher (2002) for further details. In the *agglomerative* procedure, a cluster of observations is merged into another cluster. In this sequence, the number of clusters decreases, but the clusters themselves increase in size. Thus, in an agglomerative procedure, one ends with one single cluster containing the whole data set, even though the procedure starts with m clusters.

On the other hand, in the divisive method, we start with one large cluster with all the m items and end up with m clusters containing each item as a cluster of its own. In what follows, we introduce three key methods of hierarchical clustering approach.

Single linkage (nearest neighbor method)

In this method, the distance between any two clusters is defined as the distance between the nearest pair of objects in the two clusters. If cluster A is the set of objects $x_1, x_2, \ldots, x_m$ and cluster B is the set of objects $y_1, y_2, \ldots, y_n$, then the single linkage distance between A and B is given by

$$d(A, B) = \min\{d(x_i, y_j), \text{ for } x_i \in A \text{ and } y_j \in B\} \qquad (13.2)$$

To begin with, we calculate the distance $d(A, B)$ for each pair of clusters. Then, we merge any two clusters with the smallest distance. We repeat the process with the merged clusters and calculate cluster distances again among cluster pairs. Once again, we merge two clusters with smallest distance into one single cluster. When one single cluster is obtained, the procedure comes to an end.

Complete linkage (farthest neighbor method)

In this approach, the distance between cluster A and cluster B is defined as the maximum distance between a point in cluster A and a point in cluster B. Similar to the single linkage method, the distance function between A and B is given by

$$d(A, B) = \max\{d(x_i, y_j), \text{ for } x_i \in A \text{ and } y_j \in B\} \qquad (13.3)$$

This method tends to generate clusters at the early stages that have objects within a narrow distance from each other.

Average linkage method

In this approach, the average distance between all possible pairs of objects with one object in each pair belonging to a distinct cluster gives the distance between two clusters. Thus, the distance between any two clusters A and B is defined as the average of the mn distances between the m points in A and n points in B and is given as follows:

$$d(A,B) = \frac{1}{mn} \sum_{i=1}^{m} \sum_{j=1}^{n} d(x_i, y_j) \quad (13.4)$$

There are other methods in agglomerative clustering, such as centroid, median, and Ward approaches. They are not discussed here. For further details, the reader is referred to Rencher (2002).

K-means method

The *K*-means method of cluster analysis is popular due to its simplicity. In this method, clusters are formed by specifying the number of clusters, say k, and assigning each object to one of the *K*-clusters to minimize the sum of distances from the mean of each cluster.

The algorithm for *K*-means clustering begins with an initial partition of the cases into *K*-clusters. In subsequent steps, the partition of cases is modified to reduce the sum of the distances for each case from the mean of the cluster to which the case belongs. This leads to a new partition for which the sum of distances is strictly smaller than before. This method is extremely fast, and there is a possibility that the improvement in the steps leads to less than *k*-partitions. In general, the number of clusters in the data is not known in advance. Thus, it may be a good idea to run the algorithm with different values of k to determine how the sum of distances reduces with increase in the value of k (Bacher, 2002).

In this chapter, we use the *K*-means method to demonstrate the use of cluster analysis for classifying households among different dimensions of assets, income, and food security.

Review of selected studies using cluster analysis

Cluster analysis is primarily used for poverty mapping exercises.[1] For example, households tend to cluster together in villages or other small geographic and administrative units, which are relatively homogeneous. In other words, households that are close together tend to be more similar than households far apart, allowing the researcher to specify a regression model of

---

1. It is beyond the scope of this chapter to discuss in detail the various methodologies of poverty mapping. For an extensive discussion, see Hentschel et al. (2000).

household welfare of a cluster to be dependent on household characteristics and household level error within the cluster.

The rationale behind such an analysis is threefold:

1. Aggregate (national level) indicators provide an impression that conditions within a country are uniform. However, evidence suggests otherwise. There remains significant geographic variation in the incidence of poverty, which can be attributable to differences in resource endowments, educational levels, health services, etc. As one maps the relevant administrative units, the geographic variability in the aggregate data becomes apparent.
2. Poverty maps are an important tool for targeting resources and interventions. Detailed information on the location of the target groups can facilitate resources to be used more effectively so that the most needy groups are reached effectively.
3. Maps encourage visual comparison, and it becomes easier to look at spatial trends, clusters, or other patterns in the data. Disaggregated information provides local decision-makers with the facts that are necessary for local decision-making.

In addition to its use in poverty mapping, cluster analysis has also been used for examining dietary and nutrition patterns of a particular segment of the population (for example, children or the elderly). This is because the cluster analysis approach is very useful for summary and descriptive purposes, although the results depend on local situations that make it difficult to transfer to other populations. A number of studies have examined dietary patterns by estimating the usual daily intake from the food frequency questionnaires (see, for example, Wirfalt and Jeffery (1997), Greenwood et al. (2000), Millen et al. (2001)). It is important to note that these studies have been undertaken mainly for developed countries such as the United States and European countries.

The general findings of these studies indicate that there exist distinct dietary patterns within a given population. The analytical method can thus provide a rationale for developing intervention strategies aimed at preventing and managing chronic health conditions such as obesity, colon cancer, and cardiovascular disease.

The food intake pattern of elderly individuals has been studied both for southern European countries (SENECA investigators, 1996; Leite et al., 2003) and the United States (Akin et al., 1986). For Italy, the overall intake of fish, fruits, and vegetables is similar to that reported in Portugal and Spain, while the consumption of dairy and animal products for Italy, such as milk, eggs, and meats, is similar to that of France. The overall food group intakes among elderly Americans revealed some important differences. They consume a higher intake of high-fat meats and other animal fats compared with countries in Southern Europe, while the consumption of fruits and vegetables is greater in Southern Europe relative to that in the United States. We now review some of the studies that use cluster analysis to classify households on the various dimensions of poverty.

Deininger et al. (2004) examined the effects of targeting asset redistribution (land reform[2]) on poverty alleviation in Zimbabwe. The analysis was done in two main steps. First, the target groups were identified through clustering techniques by generating household groups that closely resembled each other in multiple dimensions. In the second step, the trajectories of income and asset of the groups of households were traced over time.

The main result from the analysis demonstrated that land reform could uplift the disadvantaged households significantly. However, it takes time for the benefits of land reform to accrue. Land reform can alter the lives of individuals in many dimensions, and the benefits could come from the combination of access to resources, technology, institutional support, social capital, learning opportunities, and much more. Over time, it seems that the majority of the households in the panel had made impressive gains.

Shinns and Lyne (2003) identified different dimensions of poverty that affected the current and future well-being of households within a community of land reform beneficiaries in the Midlands of Kwazulu-Natal in South Africa. The study used objectively measured variables representing the broader symptoms of poverty, namely, the quality of housing, health, income, and household wealth. A cluster analysis methodology was undertaken to classify households according to their poverty profiles. A census survey of 38 land reform beneficiary households—members of a communal property association established to purchase Clipstone (a 630 ha farmland which was sold to the beneficiary households by the owners of Clipstone)—was conducted during May 2002.

Although no explicit statements can be made about the underlying causes of poverty using cluster analysis, the study reveals dimensions of poverty that can help in distinguishing between short-term and long-term strategies in alleviating poverty. The study provided some strategies (both short and long term) and estimated the annual net costs to the government in implementing them.

Summarizing, cluster analysis is primarily done for grouping households on the basis of certain economic characteristics (such as their income or asset situations) or when certain program interventions are desirable (such as land reforms) to improve their economic conditions. The purpose of this sort of

---

2. The main objectives of the land reform program were as follows: (1) to alleviate population pressure in the communal areas; (2) to improve the base for productive agriculture for the peasant farming sector; (3) to improve the standard of living of the poorest segments of the population; (4) to bring underutilized land into full production for implementing an equitable program of land redistribution; (5) to improve the infrastructure of economic production; and (6) to achieve national stability and progress of the country that had only recently emerged from the turmoil of war.

analysis is mainly for targeted interventions by policy makers or program managers and not in understanding the underlying causes of food insecurity or poverty. In the first step, the target groups (the income poor or food insecure households) are identified through clustering techniques by generating household groups that closely resemble each other in multiple dimensions. In the second step, the trajectories of certain economic variables (such as income and assets) of the groups of households are traced over time. In the next section, we demonstrate this method using the Malawi data set.

Several studies use the K-means clustering techniques to identify the zones or the subgroups that require economic assistance and policy intervention. The technique along with related methods has been applied to data from Namibia, Nigeria, China, and India.

Oparinde et al. (2017) combine K-means cluster analysis to behavioral economics, to examine why attitudes regarding cultivation of Provitamin A GM cassava vary across farmers in Nigeria. The researchers use K-means cluster to identify three different classes of attitudes toward opposition: low, medium, and high. Particularly, the application of K-means cluster analysis indicates that smallholder farmers are supportive toward GM adoption. Overall, the analysis suggests that only about 25% of the farmers have negative attitudes toward GM adoption, primarily due to ignorance, and religious norms, beliefs, and customs.

Muthoni et al. (2017) apply the K-means cluster analysis to identify locations with similar biophysical and socioeconomic characteristics, for improving sustainable implementation of crop varieties in Tanzania. The study is an interesting application of K-means clustering technique to geospatial tools that help locate development domains that possess agricultural potential, but require guidance and investment opportunities.

Sethi and Pandi (2014) apply the K-means clustering technique to account for rural—urban differences in per capita consumption expenditures in India and identify the rural and urban regions that lag behind economically. Similarly, Chamboko et al. (2017) link K-cluster methods to principal component and bivariate analysis and map various dimensions of poverty for Namibia. Likewise, Huang and Browne (2017) examine Chinese mortality data, and Wu et al. (2018) examine the differences in fertility rates across China and identify which regions require targeting.

Empirical analysis: *K*-Means clustering

In hierarchical clustering, one requires a distance of similarity matrix between all pairs of cases, which can be a huge matrix given that we have 604 households. The clustering method that does not require computation of all possible distances among cases is the K-means clustering method. In this method, one starts with an initial set of means and classifies the cases based on their distance to the means. The cluster means are computed again, using the cases that are assigned to the cluster, and then the cases are reclassified based on the new set of means. This process is repeated until the cluster means do not change much between successive steps.

In this section, we want to classify households based on their food security and poverty dimensions along with the correlates of food security (such as productivity of maize) and poverty (such as land owned and livestock possessed). The purpose is to classify households among different dimensions of assets, income, and food security.

Data description

The variables used in the present analysis are as follows:

1. FOODSEC: This is a measure of food security and is a weighted average of three components: namely the number of livestock owned (LIVSTOCKSCALE), the number of meals consumed per day (NBR), and stocks of food running out (RUNDUM). It is a continuous variable with higher values of index denoting food secure households. The index ranges between 0 and 1, with 0 denoting completely food insecure households
2. Per capita expenditure (PXTOTAL): This is a proxy for income level of the household, since consumption expenditure by households is a better indicator of lifetime welfare than income
3. Size of land owned (LANDO): This is a variable that measures assets of the household. The values range from 0 to 6, with higher values indicating that the household has greater amounts of land available
4. LIVSTOCKSCALE: This is another measure of assets of the household and is measured in tropical livestock units (described in Chapter 2). It is a continuous variable scaled from 0 to 1, with 0 indicating that the household owns no livestock.
5. PRODLMAIZ: This is a measure of yield or technology, which can affect food security in a favorable way. The ratio of total quantity harvested of local maize to household size is considered as a measure of yield (since local maize is the main crop produced and consumed in Malawi).

Initial partitions and optimum number of clusters

To choose the initial cluster means, a hierarchical method (on the lines of Bacher, 2002) was used for the starting configuration. We ran the cluster and saved the membership of the variables as Z-scores. In other words, all the variables were converted with zero mean and a standard deviation of unity. The procedure was adopted since variables were measured in different units (for example, land was measured in hectares, while livestock was measured in TLU terms). If standardized units were not undertaken, we had to worry about the variables that had large values to have a large impact on the distance compared with variables that had smaller values. We renamed the new membership variable to Cluster_ and computed the cluster centers (means) using the AGGREGATE command. We stored the means in a new data file, reread the original data, and specified the quick cluster command that reads the saved centers as the starting partition.

The procedure, however, still does not determine the optimum number of clusters. We thus undertake the *F-max* statistic, which analyzes the null hypothesis that a solution with k is not improved by a solution with $(k + 1)$ clusters. We allowed for a systematic variation of the number of clusters from $k = 2$ to $k = 8$. By the *F-max* criterion, the solution with the highest *F-max* value needs to be chosen for the optimum number of clusters. We found $k = 4$ as having the highest value for the *F-max* statistic. However, this solution was not chosen in the final specification, since we could not classify the households on all the three dimensions of asset, income, and food security. We chose the value $k = 7$, which provided the next highest value for the *F-max* statistic as the optimum number of clusters, since it gave us a better classification of the households based on the asset, income, and food security configurations.

Descriptive characteristics of the cluster of households

First, we look at the mean value of the aforementioned variables for each of the cluster of households to understand how they can be characterized on the basis of food security, assets, and income (Table 13.1). We do not report the households belonging to the seventh cluster, since there are only two members and characteristics such as food security and livestock ownership are missing.

The first group (group 1) of households can be termed as income rich but a relatively asset poor, and food-insecure group of households. The per capita

TABLE 13.1 Mean values of selected variables for clusters of households in Malawi.

	Group 1	Group 2	Group 3	Group 4	Group 5	Group 6
Variables	$n = 73$	$n = 23$	$n = 280$	$n = 92$	$n = 10$	$n = 124$
PXTOTAL	43.07	18.37	10.33	10.45	114.86	12.09
FOODSEC	0.35	0.58	0.30	0.59	0.59	0.30
LANDO	2.51	4.26	1.61	3.05	3.20	3.93
LIVSTOCKSCALE	0.06	0.28	0.05	0.85	0.66	0.07
PRODLMAIZ	136.26	554.66	66.29	83.41	268.70	88.95

expenditure exceeds the average per capita expenditure for all the households (17.27 kwacha per household size). This group of households is relatively more asset poor (own average land size and own almost no livestock units) and is, on average, food insecure. Although the yields of local maize generate some income, their average distance to the market place is quite high. Thus, it is likely that lack of market access coupled with lack of assets is not enabling these households to be food secure.

The second group of households can be significantly distinguished from the first in that they are relatively asset rich and food secure but are still relatively income poor. They are also significantly more productive in local maize (the average productivity is 106.56 per household) but are still income poor. This may be because adequate food in a region may not guarantee freedom from hunger. It may be the case that undiversified production and lack of control over land, labor, and prices all contribute to households being unable to meet their income needs.

The third group of households is poor in all dimensions (namely in income, assets, and food security) and constitutes a significant portion of the sample. These households are the ones trapped in chronic poverty and require the most attention from both government agencies and the international donor community. It is likely that constrained by the extremely low incomes, these households have not accumulated any assets and may not be able to finance adequate nutrition.

The fourth group of households is food secure and owns significant amount of assets (both livestock and land) but is income poor and less productive. What distinguishes these households from group 2 is their productivity of local maize, is below the average, and is significantly lower than group 2. It appears that producing only local maize may not generate enough income in the long run and a broader strategy of crop diversification should be conceived in generating higher incomes in the long term. Additionally, production of more food for home consumption and increasing output of marketed products that increase farm income should be carefully thought of by policy makers.

Turning to group 5, we find that this group of households is rich in all dimensions (namely income, food security, and assets). However, the number of individuals in this group being so small ($n = 10$) suggests that poverty in Malawi is both chronic and widespread.

Finally, looking at group 6, we find that this group can be termed as land rich but both income poor and food insecure. The characteristic that distinguishes these households from group 4 is that, in spite of having a significant amount of land, these households are less productive, more food insecure, and income poor. It may be the case that these households are located in a remote area where there is not much fertile land, and thus, there is less scope for income-generating activities. Additionally, we find that these households are located far from an ADMARC center (the mean SADMDIST was found to be 2.47).

Thus, lack of market access could have acted as an additional bottleneck for the households to sell their products during periods of crisis.

Cluster centers

To arrive at the ideal number of clusters in the final solution, Ward's technique was used. The Ward's method[3] joins those clusters whose combination leads to a minimum increase in within cluster sum of squares, while maximizing the between cluster sum of squares. Second, using the output from Ward's procedure as initial subgroup seeds, the K-means was applied to determine the final case location in the separate subgroups. The K-means clustering is an iterative partitioning procedure that reproduces the k number of disjoint clusters through minimizing the sum of squared distances from the cluster centroid means.

The first step in the K-means clustering is to find the k centers. This is done iteratively. The initial set of centers is reported in Table 13.2.

After the initial cluster centers have been selected, each household (or case) is assigned to the closest cluster, based on its distance from the cluster centroids. After all the cases have been assigned to clusters, the cluster centers are computed again based on all the cases in the cluster. The case assignment is done again, using the updated cluster center. The iteration stops when no cluster center changes appreciably. We look at the final cluster centers to determine how the households can be classified.

Table 13.3 reports the final cluster centers. It is evident that cluster 1 has a higher average value for per capita expenditure (in terms of Z-scores) but a lower than average value for the rest of the variables, confirming our earlier insight (from the descriptive statistics) that this cluster can be coined as income rich, asset poor, and food insecure households. Similarly, looking at cluster 3, we can identify that the households belonging to this cluster have lower than average values for all the variables suggesting that they are poor in all dimensions.

Table 13.3 thus confirms the description of Table 13.1, by grouping households in asset, income, and food security dimensions. One can also look at the distance among the cluster centroids to determine how far the cluster centers are from each other. Table 13.4 gives us the distance among the various cluster centers.

Table 13.4 provides a precise picture as to which clusters are similar to each other based on the distance between the cluster centers. We find from the

---

3. The algorithm starts with an initial partition of the cases into K-clusters. In subsequent steps, the partition of cases is modified to reduce the sum of the distances for each case from the mean of the cluster to which the case belongs. This leads to a new partition for which the sum of distances is smaller than before. It is a good idea to run the algorithm with different values of K, to determine how the sum of distances reduces with increase in the value of K.

Classifying households on food security and poverty **Chapter | 13** **511**

TABLE 13.2 Initial cluster centers.

Cluster	1	2	3	4	5	6
Z-PXTOTAL Z-score: per capita expenditure	−0.03738	0.37477	−0.20655	−0.13189	2.64974	−0.41473
Z-LIVSTOCK SCALE Z-score (LIV STOCKS CALE)	−0.15277	0.34213	−0.48009	2.40895	−0.46962	−0.51123
Z-FOODSEC Z-score (FOODSEC)	0.02057	1.26577	−0.48766	1.47719	−0.31929	−0.82172
Z-LANDO Z-score: Size of land owned	0.52405	1.00929	−0.79451	0.38942	0.49160	1.02663
Z-PRODLMAIZ Z-score (PRODLMAIZ)	0.04398	1.74494	−0.13471	0.08025	0.20773	−0.35410

TABLE 13.3 Final cluster centers.

Cluster	1	2	3	4	5	6
Z-PXTOTAL Z-score: per capita expenditure	1.32836	0.04799	−0.36927	−0.36269	5.04967	−0.27773
Z-LIVSTOCKSCALE Z-score(LIV STOCK SCALE)	−0.43871	0.26500	−0.46422	2.09280	1.47314	−0.39858
Z-FOODSEC Z-score (FOODSEC)	−0.09215	1.36556	−0.38002	1.43801	1.43283	−0.39102
Z-LANDO Z-score: Size of land owned	−0.17688	1.32855	−0.94295	0.29302	0.41803	1.04236
Z-PRODLMAIZ Z-score (PRODLMAIZ)	0.17930	2.65315	−0.23435	−0.13315	0.96240	−0.10037

TABLE 13.4 Distances between cluster centers.

Cluster	1	2	3	4	5	6
1		3.556	1.930	3.454	4.560	2.058
2	3.556		4.153	3.514	5.493	3.361
3	1.930	4.153		3.374	6.300	1.993
4	3.454	3.514	3.374		5.558	3.182
5	4.560	5.493	6.300	5.558		6.061
6	2.058	3.361	1.993	3.182	6.061	

We do not report the values of the variables for cluster number 7, since there are very few households in this cluster.

above matrix that households belonging to cluster 1 are similar to households belonging to clusters 3 and 6. This is possibly because the dimensions along which households become vulnerable are more or less similar. For cluster 3, households are income poor, asset poor, and food insecure (i.e., they are poor in all the dimensions), whereas in cluster 1, households are income rich but asset poor and food insecure. Similarly, households in cluster 6 are land rich but both income poor and food insecure. One can interpret the results as if the underlying causes of vulnerability may be similar among clusters 1, 3, and 6.

Cluster analysis in STATA

We can conduct cluster analysis in STATA using the cluster command. We use the example provided in Hamilton (2006) that contains information on living conditions across many countries. We have the following variables from Hamilton (2006) (Table 13.5):

1. country: Name of the country
2. region: Region
3. gdp: Gross domestic product
4. school: Mean years of schooling for adults
5. adfert: Adolescent fertility: births/1000 females
6. chldmort: Probability of dying before age 5
7. life: Life expectancy at birth
8. pop: Population
9. urban: Percent of population that is in urban areas
10. femlab: Female/male ratio in labor force
11. literacy: Adult literacy rate
12. CO_2: Tons of CO_2 emitted per capita

TABLE 13.5 Living conditions on 32 families.

Id	x1	x2	x3	x4	x5	x6	x7	x8
1	0	0	0	4	3	3	3	3
2	0	0	1	4	3	3	3	3
3	0	0	0	4	3	3	3	3
4	0	0	0	4	3	3	3	3
5	0	1	0	1	0	0	3	3
6	0	1	0	1	1	1	3	3
7	0	1	0	1	1	1	3	3
8	0	1	0	1	2	1	3	3
9	0	1	0	1	2	1	3	3
10	1	1	1	3	1	2	3	3
11	0	2	0	1	2	2	3	3
12	0	2	0	1	3	2	3	3
13	1	2	1	1	4	4	1	1
14	1	2	1	3	1	2	3	3
15	1	2	1	3	2	2	3	3
16	1	2	1	3	2	3	3	3
17	1	2	1	3	3	3	3	3
18	0	3	1	2	3	3	3	3
19	1	3	1	2	3	3	1	2
20	1	3	1	2	3	3	2	2
21	1	3	1	2	4	4	1	1
22	1	3	1	2	4	4	1	1
23	1	3	1	3	3	2	1	1
24	1	3	1	3	3	2	1	1
25	1	3	1	3	3	3	1	2
26	1	3	1	3	3	3	1	2
27	1	3	1	3	4	3	1	2
28	1	3	1	3	4	4	2	3
29	1	3	1	3	4	4	2	3
30	1	3	1	3	4	4	3	2

Continued

TABLE 13.5 Living conditions on 32 families.—cont'd

Id	x1	x2	x3	x4	x5	x6	x7	x8
31	1	3	1	3	4	4	1	2
32	1	3	1	3	4	4	1	2

We do not report the values of the variables for cluster number 7, since there are very few households in this cluster.

We wish to group countries using the Wards linkage method. To produce the results, we use the command lines in STATA:

- cluster wards gdp school adfert chldmort life pop urban femlab literacy CO_2, name(clward)
- cluster generate ward4 = groups(4)
- sort ward4
- label variable ward4 "country type"
- by ward4: list country

The command cluster wards is the procedure we wish to use and we name the cluster clward for later use. We create a cluster of 4 groups, and we can generate the countries that are listed in these four groups. The STATA output is produced in the following:

```
-> ward4 = 1

         country
1.       Egypt
2.       Thailand
3.       Italy
4.       Philippines
5.       Ethiopia

6.       Mexico
7.       Viet Nam
8.       Turkey
9.       Congo (Dem Rep)
10.      Iran


-> ward4 = 2

         country
1.       Russian Federation
2.       Indonesia
3.       Bangladesh
4.       Brazil
5.       Pakistan

6.       Nigeria
```

We omit the portion of the output covering group 3, because there are over 100 countries in this group. The last group has only two countries, India and China. To get a better understanding, of the current classification, we can use the following command that groups the clusters according to basic descriptive statistics:

- tabstat gdp school adfert chldmort life pop urban femlab literacy CO$_2$, by(ward4) stat(n mean sd)

The STATA output at this stage is:

```
. tabstat gdp school adfert chldmort life pop urban femlab literacy co2, by(ward4) stat(n mean sd)

Summary statistics: N, mean, sd
  by categories of: ward4 (country type)

  ward4 |      gdp    school    adfert  chldmort      life       pop     urban    femlab  literacy       co2
--------+------------------------------------------------------------------------------------------------------
      1 |       10        10        10        10        10        10        10        10        10        10
        |  8110.96  6.201667     59.06    48.875     69.53  7.73e+07  48.61667   .60836  81.07667     11.35
        | 8309.691  2.459793  53.80058  60.45088  9.957614  1.49e+07  20.90246  .2086741  21.06122  9.952638
--------+------------------------------------------------------------------------------------------------------
      2 |        6         6         6         6         6         6         6         6         6         6
        |   5228.9  6.066667     63.25      62.5  65.05555  1.71e+08  52.81944    .6024  75.23333      9.75
        | 4923.492  2.012958  34.24808  49.53206  7.669651  3.56e+07  22.61621  .2013836  20.77486   14.6451
--------+------------------------------------------------------------------------------------------------------
      3 |      108       108       108       108       108       108       108       108       108       108
        | 9417.697  6.739667  60.41759  57.87963  66.42731  1.14e+07  51.31543  .6948685  80.85162   16.7537
        | 12448.05  2.940194  47.83682  56.08115  9.883906  1.21e+07  22.58537  .191925  19.82191   29.6913
--------+------------------------------------------------------------------------------------------------------
      4 |        2         2         2         2         2         2         2         2         2         2
        |   3911.6      5.75     47.35    46.125  68.44166  1.25e+09    37.075    .6248      78.4      10.5
        | 1787.283  2.215602  55.08362  34.47146   5.99862  9.90e+07  10.94837  .3142383  22.06173  8.202438
--------+------------------------------------------------------------------------------------------------------
  Total |      126       126       126       126       126       126       126       126       126       126
        | 9027.123  6.649471   60.2373  57.19841  66.64021  4.39e+07  50.90873  .6824873  80.56303  15.89206
        | 11828.67  2.844162  47.34622  55.42751  9.709974  1.59e+08  22.20823  .1949653   19.7726  27.84679
```

We can see that the third group has a lot of countries whose mean GDP is much smaller than the other groups. This group also has lower adolescent schooling and fertility ratios. We can divide the countries into more groups using STATA, to find a sharper pattern in the data.

Conclusion and implications

The purpose of this chapter is to classify (cluster) households among the various dimensions of vulnerability. The dimensions of vulnerability chosen are assets, income, and household food security. The analysis is undertaken using a *K-means* cluster analysis. The advantage of this method over hierarchical cluster analysis is that it can provide clusters that can satisfy some optimality criteria when the number of clusters is known. While determining the optimum number of clusters is somewhat arbitrary, we used the *F-max* statistic for various values of *k*. The value of *k* is chosen, where the *F-max* statistic is the second maximum. Additionally, in a hierarchical clustering method, one requires a distance of similarity matrix between each pair of cases, which can be a huge matrix given that our sample consists of 604 households.

While no explicit statements can be made about the underlying causes of poverty or the ways in which the fundamentals can be addressed, we can understand the various dimensions of vulnerability to help distinguish the strategies that can relieve the symptoms of vulnerability. From the present analysis, we find that households belonging to clusters 1, 3, and 6 are vulnerable in different dimensions. While in the short term, improvement in market access, distribution of land, and other assets can be one set of policy measures, in the long run, crop diversification is likely to help households move out of chronic poverty. At the same time, improvement in infrastructure, such as construction of new roads, can facilitate market expansion, which can reduce input prices, raise the output prices of crops, and benefit the producers and consumers at the same time.

Cluster analysis in R

We are going to conduct a K-means cluster analysis in this chapter. We can conduct cluster analysis in R using *K-means* command. We use a data set with 32 observations and 9 variables. Before performing an analysis, we produce descriptive statistics with the *summary* command.

```
> summary(data_chapter13)
      ID              x1              x2              x3              x4
Min.   : 1.00   Min.   :0.000   Min.   :0.000   Min.   :0.0000   Min.   :1.000
1st Qu.: 8.75   1st Qu.:0.000   1st Qu.:1.000   1st Qu.:0.0000   1st Qu.:1.750
Median :16.50   Median :1.000   Median :2.000   Median :1.0000   Median :3.000
Mean   :16.50   Mean   :0.625   Mean   :2.031   Mean   :0.6875   Mean   :2.469
3rd Qu.:24.25   3rd Qu.:1.000   3rd Qu.:3.000   3rd Qu.:1.0000   3rd Qu.:3.000
Max.   :32.00   Max.   :1.000   Max.   :3.000   Max.   :1.0000   Max.   :4.000
      x5              x6              x7              x8
Min.   :0.000   Min.   :0.000   Min.   :1.000   Min.   :1.000
1st Qu.:2.000   1st Qu.:2.000   1st Qu.:1.000   1st Qu.:2.000
Median :3.000   Median :3.000   Median :3.000   Median :3.000
Mean   :2.781   Mean   :2.688   Mean   :2.219   Mean   :2.438
3rd Qu.:4.000   3rd Qu.:3.250   3rd Qu.:3.000   3rd Qu.:3.000
Max.   :4.000   Max.   :4.000   Max.   :3.000   Max.   :3.000
```

Among these nine variables, we only use variables *x1* through *x4* to perform a cluster analysis with four groups. The *K-means* command in R is undertaken for the analysis, and we can print output as in the following.

The *K*-means clustering with four clusters has the number of observations: 4, 7, 15, and 6 for each group. Cluster means and vector are also produced by the *K-means* command. Type the command *set.seed(123)* prior to the *K-means* command and run both the commands simultaneously so that the cluster sizes do not vary every time we run the *K-means* command.

518 SECTION | III Special topics on poverty, nutrition, and food policy analysis

```
> kmeans(data_chapter13[,2:5],4)
K-means clustering with 4 clusters of sizes 4, 7, 15, 6

Cluster means:
        x1       x2    x3    x4
1 0.0000000 0.000000 0.25 4.000000
2 0.0000000 1.285714 0.00 1.000000
3 1.0000000 2.600000 1.00 3.000000
4 0.8333333 2.833333 1.00 1.833333

Clustering vector:
 [1] 1 1 1 1 2 2 2 2 2 3 2 2 4 3 3 3 3 4 4 4 4 3 3 3 3 3 3 3
[30] 3 3 3

Within cluster sum of squares by cluster:
[1] 0.750000 1.428571 5.600000 2.500000
 (between_SS / total_SS =  87.7 %)

Available components:

[1] "cluster"     "centers"    "totss"    "withinss"
[5] "tot.withinss" "betweenss"  "size"     "iter"
[9] "ifault"
```

Now we plot the graph by clusters. By using the *plot* command and *cluster* argument to present different colors within groups. The plots on each variable are as follows:

```
> plot(data_chapter13[,2:5],col=cluster1$cluster)
```

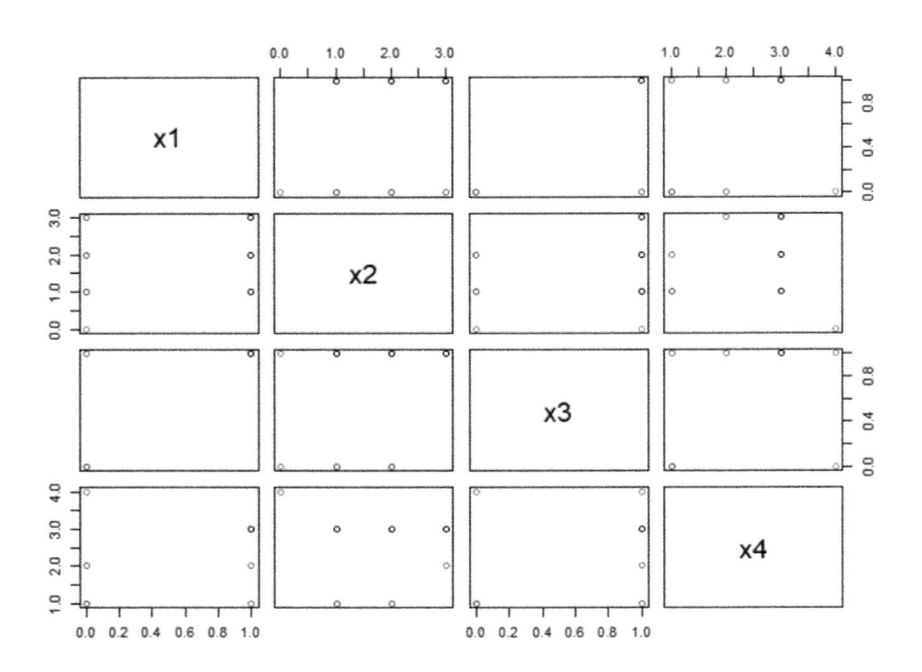

The aforementioned graphs present for clusters between two variables. Since the data sets consist of categorical and dummy variables, we cannot see the grouped cluster. However, if the data set has a continuous variable, it would be grouped plot graphs.

Exercises

1. What is the purpose of cluster analysis and when is it appropriate to use instead of factor analysis? Under what condition is hierarchical or nonhierarchical cluster analysis appropriate?
2. What is meant by agglomerative and divisive method in cluster analysis?

STATA workout−1

The following table presents data on 17 of the poorer districts in the state.

District	Poverty	Foodsec	Drinkdt	Hltdt
1	0.32	2.813	2.813	2.809
2	−0.15	2.831	2.830	2.826
3	0.30	2.814	2.810	2.808
4	0.26	2.816	2.811	2.807
5	0.28	2.789	2.790	2.783
6	0.34	2.811	2.810	2.811
7	0.61	2.810	2.810	2.809
8	−0.05	2.806	2.806	2.803
9	−0.05	2.807	2.799	2.799
10	0.58	2.797	2.779	2.782
11	0.52	2.802	2.801	2.798
12	0.38	2.804	2.804	2.797
13	0.08	2.805	2.798	2.798
14	−0.15	2.835	2.834	2.829
15	0.59	2.812	2.811	2.808
16	0.08	2.826	2.823	2.819
17	0.04	2.820	2.819	2.814

The table provides information on indices of poverty, food security, distance to drinking water facilities, and distance to a health facility. A relatively poorer districts have a smaller poverty index value. Higher values of foodsec, drinkdt, and hltdt indicate greater food insecurity and lower access to public goods.

- Generate a three-group cluster in STATA using the **K-means** command, with a starting seed value = 123456.
- Generate another three-group cluster, with a starting seed value = 654321.
- Do starting values matter?

- Compare the first result with a three-group cluster using the **wards** command.

The input in STATA for the ***K*-means** procedure with starting seed value 123456 is as follows:

- cluster *K*-means poverty foodsec drinkdt hltdt, k(3) name(cluster3a) s(kr(123456))
- cluster list cluster3a
- table cluster3a
- tabstat poverty foodsec drinkdt hltdt, by(cluster3) stat(n mean min max)

The relevant STATA output under the corresponding command is as follows:

```
. cluster kmeans poverty foodsec drinkdt hltdt, k(3) name(cluster3a) s(kr(123456))

. cluster list cluster3a
cluster3a  (type: partition,  method: kmeans,  dissimilarity: L2)
     vars: cluster3a (group variable)
    other: cmd: cluster kmeans poverty foodsec drinkdt hltdt, k(3) name(cluster3a) s(kr(123456))
           varlist: poverty foodsec drinkdt hltdt
           k: 3
           start: krandom(123456)
           range: 0 .

. table cluster3a
```

cluster3a	Freq.
1	2
2	5
3	10

As mentioned in the previous sections, *K*-means clustering is a well-established nonhierarchical clustering technique. Under *K*-means, the number of clusters is provided *apriori*, and the program searches for the best solution with that specified number of clusters. The procedure computes the means of each cluster. Each observation's centroid is placed in a relevant cluster, given its position with respect to the mean.

A starting point is needed for the **cluster** command, and **prandom** option allows the program to pick a random seed value. However, it is better to provide a random starting value, which helps researchers recheck and recalibrate the results. The **table cluster3a** produces a frequency table that indicates the number of districts, or observations in each group. The **tabstat** command organizes the groups by key features:

```
. tabstat poverty foodsec drinkdt hltdt, by(cluster3a) stat(n mean min max)

Summary statistics: N, mean, min, max
  by categories of: cluster3a

  cluster3a |   poverty    foodsec    drinkdt      hltdt
  ----------+--------------------------------------------
          1 |         2          2          2          2
            |      -.15      2.833      2.832     2.8275
            |      -.15      2.831       2.83      2.826
            |      -.15      2.835      2.834      2.829
  ----------+--------------------------------------------
          2 |         5          5          5          5
            |       .02     2.8128      2.809     2.8066
            |      -.05      2.805      2.798      2.798
            |       .08      2.826      2.823      2.819
  ----------+--------------------------------------------
          3 |        10         10         10         10
            |      .418     2.8068     2.8039     2.8012
            |       .26      2.789      2.779      2.782
            |       .61      2.816      2.813      2.811
  ----------+--------------------------------------------
      Total |        17         17         17         17
            |  .2341176   2.811647   2.808706   2.805882
            |      -.15      2.789      2.779      2.782
            |       .61      2.835      2.834      2.829
```

The first two group with two districts represent the poorest sectors. The mean values in the foodsec, drinkdt, and hltdt are above the centers of groups 2 and 3. To check if starting values make a difference, the following input lines are used:

- cluster *K*-means poverty foodsec drinkdt hltdt, k(3) name(cluster3b) s(kr(654321))
- cluster list cluster3b
- table cluster3b
- tabstat poverty foodsec drinkdt hltdt, by(cluster3b) stat(n mean min max)

The corresponding output from STATA is given in the following. It is possible that clusters with different starting values are different:

```
. cluster kmeans poverty foodsec drinkdt hltdt, k(3) name(cluster3b) s(kr(654321))

. cluster list cluster3b
cluster3b (type: partition, method: kmeans, dissimilarity: L2)
      vars: cluster3b (group variable)
      other: cmd: cluster kmeans poverty foodsec drinkdt hltdt, k(3) name(cluster3b) s(kr(654321))
             varlist: poverty foodsec drinkdt hltdt
             k: 3
             start: krandom(654321)
             range: 0 .

. table cluster3b

  cluster3b |     Freq.
  ----------+----------
          1 |         6
          2 |         4
          3 |         7
```

The three groups now have different number of observations as shown in the frequency table. The **tabstat** command provides a clearer picture:

```
. tabstat poverty foodsec drinkdt hltdt, by(cluster3b) stat(n mean min max)

Summary statistics: N, mean, min, max
  by categories of: cluster3b

cluster3b |   poverty   foodsec   drinkdt     hltdt
----------+--------------------------------------------
        1 |         6         6         6         6
          |  .3133333  2.807833  2.806333    2.8025
          |       .26     2.789      2.79     2.783
          |       .38     2.816     2.813     2.811
----------+--------------------------------------------
        2 |         4         4         4         4
          |      .575   2.80525   2.80025   2.79925
          |       .52     2.797     2.779     2.782
          |       .61     2.812     2.811     2.809
----------+--------------------------------------------
        3 |         7         7         7         7
          | -.0285714  2.818571  2.815571  2.812571
          |      -.15     2.805     2.798     2.798
          |       .08     2.835     2.834     2.829
----------+--------------------------------------------
    Total |        17        17        17        17
          | .2341176  2.811647  2.808706  2.805882
          |      -.15     2.789     2.779     2.782
          |       .61     2.835     2.834     2.829
----------+--------------------------------------------
```

Certainly, there is a difference in the way the groups are organized, with group 3 being the most poor, but with 7 observations. The next command, **table cluster3a cluster3b, col** presents a cross-tab between the two results, indicating a comparison between the two outputs:

```
. table cluster3a cluster3b, col

          |       cluster3b
cluster3a |   1     2     3    Total
----------+-------------------------
        1 |                2     2
        2 |                5     5
        3 |   6     4           10
```

The same strategy is carried out using the **wards** command and the relevant **tabstat** results are as follows:

- cluster ward poverty foodsec drinkdt hltdt, name(clward)
- cluster generate ward3 = groups(3)
- sort ward3

```
. tabstat poverty foodsec drinkdt hltdt, by(ward3) stat(n mean min max)

Summary statistics: N, mean, min, max
  by categories of: ward3
```

ward3	poverty	foodsec	drinkdt	hltdt
1	6 .3133333 .26 .38	6 2.807833 2.789 2.816	6 2.806333 2.79 2.813	6 2.8025 2.783 2.811
2	4 .575 .52 .61	4 2.80525 2.797 2.812	4 2.80025 2.779 2.811	4 2.79925 2.782 2.809
3	7 -.0285714 -.15 .08	7 2.818571 2.805 2.835	7 2.815571 2.798 2.834	7 2.812571 2.798 2.829
Total	17 .2341176 -.15 .61	17 2.811647 2.789 2.835	17 2.808706 2.779 2.834	17 2.805882 2.782 2.829

The **wards** command is also very similar to the previous results with *k*-**means** cluster. The same number of districts are identified in both procedures. The cross-tab also verifies this as follows:

```
. table cluster3a ward3
```

cluster3a	ward3 1	2	3
1			2
2			5
3	6	4	

STATA workout–2

The table has information on poverty status on 32 families, along with key indicators. Is it possible to identify relevant clusters using ***K*-means**, for clusters with groups 3 and 4? The indicators are as follows:

Poverty − Poverty Status Index (0 for self-reported as, "not poor", 1 self-reported as "poor")

Foodsec − Food Insecurity Index (higher value of the index reflects greater insecurity)

Child − Number of children below age 5

DEP − Number of dependents

drinkdt — Distance to drinking water (higher values represent lesser access)

Hltdt —Distance to a health facility (higher values represent lesser access)

Mktdt — Distance to a market (higher values represent lesser access

HH	Poverty	Foodsec	School	Child	dep	Drinkdt	Hltdt	mktdt
1	0	0	0	4	3	3	3	3
2	0	0	1	4	3	3	3	3
3	0	0	0	4	3	3	3	3
4	0	0	0	4	3	3	3	3
5	0	1	0	1	0	0	3	3
6	0	1	0	1	1	1	3	3
7	0	1	0	1	1	1	3	3
8	0	1	0	1	2	1	3	3
9	0	1	0	1	2	1	3	3
10	1	1	1	3	1	2	3	3
11	0	2	0	1	2	2	3	3
12	0	2	0	1	3	2	3	3
13	1	2	1	1	4	4	1	1
14	1	2	1	3	1	2	3	3
15	1	2	1	3	2	2	3	3
16	1	2	1	3	2	3	3	3
17	1	2	1	3	3	3	3	3
18	0	3	1	2	3	3	3	3
19	1	3	1	2	3	3	1	2
20	1	3	1	2	3	3	2	2
21	1	3	1	2	4	4	1	1
22	1	3	1	2	4	4	1	1
23	1	3	1	3	3	2	1	1
24	1	3	1	3	3	2	1	1
25	1	3	1	3	3	3	1	2
26	1	3	1	3	3	3	1	2
27	1	3	1	3	4	3	1	2
28	1	3	1	3	4	4	2	3
29	1	3	1	3	4	4	2	3
30	1	3	1	3	4	4	3	2
31	1	3	1	3	4	4	1	2
32	1	3	1	3	4	4	1	2

First, the ***K*-means** procedure is used to perform a cluster for 3 groups:

```
. cluster kmeans poverty foodsec school child dep drinkdt hltdt mktdt, k(3) name(cHH) s(kr(123456))
. cluster list cHH
cHH (type: partition, method: kmeans, dissimilarity: L2)
    vars: cHH (group variable)
   other: cmd: cluster kmeans poverty foodsec school child dep drinkdt hltdt mktdt, k(3) name(cHH) s(kr(123456))
          varlist: poverty foodsec school child dep drinkdt hltdt mktdt
          k: 3
          start: krandom(123456)
          range: 0 .

. table cHH
```

cHH	Freq.
1	8
2	8
3	16

The frequency table indicates that 16 families in group 3, with 8 families in the first two group. The **tabstat** command produces the necessary classification:

```
. tabstat poverty foodsec school child dep drinkdt hltdt mktdt, by(cHH) stat(n mean min max)
Summary statistics: N, mean, min, max
  by categories of: cHH
```

cHH	poverty	foodsec	school	child	dep	drinkdt	hltdt	mktdt
1	8	8	8	8	8	8	8	8
	.875	3	1	2.625	3.125	2.75	1.375	1.875
	0	3	1	2	3	2	1	1
	1	3	1	3	4	3	3	3
2	8	8	8	8	8	8	8	8
	1	2.875	1	2.5	4	4	1.5	1.875
	1	2	1	1	4	4	1	1
	1	3	1	3	4	4	3	3
3	16	16	16	16	16	16	16	16
	.3125	1.125	.375	2.375	2	2	3	3
	0	0	0	1	0	0	3	3
	1	2	1	4	3	3	3	3
Total	32	32	32	32	32	32	32	32
	.625	2.03125	.6875	2.46875	2.78125	2.6875	2.21875	2.4375
	0	0	0	1	0	0	1	1
	1	3	1	4	4	4	3	3

The aforementioned classification shows that group 3 is relatively well-off, given that the mean poverty index is closer the lowest. However, groups 2 and 3 are relatively worse off. Both groups 2 and 3 have higher food insecurity and lower access to drinking water.

For comparison, a second clustering with 4 groups is provided in the following. The **tabstat** 4 groups are more specific with the grouping, as they narrow the field of relatively well-off families. Furthermore, the access to drinking water, health, and marketing is sharply different with this cluster group.

```
. cluster list cHH
cHH  (type: partition,  method: kmeans,  dissimilarity: L2)
     vars: cHH (group variable)
     other: cmd: cluster kmeans poverty foodsec school child dep drinkdt hltdt mktdt, k(4) name(cHH) s(kr(123456))
            varlist: poverty foodsec school child dep drinkdt hltdt mktdt
            k: 4
            start: krandom(123456)
            range: 0 .

. table cHH
```

cHH	Freq.
1	8
2	8
3	9
4	7

```
. tabstat poverty foodsec school child dep drinkdt hltdt mktdt, by(cHH) stat(n mean min max)

Summary statistics: N, mean, min, max
  by categories of: cHH
```

cHH	poverty	foodsec	school	child	dep	drinkdt	hltdt	mktdt
1	8	8	8	8	8	8	8	8
	.875	3	1	2.625	3.125	2.75	1.375	1.875
	0	3	1	2	3	2	1	1
	1	3	1	3	4	3	3	3
2	8	8	8	8	8	8	8	8
	1	2.875	1	2.5	4	4	1.5	1.875
	1	2	1	1	4	4	1	1
	1	3	1	3	4	4	3	3
3	9	9	9	9	9	9	9	9
	.5555556	1	.6666667	3.444444	2.333333	2.666667	3	3
	0	0	0	3	1	2	3	3
	1	2	1	4	3	3	3	3
4	7	7	7	7	7	7	7	7
	0	1.285714	0	1	1.571429	1.142857	3	3
	0	1	0	1	0	0	3	3
	0	2	0	1	3	2	3	3
Total	32	32	32	32	32	32	32	32
	.625	2.03125	.6875	2.46875	2.78125	2.6875	2.21875	2.4375
	0	0	0	1	0	0	1	1
	1	3	1	4	4	4	3	3

Chapter 14

Household care as a determinant of nutritional status—application of instrumental variable estimation

Chapter outline

Introduction	527	Stage 2: estimating the determinants of child health (weight-for-age Z-Scores)	541
Review of selected studies	529		
Federal nutrition programs and children's health in United States	534	IV estimation using STATA	543
Parental unemployment and children's health in Germany	536	**Conclusions**	**545**
		Instrumental variable estimation using R	546
Food security using the Gallup World Poll	538	**Exercises**	**547**
Empirical analysis	**539**	STATA workout 1	548
Stage 1: estimating child-care practices	540	STATA workout 2	552

You cannot achieve environmental security and human development without addressing the basic issues of health and nutrition

Gro Harlem Brundtland.

Introduction

Using case studies from the child nutrition literature, this chapter introduces the instrumental variable (IV) model. The analysis is shown in two steps. Estimation of the determinants of child-care practices is carried out in the first stage, while, in the second stage, the predicted value of child-care practices along with other control varieties is used to determine the impact of child nutrition outcomes. We illustrate the usefulness of the IV method for the federally funded nutrition programs in the United States. We relate these issues to children's health in Germany and to the one child—only policy in China.

Provision of care to mothers and children has profound implication for health and nutrition outcomes (Rasaily et al., 2020). One of the strongest and most consistent findings in the health economics literature is the positive relationship between maternal schooling and child health. This empirical relationship has been confirmed across different time periods, countries, and measures of child health (Behrman and Deolalikar, 1988; Strauss and Thomas, 1995). Furthermore, the international literature clearly shows that inadequacy of early care and feeding practices and the presence of infections are major determinants of undernutrition after birth (Haddad et al., 1996; Ruel, 2001). Thus, adequate child-care practices, timely availability of food, and well-placed environmental and health conditions could reverse undernutrition.

For the past 20 years or so, child health and nutrition interventions have benefited from the advances of the new approaches in dealing with the prevalence of infectious diseases and the newly identified role of micronutrients in the nutritional status and in conceptualizing child-care practices (Engle et al., 1999). In the fourth report on the World Nutrition Situation (UN, ACC/SCN, 2000), the frequent nutritional problems in developing countries were identified by the following factors: fetal undernutrition, stunting, wasting, and underweight in children less than 5 years of age. This report also addressed micronutrient deficiency and showed that, worldwide, approximately 11.7 million newborns have weights below 2500 g, which reflects fetal undernutrition. Additionally, micronutrient deficiencies coexist with poor nutritional status. This affects a large number of people in developing countries. Around 250 million children under the age of 5 are clinically deficient in vitamin A and almost 740 million people are iodine deficient.

The issue is clearly important since there is an increasing recognition that feeding practices (such as breastfeeding and complementary feeding practices) can serve as examples of care practices that are essential for improving child nutrition (Engle et al., 2000). First, there is a general consensus that increasing income alone is not sufficient for improving children's nutritional outcomes. Development projects that increase men's income relative to women have demonstrated that its impact on child nutritional outcomes is small and often negligible (Kennedy and Garcia, 1993). Second, several studies (Christian et al., 1988; Ruel et al., 1992) have convincingly established that behavioral factors, such as a mother's ability to plan and organize her time, can have a significant impact on child nutritional status. In other words, if the time constraints on the mother are reduced, there can be significant improvement in child health outcomes. Finally, international agencies, such as the UNICEF, have taken the lead in advocating the role of child-care for child nutrition.

During the 1990s, the UNICEF (1990) proposed a conceptual framework that emphasized that care for women and children was equally important for child survival as are food security and healthcare services. The proposed framework argued that food, health, and care were all necessary for child health outcomes, but none of them alone was sufficient for healthy growth and development. For example, breastfeeding is a practice that provides food, health, and care simultaneously. While care was not defined till 1990, its importance was obvious. The initial definition consisted of actions of caregivers that converted food and health services to positive health outcomes for the child. While originally the outcome measure of interest was child survival, later outcomes of growth and development were also included.

In this chapter, we use an IV estimation technique since child-care practices are essentially endogenous. In other words, child-care practices are determined by a set of independent variables and the IVs. This approach is appropriate since child-care practices need to be instrumented out by a set of exogenous variables that are correlated with child-care practices but are not correlated with the child outcome variable, namely weight for age. Then, in the second stage, we regress the outcome variable (namely weight for age) on the predicted child-care variable along with other control variables. Ordinary least squares (OLS) estimation of the regression with child nutritional status as an outcome variable and food, health, and care, as the proximate variables could be biased for two reasons. First, there may be unobserved variables that are part of the error term but are correlated with the variables included on the right side. Second, explanatory variables may exist that are endogenous or jointly determined with the outcome variable and hence are correlated with the error term. The approach to address the aforementioned problems is to use the IV approach. The credibility of this approach will rest on the ability to find variables that are correlated with the suspected endogenous explanatory variables but are not correlated with the outcome variable.

This chapter is organized as follows: in the next section, we briefly discuss the literature of how child-care and maternal nutritional knowledge are critically important in determining children nutritional outcomes such as weight for age. The third section presents the empirical model and discusses the results, while the final section makes some concluding remarks and research agenda on child-care practices on child nutritional outcomes.

Review of selected studies

Since the conceptual and measurement issues on child-care practices have been illustrated in Chapter 7, we will not reiterate them here. We now review some of the case studies that emphasize the role of child-care and maternal nutritional knowledge (in addition to maternal schooling) on child health outcomes.

Garrett and Ruel (1999) explored in depth the question of whether the factors that determine food and nutrition security are different between the rural and urban areas and the implications of these differences for the design and operation of food and nutrition programs. The study answered these questions using a data set from a 1996–97 national household survey of living conditions in Mozambique. The information was collected at the level of the province and capital, Maputo, for 8274 households. Anthropometric measures of height, along with age in months, were also collected for children below 5 years.

Following the standard household utility maximization model, a demand function for calories and a production function for child nutritional status were specified. The demand function for calories was a function of prices, income (instrumented out by assets), and other exogenous factors including demographic characteristics. The nutrition function was determined by caregiving behaviors, health status, and household characteristics and calorie availability at the household level. Since income was endogenous, using least squares estimates would produce biased and inconsistent estimates. Thus, IV estimation was used to eliminate the correlation between the explanatory variable and the error term. A two-stage least squares (2SLS) approach was used to control for this endogeneity with an index of household assets as the identifying instrument, in combination with other exogenous variables in the first stage equation.

The study demonstrated that income was an essential determinant of calorie availability and child nutritional outcomes in both rural and urban areas. Thus, investment in education, increasing agricultural productivity, and investment in rural infrastructure were the key determinants of poverty reduction strategies. Women's education was also important in improving child nutritional status. In the long run, improving girls' formal education and women's literacy and job skills would raise household incomes. In Mozambique, not only did maternal education have a positive effect on young children's nutritional status above and beyond the income effect, but it enhanced the positive effect for the young children (less than 2 years of age). It is likely that maternal education affected child nutrition through its effect on greater nutrition knowledge and improved caregiving practices.

Overall, the results indicated that the determinants of food security and nutritional status were not very different between the rural and urban areas. However, policy makers and program administrators should not simply transfer programs from the rural to the urban areas, since community-level specific conditions need to be identified for a program to be successful.

In a comprehensive study, Glewwe (1999) explored the mechanisms through which mothers' education raised child health. The study analyzed three channels through which mothers' education influences child health:

1. Direct acquisition of basic health knowledge in school.
2. Literacy and numeracy skills learned in school could enhance mothers' abilities to treat child illnesses and could also help mothers increase their stock of knowledge after leaving school.

3. Exposure to modern society via schooling could change women's attitudes toward traditional methods of raising children and treating their health problems.

The study used data from the 1990—91 Moroccan LSMS household surveys conducted by the World Bank to assess the relative importance of these three mechanisms through which mothers' education affects child health. An important aspect of the data related to information from mothers regarding their amount of knowledge about health conditions. The tests for health knowledge included were

1. 5 questions on health knowledge,
2. 12 questions on general knowledge,
3. an oral mathematics test of 10 questions,
4. a set of written mathematics tests with varying degrees of difficulty,
5. a set of Arabic reading and writing tests, and
6. a set of French reading and writing tests.

A sample of 2171 households between the ages of 9 and 69 were surveyed, with the final sample of children being 1495.

The main conclusions of the study can be summarized as follows:

1. Health knowledge was the most important skill through which mothers are better prepared to improve children's health.
2. Schooling affected mothers' health knowledge in Morocco indirectly—health knowledge was learned using literacy and numeracy skills acquired in school.

The aforementioned conclusions have direct policy implications for Morocco. First, health knowledge should be directly taught in the schools. They should be taught at an early age since girls dropping out early can never acquire sufficient numeracy and literacy skills that will help them in acquiring health knowledge. Second, school quality should not be neglected since women will be unable to raise their level of health knowledge if they leave school without basic literacy and numeracy skills.

Block (2003)[1] addressed the following set of questions:

1. How does nutrition knowledge affect household budget allocation between food and nonfood expenditures?
2. Within the food budget, does nutrition knowledge affect the allocation of spending on micronutrient rich foods[2] versus staples?

---

1. While this study does not examine the impact of nutritional knowledge on anthropometric outcomes, it is important as it studies how nutritional knowledge affects the demand for micronutrient-rich foods.
2. "Micronutrient-rich foods" refers to a composite commodity constructed from the household survey data. This composite commodity consists of beef, fish, chicken, vegetables, fruits, milk, and eggs.

3. How do key demand parameters differ as a function of maternal nutrition knowledge?

The motivation for this study lies in the fact that nutrition knowledge might operate in increasing the demand for micronutrient-rich foods. The critical demand parameters included were budget shares, as well as income and own price elasticities of demand for micronutrient-rich foods. The data were obtained from a detailed survey by the Helen Keller International (an NGO that undertakes social marketing campaigns) and covered the entire province of Central Java. The survey began in December 1995 and involved regular collection of information on dietary diversity, expenditures, asset ownership, demographics, and nutritional status. For each round, a random sample of 7200 households was chosen, and each time a total of 30 villages was selected from each of the province's six agroecological zones.

The study demonstrated that the households' inclination to reduce expenditures on high-quality foods was a function of nutritional knowledge of the mother. The estimated cross-price elasticity between micronutrients and eggs was substantial. It was negative for households lacking nutrition knowledge and was zero for households with nutrition knowledge. Thus, maternal nutrition knowledge emerged as the most important factor for coping with the consequences of macroeconomic crises.

Blunch (2005) examined the impact of maternal literacy and numeracy skills on the production of children's health in Ghana. Ghana is an ideal candidate for investigating these issues, as its education system is one of the most developed in sub-Saharan Africa. Second, an important priority of the Government of Ghana has been to provide basic literacy and numeracy skills through adult literacy programs for individuals who never attended school. Multiple paths to literacy and numeracy skills can thus be studied. The analysis considered both child health inputs and outputs and examined the determinants of vaccinations, postnatal care, and mortality. The Ghana Living Standards Survey (GLSS) was a nationally representative household survey, which was conducted over four cross sections. These surveys were conducted in 1987/88, 1988/89, 1991/92, and 1998/99. The most recent round was used for analysis in the current study. The surveys contained information on educational attainment, participation in adult literacy courses, literacy and numeracy skills, and information on background variables such as age, gender, ethnicity, etc., which were important determinants of human capital formation. The community questionnaire contained information on access to facilities, including schools and adult literacy programs.

IV estimation was undertaken to account for the potential endogeneity of skills, schooling, and adult literacy participation rates. For comparison purposes, least squares estimates were also presented, where skills and schooling were taken as given. The models were estimated for the full sample and for three different subsamples, namely rural and urban areas and mothers who did not complete any formal schooling.

The study found a positive impact of maternal formal schooling on child health input demand and child mortality, which was consistent with the previous literature. Additionally, the study found a substantial impact from adult literacy course participation and some impact from literacy and numeracy skills. The estimated impacts from maternal adult literacy program participation were often substantial compared with the estimated impacts from formal maternal education.

The implication of the findings was the potentially important role of adult literacy programs in promoting child health, possibly through the inclusion of health topics in the curriculum. One reason for the differential impact of formal and nonformal education is that adult literacy programs could effectively provide knowledge by introducing topics to mothers such as immunization, safe motherhood and child care, and safe drinking water. Thus, promoting more adult literacy courses could help in improving child health conditions in the future.

From the aforementioned studies, it is reasonable to conclude that maternal nutrition knowledge (working through both formal and nonformal education) and child-care practices (such as breastfeeding and complementary feeding practices) could effectively improve child health outcomes in the long run. Since both maternal nutrition knowledge and child-care practices are essentially endogenous, running least squares estimates on the health outcome variable would produce biased and inconsistent estimates. Thus, it is appropriate to control for endogeneity by undertaking IV estimation.

Several studies in recent years have identified the crucial role of maternal literacy on child nutritional status. Basnet et al. (2020) apply the Alive & Thrive database from Bangladesh, Vietnam, and Ethiopia and relate maternal education to overall household care and childhood nutritional status. Maternal education, along with maternal height, weight, mental wellness, autonomy with decision making, employment, and social support, are all complementary factors that influence breastfeeding, dietary diversity, hygiene, and preventive immunization.

Similarly, Miller et al. (2020) and Cunningham et al. (2017) also undertake a detailed study for Nepali children and demonstrate that home environmental quality, measured via household wealth and maternal education, strongly influences consumption of eggs and dairy products, and incidentally childhood health positively. Headey and Martin (2016), Headey et al. (2020), Nguyen et al. (2016, 2018), and Young et al. (2018) also examine the same issue for Bangladesh, India, Nepal, and Pakistan and conclude that improvements in maternal education and sanitation facilities are key drivers of positive of height-for-age Z-scores.

As noted in Chapter 7, and in later sections, substantive evidence spanning Asia, Africa, and Latin America link maternal education and home environment factors to childhood health status. Favara (2018) for Peru, Dasgupta et al. (2017) and Hoddinott et al. (2018) for Bangladesh, Jansen et al. (2015) for Columbia, Ervin and Bubak (2019) for Paraguay, and Masters et al. (2018), for Ethiopia,

Nigeria, and India, all examine different data and circumstances and arrive at the following consensus: maternal education, maternal group participation, female community organizations, alongside rights, and empowerment, coupled with access to health, water, and related public health infrastructure, all influence the home care environment and positively influence dietary diversity and foster decent height-for-age Z-scores.

Federal nutrition programs and children's health in United States

In the United States, there has been a rapid increase of national and federal healthcare spending, and consequently, the effectiveness of many public health programs is under close scrutiny. The largest federal nutrition program in the United States is the Special Supplemental Nutrition Program for Women, Infants, and Children (WIC). The program directly benefits pregnant, postpartum, and breastfeeding women, as well as infants and children up to 5 years of age who have low income and are nutritionally at risk. Overall, the program is cited as a success, because the General Accounting Office reports that for every one dollar spent on WIC, the government saves $3.5 in medical care for premature birth and low-birth-weight birth. Many researchers have attempted to quantify and verify these claims.

One of the main problems from these empirical studies seems to be riddled with selection bias. In the context of this chapter, Gai and Feng (2012) have controlled for selection bias using IVs and latent variable estimation. WIC participants are not randomly selected into the program, and hence, women in the WIC program may be very different from other eligible women who choose not to participate. Consequently, choice of participation becomes endogenous and has to be accounted for in the estimation. Women in the program may have some advantages in terms of social network, family support, motivation, and awareness. Consequently, these women are more than likely to succeed in the program, given their characteristics. A usual strategy to control for selection bias is to use IV method.

Gai and Feng (2012) use several socioeconomic and health behavior information of the mothers as IVs. Even after controlling for selection bias and endogeneity, Gai and Feng (2012) find that WIC has a strong effect for most disadvantaged women and infants. Gai and Feng (2012) conclude that it is important to quantify the average WIC's effect and, more importantly, to identify areas and subgroups that can benefit most from the program.

More resources within WIC should be directed toward women and infants who are most likely to benefit from the program. For instance, Gai and Feng (2012) estimation results indicate that participants at the low end of the birth weight distribution appear to benefit more from the WIC program. Minority women and infants are more likely to be in this category because of their limited access to social service and public health resources.

Several studies note the value of the WIC program. For instance, Kreider et al. (2016) note how WIC reduces child food insecurity by at least 3.6%. The WIC program has also provided more access to those in food deserts (Wu et al., 2017). Most importantly Fang et al. (2019) note how WIC participation positively influences the choice and quality of food purchases among the participants.

A lot of evidence is now available to support the positive effects of prenatal WIC participation on better childhood outcomes, showing lower incidence of ADHD and academic performance as in Chorniy et al. (2018).

On the other hand, Bullinger and Gurley-Calvez (2016) note how WIC participation reduces breastfeeding by 50% and work leave duration by 20%. Relatedly, Topolyan and Xu (2017) note the differential effect of WIC participation by infants versus mothers on breastfeeding decisions: while no significant relationship exists between mother's participation and reduced breastfeeding, whereas a significant relationship exists between infant's participation and a reduction in postnatal breastfeeding.

The aforementioned differential impact calls into question, the efficacy of WIC policy that requires infants, but not mothers to be enrolled for free formula. Cakir et al. (2018) and Bronchetti et al. (2018) identify another differential of the WIC program, which is brought about by inflation across different regions: participants residing in higher-cost areas buy fewer fruits and vegetables (than similar participants in lower-cost areas), leading to lower use of preventive healthcare and more missed-school days among children. Likewise, Robinson (2016) notes similar restrictions on the program's health benefits coverage to siblings.

Haeck and Lefebvre (2016) note that the Canadian "oeuf-lait-orange" (eggs−milk−oranges) prenatal program significantly increased birth weights by 70 g reduced the probability of low birth weights. Interestingly, the prenatal nutrition program produces better results in Canada than the SNAP and the WIC programs in the United States, in terms of cost, simplicity, and outcomes. Relatedly, WIC's removal of 100% juices and whole milk from food packages, in 2009, led to reductions in the intakes of both desirable and undesirable nutrients from milk.

Further limitations within WIC are generated due to certain provisions and requirements under the Affordable Care Act as noted by Chatterji et al. (2019), Robinson (2018), Liu (2018), and Leguizamon and Leguizamon (2018):

- Reductions in the likelihood of being married, and cohabitating
- Increases in the likelihood of being single
- Reductions in the likelihood of being a single parent
- Reductions in young adult's participation in all nutrition assistance programs
- Increased likelihood of young adults living with their parents
- Reductions in the likelihoods of serving in the military

- Reduced probabilities of being privately insured
- Improvements in the overall health of male children, but not for female children

It is well known that infants of normal birth weight outperform infants of low birth weights. Infants with low birth weights have lower education achievements and lower productivity and earnings in their adult lives. Consequently, WIC should focus more on the most disadvantaged women and infants. Using that strategy, we can adapt toward healthcare cost savings in the long run.

Parental unemployment and children's health in Germany

An interesting stylized fact concerning the heights of children has occurred in East Germany. Between 1990 and 1995, there was a substantial height increase of school starters in East Germany. However, this upward trend in children's height suddenly stopped and even developed into a downturn between 1997 and 2000.

Baten and Bohm (2009) investigate the impact of unemployment rates in East Germany on the observed trends in children's height. The researchers use a large data set of over 250,000 children and collect information on all anthropometric measures.

The researchers are motivated by the simple idea that if both parents are unemployed, the quality of nutrition, medical resources, and family life could decline in many cases. Thus, frustration and psychological stress might pave the way for reduced care or other compensating behavior, resulting in parents allowing more unhealthy behavior than before.

Baten and Bohm (2009) demonstrate that unemployment mattered for the height of young children in Eastern Germany, 1994–2006. The result that parental unemployment renders children shorter is robust even after controlling for a number of other factors. Obviously, this result has substantial economic, political, and social implications. The researchers suggest that health-related goods should be made accessible to the unemployed, for free. This policy is particularly relevant for families with many children, since children's height is very sensitive to the number of siblings. The researchers also suggest the use of socioeconomic data to understand secular changes of height.

> **Section highlights: boys are better under a one child–only policy in China**
>
> The one-child policy in China has been implemented for three decades. Under this policy, most couples are allowed to have only one child, while ethnic minorities and many rural residents have been granted the freedom to have more than one child. Zhai and Gao (2010) note that since many families prefer to have

> **Section highlights: boys are better under a one child–only policy in China—cont'd**
>
> many children, especially boys, it is of interest to examine whether a child's gender and the number of siblings influence a child's enrollment in center-based care.
>
> Studies indicate that there are long-term positive effects of high-quality center-based care programs, such as improvement in school achievement, social skills, college attendance, health, and future earnings. Furthermore, center-based care is also associated with a reduction in grade retention, high-school dropout rates, teen pregnancies, delinquency, and crime. Zhai and Gao (2010) also note that center-based care is also in many cases substituted for parental and grandparental care in many rural areas. Consequently, in recent years, the demand for center-based care has increased.
>
> The economic problem arises because of potential conflicts between the one-child policy and the traditional preference to have many children, especially sons. Parents are now forced to optimally allocate the investment in their children, including child-care arrangements. If family resources are scarce, then it is possible that parental investments are placed on the eldest son even if he had younger brothers. In that case, without competition from siblings, children without siblings, including only girls, may be more likely to receive better child care and education compared with those with siblings. It is also possible that the one-child policy might lead to discrimination against girls, especially in rural areas.
>
> Zhai and Gao (2010) examine the effects of child gender and siblings on center-based care enrollment. They study the effects of child gender and siblings on center-based care enrollment in the context of one-child policy, in a context where there is a preference to have many children, especially sons. Specifically, Zhai and Gao (2010) test whether the chance of the focal child's enrollment in center-based care is significantly affected by the child's gender, the presence of siblings, the gender of the siblings, and other characteristics.
>
> They do not find any gender-based discrimination regarding the odds of receiving center-based care. However, they find that having one or more siblings, especially male siblings, siblings older than the focal children, or school-age siblings, tends to reduce focal children's odds of receiving center-based care. Moreover, children without siblings were more likely to receive center-based care than their peers who had siblings.
>
> Interestingly, there is no evidence that a child's gender by itself played a role in the center-based care enrollment. However, the presence of siblings and siblings' gender mattered. Zhai and Gao (2010) also show that children from low-income households had significantly lower odds of receiving center-based care.
>
> Most interestingly, the presence of male, older, or school-aged siblings, rather than female, younger, or preschool-aged siblings, significantly reduced children's odds of receiving center-based care. That is, gender still plays an important role in families' investment, and a school-age *male* sibling may spend a proportion of family resources that is large enough to significantly reduce his preschool-age sibling's chance to attend child-care centers.

Continued

> **Section highlights: boys are better under a one child–only policy in China—cont'd**
>
> The findings of Zhai and Gao (2010) provide important implications for policy on child-care and education in the context of the one-child policy in China. Policy makers should make efforts to improve the enrollment and quality of center-based care and to increase the equity of education for both boys and girls as well as for children with and without siblings.

Food security using the Gallup World Poll

The global food crisis of 2007−2008 involved approximately a doubling of international wheat and maize prices in the space of 2 years and a tripling of international rice prices in the space of just a few months. Estimates from various organizations indicate that about 1 billion people went hungry during this time period, with every indication that global food security situation had worsened in 2008.

Heady (2013) questions these conclusions, because the simulations that were used to make predictions had assumed rates of international price transmission to domestic markets rather than using observed price changes. Secondly, it is possible that wages might have adjusted to higher food prices. Thirdly, strong income or wage growth may have cushioned some of the inflationary effects. Finally, households could have used many coping and adjustment strategies to work through the tough time period.

To account for these possibilities, Heady (2013) undertakes an ex-post survey analysis collected before, during, and shortly after the 2008 food crisis across a large number of countries. Specifically, Heady (2013) uses the results from an indicator of self-assessed problems affording sufficient amounts of food, which was collected as part of the Gallup World Poll (GWP).

Surprisingly, Heady's Gallup data show that at the peak of the crisis, global food insecurity was either not higher or even substantially lower than it was before the crisis. However, these surprisingly optimistic global trends mask large regional variations.

Heady (2013) shows that global trends were clearly driven by declining food insecurity in India and several other large developing countries. Food insecurity increased in many African countries and most Latin American countries. It decreased somewhat in Eastern Europe and Central Asia, but it probably rose in the Middle East. Overall, Heady (2013) finds that strong real income growth largely offsets the adverse impacts of food inflation in many developing countries, including those with the largest poor populations.

The GWP question on food security: "Have there been times in the past 12 months when you did not have enough money to buy the food that you or your family needed?"

A simple yes or no answer is recorded and is referred to as the "food insecurity" indicator.

Heady (2013) notes that the indicator exhibits large variations across countries. Food insecurity is highest in sub-Saharan Africa, which is by far the poorest region in the world in monetary terms. Food insecurity in South Asia is higher than in East Asia, and the indicator is also high in Latin America. Low-income countries have food insecurity rates that are 17% points higher than middle-income countries, and the same difference is observed between middle- and upper-income countries.

Anthropometric indicators are also highly associated with nonfood factors, such as health, education, family planning, and cultural norms. Relatedly, Heady (2013) finds that GDP per capita, mean household income, poverty rates, hunger rates, and anthropometric indicators are significantly correlated with GWP.

In addition, the study finds no evidence that global food insecurity was higher in 2008 than it was in previous years, and the study does indicate that there are significant regional effects, particularly in Africa and Latin America. Thus, there are strong reasons for making a large push to improve the measurement of food security. Importantly, the global food crisis of 2007–2008 has revealed some significant failures in the global scene with regard to developing coping strategies and welfare policies in an acceptable timeframe.

The study finds that if strong economic growth had not been prevalent in substantial parts of the developing countries, prior to and during the food crisis, the impacts of higher food prices on poorer households would have far negative impacts. This observation should serve as a crucial warning for the years to come.

Recently, studies use the FAO Food Insecurity Experience Scale (FIES) module within the GWP, as in Smith et al. (2017), Grimaccia and Naccarato (2019), who demonstrate the significant influence, across the globe, of low levels of education, composition and number of children in the household, location of dwellings, weak social networks, less social capital, low household income, and being unemployed on food insecurity.

Empirical analysis

We undertake the analysis in two steps. In the first stage, we estimate the determinant of child-care practices by using the IV technique, while, in the second stage, the predicted value of child-care practices along with other controls (such as household and community characteristics) is included to determine the impact on child health outcomes (weight-for-age Z-scores).

Stage 1: estimating child-care practices

The first stage requires instruments to predict child-care practices. We include breastfeeding practices as an instrument of child-care practices. This is a dichotomous variable indicating whether the child is breastfed or not during his or her infanthood. The explanatory variable is how many times the child attends clinic (ATTCLINI). This is a continuous variable denoting how many times the mother took the child to a clinic during his or her sickness. The instruments that are used are as follows:

1. MKTGT5: a dichotomous variable assuming a value of 1 if the local market is at a distance of greater than 5 km and is zero otherwise.
2. AGEMNTH and AGESQ: the age of the child in months and its square. It can be expected that the younger the child (below 24 months), the greater is the need for breastfeeding.
3. FOODAVAI: a categorical variable assuming values from 1 to 3, with a value of 1 indicating that adequate food is available, 2 denoting food is not adequate, while 3 denotes when the respondent was not sure whether food was adequate or not.
4. SELPOINT: a categorical variable assuming values from 0 to 5, representing the distance of the household to a private trader's selling point. While this variable determines food security, it is exogenous in determining the child health outcome.
5. STAPLEFT: a dichotomous variable assuming two values 0 and 1. The variable measures whether any staple food was left for the household or not. A value of 1 denotes some staple food was left over, while a value of 0 denotes no staple food being left over. This variable can also be considered as exogenous in the determination of child health outcome.
6. Child0_5: a continuous variable denoting the number of children in the household. This variable indirectly measures birth spacing. If there are more children in the household, it signifies that the mother is time constrained and may not be able to provide good child-care practices.
7. LOCMKT: a categorical variable, assuming values from 1 to 5. This variable represents the distance of the household to the local market and is a measure of market access.
8. DIARRHEA: a dichotomous variable that indicates the household environment of sanitation conditions indirectly. A value of 1 denotes the presence of diarrhea, while a value of 0 indicates absence of diarrhea. This variable can negatively influence child-care practices since, with a poorer household/community environment, it is more likely that the mother will have less time to care for the child.

Thus, to obtain the two-stage least squares solution, we choose from the menus:

Analyze
Regression
Two-Stage Least Squares

- Dependent: BFEEDNEW
- Explanatory: ATTCLINI
- Instrumental: MKTGT5, AGEMNTH, AGESQ, FOODAVAI, SELPOINT, STAPLEFT, Child0_5, LOCMKT, DIARRHEA

Table 14.1 shows the two-stage least squares results.

From Table 14.1, we find that the number of times the mother took the child to a clinic was a significant predictor of child-care practices as measured by breastfeeding practices. Additionally, from the Hausman exogeneity test, which is distributed as $F(J, n - k)$ degrees of freedom, where J is the set of linear restrictions in this regression model (in this example is just 1), n is the number of observations, and k is the number of instruments in the model. The critical value of $F(1, 290)$ at the 1% level is 6.63. Since the computed value of $F = 41.88$ is much greater than the critical value, we can reject the null hypothesis of exogeneity and conclude that IV estimation should be used instead of OLS. Let us call this predicted value of child-care practices as FIT_1.

Stage 2: estimating the determinants of child health (weight-for-age Z-Scores)

In this stage, we run a reduced form demand function for child nutritional status based on weight for age as the dependent variables (see Eq. (10.9) of Chapter 10 for how the reduced form specification is undertaken). The coefficients are estimated using least squares, which results in the smallest sum of squares differences between the observed and the predicted values of the dependent variable. Table 14.2 gives the estimated coefficients. We compare these results with the OLS estimates from Chapter 10, so as to provide a comparative picture of the difference in these estimates.

Table 14.2 compares the OLS estimates with the IV estimates (question: why is the estimated coefficient of ATTCLINI not reported in the IV estimate?)

TABLE 14.1 Two-stage least squares regressions on child-care practices.

Variable	Coefficient	t-Stat	P value	Hausman endogeneity test (F-Test)	R^2
Constant	−0.193	−0.703	0.482	$F(1, 290) = 41.88$	0.126
ATTCLINI	1.348	6.472[a]	0.00		

Notes: Hausman (1978) endogeneity test; H_0, accept exogeneity, i.e., OLS should be used; H_1, reject exogeneity, i.e., IV should be used.
[a]Denotes statistically significant at the 1% level.

TABLE 14.2 Using predicted breastfeeding practices on weight-for-age Z-scores.

Variables	OLS	IV
Constant	−2.93[a](−5.95)	−2.65[a](−5.43)
EDUCSPOUS	0.156[a](2.70)	0.17[a](2.93)
ATTCLINI	0.439[a](3.10)	
DRINKDST	−0.147[a](−2.938)	−0.133[a](−2.63)
LATERINE	0.152(1.03)	0.141(0.944)
AGEMNTH	−0.08[a](−4.28)	−0.059[a](−3.27)
AGESQ	0.001[a](3.386)	0.001[b](2.60)
CLINFEED	0.699[a](3.91)	0.723[a](3.98)
DIARRHEA	−0.62[a](−3.126)	−0.682[a](−3.41)
BFEEDNEW or FIT_1	0.625[a](2.79)	0.364[c,a] (3.44)
HEALTDST	−0.06(−0.85)	−0.03(−0.527)
R^2	0.256	0.232
F	7.47	7.23

[a]Denotes at 1% level of significance.
[b]Denotes at 5% level of significance.
[c]Denotes the endogenous variable, the fitted value of breastfeeding. The terms in parentheses denote t-statistic.

of the individual, household, and community characteristics on child health outcomes as measured by weight-for-age Z-score. The estimated coefficients are of the correct sign, and the IV estimates are measured with more precision. However, the magnitude of the estimated impact significantly differs between the two specifications. For the IV estimates, the impact of maternal education on 1 standard deviation weight-for-age Z-scores increases from 0.156 to 0.17. This is possibly because maternal health endowments such as maternal age or maternal height are not controlled for. Similarly, the estimated impact of diarrhea increases in absolute magnitude. A one unit increase in the prevalence of diarrhea reduces the weight-for-age Z-score by 0.68 standard deviation points, whereas in the OLS estimate, the impact was 0.62 standard deviation points. One possible reason for this result can be that the prevalence of diarrhea affects child-care practices initially (in the first stage regression), and this reduction of child-care feeds back into lowering the levels of child health outcomes. Similarly, the estimated impact of feeding the child in clinic (CLINFEED) also increases the weight-for-age Z-score more in the IV estimate compared with the OLS—0.72 standard deviation units in the former versus 0.69 standard deviation units in the latter.

The most interesting result, however, is that the estimated impact of breastfeeding practices substantially differs in the two specifications. In the OLS specification, the estimated impact of breastfeeding practices was 0.625, whereas the predicted breastfeeding practices increase weight-for-age Z-scores by 0.36 standard deviation units. This substantial difference can be attributed to the fact that OLS was overestimating the impact of breastfeeding practices on child health outcomes. Once breastfeeding practices were instrumented out, by the relevant instrument variables, the effect is much more precisely estimated. Overall, one can infer that the IV estimates are more precisely estimated than the OLS specification.

From the aforementioned results, it is reasonable to conclude that both maternal education and child-care practices are important determinants of child health outcomes. This is consistent with the empirical literature as we have seen before. However, child-care practices are essentially endogenous, and one needs to instrument out child-care practices as done in this chapter. In addition, we can also infer that individual characteristics (such as age of the child) and community characteristics (such as distance to a protected water source and prevalence of diarrhea) are extremely important determinants of child health outcome.

IV estimation using STATA

In STATA, the IV regress procedure implements the IV regression using the 2SLS procedure, and Durbin–Wu–Hausman (DWH) test of endogeneity using the estat command. We illustrate this procedure from Gajanan et al. (2014) example seen in the last chapter on willingness to pay for reliable electricity.

Example: Recall from the last chapter that we have the following variables from 450 farmers randomly selected from two villages in Tamil Nadu:

Variable	Variable description
Y	Dependent variable that tracks the willingness to pay for a reliable package of electricity (1 = "yes"; 0 = "no")
EDUC	Educational status of the respondent (head of the household): is a dummy variable = 1 if the head of the household is a literate; 0, if not
Household	Household type (1 = single, 2 = nuclear, 3 = joint family)
Income	The net annual household income in rupees
Debt_level	Debt (1 = the household has debt; 0 if not)
Irrigation	The source of irrigation before getting an electric pump set connection (1 = tank irrigation; 2 = canal irrigation; 3 = nonelectrical methods)
Farm Size	Farm size (1 = marginal; 2 = small; 3 = medium; 4 = large)
Pumps	Number of electric pump set connections (1, 2; 3 = more than 2)

Continued

—cont'd	
Variable	Variable description
Pump Use_1	Use of electric pumps everyday in the last month (yes = 1; no = 0)
Pump Use_2	Number of times the pump set was switched off yesterday
Pump Age	Age of the first electric pump set (in years)
Crop Pattern	Factors influencing the cropping pattern (1 = availability of water; 2 = power tariff; 3 = market conditions; 4 = others)
Well Quality	Quality of the well water at present (1 = excellent; 2 = good; 3 = satisfactory; 4 = bad; 5 = very bad)
Water Level	Sufficient water available for irrigation (1 = yes; 0 = no)
Power cut	Was there power cut yesterday (1 = yes; 0 = no)
Burnout_1	Aware that voltage fluctuations cause motor burnouts (1 = yes; 0 = no)
Burnout_2	Motor burn out during the last year due to voltage fluctuation (1 = yes; 0 = no)
Reliability	Rate the reliability of power supply to agriculture (1 = very good; 2 = good; 3 = satisfactory; 4 = bad; 5 = very bad)
Power Need	Quality of power is indispensable to cheap irrigation (1 = yes; 0 = no)

Whitehead (2006) has shown that the respondent's perception of water quality and their WTP are endogenous. We can test this hypothesis using the data from Gajanan et al. (2014).

In the first stage, we instrument the quality of water, *Well Quality* on *Farm Size* and all the other exogenous variables. We then use the predicted value of *Well Quality* to produce the second-stage regression. We can implement all this in STATA with the following command:

.ivregress 2sls y (qwelpres = tarow1) $xlist5, vce(robust) first
estat endogenous

The ivregress command is used for the IV procedure, and the option first produces the first-stage regression results in the output. The outputs from both the procedures and the DWH statistic are produced in the following:

Note that *qwelpres* or *Well Quality* is the dependent variable in this case. It is interesting to note that Farm Size (or *tarow1*) is significant and positively related to well quality perception. In the second stage, we get:

Interestingly, we find that the WTP decreases with increases in well-water quality perception. This feature indicates that even if the perception of well-water quality is endogenous to farm size, it has a big effect on WTP. The endogeneity result is given in the following:

As the chi-square value indicates, the null hypothesis of no-endogeneity can be rejected. Thus, perception of well-water quality is endogenous and can be instrumented via Farm Size.

Conclusions

Numerous studies (Behrman and Deolalikar, 1988; Strauss and Thomas, 1995; Glewwe, 1999; Blunch, 2005) have demonstrated that both maternal education and nutrition knowledge generated mostly through nonformal education, such as adult literacy programs, can improve child nutritional outcomes through the mediating effect of improved child-care practices.

This chapter demonstrates the importance of maternal education and child-care practices using child care as an endogenous variable. This is done by pursuing an IV-based (2SLS) estimation strategy in the spirit of Garrett and Ruel (1999) and Blunch (2005). Our analysis demonstrates that child-care practices are an important determinant of child nutritional outcomes. However, the estimated impact of child-care practices on child nutritional outcomes substantially declines in magnitude in the IV estimate relative to the OLS specification. The result indicates that child-care practices are overestimated in the latter specification, and thus, one needs to control for endogeneity of child-care practices. Additionally, we found women's education was critical in improving children's nutritional status, which is consistent with previous findings. The IV estimates showed that the impact of maternal education on one standard deviation weight-for-age Z-scores was 0.17. Thus, in the long run, it is extremely important to improve women's formal education and women's literacy and job skills that can raise household income. Higher levels of women's education in the long run may also lead to reductions in fertility and lengthen the time between births, which will result in lower household sizes.

Furthermore, we found that younger children were more prone to malnutrition, and more attention should be directed to attenuate these conditions. Community characteristics, such as distance to a water source and the prevalence of diarrhea, were also important determinants of child nutritional outcomes. Thus, programs should concentrate on providing sanitation and clean water to households, especially those with children below the age of 5.

In conclusion, the results of this chapter can provide general guidelines to program managers, government officials, as well as researchers, as to what sort of interventions are necessary and when to undertake them in improving child health and nutritional outcomes in the short and long term. Creating programs and making policies that are flexible and sustainable, given the needs and resources of the community, can be extremely beneficial for administrators to reduce malnutrition.

Instrumental variable estimation using R

With given information on weight-for-age Z-scores (Z), breastfeeding practices (BF), clinical attendance (ATCL), age in months (AGEM), latrine facilities (L), and food availability (FA), we are going to use these data set to check if BF can be instrumented using ATCL and whether the null hypothesis of no-endogeneity status. The breastfeeding practices variable is correlated with weight-for-age Z-scores. Thus, the error terms cannot be met with the independence assumption. OLS model will not be valid for endogenous variables in it. We undertake the analysis in two steps. In the first stage, we estimate the determinant of breastfeeding practices by using the IV technique. In the second stage, the predicted value of breastfeeding practices along with other exogenous variables is included in determining the impact on weight-for-age Z-scores.

Stage 1: Estimating Breastfeeding Practices

The first stage requires instruments or clinical attendance variables to predict breastfeeding practices. Therefore, we are going to run a regression model on breastfeeding with instrument variable ATCL and other exogenous variables.

Stage 2: Estimating the Determinants of Weight-for-Age Z-Scores

In this stage, we then use the predicted value of breastfeeding to produce the second-stage regression. We can implement all this in R with the following command.

We first classify variables with the names. Y is a dependent variable that we want to see how other variables affect it. The breastfeeding variable is an endogenous one. ATCL is an instrument variable, and AGEM, EDS, L, and FA variables are exogenous variables.

```
> Y<-cbind(Z)
> Endogenous<-cbind(BF)
> Instrument<-cbind(ATCL)
> Exogenous<-cbind(AGEM,EDS,L,FA)
```

We can use the *ivreg* command to undertake the IV estimation in R. We conduct the analysis using the *ivreg* command; we conduct a regression with endogenous and exogenous variables on the dependent variable. At the same time, we also set up the first regression with exogenous and IVs after |. By using the *ivreg*, we can conduct two-stage least square analysis. Note that adding diagnostics in the summary command will print specification tests with results.

```
> iv2<-ivreg(Y~Endogenous+Exogenous|Exogenous+Instrument)
```

From the result table, endogenous is the predicted value of the first regression on breastfeeding, where the independent variables are the ATCL and other exogenous variables. First, it is the test for weak instruments with several instruments. The null hypothesis is H_0: "All instruments are weak." Second test is the Hausman test for endogeneity, where the null hypothesis is H_0: $Cov(x, e) = 0$. Thus, rejecting the null hypothesis indicates the existence of endogeneity and the need for IVs. The last test is the validity of instruments. The Sargan test is also called a test for overidentifying restrictions. The null hypothesis is that the covariance between the instrument and the error term is zero, that is, H_0: $Cov(z, e) = 0$. Thus, rejecting the null hypothesis indicates that at least one of the extra instruments is not valid. In this case, we cannot conduct the Sargan test because there is only one instrument variable.

```
> summary(iv2,diagnostics=TRUE)

Call:
ivreg(formula = Y ~ Endogenous + Exogenous | Exogenous + Instrument)

Residuals:
    Min      1Q  Median      3Q     Max
-1.73268 -0.71510 -0.09236  0.67750  2.59196

Coefficients:
              Estimate Std. Error t value Pr(>|t|)
(Intercept)    7.82173    5.80395   1.348    0.191
Endogenous    -0.45553    4.20251  -0.108    0.915
ExogenousAGEM -0.01979    0.08493  -0.233    0.818
ExogenousEDS  -0.04689    0.16445  -0.285    0.778
ExogenousL     1.51553    2.10001   0.722    0.478
ExogenousFA   -0.47694    1.20782  -0.395    0.697

Diagnostic tests:
                 df1 df2 statistic p-value
Weak instruments   1  22     0.354   0.558
Wu-Hausman         1  21     0.092   0.765
Sargan             0  NA        NA      NA

Residual standard error: 1.073 on 22 degrees of freedom
Multiple R-Squared: 0.09836,    Adjusted R-squared: -0.1066
Wald test: 1.079 on 5 and 22 DF,  p-value: 0.3989
```

As a result, there are no significant variables in the analysis, and the Hausman test cannot be rejected. Therefore, in the data, there is no need to use instrument variables for addressing endogeneity. However, in other research, instrument variables are commonly employed to address the endogenous problem.

Exercises

1. Define the following terms without reference to this chapter:
    a. endogenous variables,
    b. simultaneity bias,
    c. two-stage least squares, and
    d. identification.

2. What are the properties of the IV estimate? When is IV estimation useful? (Answer them in your own words.)
3. Which of the equations in the following systems is simultaneous? Be sure to specify the variables that are endogenous and the ones that are exogenous.

   a. $Y_{1t} = f_1(Y_{2t}, X_{1t}, X_{2t-1})$
   $Y_{2t} = f_2(Y_{3t}, X_{3t}, X_{4t})$
   $Y_{3t} = f_3(Y_{1t}, X_{1t-1}, X_{4t-1})$

   b. $Y_{1t} = f_1(Y_{2t}, X_{1t}, X_{2t})$
   $Y_{2t} = f_2(Y_{3t}, X_{5t})$

4. From the existing literature as explained in this chapter, describe in your own words how child-care practices can influence child nutritional outcomes. What are the channels through which maternal formal schooling and nutritional knowledge influence child nutritional outcomes?
5. Suppose that your colleague recently estimated a simultaneous equation and found that the OLS results were almost identical to the 2SLS results.
   a. What is the value of 2SLS in such a case?
   b. Does the similarity between the 2SLS and OLS estimates indicate a lack of bias?

STATA workout 1

The following table has information on weight-for-age Z-scores (Z) from 28 families. Some key indicators of Z: breastfeeding practices (BF = 0 indicating limited or no breastfeeding, and BF = 1 indicating adequate postnatal feeding for 24 months), clinical attendance (ATCL), age in months (AGEM), latrine facilities (L), food availability (FA), and WIC participation (0 for nonparticipation, 1 for receiving WIC benefits).

Obs	Z	BF	ATCL	AGEM	EDS	L	FA	WIC
1	5.17	0	1	80	12	0	1	1
2	5.8	0	0.537	91	16	0	0	1
3	6.42	0	1	90	12	1	1	1
4	6.55	0	1	79	12	1	1	1
5	7.19	0	0.543	90	12	1	0	1
6	5.16	0	0.543	71	12	1	1	1
7	5.41	0	1	89	12	1	1	1
8	6.9	0	0.966	71	8	0	0	1

Household care as a determinant of nutritional Chapter | 14 **549**

—cont'd

Obs	Z	BF	ATCL	AGEM	EDS	L	FA	WIC
9	6.03	0	0.586	80	14	1	1	1
10	4.3	0	0.653	71	6	0	1	1
11	6.46	0	0.652	68	12	1	1	0
12	6.58	0	0.15	69	17	1	1	0
13	6.39	0	0.094	71	6	1	0	0
14	7.4	0	0.51	90	8	1	1	0
15	5.58	1	0.104	81	16	1	0	0
16	6.67	1	0.955	81	12	1	0	0
17	7.45	1	0.089	66	17	1	0	0
18	6.58	1	0.336	70	17	1	0	0
19	5.35	1	0.336	71	15	1	1	0
20	6.11	1	0.614	71	14	1	0	0
21	7.13	1	0	85	12	1	0	0
22	7.44	1	0	81	12	1	1	0
23	6.17	1	0.388	67	14	1	0	1
24	7.64	1	0	66	12	1	1	1
25	6.27	1	0.516	75	12	1	1	1
26	8.06	1	0.306	67	12	1	0	1
27	7.44	1	0.722	70	16	1	1	1
28	9.15	1	0.975	72	9	1	1	1

Use these data to test if BF can be instrumented with ATCL. Compare these results to the OLS results using the **regress** command, with all the key indicators as drivers of Z.

The input lines for the IVs procedure, with ATCL as the instrument for BF, along with the robust standard errors to control for heteroscedasticity are as follows:

. ivregress 2sls Z (BF = ATCL) AGEM EDS L FA WIC, first vce(robust)
. estat endogenous
. regress Z BF ATCL AGEM EDS L FA WIC, first vce(robust)

The STATA output portion for this portion is as follows:

```
. ivregress 2sls Z (BF = ATCL) AGEM EDS L FA WIC, first vce(robust)
```

First-stage regressions

```
                                    Number of obs   =       28
                                    F(  6,    21)   =    12.75
                                    Prob > F        =   0.0000
                                    R-squared       =   0.4212
                                    Adj R-squared   =   0.2558
                                    Root MSE        =   0.4392
```

		Robust				
BF	Coef.	Std. Err.	t	P>|t|	[95% Conf. Interval]	
AGEM	-.017727	.0089031	-1.99	0.060	-.036242	.000788
EDS	.0328279	.0349549	0.94	0.358	-.0398648	.1055207
L	.4363517	.188768	2.31	0.031	.043787	.8289163
FA	-.2406567	.1962791	-1.23	0.234	-.6488415	.1675281
WIC	.0091756	.208422	0.04	0.965	-.4242617	.4426129
ATCL	-.1709827	.3329596	-0.51	0.613	-.8634101	.5214448
_cons	1.288503	.7997287	1.61	0.122	-.374624	2.95163

Interestingly, AGEM and L are significant in the first-stage regression results. While the R-squared and the adjusted R-squared values are low, the F (6,21) indicates that the overall model is significant. However, after controlling for endogeneity, the IV estimation indicates a poor fit, since none of the variables is significant.

Instrumental variables (2SLS) regression

```
                                    Number of obs   =       28
                                    Wald chi2(6)    =    16.43
                                    Prob > chi2     =   0.0116
                                    R-squared       =   0.3114
                                    Root MSE        =   .83122
```

		Robust				
Z	Coef.	Std. Err.	z	P>|z|	[95% Conf. Interval]	
BF	.7147629	2.895962	0.25	0.805	-4.961219	6.390745
AGEM	.0020237	.0596443	0.03	0.973	-.1148769	.1189244
EDS	-.0846578	.1041817	-0.81	0.416	-.2888502	.1195345
L	1.120881	1.562643	0.72	0.473	-1.941843	4.183606
FA	-.2551597	.8461757	-0.30	0.763	-1.913634	1.403314
WIC	.3344966	.3422542	0.98	0.328	-.3363093	1.005303
_cons	6.066132	4.297671	1.41	0.158	-2.357147	14.48941

Instrumented: BF
Instruments: AGEM EDS L FA WIC ATCL

Household care as a determinant of nutritional Chapter | 14 **551**

As mentioned before, the **estat endogenous** command is used to test for endogeneity, in this case, BF with ATCL. STATA reports two statistics: a robust VCE, and a robust regression-based test results are reported. In both cases, the test statistic is insignificant, and the null hypothesis of no-endogeneity cannot be rejected. Hence, then the variables being tested cannot be treated as endogenous.

```
. estat endogenous

Tests of endogeneity
Ho: variables are exogenous

Robust score chi2(1)        =    .000014   (p = 0.9970)
Robust regression F(1,20)   =    9.9e-06   (p = 0.9975)
```

In comparison, the OLS model with the **regress** command also does not produce a good fit. However, variable L is the only significant factor in the equation.

```
. regress Z BF ATCL AGEM EDS L FA WIC, first vce(robust)
```

Linear regression

Number of obs = 28
F(7, 20) = 1.78
Prob > F = 0.1463
R-squared = 0.3115
Root MSE = .9835

Z	Coef.	Robust Std. Err.	t	P>|t|	[95% Conf. Interval]
BF	.7033597	.532688	1.32	0.202	-.4078079 1.814527
ATCL	-.0019497	.6210001	-0.00	0.998	-1.297333 1.293434
AGEM	.0018216	.0225996	0.08	0.937	-.0453203 .0489635
EDS	-.0842835	.0742746	-1.13	0.270	-.2392176 .0706506
L	1.125857	.646801	1.74	0.097	-.2233461 2.47506
FA	-.2579039	.4150392	-0.62	0.541	-1.12366 .6078526
WIC	.3346013	.4062347	0.82	0.420	-.5127894 1.181992
_cons	6.080826	2.179677	2.79	0.011	1.534099 10.62755

The IV technique in this example does not yield a significant model. With more observations, other key explanatory variables can usually generate different results. The second workout explores another alternative specification for the same data.

STATA workout 2

How is the participation in the WIC program affected by the other variables listed before? Assume that BF is endogenously related to ATCL. Test for the endogeneity between BF and ATCL using the **estat endogenous** command. Check for endogeneity using the Hausman and the Wu tests.

The **ivregress** and the **estat** commands are implemented as follows:

. ivregress 2sls WIC (BF = ATCL) Z AGEM EDS L FA, first vce(robust)
. estat endogenous

. ivregress 2sls WIC (BF = ATCL) Z AGEM EDS L FA, first vce(robust)

First-stage regressions

```
                                    Number of obs   =        28
                                    F(  6,    21)   =      8.23
                                    Prob > F        =    0.0001
                                    R-squared       =    0.4747
                                    Adj R-squared   =    0.3246
                                    Root MSE        =    0.4184
```

BF	Coef.	Robust Std. Err.	t	P>|t|	[95% Conf. Interval]
Z	.1308818	.080722	1.62	0.120	-.0369887 .2987524
AGEM	-.016224	.0086374	-1.88	0.074	-.0341864 .0017383
EDS	.0409041	.0329202	1.24	0.228	-.0275572 .1093653
L	.2607451	.219322	1.19	0.248	-.19536 .7168501
FA	-.190896	.1963531	-0.97	0.342	-.5992346 .2174427
ATCL	-.1755414	.2828505	-0.62	0.542	-.7637612 .4126783
_cons	.3485764	1.045734	0.33	0.742	-1.826146 2.523299

While Z and AGEM are significantly related to BF. The predicted values of BF are now used to derive the second stage regression. The IV results are omitted, and the results of the **estat** command are given as follows:

. estat endogenous

Tests of endogeneity
Ho: variables are exogenous

Robust score chi2(1)   =   4.2076   (p = 0.0402)
Robust regression F(1,20)   =   3.50626 (p = 0.0758)

Both statistics for endogeneity indicate the null hypothesis of no-endogeneity can be rejected. To further explore this via Hausman and Wu tests, we implement the following commands in STATA:

. quietly ivregress 2sls WIC (BF = ATCL) Z AGEM EDS L FA
. est store ivregress
. regress WIC Z AGEM EDS L FA BF
. hausman ivregress .,constant sigmamore

The first input line produces the first and second stage regressions of the 2sls procedure, and the output is suppressed using the **quietly** feature. The second line helps to store the results of the **ivregress** procedure. The third line estimates the same equation as an OLS using the **regress** command. The last line generates the results from the Hausman test. The results are given as follows:

. regress WIC Z AGEM EDS L FA BF

Source	SS	df	MS		
Model	1.64913112	6	.274855187		
Residual	5.20801174	21	.248000559		
Total	6.85714286	27	.253968254		

Number of obs = 28
$F_{(6, 21)}$ = 1.11
Prob > F = 0.3906
R-squared = 0.2405
Adj R-squared = 0.0235
Root MSE = .498

WIC	Coef.	Std. Err.	t	P>|t|	[95% Conf. Interval]
Z	.0899451	.1115091	0.81	0.429	-.1419507 .321841
AGEM	.0013198	.012475	0.11	0.917	-.0246234 .027263
EDS	-.0009236	.0356675	-0.03	0.980	-.0750982 .073251
L	-.5593119	.3237504	-1.73	0.099	-1.232588 .1139638
FA	.2835545	.2054959	1.38	0.182	-.1437975 .7109065
BF	-.1071187	.2571556	-0.42	0.681	-.641903 .4276656
_cons	.2661262	1.285727	0.21	0.838	-2.40769 2.939942

The only significant variable in the OLS regression is L, which is negative, indicating that better placed households with latrine facilities opt out of the WIC program. STATA also generates the difference in the coefficients between the 2SLS and OLS and the computations involving the variance—covariance matrix.

| | Coefficients | | | |
| | (b) | (B) | (b-B) | sqrt(diag(V_b-V_B)) |
	ivregress	.	Difference	S.E.
BF	-3.271121	-.1071187	-3.164002	1.823104
Z	.499508	.0899451	.4095629	.2359909
AGEM	-.055323	.0013198	-.0566428	.0326377
EDS	.1363271	-.0009236	.1372507	.0790841
L	.4227981	-.5593119	.98211	.5658935
FA	-.4247974	.2835545	-.7083519	.4081536
_cons	1.341463	.2661262	1.075337	.619611

b = consistent under Ho and Ha; obtained from ivregress
B = inconsistent under Ha, efficient under Ho; obtained from regress

Test: Ho: difference in coefficients not systematic

chi2(1) = (b-B)'[(V_b-V_B)^(-1)](b-B)
        =   3.01
Prob>chi2 =   0.0827

The null hypothesis assumes that the OLS estimator is consistent. If the null hypothesis is rejected, then 2SLS is preferred to OLS. Since the *P*-value is 0.08, we can reject the null at 0.1%. Hence, there is some evidence of endogeneity between the chosen variables. Another way to track this is to explicitly incorporate the predicted values of BF from the first-stage regression into the second stage and check for the significance of that variable. The input lines that establish this Wu-test version are as follows:

. quietly reg BF ATCL Z AGEM EDS L FA
. predict BFhat, xb
. reg WIC Z AGEM EDS L FA BFhat

The corresponding output is shown in the following. Note that the t-value of BFhat is 1.92, which implies that we can reject the null and conclude that BFhat is a significant variable, and thus note the compatibility between the alternative versions of the endogeneity tests. Note that WIC is dichotomous variable, and the aforementioned methods can be incorporated into the logistic model explored in the previous chapter. Similar exercises that check for the endogeneity of WIC and BF and other combinations are also possible.

```
. quietly reg BF ATCL Z AGEM EDS L FA

. predict BFhat, xb

. reg WIC Z AGEM EDS L FA BFhat
```

Source	SS	df	MS		Number of obs	=	28
					F(6, 21)	=	1.87
Model	2.38892882	6	.398154804		Prob > F	=	0.1335
Residual	4.46821403	21	.212772097		R-squared	=	0.3484
					Adj R-squared	=	0.1622
Total	6.85714286	27	.253968254		Root MSE	=	.46127

WIC	Coef.	Std. Err.	t	P>|t|	[95% Conf.	Interval]
Z	.4995081	.2417616	2.07	0.051	-.0032627	1.002279
AGEM	-.055323	.0323639	-1.71	0.102	-.1226274	.0119814
EDS	.1363271	.0803575	1.70	0.105	-.0307855	.3034397
L	.4227982	.6038804	0.70	0.492	-.8330397	1.678636
FA	-.4247975	.4232675	-1.00	0.327	-1.30503	.4554354
BFhat	-3.271121	1.705377	-1.92	0.069	-6.817647	.2754039
_cons	1.341463	1.321989	1.01	0.322	-1.407763	4.09069

Chapter 15

Achieving an ideal diet—modeling with linear programming

Chapter outline

Introduction	557	Using solver in excel to obtain an	
Review of the literature	559	LP solution	568
Linear programming model	563	Step 1: Setting the problem in excel	568
Solution procedures	565	Step 2: Solving the parameters of	
Graphical solution approach	565	the model	570
Some qualifications about the		Step 3: Deriving the results	570
optimum	567	**Summary**	**572**
		Exercises	**573**

Nothing will benefit human health and increase the chances for survival of life on Earth as much as the evolution to a vegetarian diet.

Albert Einstein

Introduction

Nutrition planners face several programmatic challenges in designing appropriate interventions. Frequently asked field questions include the following: Is it possible to design a diet with locally available food that meets the daily recommended allowances? How do we minimize the cost of achieving such diets? These questions could be answered through a linear programming approach that helps to examine the compatibility of different mathematical inequalities in using simple mathematics to address the aforementioned questions. This chapter introduces elements of linear programming and demonstrates how food and nutrition problems could be solved using an Excel spreadsheet. Interpretations of the results of linear programming problems for policy applications are also given.

Although finding a combination of foods that can provide nutritional sufficiency to different types of households has been a subject of investigation for more than 70 years in the developed world (Stigler, 1945) and has also been used in developing countries for several decades (Calkins, 1981; Babu and Hallam, 1989), the idea of using least cost diets to design nutrition interventions in developing countries has gained momentum recently (Vossenaar et al., 2017; Hirvonen et al., 2019; Bai et al., 2021; WFP, 2019).

Micronutrient deficiency continues to be a serious development challenge in many developing countries. The diets of poor people in these countries are usually deficient in key nutrients such as iron, zinc, calcium, vitamin A, and vitamin C. This deficiency can be explained by either a shortage of micronutrient-dense foods (foods with a high concentration of nutrients in relation to energy) in their diets or an inappropriate selection of local foods. The aforementioned two possibilities have different programmatic implications. If deficiency is caused by shortage of micronutrients, it can be improved by increasing the availability of nutrient-rich foods—either via food fortification or through agricultural programs that introduce new crop varieties. On the other hand, if micronutrient deficiency is caused by inappropriate selection of foods, nutrition education programs that emphasize the best use of locally available nutrient-rich foods should be given priority (Darmon et al., 2002).

The issues that underline the aforementioned alternative programmatic possibilities can be addressed as follows:

1. Is it possible to design a diet that fulfills the nutritional recommendations through the use of locally available foods?
2. If such a diet is feasible, what is the best combination of these foods (the minimum cost) that can achieve a nutrient-dense diet?

The aforementioned issues can be answered using a "trial-and-error" approach or by "expert" guessing. However, an efficient and rigorous method based on linear programming can address the aforementioned questions. Linear programming is an approach that examines the compatibility of different mathematical inequalities using simple mathematics to solve the problem of the type discussed earlier (Ferguson et al., 2006).

The importance of this method can be understood from a developing country perspective, where households have limited income but need to meet the nutritional requirements with that income. Thus, they have to minimize their total expenditure, while, at the same time, they attain minimum nutritional requirements. Thus, determining the optimal diet (at the minimum cost) is important so that individuals can attain the daily nutritional requirements of their body with their limited income.

In this chapter, we introduce the basic elements of linear programming, as it is applied to solve the diet/nutrition problem. It is a powerful tool for analyzing the cost of a nutritionally adequate ration prepared from different locally available foods. The method could be further refined by taking into account costs that were not included in the present study, such as the cost of targeting food distribution, of administrative overheads, or of training food aid staff.

The sensitivity of linear programming to selected constraints is its major weakness. This approach should not be used in isolation, and the validity of the conclusions should always be field-tested. The chosen set of nutritional constraints should be based on internationally accepted nutritional recommendations, such as ones published by international organizations. The food consumption constraints should be derived from the food consumption data collected in the community of interest. Building up an international database of food consumption constraints for different age groups, especially for nutrient-dense foods, would facilitate the application of this method. The validity of these constraints could then be confirmed and, if necessary, adjusted on the basis of a series of simple observations.

This chapter is organized as follows: In the next section, we review some of the recent case studies that use this approach for achieving an optimal diet. The third section illustrates the principles of the linear programming model along with its underlying assumptions. We then demonstrate the use of the Excel solver to determine the least expensive food combination that respects multiple nutritional constraints. The method shows how modern computer applications can be used to answer very practical questions. The final section provides a summary of the linear programming applications for nutrition programming.

Review of the literature

The application of linear programming to determine the minimum cost of achieving the recommended daily nutrients is not a new idea. This tool started with the seminal work of Stigler (1945). Stigler posed the following question: For a moderately active individual weighing 154 pounds, how much of each of the 77 foods should be eaten on a daily basis so that the individual's nutrient intakes will be at least equal to the recommended dietary allowances (RDAs)?[1] Stigler's RDAs of interest were calories, protein, calcium, iron, vitamin A, thiamine, riboflavin, niacin, and ascorbic acid. The nutrient contents of the 77 foods were obtained from the 1940 publication of the US Department of Agriculture (USDA).

The study used a trial and error method to solve the 9 × 77 set of inequalities. Based on the cost and nutrient content of foods, the original 77 foods were brought down to 15. These 15 food items had no meat except beef liver and excluded all sugars, beverages, and planted cereals. The minimum cost of the diet during 1939 was found to be $39.93 per year and included varying amounts of wheat flour, evaporated milk, cabbage, spinach, and dried navy beans. The optimum diet consisted of foods that most individuals would

---

1. RDAs are the levels of intake of essential nutrients that, on the basis of scientific knowledge, are considered by the Food and Nutrition Board to be adequate to meet the known nutrient needs of all healthy individuals.

find unappetizing such as pork liver, spinach, dried beans, evaporated milk, and wheat flour. The aforementioned diet could be considered as the combination that makes up the human dog biscuit. However, human nutritionists did not seriously investigate the application of linear programming to determine an optimum diet till the late 1990s. We present some of the case studies from the economics and nutrition literature that apply linear programming to study food and nutrition problems.

Silberberg (1985) developed a hypothesis of tastes by humans with regard to food consumption using the law of diminishing marginal product. Since dog foods cannot be consumed by humans due to taste considerations, Silberberg introduced taste in the linear programming (LP) models. The main hypothesis postulated was as follows: As income rises, expenditure on "pure nutrition" declines as a percentage of overall food expenditure and the importance of taste increases.

The analysis starts with the assumption that foods purchased in the market were inputs, which could be used in the production of meals that provide both pleasing taste and nutrition. The data were obtained from the USDA's food consumption surveys and consisted of approximately 15,000 households for the period 1977 and 1978.

The minimum cost level diets found in the study were such that it could be given to slaves, although the calorie levels could be elevated. The various aspects of consumer behavior that are amenable in this framework could be the automobile or housing markets. Cars, in general, provide varying degrees of styles and comfort. Thus, one could expect diminishing returns to the pure transport function of cars. In other words, as income increased, a greater fraction of the price of a car could be attributable to style and comfort and less to pure transportation.

For housing, one could expect rapidly diminishing returns to pure shelter component as income increased. As income increases, a greater proportion of housing expenses could be directed toward more space and less toward pure shelter. Applying theory in the aforementioned manner could provide interesting implications of consumer behavior in general and analyzing nutrition behavior of households.

Athanasios et al. (1994) assessed the impact of food for work program (FFW) on consumption and nutrition of the households. A two-step procedure was undertaken. First, a linear programming model was used to estimate the marginal nutrient prices of four nutrients, namely calories, protein, fat, and carbohydrates. The main hypothesis was that changes in food commodity prices would affect the nutrient demand by affecting the nutrient shadow prices. Second, an econometric model was specified to estimate the own and cross-price elasticities and income elasticity of demand for each of the four nutrients. The data used in the study were collected from Baringo district in the Rift valley province of Kenya for the period August 1983 to February 1984. A random sample of 252 households was selected, out of which 100 were found to be participants in the FFW projects. The data included all the production and consumption activities of households and the data on FFW included beans, corn, and vegetable oil.

In the first step, a linear programming model was specified and run for each household. Every household had a different set of food items consumed and paid a different set of food prices. In the second step, a household's nutrient demand was specified as a function of nutrient shadow prices and other household characteristics and was estimated econometrically.

The study found that there were significant nutritional gains to FFW participants via food transfers compared with an equivalent net income transfer to the participating households. While previous research on the impact of FFW on nutrition suggested that lower-income households had significant income gains due to participation in the FFW projects, the study demonstrated that even the poorest participant households had nutritional gains of 32.46% more than the gain by all participants. The implication of these results is that significant nutritional gains can occur through food transfers as compared with an equivalent net income transfer to the participant households. Thus, the food for work program could be used as an instrument to improve the general nutrition conditions of the population.

Conforti and D'Amicis (2000) assessed the effect of adoption of a nutritionally correct food behavior on average food patterns and related expenditure. In other words, the basic question addressed was what would happen if households (Italian households) switched from their actual food pattern to one that met the nutritional requirements as defined by nutritionists. The main objective of this paper was to describe in detail what the average food pattern looks like and how much it would cost if the population followed the recommended dietary allowances (RDAs) rather than to indicate how RDAs should be met. A linear programming model was formulated by minimizing food expenditures subject to the intake of given maximum and/or minimum levels of nutrients, vector of prices, vectors of maximum and minimum requirements, and a food composition matrix.

Some important conclusions were derived from this study. First, the LP exercise supported the idea that consumers normally chose foods rather than nutrients. In other words, most food expenditure paid for tastes and habits in consumption in the Italian case. This was the reason for including the food habit constraints. Second, the pattern generated by including both the RDAs and food habit constraints showed several patterns of consumption trends that matched reality. The generated pattern with the actual trend in behavior suggested that consumption by Italians has been moving toward a healthier pattern over the years. Third, a trend toward a healthier diet does not imply per se an increase in food expenditure. Finally, the model suggests that the costs associated with an imbalanced average food pattern could have far-reaching implications, which go beyond consumption. From the epidemiological studies, the association between food intake and health status has demonstrated that even a small decrease in the risk of contracting food-related diseases could significantly affect both healthcare expenditures and labor productivity.

Briend et al. (2001) demonstrated how the linear programming technique could be applied in estimating the economic benefits from the introduction of different fortified foods, using local food prices. The paper compared the economic value of a classical blended food with that of a nutrient-dense spread known as "foodlet," a highly fortified food that can be regarded as a big tablet for childhood diets in Chad.

The economic value of two food supplements (a traditionally blended flour and a highly nutrient-dense spread known as "foodlet") was used in the study to illustrate the application of linear programming. For the nutrient-dense spread, fortification levels were chosen from previous research in Algeria. High fortification levels were made possible in this spread by the attractive taste of peanut, which could easily hide high levels of unpalatable vitamins and minerals. The flour was a blend of maize and cowpea flours with sugar, fortified with a standard mineral and vitamin mix.

The analysis also showed that a proposed program has a ratio of amount saved to amount spent less than 1, which implied that the money saved by the families will be below the amount spent by the donor. This is likely to be the case for unfortified blended flour prepared from locally available foods. These foods were more expensive than the sum of the basic ingredients used in their composition and had no superior nutritional value compared with the meal a mother would prepare at home with the same ingredients.

Garille and Gass (2001) first took the original data in 1939 used by Stigler and replicated the LP results. The data were then updated to include price changes, revised values of the RDAs, and current evaluations of the nutrient content of the 77 foods. In extending Stigler's original problem, the upper bounds of these nutrients were set that were known to be toxic or to have other undesirable properties.

The purpose of this study was to solve the diet problem for five different sets of data:

1. Stigler's original problem was updated for a 25- to 50-year-old man,
2. Stigler's original problem was updated for a 25- to 50-year-old woman.
3. The extended Stigler problem, where the constraints were incorporated for all of the current RDAs for a 25- to 50-year-old man.
4. The extended Stigler problem, where the constraints were incorporated for all of the current RDAs for a 25- to 50-year-old woman.
5. Stigler's problem using current RDAs and food nutrient contents with 1939 prices.

Additionally, for each of the problems, the minimum cost diet was solved where no excess nutrients were allowed.

The study used the nutritive content of foods in the form that would most likely be eaten, to make the problem more realistic. The 77 food items were all included in the study. However, the Bureau of Labor Statistics (BLS) list only includes about 30 of the original 77 foods. For the data set to be consistent, the prices of the food items were included from the Giant food supermarket chain in the Washington, DC area for April 1998.

The optimal solution to this excess Stigler diet problem was $40.92 annually during 1939 and was $481.16 in April 1998. The updated excess diet included wheat flour, evaporated milk, cabbage, and sweet potatoes and was a different diet than Stigler's original model (which only included spinach and navy beans but no sweet potatoes). The following minimum requirements were not satisfied with this diet: polyunsaturated fatty acids, vitamin B_6, vitamin B_{12}, pantothenic acid, sodium, potassium, magnesium, zinc, copper, iodine, vitamin D, and vitamin E. The upper limit for manganese was exceeded. All the foods in this diet were also on Stigler's reduced list of foods, and the annual cost of this diet was only $0.99 more than the annual cost of Stigler's 1939 diet.

The trial-and-error method of Stigler's original (9 × 77) problem proved that the simplex method works in practice. The concept of a diet problem led the way to many minimum cost applications, such as cattle and chicken feed, chemical, and fertilizer blending. However, the inadequacy of the Stigler's diet problem to produce a nutritious and palatable human diet caused researchers to extend the analysis to menu planning, which, in turn, raised new research questions in integer and goal programming. The advantage of the LP model was that the assumptions and concepts (such as additivity, proportionality, nonnegativity, duality) could be easily explained. Stigler's prelinear programming approach to modeling human diet can now be used for the following purposes:

1. Evaluate the nutritional content of diets for school children.
2. Plan menus for institutions (such as hospitals, jails).
3. Manage food systems.

Linear programming model

The first solution of a diet problem using linear programming was by Smith (1959). An LP problem consists of several essential elements. First, there are *decision variables* (X_j), the level of which denotes the amount undertaken of the respective unknowns, of which there are n ($j = 1, 2, \ldots, n$). Next is the *linear objective function* whose total objective value (Z) equals $c_1X_1 + c_2X_2 + \cdots + c_nX_n$. Here, c_j is the contribution of each unit of X_j to the objective function.

There are *m linear constraints*. An algebraic expression of the ith constraint is given by the following expression: $a_{i1}X_1 + a_{i2}X_2 + \cdots + a_{in}X_n \neq b_i$ ($i = 1, 2, \ldots, m$), where b_i denotes the upper limit or the right-hand side imposed by the constraint and a_{ij} is the use of the constraint by one unit of X_j. The c_j, b_i, and a_{ij} are the exogenous parameters of the model. The LP problem then is to choose $X_1, X_2, \ldots, X_n$ so as to maximize the following function (Hazell and Norton, 1986):

$$\text{Max } c_1X_1 + c_2X_2 + \ldots + c_nX_n$$
$$a_{11}X_1 + a_{12}X_2 + \ldots + a_{1n}X_n \leq b_1$$
$$a_{21}X_1 + a_{22}X_2 + \ldots + a_{2n}X_n \leq b_2$$
$$\text{s.t.}$$
$$a_{m1}X_1 + a_{m2}X_2 + \ldots + a_{mn}X_n \leq b_m$$
$$X_j \geq 0$$
(15.1)

The aforementioned formulation can be expressed in matrix notation as follows:

$$\text{Max } cX$$
$$\text{s.t. } AX \leq b$$
$$X \geq 0$$
(15.2)

The model described either by Eq. (15.1) or by Eq. (15.2) has some important underlying assumptions. They are as follows:

1. **Optimization**: It is assumed that an appropriate objective function is either maximized or minimized.
2. **Fixedness**: At least one constraint has a nonzero right-hand side coefficient.
3. **Finiteness**: There are only a finite number of activities and constraints so that a solution can be found.
4. **Determinism**: All the parameters c_j, b_i, and a_{ij} are assumed to be constants.
5. **Continuity**: Resources can be used and activities produced in quantities that are fractional units.
6. **Homogeneity**: All units of the same resource or activity are identical.
7. **Additivity**: The activities are assumed to be additive in the sense that when two or more are used, their total product is the sum of their individual products. In other words, interaction between activities is not permitted.
8. **Proportionality**: The gross margin and resource requirements per unit of activity are assumed to be constant regardless of the level of activity. For example, a constant gross margin per unit of activity assumes a perfectly elastic demand curve for the product. On the other hand, constant resource requirements per unit of activity are equivalent to a Leontief production function.

The assumptions of linearity and proportionality together define the linearity in activities, thereby giving rise to the name linear programming. However, the assumptions underlying the LP model are stringent, since both the objective function and the constraints are assumed to be linear. This only implies that the optimum solution will lie in one of the corners of the constraint. We will demonstrate in the next section the graphical procedure of obtaining a solution.

Solution procedures

This section illustrates the solution procedure of the linear programming by using both the graphical and the simplex method. The characteristic that makes linear programs easy to solve is their simple geometric structure. A solution for a linear program is any set of numerical values for the variables. These values need not be the best values and do not even have to satisfy the constraints. A "feasible solution" is a solution that satisfies all of the constraints. The "feasible set" is the set of all feasible solutions. An "optimal solution" is the feasible solution that produces the best objective function value.

Graphical solution approach

Consider the data in Table 15.1.

The problem is to select the least cost combination of potato and beef that will supply 2000 calories or more, less than, or equal to 1500 mg of calcium and exactly 50 g of protein. Thus, in the LP formulation, the problem can be rewritten as follows:

$$Minimize\ 0.1\ X_1 + 1.3\ X_2$$
$$s.t.\ 76\ X_1 + 242\ X_2 \geq 2000$$
$$0.3\ X_1 + 3\ X_2 = 50$$
$$7\ X_1 + 11\ X_2 \leq 1500$$
$$and\ X_1, X_2 \geq 0$$

TABLE 15.1 Summary of data for linear programming.

	X_1 Potato	X_2 Beef	Constraint
Calories (kcal/100 g)	76	242	≥2000
Protein (g/100 g)	0.3	3.0	= 50
Calcium (mg/100	7.0	11.0	≤1500
Cost ($/100 g)	0.1	1.3	

The constraints $X_1, X_2 \geq 0$ restrict us to the points on or to the right of the vertical axis and to the points on or above the horizontal axis. Next, we draw the individual constraints. To find the points that satisfy the first constraint ($76 X_1 + 242 X_2 \geq 2000$), we construct the line $76 X_1 + 242 X_2 = 2000$ by finding the two points that lie on the line and then constructing a line through these points. First, set $X_1 = 0$ and solve for X_2. This yields the point ($X_1 = 0$ and $X_2 = 8.264$). Then we set $X_2 = 0$ and solve for X_1. This yields the point ($X_1 = 26.315$ and $X_2 = 0$). This line is plotted in Fig. 15.1. Now, we have to determine which side of the line the point satisfies the constraint. If one point satisfies the constraint, then all the points on the same side of the line satisfy the constraint. Analogously, if one point does not satisfy the constraint, then no point on that side of the line satisfies the constraint. Suppose we choose the point ($X_1 = 0$ and $X_2 = 0$). This point does not satisfy the constraint $76 X_1 + 242 X_2 \geq 2000$. Hence, all the points to the lower left will also not satisfy the constraint. Thus, points above the upper right of this line will satisfy the constraint.

We do the same thing for the protein constraint $0.3 X_1 + 3 X_2 = 50$. Since this is an equality, we first set $X_1 = 0$ and solve for X_2. This yields the point ($X_1 = 0$ and $X_2 = 16.67$). Next, we set $X_2 = 0$ and solve for X_1. This yields the point ($X_1 = 166.67$ and $X_2 = 0$). Now, if we choose a point such as ($X_1 = 0$ and $X_2 = 0$), we find that the constraint is not satisfied. Hence, the solution

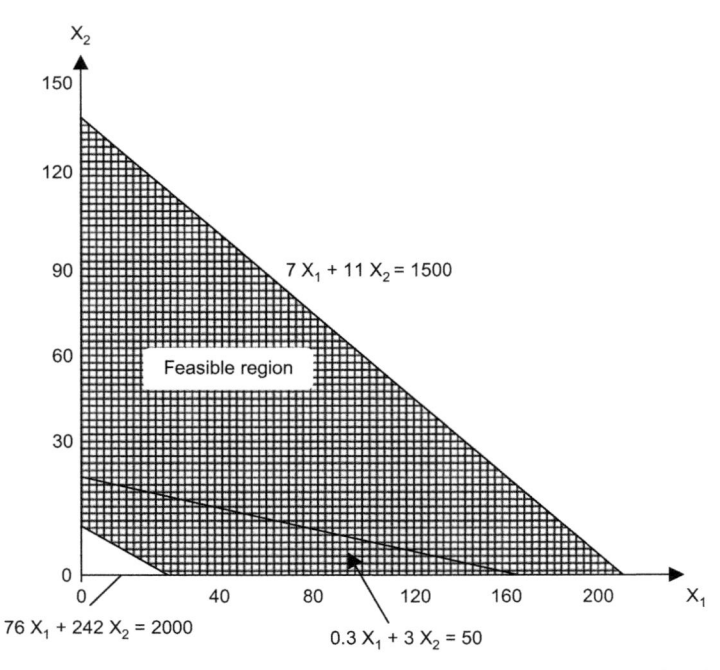

FIGURE 15.1 Feasible region of the above LP problem. *Source: from Anderson and Earle, 1983.*

must lie on the line, and neither above nor below it. Finally, we choose the calcium constraint $7 X_1 + 11 X_2 \leq 1500$. We find two points on the line $7 X_1 + 11 X_2 = 1500$. We first set $X_1 = 0$ and solve for X_2. This yields the point ($X_1 = 0$ and $X_2 = 136.36$). Next, we set $X_2 = 0$ and solve for X_1. This yields the point ($X_1 = 214.28$ and $X_2 = 0$). This gives us the line on the upper right-hand side. Now, if we choose a point such as ($X_1 = 0$ and $X_2 = 0$), we find that the constraint is satisfied. Therefore, all the points to the lower left also do. The shaded area represents the feasible set. The feasible set for a linear program will always have a shape like the one in Fig. 15.1, with edges being straight lines and corners where the edges meet. The corners of the feasible set are called *extreme points*. Note that each extreme point is formed by the intersection of two or more constraints.

Thus, the fundamental theorem of linear programming can be stated as follows: If a finite optimal solution exists, then at least one extreme point is optimal. In this rather simple example, we know that the solution must lie at the point where the line $0.3 X_1 + 3 X_2 = 50$ has an intercept on the X_1 axis. We want to find the optimum values of X_1 and X_2 such that the objective function $0.1 X_1 + 1.3 X_2$ is as small as possible. This is obtained if we set $X_2 = 0$. In this case, $X_1 = 166.67$. Substituting these values of X_1 and X_2 in the constraint functions, we obtain the optimum number of calories, protein, and calcium to be 12,666.67, 50.001, and 1166.67 respectively.

Some qualifications about the optimum

The fundamental theorem of linear programming states that if a finite optimum exists, there exists an extreme point which is optimal. However, this optimal solution may not be unique. Two or more adjacent extreme points (that share a common edge) may tie for the best solution. In this case, not only are extreme points optimal, but all the points on the edge connecting them are also optimal. In such a case, we have a situation of *multiple optima*.

In addition, sometimes a linear program can have an unbounded solution. In such a situation, the objective function can achieve a value of positive infinity for a maximization problem and negative infinity for a minimization problem. Consider the following problem:

$$\text{Maximize } X_1 + 2 X_2$$
$$\text{s.t. } X_1 \leq 10$$
$$2 X_1 + X_2 \geq 5$$
$$\text{and } X_1, X_2 \geq 0$$

As long as X_1 is kept less than or equal to 10, X_2 can be increased without limit, and the objective function will increase without bound. Thus, in this example, there is no finite optimum. We call such a solution an *unbounded solution*. Unboundedness refers to the objective function value and not the constraint set. When an unbounded problem occurs, the modeler should carefully study the situation to determine the limitations that exist and which are not explicitly stated in the constraints. We now demonstrate the aforementioned results using Microsoft Excel solver program to show how computer software can be used to solve an LP problem.

Using solver in excel to obtain an LP solution

There are numerous software packages that can be used to solve linear programming problems—LINDO and GAMS being the most popular ones. All these packages are usually DOS based and intended for a niche market.

In recent years, however, several business packages such as spreadsheets have started to include an LP solving option. The inclusion of an LP solving capability in a program such as Excel is attractive for two reasons. First, Excel is perhaps the most popular spreadsheet that is used in business organizations and universities and thus is accessible. Second, the spreadsheet offers very convenient data entry and editing features that can allow the student to gain better understanding of how to construct linear programs. We demonstrate in this section how to use the Solver function to solve the diet problem in the aforementioned section.

First, to use Excel, the Solver Add-In must be included. To add this facility to your Tools menu, one needs to carry out the following steps:

1. Select the menu option Tools and then hit Add-Ins (which will take a few moments).
2. From the dialog box presented, check the box for Solver Add-In.
3. On clicking OK, you can then access the Solver option from the new menu option Tools and then hit on Solver.

We now illustrate the steps that are necessary to solve the diet problem.

The best approach to entering the problem into Excel is first to list in a column the names of the objective function, decision variables, and constraints. Then enter some arbitrary starting values in the cells for the decision variables, usually zero. Excel will vary the values of the cells as it determines the optimal solutions. Having assigned the decision variables with some arbitrary starting values, one can use cell references explicitly in writing the formulae for the objective function and the constraints.

Step 1: Setting the problem in excel

Entering the formulae for the objective and constraints, the objective function in B5 is given by 0.1*B9 + 1.3 B10. The constraints will be given by putting the right-hand side values in adjacent cells (Fig. 15.2).

Calories (B14) = 76*B9 + 242*B10
Protein (B15) = 0.3*B9 + 3*B10
Calcium (B16) = 7*B9 + 11*B10

Achieving an ideal Chapter | 15 **569**

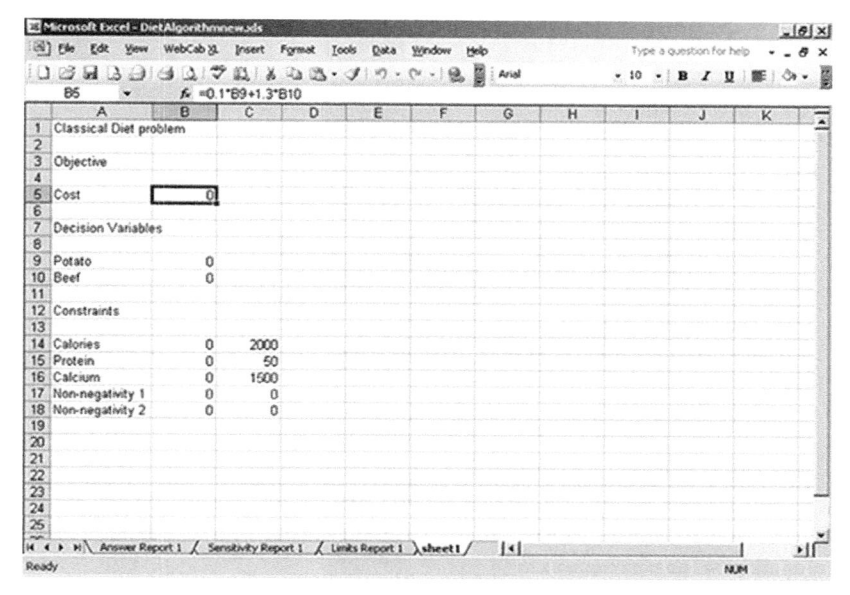

FIGURE 15.2 Excel box.

Nonnegativity 1 (B17) = B9
Nonnegativity 2 (B18) = B10

Now selecting the menu option tools and hitting on Solver, Fig. 15.3 is revealed.

FIGURE 15.3 Solver parameters box.

Step 2: Solving the parameters of the model

Select whether you wish to maximize or minimize the problem. In this example, set the target cell B5 to a Min. Next enter the range of cell that you want Solver to vary—the decision variables. Click on the white box (by changing cells) and select cells B9 and B10. Next, you can enter the constraints by first clicking the "Add" button and add each constraint of the equation. Having added all the constraints, click the "OK" button and the Solver dialog box would look like the one shown in Fig. 15.3. Before clicking "Solve," it is important to go to the Options button and check the "Assume Linear Model" to make the model linear. This step is necessary for a solution to exist and for generating the relevant sensitivity report. After selecting this option, click "Solve" and the Solver will find the optimal values of the decision variables. Observe that Solver has altered all the values in your spreadsheet and replaced them with the optimal results.

One can use the Solver Results dialog box to generate three reports. To select all the three at once, hold down the CTRL button of the keyboard, and drag the mouse over all three. Fig. 15.4 should appear.

Step 3: Deriving the results

It is a good practice to get Solver to restore the original values in the spreadsheet so that adjustments can be made to the model. Three reports are now generated in new sheets in the current workbook of Excel. The Answer report (Fig. 15.5) gives details of the solution. In our case, the cost of the diet is minimized at 16.67 when 0 units of beef and 166.67 units of potatoes are consumed. Recall, this was the solution[2] obtained graphically in

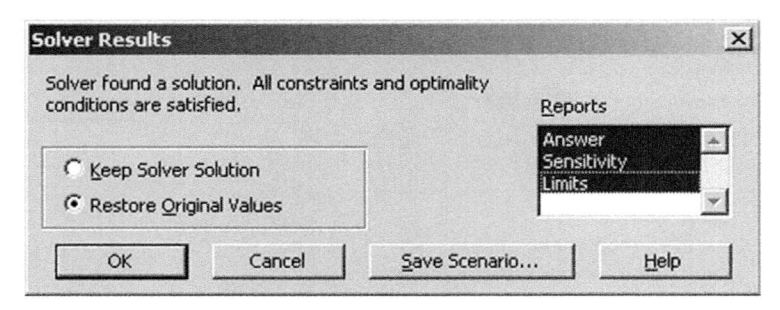

FIGURE 15.4 Solver results box.

---

2. The method used by Excel to solve the LP problem is called the simplex method. For an excellent discussion of this method, see the following website: http://cba.fiu.edu/dsis/davidsoj/Supplement%20B.pdf.

Microsoft Excel 11.0 Answer Report

Worksheet: [DietAlgorithmnew.xls]sheet1

Report Created: 4/28/2005 10:44:43 AM

Target Cell (Min)

Cell	Name	Original Value	Final Value
$B$5	Cost	0	16.66666667

Adjustable Cells

Cell	Name	Original Value	Final Value
$B$9	Potato	0	166.6666667
$B$10	Beef	0	0

Constraints

Cell	Name	Cell Value	Formula	Status	Slack
$B$14	Calories	12666.66667	$B$14>=$C$14	Not Binding	10666.66667
$B$15	Protein	50	$B$15=$C$15	Not Binding	0
$B$16	Calcium	1166.666667	$B$16<=$C$16	Not Binding	333.3333333
$B$17	Non-negativity 1	166.6666667	$B$17>=$C$17	Not Binding	166.6666667
$B$18	Non-negativity 2	0	$B$18>=C$18	Binding	0

FIGURE 15.5 Optimized values of decision variables and constraints.

the previous section. Finally, the sensitivity report provides information about how sensitive the solution is to changes in the constraints given in Fig. 15.6.

The sensitivity report given in Fig. 15.6 is fairly standard and provides information on shadow prices, reduced cost, and the upper and lower limits for the decision variables and constraints. In conclusion, Excel Solver is a simple but effective tool for allowing users to explore linear programs. It can be used for large problems with hundreds of variables and constraints and can be done relatively quickly. However, for teaching purposes, a small problem can provide important insights about the structure of an LP. One of the main limitations of this method is that the tableaus generated at each iteration cannot be found. Other programs such as Lindo allow this.

Microsoft Excel 11.0 Sensitivity Report

Worksheet: [DietAlgorithmnew.xls]sheet1

Report Created: 4/28/2005 10:44:43 AM

Adjustable Cells

Cell	Name	Final Value	Reduced Cost	Objective Coefficient	Allowable Increase	Allowable Decrease
$B$9	Potato	166.6666667	0	0.1	0.03	1E+30
$B$10	Beef	0	0	1.3	1E+30	0.3

Constraints

Cell	Name	Final Value	Shadow Price	Constraint R.H. Side	Allowable Increase	Allowable Decrease
$B$14	Calories	12666.66667	0	2000	10666.66667	1E+30
$B$15	Protein	50	0.333333333	50	14.28571429	42.10526316
$B$16	Calcium	1166.666667	0	1500	1E+30	333.3333333
$B$17	Non-negativity 1	166.6666667	0	0	166.6666667	1E+30
$B$18	Non-negativity 2	0	0.3	0	16.66666667	5.649717514

FIGURE 15.6 Sensitivity report and shadow prices.

Summary

Since the seminal work of Stigler in 1945, the concept of a diet or blending problem led the way to many minimum cost applications, such as cattle and chicken feed, and chemical and fertilizer blending. The inadequacy of the Stigler's diet problem to produce a palatable and a nutritious human diet caused researchers to extend the approach to menu plans in schools, which, in turn, raised new research questions in integer and goal programming. The LP model of the basic diet problem can be used to explain all the concepts of linear programming. As pointed out by Garille and Gass (2001), it is a classic example of a correct mathematical model of a real-world problem that does not produce a valid solution.

Linear programming is a tool for analyzing the cost of a nutritionally adequate ratio prepared from different locally available foods. The general principles and a selected application of this approach were presented in this chapter. From a nutritionist perspective, very practical questions on complementary feeding have been debated over many years with no clear solution. Without the help of linear programming, it is impossible to use intuition and trial-and-error approaches to arrive at solutions to problems that required solving hundreds of equations simultaneously. Linear programming should clarify these important issues. Additionally, everyone agrees that nutrient recommendations by different expert committees are difficult to implement in practice. In fact, few practitioners have been successful in providing recommendations that are realistic and consistent with the recommended nutrient intakes. It is often the case that diets based on current food recommendations do not provide the recommended nutrients. Linear programming can make major progress in this area.

In conclusion, linear programming could be a powerful tool for formulating sound nutritional advice, especially in the context of complementary feeding practices in poor countries. It has the potential to improve infant and young children's nutrition with behavior change and effective communication strategies. This approach to human nutrition is long overdue.[3] This method has wide applications for different types of nutrition intervention programs, including supplementation, fortification, and agriculture programs. Despite its limitations due to the underlying assumptions, linear programming clearly provides useful information for evaluating the economic benefits of different nutrition intervention programs for the poor.

Exercises

1. What are the three primary components of a constrained optimization model? Explain the difference between a parameter and a decision variable. What are the primary assumptions underlying a linear programming model?
2. What does it mean when a problem has an unbounded solution? What does it mean to perform sensitivity analysis?
3. You are given the following diet problem:
   Min. $C = 0.6 X_1 + X_2$
   s.t. $10X_1 + 4X_2 \geq 20$ (calcium constraint)
   $5X_1 + 5X_2 \geq 20$ (protein constraint)
   $2X_1 + 6X_2 \geq 12$ (vitamin A constraint)
   and $X_1, X_2 \geq 0$.

   Solve the problem using Excel Solver. What are the optimum quantities of X_1 and X_2 that you obtain? What is the minimum cost of the aforementioned diet?

   - Solve the following linear program problem graphically:
     Maximize $z = X_1 + 2X_2$
     s.t. $6X_1 + 3X_2 \leq 15$
     $2X_1 - X_2 \geq 4$
     and $X_1, X_2 \geq 0$.

---

3. Unfortunately, this technique has not been part of the curriculum in most of the universities specializing in nutrition sciences, and thus, nutritionists have not used this technique to determine the minimum cost of achieving an optimal diet.

Chapter 16

Food and nutrition program evaluation

Chapter outline

Introduction	575	Randomization and	
Recent developments	577	development policy: applying	
Randomization	577	the methods	588
Instrumental variables	579	**Summary and conclusions**	**591**
Difference-in-difference	583	Section highlights: nobelprize	
Regression discontinuity design	585	worthy	591
Propensity score matching and		**STATA workout 1**	**592**
pipeline comparisons	587	**STATA workout 2**	**594**

Randomized controlled trials have been used in economics for 50 years, and intensively in economic development for more than 20. There has been a great deal of useful work, but RCTs have no unique advantages or disadvantages over other empirical methods in economics. They do not simplify inference, nor can an RCT establish causality. Many of the difficulties were recognized and explored in economics 30 years ago but are sometimes forgotten.

Angus Deaton, 2020

Introduction

Addressing problems of food security and malnutrition requires designing and implementing cost-effective intervention programs. However, much of the programs implemented to do so go beyond the pilot stage. Projects that are not evaluated before scaling up to a large intervention program can fail and waste resources. Evaluating food security and nutrition interventions requires a good understanding of the methods involved and appropriate application of these methods. In this chapter, we review the methods for program evaluation and introduce the reader to the latest development in the field including the randomized control trials (RCTs) that have come to dominate the program evaluation literature.

Evaluating the impact of the food and nutrition programs on the beneficiaries is a key activity to save resources and to redefine the program to meet its welfare goals. A sad truth in development practice is that the pilot programs are implemented poorly and some of them are scaled up even before they are evaluated for their impact on the targeted population. This is partly because the capacity for program evaluation is weak. In this chapter, we review the methods of program evaluation and present the examples of the methods.

Suppose we want to answer questions such as "What is the effect of providing free lunch at schools on attendance and grades?" To correctly answer this question, we should be able to find out how would students who had the free lunch have performed in the absence of the program. "What would have happened to these students in the absence of the free lunch program?" is known as a counterfactual question. This question is very difficult to answer. At any point of time, a student is either exposed to this program or not.

To answer the counterfactual question, we need information on a comparison group, or a control group that is not exposed to the program, but whose performance in attendance and exam grades are very similar to those of the students who availed themselves of the free lunch. While we may be able to find a group that did not participate in the free lunch program, it is not reasonable to compare this group's performance with those who received the treatment. Such a comparison is not reasonable, because the final performance on attendance and exam grades may be either due to the program, or due to preexisting differences between these two groups.

In other words, programs and treatments such as the introduction of free lunches, or textbooks, medication, and information brochures are placed in specific locations (such as poorer or richer neighborhoods). Usually individuals are screened for participation (on basis of height, weight, gender, or other measures). Furthermore, individuals are often asked to participate in these programs on a voluntary basis, creating a problem of "self-selection." Families chose to send their children to school based on location or on the school's reputation. For many of these and other reasons, the groups that do not participate in a given treatment are not a good comparison for those that are exposed to the said treatment. Any difference in the performance between these groups could be due to both the program itself, or to preexisting differences, which creates the problem known as "selection bias."

In recent years, development economists have designed many new procedures to tackle the selection bias and decompose the overall difference into a "treatment effect" and a bias term. The newer methods and research design typically compare outcomes between beneficiaries and a control group, both before and after a project has been implemented. The recent developments in the economics of impact evaluation are as follows:

1. Randomization
2. Instrumental variables
3. Difference-in-difference

4. Regression discontinuity (RD) design
5. Propensity score matching (PSM) and pipeline comparisons

Each of these methods seeks to find the true causal impact of a program by not only observing the participants' outcomes before and after the project but also answering the question "what would have been the participants' outcomes in the absence of the project?" both before and after the project. Finally, which method should be used for a particular program or a policy depends upon a combination of factors: program's design, implementation, data collection, and feasibility. We will provide a short description of each of these methods in the next section. We also illustrate the implementation of some of these methods through a simple example using STATA.

Recent developments

Randomization

Recent research in Development Economics, by Duflo et al. (2011); Banerjee and Duflo (2007, 2008, 2009), indicates the importance of randomized trials and natural experiments. Duflo et al. (2007) provide an excellent introduction to randomization and to all the recent econometric issues involved in impact evaluation. Randomization is a technique that randomly assigns a predetermined fraction of the eligible beneficiaries to the project, creating the treatment group. The remaining eligible beneficiaries make up the control group. The difference in outcomes between the treatment and control group is the impact of the project. For a recent treatment of introduction to the randomization, see Duflo et al. (2011).

We can use the example in Duflo et al. (2007) as a good starting point to illustrate the workings of randomization and all the other methods. Suppose we decide to provide free textbooks to children in schools. What effect will this treatment have on average grades? That is, suppose we are interested in measuring the impact of textbooks on learning. We can explore this issue by thinking about the average test scores of students in a given school i in two ways:

- Y_i^T = the average test score if school i has textbooks, and
- Y_i^C = the average test score of the *same* school i if it has no textbooks

We are interested in the outcome: $Y_i^T - Y_i^C$.

While theoretically every school has these two *potential* outcomes, only one is observed. Now suppose in the population of schools, there are several with textbooks, while others without. One easy method to evaluate the impact is to compute the averages from both groups and then calculate the difference D, i.e.,

$$D = E\left[Y_i^T / \text{school has textbooks}\right] - E\left[Y_i^C / \text{schools has no textbooks}\right]$$

$$= E\left[Y_i^T / T\right] - E\left[Y_i^C / C\right]$$

Now let us think of $E[Y_i^C/T]$ as the expected outcome of a student in the treatment group, who was not treated, which is a theoretical possibility. Suppose we add and subtract this to D, then we can decompose the effects as follows:

$$= E[Y_i^T/T] - E[Y_i^C/T] - E[Y_i^C/C] + E[Y_i^C/T]$$

$$= E[Y_i^T - Y_i^C/T] + E[Y_i^C - Y_i^C/T] \text{ ``the treatment effect'' ``the selection bias''}$$

The first term is the *treatment effect*, which indicates the effect of treatment on the treated. It captures the average effect of textbooks in the schools where textbooks were offered. The second term is the *selection bias*, which captures the potential untreated scores between treatment and control groups, i.e., treatment schools may have had different test scores on average, *even if they had not been treated*.

Suppose parents consider education as a high priority, and push the children to do their homework, and undertake private tuitions. In this case, $E[Y_i^C/T]$ would exceed $E[Y_i^C/C]$. In this case, the actual impact D would be biased upwards. The impact could also be biased downward if textbooks were given to schools in disadvantaged areas. As mentioned before, the entity, $E[Y_i^C/T]$, is not observed, and hence, it is not possible for researchers to account for selection bias. Duflo et al. (2007) provide a good summary of the current research that is underway to correct this empirical problem. According to Duflo et al. (2007), randomization solves the selection bias issue.

Randomization, or assigning the treatment randomly, is often the fairest way to allocate beneficiaries to a project. This is particularly useful, if the resources needed to roll out the project to all beneficiaries are not sufficient. By this method, each beneficiary has an equal chance of receiving the treatment. The regression counterpart to D is

$$Y_i = a + bT + e_i \tag{16.1}$$

where T is a dummy variable indicating the assignment to the treatment group. The estimated coefficient $\hat{b}$ via OLS captures the impact of the policy. Duflo et al. (2007) point out how this estimation procedure solves the problem of selection bias and further point out various strategies to correctly implement randomization. We end this section with a simple example to implement (Eq. 16.1).

Example: Test Scores and Free Lunch

Assume we have data on the test scores of students some of whom have been treated to a free lunch program in their schools. In particular, suppose 1043 students were randomly treated to the free lunch program and 1346 were not, for a total of 2389 students in the sample. Then a regression counterpart of Eq. (16.1) in this context is

$$Score_i = a + bD + e_i \tag{16.1.1}$$

where $Score_i$ is the score of the ith student in the sample, and D is a dummy variable taking the value 1 if the student was treated. The regression implemented in STATA produces the following output:

```
. regress score d

      Source |       SS       df       MS              Number of obs =    4778
-------------+------------------------------           F(  1,  4776) =    6.79
       Model |   45277.3565     1   45277.3565         Prob > F      =  0.0092
    Residual |   31855914.9  4776   6669.99893         R-squared     =  0.0014
-------------+------------------------------           Adj R-squared =  0.0012
       Total |   31901192.2  4777   6678.08085         Root MSE      =   81.67

       score |      Coef.   Std. Err.      t    P>|t|     [95% Conf. Interval]
           d |   6.206817   2.382272     2.61   0.009     1.536467    10.87717
       _cons |   1220.483   1.574075   775.37   0.000     1217.397    1223.568
```

The first line (**.regress score d**) is the command line in STATA that produces the OLS estimates of Eq. (16.1.1).

We observe each student for two time periods, and hence, we have a total of 4778 observations. The important portion of the output is the regression results in the last two lines. Based on the output, we can write the estimated regression as follows:

$$Score_i = 1220.5 + 6.2D + e_i$$

From this simple example, we see that participation improves the score by 6.2 points, hence indicating that the free lunch programs have some impact on test scores. The overall R-squared is 0.001 meaning that the model itself does not account for all variations in scores. This means that there are other factors not included in our model that may explain the variation in *Score*. The goal of this example is not to generate the best explanatory model, but to illustrate in simplest terms the implementation of randomization with a simplistic mock data on free lunch and scores.

Instrumental variables

As described in Chapter 14, instrumental variable (IV) is a technique that identifies variables that determine program participation or the receipt of a project, but which do not influence project outcomes. The IV method is used in statistical analysis to control for selection bias, due to unobservables. The IV method identifies the exogenous variation in outcomes attributable to the program, recognizing, for example, that the participant placement may not be random but purposive. The chosen instruments are first used to predict program participation. In the second step, we examine how the outcome indicator varies with the predicted values. That is, in the first step, the chosen instruments are used to simulate who would have been in the treatment group and who would have been in the control group, given that the participation was based on the instruments. The difference in outcomes between these simulated treatment and control groups is then the impact of the project. Often, one can use geographic variation in program availability and program characteristics as instruments especially when endogenous program placement seems to be a source of bias.

The efficiency of the estimator in Eq. (16.2) can be improved if we add some independent variables, say x_1 and x_2, such that we have

$$Y_i = a + bT + \alpha_1 x_{1i} + \alpha_2 x_{2i} + e_i \qquad (16.2)$$

Suppose we want to measure the weights of children who are in families that receive special food stamps that are valid only for healthy diet, such as vegetables and fruits, and not valid for unhealthy nutrition such as saturated fat in carbonated drinks and excess sugar as in fruit loops. Suppose our variable of interest in this outcome is Y_i for a given child in the observation. As usual, T reflects the time period of the treatment, and let us say that x_1 tracks the participation status ("received food stamps"). Unfortunately, in the real world, people do not follow the rules and find ways to hoodwink the system. In this selective food program assistance for instance, families may receive the food stamps but may simply exchange them with their neighbors or friends or others for cigarettes or even cash. Furthermore, to complicate the process, the neighbors or friends receiving the said food stamps, and who might end up using them, might unfortunately have been placed in our control group. This problem of *partial compliance* makes x_1 correlate with e_i, leading to a bias in the OLS estimator.

The IV estimation is very useful if we have data on the treatment received (x_1) and on another control variable, say z_1, which can be used as an instrument for the treatment actually received (x_1). For z_1 to qualify as a valid instrument for x_1, two conditions must be satisfied. First, it must be relevant, and this measure of relevance is usually given by a positive correlation between z_1 and x_1, or Corr $(z_1, x_1) \neq 0$. Second, the instrument must not be correlated with the random term, or the instrument must satisfy the exogeneity condition Corr $(z_1, e_i) = 0$. Thus, an instrument that is relevant and exogenous can capture movements in x_1 that are exogenous. If the initial assignment of the treatment is random, then z_1 is distributed independent of e_i, and the problem of partial compliance can be addressed successfully. In practice, the IV estimator is implemented via a 2SLS procedure, where in the first stage the originally designed independent variable x_1 is regressed on all the exogenous variables and z_1. That is, in the first-stage regression, we have

$$x_{1i} = \gamma + \delta T + \beta_1 x_{2i} + \beta_2 z_{1i} + \acute{o}_i \qquad (16.2.1)$$

In the second stage, we go back and estimate Eq. (16.2), after replacing x_{1i} with $\widehat{x}_{1i}$, the predicted value of x_{1i} obtained after the first-stage estimation in Eq. (16.2.1). That is, in the second stage, we estimate

$$Y_i = a + bT + \alpha_1 x_{1i} + \alpha_2 x_{2i} + e_i \qquad (16.2.2)$$

We end this section with a simple illustration in STATA that implements the issues discussed in Eqs. (16.2), (16.2.1), and (16.2.2). We continue with our previous example on test scores and free lunch.

Example: Test Scores and Free Lunch

Recall that we have data on the test scores of students some of whom have been treated to a free lunch program in their schools. We extend this example by adding observations on the same students for two time periods before treatment. Then a regression counterpart of Eq. (16.2) in this context is

$$Score_{it} = \alpha + \beta_1 D + \beta_2 Treat + \delta D_1 \times Treat + \varepsilon_{it} \qquad (16.2.3)$$

In Eq. (16.2.3), the variable D is a dummy variable that tracks whether the given student i in the sample "received the treatment" (or $D = 1$), and the *Treat* is a dummy variable that takes the value 1 during the posttreatment period. As discussed earlier, suppose we wish to use the IV method and attempt to use an instrument for D, and let us say we have information on the family income level of each student.

Consequently, we can regress D on all the variables in the system, including income, as indicated in Eq. (16.2.1) and obtain the first-stage estimates. We can then derive the second-stage results using the steps outlined in Eq. (16.2.2). We reproduce this portion of the STATA output as follows.

```
. ivregress 2sls score treat td (d = income), vce(robust) first
```

First-stage regressions

```
                              Number of obs   =      7074
                              F( 3,   7070)   = 953307.04
                              Prob > F        =    0.0000
                              R-squared       =    0.3342
                              Adj R-squared   =    0.3339
                              Root MSE        =    0.4047
```

d	Coef.	Robust Std. Err.	t	P>|t|	[95% Conf. Interval]
treat	-.4349979	.007227	-60.19	0.000	-.4491649 -.4208308
td	.9987585	.0007235	1380.54	0.000	.9973403 1.000177
income	-.0008559	.0002877	-2.98	0.003	-.0014198 -.000292
_cons	.4763401	.0155712	30.59	0.000	.4458159 .5068643

Instrumental variables (2SLS) regression

```
                              Number of obs   =    7074
                              Wald chi2(3)    = 8030.66
                              Prob > chi2     =  0.0000
                              R-squared       =  0.2040
                              Root MSE        =  131.32
```

score	Coef.	Robust Std. Err.	z	P>|z|	[95% Conf. Interval]
d	-213.407	109.9931	-1.94	0.052	-428.9896 2.175474
treat	143.5663	47.93992	2.99	0.003	49.60583 237.5269
td	218.8511	110.0504	1.99	0.047	3.156234 434.5459
_cons	1107.694	47.9302	23.11	0.000	1013.752 1201.635

Instrumented: d
Instruments: treat td income

The STATA command in the first line (ivregress 2sls score treat td (d = income), vce(robust) first) is entered at the command line, which runs the IV regression using the 2SLS procedure. In addition, we specify that *income* is an instrument for D. The option vce(robust) is used to produce efficient standard errors to control for heteroscedasticity, and finally, the option first produces the first stage estimates.

The first part of the output presents the first-stage regression of the exogenous variable D_i on all the exogenous variables, including *income*. The first-stage regression produces a decent fit of the model, and the coefficient of income is negative, and highly, statistically significant.

The second-stage regression results are in the second portion of the output. These results are key to the example, where $Score_i$ is regressed on all the exogenous variables. Note that the variable D is negative and significant at 10% level. Consequently, participation in the program has a significant effect on scores, even after controlling for income levels. Finally, we note that Treat is still positive and significant.

Difference-in-difference

The difference-in-difference (DD) is a quasi-experimental technique that measures the causal effect of some nonrandom intervention (Angrist and Kruegger, 1991; Wooldridge, 2002; Stock and Watson, 2011). It has been widely used in economics, education, and law to test the effectiveness of policy intervention. In the simplest quasi-experiment, an outcome variable is observed for one group before and after it is exposed to a treatment. The same outcome is observed for a second group (control group) that is not exposed to the treatment. The change in the outcome variable in the treatment group compared with the change in the outcome in the control group gives a measure for the treatment effect.

The DD procedure tells us whether a particular policy has an impact on those for whom the policy is intended. The DD method answers the counterfactual question: What would have happened to the outcome, if the said intervention had not taken place? If we can answer the counterfactual question, then we can compare this answer to the factual situation, wherein the policy intervention was initiated. The true impact of the treatment is the difference between the factual values and the answer to the counterfactual question.

The DD methodology is useful when we have natural experiments. A natural experiment arises when only a subset of participants is exposed to a policy intervention. The group that receives the treatment is called the treatment group, and the group that does not receive the treatment is known as the control group. The policy intervention is expected to have some outcomes, exposing the intervention to the treatment group, while leaving the control group out of the intervention. This methodology provides a good framework, to study the true impact of the intervention. After the intervention, one can examine the differences in the outcomes from both the groups. Significant differences in outcomes between the treatment and control group, holding other things equal, indicate the necessary and the intended effects of the said intervention.

The difference in a given outcome between recipients of the project (the treatment group) and a comparison or control group is computed before the project is implemented. This difference is called the "first difference." The difference in outcomes between treatment and control groups is again computed sometime after the project is implemented, and this is called the "second difference." Under the difference-in-difference technique, the impact of the project is the second difference less the first difference. The logic is that the impact of the project is the difference in outcomes for treatment and control groups after the project is implemented, net of any preexisting differences in outcomes between treatment and control groups that predate the project.

It is usual to define a treatment as a form of policy intervention. The outcome variable can be defined as a child's height or weight or a related index to denote the nutritional status, which is observed for pre- and postpromotion time periods. We also have a control group, which received no intervention during the entire time period. Let y be an outcome variable such as the BMI of a child. Then the effect δ_d of a policy intervention on BMI is given by

$$\delta_d = \left(\overline{y}_{at}^d - \overline{y}_{bt}^d\right) - \left(\overline{y}_{at}^n - \overline{y}_{bt}^n\right)$$

where the superscript d denotes the child that receives the treatment; the superscript n denotes the child that did not receive any treatment i.e., the control group; the subscript bt denotes the time period before the treatment started, while the subscript at denotes the period after the treatment started.

The expression $\overline{y}_{at}^d$ denotes the average BMI, in the treatment group after the treatment and the expression $\overline{y}_{bt}^d$ is the average BMI in the treatment group before the treatment; the expressions $\overline{y}_{at}^n$ and $\overline{y}_{bt}^n$ are the corresponding averages of the control group. The effect δ_d is the difference of the two differences between the treatment and control groups.

The first difference $\left(\overline{y}_{at}^d - \overline{y}_{bt}^d\right)$ captures the difference in mean BMI before and after the implementation of the policy in the *treatment* group, while the second difference $\left(\overline{y}_{at}^n - \overline{y}_{bt}^n\right)$ captures the difference in mean BMI between the two periods in the *control* group. The DD method cancels out the common trends in the control and the treatment group, and hence, the resulting difference between the two differences accounts for the effect of the promotion.

In practice, the DD estimator is implemented as a regression equation, where the level of the outcome variable, s, is used to estimate the model (see Stock and Watson, 2011; page 493):

$$s_{it} = \alpha + \beta_1 X_i + \beta_2 T + \delta X_i * T + \varepsilon_{it} \qquad (16.4)$$

where s_{it} is the BMI of, say, a given child i at time t, and X_i is a dummy variable taking the value 1 if the child is in the treatment group and 0 if it is in the control group, and T is a dummy variable which takes the value 1 in the postpromotion period and 0 otherwise. The DD estimator is δ, the coefficient of the interaction between X_i and T. Note this interaction term is a dummy variable that takes the value 1 only for the treatment group in the postpromotion period. ε_{it} is the error term assumed i.i.d Normal.

Within this framework, additional treatment effects can be easily added, which is very useful for policy analysis. We provide an example in the following to illustrate this method. It is possible to extend the aforementioned equation accordingly and estimate many general models to illustrate the importance of the DD method. We provide an example of the DD method continuing with the mock data from the school free lunch program and test scores.

Example: Test Scores and Free Lunch Continued

Recall that we have data on the test scores of students some of whom have been treated to a free lunch program in their schools, in which 1043 students were randomly treated to the free lunch program, while 1346 were not. We extend this example by adding observations on the same students for two time periods before treatment, for a total of 9556 students in the sample. Then a regression counterpart of Eq. (16.4) in this context is

$$Score_{it} = \alpha + \beta_1 D_i + \beta_2 Treat + \delta D_i * Treat + \varepsilon_{it} \qquad (16.4.1)$$

We reproduce this portion of the STATA output in the following. The estimated regression from the output is

$$Score_{it} = 1009.2 + 12.14 D_i + 211.2 Treat - 5.9 D_i * Treat + \varepsilon_{it}$$

The results tell us that being in the treatment group ($D = 1$) or in the treatment period ($Treat = 1$) increases the score of the ith student in the sample within the said groups. However, our coefficient of interest ($\delta = -5.9$) is negative and insignificant. Note that the variable $D_i \times$Treat (variable td in the STATA output) is 1 only in the posttreatment period for the participants in the sample.

Regression discontinuity design

This is used when a cutoff point on a continuous variable such as a poverty index is used to determine who receives benefits from a given project. The impact of the project can then be estimated by comparing outcomes for beneficiaries who just qualify for the project on this score, with outcomes for individuals who just fail to qualify for the project given their score. Stock and Watson (2011) provide a good example to illustrate the application of the RD estimators. Suppose students are mandated to enroll in a summer school for intensive study, if their grades from the previous year fall below a certain threshold. Suppose all the students whose grades were below the threshold attend the summer school. As a consequence, their grades in the next year should improve. Thus, we can define an outcome variable, which is basically the next year's grade of all students, including all nonparticipants. If the minimum grade threshold was the only requirement to attend the summer school program, then it is reasonable to conclude that any jump in the next year's grade can be attributed to summer school attendance. Since the cutoff occurs at some arbitrary threshold, the RD can be estimated as

$$GPA_i = a_0 + b_1 Program_i + b_2 Grade_i + e_i \tag{16.5}$$

where GPA_i is the next year's GPA of the ith student in the sample. $Grade_i$ is the last year's grade of the ith student. $Program_i$ is a dummy variable that takes the value 1, if the student attended the summer program, and this depends on the mandated threshold ($Program_i = 1$, if $Grade_i < g_0$). Many alternative specifications and interactions could be included in Eq. (16.5) to capture the effect of the treatment. Finally, it is possible that many students who were supposed to attend the program, and further, many students who were above the cutoff attended the program anyway. Stock and Watson (2011, page 495) discuss ways of dealing with these types of fuzzy situations.

Lee and Munk (2008) and McEwan and Shapiro (2008) provide many examples, wherein treatments are administered after establishing appropriate cutoff points. Some useful policy interventions that have been analyzed using the RD methods are for remedial reading and math classes for underprepared students, and for the effect of delayed school entry on academic outcomes.

Using the example from the previous section, a program can be designed such that free meals are provided to children, whose family income is below a certain threshold. Suppose the relationship between the outcome variable (say exam scores, Y) and an independent variable, say X, is given by the following simple regression:

$$Y = a + bX + \varepsilon$$

After receiving free meals, some of the treated students may end up with higher scores, which may shift the regression line for these individuals by a factor d, so as to affect the aforementioned regression equation in the following manner:

$$Y = a + dT + bX + \varepsilon$$

where T is a dummy variable, with the value equal to 1, if the student gets free meals. The idea behind RD is to estimate the program's effect or arrive at d. The following graph from Lee and Lemieux (2010) shows the jump at the cutoff point given by c on the horizontal axis.[1] The effect of the treatment effect is given by τ, and (A'' & B') are observations such as close to the cutoff at c, which would yield the value of τ, as c is also assumed to be close to (c'' & c').

---

1. As Lee and Lemieux (2010) provide an excellent survey of the RD method, its implementation, and limitations.

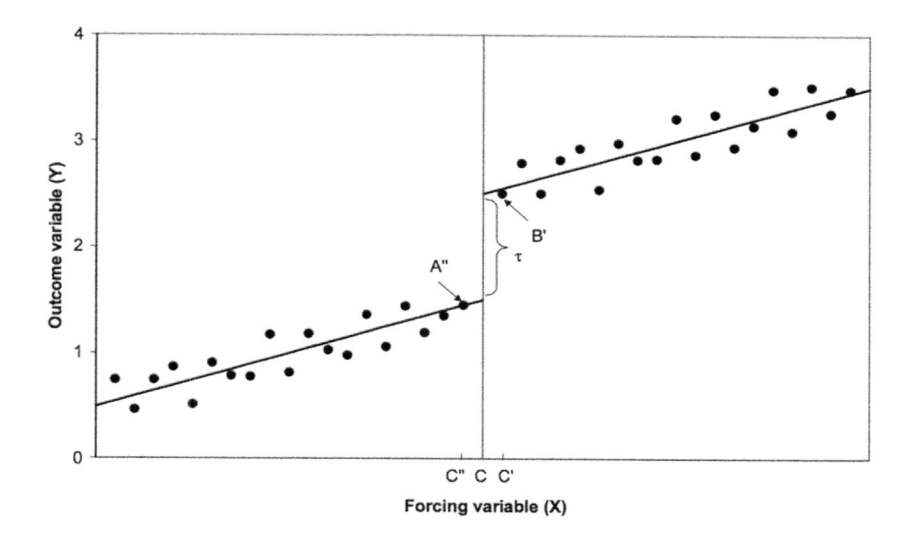

[Figure: RD plot with outcome variable (Y) vs forcing variable (X), showing points A", B', τ, and C", C, C' on x-axis]

The important issue in RD is once again, the answer to the counterfactual: An individual i in the sample is either exposed to the treatment and has an outcome $Y_i(1)$, or is not exposed to the treatment and receives. The basic idea is in RD, as in DD is to estimate τ, by computing $Y_i(1) - Y_i(0)$. Several refined methods and techniques have been developed by researchers studying program evaluation and intervention efforts. For example, Lee and Lemieux (2010) graph the data, like the one shown before, and place the variable with the cutoff value into different bins. The average value of the outcome variable is computed for each bin, and this average is graphed against the midpoints of the bins. This allows them to obtain a polynomial regression specification.

Ludwig and Miller (2007), Lee and Lemieux (2010), Imbens-Kalyanaraman (2012), Calonico et al. (2014a), or CCT (2014a) are some of the major econometric studies that provide alternative methods and techniques to derive RD estimators for τ. Calonico et al. (2014b) also produce newer methods that generate globally robust estimates, and confidence bands in RD plots. These methods are illustrated in STATA workout section at the end of the chapter. Also see Babu Gajanan and Hallam (2017, Chapter 13) for recent examples relating school lunch provisions on academic performance and welfare, and also STATA program code and output that apply RD estimation, following CCT (2014a,b) and Calonico et al. (2016).

Propensity score matching and pipeline comparisons

PSM is another useful procedure that identifies a suitable control group that is comparable to the treatment group. PSM identifies a group of individuals, who are not exposed to the treatment, but who, with their given observable characteristics, have the same probability of receiving the treatment, as individuals in the treatment group.

Finally, the project's impact is computed through the difference in outcomes between the treatment and comparison group, based on the assumption that the treatment group members are similar to those who are about to receive the project. This assumption allows the PSM method to produce matching scores for the two groups, in the way the treatment was initially administered. Rosenbaum and Rubin (1983) and Becker and Ichino (2002) have produced extensive methods and procedures to compute the average effect of treatment on the treated (ATT). A brief implementation of these methods following the procedures developed by Becker and Ichino (2002) is provided later in the chapter. Additional examples following these STATA methods are in Babu, Gajanan, and Hallam (2017, Chapter 13).

Randomization and development policy: applying the methods

Randomization along with the associate econometric techniques has become an attractive way to model program evaluation and is actively used by development economists.[2] Several interesting examples are available in the literature, which demonstrate the application of the methods discussed in this chapter: DD, RD, and PSM alongside randomization and related instrumental variables. For example, Warren et al. (2020) use DD method to examine whether infant and child feeding improves, when households can gain knowledge, and as a consequence, undertake changes in their behavior. The researchers use the Alive and Thrive data from Bangladesh and note that households exposed to intensive intervention increased their expenditures on important food items for mothers and children. Households that received intensive intervention were exposed to campaigns and mass media promotions pertaining to the advantages of breastfeeding and related feeding. Intensive intervention also involved interpersonal counseling and community mobilization. The DD results indicate that, compared with nonintensive groups, households that received intensive interventions had higher expenditures on eggs and flesh foods, higher women's employment and greater autonomy over income, and higher consumption of key food groups among women and children. Furthermore, in the targeted areas, jewelry ownership decreased significantly.[3]

Likewise, using an interesting data set covering Muslim and non-Muslim immigrants in Denmark, Greve et al. (2017) show that the Muslim students, particularly females and those from low socioeconomic background, score poorly in Danish, English, Mathematics, and Science. The researchers use DD

---

2. See Duflo et al. (2011) and Duflo et al. (2007) for the theoretical developments, methodological foundations, and limitations of randomization. For other useful applications, see the World Bank Africa Program.
3. Chowhan and Stewart (2014) and You et al. (2016) are good examples of applications of IV toward maternal employment, nutrition, and China's paradoxical situation where economic growth accompanies a decline in nutrient intake.

and demonstrate that the lower scores of the treated Muslim immigrant children's can be attributed to fetal malnutrition that they could have been subjected to, due to their mothers' fasting during Ramadan.

Headey and Palloni, 2019 conduct a DD analysis for a big panel across 59 countries and show how access to piped water helps in reducing diarrhea and child mortality but has insignificant effects with stunting and wasting. Singh (2017) also provides evidence of positive spillover externalities derived from nutritional health information programs within urban slums in Chandigarh, India, using a DD approach.

Chakrabarti et al. (2019) apply the DD method to show how fortified wheat programs in India were partially successful in combating anemia. The idea that commuting to work has a positive influence on dietary diversity is evidenced for rural India by Sharma and Chandrasekhar (2016), where the DD method shows that households with rural—urban commuters enjoy higher dietary diversity, compared with the control groups with no commuters.

Similarly, Scarlato et al. (2016) combine DD and RD methods and show the positive effect of Chile Solidario cash transfer program on women employment. Another interesting study that adopts RD methodology is by Guarnieri and Rainer (2018), which examines the impact of women's empowerment and domestic violence. The researchers use a novel data set from Cameroon that has information from two distinct colonial regimes: France and the United Kingdom. The RD results indicate that women from the British areas benefitted from a universal education system and also enjoyed better labor outcomes, compared with their counterparts from the French territories. Ironically, compared with the women from French territories, women from the British territories were 30% more likely to be victims of domestic violence.

In a detailed study linking DD and PSM, Aurino et al. (2019) examine the impact of emergency school feeding and food distribution program on academic performance, during conflict in Mali. School feeding has a positive impact on enrollment, additional time in school, and lesser time spent in work among girls, particularly in those areas with high conflict intensity. While some areas also showed a decline in attendance, particularly among boys, school feeding raised outcomes for children who were not close to the conflict-infested areas.

Tranchant et al. (2019) also examine the impact of conflict on food security in Mali. Combining DD and PSM methods, the researchers find that food assistance has a positive influence on nonfood and food expenditures, caloric and micronutrient consumption, children's height, and improved zinc consumption, particularly among the households located in close proximity to armed groups.

In this context, Alemu et al. (2018) apply PSM estimator and show that self-help groups (SHGs) in apple production in Ethiopia help toward empowering women at the community level. However, the treated group exposed to SHG also suffer major "backlash" from husbands, indicating systemic gendered inequalities. Relatedly, Deininger and Liu (2009) and Swain and Verghese (2014) also examine the same issue using PSM.

Bose and Das (2017) also indicate using a DD approach that the amendment to the Hindu Succession Act in India significantly increased women's education but, at the same time, decreased educational attainment of children, particularly boys, of the treated mothers. Grillos (2018) finds similar results using DD, with increased women empowerment at the community level, based on interventions to increase drought preparedness in Kenya.

Kabunga et al. (2017) apply PSM methods and show that households adopting dairy cows exhibit higher milk sales, intakes along with lower child stunting, indicating the importance of livestock promotion. In another setting, Kumar et al. (2016) apply PSM and demonstrate the success of the public distribution system (PDS) in India.

The impact of maternal employment on child nutritional status has been discussed in several parts of the book. Interestingly, Rashad and Sharaf (2019) conduct PSM and IV for data from Egypt and show that maternal employment has adverse effects on child health, once again stressing the importance of the underlying pathways that lie hidden in the nutrition transmission process. Employment and education of women are important drivers of development pathways, and Samad and Zhang (2019) use PSM to show that access to electricity positively influences women's empowerment in rural India. Relatedly, Melesse et al. (2018) use PSM to show how joint land certification has a significant and a positive impact on women's empowerment in Ethiopia.

Verwimp and Munoz-Mora (2018) also adopt PSM and IV methods to an interesting data set from Burundi, which covers food security of individuals and households who return home after being displaced due to civil war and internal conflicts. When compared with the control group, namely the non-displaced or those that had returned many years ago, the treated groups covering recent returnees display greater food insecurity.

A host of studies employ these methods to interesting data and examine all the issues that have been discussed in this book, dealing with nutrition, health, women empowerment, and poverty, alongside cash transfer programs and related public interventions. Averett and Smith (2014) use the PSM method to investigate whether debt leads to obesity and note that the effect of "credit card debt" has significant influence on women's weight.

Chase and Sherburne-Benz (2001) adopt the pipeline comparison, asking the question "What is the effect of the social fund on education and health outcomes?" in Zambia. The study concludes that the social fund positively influences school attendance and the households' share of expenditure on education.

Likewise, Novak (2014) dealing with clean water in Senegal and the "Accelerated Electricity Access Expansion" for Ethiopia are all good examples of applying PSM techniques to address key questions in development policy. The World Bank website also presents other interesting applications of this method. Rosenbaum and Rubin (1983), Imbens (2010), Dehejia and Wahba (2002), and Babu, Gajanan and Hallam (2017) all provide detailed descriptions and methods associated with the development and application of PSM.

Summary and conclusions

Evaluating food and nutrition interventions has been a key area for interest to policy analysts for sometime (Babu, 2002b). In recent years, economists have started looking at policy and the impact of development programs on targets and the sample population, using different econometric techniques, to control for self-selection and related problems with design. In this chapter, we examined five popular techniques: randomization, instrumental variables, DD, RD, PSM, and pipeline comparison. We also illustrated the implementation of the first three methods through an illustrative example using STATA.

Section highlights: nobelprize worthy

Professors Abhijit Banerjee, Esther Duflo, and Michael Kremer (BDK) were awarded the 2019 Nobel Prize in Economics, for their innovative methods involving experimental methods and RCTs. BDK's contributions span several key areas in Development Economics: women empowerment, gender inequality, schooling, health, nutrition, environment, and poverty. BDK's influence with randomization and experimental techniques has spawned countless studies in developing countries, such as those using Alive and Thrive data for Bangladesh, Ethiopia, and Vietnam, or the PROGRESA scheme for Mexico.[4]

Duflo (2012) and related papers demonstrate that while women's opportunities expand with policies that promote overall development and target specific women empowerment facets, such policies are not sufficient to combat systemic gender inequalities. Rather, Duflo (2012) argues that for a long time to come, it behooves agencies to promote women in favor of men.

Miguel and Michael's (2004) seminal work and follow-up research on deworming studies in Kenyan schools, and the role of water and sanitation has significantly influenced research and application of RCTs in health economics, chartering new areas that stress the role of externalities and careful data gathering.[5]

The experimental designs and RCTs have also helped researchers understand the subtleties inherent in credit markets, public school programs,

---

4. See de Brauw and Hoffman (2020), Quattrochi et al. (2020), Rodgers et al. (2020), and Das (2020) for an extensive discussion of these applications.
5. Subramanian (2007) explores the contributions of Michael Kramer's work Development Economics and related areas in great detail.

nutrition fortification, marriage markets, and the difficulties that arise with program implementation. Banerjee and Duflo's *Poor Economics* has generated several conversations and studies in the field and is currently a standard text in economics departments across the globe, for classes in Poverty, Policy, and Development Economics. BDK's methods have also been successfully applied to examine policies toward corruption, drunk-driving, trade sanctions on dictatorial regimes, and the effectiveness of humanitarian aid.

Although the RCTs and the techniques described earlier have become very popular among development economists, these have also been subject to many criticisms. Pinstrup-Andersen (2012), Heckman and Sergio (2009), and Deaton (2010) have expressed doubt regarding the theoretical foundations underlying the econometrics. Deaton (2010) has also questioned the validity of the results of Imbens and Angrist (1994), concerning the local average treatment effect (LATE). Rosenzweig (2012) and Ravallion (2012) have noted related criticisms and qualifications pertaining to estimation methods and scalability of experimental designs. Dehejia (2013) and White (2014) have evaluated all the methods and have provided insights about the internal consistencies across these applications. Menon et al. (2020) demonstrate how RCTs and experimental methods can combine other related information to produce useful results for next-generation policy proposals. Babu, Gajanan and Hallam (2017) provide a summary of the criticisms and practical tips to pursue intervention methods.

Besides these methods, there are other methods that have also been used, such as PSM and pipeline comparison. PSM is a tool for identifying a suitable comparison group to compare to the recipients of the project (the treatment group). Essentially, PSM finds a comparison group comprising individuals who did not in fact receive the benefits of the project, but who, given their observable characteristics, had the same probability of receiving the benefits of the project as individuals in the treatment group. The project's impact is then the difference in outcomes between the treatment and comparison group. Similarly, pipeline comparison method compares outcomes of beneficiaries who have already received the benefits of the project, with those who have not yet received the benefits of the project but are about to. This method relies on the assumption that the beneficiaries who have already received the benefits of the project are similar to those who are about to receive the benefits of the project.

STATA workout 1

The following table presents observations on 32 students based on the example on textbooks and score: $D = 1$ indicates whether the student received the treatment, and the variable $Treat = 1$ indicates that the treatment was administered. Income and score represent family income and the test scores.

Use these data to examine the effect of this RCT. Also check the results based on the DD estimator.

Obs	I	D	Treat	Income	Score
1	1	0	0	3276.009	45
2	1	0	1	5992.015	45.5
3	2	0	0	7304.947	55
4	2	0	1	9454.144	55.8
5	3	0	0	7188.219	63
6	3	0	1	2268.913	60.2
7	4	0	0	9344.465	55.5
8	4	0	1	2927.189	65.3
9	5	0	0	6691.779	45.6
10	5	0	1	3209.162	48.7
11	6	0	0	4107.006	54.7
12	6	0	1	7766.274	55.8
13	7	0	0	1546.991	60.2
14	7	0	1	7514.323	63.2
15	8	0	0	1737.605	55.7
16	8	0	1	4001.105	65.3
17	9	1	0	4410.61	55.6
18	9	1	1	8482.669	58.7
19	10	1	0	8353.924	64.7
20	10	1	1	3675.96	65.8
21	11	1	0	2179.76	70.2
22	11	1	1	2280.997	73.2
23	12	1	0	10,067.33	65.7
24	12	1	1	2783.191	75.3
25	13	1	0	3060.225	65.6
26	13	1	1	3036.27	68.7
27	14	1	0	2168.629	74.7
28	14	1	1	2626.472	75.8
29	15	1	0	2570.849	80.2
30	15	1	1	2837.423	83.2
31	16	1	0	7736.791	75.7
32	16	1	1	9037.849	85.3

As mentioned in the discussion surrounding RCTs, the following command line implements the effect RCT:

```
. regress score d
```

Source	SS	df	MS		Number of obs	=	32
					F(1, 30)	=	32.00
Model	1858.97493	1	1858.97493		Prob > F	=	0.0000
Residual	1742.88959	30	58.0963196		R-squared	=	0.5161
					Adj R-squared	=	0.5000
Total	3601.86452	31	116.189178		Root MSE	=	7.6221

score	Coef.	Std. Err.	t	P>|t|	[95% Conf. Interval]
d	15.24375	2.694817	5.66	0.000	9.740197 20.7473
_cons	55.90625	1.905524	29.34	0.000	52.01465 59.79785

Based on the output, the estimated equation is

$$Score_i = 55.9 + 15.24D + e_i$$

In this example, participation improves scores by 15.24 points. The coefficient of D is significant, and because of a single variable regression, the R-squared is only around 0.51. The results can be compared with the following DD estimator executed:

```
. gen td = treat*d

. regress score d treat td
```

Source	SS	df	MS			
Model	1968.91068	3	656.30356	Number of obs	=	32
Residual	1632.95384	28	58.3197799	F(3, 28)	=	11.25
				Prob > F	=	0.0001
				R-squared	=	0.5466
				Adj R-squared	=	0.4981
Total	3601.86452	31	116.189178	Root MSE	=	7.6367

score	Coef.	Std. Err.	t	P>|t|	[95% Conf. Interval]
d	14.7125	3.818369	3.85	0.001	6.890922 22.53407
treat	3.137501	3.818369	0.82	0.418	-4.684074 10.95908
td	1.062502	5.39999	0.20	0.845	-9.998875 12.12388
_cons	54.3375	2.699995	20.13	0.000	48.80681 59.86819

The estimated regression in this case is

$$Score_{it} = 5433 + 14.71D_i + 3.13Treat + 1.06td + e_{it}$$

While the coefficient of *td* is positive, the interaction along with *Treat* is insignificant. The only significant variable is D, which indicates that being in the treatment group is relatively more important than the actual treatment period. The next exercise compares these outcomes to pipeline comparison.

STATA workout 2

Use the same data from before and the STATA algorithm developed by Becker and Ichino (2002) to produce ATT using stratification matching.

The **pscore** command: This estimates the propensity score and tests whether the balancing property holds. The propensity score is the probability of getting a treatment for each student, and the balancing property tests the assumption that the observations with the same propensity score has the same distribution of characteristics (income) independent of whether the student receives a textbook.

The option **pscore(ps1)**: This specifies that the variable name for the estimated propensity score is *ps1*.

Comsup: This specifies that the analysis be restricted to the region of common support within the distributions of the treated and the control groups.

The input line and the output from the following STATA show that the regions of common support are [0.406, 0.506]. The results from the probit estimation are in the first step:

```
pscore d treat income, pscore(ps1) blockid(blockf1) comsup level(0.001)
```

```
***************************************************
Algorithm to estimate the propensity score
***************************************************

The treatment is d
```

D	Freq.	Percent	Cum.
0	16	50.00	50.00
1	16	50.00	100.00
Total	32	100.00	

```
Estimation of the propensity score

Iteration 0:    log likelihood = -22.18071
Iteration 1:    log likelihood = -22.008047
Iteration 2:    log likelihood = -22.008047

Probit regression                                Number of obs   =      32
                                                 LR chi2(2)      =    0.35
                                                 Prob > chi2     =  0.8414
Log likelihood = -22.008047                      Pseudo R2       =  0.0078
```

d	Coef.	Std. Err.	z	P>|z|	[95% Conf. Interval]
treat	-.0117745	.4449567	-0.03	0.979	-.8838737 .8603246
income	-.000048	.0000816	-0.59	0.557	-.000208 .000112
_cons	.2453721	.522652	0.47	0.639	-.779007 1.269751

```
Note: the common support option has been selected
The region of common support is [.40610678, .55619935]
```

Becker and Ichino (2002) present all the details of PSM implementation in STATA. The algorithm computes the propensity score, and sorts the data and places the propensity scores in matching intervals. The next step is to compute the ATT. Becker and Ichino (2002) present four methods of sorting the data: stratification matching, radius matching, kernel matching, and nearest-neighbor matching. The following example presents the results of stratification matching. Stratification matching splits the variation in the propensity score into different intervals, such that the treated and the control units within each interval have the same propensity score on average.

```
Note: the common support option has been selected
The region of common support is [.40610678, .55619935]


Description of the estimated propensity score
in region of common support

                    Estimated propensity score
      ─────────────────────────────────────────────────────────
            Percentiles      Smallest
       1%     .4061068       .4061068
       5%     .4129626       .4129626
      10%     .4201823       .4196087     Obs                30
      25%     .4495198       .4207558     Sum of Wgt.        30

      50%     .5179488                    Mean         .4957299
                              Largest     Std. Dev.    .0512459
      75%     .5392598       .5494117
      90%     .5495264       .5496412     Variance     .0026261
      95%     .5559884       .5559884     Skewness    -.3949527
      99%     .5561993       .5561993     Kurtosis     1.562483


************************************************************
Step 1: Identification of the optimal number of blocks
Use option detail if you want more detailed output
************************************************************

The final number of blocks is 3

This number of blocks ensures that the mean propensity score
is not different for treated and controls in each blocks
```

The output indicates the number of blocks and presents the implementation and shows how the algorithm satisfies the definition and balancing (wherein, within each interval, the means of each characteristic is different between the treated and control groups).

The balancing property is satisfied, and the algorithm has identified three blocks to match the scores. Of interest is to see if the textbooks affect the scores, between the matched treatment and control groups.

The balancing property is satisfied

This table shows the inferior bound, the number of treated
and the number of controls for each block

Inferior of block of pscore	D 0	1	Total
.4	14	16	30
Total	14	16	30

Note: the common support option has been selected

```
*******************************************
End of the algorithm to estimate the pscore
*******************************************
```

The command **atts score d treat dt income, pscore(ps1) blockid(blockf1) comsup** generates the ATT using the stratification method. The following result indicates that the ATT in this case is 15.5 points:

ATT estimation with the Stratification method
Analytical standard errors

n. treat.	n. contr.	ATT	Std. Err.	t
16	14	15.536	2.844	5.462

See Babu, Gajanan and Hallam (2017, Chapter 11) for the other methods (stratification, radius matching, and kernel matching) implemented in STATA for this example, following Becker and Ichino (2002).

Chapter 17

Multidimensional poverty and policy

Chapter outline

Multidimensional child poverty and gender inequalities	602	The Alkire—Foster method	604
		STATA implementation	609
Multidimensional energy poverty	603	STATA workout	614
Financial exclusion and Multidimensional Poverty Index	604		

> *As long as poverty, injustice and gross inequality persist in our world, none of us can truly rest*
>
> Nelson Mandela.

The Alkire—Foster (AF) method to construct the Multidimensional Poverty Index (MDPI) for population and subgroups has become very influential in poverty analysis and policy. In a series of papers, Professors Sabina Alkire and James Foster (2011, 2013)[1] demonstrate that poverty analysis cannot be confined narrowly to a unidimensional metric, usually based on income. Rather, the construction of a poverty index must include a whole range of deprivations, such as access to health, food, clean water, sanitation, schooling, housing, and empowerment. Indeed, throughout this book, there has been an emphasis on several of these topics, and this chapter provides a natural transition toward the measurement of poverty via MDPI.

The AF method has also attracted researchers because the ensuing MDPI satisfies several theoretical properties that are required of measurement indices: symmetry, scale invariance, transferability, and monotonicity. An attractive feature of the AF method is that it allows for the decomposition of the population among different subgroups and across different dimensions.

Studies that have applied and extended MDPI to different settings have proliferated in recent years. For instance, Alkire et al. (2017b) illustrate the robustness of MDPI for an extensive data set covering 34 counties covering 2.5 billion people. The study identifies those nations where poverty reduction has

---

1. For related work by AF and relevant history, see Akire et al. (2015).

Food Security, Poverty and Nutrition Policy Analysis. https://doi.org/10.1016/B978-0-12-820477-1.00025-5
Copyright © 2022 Elsevier Inc. All rights reserved.

been fastest, particularly among the poorest. The study also explores detailed decomposition analysis to identify differences between urban—rural locations, subnational regions, and ethnic groups.

In another study using data from China, Alkire and Fang (2019) note that the mismatch between MDPI and income-based metrics. Furthermore, Alkire and Fang (2019) also show that MDPI is much more stable than income-based measures. Alkire et al. (2017) also use data from Chile to illustrate how the construction of MDPI is sensitive to the share of dimensions in which households are deprived and to the duration of poverty.

Using Chinese panel data, Shen et al. (2019) note that the level of multidimensional poverty in China is relatively low and has decreased over time. Although the poor are deprived in key dimensions, as in health and education, overall, the researchers conclude that China's development-oriented commitment has greatly improved the capabilities of the poor and could serve as a model for other economies.

Several studies have explored the MDPI measures for Africa. Ogutu et al. (2020) examine the influence of having contracts with supermarkets on the poverty status of smallholder farmers in Kenya. Using panel data of small vegetable farmers from Kenya, the researchers found that significant reductions in income and multidimensional poverty accompany supermarket contracts among the poorest households.

In a detailed study for Ethiopia, Abeje et al. (2020) construct MDPI using the AF method and compare their results to Correlation Sensitive Poverty Index (CSPI). Their results show a large divergence between MDPI and CSPI estimates, in terms of identification of the poor. The importance of multiple dimensions such as living standards, land ownership, livestock ownership, and locational status together influence deprivations, and the study stresses this aspect for policy interventions.

Likewise, Berenger (2019) adopts the AF methodology and demonstrates the advantages of constructing an MDPI for four African countries, Malawi, Mozambique, Tanzania, and Zimbabwe and tracks the usefulness of Demographic and Health measures in subgroup decomposition.

MDPI studies have also been conducted for Asian economies. Pham et al. (2020) apply the AF method to construct and apply MDPI toward effective targeting strategies using Vietnam's Household Living Standard Survey data from 2014. The study examines the efficacy of the government's extant poverty fund-disbursement program and illustrates how the program fails to capture the complex reality behind the incidence of poverty. The study uses seven dimensions to capture poverty: income, health, education, housing, assets, basic services, and economic status. These dimensions are overlaid on spatial maps to arrive at decompositions based on commune/village, district, and province, yielding a much richer understanding of the diversity of poverty and the need to target the most depressed regions more effectively.

Iqbal et al. (2020) also find that an MDPI performs much better than a unidimensional headcount measure of economic deprivation for data from Punjab, Pakistan. Using the MDPI measures developed by AF, the research identifies the larger household size, inadequate wealth, low education level, and the geographical location as important dimensions that determine a household's poverty status.[2]

In an interesting study, Mahadevan and Jayasinghe (2020) examine the after-effects of a postwar conflict on multidimensional poverty in Sri Lanka. The study notes a decline in poverty in war-affected and nonaffected areas, as well as among all ethnic groups. The researchers attribute this decline in poverty to targeted policies in all regions. However, there are still lingering effects of war on the poverty status of Tamil minorities from the war-affected regions.

Chen et al. (2019) examine data from Taiwan and find that MDPI measures correlate significantly with age, socioeconomic status, marital status, household income, and household size, along with the level of urbanization and service-manufacturing ratio.

Alkire and Seth (2015) estimate MDPI for India between 1999 and 2006, during a time of high economic growth. Although economic growth indicates a fall in poverty at the national level, the reduction has not been uniform across different subgroups based on regions, castes, and religions and for the poorest groups. Gayathri and Rajagopal (2019) also note similar regional disparities in living standards and health, particularly in rural areas in India, using more recent data for Tamilnadu.

In an interesting assessment of "Trickle-Down" policies, Mitra (2018) examines NSS data for India and observed that several trends changed when the income-based measures were replaced with multidimensional poverty measurements to access Trickle-Down effect. One such major results uncovered that Hindus are poorer than Muslims, contradicting income-based findings.

Sevinc (2020) extends the AF measure of MDPI to include polychoric factor loadings to fine-tune the weighting structure of the dimensions in the achievement matrix for data from the United Kingdom. The novel approach produces consistent measures of poverty, and the study finds that multidimensional poverty has fallen during the sample period.

Suppa (2018) applies the AF methodology to German data and finds evidence that supports using material deprivation and lack of employment as better measures to capture poverty and gaps in poverty among different subgroups and also demonstrates the limitations of a purely income-based metric.

---

2. See Mahmood et al. (2019) for an alternative approach to conceptualize poverty measurement for Pakistan.

The AF measure of MDPI has been extended to investigate other forms of deprivation and inequities, such as child poverty, women's empowerment, and access to energy and credit. Das and Suresh (2018) extend AF methods to construct a Multidimensional Food Security Index (MFSI) for Bangladesh and identify that Education, Dietary Diversity, and Undernourishment as important dimensions, alongside women empowerment as important determinants of poverty. We explore other extensions in terms of fuel-poverty and child-poverty in the next section.

Multidimensional child poverty and gender inequalities

Research has also led studies to adopt AF methodology to develop poverty measures that capture the status of poverty among children. For instance, Mahrt et al. (2018) note that about 46% of children are multidimensionally poor in Mozambique and that the intensity of deprivation is much larger than in neighboring countries.

Using data from 2006 to 2009 for a host of developing countries including Ethiopia, India, Peru, and Vietnam, Kim (2019) shows that while the headcount ratio in both monetary and MDP terms goes down, children remaining in monetary poverty continue to remain poor under MDPI.

Most importantly, even those children who escaped monetary poverty during this time period continued to remain poor under MDPI. Like the other studies, this result indicates that a successful policy needs to delve deep into the myriad complexities of poverty.

Cuesta et al. (2020) compare MDCP indices of Mexico and Uganda and note how Mexico satisfies three important conditions needed for MDCP measures to be useful, namely, consensus, capacity, and polity. While Uganda, like Mexico, has the needed capacity to adopt the MDCP index for policy-making, Uganda lacks the necessary political framework to work across multiple contexts to bring policy to fruition.[3]

In an interesting study Nicaragua's Demographic and Health Surveys, Altamirano and Teixeira (2017) finds that the MDPI to be higher for families with male-headed households than for single-mother or female-headed biparental families, contradicting the received rendition that females are more vulnerable than men. Furthermore, their MDPI results also uncover the importance of urban—rural gaps in living standards, housing, and educational attainments.[4]

---

3. Likewise, Stewart and Roberts (2019) critically examine the prevailing notions among researchers in the United Kingdom about measurement of child poverty and conclude that the government's conservative approach and political incentives support only unidimensional income-based metrics and fail to appreciate the importance of MDPI measures, thereby falling behind in policy outlook.
4. See Najera Catalan and Gordon (2020) for a detailed exposition of the limitations, the unreliability, and the invalidity of MPI, based on the index's statistical properties, as evidenced through its application to six Latin American countries.

Similarly, Espinoza-Delgado and Kalsen (2018) add dimensions such as employment, domestic work, and social protection and show increases in the intensity of feminization of poverty in Nicaragua. The study also supports the finding that female-headed households are better than those led by males. Iqbal et al. (2020) also find a similar result across geographical districts of Punjab, Pakistan.

Researchers have also combined the AF measure to other measures that capture human deprivation and status, as in Vaz et al. (2018), and Vaz et al. (2016), who construct a Relative Autonomy Index to measure women's autonomy in Chad and find that women have less autonomous motivation, compared with men.

Likewise, Maduekwe et al. (2020) use data from Malawi and link the AF measure of MDP and Castleman's Theory of Human Recognition and Economic Development to construct an index of Human Recognition Deprivation. They apply their methods to examine subgroups of female farmers and off-farm women in terms of humiliation, dehumanization, violence, and lack of autonomy within the household and the community.

Multidimensional energy poverty

Globally, millions of households are unable to purchase the required amounts of energy for their basic needs, such as heating, cooking, and lighting. A significant share of the population across rural and urban areas in Africa, Asia, and Latin America is unable to access required energy. Consequently, the UN Sustainable Development Goals lists access to modern energy forms for everyone as one of the targets. Researchers have begun to construct multidimensional energy poverty indices, based on access to fuel and related socioeconomic variables, and currently, a host of studies has uncovered the importance of fuel poverty and its connection to a wider system of inequalities.

Using French data, Charlier and Legendre (2019) combine poverty characteristics to develop and test the robustness of a multidimensional fuel poverty indicator, thereby classifying levels of households' fuel poverty. Charlier and Legendre (2019) note that between 50 million and 125 million people in the EU27 countries and in the 12 new member states suffer from this predicament.

Aristondo and Onaindia (2018) note that energy poverty has worsened in Spain between 2004 and 2015. Their multidimensional energy poverty index (MEPI) construction includes accessibility indicators (heating, arrears on utility bills, quality of roofs, windows, and walls) for two subgroups living in three different areas. This study also reveals a high incidence of energy poverty in rural and Southern Spain.

Robinson et al. (2018) include geographical location characteristics along with expenditure-related variables to construct a detailed MEPI for England. The researchers found that the standard expenditure-based indicators do not adequately capture the multidimensionality of fuel poverty. In particular, along with demographic and income characteristics, households' sociospatial vulnerabilities affect fuel poverty estimates.

Israel-Akinbo et al. (2018) also develop a MEPI and track massive energy poverty among low-income rural and urban households in South Africa. Using repeated cross-sectional data from Ghana, Crentsil et al. (2019) stress the need for intensifying the LPG promotions to reduce the incidence and intensity of multidimensional energy poverty (MEP). Ozughalu and Ogwumike (2019) develop and examine a MEPI and note one-fifth of Nigeria suffers from energy poverty, with rural households and women being the most affected.

Sadath and Acharya (2017) also find extreme MEP among households belonging to tribal and lower class communities in India, producing health and gender inequities. Likewise, Mahmood and Shah (2017) track survey data to uncover MEP among rural households in Pakistan, where 26.4% of the population suffers from MEP.

Mendoza et al. (2019) find that the MEPI scores between 2011 and 2016 have actually improved in the Philippines, implying lower levels of energy poverty. Their results also support a high correlation between MEPI and income poverty. Zhang et al. (2019) also obtain similar results for Chinese Panel data that link multidimensional energy poverty status to household's health.

In an interesting study for Japan, Chapman and Okushima (2019) note that MPEI estimates are closely connected to households' awareness of issues surrounding environmental and ecosystem protection. Consequently, a national policy geared toward low-carbon energy transitions is not likely to succeed if energy-poor households are not included in the use of solar capital and low-carbon substitutes.

Financial exclusion and Multidimensional Poverty Index

Among the multiple dimensions that are included in the achievements matrix, access to credit is also a key metric (Churchill and Marisetty, 2020). Researchers have started to incorporate this variable to construct MDPI in their studies. For example, using data from Assam, India, Das (2019) shows a large incidence of MDP among semiformal and informal borrowers. Furthermore, the study also shows that formal sources are more effective in benefiting households closest to the poverty line. Likewise, using a multidimensional approach for household data from Ghana, Koomson et al. (2020) uncover how financial inclusion reduces household poverty, reduces household's exposure to future poverty, with demonstrable effects in rural areas, particularly among female-headed households.

The Alkire—Foster method

We illustrate the workings of the AF measure with an example. Suppose we begin with the following achievements matrix for seven individuals, comprising of three males and four females, in four dimensions that measure poverty (income, the number of schooling years, access to healthcare centers, and access to clean water):

Id	Gender	Income	School	Health	Water
1	Male	15,000	14	No	No
2	Male	13,121	4	Yes	Yes
3	Male	15,345	5	Yes	No
4	Female	15,121	14	No	Yes
5	Female	14,567	11	No	No
6	Female	13,989	4	Yes	Yes
7	Female	11,000	5	No	No

The following steps are used to construct the AF measure of multidimensional poverty:

Step 1: Use the deprivation cutoff vector (z) and derive the deprivation matrix, g^0.

Step 2: Use the weights assigned to each dimension, from the weight vector, w, and derive the deprivation score vector.

Step 3: Use the cutoff score, k, and identify the persons who are poor, using the censored deprivation score vector $c(k)$.

Step 4: Compute the following indices: multidimensional headcount ratio or the incidence of poverty (H), intensity of poverty (A) and the adjusted headcount ratio (M_0).

We illustrate the workings of each step in the following:

Step 1: Assume we have a deprivation cutoff vector, which helps identify those individuals who are deprived in specific dimensions. Say that the cutoff vector z = (14,000, 6, Has access to health, Has access to water).

Based on the information in z, we construct the deprivation matrix, g^0, by assigning a deprivation status indicator equal to 1, to persons who are deprived in specific dimensions, and a deprivation status indicator equal to 0, to persons who are at or above the cutoff. The deprivation matrix, g^0 for the example is as follows:

$$g^0 = \begin{matrix} 0 & 0 & 1 & 1 \\ 1 & 1 & 0 & 0 \\ 0 & 1 & 0 & 1 \\ 0 & 0 & 1 & 0 \\ 0 & 0 & 1 & 1 \\ 1 & 1 & 0 & 0 \\ 1 & 1 & 1 & 1 \end{matrix}$$

Step 2: Assume that the dimensions are weighted equally. Then, the corresponding weights vector is $w = (0.25, 0.25, 0.25, 0.25)$. Using the weights vector and the deprivation matrix, we can derive the deprivation score vector, by multiplying each dimensional entry from g^0 with the corresponding weight.

For instance, the first individual's deprivation score is $(0 \times 0.25) + (0 \times 0.25) + (1 \times 0.25) + (1 \times 0.25) = 0.5$. Consequently, the deprivation score vector for all seven individuals is then $c = (0.5, 0.5, 0.5, 0.25, 0.5, 0.5, 1)$.

Step 3: In this step, we identify those who are poor by establishing a cutoff score k. In this example, we use the intermediate cutoff, established by AF, and set $k = 0.5$. This implies that a person is poor if the individual is deprived of 50% of all the weighted dimensions. We use the cutoff score, k, and identify the persons who are poor, using the censored deprivation score vector $c(k)$.

The deprivation vector, c, indicates that the fourth individual has a score of 0.25 and, hence, is above the cutoff threshold. The censored deprivation vector $c(k)$ uses the same deprivation score if the value is bigger than or equal to 0.5. If the score in vector $c < 0.5$, then that score in the $c(k)$ is set to 0. With these calibrations, the censored deprivation vector $c(k) = (0.5, 0.5, 0.5, 0, 0.5, 0.5, 1)$.

Step 4: We can compute multidimensional headcount ratio or the incidence of poverty (H), intensity of poverty (A), and the adjusted headcount ratio (M_0), as follows:

$H = \frac{q}{n}$, where q stands for the number of individuals who are poor, and n stands for the total sample size, and H represents the incidence of poverty in the total population.

$A = \sum_{i=1}^{q} c_i(k) / q$, where A represents the intensity of poverty among the poor.

$M_0 = H \times A$, or the adjusted headcount ratio.

Hence, for the achievements matrix for the seven individuals, we have

$$H = \frac{q}{n} = \frac{6}{7} = 0.857$$

$$A = \frac{\sum_{i=1}^{q} c_i(k)}{q} = \frac{3.5}{6} = 0.58$$

$$M_0 = H \times A = 0.857 \times 0.58 = 0.497$$

AF defines M_0 as the multidimensional headcount ratio, also known as the MDPI. M_0 is a combination of both the incidence of poverty (H) and the intensity of deprivations faced by the poor in the population (A).

The uncensored headcount ratio of each dimension (h_j) indicates the proportion of individuals who are deprived in that dimension. From the deprivation matrix, it is clear that three out of seven or 42% of individuals are deprived in the income dimension. The uncensored headcount ratio for each of the other dimensions is 57% (or 4/7).

It is also possible to partition the individuals into different subgroups and identify the share of poverty among the subgroups to overall poverty. To identify the share of subgroup poverty levels and their comparison to overall poverty, we have to compute subgroup headcount ratios and population shares. We undertake this task, for the aforementioned example, by dividing the data into two groups based on gender. A new binary variable is defined as equal to 1 for males and 0 for females, such that the deprivation matrix is now as follows:

$$g^0 = \begin{matrix} 0 & 0 & 1 & 1 & 1 \\ 1 & 1 & 0 & 0 & 1 \\ 0 & 1 & 0 & 1 & 1 \\ 0 & 0 & 1 & 0 & 0 \\ 0 & 0 & 1 & 1 & 0 \\ 1 & 1 & 0 & 0 & 0 \\ 1 & 1 & 1 & 1 & 0 \end{matrix}$$

The last column in g^0 indicates that the sample has three males and four females. Assume that we now have different weights for the dimensions, with $w = (0.4, 0.25, 0.25, 0.1)$, with income receiving the highest weight.

As before, we can derive the deprivation score vector, by multiplying each dimensional entry from g^0 with the corresponding weight. For instance, the first individual's deprivation score is $(0 \times 0.40) + (0 \times 0.25) + (1 \times 0.25) + (1 \times 0.1) = 0.35$. Consequently, the deprivation score vector for all seven individuals is then $c = (0.35, 0.65, 0.35 \mid 0.25, 0.35, 0.65, 1)$. Note that the first three individuals are males, who belong to the first subgroup. For notational convenience, we have used the line $\mid$, to represent the subgroup partition.

If we continue with the cutoff at $k = 0.5$, then the censored deprivation vector is now $c = (0, 0.65, 0 \mid 0, 0, 0.65, 1)$. The following table summarizes the adjusted headcount ratio for the two subgroups (M_0^1, M_0^2), and for the whole sample (M_0).

	Females (gender = 0)	Males (gender = 1)	Total
$H = \frac{q}{n}$	2/4 = 0.5	1/3 = 0.33	3/7 = 0.428
$A = \frac{\sum_{i=1}^{q} c_i(k)}{q}$	1.65/2 = 0.825	0.65	$A = \frac{0.65+0.65+1}{7}$ = 0.766
$M_0 = H \times A$	0.5 × 0.825 = 0.4125	0.33 × 0.65 = 0.214	0.428 × 0.766 = 0.327
Population share	4/7 = 0.57	3/7 = 0.43	100%

608 SECTION | III Special topics on poverty, nutrition, and food policy analysis

Note that the adjusted headcount ratio for females is larger than that for the males. This aspect is also highlighted by this group's contribution to the overall adjusted headcount ratio. The contribution of each subgroup to overall M_0, depends upon the subgroup's adjusted headcount ratio weighted by its population share (v_i) as follows:

$$D_0 = \frac{M_0^1 \times v_0}{M_0} = \frac{0.4125 \times 0.57}{0.327} = 0.72$$

$$D_1 = \frac{M_1^1 \times v_1}{M_0} = \frac{0.214 \times 0.43}{0.327} = 0.28$$

Recall that each individual's deprivation score vector c = (0.35, 0.65, 0.35 | 0.25, 0.35, 0.65, 1), and the cutoff value is k = 0.5. Hence, the first, third, fourth, and fifth individuals are not considered poor because their deprivation scores are less than the cutoff. Using this consideration, we transform the original uncensored deprivation matrix, g^0, to obtain a censored deprivation matrix, $g^0(k)$:

$g^0(k)=$

Income	School	Health	Water
0	0	0	0
1	1	0	0
0	0	0	0
0	0	0	0
0	0	0	0
1	1	0	0
1	1	1	1

Given the new censored deprivation matrix, $g^0(k)$, we get a censored headcount ratio, $h_i(k)$, in each dimension. The following table summarizes the values of the uncensored and the censored headcount ratios, h_i and $h_i(k)$, for each dimension.

	Income	School	Health	Water
h_i	0.42	0.57	0.57	0.57
$h_i(k)$	0.42	0.42	0.14	0.14
w	0.4	0.25	0.25	0.1
$\varphi^0(k)$	0.51	0.32	0.10	0.04

The first row presents the uncensored headcount ratio (h_i) in each dimension, which is derived from the original achievements matrix. Recall that from the uncensored deprivation matrix g^0, 42% of individuals are deprived of the income dimension.

The censored headcount ratio in each dimension, $h_i(k)$, is derived from the censored deprivation matrix, $g^0(k)$. The censored headcount ratios, along with the weights, w, provide us, $\varphi^0(k)$, or the percentage contribution of each dimension to the overall adjusted headcount ratio.

Consider the income dimension, with a censored headcount ratio of 0.42, and with a weight of 0.4. The contribution of the income dimension to the adjusted headcount ratio is $(0.42 \times 0.4)/0.327 = 0.51$, or 51%.

Likewise, the contribution of the schooling dimension is $(0.42 \times 0.25)/0.327 = 0.32 = 32\%$. Note that while the censored headcount ratios for health and water are the same, the contribution of water to the overall adjusted headcount ratio is only 4%, which is much smaller than the role of income, a natural consequence of the relative weights attached to each of these corresponding dimensions. In the following section, we develop and implement the MDPI calculation using STATA.

STATA implementation

Pacifico and Poege (2017) have developed an implementation procedure in STATA that computes the desired indices, including subgroup decompositions. We illustrate the workings of the Pacifico—Poege **mpi** command, for the same example in the previous section, using the following input lines:

- generate inc_i = (income < 14000)
- generate sch_i = (school < 6)
- mpi d1(inc_i) d2(sch_i) d3(health) d4(water) [pw=weights], cutoff(0.5)

We begin with the achievements matrix for the seven individuals, in income, schooling, access to health, and water. The first three input commands using **generate** create the two new binary variables inc_i, sch_i, based on the cutoff for income levels, and the number of schooling years. These commands establish the deprivation matrix, g^0, that we derived before.

The third line estimates the AF poverty measures using the **mpi** command: mpi d1(inc_i) d2(sch_i) d3(health) d4(water) [pw=weights], cutoff(0.5)

In this example, there are four distinct deprivation domains, and these are listed as d1(inc_i), d2(sch_i), d3(health), and d4(water). The syntax [pw=weights] indicates that the data will be weighted. However, since we have not specifically provided the weights, the four domains will be weighted equally, by default. The cutoff(0.5) helps to establish the cutoff based on the weighted sum of the indicators from the deprivation matrix. The corresponding portion of the STATA output is reproduced in the following:

Indicator	Type	Weight	Deprived
Domain 1			
inc_i	Binary	.25	42.857 %
Domain 2			
sch_i	Binary	.25	57.143 %
Domain 3			
health	Binary	.25	57.143 %
Domain 4			
water	Binary	.25	57.143 %

Deprived: Percentage of individuals whose indicator values are below the threshold.

Main results   N = 7

		Coef.	Std. Err.	[95% Conf. Interval]
Main				
	H	0.857	0.143	0.577 1.137
	M0	0.500	0.109	0.286 0.714
Additional				
	A	0.583	0.082	0.422 0.744

Note: Adjusted Multidimensional Headcount   M0 = H*A

Indicator		M0
domain 1		
	inc_i	0.214
domain 2		
	sch_i	0.286
domain 3		
	health	0.214
domain 4		
	water	0.286
	Total	1.000

Contribution of each indicator (%)

STATA produces three tables for our analysis. The first table summarizes the four deprivation domains, indicating the variable type and its calibrated weight. The last column indicates the percentage of the overall population that is deprived in the corresponding dimension.

The second table summarizes the key results for the AF poverty measures: H, M_0, and A. The table also computes the standard errors and the confidence intervals for the said estimates. Note that the estimated values match those from the example in the previous section.

Additionally, STATA also produces a third table that shows the contribution to the overall M_0 accounted for by each deprivation indicator. Income deprivation accounts for roughly 21.4% of the overall value of M_0.

STATA can also decompose AF measures based on population subgroups. The STATA command line that computes AF measures by gender using different weights is as follows:

- mpi d1(inc_i) w1(0.4) d2(sch_i) w2(0.25) d3(health) w3(0.25) d4(water) w4(0.1) [pw=weights], cutoff(0.5) by(gender) postbymain

As discussed before, the weighting schemes identify the poor differently. The *postbymain* option allows the researcher to conduct different statistical tests on the estimates.[5] The corresponding STATA output for the aforementioned command line is given in the following:

Summary of mpi indicators

Indicator	Type	Weight	Deprived
Domain 1			
inc_i	Binary	.4	42.857 %
Domain 2			
sch_i	Binary	.25	57.143 %
Domain 3			
health	Binary	.25	57.143 %
Domain 4			
water	Binary	.1	57.143 %

Deprived: Percentage of individuals whose indicator values are below the threshold.

The first table summarizes the details of the mpi indicators, including the percentage of people deprived in each indicator, in the pooled data. The second table presents the main results from the analysis, H, A, and M_0. The third table produces the censored headcount ratios in each dimensions, $h_i(k)$. These two

---

5. See Pacifico and Poege (2017) for details.

tables summarize the main results from the data, which match the values derived in the example.

The last three tables produce the decomposition by gender subgroups in the population. The subgroup decomposition portion of STATA produces the absolute and relative value of the indices, absolute value of the indices in each subgroup, and related population shares.

The first table under subgroup decomposition provides the absolute values of H, A, and M_0 for each subgroup and the whole sample, which have been computed before: Males comprise 42.9% of the sample, and the incidence of poverty is 33.3%, against 50% for the females.

The second table shows the proportional contributions of each subgroup of the population to the overall index. In this example, 71% of the overall M_0 is attributable to females and is computed by weighted absolute index divided by the overall index, with weights being the population shares: $(0.42 \times 0.571)/0.329 = 0.717$ and $(0.217 \times 0.429)/0.329 = 0.283$.

```
Main results                                              N = 7

                    Coef.     Std. Err.    [95% Conf. Interval]

Main
         H          0.429     0.202         0.033      0.825
         M0         0.329     0.161         0.013      0.644

Additional
         A          0.767     0.103         0.565      0.968

Note: Adjusted Multidimensional Headcount           M0 = H*A
```

```
         Indicator          M0

domain 1
            inc_i           0.522
domain 2
            sch_i           0.326
domain 3
            health          0.109
domain 4
            water           0.043

            Total           1.000
```

Contribution of each indicator (%)

Decomposition by subgroups

MPI by: gender

	gender_0	gender_1	Total
H	0.500	0.333	0.429
M0	0.412	0.217	0.329
pop share	0.571	0.429	1.000

Indices by subgroup (absolute)

	gender_0	gender_1	Total
H	0.667	0.333	1.000
M0	0.717	0.283	1.000

Contribution of subgroups to indices (%)

		gender_0	gender_1	Total
M0	inc_i	0.485	0.615	0.522
	sch_i	0.303	0.385	0.326
	health	0.152	0.000	0.109
	water	0.061	0.000	0.043
	Total	1.000	1.000	1.000

Contribution of each indicator (%)

STATA also produces a third table that shows the contribution to the overall M_0 accounted for by each deprivation indicator, across gender decomposition. Income deprivation accounts for roughly 21.4% of the overall value of M_0 for females, while within the male subgroup, income deprivation accounts for nearly 61.5 of their overall M_0. Likewise, while the schooling dimension accounts for roughly 30.3% of the overall M_0, within females, the same dimension accounts for nearly 38.5% of the overall M_0, within the male subgroup.

STATA workout

1. Use the mpi command and derive the AF measures for the following achievements matrix from Alkire et al. (2015), Chapter 5. Define binary variables for malnourished and access to sanitation (1 = Yes, 0 = No). Define the cutoff vector as (500, 5, not malnourished, has access), and set $k = 0.5$. Use the default weights in STATA. Compare your results with Alkire et al. (2015, Chapter 5).

Person	Income	Years of schooling	Malnourished	Access to sanitation
1	700	14	No	Yes
2	300	13	No	No
3	800	1	No	Yes
4	400	3	Yes	No

2. Reproduce the same results for the following achievements matrix from Alkire et al. (2015), using the same cutoff, as in the previous problem. Compare your results with the previous problem. What has happened to the headcount ratio, and the intensity index? What does that indicate about the status of the poor in the previous problem? For a full discussion, compare your results with Alkire et al. (2015, Chapter 5).

Person	Income	Years of schooling	Malnourished	Access to sanitation
1	700	14	No	Yes
2	300	13	No	No
3	800	1	Yes	Yes
4	400	3	No	Yes

3. Divide the data in problem 1 into two subgroups: Persons 1, 2, and 3 in group 1, and Person 4 in group 2. Use the following weight vector $w = (0.4, 0.25, 0.25, 0.10)$ and set $k = 0.4$. Compute the censored headcount ratio and the adjusted headcount ratio using the mpi command in STATA. Compare your results with Alkire et al. (2015, Chapter 5).

4. Suppose $z = (13, 12, 3, 1)$ and $k = 2$, for the achievement matrix given in the following for 4 dimensions:

Person	D1	D2	D3	D4
1	13.1	14	4	1
2	15.2	7	5	0
3	12.5	10	1	0
4	20	11	3	1

Use STATA to derive the deprivation matrix, the censored deprivation matrix, and the adjusted headcount ratio. Compare your results with Alkire and Forter (2013).

Section IV

Technical appendices

Appendix 1

Introduction to software access and use

This third edition of this book has been written, as with the first two editions, to address a critical need faced by policy analysts, development professionals, and program managers in carrying out a preliminary descriptive analysis of the regular baseline and end line data they collected from the project and program implementation as well as to learn from the monitoring and evaluation data collected throughout the period of development projects. Often such collected data do not see their analysis due to lack of capacity of the development professionals. The book tries to such capacity by introducing methods of data analysis that are simple, practical, and communicable. This book aims to share the basic descriptive data analysis methods with respect to food security, poverty, and nutrition and nutrition programming and policymaking. Strong emphasis is placed on the most common descriptive analysis methods—means, cross-tabulation, correlation, ranking, t-test, and one-way analysis of variance. Simple linear regression, factor analysis, and multivariate regression analysis have also been included in this manual as examples of predictive analysis. This book also tries to address how food security and nutrition variables can be created, grouped, etc., for analysis. This is not a comprehensive statistical textbook but, rather, we have tried to highlight, by giving examples, how basic descriptive and predictive analysis could be used to study issues related to food security, poverty, and nutrition.

Use of computers for data analysis has become commonplace. The statistics discussed in this book could be carried out with one of the most widely used and comprehensive statistical programs in the social sciences—SPSS/PC+ Windows 2000 version 12.0, and the implementation via STATA. In this revised third edition, we have introduced the codes for analysis using "R" language at the end of most of the chapters, leaving some room for the reader to develop their own codes for other chapters.

This edition of the book outlines how to conduct some of the common statistical procedures and tests. Each test and procedure is described in the form of sample programs using a specific policy issue as an example. Fifteen such policy issues are presented. The programs are presented at the end of each chapter. These sample programs and data files are available on the Elsevier website. Please note that there may be several alternative descriptive analysis techniques or programs for any given issue.

Appendix 2

Software information

SPSS/PC+ is a computer software package—a set of computer programs written by experts for other individuals to use with their own data. It was developed specifically for analyzing data from social science research. This package is commonly used in various disciplines such as economics, marketing, psychology, and sociology. The abbreviation SPSS stands for Statistical Package for the Social Sciences. SPSS has a user-friendly graphical interface and also allows programming. These programs are being continuously updated and so there are various versions in existence. Although we use SPSS 12.0 in developing the present chapters, recently, the package has been extended to include a newer version SPSS 16.0. SPSS/PC commands are generally executed in a batch mode where the user creates a text file with the programs and submits it for execution in a batch mode. The text file can be written within the SPSS/PC+ by using Syntax window text editor or using dos EDLIN, NORTON EDITOR, or any other word processing programs (saving the file as dos ASCII text). The program can also run interactively, but this is not recommended since it is more difficult to recall what you have done with interactive processing, as opposed to batch processing where the complete file is run. The third method for executing SPSS/PC+ commands is a user-friendly menu system for building and executing commands, which is interactive as well. A list of interacting commands is presented in Appendix 3.

STATA statistical data analysis

STATA statistical software is a complete, integrated statistical software package that provides everything you need for data analysis, data management, and graphics. The software is developed by StataCorp, 4905 Lakeway Drive, College Station, Texas 77845, United States. STATA statistical software is a comprehensive statistical software package that provides everything you need for data analysis, data management, and graphics. The software has many features such as treatment effects, multilevel generalized linear model, power and sample size, generalized SEM, and so on. STATA is also easy to use, using both a point-and-click interface and an intuitive command syntax. STATA's statistical tools include survival models with frailty, dynamic panel data regressions, generalized estimating equations, multilevel mixed models, models with sample selection, multiple imputation, ARCH, and estimation with complex survey samples. In our chapters, we also implemented such as linear and generalized linear models, regressions with count or binary outcomes, ANOVA/MANOVA, cluster analysis, and summary statistics. STATA's data management features give you complete control of all types of data: you can combine and reshape data sets, manage variables, and collect statistics across groups or replicates. You can work with byte, integer, long, float, double, and string variables. With STATA's Internet capabilities, new features and official updates can be installed over the Internet with a single click. Many new features and informative articles are published quarterly in the refereed Stata Journal. STATA will run on Windows, Mac, and Linux/Unix computers.

"R" package for statistical analysis

In this edition of this book, we introduce the R package for statistical analysis. The major advantage of this package is that it is an open-source software package, which does not require any licensing for its use. Given that several developing country institutions are still behind in procuring latest version of the proprietary statistical packages, largely due to budget constraint, training next generation of policy analysts in the open-source software packages becomes imperative. R provides a suite of software for data management, analysis, graphing, and display of results. The programming language is easily learned by readers new to programming methods. the user-friendly nature of the programming approaches for data input and output makes it a versatile environment for data analysis. R package includes a wide range of statistical and graphical techniques. R complies and runs on UNIX, Linux, Windows, and MacOS systems. The following link provides a set of webpages that could be used to explore a wide range of options that R provides including instruction to downloading and installing the R package.

Installing and downloading R

Although R is an open-source software package developed and managed by a community of researcher, to use the software you need to install it in your computer. Visit the homepage of R software: https://www.r-project.org/, and then follow the instructions there to download the latest version of R. Installation process requires careful selection and acceptance of the defaults options. There are a number of videos available on the Internet for the beginners of R software, which should be useful to get started. The one that is viewed by a larger number of viewers normally gives the hint that it is the one most sought after and recommended by peers to others. Going through videos for beginners of R will help to get a handle of initial commands, and practicing them will bring you up to speed for the basic exercises that are described in the chapters of this book.

Appendix 3

SPSS/PC+ environment and commands

Although we have introduced the STATA and "R" languages in the last two editions, respectively, moving away from SPSS/PC+, we are retaining the basics of the SPSS programming in this edition as well for the benefit of the readers who may be using SPSS as their main programming language. As we have discussed the computation and data management techniques in the beginning of the manual, we will concentrate our attention in this section on the SPSS environment and the different menu options. First, to start an SPSS/PC, double-click on the SPSS icon from the desktop or wherever the icon is located. This might take a minute or so to load. Once started, the SPSS Data Editor window is displayed. Each SPSS session can only display one data editor at a time. However, multiple output windows can be open at any time. Multiple sessions can be open, with each displaying a different data set, if there is enough memory available.

The main SPSS window

The main SPSS window contains all the activities of an SPSS session. The window contains a toolbar, a collection of menus, and a status bar. If one of the subwindows becomes larger than the main window, or if it shifts outside the area of the main window, then the main window will develop scroll bars (see Fig. A3.1).

Menus

SPSS has 11 menus available for viewing a data file, output file, or syntax file. They are as follows: File, Edit, View, Data, Transform, Analyze, Graphs, Utilities, Add-ons, Window, and Help. The following toolbars are available in SPSS.

FIGURE A3.1 Data toolbar.

Data editor toolbar

The first button is open a file. To open an existing data file, select the data window. Then select file, and click on open. The second button indicates save file. To save the contents of a data window, select file and then click on save. If the file is being saved for the first time, enter the filename and location, and click on save. If no extension is added to the filename, SPSS will automatically add a .sav extension.

The third button is the print. To print a document, just go to file, and then click on print. You have the choice of printing only what is highlighted from the active window or printing the entire contents of the window. The next box is the dialog recall and contains the submenus (see Fig. A3.2).

The next box in the toolbar is the go to case number, with which one can locate a particular case number in the data file. The next icon is the variables. After clicking this icon, one can see what the variable name is, its measurement level (whether it is categorical or continuous), and the value labels of the variable. The next icon is the insert variable icon with which one can insert an additional variable. The next icon is the split file, with which the following options are available (see Fig. A3.3):

1. One can analyze all the cases.
2. One can compare groups (for example, female-headed households versus male-headed households).
3. One can also organize output by groups.

FIGURE A3.2 Dialog recall box.

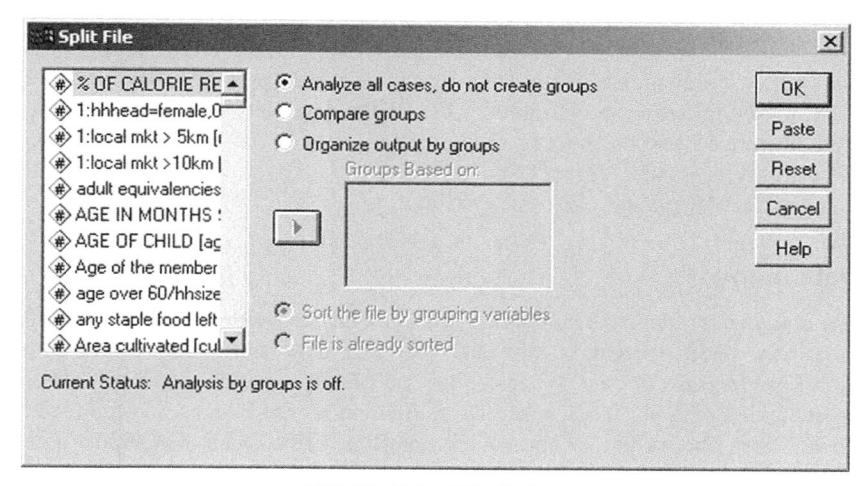

FIGURE A3.3 Split file box.

The next icon is the weight cases icon. This is useful when one needs to weight the cases by one of the frequency variables. For example, it may be interesting to weight the households proportional to the sample of households in each region, or the weight can be inversely proportional to the probability of a household being selected in the sample.

The next icon is the select cases icon. The following options are available with this icon. One can select all the cases, or certain cases can be selected based on the If condition being satisfied. Additionally, one can select a random sample of cases or cases based on a certain range (for example, for only one region, such as Mzuzu). One can also use a filter variable to select only the filtered cases. The next icon is the value labels of the variable and shows what values the variable takes. The next icon is the use sets icon, with which one can use all the existing variables or a new set of variables can be created.

File menu

We have already discussed aspects of the file menu before (for example, the data editor toolbar). This menu deals with all the file handling aspects of SPSS. In this menu, one can also read text data from an existing file. The command to exit SPSS once you are done is also located in this menu.

Edit menu

This menu also shares common applications such as in Microsoft windows applications. This menu contains commands for undoing the last action, cut, copy, and paste actions. It also has some special copy options for certain elements and the commands for find and replace.

View menu

Two views are available in the Data Editor—variable and data. The variable view is used to define the variables, whereas the data view is used to enter the data. To move between the data and variable views in the Data Editor:
    Select View, and click on Data.
    Or Select View, and click on Variable.

Data menu

The data menu affects the contents of the Data Editor window and can be used to correct, select, weight, or sort data. One can also merge files or split data files. For merging data sets, one can either add cases or variables. To add cases, open the first data file. Then select Data, click on Merge File, and then hit Add Cases. Then choose the second source data file. Click Open. The Add Cases dialog box is opened. Identify and match any wanted variables and then click on OK. When two sets of data have the same number of cases, variables from one can be appended to the other. To add variables, resort to the following steps: Open the first data file. Select Data and then click on Merge File and hit Add Variables. Then choose the second source of the data file and click open. The Add Variable dialog box will be opened. Select the options as required, and click on OK.

    Additionally, one can insert new variables and new cases, or cases can be sorted using a particular variable. The changes will be temporary unless you specifically save the file with the changes.

Transform menu

This menu is used to make changes to selected variables in the data file. One can compute new variables based on the values of existing ones. Random numbers can be generated and ranking or recoding of the data can also be done.

Analyze menu

This menu contains the crux of the whole package. It contains all the categories of statistics that SPSS is capable of performing. The little arrows to the right of all menu options indicate that there is a submenu of specific tests or groups of tests. The submenus are too detailed so that explanations cannot be provided here. We will discuss some of these issues in Appendix 5.

Graphs menu

This menu enables one to plot graphs from the data. Some of these graphs can be plotted from the dialog box options. SPSS can plot various kinds of graphs such as bar, line, area, and pie. It can also produce histograms, scatterplots up to three dimensions, and tests of normal curves (such as the P–P and the Q–Q).

Utilities menu

This menu contains some useful tools such as index of commands with brief descriptions, information on the current variables, information on files available, and a facility for changing fonts.

Add-ons menu

Several SPSS products are additionally available with SPSS 12.0. They are described in the following.

SPSS Categories performs correspondence analysis, nonlinear principal components, canonical correlation, and regression with nominal, ordinal, and continuous variables.

SPSS Conjoint analyzes designs in which subjects consider jointly two stimuli at a time. It generates the design, prints the stimulus combinations, and analyzes the data.

SPSS Maps does statistical data maps called thematic maps. They can display simple information such as average income of a country or state. They can also display how well a model such as regression analysis fits in different regions.

SPSS Exact Tests calculates the exact probability values for statistical tests when small or very unevenly distributed samples could make the usual tests inaccurate.

It simply adds new options to many of SPSS categorical and nonparametric analyses.

SPSS Complex Samples accurately analyzes and predicts numerical outcomes from complex sample designs. It uses the new complex samples general linear model (CSGLM) to build linear regression, ANOVA, and ANCOVA models to predict numerical outcomes.

SPSS Missing Value Analysis is a new innovation in SPSS 16.0. Missing data can be very treacherous since it is difficult to identify the problem. Missing data can cause very serious problems. First, most statistical procedures automatically eliminate cases with missing data, which implies that in the end you may not have enough data to perform the analysis. Second, the analysis might run, but the results may not be statistically significant. Third, the results can be misleading if the analysis is not based on a random sample of all cases. The SPSS Missing Value Analysis provides two methods for maximum likelihood estimation and imputation. First, the Expectation Maximization (EM) algorithm is an iterative algorithm that can provide estimates of statistical measures such as correlations or imputed values for missing values in the presence of a general pattern of missing data. Second, regression imputation relies on the fact that the EM approach is mathematically similar to using regression to fill in the missing values using predicted values from a regression of a given variable on other variables in the analysis.

Either or both of these methods can be tried on the data using SPSS Missing Value Analysis. In general, either of these methods is superior to naïve approaches such as listwise deletion, pairwise deletion, or mean substitution.

Window menu

This menu is found in many MS Windows programs and is simply a series of commands to move, select, and resize the subwindows in SPSS. Most of the commands here can simply be done with a mouse.

Help menu

The Help menu is SPSS that provides a variety of ways of obtaining help. It mainly consists of four tabs below the menu bar. They are contents, index, search, and favorites. The Contents tab displays a list of available help topics organized by categories.

To display a topic using the Contents tab, select Help, Topics, and the Contents tab. The Index tab displays an alphabetic list of available help topics. To display a topic using this tab, select Help, Topics, and the Index tab. Type in a word into the text box to move to that term in the list or you can scroll through the list of topics, and select the topic of interest in the list box. Click Display to display the topic.

To obtain help about a specific command while in a dialog box, open the dialog box and click Help. Additionally, SPSS has a useful tutorial system to enable the user to work through a series of tasks at their own speed. To start the tutorial, select Help and click on Tutorial. Work through each lesson in the Help window is provided.

Output window

The output window is where SPSS stores the results of the analysis that you have performed. This window is a straightforward text window, and the files you save from it are files with an extension spo. To save the contents of an output window, select File and click on Save. If the file is being saved for the first time, enter the filename and the location. You cannot enter any additional information in this window manually. However, you can edit the contents of the window if you wish. To close an output window, select File and click on Close. Then click Yes to save the contents of the output window when prompted. One of the open output windows is known as the designated window. When a command is executed and output is produced, the output is placed into the designated output window. To specify an output window as the designated window, select the Utilities, and click on Designate Window.

Syntax window

The Syntax window is very similar to the output window in its appearance. It is a text window and has a toolbar. However, the purpose of this window is to type in SPSS commands to analyze the data set. A list of SPSS commands is also given in this appendix. There are certain analyses that SPSS will only perform through the command language. Let us now look at some of the SPSS/PC+ commands that have been done throughout the chapters.

SPSS/PC+ commands

This section describes a subset of SPSS/PC+ commands that are essential to understand the programs. There are three main categories of SPSS/PC+ commands: data definition and manipulation, procedure, and operation.

Data definition and manipulation

Get File "finalmainnew.sav." This command retrieves the SPSS system file into the active file.

Save File "finalmainnew.sav." This command saves the active file as SPSS system file with the relevant filename.

Write. This command writes the active file into an ASCII file.

Compute var 3=var1/var 2. This command creates a new variable called var 3, which is the ratio of var 1 and var 2.

Recode var 3 (1=2). This command changes all code in var 3 from a value of 1−2.

Process IF (var 3=2). This command temporarily selects cases where var 3 is equal to 2 for the subsequent procedure.

Select IF (var 3=2). This command permanently selects cases where var 3 is equal to 2 for all subsequent procedures.

Procedure

Rank var 3. This command creates a new variable called Rvar 3, which assigns rank to the variable.

List var 3. This command lists the value of variable 3 for all the cases.
Aggregate file = "temp.sav."
/Break dummy
/ncase = NUMISS (var3). This command creates a new system file temp.sav, which contains as many cases as there are values of dummy. Each case includes two variables, dummy and ncase (which is the unweighted number of cases in the break group).

Operations

Descriptives
    variables = var 1 var 2 var 3
    /STATISTICS=MEAN SUM STDDEV VARIANCE RANGE MIN MAX SEMEAN KURTOSIS SKEWNESS.

The aforementioned command computes the descriptive statistics such as mean, sum, standard deviation, variance, range, minimum, maximum, standard error, kurtosis, and skewness for variables 1, 2, and 3.

Frequencies
Variables=var 1 var 2 var 3
/Order=Analysis

This command finds the frequency distribution of the number of cases into different groups.

Crosstabs
/Tables=var 3 by var 2

This command is used to examine the statistical relationship between variables 3 and 2. In addition, the degree of association between the variables can also be determined through the use of statistical tests such as chi-square and phi.

T-TEST
GROUPS = var 1 (0 1)
/MISSING = ANALYSIS
/VARIABLES = var 2

T-test computes the independent sample Student's t-test. This is used to examine the effects of one independent variable var 1 on one dependent variable var 2. The result of this test enables us to determine if the two means differ significantly.

Oneway
var 1 by var 2
/Missing Analysis

The aforementioned command computes the analysis of variance for between subjects which contain only one independent variable var 2. This is used to determine if two or more group means differ significantly.

Correlations
/Variables = var 1 var 2 var 3
/print = twotail nosig
/missing = pairwise

The aforementioned command is used to obtain Pearson's product moment correlation coefficient for pairs of variables (such as variables 1, 2, and 3), and to determine the significance of these coefficients.

REGRESSION
/MISSING LISTWISE
/CRITERIA=PIN(.05) POUT(.10)

/NOORIGIN
/DEPENDENT var 3
/METHOD=ENTER var 1 var 2

The aforementioned command runs a regression model with var 3 as the dependent variable and var 1 and var 2 as the independent variables. The criterion for removal of a variable is based on the probability of the F-distribution.

Appendix 4

Data handling

This appendix will discuss the basic concepts of how to import and export data from different programs such as an ASCII text editor or Microsoft Excel into SPSS. This can be extended to STATA and R programming languages as well. It will then discuss the data structure of the file Finalmain.sav that was used for developing the current chapters. Finally, this appendix will define all the variables that were used for developing the analysis in the previous chapters.

The example data used for illustrative purposes in the previous editions of the book were collected through the Food Security and Nutrition Monitoring Survey (FSNM) conducted with the technical help of Cornell University and UNICEF funding during May 1992. Food Security and Nutrition Monitoring Surveys conducted in Malawi provided information to policy decision-makers on the food security and nutrition situation in various parts of the country. They involved use of four different chapters: food security; markets and prices; household income and expenditure; and nutrition monitoring. The surveys were conducted twice a year to capture the effects of seasonality on the food availability and nutritional status of the population.

The expenditure chapter used the 1-month recall approach to collect information on the monthly expenditure of 15 food and 12 nonfood commodities. Details on the quantity of purchase, value of food consumed from own production, and value of purchased food using cash and kind were recorded. The food security chapter generated information on the demographic characteristics, cultivation characteristics, livestock ownership, food production, marketing and storage, food purchasing behavior, food intake pattern, composition of meals, and nature and availability of employments. The markets and prices chapter implemented at the village level generated information on food markets, accessibility of organized markets, private trader operations and availability of food in the markets, and prices of several food and nonfood commodities. Details on the sampling methodology of FSNM have been published elsewhere.

In this book, we used the data collected in the Mzuzu, Salima, and Ngabu Agricultural Development Districts (ADD) during 1991−1992 as our sample data. The sample consists of 604 households out of which 197 households had information on 304 children (their height, age, weight) below the age of 5. The sample consisted of 733 children of all ages. The name of this household level file is FINALMAIN.SAV. In the multiple regression chapter, influential observations or outliers have been removed, since multivariate models are extremely sensitive to the presence of outliers. We call this file FINALMAINNEW.SAV. Both these files are provided together with the syntaxes in the CD-ROM.

505Importing ASCII data into SPSS

Here, we show how to import data into SPSS. Readers should be able to practice such procedures in STATA and R programming languages. Many times, data are made available in text format, also called "ASCII," which stands for American Standard Code for Information Interchange. The researcher needs to read the file into a package such as SPSS, ensuring that the variables included in the file are defined according to specifications provided with the data file. Text files are convenient for two main reasons:

1. They can be read by different programs and they are usually small.
2. Data files in binary format created by a program (such as SPSS or Excel) are larger due to the "overhead" in the space that is created by the data. This is because SPSS data files will include not only the values of each variable but also the names of the variables, the type (numeric or string) of the variable, and any associated variable or value labels. The two most frequent types of text data files are files in "fixed" and "delimited" formats. For the sake of brevity, we will only discuss the fixed format. The delimited format is similar in nature, when the data are exported to SPSS. In a fixed format, each variable takes a certain number of spaces in the file. For example, a variable that takes values between 0 and 10,000 would take five spaces or columns in the file. In the delimited case, the values of the different variables are separated by a specific delimiter, such as comma or semicolon.

To read a fixed ASCII formatted file, the first step is to open a blank SPSS data editor. Then from the file menu, select Read Text Data. Locate the text file from the selected directory. Once the file is selected, SPSS will open the Text Import Wizard, which will walk you through the steps in reading the data in SPSS format. The next step in the SPSS text import wizard will ask, "How are your variables arranged?" and you should select Fixed. Then it will ask, "Are variables names included at the top of your file?" and select No and click on Next. In step 3 of the wizard, the default options should be to start at row 1 as the first observation, with each line representing one case. Then click on Next. In step 4 of the wizard, define the variables by clicking between them, which will insert a line between any two columns. If there is mistake, point to the arrow in the preview and drag it beyond the ruler and release the mouse button. Then click on Next to go to the next step (step number 5). In this step, click on each of the columns to name the variables: add-code, epa-code, MBRNBR, etc. The variable names can be up to eight characters in length. Click again on

Next and click on Finish to import the data. The data will be displayed as a spreadsheet. From the File menu, select Save as and save the file as an SPSS data file (call it Finalmain.sav).

Importing data from excel

To import data from an Excel spreadsheet, select Open Data from the File menu. In the dialog box, select Excel (*.xls) under Files of Type. The pop-up box shown in Fig. A4.1 appears.

Among the sample files, point to finalmain.xls (Fig. A4.2). Once you click on the Open button, SPSS will prompt to select the worksheet within the file, as well as to read the variables in the first row of the data. The researcher can also specify a particular range of cells within the worksheet. For example, one could retrieve only the data included in "A1: IV1601." SPSS will read the variables in the first row relative to the range selected.

It is important to note that SPSS expects cases or observations to be represented in rows, while the columns are variables.

Data structure

The data structure of the chapters is given in Table A4.1. For the food security chapters, information on 604 households is utilized, with household as the unit of analysis. On the other hand, for determinants of child nutrition, information on 197 households is utilized with children below the age of 5 as the unit of analysis. The households come from three main regions or add-codes (region 2 = Mzuzu, region 4 = Salima, and region 8 = Ngabu). Each household is uniquely identified from the combination of the region (add-code) and the subregion (epa-code).

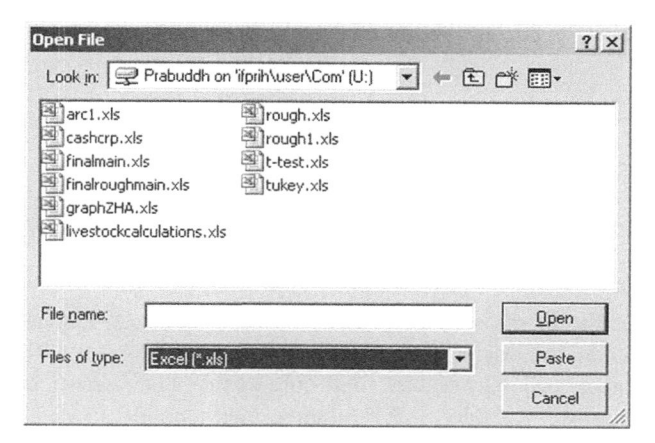

FIGURE A4.1 Opening excel file into SPSS.

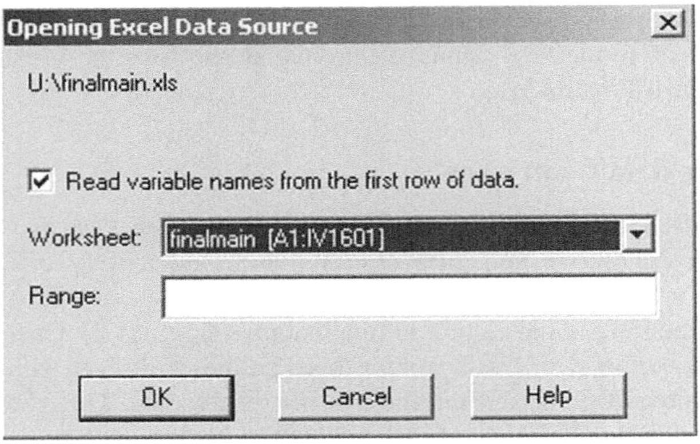

FIGURE A4.2 Opening the data source.

TABLE A4.1 Data structure of Finalmain.sav

Households	Food security Variables 12 ....N1	Both Variable both	Nutrition Variables 12 ....N2
1	1		
2	1		
•	Yes •	Yes	
•	•		
•	•		
197	1		
1	0		
2	0		
•	Yes •	No	
•	•		
•	•		
407	0		

From the aforementioned data structure, it is immediately apparent that the 197 households (or cases) have information pertaining to food security, nutrition, market access variables, and community characteristics variables. Thus, the dummy variable BOTH assumes a value of 1 for this set of households. The various characteristics of the household such as member's age, sex, and other demographic characteristics such as child age are also available for this set of households. On the other hand, for the remaining 407 households, we have information on only variables related to food security. Child-related information, community characteristics, and market access variables are missing for this set of households. Thus, the dummy variable BOTH assumes a value of 0 for this set of households. Additionally, the number of variables related to food security N1 is greater than the number of variables related to child nutrition N2. In other words, N1 > N2. In the following, we define some terminology related to this data structure.

List of useful terms

Records (or cases or observations): These are individual observations such as individuals, farm plots, countries, industries, villages, etc. They are usually considered to be the rows of the data file.

Variables: They are characteristics, location, or dimensions of each record. They are considered as columns of the data file. "Key Variables" are ones that are needed to identify a record in the data. Thus, for the food security chapters, MBRNBR = 1 or the household is the key variable. On the other hand, for the nutrition chapters, AGEMNTH ≤5 or the child below the age of 5 is the unit of analysis and thus is the key variable.

Discrete (or categorical) variables: These are variables that have only a limited number of different values. Examples may include region (or add-code), sex, income category, and educational level of the household head.

Binary (or dichotomous) variables: These variables are a type of discrete variable that assumes only two values. They may represent yes/no, educated/noneducated, or other variables with only two values.

Continuous variables: These are variables whose values are not limited and can vary over a much wider range. Examples include income, rice production, expenditures on cassava, calories per adult equivalent, and so on. Unlike discrete variables, they are usually expressed in units such as Malawian Kwacha, hectares, kilograms, etc., and may assume fractional values.

Variable labels: These are longer names associated with each variable to explain them in tables and graphs. For example, the variable label for MBRNBR might be "Member Number," and the label for LANDO might be "Size of Land Owned."

Value labels: These are longer names attached to each value of a variable. For example, if the variable LANDO has five values, each value is associated with a name. 0 could indicate "landless," while 1 could indicate "holding less than 0.5 hectares," and so on.

We next define the variables that were associated in the development of this manual and the nature of the variables (whether they are continuous, discrete, or dichotomous; Table A4.2). We also indicate where the variable might be used (whether in the food security (FS) or nutrition (N) or both) for subsequent analysis.

TABLE A4.2 Labels and summary descriptive statistics of variables used in the chapters.

Variable	Labels	Value Labels1	Nature	Mean	Std. Dev.	N
MBRSEXNEW	Sex of the member	Male = 1 Female = 0	Dichotomous	0.81	0.39	197
MBRAGE (household head)	Age of the member		Continuous	39.12	10.98	197
EDUCHEAD	Education of the household head	(1–7)	Discrete	2.99	1.29	197
EDUCSPOUS	Education of the spouse or mother	(1–7)	Discrete	2.19	1.31	179
LANDO	Size of land owned	(0–5)	Discrete	2.55	1.17	604
C0_5P (RATIO)	Child 0–5/hh size		Continuous	0.267	0.12	604
A60p	Age over 60/hh size		Continuous	0.037	0.20	604
XTOTAL	Total household expenditure		Continuous	97.88	123.60	604
XFD	Total food expenditure		Continuous	47.92	55.12	604
RUNDUMCAS	Dummy for stock of cassava running out	(0–3)	Discrete	1.17	1.04	140
RUNDUMRICE	Dummy for stock of rice running out	(0–3)	Discrete	0.73	0.71	109
UNDUMSORG	Dummy for stock of sorghum running out	(0–3)	Discrete	0.52	0.72	77
RUNDUMMILL	Dummy for stock of millet running out	(0–3)	Discrete	0.65	0.61	113
RUNDUMBEAN	Dummy for stock of beans running out	(0–3)	Discrete	0.75	0.68	96
RUNDUMPIGPEA	Dummy for stock of pigpeas running out	(0–3)	Discrete	0.80	0.92	10
RUNDUMCOWPEA	Dummy for stock of cowpeas running out	(0–3)	Discrete	0.56	0.73	16

RUNDUMGNUT	Dummy for stock of groundnuts running out	(0–3)	Discrete	0.73	0.44	90
RUNDUMSPOTA	Dummy for stock of small potato running out	(0–3)	Discrete	0.29	0.57	112
SADMDIST	Nearest ADMARC market to sell	(1–5)	Discrete	2.18	1.28	304
PSTRADER	Private traders buy staple?	(1–2)	Dichotomous	1.91	0.28	304
PBTRADER	Private traders sell staple?	(1–2)	Dichotomous	1.91	0.29	304
SELPOINT	Private traders' selling point		Continuous	2.53	2.14	304
LOCMKT	Distance of the local produce market		Continuous	2.38	1.14	304
DRINKDST	Distance to protected drinking source	(1–5)	Discrete	2.07	1.40	304
LATERINE	Latrine in house	Yes = 1 No = 0	Dichotomous	0.39	0.48	304
HEALTSRC	Source of healthcare	(1–3)	Discrete	1.03	0.179	304
HEALTDST	Distance to a health center in km	(1–4)	Discrete	2.40	0.987	304
AGEMNTH	Age of child		Continuous	27.56	16.13	298
CLINFEED	Clinic feeding	Yes = 1 No = 2	Dichotomous	1.32	0.59	294
ZWA	Weight-for-age Z-score		Continuous	−1.47	1.32	282
ZHA	Height-for-age Z-score		Continuous	−1.76	1.51	240
ZWH	Weight-for-height Z-score		Continuous	−0.38	1.27	254
SICKFEED	Number of times child fed during sickness		Continuous	1.42	1.21	295

Continued

TABLE A4.2 Labels and summary descriptive statistics of variables used in the chapters.—cont'd

Variable	Labels	Value Labels1	Nature	Mean	Std. Dev.	N
CHILD0_5	No. of children between 0 and 5 years		Continuous	1.70	0.618	304
CASHCROP	Cash crop growers	Yes = 1 No = 0	Dichotomous	0.347	0.476	604
PCALMZ	Share of calories from maize		Continuous	63.66	38.71	600
PCALCR	Share of calories from other grains		Continuous	9.76	21.91	600
PCALRT	Share of calories from roots/tubers		Continuous	16.54	27.01	600
PCALMT	Share of calories from meat, fish, eggs		Continuous	4.91	16.43	600
PCALVEG	Share of calories from vegetables		Continuous	1.98	9.92	600
PCALFAT	Share of calories from fat		Continuous	0.87	6.20	600
PCALPUL	Share of calories from pulses		Continuous	0.76	5.24	600
PXFD	Per capita food expenditure		Continuous	8.47	10.62	604
WATER	Drinking water source	Protected = 1 Unprotected = 0	Dichotomous	0.46	0.499	304
HOSPITAL	Health source dummy	Yes = 1 No = 0	Dichotomous	0.967	0.18	304
BFEEDNEW	Breastfeeding	Yes = 1 No = 2	Dichotomous	1.57	0.49	294

Appendix 5

SPSS programming basics

The current edition of this book demonstrates the statistical methods in each chapter using STATA and R codes. However, SPSS remains a popular programming package among social scientists in developing and developed countries. SPSS is also a popular introductory package for courses in statistics. Readers of the earlier editions of the book also appreciated introduction of the programming basics using SPSS.

The purpose of this appendix is to explain how to use SPSS syntax, submitting some elementary statistical procedures, transforming data using various commands, sorting data, and saving outfiles. We will use a few examples to demonstrate each of the aforementioned.

Using SPSS syntax

Writing programs in SPSS (using SPSS syntax) allows a researcher to use additional features, which cannot be used in menu and dialog windows. It also allows replication of the steps in the analysis quickly and efficiently. The following steps may be done from the syntax editor:

Step 1: First open a syntax window. To do this, click on

▶ From the menus, choose File

New
Syntax
Select the file and click OK. An example of a syntax editor may be as in Fig. A5.1.

Step 2: Save the file under a name (for example, wissue5new.sps). To do this, the following should be done:

▶ From the menus, choose File

Save
Name of the file
Be sure to put periods at the end of each SPSS statement
Step 3: To run all the commands in the syntax file, do the following:

642 SECTION | IV Technical appendices

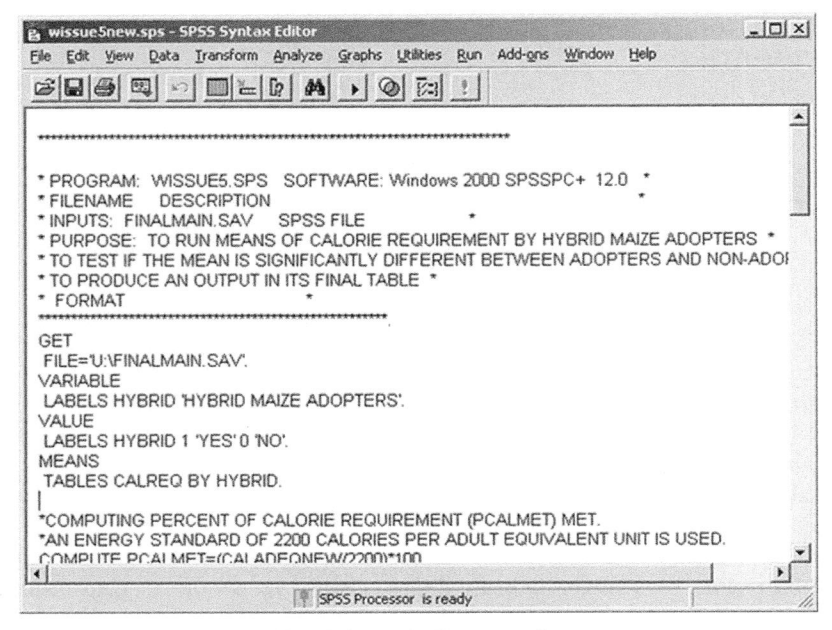

FIGURE A5.1 SPSS syntax editor.

▶ From the menus, choose Run

Select "All" under the Run menu

Alternatively, one can highlight the command to be run (Ctrl-A) and run the command by pressing Ctrl-R. Once the program has been submitted, an active file is generated from the data, and the variable names and the output automatically show up in the SPSS output viewer window.

If the data set has been read without any errors, one can run any number of additional procedures against it during the session. In the syntax window, move the cursor to the next line by pressing the Enter key. To see what variables are in the active file, type:

display.

Then, with the cursor still on the line typed, press the Run button to submit the procedure. To see all the cases, submit the following command in a similar fashion:

list.

The following rules for writing SPSS syntax may be taken into consideration by the researcher:

- Commands are not case sensitive. In other words, upper- and lowercases do not matter.

- Command options start with a slash (/).
- If multiple lines are used for a command, column 1 of each continuation line must be blank.
- Commands should be terminated with a period.
- Comments are indicated by an asterisk (*) in the first column.

When a syntax file is created, SPSS does not carry out certain commands until it is asked to do so. Thus the "Get File" command is necessary for SPSS to execute the commands. A very useful resource for SPSS commands can be found on the web by visiting Rayland Levesque's website at http://pages.infinit.net/rlevesqu/spss.htm.

Submitting statistical procedures

One can submit different statistical procedures one at a time. They are given in the following.

SHOW: This command displays the status of various running options.
CROSSTABS tables (contingency tables):
CROSSTABS TABLES CALREQ BY cashcrop
/cells=count row column
/STATISTICS=ALL.
T-test Groups = HYBRID (0,1)/VAR=FOODSEC.

(where the group HYBRID contains cases such that the first group has values equal to 0, while the second group has values equal to 1. The groups are compared on their mean scores for the variable FOODSEC).

DESCRIPTIVES
VARIABLES=pcalmz pcalcr pcalrt pcalmt pcalmlk pcalveg pcalpul
/STATISTICS=MEAN

(produces descriptive statistics such as the mean of the continuous variables per capita calorie of maize, per capita calories from other grains, per capita calories from roots and tubers, per capita calories from meat, fish, and eggs, per capita calories from milk, per capita calories from vegetables, and per capita calories from the pulses).

CORRELATIONS VARIABLES=ZHA ZWH ZWA

(produces correlation coefficient for height-for-age, weight-for-height, and weight-for-age Z-scores).

Data transformation techniques

We now look at a few techniques to transform data using the data set collected and entered. This can include data modifications, such as the creation of new variables or recoding the existing variables.

Compute statement

The Compute statement is used mainly to create new variables. Various arithmetic and statistical functions are available to use. The complete list can be found in the Command syntax reference from the Help menu. We provide a few examples of the Compute statement.

COMPUTE PCALMET=(CALADEQNEW/2200)*100
EXECUTE

This statement will compute the percent of calorie requirements that are met by the household with the energy standard of 2200 calories per adult equivalent unit being used as the basis.

COMPUTE FOODSEC=0.279762*NBR +0.482143*RUNDUM1 + 0.238095*LIVSTOCKSCALE
EXECUTE

This statement will compute a food security indicator as a weighted average of the number of meals consumed by the household, stocks of food running out, the number of livestock owned, and the household normalized.

COMPUTE
pcalmz=(calmzg+calmzf)/cal*100

This statement will compute the per capita calorie consumption of maize by the household as the ratio of the sum of the calories from maize grain and maize flour to the total calories consumed by the household times 100.

Recode statement

The Recode statement is used in SPSS to classify continuous variables into discrete ones. The general format is as follows:

Recode varname or varlist (old value=new value)/varname (old value=new value). An example is given in the following.

*RECODING MEMBER SEX
RECODE MBRSEX (1=1) (2=0) INTO MBRSEXNEW. VARIABLE LABELS MBRSEXNEW 'SEX OF THE MEMBER'. EXECUTE

This command recodes the member sex, with the cases of the MBRSEX variable being 1 remaining unchanged, while the cases of the MBRSEX being 2 changes to a value of 0. This new variable is then termed MBRSEXNEW.

The If statement

The If statement is used in SPSS to create new variables based on logical conditions. Various relational operators (example, $>$ $\leq$) and logical operators (such as AND, OR, NOT) are used for this purpose. The following examples demonstrate how this statement is used to compute a new variable in SPSS.

IF (DRINKSRC=2 OR DRINKSRC=6 OR DRINKSRC=7) WATER=1. IF (DRINKSRC=1 OR DRINKSRC=3 OR DRINKSRC=5 OR DRINKSRC=0) WATER=0

In the aforementioned example, we categorize drinking water source into two categories: protected and unprotected. When the cases of the DRINKSRC variable attain values of 2, 6, and 7, then the variable WATER attains a value of 1, indicating protected water source. Similarly, when the cases of the DRINKSRC variable attain values of 0, 1, 3, and 5, then the variable WATER attains a value of 0, indicating unprotected water source.

IF (PCALMET LT 80) CALREQ=0.
IF (PCALMET GE 80) CALREQ=1

In this example, if the value of the variable PCALMET is less than 80, then the CALREQ variable attains a value of 0, indicating that calorie requirements for the particular member of the household are not met. On the other hand, if the value of the variable PCALMET is greater than or equal to 80, then the CALREQ variable attains a value of 1, indicating that calorie requirements for the particular member of the household are attained.

Data selection statements

The Sample statement selects a random sample. Using the temporary statement keeps the full sample in the active file. The general format is given as follows:

Sample factor. or
Sample n from m

The following Sample statement selects approximately 25% of a random sample.

temporary. sample 0.25. list id

We next look how data can be sorted and selected number of observations selected from a particular data set.

Sorting and selecting data

Data files are not always organized in an ideal manner for specific needs of the researcher. To undertake data analysis in an organized fashion, one can sort a data set based on the values of one or more variables. Additionally, one can restrict the analysis to a subset of cases to analyze the subset of cases.

Sorting cases is often necessary for certain kinds of analysis. From the menu option, choose Data and then sort cases. This will open the Sort Cases dialog box (Fig. A5.2).

For example, one may be interested in sorting cases by per capita expenditure of the household and educational status of the spouse or the mother. The order in which the variables are sorted determines the order in which cases are sorted. In the aforementioned example, cases will be sorted by the per capita expenditure (in Kwacha) within categories of educational status of the mother.

It is also possible to select a subset of cases based on a specific subgroup. This is important since the researcher may be only interested in a particular subset of the original sample (e.g., education of the household head to be more than the

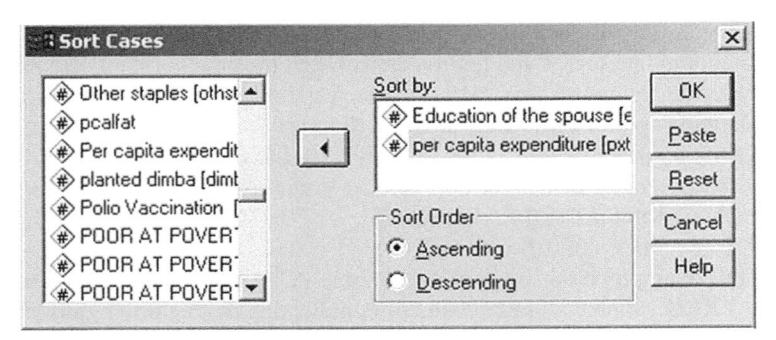

FIGURE A5.2 Sort cases dialog box.

secondary level or per capita expenditure above a certain threshold value). We will use the following criteria to define a subgroup that includes the following:

1. Variable values and ranges
2. Logical expressions

For the first case, the example is as follows: USE ALL
COMPUTE filter_$=(MBRNBR=1)
VARIABLE LABEL filter_$ 'MBRNBR=1 (FILTER)'. VALUE LABELS filter_$ 0 'Not Selected' 1 'Selected'. FORMAT filter_$ (f1.0)
FILTER BY filter_$. EXECUTE

In the aforementioned example, the researcher selects only the observations for member number 1 from the whole sample. Thus, the household becomes the unit of observation for the analysis.

For selecting a subgroup based on a conditional or logical expression, an example is given in the following:
COMPUTE filter_$=(MBRREL=3 and educhead>=4)
VARIABLE LABEL filter_$ 'MBRREL=3 and educhead>=4 (FILTER)'.
VALUE LABELS filter_$ 0 'Not Selected' 1 'Selected'
FORMAT filter_$ (f1.0). FILTER BY filter_$. EXECUTE

In this example, the researcher selects child-level observations and the households for which education of the household head is greater than primary level. Thus, the aforementioned conditional expression selects only the cases for which the aforementioned statement is true.

Saving files

SPSS-formatted data sets are also referred to as system files or save files and the extension name associated with them is .sav. A system file contains information such as variable names, formats, variable and value labels, and missing value indicators stored in a binary format. Thus, the Save Outfile statement can be used to save a permanent data set (as a .sav file). The general format for saving is as follows:

Save Outfile='filename'/options

To check the directory for determining the variables in the active file, the following command may be undertaken:
display

Additionally, one may want to save a system file that contains only a subset of the variables in the active file. The drop option is especially important if the data set contains a large number of cases or variables. For example, you might want to save the outfile finalmain.sav as finalmainrecent.sav by dropping some variables such as quantity harvested of various crops.

The Get File statement, on the other hand, is used to access a permanent SPSS data set (a .sav file). The general format is as follows: Get File='filename'/options.

Thus, to access the system file that you just created, submit the command:
Get File='finalmain.sav'

This command will generate the file much faster than processing a syntax file with the data list input commands.

Working with output

The results from running a statistical procedure are displayed in the viewer. We use the file finalmain.sav and chapter8.spo files to demonstrate how to use the viewer.

The viewer window is divided into two panes: (1) the outline pane contains the outline for all the information stored after the syntax is run; and (b) the contents pane consists of all the tables, charts, and the output (Fig. A5.3). To hide a table or chart, double-click its icon in the outline pane. The open book icon will change to a closed book icon, implying that the information associated with it is now hidden. To redisplay the hidden output, just double-click the closed book icon.

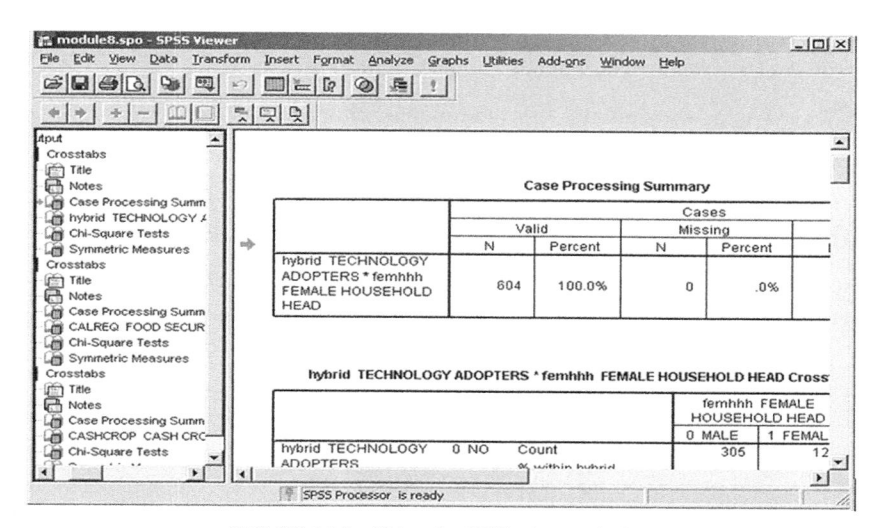

FIGURE A5.3 Using the SPSS viewer window.

Default tables produced in the viewer window may not display the information in a clear fashion. Pivoting the tables allows one to transpose the rows to columns and the columns to rows. To transpose the rows into columns and vice versa, the following steps can be taken:

▶ Double-click the CALREQ* FEMHHH Cross-tabulation table.
▶ From the Pivot menu, choose

Pivot
Transpose Rows and Columns

To have a custom table look, pivot tables are usually created with standard formatting. To change the formatting of any text within a table such as font name, font size, font style, and color, the following can be done:

▶ Double-click the CALREQ* FEMHHH Cross-tabulation table.
▶ From the menus, choose

View
Toolbar

▶ Click the text CALREQ Food Security.
▶ From the drop-down list of font sizes on the toolbar, select 12, and to change the color of the text, click on text color and select any color.

To edit the contents of the table (such as the title), double-click on the title, and change the contents. Hiding rows and columns that are necessary can also be extremely useful, since some of them may be unnecessary. The following procedure hides the unnecessary columns from the Case Processing Summary Table.

▶ Double-click the Table.
▶ Ctrl-Alt and click on valid percent column.
▶ Right-click on the highlighted column and select Hide Category from the drop down menu. The column is hidden but not deleted.

To redisplay the column, from the menus, choose
View
Show All

Table looks

The look of a table is extremely important in providing a clear, concise, and meaningful report. To customize a format to fit the specific needs, such as changing background colors, border styles, etc., the following procedure can be done:

▶ Double-click the CALREQ* FEMHHH Cross-tabulation table.
▶ From the menus, choose

Format
TableLooks

▶ Select Academic (VGA Narrow), for example, and click Edit Look.
▶ Next, click on Cell Formats tab to view the formatting options.

The options include font size, style, and color. Other options include alignment, shading, background colors, and margin sizes. The new style can be saved in an appropriate format. To do this:

▶ Click Save As.
▶ Navigate to the desired directory and enter a name for the new style in the File Name text box.
▶ Click on Save and click OK to apply the changes before returning to the viewer.

Using results in other applications

The output from the viewer window can be used in many applications. The following is an example of how to use the results in Microsoft Word. To paste the results as rich text, the following procedure needs to be done:

▶ Click the CALREQ* FEMHHH Cross-tabulation table.
▶ From the menus, choose

Edit
Copy

▶ Open Microsoft Word. From the Word menu, choose

Edit
Paste Special

▶ Select Formatted Text (RTF) in the Paste Special box. The diagram shown in Fig. A5.4 pops up.

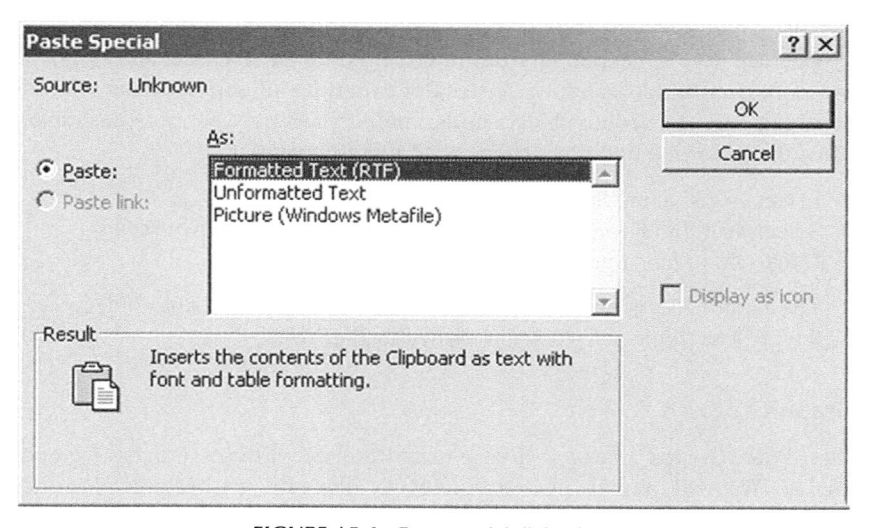

FIGURE A5.4 Paste special dialog box.

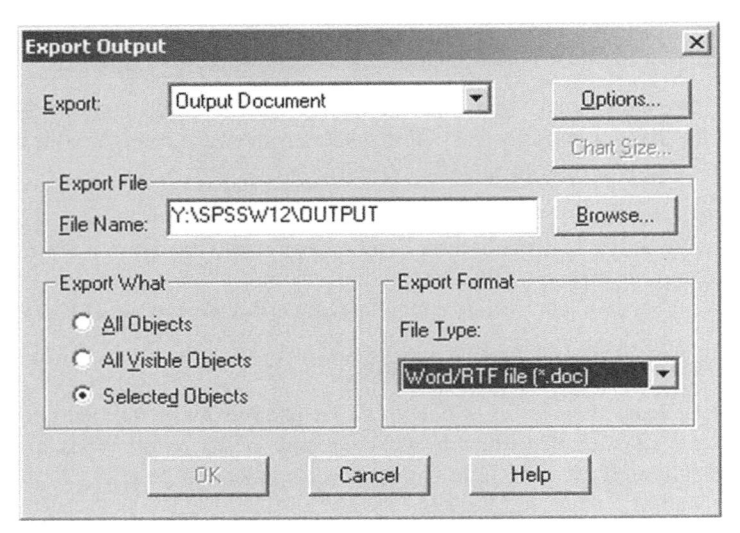

FIGURE A5.5 Export output dialog box.

▶ Click OK to paste the results into the current document. The table will now be displayed in the document.

SPSS also allows exporting results into Word or Excel files. In the viewer outline pane, one can select the specific items that need to be exported. This can be done as follows:

▶ From the viewer menus, choose

File
Export
The export dialog box (Fig. A5.5) pops up as follows:
One can save the exported Word or Excel file to any location and assign it any name that Windows allows. Instead of exporting all objects in the viewer, exporting only the visible objects in the outline pane may be more desirable. Thus, the following additional steps need to be undertaken:

▶ Select All Visible Objects in the Export Group.
▶ Select Word/RTF file (*.doc) from the File Type drop-down menu.
▶ Click OK to generate the word file.

The output file is named outputvis.doc, and only the visible objects are displayed. The output file looks as shown in Fig. A5.6.

Graphics

The SPSS "Graphs" menu is a very useful tool for different kinds of graph display. We will provide two examples to illustrate the features that are

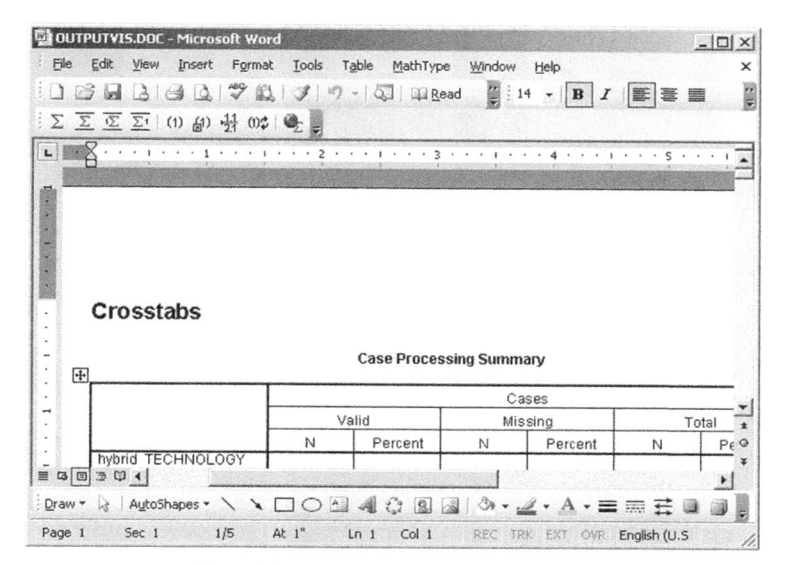

FIGURE A5.6 Output file exported in Word.

available with SPSS. There are various formatting options (such as changing the color of the chart elements, including value labels, changing titles, etc.) in the SPSS output viewer similar to what is described before.

Scatterplots

A useful way of visualizing relationship between any two variables is by creating a matrix of scatterplots. To do this, the following procedure may be selected:

▶ From the graph menu, choose

Scatter
Simple

▶ Then click on the "Define" button to open the scatterplot dialog window.
▶ In the "Scatterplot Simple" dialog window, select ZWA and DRINKDST.

The window shown in Fig. A5.7 pops up.

The "Options" button specifies how the missing values are treated. The default is "listwise." This means that if two variables are being analyzed, if the value of one variable is missing for any of the cases, the case is dropped from the analysis. Another option is to exclude cases variable by variable. In this case, cases are dropped depending on the variables necessary to compute a specific statistic.

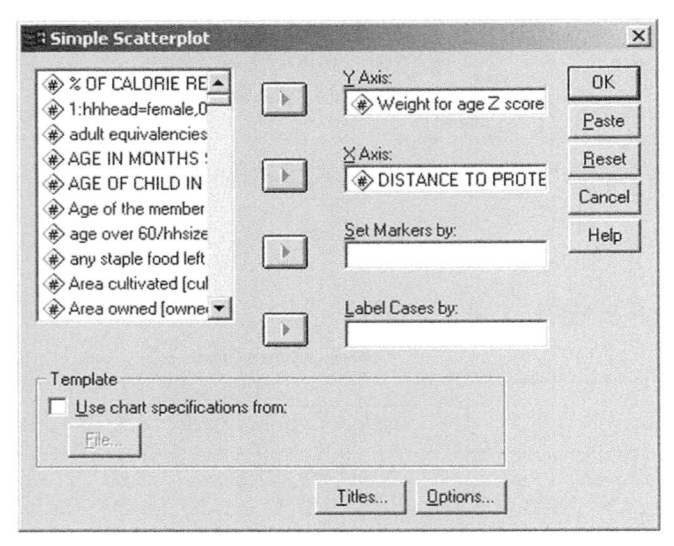

FIGURE A5.7 Simple scatterplot.

▶ Click on the "OK" button to produce the scatterplot.

Histograms

To create a histogram, the following procedure needs to be followed:

▶ From the graph menu, choose Histogram.

The window shown in Fig. A5.8 pops up.

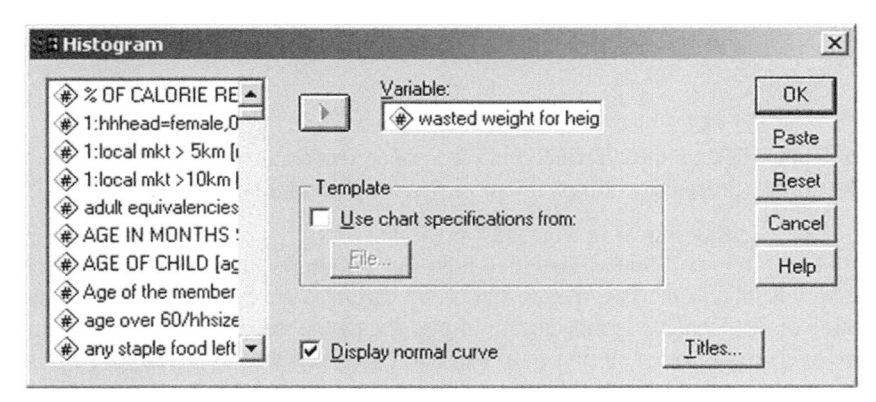

FIGURE A5.8 Histogram plot of weight-for-height Z-**score**.

SPSS programming basics Appendix | 5 **653**

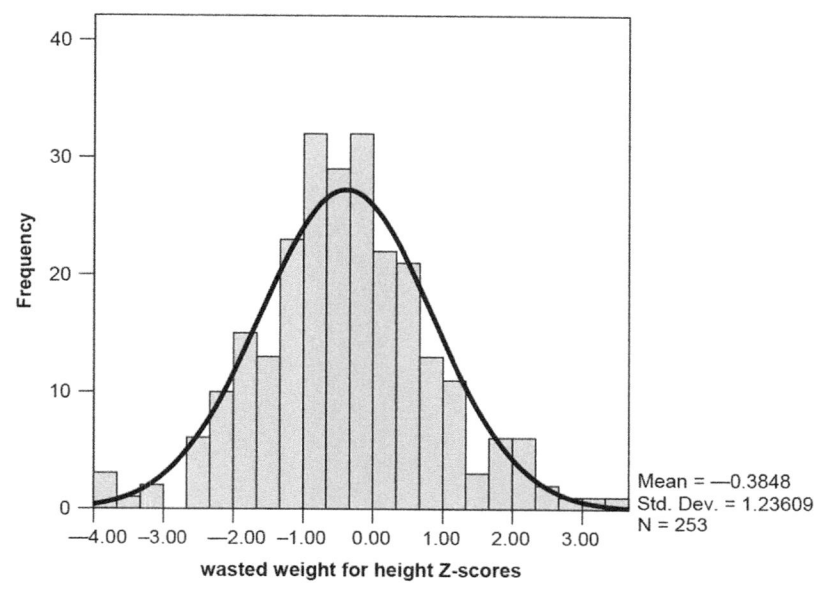

FIGURE A5.9 Histogram of weight-for-height Z-score.

▶ Move the ZWH variable to the "Variable" field.
▶ Click on "Display Normal Curve" box.

The normal curve overlay allows us to visualize whether the variable under analysis is close or far from resembling a normal distribution.

▶ Click on the "OK" button to produce the histogram. Fig. A5.9 gives the histogram of weight-for-height Z-scores.

Appendix 6

STATA—a basic tutorial

The second edition of this book introduced STATA. We continue to demonstrate STATA codes and programming in this edition as well. As mentioned in Appendix 2, STATA is a software that is very powerful and flexible, which has become very popular in recent years among researchers, students, and policy analysts. In this book, we have used STATA in many chapters and have shown the implementation using commands and output from STATA. To summarize, we have illustrated the following applications in this revised edition:

Chapter	Concept	STATA Command
2	t-tests	ttest
3	Chi-square tests	tabulate
4	Cramer's V	tab x y, column nokey chi2V
5	One-way ANOVA	oneway
6	Factor analysis	factor
7	Two-way ANOVA	anova
8	Correlation	correlate or pwcorr
9	Simple regression	regress
10	Multiple regression	regress
11	Discriminant analysis	manova and candisc
12	Logistic regression	logit
13	K-means clustering	tabstat

There are lots of resources on the web that provide interesting and useful examples and suggestions. For instance, the STATA website (www.stata.com) itself has a host of helpful services and programs. There are helpful guides for different users (in particular, see http://www.stata.com/why-use-stata/). In addition, there are tutorials set up by different user groups that give specific line commands and output for different texts in Econometrics and Statistical Modeling. In particular, see http://www.ats.ucla.edu/stat/stata/ and http://data.princeton.edu/stata/. We highly recommend the resources page in STATA itself http://www.stata.com/links/resources-for-learning-stata/, where one can find information and articles from *The Stata Journal*, among many other useful links.

In this revised edition, we have provided a lot of hands-on examples in all the chapters to cover many basic and important programs. Please note that STATA examples in the text are meant for illustrative purposes. In many instances, we note how certain key statistical properties such as independence, normality, and sampling adequacy fail for our mock data. We note those failures and indicate how the reader can cross-check the validity of the data through adequate tests. It is important for students to understand these diagnostic aspects, which are crucial steps in research.

We provide a basic tutorial in the following, where we begin by typing a small dataset in EXCEL, and bring that data into STATA, and show its implementation for one small example.

We begin with a small data typed in EXCEL with five observations and four columns: observations, corn output, fertilizer usage, and pesticide usage. The data typed in EXCEL is given in the following, with a portion of the EXCEL in the display window:

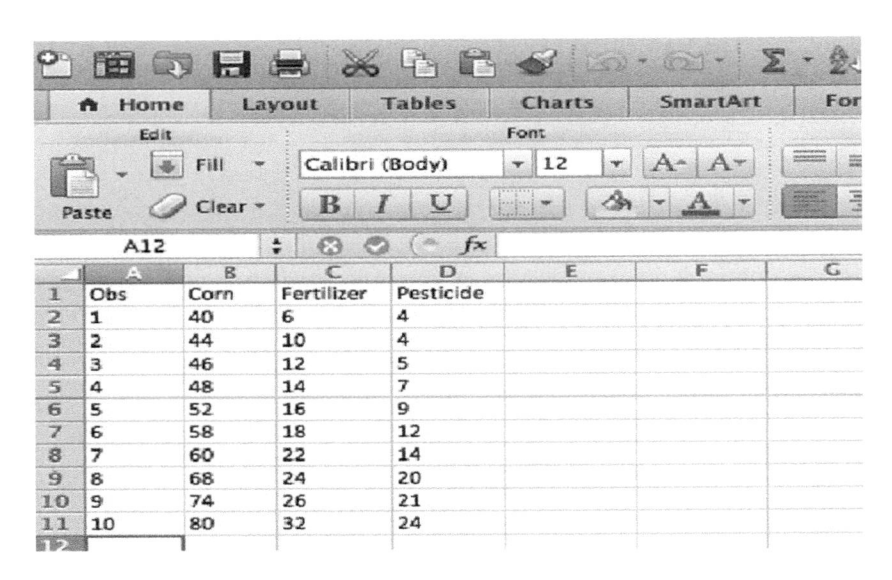

Obs	Corn	Fertilizer	Pesticide
1	40	6	4
2	44	10	4
3	46	12	5
4	48	14	7
5	52	16	9
6	58	18	12
7	60	22	14
8	68	24	20
9	74	26	21
10	80	32	24

To use this in STATA, we first save it as a .csv file. For the time being, let us save it in our desktop and call it a.csv. We now open STATA and bring this file into STATA for later use. When we open STATA, the window display looks like this:

STATA—a basic tutorial Appendix | 6

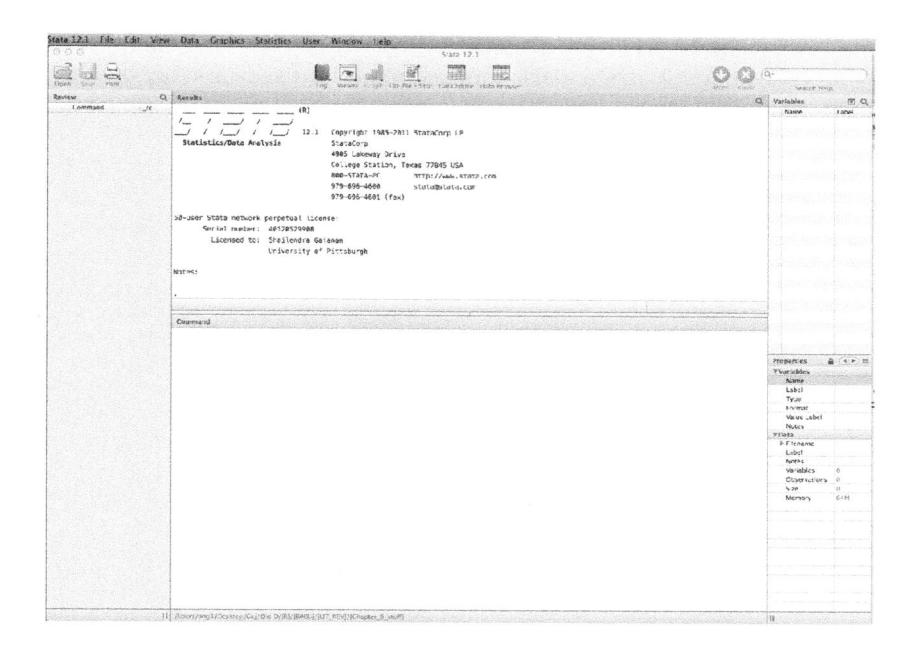

We type our commands in the second half of the screen, and hit return. The result will be placed in the top portion of the window. We first type a command letting STATA use the a.csv file. The command is the .insheet using command, and we present the input line and STATA's results in the next window display as follows:

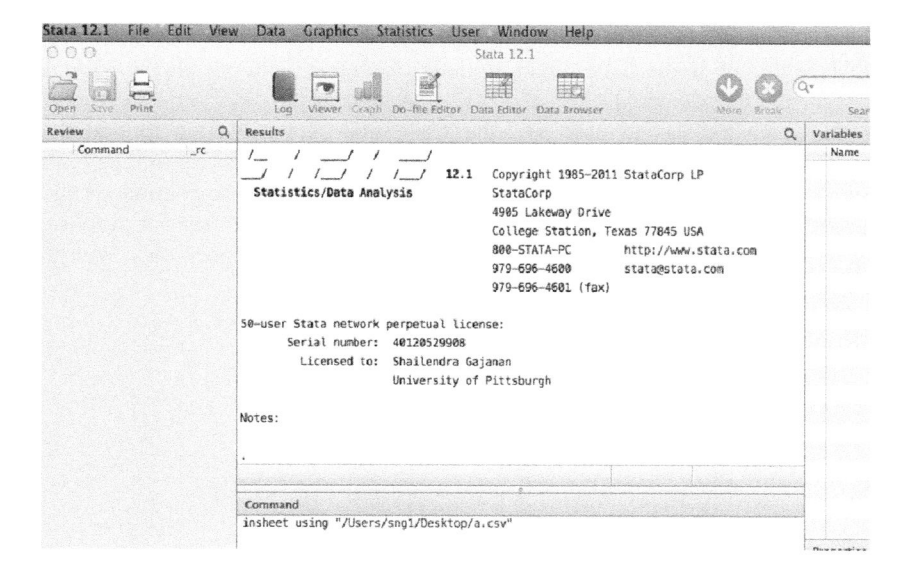

After typing the following command,

.insheet using "/Users/sng1/Desktop/a.csv" we hit the return key, and the command is pushed up and the STATA response is as follows:

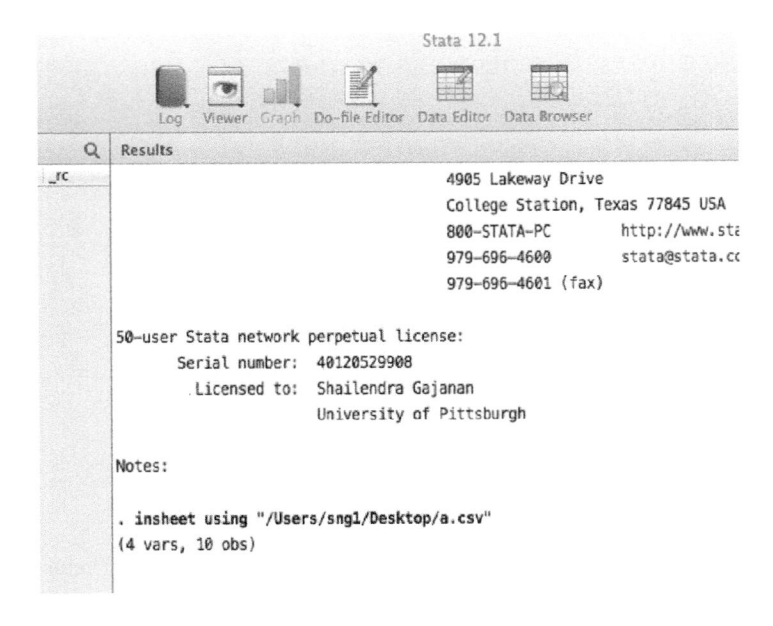

```
Stata 12.1

                                4905 Lakeway Drive
                                College Station, Texas 77845 USA
                                800-STATA-PC          http://www.sta
                                979-696-4600          stata@stata.cc
                                979-696-4601 (fax)

50-user Stata network perpetual license:
      Serial number:  40120529908
      Licensed to:    Shailendra Gajanan
                      University of Pittsburgh

Notes:

. insheet using "/Users/sng1/Desktop/a.csv"
(4 vars, 10 obs)
```

The top panel indicates that the a.csv file can be now read in STATA and it indicates that there are 4 variables and 10 observations, which is what we started out with. Before quitting, we have to save our work. So, at this stage, we save the file in the same directory under the name a.dta (all files in STATA will have .dta extension).

Now our goal is to use the a.dta file. Suppose we wanted to find the descriptive statistics and run some regressions. It is not necessary to type each command sequentially. Rather, we can create a file with all the commands and save it in the same directory. This file has a .do extension. So, for our purpose, we will open STATA and create a new file called a.do, and type 5 lines. We can then execute the program and see the output.

To create a new .do file, we open STATA and under the File option, we click on New Do-File. A Blank screen opens, and we type our commands and save the file as a.do:

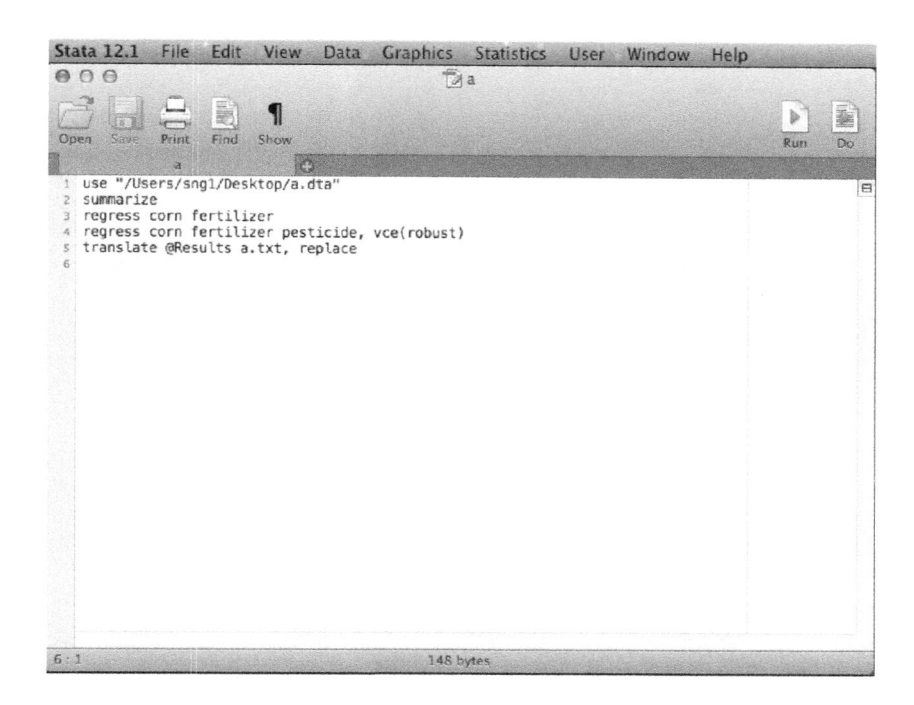

```
use "/Users/sng1/Desktop/a.dta"
summarize
regress corn fertilizer
regress corn fertilizer pesticide, vce(robust)
translate @Results a.txt, replace
```

We have typed five lines in the blank screen, and we can save it in the same directory as a.do. The lines in the .do file are seen in the aforementioned display.

The first line:
.use "/Users/sng1/Desktop/a.dta"
indictates the file that we have to use in the given directory.

The next line is the .summarize command which produces the descriptive statistics.

The first .regress command in the third line produces the output for the simple regression, while the second is the multiple regression command which has the vce(robust) option, for producing robust standard errors.

The last line:
.translate @Results a.txt, replacetakes all the results from the a.do file and places them in a txt file called a.txt in the same directory. The .replace option rewrites this file automatically as newer commands are added.

660 SECTION | IV Technical appendices

To execute the file, first open STATA, and click on the File tab on the top, and open the do file from y our directory. The a.do file in this case can be opened and placed alongside the STATA window.

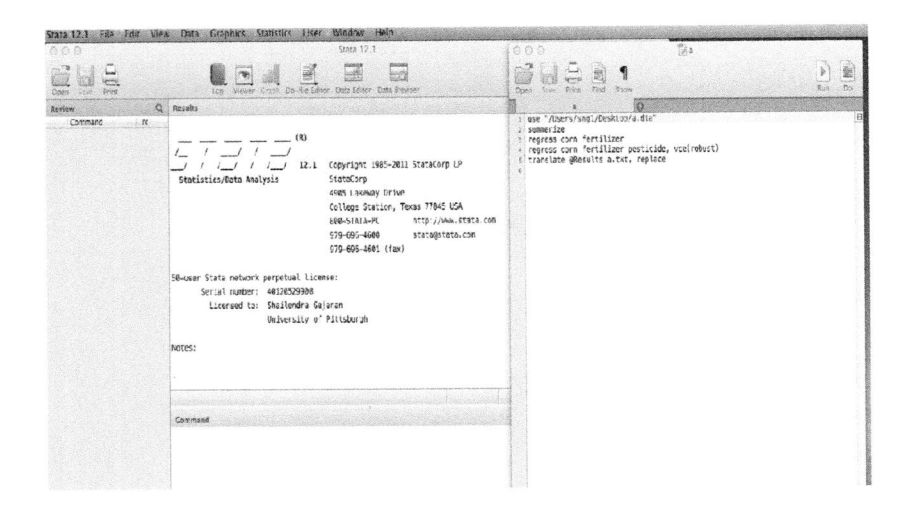

To execute the file, just click on the Do tab on the top right-hand corner in the a.do file. We show the results in the top-panel of the STATA window, as follows.

As you can see, each command line has been executed and a text file with all the results is also produced from the .translate command.

This simple tutorial is meant to give you a starting point for using STATA. As mentioned in the beginning of the appendix, you can find lots of resources in the web and also from many helpful guides that are available.

Hamilton (2006), Cameron and Trivedi (2010), and Acock (2012) contain lots of information on programming and interpretation.

STATA and R are software that is very powerful and flexible, which has become very popular in recent years among researchers, students, and policy analysts. In this book, we have updated R commands in many chapters and have shown the implementation using commands and output from R, which are corresponding the implementation from STATA. To summarize, we have illustrated the following applications in this revised edition:

Chapter	Concept	STATA command	R command
2	t-tests	ttest	t.test
3	Chi-squared tests	tabulate	chisq.test
4	Cramer's V	tab x y, column nokey chi2 V	cramersV
5	One-way ANOVA	oneway	oneway.test
6	Factor analysis	factor	factanal
7	Two-way ANOVA	anova	aov
8	Correlation	correlate or pwcorr	cor or rcorr
9	Simple regression	regress	lm
10	Multiple regression	regress	lm
11	Discriminant analysis	manova and candisc	manova and candisc
12	Logistic regression	logit	glm
13	K-means clustering	tabstat	kmeans
14	Instrumental analysis	ivregress	ivreg

Appendix 7

Anthropometric indicators—computation and use

In this edition, we continue to use nutritional indicators including anthropometric indicators for the studying nutrition policy and programs. Anthropometry[1] is a very useful tool for assessing and predicting performance, health, and survival of individuals and reflects economic and social well-being of a population. It is a widely used technique and is inexpensive. It generally measures the nutritional status of an individual or a population group. The four building blocks used to undertake anthropometric assessment are age, sex, height, and weight. Anthropometry can be used for various purposes, depending on the indicators selected. For example, wasting (low weight for height) can be useful for screening children at risk and for measuring short-term changes in nutritional status. However, this indicator is not useful for evaluating changes in a population over a longer time period. For assessing the nutritional status of children, three most common measures used are as follows:

- Weight for height
- Height for age
- Weight for age

What are the indices of nutritional status of children?

Low weight for height: Wasting or thinness indicates a recent or severe process of weight loss, which is often associated with acute starvation and/or severe disease. Additionally, it can be also caused by chronic unfavorable conditions such as severe food shortage, disease, or infection. The prevalence of wasting is usually below 5% even in poor countries. The Indian subcontinent, where

---

1. This appendix uses the materials from Food and Nutrition Assessment Technical Assistance (FANTA) (2003) Anthropometric Indicators Measurement Guide.

higher prevalence is found, is an important exception. A prevalence rate above 5% is alarming, since it can lead to mortality. If prevalence is between 10% and 14%, it can be regarded as serious, and if prevalence is above 15%, it is considered critical.

High weight for height: This is often referred to as overweight. Even though there is a strong correlation between high weight for height and obesity, greater lean body mass can also contribute to high weight for height. At the individual level, obesity should not be used to describe high weight for height. However, at the population level, high weight for height can be considered as an indicator for obesity, since the majority of individuals are obese.

Low height for age: Low height for age (often referred to as stunting) stems from a slowing in the growth of the fetus and results in a failure to achieve expected length as compared with a healthy well-nourished child of the same age. It is an indicator of past growth failure and is associated with a number of long-term factors such as insufficient protein and energy intake, frequent infection, inappropriate feeding practices, and poverty. The worldwide variation of the prevalence of stunting is considerable and ranges from 5% to 65% (see, for example, the Introduction section of the WHO Global Database on Child Growth and Malnutrition). For the purpose of evaluation, it is preferable to use children below 2 years of age. This is because the prevalence of stunting in children of this age is likely to be more responsive to the impact of interventions than in older children. Information on stunting for individual children can be clinically useful as a tool for diagnosis. It is useful for evaluation purposes but is not recommended for monitoring, since it does not change in the short term (e.g., between 6 and 12 months).

Low weight for age: This is often referred to as underweight and is a composite measure of stunting and wasting. However, due to its composite nature, the interpretation becomes more complex. For example, weight for age fails to distinguish between short children of adequate body weight and tall, thin children. However, in the absence of significant wasting in a community, similar information can be obtained from weight for age and height for age, as both reflect long-term health and nutritional experience of the individual or population.

In addition to the aforementioned indicators of child nutritional status, two recommended annual monitoring indicators are also used. They are as follows:

Percent of eligible children in growth monitoring and promotion programs

This measure supports program management and provides information on coverage and targeting and may provide a useful basis for supervision of field staff. This measure also provides an indication of trends in service delivery and use. Thus, it has the potential to demonstrate successes of efforts to achieve specified project results.

Percent of children in growth monitoring and promotion programs gaining weight during past 3 months

This measure is used to increase growth promotion and health education counseling. The information can bring about a positive communication between the health worker and caregiver in impacting the health of the child. The measure is most useful when added with other information such as food availability and presence of infection. This indicator is also used as a surveillance tool and may be useful in a community facing severe food or health-related stress.

Comparison of anthropometric data to reference standards and computation of Z-scores

The reference standards most commonly used to standardize measurements were developed by the US National Center for Health Statistics (NCHS) and are recommended for international use by the World Health Organization (WHO). The data and methodology can be found at the following website: http://www.who.int/nutgrowthdb/. The reference population chosen by NCHS was a statistically valid random population of healthy infants and children. This method was based on the assumption that, until the age of 10, children from well-nourished and healthy families throughout the world grow at approximately the same rate and attain the same height and weight as children from industrialized countries. The NCHS/WHO international reference tables can be used for standardizing anthropometric data from around the world and can be obtained on FANTA's website at http://www.fantaproject.org/publications/ anthropom.shtml.

References are used to standardize a child's measurement by comparing the child with the median or average measure for children at the same age and sex. For example, if the length of a 5-month-old boy is 59 cm, it would be difficult to know if he was healthy without comparison to a reference standard. The reference or median length for a population of 5-month-old boys is 65.9 cm, and a simple comparison of lengths would indicate that the child was almost 7 cm shorter than expected. Thus, difference in measurements can be expressed by taking into account age and sex considerations. This can be done in two main ways:

1. Standard deviation units or Z-scores
2. Percentage of the median

Computation of Z-scores

Z-scores are usually used by the international nutrition community since they offer two main advantages. First, using Z-scores allows one to identify a fixed point in the distributions of different indices and across different ages. Second, this approach allows the mean and standard deviation to be calculated for the Z-scores for a group of children. It is a statistic recommended for use when reporting results of nutritional assessments.

The Z-score is defined as the difference between the value for an individual and the median value of the reference population for the same age or height, divided by the standard deviation of the reference population. This can be written as follows:

$$Z-\text{score}(\text{or SD score}) = \frac{(\text{observed value}) - (\text{median reference value})}{\text{standard deviation of reference population}} \quad (A7.1)$$

Let us give an example to illustrate how Z-scores are computed. This example is borrowed from the Anthropometric Indicators Measurement Guide (FANTA, 2003). Suppose that a 19-month-old boy weighs 9.8 kg. By looking at the reference standards, we find that the average weight of a healthy boy of 19 months is 11.7 kg. Since the child is below the median, we want the use the lower standard deviation value, which in this case is 1.2 kg. Then, using Eq. (A7.1), we find that the Z-score for this child is -1.58 SD units.

The distribution of Z-scores follows a normal distribution. The commonly used cutoffs of -3, -2, and -1 Z-scores are, respectively, the 0.13th, 2.28th, and 15.8th percentiles. The percentiles can be thought of as the percentage of children in the reference population below the equivalent cutoff. The cutoffs enable the different individual measurements to be converted in a prevalent statistic. The most common cutoff with Z-scores is -2 standard deviations. This implies that children with a Z-score for wasting, stunting, and underweight below -2 SD are considered severely malnourished. The WHO uses the guidelines shown in Table A7.1 for classifying malnutrition.

Percentage of the median

The percentage of the median is defined as the ratio of observed value of the individual to the median value of the reference data for the same age or height for the specific sex, expressed in percentage. This can be written as follows:

$$\text{Percent of median} = \frac{\text{observed value}}{\text{median value of reference population}} \times 100 \quad (A7.2)$$

TABLE A7.1 Malnutrition classification system.

Cut-Off	Malnutrition classification
<−1 to >−2 Z-score	Mild
<−2 to >−3 Z-score	Moderate
<-3 Z-score	Severe

The median of a distribution is the value at exactly the midpoint between the largest and the smallest. For example, if a child's measurement is exactly the same as the median of the reference population, we say that they are "100% of the median."

Example: Let us assume a girl who is 26 months old and is 70 cm long. We want to use percentages of the median to compare her to the reference standards. From the table of reference values, the 50th percentile length measurement for a 26-month-old girl is 86.20. Then, using Eq. (A7.2), we find that a 26-month-old girl who is 70 cm long is 81.2% of the median.

The main disadvantage of this approach is the lack of exact correspondence with a fixed point of the distribution across age or height status. For example, depending on a child's age, 75% of the median weight for age might be above or below −2 Z-scores. Additionally, typical cutoffs for percent of the median are different for different anthropometric indicators.

Software programs used for computing anthropometric measures

When using computer software programs for anthropometric computations, three separate procedures must be undertaken. First, the raw measurement data should be entered into a computer. Second, the program should combine the raw data on the variables (such as age, sex, length, and weight) to compute the weight for age, height for age, and weight for height. Third, the software should transform these data into Z-scores so that the prevalence of nutritional conditions, such as wasting and stunting, can be computed.

There are different software programs that are used to compare a child to the reference standards. The Centers for Disease Control and Prevention (CDC) have developed a free software package called Epi Info that can handle all anthropometric computations. This is available at the following site: www.cdc.gov/epiinfo. The software enables raw anthropometric data to be converted into indices and Z-scores. It also determines the outliers of the data, which are usually the result of incorrect measurements and coding errors.

Another software, called ANTHRO, analyzes anthropometric data and can be downloaded from the WHO Global Database on Child Growth and Malnutrition at the following website: http://www.who.int/childgrowth/software/en/.

A useful Windows-based software package is available from www.nutrisurvey.de. This program was designed specifically for nutrition surveys by the Work Group on International Nutrition of the University of Hohenheim/Stuttgart in cooperation with the German agency for Technical Cooperation (GTZ). The purpose of this program was to integrate all the steps of a nutrition baseline survey into a single program. The program consists of a standard nutrition baseline questionnaire, which can be customized for the specific site, a function for printing out the questionnaire, a data entry unit that controls the data being entered, a report function, and a graphics section. The report function produces the full set of descriptive statistics of the baseline survey.

The graphics section contains standard graphs and graphics for anthropometric indices. The anthropometric indices are automatically computed using this software.

Uses of anthropometric Data

Anthropometric data can be used for a variety of purposes. They are as follows:

1. Identification of individuals or populations at risk: Depending on the specific objective, anthropometric measures can be used to reflect past or present risk and to predict future risk. Any indicator can reflect both present and future risk. For example, an indicator of present malnutrition can be an important predictor of increased risk of mortality in the future. On the other hand, a reflective indicator of past problems may have no value on predicting future risk. For example, stunting of growth in early childhood as a result of malnutrition can persist throughout the life of an individual. However, with age, the effect may fade out.
2. Selection of individuals or populations for an intervention: A good indicator should predict the benefits that can be derived from an intervention. The crucial distinction between indicators of risk and indicators of benefit is important for developing and targeting interventions. For example, low maternal height can predict low birthweight. However, low maternal weight in the population does not predict any benefits of providing an improved diet to pregnant women. Anthropometric indicators thus reflect the overall socioeconomic development among the poorest members of a population. For example, data on stunting in children and adults can show socioeconomic conditions that are not conducive to good health and nutrition. Hence, stunting can be used effectively for targeting development programs.
3. Evaluation of the effects of changing nutritional or socioeconomic variables, influencing interventions: Indicators should reflect responses for past and present interventions. Changes in wasting are a good example of a short-term indicator of malnutrition, whereas a decrease in the prevalence of stunting at the population level is a long-term indicator showing that social development is benefiting both the rich and the poor. On the contrary, a decrease in the prevalence of low birthweight can be used to indicate success in activities such as controlling malaria during pregnancy.
4. Achieving normative standards: indicators should reflect normality of individuals' health and nutrition situation. For example, some researchers have argued that moderate obesity among the elderly is not associated with poor health or increased risk of mortality. If the aforementioned statement is true, then advocacy for the need for weight control for this age group would be based only on normative considerations.

Appendix 8

Elements of matrix algebra

Matrix algebra is a mathematical framework for manipulating arrays of numbers or algebraic symbols. It is very useful for representing methods for modeling and analyzing multivariate data.

Definitions

Matrix: A matrix is a rectangular array of numbers arranged in rows and columns and looks as follows:

$$A = \begin{pmatrix} a_{11} & \cdots & a_{1n} \\ \vdots & \ddots & \vdots \\ a_{m1} & \cdots & a_{mn} \end{pmatrix} = \begin{pmatrix} 7 & & 0 \\ & \ddots & \\ 0 & & 6 \end{pmatrix}$$

where a_{ij} is an element of the ith row and the jth column. The upper case letter A denotes the matrix and stands for the whole array of elements.

Order: This is the size of the matrix. A matrix with p rows and q columns is of the order $p \times q$.

Square matrix: This is a matrix where the number of rows is equal to the number of columns.

Vector: This is a matrix with a single column (column vector) or a single row (row vector). For example:

$$A = \begin{bmatrix} a_1 \\ a_2 \\ \vdots \\ a_m \end{bmatrix} \text{ is a } m \times 1 \text{ column vector}$$

whereas $A' = (a_1 \, a_2 \ldots a_m)$ is a $1 \times m$ row vector.

Transpose of a matrix: A transposed matrix A' is a new matrix that is formed by writing the rows of A as the columns of A'. For example, if:

$$A = \begin{pmatrix} 1 & 3 \\ 2 & 4 \end{pmatrix} \text{ then } A' = \begin{pmatrix} 1 & 2 \\ 3 & 4 \end{pmatrix}$$

Equality of matrices: Two matrices are equal if they are of the same order and the corresponding elements are equal.

OPERATIONS WITH MATRICES

Addition and subtraction

Addition and subtraction of matrices are very similar to arithmetic operations in general. In matrix algebra, one needs to add or subtract the corresponding elements. For example, $C = A + B$ is formed by summing the corresponding elements of A and B to form the matrix C. The laws regarding addition and subtraction of matrices are as follows:

1. commutative law: $A + B = B + A$,
2. associative law: $(A + B) + C = A + (B + C)$, and
3. transpose law: $(A + B)' = A' + B'$.

The transpose law states that the sum of matrices equals the sum of the transposes.

Multiplication of two matrices

Two matrices A and B can be multiplied to form another matrix C. The elements of C are the sum of the cross-products of the elements in the ith row of A with the corresponding elements in the jth column of B. Thus:

$$C = AB = \sum_{k=1}^{p} a_{ik} b_{kj} \qquad (A8.1)$$

For example, if:

$$A = \begin{pmatrix} 2 & 0 \\ 0 & 3 \end{pmatrix} \quad B = \begin{pmatrix} 5 & 7 \\ 9 & 4 \end{pmatrix}$$

$$C = AB = \begin{pmatrix} 2 \times 5 + 0 \times 9 & 2 \times 7 + 0 \times 4 \\ 0 \times 5 + 3 \times 9 & 0 \times 7 + 3 \times 4 \end{pmatrix} = \begin{pmatrix} 10 & 14 \\ 27 & 12 \end{pmatrix}$$

The two matrices must be conformable for multiplication. In other words, the number of columns in the first matrix (A) must be equal to the number of rows in the second matrix. The following laws hold for multiplication of matrices:

1. associative: $A(BC) = (AB)C$,
2. commutative law does not hold: in other words, $AB \neq BA$,

3. distributive: $A(B + C) = AB + AC$, and
4. transpose of a product: $(A + B)' = B'A'$.

The transpose of a product of matrices is equal to the product of the transposes in the reverse order.

Types of matrices

1. *Diagonal matrix*: This is a square matrix with all the off-diagonal elements equal to zero. The following is an example of a diagonal matrix:

$$A = \begin{pmatrix} 5 & 0 & 0 \\ 0 & 7 & 0 \\ 0 & 0 & 9 \end{pmatrix}$$

2. *Identity matrix*: This is a diagonal matrix with all the diagonal elements being 1. Multiplying a matrix by an identity matrix does not change the matrix. Thus, if A is any matrix, then $AI = IA = A$.
3. *Symmetric matrix*: A square matrix A is symmetric if it is equal to its own transpose. In other words, $A = A'$.
4. *Orthogonal matrix*: A matrix A is orthogonal if $AA' = I$. For example,

$$A = \frac{1}{\sqrt{2}} \begin{pmatrix} 1 & 1 \\ 1 & -1 \end{pmatrix} \text{ is orthogonal.}$$

FUNCTIONS OF MATRICES

Determinant

Determinants are mathematical objects that are very useful in the analysis and solutions of a system of linear equations. The determinant of a matrix A of the form:

$$A = \begin{pmatrix} a_{11} & a_{12} & a_{13} \\ a_{21} & a_{22} & a_{23} \\ a_{31} & a_{32} & a_{33} \end{pmatrix} \text{ has a value}$$

$$|A| = a_{11}(a_{22}a_{23} - a_{23}a_{32}) - a_{12}(a_{21}a_{33} - a_{23}a_{31}) + a_{13}(a_{21}a_{32} - a_{22}a_{31})$$

A general $(k \times k)$ determinant can be expanded by minors to obtain

$$= a_{11} \begin{vmatrix} a_{22} & a_{23} & \cdots & a_{2k} \\ \vdots & \vdots & \ddots & \vdots \\ a_{k2} & a_{k3} & \cdots & a_{kk} \end{vmatrix} - a_{12} \begin{vmatrix} a_{21} & a_{23} & \cdots & a_{2k} \\ \vdots & \vdots & \ddots & \vdots \\ a_{k1} & a_{k3} & \cdots & a_{kk} \end{vmatrix}$$

$$+ \ldots \pm a_{1k} \begin{vmatrix} a_{21} & a_{22} & \cdots & a_{2(k-1)} \\ \vdots & \vdots & \ddots & \vdots \\ a_{k1} & a_{k2} & \cdots & a_{k(k-1)} \end{vmatrix} \quad (A8.2)$$

A general determinant for a matrix A has a value

$$|A| = \sum_{i=1}^{k} a_{ij} C_{ij} \quad (A8.3)$$

where C_{ij} is the cofactor of a_{ij} defined as follows:

$$C_{ij} = (-1)^{i+j} M_{ij} \quad (A8.4)$$

M_{ij} is called the minor of the matrix A formed by eliminating the ith row and the jth column. Important properties of the determinant are as follows:

1. switching any two rows or columns changes the signs,
2. multiples of rows and columns can be added together without changing the determinant's value,
3. scalar multiplication of a row by a constant c multiplies the determinant by c, and
4. any determinant with two rows or columns equal has value 0.

A matrix with a determinant of zero is called *singular*, implying that there is a linear dependency among its rows (or columns).

Computing inverse of a matrix

If A is a square matrix and $A|I=0$, then A has a unique inverse A^{-1} such that $AA^{-1} = A^{-1}A = I$. A square matrix A possessing an inverse is called nonsingular or invertible. For example, for a 3×3 matrix of the form:

$$A = \begin{pmatrix} a_{11} & a_{12} & a_{13} \\ a_{21} & a_{22} & a_{23} \\ a_{31} & a_{32} & a_{33} \end{pmatrix}$$

$$A^{-1} = \frac{1}{|A|} \begin{bmatrix} a_{22} & a_{23} & a_{13} & a_{12} & a_{12} & a_{13} \\ a_{32} & a_{33} & a_{33} & a_{32} & a_{22} & a_{23} \\ a_{23} & a_{21} & a_{11} & a_{13} & a_{13} & a_{11} \\ a_{33} & a_{31} & a_{31} & a_{33} & a_{23} & a_{21} \\ a_{21} & a_{22} & a_{12} & a_{11} & a_{11} & a_{12} \\ a_{31} & a_{32} & a_{32} & a_{31} & a_{21} & a_{22} \end{bmatrix} \qquad (A8.5)$$

Properties of an inverse matrix

1. Only square matrices have inverses.
2. The inverse of a product of matrices A and B is the product of the inverses in the reverse order. In other words:

$$(AB^{-1}) = B^{-1}A^{-1} \qquad (A8.6)$$

Proof:

- Let $C \equiv AB$.
- Then, $B = A^{-1}AB = A^{-1}C$ and $A = ABB^{-1} = CB^{-1}$.
- Therefore, $C = AB = (CB^{-1})(A^{-1}C) = CB^{-1}A^{-1}C$.
- Thus, $CB^{-1}A^{-1} = I$, where I is the identity matrix.
- Thus, $B^{-1}A^{-1} = C^{-1} = (AB)^{-1}$

Solutions of linear systems of equations

Linear systems can be represented in matrix form as follows:

$$AX = B$$

where X is unknown and A and B are known. A is the matrix of coefficients, X is the column vector of variables, and B is the column vector of solutions. The matrix A is square and nonsingular, and X needs to be solved.

Premultiplying the aforementioned equation on both sides by A^{-1}, We have

$$A^{-1}AX = A^{-1}B$$

Thus,

$$IX = A^{-1}B$$

Appendix 9

Some preliminary statistical concepts

All statistical tests produce a p value (as discussed in Chapters 2 and 3), and this is equal to the probability of obtaining the observed difference if the null hypothesis is true. To put it differently, if the null hypothesis is true, the p value is the probability of obtaining a difference at least as large as that observed due to sampling variation. Thus, if the p value is small, the data support the alternative hypothesis. On the other hand, if the p value is large the data support the null hypothesis. The question that naturally arises is how small is "small" and how large is "large"? Conventionally, a p value of 0.05 (5%) is generally regarded as sufficiently small to reject the null hypothesis. If the p value is larger than 0.05, we fail to reject the null hypothesis. The 5% value is called the "significance level" (often denoted by α) of the test. Other significance levels that are commonly used are 1% and 0.1%. In this appendix, we will introduce some preliminary statistical concepts that check the underlying assumptions of the data.

Principles of specification tests

A correct statistical method depends in part on the nature of the data being used. More specifically, it depends on what assumptions about the data are true or not. Thus, it is important to check the data to determine which assumptions are consistent with the data and which are violated by the data. Methods of checking assumptions are also known as *specification tests*. Almost all statistical assumptions are hypotheses about the data, i.e., about the *data-generating process* (*DGP*). Let us denote that an assumption being claimed as true as H_0. This is often called the *null hypothesis*. In other words,

H_0: assumption is true

The other possibility is that the assumption is not true. Let us denote this by H_1. We will refer to this as the *alternative hypothesis*. In other words:

H_1: Assumption is false

Based on the data, one can do the following:

1. not reject H_0,
2. reject H_0 (which implies accepting H_1).

Specification tests are functions of the data that are used to decide whether to accept or reject H_0. To do this, plug the data into a function called a test statistic. Let us denote this *test statistic* as T. The test statistic satisfies two criteria:

1. the distribution of T when H_0 is true must be known (either exactly or asymptotically),
2. larger values of T imply that H_0 is less likely to hold. To use the specification test, one needs to compute the T value using the data and compare it to some critical value denoted by T_c. The following rules hold:
   a. If $T > T_c$, then T is in the "rejection region" and one rejects H_0.
   b. If $T \leq T_c$, then T is in the "acceptance region" and thus one does not reject H_0.

Table A9.1 provides a visual depiction of the decisions based on T.

Type I error is the decision to reject the null hypothesis when it is true. On the other hand, *type II error* is the decision of not rejecting the null hypothesis when it is false.

The size of a test is defined as the probability of committing *type I error*. This is also called the significance level (the p value or α) as discussed earlier. The *power* of a test is defined as the probability of rejecting the null hypothesis when it is false. It is defined as follows:

Power = 1 − Prob (type II error). If β denotes the probability of type II error, then the power of a test is denoted by $1 - \beta$. Ideally, one would want the probabilities of both types of errors to be as small as possible. However, in practice, there is a trade-off between the two. The smaller the value of α, the lower is the power of the test and, thus, the higher is the probability of type II error.

Many specification tests are parametric in nature, such that the DGP is governed by a set of parameters μ and the null hypothesis being true implies that μ takes values that fall within a certain range, denoted by μ_0. Thus, one can express the aforementioned statement as follows:

$$H_0 = \mu \, \varepsilon \, \mu_0 \text{ and } H_1: \mu \notin \mu_0$$

TABLE A9.1 Decisions based on the computed T.

Decision	H_0 is True	H_0 is False
Do not reject H_0 ($T \leq T_c$)	Correct	Type II error
Reject H_0 ($T > T_c$)	Type I error	Correct

The significance level and the power of the test can then be mathematically defined as follows:

$$p \text{ value or } \alpha = \text{Prob}(T > T_C | \mu \in \mu_0)$$
$$\text{Power} = (1 - \beta) = \text{Prob}(T > T_C | \mu \notin \mu_0) \quad \text{(A9.1)}$$

An important characteristic of a specification test is that it should be *consistent*. Formally, a specification test is *consistent* if its power $(1 - \beta)$ goes to 1 as the sample size goes to infinity. Another characteristic of a good specification test is that it should have very small size (low probability of type I error) and high power (low probability of type II error). Ideally, a researcher would like to have the most powerful test, which is a specification test that has greater power than any other possible test of the same size. However, in practice, this is hardly met, and statisticians compare two tests by examining the power functions. Power functions are function of the size (significance level), the sample size, and some data-generating process.

How does one differentiate between a good and a bad specification test?

Suppose a test ignores the data completely and randomly rejects H_0 $x\%$ of the time and does not reject H_0 $(1-x)\%$ of the time. Clearly, this test is worthless. However, it is important to examine the relationship between the size of the test and its power:

$$\text{Size}(p \text{ value or } \alpha) = \text{Prob}(\text{Rejecting } H_0 \text{ when } H_0 \text{ is true}) = x$$
$$\text{Power} = (1 - \beta = 1) - \text{Prob}(\text{Not rejecting } H_0 \text{ when } H_0 \text{ is false})$$
$$= 1 - (1 - x) = x$$

Thus, for this worthless test, we have Size = Power. Since, for a given size, we want a higher power than this, a useful or a good specification test must satisfy Size < Power. This is often called *unbiasedness*.

Formally, a specification test is said to be *unbiased* if its power is greater than or equal to its size. In other words, we want the following to hold true:

$$\text{Power}(1 - \beta) \geq \text{Size}(p - \text{value or } \alpha). \quad \text{(A9.2)}$$

Appendix 10

Instrumental variable estimation

This technical appendix uses notations as given in Greene (2002). For proof of some of the statements not derived here, the reader can consult the book.

In the classical linear regression model (as discussed in Chapter 10), a disturbing assumption was that $E(\varepsilon) = 0$. The reason for this trouble is as follows:

1. It is likely to be violated.
2. If violated, the least squares estimate of β will be biased and inconsistent.

The general and common procedure to get around this problem is instrumental variable estimation.

Problem of bias in the least squares estimate

Let us start with the least squares regression model of Eq. (10.12), $Y = X\beta + \varepsilon$ and the least squares estimator of it (given by Eq. (10.22)), $\widehat{\beta} = (X'X)^{-1}X'Y$. Let us denote this estimator as $\widehat{\beta}_{\text{OLS}}$. Taking the expectations of Eq. (10.22) on both sides, we obtain:

$$E\left(\widehat{\beta}_{\text{OLS}}\right) = E\left[(X'X)^{-1}X'(X\beta + \varepsilon)\right] = E\left[(X'X)^{-1}X'X\beta + (X'X)^{-1}X'\varepsilon\right]$$
$$= \beta + E\left[(X'X)^{-1}X'\varepsilon\right]$$
$$= \beta + E_X\left[(X'X)^{-1}X'E[\varepsilon|X]\right]$$

(A10.1)

The problem with the aforementioned equation is that when X and ε are correlated, $E[\varepsilon|X]$ is not equal to zero, which implies that $\widehat{\beta}_{\text{OLS}}$ will be biased and inconsistent. Now, let us suppose that we have some variables Z that are not correlated with ε. In other words, $E[\varepsilon|Z] = 0$. Then solving for β in this equation yields the following equation:

$$\widehat{\beta}_{IV} = (Z'X)^{-1}Z'Y \qquad (A10.2)$$

679

This is simply the instrumental variable (IV) estimate of β. Asymptotically, $\widehat{\beta}_{IV}$ is a consistent estimator of β and has the following asymptotic distribution:

$$\widehat{\beta}_{IV} \sim N\left(\beta, (\sigma^2/n) Q_{ZX}^{-1} Q_{ZZ} Q_{XZ}^{-1}\right) \quad (A10.3)$$

where $Q_{ZX} = \text{plim}(Z'X/n)$, $Q_{XZ} = \text{plim}(X'Z/n)$, and $Q_{ZZ} = \text{plim}(Z'Z/n)$. The important issue in IV estimation is what variables to use for Z. The more Z is correlated with X, the larger will be $Z'X$ and the smaller will be $(Z'X^{-1})$. Thus, we have a smaller variance of $\widehat{\beta}_{OLS}$. Hence, the more Z is correlated with X, the more precise the estimate of $\widehat{\beta}_{IV}$. Often, there are more IVs than there are variables in X. In such cases, $Z'X$ is not a square matrix and cannot be inverted. The solution in such cases is to make a predicted matrix of X from Z by regressing X on Z. In other words, $\widehat{X}$ can be constructed as follows:

$$\widehat{X} = Z(Z'Z)^{-1}Z'X \quad (A10.4)$$

Then substituting $\widehat{X}$ in place of Z in Eq. (A5.2), we obtain the IV estimate as follows:

$$\widehat{\beta}_{IV} = (Z'X)^{-1}Z'Y = \left(X'Z(Z'Z)^{-1}Z'X\right)^{-1} X'Z(Z'Z)^{-1}Z'Y \quad (A10.5)$$

Properties of the independent variable estimator

1. The estimator can be computed using ordinary least squares by first regressing each variable in X on all the variables in Z and then using these predicted values by regressing Y on them. This is why the IV estimator is sometimes called the *two-stage least squares*.
2. It is always important to include all the exogenous variables in X as instruments for Z.
3. It can be demonstrated that if there are some extra IVs to choose from, the linear combination of the Z variables represented by $\widehat{X}$ is optimal over all other possible linear combinations in the sense that the covariance matrix is minimized when $\widehat{X}$ is used. (For details of this proof, refer to Greene, 2002, pp. 622–624.)
4. Although $\widehat{\beta}_{IV}$ is not unbiased, it is consistent.

Choice of Z

1. The more Z is correlated with X, the better is the estimate $\widehat{\beta}_{IV}$, since the covariance matrix of $\widehat{\beta}_{IV}$ is smaller.
2. The correlation between Z and ε should be zero.
3. From a finite sample viewpoint, the number of variables in Z should be at least greater than the number of variables in X.

Specification tests

As discussed in the previous appendix, a general form of the specification test is given by $H_0: \mu \in \mu_0$ and $H_1: \mu \in /\mu_0$.

One can think of it as follows:

H_0: assumption about the data generating process (DGP) is correct, against the alternative.

H_1: assumption about the data-generating process is not correct.

Hausman test

Let there be two estimators of μ ($\hat{\mu}$ and $\tilde{\mu}$), which need to be compared. $\hat{\mu}$ is consistent and asymptotically efficient when H_0 holds but is inconsistent when H_0 does not hold. $\tilde{\mu}$ is consistent but asymptotically inefficient when H_0 holds and consistent even when H_0 does not hold. Thus, if H_0 holds, then $\hat{\mu} = \tilde{\mu}$. However, if H_0 does not hold, then they are probably not equal. The *Hausman test* is a test to see whether these two estimators are equal to each other. Let the variance–covariance matrices of the two estimators be given as follows:

$$\text{Var}(\hat{\mu}) = \Sigma_{\hat{\mu}}$$
$$\text{Var}(\tilde{\mu}) = \Sigma_{\sim\sim} \qquad (A10.6)$$

To test the null hypothesis that $\hat{\mu} = \tilde{\mu}$, one needs an estimate of the variance–covariance matrix $\hat{\mu} - \tilde{\mu}$, which is given by the following expression:

$$\text{Var}(\hat{\mu} - \tilde{\mu}) = \Sigma_{\hat{\mu}} + \Sigma_{\sim\sim} - (\Sigma_{\hat{\mu}\sim} + \Sigma_{\sim\hat{\mu}}) = \Sigma_{\sim\sim} - \Sigma_{\hat{\mu}} \qquad (A10.7)$$

where

$$\Sigma_{\hat{\mu}\sim} = \text{Cov}(\hat{\mu}, \tilde{\mu}) \text{ and } \Sigma_{\sim\hat{\mu}} = \text{Cov}(\tilde{\mu}, \hat{\mu}).$$

The Hausman test statistic is given by

$$H = (\hat{\mu} - \tilde{\mu})'[\hat{\Sigma}_{\sim\sim} - \tilde{\Sigma}_{\hat{\mu}}]^{-1}(\hat{\mu} - \tilde{\mu}) \qquad (A10.8)$$

If the null hypothesis is true, then H is distributed as $\chi^2(r)$, where r is the number of elements in μ. Thus, large values for H indicate that the null hypothesis does not hold.

Example

If all the assumptions of the classical multiple regression model hold, then $\hat{\beta}_{\text{OLS}}$ is an efficient and consistent estimator of β. However, if X and ε are correlated, then $E[\varepsilon|X]$ is not equal to zero, which implies that $\hat{\beta}_{\text{OLS}}$ will be biased and inconsistent. In this, the two-stage least squares estimate is consistent. Thus, the Hausman test can be used to compare $\widehat{\beta}_{OLS}$ with $\widehat{\beta}_{2SLS}$. In general, this specification test will be more powerful if one focuses on variables that are potentially endogenous. If one estimates the following regression equation:

$$Y = X\beta + \widehat{X}^*\gamma + \varepsilon^* \qquad (A10.9)$$

where $\widehat{X}^*$ is the predicted value of the variables that are endogenous, using the instrument variable set Z as the regressors for those variables. The joint significance of the elements in γ can be checked using an F-test, where the degrees of freedom of the F-test are the number of endogenous variables (let us denote it by K^*) and $(N - K - K^*)$.

Overidentification test

This test can be used to determine if the instruments satisfy the criteria $E[\varepsilon|Z] = 0$. In other words, the variables belonging to Z that are not in X, which are also referred to as the excluded instruments, must satisfy the property that $E[\varepsilon|Z] = 0$. The choice of Z variables to be the excluded instruments is also referred to as the "exclusion restrictions." This test can only be implemented when the number of instruments (variables in Z) exceeds the number of variables in X. In such a case, the model is said to be *overidentified*.

Suppose that there is some variable in Z, which is not in X, correlated with ε. If this is the case, entering into the equation of interest (i.e., adding it as another variable in the regression of Y on X) would yield a coefficient on the variable significantly different from zero. The test of this is very simple. First, compute the residuals from the IV estimates as follows:

$$e = Y - X\widehat{\beta}_{IV} \qquad (A10.10)$$

In step 2, regress these residuals on all the IVs (including the variables in X assumed exogenous) in Z. Then, if one multiplies the R_2 of this regression by the sample size, it is distributed as χ^2 $(I - K)$, where I is the number of variables in Z and K is the number of variables in X. In other words, the following statement holds:

$$nR^2 \sim \chi^2(I - K) \qquad (A10.11)$$

Comments about this test

1. The aforementioned test is valid only under the assumption that at least K elements of Z are uncorrelated with ε.
2. The power of the test to reject the null hypothesis that $E[\varepsilon|Z] = 0$ when it is false may not be very high.
3. If the test rejects the null hypothesis, then there is clearly something wrong.

If the test does not reject the null, then it might be all right to carry out the test. However, there is always the possibility that the power of the test to reject the null hypothesis is weak.

When is independent variable estimation useful?

In this section, we discuss the situations under which *IV* estimation might be useful. We will not get into the technical details in sufficient length, since it is beyond the scope of this book.

Measurement error

What kind of instruments does one need? In measurement of the same thing, one would like two good proxies of the same thing. Suppose that we are interested in regressing household income. Then, household consumption expenditures or household wealth can be treated as separate proxies for income. Hence, one or the other could be used as instruments for income. It does not matter if the instrument is measured with error. However, it must be the case that measurement error in the instrument is uncorrelated with the measurement error in the variable of interest. A second example of measurement error from the nutrition literature will illustrate the point further. Suppose one wants to examine the impact of child malnutrition on school attendance. Child height for age and weight for height are two indicators of child nutritional status. However, they are only approximate and thus contain measurement error. In such a case, it is not a good idea to use height for age as an instrument for weight for height, since they indicate different information on child nutritional status. While low height for age is a measure of stunting and is a long-term phenomenon, low weight for height is a measure of wasting and pertains mainly to the short-term nutritional status.

Simultaneous equations

In a market economy, both quantity demanded and supplied of a commodity are determined by the price. The supply and demand curves are functions of the price and other variables, and at equilibrium, both the quantity sold and the price are determined. Since price is determined simultaneously with quantity, both are endogenous, and if the least squares estimate is used to estimate the demand or supply curve as a function of the price, one will get biased estimates. (For additional technical details, the reader can refer to Greene, 2002, pp. 592−594).

Endogeneity/omitted variable bias

Suppose we regress wages on schooling and work experience, and let us assume that schooling is also determined by the price of schooling, parents' education, and genetic intelligence. Suppose further that genetic intelligence also determines wages above and beyond its impact on schooling. Then it becomes part of the error term in the wage regression equation. However, from the schooling regression equation, we see that schooling is also correlated with genetic intelligence and, thus, with the error term in the second equation. Thus, one can use parental education and the price of schooling as the instruments for schooling, as long as they are uncorrelated with the error term.

Another example from the health economics literature may be assessing the impact of health programs on health outcomes. If health programs are deliberately targeted to areas where the health situation is poor, the outcome can lead to bad health.

Appendix 11

Statistical tables

Tables of the normal distribution

Probability content from $-\infty$ to Z

Z	0.00	0.01	0.02	0.03	0.04	0.05	0.06	0.07	0.08	0.09
0.0	0.5000	0.5040	0.5080	0.5120	0.5160	0.5199	0.5239	0.5279	0.5319	0.5359
0.1	0.5398	0.5438	0.5478	0.5517	0.5557	0.5596	0.5636	0.5675	0.5714	0.5753
0.2	0.5793	0.5832	0.5871	0.5910	0.5948	0.5987	0.6026	0.6064	0.6103	0.6141
0.3	0.6179	0.6217	0.6255	0.6293	0.6331	0.6368	0.6406	0.6443	0.6480	0.6517
0.4	0.6554	0.6591	0.6628	0.6664	0.6700	0.6736	0.6772	0.6808	0.6844	0.6879
0.5	0.6915	0.6950	0.6985	0.7019	0.7054	0.7088	0.7123	0.7157	0.7190	0.7224
0.6	0.7257	0.7291	0.7324	0.7357	0.7389	0.7422	0.7454	0.7486	0.7517	0.7549
0.7	0.7580	0.7611	0.7642	0.7673	0.7704	0.7734	0.7764	0.7794	0.7823	0.7852
0.8	0.7881	0.7910	0.7939	0.7967	0.7995	0.8023	0.8051	0.8078	0.8106	0.8133
0.9	0.8159	0.8186	0.8212	0.8238	0.8264	0.8289	0.8315	0.8340	0.8365	0.8389
1.0	0.8413	0.8438	0.8461	0.8485	0.8508	0.8531	0.8554	0.8577	0.8599	0.8621
1.1	0.8643	0.8665	0.8686	0.8708	0.8729	0.8749	0.8770	0.8790	0.8810	0.8830
1.2	0.8849	0.8869	0.8888	0.8907	0.8925	0.8944	0.8962	0.8980	0.8997	0.9015
1.3	0.9032	0.9049	0.9066	0.9082	0.9099	0.9115	0.9131	0.9147	0.9162	0.9177
1.4	0.9192	0.9207	0.9222	0.9236	0.9251	0.9265	0.9279	0.9292	0.9306	0.9319
1.5	0.9332	0.9345	0.9357	0.9370	0.9382	0.9394	0.9406	0.9418	0.9429	0.9441
1.6	0.9452	0.9463	0.9474	0.9484	0.9495	0.9505	0.9515	0.9525	0.9535	0.9545
1.7	0.9554	0.9564	0.9573	0.9582	0.9591	0.9599	0.9608	0.9616	0.9625	0.9633
1.8	0.9641	0.9649	0.9656	0.9664	0.9671	0.9678	0.9686	0.9693	0.9699	0.9706
1.9	0.9713	0.9719	0.9726	0.9732	0.9738	0.9744	0.9750	0.9756	0.9761	0.9767
2.0	0.9772	0.9778	0.9783	0.9788	0.9793	0.9798	0.9803	0.9808	0.9812	0.9817
2.1	0.9821	0.9826	0.9830	0.9834	0.9838	0.9842	0.9846	0.9850	0.9854	0.9857
2.2	0.9861	0.9864	0.9868	0.9871	0.9875	0.9878	0.9881	0.9884	0.9887	0.9890
2.3	0.9893	0.9896	0.9898	0.9901	0.9904	0.9906	0.9909	0.9911	0.9913	0.9916
2.4	0.9918	0.9920	0.9922	0.9925	0.9927	0.9929	0.9931	0.9932	0.9934	0.9936
2.5	0.9938	0.9940	0.9941	0.9943	0.9945	0.9946	0.9948	0.9949	0.9951	0.9952
2.6	0.9953	0.9955	0.9956	0.9957	0.9959	0.9960	0.9961	0.9962	0.9963	0.9964
2.7	0.9965	0.9966	0.9967	0.9968	0.9969	0.9970	0.9971	0.9972	0.9973	0.9974
2.8	0.9974	0.9975	0.9976	0.9977	0.9977	0.9978	0.9979	0.9979	0.9980	0.9981
2.9	0.9981	0.9982	0.9982	0.9983	0.9984	0.9984	0.9985	0.9985	0.9986	0.9986
3.0	0.9987	0.9987	0.9987	0.9988	0.9988	0.9989	0.9989	0.9989	0.9990	0.9990

Far-right tail probabilities.

Z	P{Z to oo}	Z	P{Z to oo}	Z	P{Z to oo}	Z	P{Z to oo}
2.0	0.02275	3.0	0.001350	4.0	0.00003167	5.0	2.867 E-7
2.1	0.01786	3.1	0.0009676	4.1	0.00002066	5.5	1.899 E-8
2.2	0.01390	3.2	0.0006871	4.2	0.00001335	6.0	9.866 E-10
2.3	0.01072	3.3	0.0004834	4.3	0.00000854	6.5	4.016 E-11
2.4	0.00820	3.4	0.0003369	4.4	0.000005413	7.0	1.280 E-12
2.5	0.00621	3.5	0.0002326	4.5	0.000003398	7.5	3.191 E-14
2.6	0.004661	3.6	0.0001591	4.6	0.000002112	8.0	6.221 E-16
2.7	0.003467	3.7	0.0001078	4.7	0.000001300	8.5	9.480 E-18
2.8	0.002555	3.8	0.00007235	4.8	7.933 E-7	9.0	1.129 E-19
2.9	0.001866	3.9	0.00004810	4.9	4.792 E-7	9.5	1.049 E-21

T-distribution table

df	$\alpha = 0.1$	0.05	0.025	0.01	0.005	0.001	0.0005
∞	$t_\alpha = 1.282$	1.645	1.960	2.326	2.576	3.091	3.291
1	3.078	6.314	12.706	31.821	63.656	318.289	636.578
2	1.886	2.920	4.303	6.965	9.925	22.328	31.600
3	1.638	2.353	3.182	4.541	5.841	10.214	12.924
4	1.533	2.132	2.776	3.747	4.604	7.173	8.610
5	1.476	2.015	2.571	3.365	4.032	5.894	6.869
6	1.440	1.943	2.447	3.143	3.707	5.208	5.959
7	1.415	1.895	2.365	2.998	3.499	4.785	5.408
8	1.397	1.860	2.306	2.896	3.355	4.501	5.041
9	1.383	1.833	2.262	2.821	3.250	4.297	4.781
10	1.372	1.812	2.228	2.764	3.169	4.144	4.587
11	1.363	1.796	2.201	2.718	3.106	4.025	4.437
12	1.356	1.782	2.179	2.681	3.055	3.930	4.318
13	1.350	1.771	2.160	2.650	3.012	3.852	4.221
14	1.345	1.761	2.145	2.624	2.977	3.787	4.140
15	1.341	1.753	2.131	2.602	2.947	3.733	4.073
16	1.337	1.746	2.120	2.583	2.921	3.686	4.015
17	1.333	1.740	2.110	2.567	2.898	3.646	3.965
18	1.330	1.734	2.101	2.552	2.878	3.610	3.922
19	1.328	1.729	2.093	2.539	2.861	3.579	3.883
20	1.325	1.725	2.086	2.528	2.845	3.552	3.850
21	1.323	1.721	2.080	2.518	2.831	3.527	3.819
22	1.321	1.717	2.074	2.50Ba8	2.819	3.505	3.792
23	1.319	1.714	2.069	2.500	2.807	3.485	3.768
24	1.318	1.711	2.064	2.492	2.797	3.467	3.745
25	1.316	1.708	2.060	2.485	2.787	3.450	3.725
26	1.315	1.706	2.056	2.479	2.779	3.435	3.707
27	1.314	1.703	2.052	2.473	2.771	3.421	3.689
28	1.313	1.701	2.048	2.467	2.763	3.408	3.674

Critical points of the chi-square distribution

Cumulative probability

df	0.005	0.010	0.025	0.05	0.10	0.25	0.50	0.75	0.90	0.95	0.975	0.99	0.995
1	0.39E-4	0.00016	0.00098	0.0039	0.0158	0.102	0.455	1.32	2.71	3.84	5.02	6.63	7.88
2	0.0100	0.0201	0.0506	0.103	0.211	0.575	1.39	2.77	4.61	5.99	7.38	9.21	10.6
3	0.0717	0.115	0.216	0.352	0.584	1.21	2.37	4.11	6.25	7.81	9.35	11.3	12.8
4	0.207	0.297	0.484	0.711	1.06	1.92	3.36	5.39	7.78	9.49	11.1	13.3	14.9
5	0.412	0.554	0.831	1.15	1.61	2.67	4.35	6.63	9.24	11.1	12.8	15.1	16.7
6	0.676	0.872	1.24	1.64	2.20	3.45	5.35	7.84	10.6	12.6	14.4	16.8	18.5
7	0.989	1.24	1.69	2.17	2.83	4.25	6.35	9.04	12.0	14.1	16.0	18.5	20.3
8	1.34	1.65	2.18	2.73	3.49	5.07	7.34	10.2	13.4	15.5	17.5	20.1	22.0
9	1.73	2.09	2.70	3.33	4.17	5.9	8.34	11.4	14.7	16.9	19.0	21.7	23.6
10	2.16	2.56	3.25	3.94	4.87	6.74	9.34	12.5	16.0	18.3	20.5	23.2	25.2
11	2.60	3.05	3.82	4.57	5.58	7.58	10.3	13.7	17.3	19.7	21.9	24.7	26.8
12	3.07	3.57	4.40	5.23	6.30	8.44	11.3	14.8	18.5	21.0	23.3	26.2	28.3
13	3.57	4.11	5.01	5.89	7.04	9.3	12.3	16.0	19.8	22.4	24.7	27.7	29.8
14	4.07	4.66	5.63	6.57	7.79	10.2	13.3	17.1	21.1	23.7	26.1	29.1	31.3
15	4.60	5.23	6.26	7.26	8.55	11.0	14.3	18.2	22.3	25.0	27.5	30.6	32.8
16	5.14	5.81	6.91	7.96	9.31	11.9	15.3	19.4	23.5	26.3	28.8	32.0	34.3
17	5.70	6.41	7.56	8.67	10.1	12.8	16.3	20.5	24.8	27.6	30.2	33.4	35.7
18	6.26	7.01	8.23	9.39	10.9	13.7	17.3	21.6	26.0	28.9	31.5	34.8	37.2
19	6.84	7.63	8.91	10.1	11.7	14.6	18.3	22.7	27.2	30.1	32.9	36.2	38.6
20	7.43	8.26	9.59	10.9	12.4	15.5	19.3	23.8	28.4	31.4	34.2	37.6	40.0
21	8.03	8.90	10.3	11.6	13.2	16.3	20.3	24.9	29.6	32.7	35.5	38.9	41.4
22	8.64	9.54	11.0	12.3	14.0	17.2	21.3	26.0	30.8	33.9	36.8	40.3	42.8
23	9.26	10.2	11.7	13.1	14.8	18.1	22.3	27.1	32.0	35.2	38.1	41.6	44.2
24	9.89	10.9	12.4	13.8	15.7	19.0	23.3	28.2	33.2	36.4	39.4	43.0	45.6
25	10.5	11.5	13.1	14.6	16.5	19.9	24.3	29.3	34.4	37.7	40.6	44.3	46.9
26	11.2	12.2	13.8	15.4	17.3	20.8	25.3	30.4	35.6	38.9	41.9	45.6	48.3
27	11.8	12.9	14.6	16.2	18.1	21.7	26.3	31.5	36.7	40.1	43.2	47.0	49.6
28	12.5	13.6	15.3	16.9	18.9	22.7	27.3	32.6	37.9	41.3	44.5	48.3	51.0
29	13.1	14.3	16.0	17.7	19.8	23.6	28.3	33.7	39.1	42.6	45.7	49.6	52.3
30	13.8	15.0	16.8	18.5	20.6	24.5	29.3	34.8	40.3	43.8	47.0	50.9	53.7
31	14.5	15.7	17.5	19.3	21.4	25.4	30.3	35.9	41.4	45.0	48.2	52.2	55.0
32	15.1	16.4	18.3	20.1	22.3	26.3	31.3	37.0	42.6	46.2	49.5	53.5	56.3
33	15.8	17.1	19.0	20.9	23.1	27.2	32.3	38.1	43.7	47.4	50.7	54.8	57.6
34	16.5	17.8	19.8	21.7	24.0	28.1	33.3	39.1	44.9	48.6	52.0	56.1	59.0
35	17.2	18.5	20.6	22.5	24.8	29.1	34.3	40.2	46.1	49.8	53.2	57.3	60.3
36	17.9	19.2	21.3	23.3	25.6	30.0	35.3	41.3	47.2	51.0	54.4	58.6	61.6
37	18.6	20.0	22.1	24.1	26.5	30.9	36.3	42.4	48.4	52.2	55.7	59.9	62.9
38	19.3	20.7	22.9	24.9	27.3	31.8	37.3	43.5	49.5	53.4	56.9	61.2	64.2
39	20.0	21.4	23.7	25.7	28.2	32.7	38.3	44.5	50.7	54.6	58.1	62.4	65.5
40	20.7	22.2	24.4	26.5	29.1	33.7	39.3	45.6	51.8	55.8	59.3	63.7	66.8
41	21.4	22.9	25.2	27.3	29.9	34.6	40.3	46.7	52.9	56.9	60.6	65.0	68.1
42	22.1	23.7	26.0	28.1	30.8	35.5	41.3	47.8	54.1	58.1	61.8	66.2	69.3
43	22.9	24.4	26.8	29.0	31.6	36.4	42.3	48.8	55.2	59.3	63.0	67.5	70.6
44	23.6	25.1	27.6	29.8	32.5	37.4	43.3	49.9	56.4	60.5	64.2	68.7	71.9
45	24.3	25.9	28.4	30.6	33.4	38.3	44.3	51.0	57.5	61.7	65.4	70.0	73.2
	0.005	0.010	0.025	0.05	0.10	0.25	0.50	0.75	0.90	0.95	0.975	0.99	0.995

95% points for the F distribution
Numerator degrees of freedom

	*	1	2	3	4	5	6	7	8	9	10	*
Denominator degrees of freedom	1	161	199	216	225	230	234	237	239	241	242	1
	2	18.5	19.0	19.2	19.2	19.3	19.3	19.4	19.4	19.4	19.4	2
	3	10.1	9.55	9.28	9.12	9.01	8.94	8.89	8.85	8.81	8.79	3
	4	7.71	6.94	6.59	6.39	6.26	6.16	6.09	6.04	6.00	5.96	4
	5	6.61	5.79	5.41	5.19	5.05	4.95	4.88	4.82	4.77	4.74	5
	6	5.99	5.14	4.76	4.53	4.39	4.28	4.21	4.15	4.10	4.06	6
	7	5.59	4.74	4.35	4.12	3.97	3.87	3.79	3.73	3.68	3.64	7
	8	5.32	4.46	4.07	3.84	3.69	3.58	3.50	3.44	3.39	3.35	8
	9	5.12	4.26	3.86	3.63	3.48	3.37	3.29	3.23	3.18	3.14	9
	10	4.96	4.10	3.71	3.48	3.33	3.22	3.14	3.07	3.02	2.98	10
	11	4.84	3.98	3.59	3.36	3.20	3.09	3.01	2.95	2.90	2.85	11
	12	4.75	3.89	3.49	3.26	3.11	3.00	2.91	2.85	2.80	2.75	12
	13	4.67	3.81	3.41	3.18	3.03	2.92	2.83	2.77	2.71	2.67	13
	14	4.60	3.74	3.34	3.11	2.96	2.85	2.76	2.70	2.65	2.60	14
	15	4.54	3.68	3.29	3.06	2.90	2.79	2.71	2.64	2.59	2.54	15
	16	4.49	3.63	3.24	3.01	2.85	2.74	2.66	2.59	2.54	2.49	16
	17	4.45	3.59	3.20	2.96	2.81	2.70	2.61	2.55	2.49	2.45	17
	18	4.41	3.55	3.16	2.93	2.77	2.66	2.58	2.51	2.46	2.41	18
	19	4.38	3.52	3.13	2.90	2.74	2.63	2.54	2.48	2.42	2.38	19
	20	4.35	3.49	3.10	2.87	2.71	2.60	2.51	2.45	2.39	2.35	20
	21	4.32	3.47	3.07	2.84	2.68	2.57	2.49	2.42	2.37	2.32	21
	22	4.30	3.44	3.05	2.82	2.66	2.55	2.46	2.40	2.34	2.30	22
	23	4.28	3.42	3.03	2.80	2.64	2.53	2.44	2.37	2.32	2.27	23
	24	4.26	3.40	3.01	2.78	2.62	2.51	2.42	2.36	2.30	2.25	24
	25	4.24	3.39	2.99	2.76	2.60	2.49	2.40	2.34	2.28	2.24	25
	26	4.23	3.37	2.98	2.74	2.59	2.47	2.39	2.32	2.27	2.22	26
	27	4.21	3.35	2.96	2.73	2.57	2.46	2.37	2.31	2.25	2.20	27
	28	4.20	3.34	2.95	2.71	2.56	2.45	2.36	2.29	2.24	2.19	28
	29	4.18	3.33	2.93	2.70	2.55	2.43	2.35	2.28	2.22	2.18	29
	30	4.17	3.32	2.92	2.69	2.53	2.42	2.33	2.27	2.21	2.16	30
	35	4.12	3.27	2.87	2.64	2.49	2.37	2.29	2.22	2.16	2.11	35
	40	4.08	3.23	2.84	2.61	2.45	2.34	2.25	2.18	2.12	2.08	40
	50	4.03	3.18	2.79	2.56	2.40	2.29	2.20	2.13	2.07	2.03	50
	60	4.00	3.15	2.76	2.53	2.37	2.25	2.17	2.10	2.04	1.99	60
	70	3.98	3.13	2.74	2.50	2.35	2.23	2.14	2.07	2.02	1.97	70
	80	3.96	3.11	2.72	2.49	2.33	2.21	2.13	2.06	2.00	1.95	80
	100	3.94	3.09	2.70	2.46	2.31	2.19	2.10	2.03	1.97	1.93	100
	150	3.90	3.06	2.66	2.43	2.27	2.16	2.07	2.00	1.94	1.89	150
	300	3.87	3.03	2.63	2.40	2.24	2.13	2.04	1.97	1.91	1.86	300
	1000	3.85	3.00	2.61	2.38	2.22	2.11	2.02	1.95	1.89	1.84	1000

Numerator degrees of freedom

	*	11	12	13	14	15	16	17	18	19	20	*
Denominator degrees of freedom	1	243	244	245	245	246	246	247	247	248	248	1
	2	19.4	19.4	19.4	19.4	19.4	19.4	19.4	19.4	19.4	19.4	2
	3	8.76	8.74	8.73	8.71	8.70	8.69	8.68	8.67	8.67	8.66	3
	4	5.94	5.91	5.89	5.87	5.86	5.84	5.83	5.82	5.81	5.80	4
	5	4.70	4.68	4.66	4.64	4.62	4.60	4.59	4.58	4.57	4.56	5
	6	4.03	4.00	3.98	3.96	3.94	3.92	3.91	3.90	3.88	3.87	6
	7	3.60	3.57	3.55	3.53	3.51	3.49	3.48	3.47	3.46	3.44	7

	8	3.31	3.28	3.26	3.24	3.22	3.20	3.19	3.17	3.16	3.15	8
	9	3.10	3.07	3.05	3.03	3.01	2.99	2.97	2.96	2.95	2.94	9
	10	2.94	2.91	2.89	2.86	2.85	2.83	2.81	2.80	2.79	2.77	10
	11	2.82	2.79	2.76	2.74	2.72	2.70	2.69	2.67	2.66	2.65	11
	12	2.72	2.69	2.66	2.64	2.62	2.60	2.58	2.57	2.56	2.54	12
	13	2.63	2.60	2.58	2.55	2.53	2.51	2.50	2.48	2.47	2.46	13
	14	2.57	2.53	2.51	2.48	2.46	2.44	2.43	2.41	2.40	2.39	14
	15	2.51	2.48	2.45	2.42	2.40	2.38	2.37	2.35	2.34	2.33	15
	16	2.46	2.42	2.40	2.37	2.35	2.33	2.32	2.30	2.29	2.28	16
	17	2.41	2.38	2.35	2.33	2.31	2.29	2.27	2.26	2.24	2.23	17
	18	2.37	2.34	2.31	2.29	2.27	2.25	2.23	2.22	2.20	2.19	18
	19	2.34	2.31	2.28	2.26	2.23	2.21	2.20	2.18	2.17	2.16	19
	20	2.31	2.28	2.25	2.22	2.20	2.18	2.17	2.15	2.14	2.12	20
	21	2.28	2.25	2.22	2.20	2.18	2.16	2.14	2.12	2.11	2.10	21
	22	2.26	2.23	2.20	2.17	2.15	2.13	2.11	2.10	2.08	2.07	22
	23	2.24	2.20	2.18	2.15	2.13	2.11	2.09	2.08	2.06	2.05	23
	24	2.22	2.18	2.15	2.13	2.11	2.09	2.07	2.05	2.04	2.03	24
	25	2.20	2.16	2.14	2.11	2.09	2.07	2.05	2.04	2.02	2.01	25
	26	2.18	2.15	2.12	2.09	2.07	2.05	2.03	2.02	2.00	1.99	26
	27	2.17	2.13	2.10	2.08	2.06	2.04	2.02	2.00	1.99	1.97	27
	28	2.15	2.12	2.09	2.06	2.04	2.02	2.00	1.99	1.97	1.96	28
	29	2.14	2.10	2.08	2.05	2.03	2.01	1.99	1.97	1.96	1.94	29
	30	2.13	2.09	2.06	2.04	2.01	1.99	1.98	1.96	1.95	1.93	30
	35	2.07	2.04	2.01	1.99	1.96	1.94	1.92	1.91	1.89	1.88	35
	40	2.04	2.00	1.97	1.95	1.92	1.90	1.89	1.87	1.85	1.84	40
	50	1.99	1.95	1.92	1.89	1.87	1.85	1.83	1.81	1.80	1.78	50
	60	1.95	1.92	1.89	1.86	1.84	1.82	1.80	1.78	1.76	1.75	60
	70	1.93	1.89	1.86	1.84	1.81	1.79	1.77	1.75	1.74	1.72	70
	80	1.91	1.88	1.84	1.82	1.79	1.77	1.75	1.73	1.72	1.70	80
	100	1.89	1.85	1.82	1.79	1.77	1.75	1.73	1.71	1.69	1.68	100
	150	1.85	1.82	1.79	1.76	1.73	1.71	1.69	1.67	1.66	1.64	150
	300	1.82	1.78	1.75	1.72	1.70	1.68	1.66	1.64	1.62	1.61	300
	1000	1.80	1.76	1.73	1.70	1.68	1.65	1.63	1.61	1.60	1.58	1000

Numerator degrees of freedom

	*	21	22	23	24	25	26	27	28	29	30	*
Denominator	1	248	249	249	249	249	249	250	250	250	250	1
degrees of	2	19.4	19.5	19.5	19.5	19.5	19.5	19.5	19.5	19.5	19.5	2
freedom	3	8.65	8.65	8.64	8.64	8.63	8.63	8.63	8.62	8.62	8.62	3
	4	5.79	5.79	5.78	5.77	5.77	5.76	5.76	5.75	5.75	5.75	4
	5	4.55	4.54	4.53	4.53	4.52	4.52	4.51	4.50	4.50	4.50	5
	6	3.86	3.86	3.85	3.84	3.83	3.83	3.82	3.82	3.81	3.81	6
	7	3.43	3.43	3.42	3.41	3.40	3.40	3.39	3.39	3.38	3.38	7
	8	3.14	3.13	3.12	3.12	3.11	3.10	3.10	3.09	3.08	3.08	8
	9	2.93	2.92	2.91	2.90	2.89	2.89	2.88	2.87	2.87	2.86	9
	10	2.76	2.75	2.75	2.74	2.73	2.72	2.72	2.71	2.70	2.70	10
	11	2.64	2.63	2.62	2.61	2.60	2.59	2.59	2.58	2.58	2.57	11
	12	2.53	2.52	2.51	2.51	2.50	2.49	2.48	2.48	2.47	2.47	12
	13	2.45	2.44	2.43	2.42	2.41	2.41	2.40	2.39	2.39	2.38	13
	14	2.38	2.37	2.36	2.35	2.34	2.33	2.33	2.32	2.31	2.31	14
	15	2.32	2.31	2.30	2.29	2.28	2.27	2.27	2.26	2.25	2.25	15
	16	2.26	2.25	2.24	2.24	2.23	2.22	2.21	2.21	2.20	2.19	16
	17	2.22	2.21	2.20	2.19	2.18	2.17	2.17	2.16	2.15	2.15	17
	18	2.18	2.17	2.16	2.15	2.14	2.13	2.13	2.12	2.11	2.11	18
	19	2.14	2.13	2.12	2.11	2.11	2.10	2.09	2.08	2.08	2.07	19

20	2.11	2.10	2.09	2.08	2.07	2.07	2.06	2.05	2.05	2.04	20
21	2.08	2.07	2.06	2.05	2.05	2.04	2.03	2.02	2.02	2.01	21
22	2.06	2.05	2.04	2.03	2.02	2.01	2.00	2.00	1.99	1.98	22
23	2.04	2.02	2.01	2.01	2.00	1.99	1.98	1.97	1.97	1.96	23
24	2.01	2.00	1.99	1.98	1.97	1.97	1.96	1.95	1.95	1.94	24
25	2.00	1.98	1.97	1.96	1.96	1.95	1.94	1.93	1.93	1.92	25
26	1.98	1.97	1.96	1.95	1.94	1.93	1.92	1.91	1.91	1.90	26
27	1.96	1.95	1.94	1.93	1.92	1.91	1.90	1.90	1.89	1.88	27
28	1.95	1.93	1.92	1.91	1.91	1.90	1.89	1.88	1.88	1.87	28
29	1.93	1.92	1.91	1.90	1.89	1.88	1.88	1.87	1.86	1.85	29
30	1.92	1.91	1.90	1.89	1.88	1.87	1.86	1.85	1.85	1.84	30
35	1.87	1.85	1.84	1.83	1.82	1.82	1.81	1.80	1.79	1.79	35
40	1.83	1.81	1.80	1.79	1.78	1.77	1.77	1.76	1.75	1.74	40
50	1.77	1.76	1.75	1.74	1.73	1.72	1.71	1.70	1.69	1.69	50
60	1.73	1.72	1.71	1.70	1.69	1.68	1.67	1.66	1.66	1.65	60
70	1.71	1.70	1.68	1.67	1.66	1.65	1.65	1.64	1.63	1.62	70
80	1.69	1.68	1.67	1.65	1.64	1.63	1.63	1.62	1.61	1.60	80
100	1.66	1.65	1.64	1.63	1.62	1.61	1.60	1.59	1.58	1.57	100
150	1.63	1.61	1.60	1.59	1.58	1.57	1.56	1.55	1.54	1.54	150
300	1.59	1.58	1.57	1.55	1.54	1.53	1.52	1.51	1.51	1.50	300
1000	1.57	1.55	1.54	1.53	1.52	1.51	1.50	1.49	1.48	1.47	1000

Numerator degrees of freedom

	*	31	32	33	34	35	36	37	38	39	40	*
Denominator degrees of freedom	1	250	250	250	251	251	251	251	251	251	251	1
	2	19.5	19.5	19.5	19.5	19.5	19.5	19.5	19.5	19.5	19.5	2
	3	8.61	8.61	8.61	8.61	8.60	8.60	8.60	8.60	8.60	8.59	3
	4	5.74	5.74	5.74	5.73	5.73	5.73	5.72	5.72	5.72	5.72	4
	5	4.49	4.49	4.48	4.48	4.48	4.47	4.47	4.47	4.47	4.46	5
	6	3.80	3.80	3.80	3.79	3.79	3.79	3.78	3.78	3.78	3.77	6
	7	3.37	3.37	3.36	3.36	3.36	3.35	3.35	3.35	3.34	3.34	7
	8	3.07	3.07	3.07	3.06	3.06	3.06	3.05	3.05	3.05	3.04	8
	9	2.86	2.85	2.85	2.85	2.84	2.84	2.84	2.83	2.83	2.83	9
	10	2.69	2.69	2.69	2.68	2.68	2.67	2.67	2.67	2.66	2.66	10
	11	2.57	2.56	2.56	2.55	2.55	2.54	2.54	2.54	2.53	2.53	11
	12	2.46	2.46	2.45	2.45	2.44	2.44	2.44	2.43	2.43	2.43	12
	13	2.38	2.37	2.37	2.36	2.36	2.35	2.35	2.35	2.34	2.34	13
	14	2.30	2.30	2.29	2.29	2.28	2.28	2.28	2.27	2.27	2.27	14
	15	2.24	2.24	2.23	2.23	2.22	2.22	2.21	2.21	2.21	2.20	15
	16	2.19	2.18	2.18	2.17	2.17	2.17	2.16	2.16	2.15	2.15	16
	17	2.14	2.14	2.13	2.13	2.12	2.12	2.11	2.11	2.11	2.10	17
	18	2.10	2.10	2.09	2.09	2.08	2.08	2.07	2.07	2.07	2.06	18
	19	2.07	2.06	2.06	2.05	2.05	2.04	2.04	2.03	2.03	2.03	19
	20	2.03	2.03	2.02	2.02	2.01	2.01	2.01	2.00	2.00	1.99	20
	21	2.00	2.00	1.99	1.99	1.98	1.98	1.98	1.97	1.97	1.96	21
	22	1.98	1.97	1.97	1.96	1.96	1.95	1.95	1.95	1.94	1.94	22
	23	1.95	1.95	1.94	1.94	1.93	1.93	1.93	1.92	1.92	1.91	23
	24	1.93	1.93	1.92	1.92	1.91	1.91	1.90	1.90	1.90	1.89	24
	25	1.91	1.91	1.90	1.90	1.89	1.89	1.88	1.88	1.88	1.87	25
	26	1.89	1.89	1.88	1.88	1.87	1.87	1.87	1.86	1.86	1.85	26
	27	1.88	1.87	1.87	1.86	1.86	1.85	1.85	1.84	1.84	1.84	27
	28	1.86	1.86	1.85	1.85	1.84	1.84	1.83	1.83	1.82	1.82	28
	29	1.85	1.84	1.84	1.83	1.83	1.82	1.82	1.81	1.81	1.81	29

	30	1.83	1.83	1.82	1.82	1.81	1.81	1.80	1.80	1.80	1.79	30
	35	1.78	1.77	1.77	1.76	1.76	1.75	1.75	1.74	1.74	1.74	35
	40	1.74	1.73	1.73	1.72	1.72	1.71	1.71	1.70	1.70	1.69	40
	50	1.68	1.67	1.67	1.66	1.66	1.65	1.65	1.64	1.64	1.63	50
	60	1.64	1.64	1.63	1.62	1.62	1.61	1.61	1.60	1.60	1.59	60
	70	1.62	1.61	1.60	1.60	1.59	1.59	1.58	1.58	1.57	1.57	70
	80	1.59	1.59	1.58	1.58	1.57	1.56	1.56	1.55	1.55	1.54	80
	100	1.57	1.56	1.55	1.55	1.54	1.54	1.53	1.52	1.52	1.52	100
	150	1.53	1.52	1.51	1.51	1.50	1.50	1.49	1.49	1.48	1.48	150
	300	1.49	1.48	1.48	1.47	1.46	1.46	1.45	1.45	1.44	1.43	300
	1000	1.46	1.46	1.45	1.44	1.43	1.43	1.42	1.42	1.41	1.41	1000

Numerator degrees of freedom

		*	45	50	60	70	80	100	120	150	300	1000	*
Denominator	1	251	252	252	252	253	253	253	253	254	254	1
degrees of	2	19.5	19.5	19.5	19.5	19.5	19.5	19.5	19.5	19.5	19.5	2
freedom	3	8.59	8.58	8.57	8.57	8.56	8.55	8.55	8.54	8.54	8.53	3
	4	5.71	5.70	5.69	5.68	5.67	5.66	5.66	5.65	5.64	5.63	4
	5	4.45	4.44	4.43	4.42	4.41	4.41	4.40	4.39	4.38	4.37	5
	6	3.76	3.75	3.74	3.73	3.72	3.71	3.70	3.70	3.68	3.67	6
	7	3.33	3.32	3.30	3.29	3.29	3.27	3.27	3.26	3.24	3.23	7
	8	3.03	3.02	3.01	2.99	2.99	2.97	2.97	2.96	2.94	2.93	8
	9	2.81	2.80	2.79	2.78	2.77	2.76	2.75	2.74	2.72	2.71	9
	10	2.65	2.64	2.62	2.61	2.60	2.59	2.58	2.57	2.55	2.54	10
	11	2.52	2.51	2.49	2.48	2.47	2.46	2.45	2.44	2.42	2.41	11
	12	2.41	2.40	2.38	2.37	2.36	2.35	2.34	2.33	2.31	2.30	12
	13	2.33	2.31	2.30	2.28	2.27	2.26	2.25	2.24	2.23	2.21	13
	14	2.25	2.24	2.22	2.21	2.20	2.19	2.18	2.17	2.15	2.14	14
	15	2.19	2.18	2.16	2.15	2.14	2.12	2.11	2.10	2.09	2.07	15
	16	2.14	2.12	2.11	2.09	2.08	2.07	2.06	2.05	2.03	2.02	16
	17	2.09	2.08	2.06	2.05	2.03	2.02	2.01	2.00	1.98	1.97	17
	18	2.05	2.04	2.02	2.00	1.99	1.98	1.97	1.96	1.94	1.92	18
	19	2.01	2.00	1.98	1.97	1.96	1.94	1.93	1.92	1.90	1.88	19
	20	1.98	1.97	1.95	1.93	1.92	1.91	1.90	1.89	1.86	1.85	20
	21	1.95	1.94	1.92	1.90	1.89	1.88	1.87	1.86	1.83	1.82	21
	22	1.92	1.91	1.89	1.88	1.86	1.85	1.84	1.83	1.81	1.79	22
	23	1.90	1.88	1.86	1.85	1.84	1.82	1.81	1.80	1.78	1.76	23
	24	1.88	1.86	1.84	1.83	1.82	1.80	1.79	1.78	1.76	1.74	24
	25	1.86	1.84	1.82	1.81	1.80	1.78	1.77	1.76	1.73	1.72	25
	26	1.84	1.82	1.80	1.79	1.78	1.76	1.75	1.74	1.71	1.70	26
	27	1.82	1.81	1.79	1.77	1.76	1.74	1.73	1.72	1.70	1.68	27
	28	1.80	1.79	1.77	1.75	1.74	1.73	1.71	1.70	1.68	1.66	28
	29	1.79	1.77	1.75	1.74	1.73	1.71	1.70	1.69	1.66	1.65	29
	30	1.77	1.76	1.74	1.72	1.71	1.70	1.68	1.67	1.65	1.63	30
	35	1.72	1.70	1.68	1.66	1.65	1.63	1.62	1.61	1.58	1.57	35
	40	1.67	1.66	1.64	1.62	1.61	1.59	1.58	1.56	1.54	1.52	40
	50	1.61	1.60	1.58	1.56	1.54	1.52	1.51	1.50	1.47	1.45	50
	60	1.57	1.56	1.53	1.52	1.50	1.48	1.47	1.45	1.42	1.40	60
	70	1.55	1.53	1.50	1.49	1.47	1.45	1.44	1.42	1.39	1.36	70
	80	1.52	1.51	1.48	1.46	1.45	1.43	1.41	1.39	1.36	1.34	80
	100	1.49	1.48	1.45	1.43	1.41	1.39	1.38	1.36	1.32	1.30	100
	150	1.45	1.44	1.41	1.39	1.37	1.34	1.33	1.31	1.27	1.24	150
	300	1.41	1.39	1.36	1.34	1.32	1.30	1.28	1.26	1.21	1.17	300
	1000	1.38	1.36	1.33	1.31	1.29	1.26	1.24	1.22	1.16	1.110	1000

TABLE C.6 Critical values for F_{MAX} (S^2_{MAX}/S^2_{MIN}) distribution for $\alpha = 0.05$ and 0.01.

$\alpha = 0.05$

df	k=2	3	4	5	6	7	8	9	10	11	12
4	9.60	15.5	20.6	25.2	29.5	33.6	37.5	41.1	44.6	48.0	51.4
5	7.15	10.8	13.7	16.3	18.7	20.8	22.9	24.7	26.5	28.2	29.9
6	5.82	8.38	10.4	12.1	13.7	15.0	16.3	17.5	18.6	19.7	20.7
7	4.99	6.94	8.44	9.70	10.8	11.8	12.7	13.5	14.3	15.1	15.8
8	4.43	6.00	7.18	8.12	9.03	9.78	10.5	11.1	11.7	12.2	12.7
9	4.03	5.34	6.31	7.11	7.80	8.41	8.95	9.45	9.91	10.3	10.7
10	3.72	4.85	5.67	6.34	6.92	7.42	7.87	8.28	8.66	9.01	9.34
12	3.28	4.16	4.79	5.30	5.72	6.09	6.42	6.72	7.00	7.25	7.48
15	2.86	3.54	4.01	4.37	4.68	4.95	5.19	5.40	5.59	5.77	5.93
20	2.46	2.95	3.29	3.54	3.76	3.94	4.10	4.24	4.37	4.49	4.59
30	2.07	2.40	2.61	2.78	2.91	3.02	3.12	3.21	3.29	3.36	3.39
60	1.67	1.85	1.96	2.04	2.11	2.17	2.22	2.26	2.30	2.33	2.36
∞	1.00	1.00	1.00	1.00	1.00	1.00	1.00	1.00	1.00	1.00	1.00

$\alpha = 0.01$

df	k=2	3	4	5	6	7	8	9	10	11	12
4	23.2	37	49	59	69	79	89	97	106	113	120
5	14.9	22	28	33	38	42	46	50	54	57	60
6	11.1	15.5	19.1	22	25	27	30	32	34	36	37
7	8.89	12.1	14.5	16.5	18.4	20	22	23	24	26	27
8	7.50	9.9	11.7	13.2	14.5	15.8	16.9	17.9	18.9	19.8	21
9	6.54	8.5	9.9	11.1	12.1	13.1	13.9	14.7	15.3	16.0	16.6
10	5.85	7.4	8.6	9.6	10.4	11.1	11.8	12.4	12.9	13.4	13.9
12	4.91	6.1	6.9	7.6	8.2	8.7	9.1	9.5	9.9	10.2	10.6
15	4.07	4.9	5.5	6.0	6.4	6.7	7.1	7.3	7.5	7.8	8.0
20	3.32	3.8	4.3	4.6	4.9	5.1	5.3	5.5	5.6	5.8	5.9
30	2.63	3.0	3.3	3.4	3.6	3.7	3.8	3.9	4.0	4.1	4.2
60	1.96	2.2	2.3	2.4	2.4	2.5	2.5	2.6	2.6	2.7	2.7
∞	1.00	1.0	1.0	1.0	1.0	1.0	1.0	1.0	1.0	1.0	1.0

TABLE A.14 Test for equal covariance matrices, $\alpha = 0.05$.

V	k=2	k=3	k=4	k=5	k=6	k=7	k=8	k=9	k=10
p=2									
3	12.18	18.70	24.55	30.09	35.45	40.68	45.81	50.87	55.86
4	10.70	16.65	22.00	27.07	31.97	36.75	41.45	46.07	50.64
5	9.97	15.63	20.73	25.57	30.23	34.79	39.26	43.67	48.02
6	9.53	15.02	19.97	24.66	29.19	33.61	37.95	42.22	46.45
7	9.24	14.62	19.46	24.05	28.49	32.83	37.08	41.26	45.40

8	9.04	14.33	19.10	23.62	27.99	32.26	36.44	40.57	44.64
9	8.88	14.11	18.83	23.30	27.62	31.84	35.98	40.05	44.08
10	8.76	13.97	18.61	23.05	27.33	31.51	35.61	39.65	43.64
11	8.67	13.81	18.44	22.85	27.10	31.25	35.32	39.33	43.29
12	8.59	13.70	18.30	22.68	26.90	31.03	35.08	39.07	43.00
13	8.52	13.60	18.19	22.54	26.75	30.85	34.87	38.84	42.76
14	8.47	13.53	18.10	22.42	26.61	30.70	34.71	38.66	42.56
15	8.42	13.46	18.01	22.33	26.50	30.57	34.57	38.50	42.38
16	8.38	13.40	17.94	22.24	26.40	30.45	34.43	38.36	42.23
17	8.35	13.35	17.87	22.17	26.31	30.35	34.32	38.24	42.10
18	8.32	13.30	17.82	22.10	26.23	30.27	34.23	38.13	41.99
19	8.28	13.26	17.77	22.04	26.16	30.19	34.14	38.04	41.88
20	8.26	13.23	17.72	21.98	26.10	30.12	34.07	37.95	41.79
25	8.17	13.10	17.55	2179	25.87	29.86	33.78	37.63	41.44
30	8.11	13.01	17.44	21.65	25.72	29.69	33.59	37.42	41.21

$p = 3$

4	22.41	35.00	46.58	57.68	68.50	79.11	89.60	99.94	110.21
5	19.19	30.52	40.95	50.95	60.69	70.26	79.69	89.03	98.27
6	17.57	28.24	38.06	47.49	56.67	65.69	74.58	83.39	92.09
7	16.59	26.84	36.29	45.37	54.20	62.89	71.44	79.90	88.30
8	15.93	25.90	35.10	43.93	52.54	60.99	69.32	77.57	85.73
9	15.46	25.22	34.24	42.90	51.33	59.62	67.78	75.86	83.87
10	15.11	24.71	33.59	42.11	50.42	58.57	66.62	74.58	82.46
11	14.83	24.31	33.08	41.50	49.71	57.76	65.71	73.57	81.36
12	14.61	23.99	32.67	41.00	49.13	57.11	64.97	72.75	80.45
13	14.43	23.73	32.33	40.60	48.65	56.56	64.36	72.09	79.72
14	14.28	23.50	32.05	40.26	48.26	56.11	63.86	71.53	79.11
15	14.15	23.32	31.81	39.97	47.92	55.73	63.43	71.05	78.60
16	14.04	23.16	31.60	39.72	47.63	55.40	63.06	70.64	78.14
17	13.94	23.02	31.43	39.50	47.38	55.11	62.73	70.27	77.76
18	13.86	22.89	31.26	39.31	47.16	54.86	62.45	69.97	77.41
19	13.79	22.78	31.13	39.15	46.96	54.64	62.21	69.69	77.11
20	13.72	22.69	31.01	39.00	46.79	54.44	61.98	69.45	76.84
25	13.48	22.33	30.55	38.44	46.15	53.70	61.16	68.54	75.84
30	13.32	22.10	30.25	38.09	45.73	53.22	60.62	67.94	75.18

$p = 4$

5	35.39	56.10	75.36	93.97	112.17	130.11	147.81	165.39	182.80
6	30.06	48.62	65.90	82.60	98.93	115.03	130.94	146.69	162.34
7	27.31	44.69	60.89	76.56	91.88	106.98	121.90	136.71	151.39
8	25.61	42.24	57.77	72.77	87.46	101.94	116.23	130.43	144.50
9	24.45	40.57	55.62	70.17	84.42	98.46	112.32	126.08	139.74
10	23.62	39.34	54.04	68.26	82.19	95.90	109.46	122.91	136.24
11	22.98	38.41	52.84	66.81	80.48	93.95	107.27	120.46	133.57
12	22.48	37.67	51.90	65.66	79.14	92.41	105.54	118.55	131.45
13	22.08	37.08	51.13	64.73	78.04	91.15	104.12	116.68	129.74
14	21.75	36.59	50.50	63.95	77.13	90.12	102.97	115.69	128.32
15	21.47	36.17	49.97	63.30	76.37	89.26	101.99	114.59	127.14
16	21.24	35.82	49.51	62.76	75.73	88.51	101.14	113.67	126.10
17	21.03	35.52	49.12	62.28	75.16	87.87	100.42	112.87	125.22
18	20.86	35.26	48.78	61.86	74.68	87.31	99.80	112.17	124.46
19	20.70	35.02	48.47	61.50	74.25	86.82	99.25	111.56	123.79

20	20.56	34.82	48.21	61.17	73.87	86.38	98.75	111.02	123.18
25	20.06	34.06	47.23	59.98	72.47	84.78	96.95	109.01	120.99
30	19.74	33.59	46.61	59.21	71.58	83.74	95.79	107.71	119.57

$p = 5$

6	51.11	81.99	110.92	138.98	166.54	193.71	220.66	247.37	273.88
7	43.40	71.06	97.03	122.22	146.95	171.34	195.49	219.47	243.30
8	39.29	65.15	89.45	113.03	136.18	159.04	181.65	204.14	226.48
9	36.71	61.39	84.62	107.17	129.30	151.17	172.80	194.27	215.64
10	34.93	58.78	81.25	103.06	124.48	145.64	166.56	187.37	208.02
11	33.62	56.85	78.75	100.02	120.92	141.54	161.98	182.24	202.37
12	32.62	55.37	76.83	97.68	118.15	138.38	158.38	178.23	198.03
13	31.83	54.19	75.30	95.82	115.96	135.86	155.54	175.10	194.51
14	31.19	53.23	74.05	94.29	114.16	133.80	153.21	172.49	191.68
15	30.66	52.44	73.01	93.02	112.66	132.07	151.29	170.36	189.38
16	30.22	51.76	72.14	91.94	111.41	130.61	149.66	166.53	187.32
17	29.83	51.19	71.39	91.03	110.34	129.38	148.25	166.99	185.61
18	29.51	50.69	70.74	90.23	109.39	128.29	147.03	165.65	184.10
19	29.22	50.26	70.17	89.54	108.57	127.36	145.97	164.45	182.81
20	28.97	49.88	69.67	88.93	107.85	126.52	145.02	163.38	181.65
25	28.05	48.48	67.86	86.70	105.21	123.51	141.62	159.60	177.49
30	27.48	47.61	66.71	85.29	103.56	121.60	139.47	157.22	174.87

References

Abass, A.B., Ndunguru, G., Mamiro, P., Alenkhe, B., Mlingi, N., Bekunda, M., 2014. Post-harvest food losses in a maize-based farming system of semi-arid savannah area of Tanzania. J. Stored Prod. Res. 57, 49–57.

Abbott, J.C., 1994. Agricultural processing enterprises: development potentials and links to the smallholder. In: von Braun, J., Kennedy, E. (Eds.), Agricultural Commercialization, Economic Development, and Nutrition. Johns Hopkins University Press and IFPRI, Washington, DC.

Abeje, M.T., Tsunekawa, A., Haregeweyn, N., Ayalew, Z., Nigussie, Z., Berihun, D., 2020. Multidimensional poverty and inequality: insights from the Upper Blue Nile Basin, Ethiopia. Soc. Indicat. Res. 149 (2), 585–611.

Acharya, R., July 2020. Reduced Food and Diet Quality, and Need for Nutrition Services during COVID-19: Findings from Surveys in Bihar and Uttar Pradesh. https://southasia.ifpri.info/2020/07/09/15081/.

Adams Katherine, P., Lybbert, T.J., Vosti, S.A., Ayifah, E., Arimond, M., Adu-Afarwuah, S., July 2018. Unintended effects of a targeted maternal and child nutrition intervention on household expenditures, labor income, and the nutritional status of non-targeted siblings in Ghana. World Dev. 107, 138–150.

Ainembabazi, J.H., Abdoulaye, T., Feleke, S., Alene, A., Dontsop-Nguezet, P.M., Ndayisaba, P.C., August 2018. Who benefits from which agricultural research-for-development technologies? Evidence from farm household poverty analysis in Central Africa. World Dev. 108, 28–46.

Aker Jenny, C., July 2010. Information from markets near and far: mobile phones and agricultural markets in Niger. Am. Econ. J. Appl. Econ. 2 (3), 46–59.

Akin, J.S., Guilkey, D.K., Popkin, B.M., Fanelli, M.T., 1986. Cluster analysis of food consumption patterns of older Americans. J. Am. Diet Assoc. 86 (5), 616–624.

Alderman, H., 2011. No Small Matter. The Impact of Poverty, Shocks, and Human Capital Investments in Early Childhood Development. World Bank, Washington, DC.

Alderman, H., Garcia, M., 1994. Food security and health security: explaining the levels of nutritional status in Pakistan. Econ. Dev. Cult. Change 42 (3), 485–507.

Alemu, Hailu, S., et al., July 2018. Women empowerment through self-help groups: the bittersweet fruits of collective apple cultivation in highland Ethiopia. J. Hum. Dev. Capab. 19 (3), 308–330. EBSCOhost. http://www-tandfonline-com.pitt.idm.oclc.org/loi/cjhd20.

Algieri, B., Aquino, A., 2011. Key Determinants of Poverty Risk in Italy.

Ali, M., Farooq, U., 2003. Diversified Consumption to Boost Rural Labor Productivity.

Alkire, S., Fang, Y., 2019. Dynamics of multidimensional poverty and uni-dimensional income poverty: an evidence of stability analysis from China. Soc. Indicat. Res. 142 (1), 25–64.

Alkire, S., Foster, J., 2011. Counting and multidimensional poverty. J. Public Econ. 95 (7-8), 476–487.

Alkire, S., Foster, J., 2013. Understandings and Misunderstandings of Multidimensional Poverty Measurement. University of Oxford. OPHI Working Paper No. 43.

Alkire, S., Seth, S., 2015. Multidimensional poverty reduction in India between 1999 and 2006: where and how? World Dev. 72, 93–108.

Alkire, S., James, F., Suman, S., Maria Emma, S., Jose' Manuel, R., Paola, B., 2015. Multi-dimensional Poverty Measurement and Analysis. Oxford University Press, Oxford and New York.

Alkire, S., Apablaza, M., Chakravarty, S., Yalonetzky, G., 2017a. Measuring chronic multidimensional poverty. J. Pol. Model. 39 (6), 983−1006.
Alsan, M., Yang, C., 2018. Fear and the Safety Net: Evidence from Secure Communities. National Bureau of Economic Research, Inc, NBER Working Papers: 24731.
Altamirano, M.A.J., Teixeira, K.M.D., 2017. Multidimensional poverty in Nicaragua: are female-headed households better off? Soc. Indicat. Res. 132 (3), 1037−1063.
Alvi, M., Shweta, G., Ruth, M.-D., Claudia, R., July 2020. Phone Surveys to Understand Gendered Impacts of Covid-19: A Cautionary Note. https://southasia.ifpri.info/2020/07/22/phone-surveys-to-understand-gendered-impacts-of-covid-19-a-cautionary-note/.
Anderson, M.D., Cook, J.T., 1999. Community food security: practice in need of theory? Agric. Hum. Val. 16, 141−150.
Anderson, A.M., Earle, M.D., 1983. Diet planning in the third world by linear and goal programming. J. Oper. Res. Soc. 34 (1), 9−16.
Andrews, M., Bickel, G., Carlson, S., 1998a. Household food security in the United States in 1995: results from the food security management project. Fam. Econ. Nutr. Rev. 11 (1−2), 17−28.
Andrews, F.M., Kelm, L., O'Malley, P.M., Rodgers, W., Welch, K.B., Davidson, T.N., 1998b. Selecting Statistical Techniques for Social Science Data: A Guide for SAS Users. SAS Institute Inc., Cary, NC.
Angrist, J., Kruegger, A., 1991. Does compulsory school attendance affect schooling and earnings? Q. J. Econ. 106 (4), 979−1104.
Anika, S.-F., Anna, G.-P., Hill, Z., September 2017. Use of informal safety nets during the supplemental nutrition assistance program benefit cycle: how poor families cope with within-month economic instability. Soc. Serv. Rev. 91 (3), 456−487.
Anwar, S., Maria, K., Aisha, A., Zahira, B., December 2015. Impact of socioeconomic and demographic factors affecting child health in selected South Asian countries. Rev. Econ. Dev. Stud. 1 (2), 143−151.
Arif, G.M., Farooq, S., Nazir, S., Satti, M., 2014. Child malnutrition and poverty: the case of Pakistan. Pakistan Dev. Rev. 53 (2), 99−118. Summer 2014.
Arimi, K., Olajide, B.R., 2016. Comparative analysis of male and female adopters of improved rice production technology in Ogun and Ekiti States, Nigeria. Int. J. Agric. Resour. Govern. Ecol. 12 (3), 246−261.
Aristondo, O., Onaindia, E., 2018. Counting energy poverty in Spain between 2004 and 2015. Energy Pol. 113, 420−429.
Armour, B.S., Pitts, M.M., Lee, C., 2007. Cigarette Smoking and Food Insecurity Among Low-Income Families in the United States, 200. Federal Reserve Bank of Atlanta,Working Paper: 2007−19.
Arndt, C., Rob, D., Sherwin, G., Laurence, H., Konstantin, M., Boipuso, M., Sherman, R., Witness, S., Dirk, van S., Lillian, A., 2020. "Impact of Covid-19 on the South African Economy: An Initial Analysis", SA-TIED Working Paper 111. https://sa-tied.wider.unu.edu/sites/default/files/pdf/SA-TIED-WP-111.pdf.
Arrieta, A., Scott, S., Kumar, N., Menon, P., Quisumbing, A., 2020. Being social is important for women's mental health: Insights from a pre-pandemic survey in rural India. August. https://southasia.ifpri.info/2020/08/19/being-social-is-important-for-womens-mental-health-insights-from-a-pre-pandemic-survey-in-rural-india/.
Asfaw, A., 2008. Fruits and vegetables availability for human consumption in Latin American and Caribbean countries: patterns and determinants. Food Pol. 33, 444−454.
Asfaw, A., Admassie, A., 2004. The role of education on the adoption of chemical fertiliser under different socioeconomic environments in Ethiopia. Agric. Econ. 30, 215−228.
Asfaw, S., Lipper, L., Dalton, T.J., Audi, P., 2012a. Market participation, on-farm crop diversity and household welfare: micro-evidence from Kenya. Environ. Dev. Econ. 17 (5), 579−601.

Asfaw, S., Shiferaw, B., Simtowe, F., Lipper, L., 2012b. Impact of modern agricultural technologies on smallholder welfare: evidence from Tanzania and Ethiopia. Food Pol. 37 (3), 283–295.
Asian Development Bank (ADB), 2001. Participatory Poverty Assessment in Cambodia, Ch. 3. The Philippines, Manila.
Athanasios, A., Bezuneh, M., Deaton, B.J., 1994. Impacts of FFW on nutrition in rural Kenya. Agric. Econ. 11 (2–3), 301–309.
Atkins, V.J., 2004. The US farm bill of 2002: implications for Caricom's agricultural export trade. Soc. Econ. Stud. 53 (3), 61–80.
Auerbach, R., 2018. Sustainable food systems for Africa. Econ. Agro-Alimentare 20 (3), 301–320.
Aurino, E., Morrow, V., November 2018. 'Food prices were high, and the dal became watery': mixed-method evidence on household food insecurity and children's diets in India. World Dev. 111, 211–224.
Aurino, E., Tranchant, J.-P., Diallo Amadou, S., Gelli, A., 2019. School feeding or general food distribution? Quasi-experimental evidence on the educational impacts of emergency food assistance during conflict in Mali. J. Dev. Stud. 55 (Suppl. 1), 7–28.
Aurino, E., Schott, W., Behrman Jere, R., Penny, M., December 2019. Nutritional status from 1 to 15 years and adolescent learning for boys and girls in Ethiopia, India, Peru, and Vietnam. Popul. Res. Pol. Rev. 38 (6), 899–931.
Averett, L.S., Smith, K.J., December 2014. Financial hardship and obesity. Econ. Hum. Biol. 15, 201–212.
Avula, R. (Ed.), 2020. POSHAN's Abstract Digest on Maternal and Child Nutrition Research, Issue 32. International Food Policy Research Institute (IFPRI), New Delhi, India.
Avula, R., Jennifer, C., Kenda, C., Edward, F., Stormer, A., Andrea, W., July 2020. Nutrition Program Adaptations and Implementation Science during COVID-19: The Case of South Asia. https://southasia.ifpri.info/2020/07/16/nutrition-program-adaptations-and-implementation-science-during-covid-19-the-case-of-south-asia/.
Azam, M.S., Shaheen, M., 2019. Decisional factors driving farmers to adopt organic farming in India: a cross-sectional study. Int. J. Soc. Econ. 46 (4), 562–580.
Azupogo, F., Elisabetta, A., Gelli, A., Bosompem Kwabena, M., Irene, A., Saskia, O., J, M., Brouwer Inge, D., Gloria, F., 2019. Agro-ecological zone and farm diversity are factors associated with haemoglobin and anaemia among rural school-aged children and adolescents in Ghana. Matern. Child Nutr. 15 (1).
Babu, S.C., 1997a. Rethinking training in food policy analysis: how relevant is it for policy reforms. Food Pol. 22 (1), 1–9.
Babu, S.C., 1997b. Facing donor community with informed policy decisions – lessons from food security and nutrition monitoring in Malawi. Afr. Dev. 22 (2), 5–24.
Babu, S.C., 1999. Designing decentralized food security and nutrition policies- a knowledge-based system approach in Malawi. Q. J. Int. Agric. 38 (1), 78–95.
Babu, S.C., 2001. Food and nutrition policies in Africa: capacity challenges and training options. Afr. J. Food Agric Nutr. Sci. 1 (1), 19–28.
Babu, S.C., 2002a. Food systems for improved human nutrition: linking agriculture, nutrition and productivity. J. Crop Prod. 6 (1/2), 7–30.
Babu, S., 2002b. Designing nutrition interventions with food systems: planning, communication, monitoring and evaluation. J. Crop Prod. 6 (1/2), 365–373.
Babu, S.C., 2006. Achieving food security through increasing access to food – strategies and options for South Asia. In: Nutrition Security in South Asia. Nutrition Foundation, New Delhi.
Babu, S.C., 2009. Hunger and food security. In: World at Risk: A Global Issues Sourcebook, second ed. CQ Press, Washington, D.C.
Babu, S.C., 2013. Policy Process and Food Price Crisis: A Framework for Analysis and Lessons from Country Studies. WIDER Working Paper 2013/070. World Institute for Development Economics Research, Helsinki.

Babu, S.C., Chapasuka, E., 1997. Mitigating the effects of drought through food security and nutrition monitoring: lessons from Malawi. U. N. Univ. Food Nutr. Bull. 18 (1), 71–81.

Babu, S.C., Gulati, A., 2005. Economic Reforms and Food Security – The Impact of Trade and Technology in South Asia. Haworth Press, New York.

Babu, S.C., Hallam, J.A., 1989. Socio-economic impacts of school feeding programmes: empirical evidence from a south Indian village. Food Pol. 14 (1), 58–66.

Babu, S.C., Mthindi, G.B., 1995a. Costs and benefits of informed food policy decisions – a case study of food security and nutrition monitoring in Malawi. Q. J. Int. Agric. 34 (3), 292–308.

Babu, S.C., Mthindi, G.B., 1995b. Developing decentralized capacity for disaster prevention – lessons from food security and nutrition monitoring in Malawi. Disasters 19 (2), 127–139.

Babu, S.C., Rajasekaran, B., 1991a. Biotechnology for rural nutrition: an economic evaluation of algal protein supplements in South India. Food Pol. 16 (5), 405–414.

Babu, S.C., Rajasekaran, B., 1991b. Agroforestry system: attitude towards risk and nutrient availability – a case study from South Indian farming systems. Agrofor. Syst. 15, 1–15.

Babu, S.C., Rhoe, V., 2002. Agroforestry systems for food and nutrition security– potentials, pathways and policy research needs. J. Crop Prod. 6 (1/2), 177–192.

Babu, S.C., Rhoe, V., 2006. Food security and poverty in central Asia. In: Babu, S.C., Djalalov, S. (Eds.), Policy Reforms and Agriculture Development in Central Asia. Springer, New York.

Babu, S.C., Sanyal, P., 2008. Persistent food insecurity in Malawi and policy options. In: Pinstrup-Andersen, P. (Ed.), Globalization and Food Security. Cornell University Press, Ithaca.

Babu, S.C., Thirumaran, S., Mohanam, T.C., 1993. Agricultural productivity, seasonality, and gender baia in rural nutrition: empirical evidence from South India. Soc. Sci. Med. 37 (11), 128–1413.

Babu, S.C., Brown, L., McClafferty, B., 2000. Systematic client consultation in development: the case of food policy research in Ghana, India, Kenya, and Malawi. World Dev. 28 (1), 99–110.

Babu, S.C., Gajanan, S.N., Hallam, J.A., 2017. Nutrition Economics: Principles and Policy Applications. Elsevier Publishers, Academic Press, New York.

Bacher, J., 2002. Cluster Analysis. University of Erlanger Manuscript, Nuremberg, Germany.

Bai, Y., Alemu, R., Block, S., Headey, D., Masters, W., 2021. Cost and affordability of nutritious diets at retail prices: Evidence from 177 countries. Food Policy 99, 101983.

Bandiera, O., Rasul, I., 2006. Social networks and technology adoption in Northern Mozambique. Econ. J. 116, 862–902.

Banerjee, A.V., Duflo, E., 2007. The economic lives of the poor. J. Econ. Perspect. 21 (1), 141–167. Winter, EBSCOhost. http://www.aeaweb.org.pitt.idm.oclc.org/jep/.

Banerjee, A.V., Duflo, E., 2008. What is middle class about the middle classes around the world? J. Econ. Perspect. 22 (2), 3–28.

Banerjee, A.V., Duflo, E., 2009. The experimental approach to development economics. Ann. Rev. Econ. 1 (1), 151–178.

Barghouti, S., Kane, S., Sorby, K., Ali, M., 2004. Agricultural diversification for the poor: guidelines for practitioners. In: Agriculture and Rural Development Discussion Paper No.1. World Bank, Washington, DC.

Barnwal, P., Kotani, K., 2013. Climatic impacts across agricultural crop yield distributions: an application of quantile regression on rice crops in Andhra Pradesh, India. Ecol. Econ. 87, 95–109.

Barrera, A., 1990. The role of maternal schooling and its interaction with public health programs in child health production. J. Dev. Econ. 32 (1), 69–91.

Barrett, C.B., 1994. Understanding uneven agricultural liberalization in Madagascar. J. Mod. Afr. Stud. 32, 449–476.

Barros, R., Fox, L., Mendonca, R., 1997. Female-headed households, poverty, and the welfare of children in urban Brazil. Econ. Dev. Cult. Change 45 (2), 231–257.

Basnet, S., Frongillo, E.A., Nguyen, P.H., Moore, S., Arabi, M., 2020. Associations of maternal resources with care behaviours differ by resource and behaviour. Matern. Child Nutr. 16, e12977. https://doi.org/10.1111/mcn.12977.

Baten, J., Bohm, A., 2009. Children's height and parental unemployment: a large scale anthropometric study on Eastern Germany, 1994–2006. Ger. Econ. Rev. 11 (1), 1–24.

Beatty, T.K.M., 2010. Do the poor pay more for food? evidence from the United Kingdom. Am. J. Agric. Econ. 92 (3), 608–621.

Becker, S., Ichino, A., 2002. Estimation of average treatment effects based on propensity scores. Stata J. 2 (4), 358–377.

Becker, W., Helsing, B., 1991. Food and Health Data: Their Use in Nutrition Policy-Making. WHO Regional Publications, European Series No. 34, Copenhagen, Denmark.

Behrman, J.R., Deolalikar, A.B., 1988. Health and nutrition. In: Chenery, H., Srinivasan, T.N. (Eds.), Handbook of Development Economics, vol. 1. Elsevier Science Publishers, Amsterdam, pp. 631–711.

Behrman, J.R., Alderman, H., Hoddinott, J., 2004. Hunger and Malnutrition. Challenge Paper for Copenhagen Consensus 2004.

Ben, S., Yonatan, M., Robert, A., Scholz, J.K., 2011. An Assessment of the Effectiveness of Anti-poverty Programs in the United States. Economics Working Paper. The Johns Hopkins University.

Benbrook, C., 2003. What Will it Take to Change the American Food System. http://organic.insightd.net/reportfiles/Kellog_Changethesystem_April_2003.pdf.

Benefo, K., Schultz, T.P., 1996. Fertility and child mortality in Cote d'Ivoire and Ghana. World Bank Econ. Rev. 10 (1), 23–158.

Bentley, M.E., et al., 1991. Maternal feeding behavior and child acceptance of food during diarrhea, convalescence, and health in the central Sierra of Peru. Am. J. Publ. Health 81 (1), 43–47.

Berenger, V., 2019. The counting approach to multidimensional poverty: the case of four African countries. S. Afr. J. Econ. 87 (2), 200–227.

Besharov, D.J., 2003. Family and Child Well-Being after Welfare reform.Transaction, New Brunswick N.J. and London.

Beynon, J., Jones, S., Yao, S., 1992. Market reform and private trade in eastern and Southern Africa. Food Pol. 17 (6), 399–408.

Bhutta, Z.A., Bawany, F., Feroze, A., Rizvi, A., Thapa, S.J., Patel, M., Supplement 2009. Effects of the crises on child nutrition and health in East Asia and the Pacific. Global Soc. Pol. 9, 119–143.

Bickel, G., Nord, M., Price, C., Hamilton, W., Cook, J., 2000. Guide to Measuring House- Hold Food Security, Food and Nutrition Service, USDA, Alexandria, VA.

Biggeri, M., et al., April 2018. Local communities and capability evolution: the core of human development processes. J. Hum. Dev. Capab. 19 (2), 126–146.

Bigsten, A., Shimeles, A., 2004. Dynamics of Poverty in Ethiopia. WIDER Research Paper No. 2004/39. United Nations University.

Billig, P., et al., 1999. Water and Sanitation Indicators Measurement Guide. Academy for Educational Development, Washington, DC.

Bird Frances, A., Pradhan, A., Bhavani, R.V., Dangour Alan, D., January 2019. Interventions in agriculture for nutrition outcomes: a systematic review focused on South Asia. Food Pol. 82, 39–49.

Black, R.E., Allen, L.H., Bhutta, Z.A., de Onis, M., Mathers, C., Rivera, J., 2008. Maternal and child undernutrition: global and regional exposures and health consequences. Lancet 371, 243–260.

Blanken, J., von Braun, J., Haen, H.D., 1994. The triple role of potatoes as a source of cash, food, and employment: effects on nutritional improvement in Rwanda. In: von Braun, J., Kennedy, E. (Eds.), Agricultural Commercialization, Economic Development, and Nutrition. Johns Hopkins University Press and IFPRI, Washington, DC.

Blaylock, J.R., Blisard, W., Noel, 1995. Food security and health status in the United States. Appl. Econ. 27 (10), 961–966.

Block, S.A., 2003. Nutrition Knowledge, Household Coping, and the Demand for Micro- Nutrient Rich Foods. Tufts University Discussion Paper No. 20. Medford, MA.

Blondel, S., Joffre, O., Planchais, G., Simon, E., 2012. Should sustainable development rely on financial incentives? Lessons from the CTE experiment in the Loire valley. Int. Bus. Res. 5 (10), 56–64.

Blunch, N.H., 2005. Maternal Schooling and Child Health Revisited. George Washington University Working Paper. Washington, DC.

Borooah, V.K., 2002. The Role of Maternal Literacy in Reducing the Risk of Child Malnutrition in India. University of Ulster Working Paper. UK.

Bose, N., Das, S., May 2017. Women's inheritance rights, household allocation, and gender bias. Am. Econ. Rev. 107 (no. 5), 150–153. EBSCOhost. http://www.aeaweb.org.pitt.idm.oclc.org/aer/.

Boss, R., Fabrizio, B., Pradhan, M., Devesh, R., July 2020. Farmer Organizations and COVID-19 in ASEAN: Role, Impact, and Opportunities. https://southasia.ifpri.info/2020/07/27/farmer-organizations-and-covid-19-in-asean-role-impact-and-opportunities/.

Bouis, H.E., 2000. Commercial vegetable and polyculture fish production in Bangladesh: their impacts on household income and dietary quality. Food Nutr. Bull. 21 (4), 482–487.

Bouis, H., Haddad, L.J., 1994. The nutrition effects of sugarcane cropping in Southern Philippine province. In: von Braun, J., Kennedy, E. (Eds.), Agricultural Commercialization, Economic Development, and Nutrition. Johns Hopkins University Press and IFPRI, Washington, DC.

Box, G.E.P., 1949. A general distribution theory for a class of likelihood criteria. Biometrika 36, 317–346.

Box, G.E.P., 1950. Problems in the analysis of growth and wear curves. Biometrics 6, 362–389.

Briend, A., Ferguson, E., Darmon, N., 2001. Local food price analysis by linear programming: a new approach to assess the economic value of fortified food supplements. Food Nutr. Bull. 22 (2), 184–189.

Broda, C., Leibtag, E., Weinstein, D., 2009. The role of prices in measuring the poor's living standards. J. Econ. Perspect. 23 (2), 77–97.

Bronchetti, E.T., et al., 2018. Local Food Prices, SNAP Purchasing Power, and Child Health. EBSCOhost. http://www.nber.org.pitt.idm.oclc.org/papers/w24762.pdf.

Bryan, E., Garner, E., 2020. What Does Empowerment Mean to Women in Northern Ghana? Insights from Research Around a Small-Scale Irrigation Intervention. International Food Policy Research Institute (IFPRI), Washington, DC. https://doi.org/10.2499/p15738coll2.133596. Discussion Paper 1909.

Bryce, J., Victora, C.J., Habicht, J.P., Vaughan, J.P., Black, R.E., 2006. The multi-country evaluation of the integrated management of childhood illness strategy: lessons for the evaluation of public health interventions. Am. J. Publ. Health 94 (3), 406–415.

Bullinger, L.R., Gurley-Calvez, T., Jan. 2016. WIC participation and maternal behavior: breast-feeding and work leave. Contemp. Econ. Pol. vol. 34 (no. 1), 158–172.

Buvinic, M., Gupta, G.R., 1997. Female-headed households and female-maintained families: are they worth targeting to reduce poverty in developing countries? Econ. Dev. Cult. Change 45 (2), 259–280.

Caglayan, E., Dayioglu, T., 2011. Comparing the parametric and semiparametric logit models: household poverty in Turkey. Int. J. Econ. Finance 3 (5), 197–207.

Cakir, M., et al., April 2018. Spatial and temporal variation in the value of the women, infants, and children program's fruit and vegetable voucher. Am. J. Agric. Econ. vol. 100 (no. 3), 691–706.

Caldwell, J.C., 1979. Education as a factor in mortality decline: an examination of Nigerian data. Popul. Stud. 33 (3), 395–413.

Calkins, P.H., 1981. Nutritional adaptations of linear programming for planning rural development. Am. J. Agric. Econ. 63 (2), 247–254.

Calonico, S., Matias, C.D., Rocio, T., 2014a. Robust nonparametric confidence intervals for regression-discontinuity designs. Econometrica 82, 2295–2326.

Calonico, S., Matias, C.D., Rocio, T., 2014b. Robust data-driven inference in the regression-discontinuity design. STATA J. 14, 909–946. Number 4.

Calonico, S., Matias, C.D., Max, F.H., 2016. Rdrobust: software for regression discontinuity designs. STATA J. 1–30. Forthcoming, Number 2.

Cameron, L.A., Dowling, J.M., Worswick, C., 2001. Education and labor market participation of women in Asia: evidence from five countries. Econ. Dev. Cult. Change 49 (3), 461–477.

Carneiro, P., Costas, M., Matthias, P., 2007. Maternal Education, Home Environments and the Development of Children and Adolescents, C.E.P.R. Discussion Papers, CEPR Discussion Papers: 6505.

Carson, C.S., 2011. Height of female Americans in the 19th century and the ante- bellum puzzle. Econ. Hum. Biol. 9 (2), 157–164.

Cavatorta, E., Bhavani, S., Artemisa, F.-M., December 2015. Explaining cross-state disparities in child nutrition in rural India. World Dev. 76, 216–237.

Chakrabarti, S., Kishore, A., Raghunathan, K., Scott, S., 2019. Impact of subsidized fortified wheat on anaemia in pregnant Indian women. Matern. Child Nutr. 15 (1).

Chamarbagwala, R., Martin, R., Waddington, H., Howard, W., 2004. The Determinants of Child Health and Nutrition: A Meta-Analysis, OED Working Paper. World Bank, Washington DC.

Chamboko, R., et al., 2017. Mapping patterns of multiple deprivation in Namibia. Int. J. Soc. Econ. vol. 44 (no. 12), 2486–2499. EBSCOhost. search.ebscohost.com/login.aspx?direct=true&db=ecn&AN=1688247&site=ehost-live.

Chand, R., 2007. Demand for Foodgrains. Economic and Political Weekly December 29.

Chandrasekar, C.P., 2013. India's Food Conundrum. The Hindu.

Chandrasekhar, C.P., 2012. India's Triumph in Rice, in The Hindu. December 23, 2012. http://www.thehindu.com/opinion/columns/Chandrasekhar/indias-triumph-in-rice/article4231844.ece?ref=sliderNews.

Chandrasekhar, C.P., January 20, 2013. The Cost of Food Security, in The Hindu. http://www.thehindu.com/opinion/columns/Chandrasekhar/the-cost-of-food- security/article4325479.ece.

Chantarat, S., et al., Aug. 2019. Natural disasters, preferences, and behaviors: evidence from the 2011 mega flood in Cambodia. J. Asian Econ. 63, 44–74.

Chapman, A., Okushima, S., 2019. Engendering an inclusive low-carbon energy transition in Japan: considering the perspectives and awareness of the energy poor. Energy Pol. 135.

Charlier, D., Legendre, B., 2019. A multidimensional approach to measuring fuel poverty. Energy J. 40 (2), 27–53.

Chase, R.S., Sherburne-Benz, L., 2001. Household Effects of Community Education and Health Initiatives: Evaluating the Impact of the Zambia Social Fund. The World Bank, Mimeo, Washington.

Chatterji, P., Liu, X., Yoruk, B., December 2019. The effects of the 2010 affordable care act dependent care provision on family structure and public program participation among young adults. Rev. Econ. Househ. vol. 17 (no. 4), 1133–1161.

Chegere Martin Julius, May 2018. Post-harvest losses reduction by small-scale maize farmers: the role of handling practices. Food Pol. 77, 103−115.
Chen, K.-M., Leu, C.-H., Wang, T.-M., 2019. Measurement and determinants of multidimensional poverty: evidence from Taiwan. Soc. Indicat. Res. 145 (2), 459−478.
Chirwa, E., 2005. Fertilizer and hybrid seeds adoption among small-holder maize farmers in Southern Malawi. Dev. South Afr. 22 (1), 1−12.
Chorniy, A.V., et al., 2018. Does Prenatal WIC Participation Improve Child Outcomes? EBSCOhost. http://www.nber.org.pitt.idm.oclc.org/papers/w24691.pdf.
Chou, S.Y., Rashad, I., Grossman, M., 2008. Fast-food restaurant advertising on television and its influence on childhood obesity. J. Law Econ. 51 (4), 599−618.
Choudhury, S., Headey, D.D., Masters, W.A., October 2019. First foods: diet quality among infants aged 6−23 months in 42 countries. Food Pol. 88.
Chowhan, J., Jennifer, S.M., June 2014. While mothers work do children shirk? Determinants of youth obesity. Appl. Econ. Perspect. Pol. 36 (2), 287−308.
Christiaensen, L., Alderman, H., 2004. Child malnutrition in Ethiopia: can maternal knowledge augment the role of income? Econ. Dev. Cult. Change 52, 287−312.
Christian, P., Abbi, R., Gujral, S., Gopaldas, T., 1988. The role of maternal literacy and nutrition knowledge in determining children's nutritional status. Food Nutr. Bull. 10 (4), 35−40.
Churchill, S.A., Marisetty, V.B., 2020. Financial inclusion and poverty: a tale of forty-five thousand households. Appl. Econ. 52 (16), 1777−1788. http://www.tandfonline.com/loi/raec20.
Claudio, Q., Rosalia, C., Gennaro, P., 2011. Measuring poverty and living conditions in Italy through a combined analysis at a sub-national level. J. Econ. Soc. Meas. 36 (1−2), 93−118.
Clay, E., 1997. Food Security: A Status Review of the Literature. ODI Research Report.
Clements, K.W., Selvanathan, S., 1994. Understanding consumption patterns. Empir. Econ. 19, 69−110.
Cohen, M., Tirado, C., Aberman, N.L., Thompson, B., 2008. Impact of Climate Change and Bioenergy on Nutrition. Paper prepared for FAO, IFPRI.
Coleman-Jensen, Alisha, J., 2010. U.S. food insecurity status: toward a refined definition. Soc. Indicat. Res. 95 (2), 215−230.
Coleman -Jensen, A., Nord, M., Andrews, M., Carlson, S., 2011. Household Food Security in the United States in 2010 Statistical Supplement, United States Department of Agriculture. Economic Research Service Administrative Publication Number 057 September 2011 27.
Conforti, P., D'Amicis, A., 2000. What is the cost of a healthy diet in terms of achieving RDAs? Publ. Health Nutr. 3 (3), 367−373.
Constantinides, S., Blake, C.E., Frongillo, E.A., Avula, R., Thow, A.-M., 2019. Double burden of malnutrition: the role of framing in development of political priority in the context of rising diet-related non-communicable diseases in Tamil Nadu, India (P22-005-19). Current Developments in Nutrition 3 (Suppl. 1), 1929.
Conway, G., March 12, 2003. From the green revolution to the biotechnology revolution: food for poor people in the 21st century. In: Paper Presented at Woodrow Wilson International Center for Scholars.
Coromaldi, M., Zoli, M., 2012. Deriving multidimensional poverty indicators: methodological issues and an empirical analysis for Italy. Soc. Indicat. Res. 107 (1), 37−54.
Cox, D.T., Snell, D.J., 1989. The Analysis of Binary Data. Chapman and Hall, London.
Craig, L.A., Weiss, T., 1997. Nutritional Status and Agricultural Surpluses in the Antebellum United States. National Bureau of Economic Research Inc. NBER Historical Working Paper: 099.
Crentsil, A.O., Asuman, D., Fenny, A.P., 2019. Assessing the determinants and drovers of multidimensional energy poverty in Ghana. Energy Pol. 133.

Cuesta, J., Biggeri, M., Hernandez-Licona, G., Aparicio, R., Guillen-Fernandez, Y., 2020. The political economy of multidimensional child poverty measurement: a comparative analysis of Mexico and Uganda. Oxf. Dev. Stud. 48 (2), 117–134. http://www.tandfonline.com/loi/cods20.

Cunningham, K., Headey Derek, D., Singh, A., Chandni, K., Pandey, R.P., June 2017. Maternal and child nutrition in Nepal: examining drivers of progress from the mid-1990s to 2010s. Glob. Food Secur. 13, 30–37.

Curtis, S., 2012. State immigration legislation and SNAP take-up among immigrant families with children. J. Econ. Issues 46 (3), 661–681.

Custodio, E., 2010. The economic and nutrition transition in Equatorial Guinea coincided with a double burden of over- and under nutrition. Econ. Hum. Biol. 8 (1), 80–87.

Dafermos, Y., Papatheodorou, C., 2013. What drives inequality and poverty in the EU? exploring the impact of macroeconomic and institutional factors. Int. Rev. Appl. Econ. 27 (1), 1–22.

Dahlberg, K.A., 2008. Pursuing Long-Term Food and Agricultural Security in the United States: Decentralization, Diversification, and Reduction of Resource Intensity. Food, Health, and the Environment Series. MIT Press, Cambridge and London.

Dao, M.Q., 2009. Poverty, income distribution, and agriculture in developing countries. J. Econ. Stud. 36 (2), 168–183.

Dargent-Molina, P., James, S.A., Strogatz, D.S., Savitz, D.A., 1994. Association between maternal education and infant diarrhea in different household and community environments of Cebu, Philippines. Soc. Sci. Med. 38 (2), 343–350.

Darmon, N., Ferguson, E., Briend, A., 2002. Linear and nonlinear programming to optimize the nutrient density of a population's diet: an example based on diets of pre- school children in rural Malawi. Am. J. Clin. Nutr. 75 (2), 245–253.

Darshini, J.S., 2012. Food security in India – challenges ahead. Int. J. Res. Commer. Econ. Manag. 2 (12), 112–120.

Das, T., 2019. Does credit access lead to expansion of income and multidimensional poverty? A study of rural Assam. Int. J. Soc. Econ. 46 (2), 252–270.

Das, J., Mar. 2020. Zen and the art of experiments: a note on preventive healthcare and the 2019 nobel prize in economics. World Dev. vol. 127 (EBSCOhost).

Das, M., Suresh, B., 2018. Multidimensional Food Security Index: Evidence from Bangladesh Using Recent Data. IFPRI, Washington, DC. https://www.researchgate.net/publication/325062263_Multidimensional_Food_Security_Index_Recent_Evidence_from_Bangladesh.

Dasgupta, S., Golam, M.M., Paul, T., Wheeler David, J., 2017. The socioeconomics of fish consumption and child health in Bangladesh. In: The World Bank, Policy Research Working Paper Series, p. 8217.

Datt, G., Jolliffe, D., Sharma, M., 2001. A profile of poverty in Egypt. Afr. Dev. Rev./Revue Africaine de Developpement 13 (2), 202–237.

David, S., 1998. Intra-household processes and the adoption of hedgerow intercropping. Agric. Hum. Val. 15, 31–42.

David, V., Moncada, M., Ordonez, F., 2004. Private and public determinants of child nutrition in Nicaragua and Western Honduras. Econ. Hum. Biol. 2, 457–488.

De, S., Sarker, D., 2011. Women's empowerment through self-help groups and its impact on health issues: empirical evidence. J. Global Anal. 2 (1), 51–82.

de Brauw, A., Hoffmann, V., March 2020. The influence of the 2019 nobel prize winners on agricultural economics. World Dev. 127. EBSCOhost. search.ebscohost.com/login.aspx?direct=true&db=ecn&AN=1815546&site=ehost-live.

De Master, Teigen, K., Daniels, J., June 2019. Desert wonderings: reimagining food access mapping. Agric. Hum. Val. 36 (2), 241−256.

de Onis, M., Frongillo, E.A., Blossner, M., 2000. Is malnutrition declining? An analysis of changes in levels of child malnutrition since 1980. Bull. World Health Organ. 78, 1222−1233.

Deaton, A., 1997. The Analysis of Household Surveys: A Microeconomic Approach to Development Policy. Johns Hopkins University Press, Washington, DC.

Deaton, A., 2010. Instruments, randomization, and learning about development. J. Econ. Lit. 48, 424−455.

Deaton, A., Zaidi, S., 2002. Guidelines for Constructing Consumption Aggregates for Welfare Analysis. Living Standards Measurement Study Working Paper No. 135. World Bank, Washington, DC.

Deb, P., Gregory Christian, A., December 2018. Heterogeneous impacts of the supplemental nutrition assistance program on food insecurity. Econ. Lett. 173, 55−60.

Dehejia, R., 2013. The porous dialectic. WIDER Working Paper No. 2013/11. United Nations University.

Dehejia, R., Wahba, S., 2002. Propensity score matching methods for non- experimental causal studies. Rev. Econ. Stat. 84, 151−161.

Deininger, K., Hoogeveen, H., Kinsey, B.H., 2004. Economic benefits and costs of land redistribution in Zimbabwe in the early 1980s. World Dev. 32 (10), 1697−1709.

Delgado, C.L., Miller, C.P.J., 1985. Changing food patterns in West Africa. Food Pol. 10 (1), 55−62.

Demurger, S., Fournier, M., 2011. Poverty and firewood consumption: a case study of rural households in Northern China. China Econ. Rev. 22 (4), 512−523.

DePolt, R.A., Moffitt, R.A., Ribar, D.C., 2009. Food stamps, temporary assistance for needy families and food hardships in three American Cities. Pac. Econ. Rev. 14 (4), 445−473.

Devi, Y.P., Geervani, P., 1994. Determinants of nutrition status of rural preschool children in Andhra Pradesh, India. Food Nutr. Bull. 15 (4), 335−342.

Devi, G., Pagi, B., Ghabru, M., 2017. Marketable surplus and post-harvest losses of maize at producer level in Tribal area of middle Gujarat. Indian J. Econ. Dev. 13 (2), 237−239.

DFID, 2002. Better Livelihoods for Poor People: The Role of Agriculture. Department for International Development, UK.

Dharmasena, S., Capps, O., Clauson, A., 2011. Ascertaining the impact of the 2000 USDA dietary guidelines for Americans on the intake of calories, caffeine, calcium, and vitamin C from at-home consumption of nonalcoholic beverages. J. Agric. Appl. Econ. 43 (1), 13−27.

Dien, L.N., Thang, N.M., Bentley, M.E., 2004. Food consumption patterns in the economic transition in Vietnam. Asia Pac. J. Clin. Nutr. 13 (1), 40−47.

Dong, Fengxia, 2007. Food Security and Biofuels Development: The Case of China. Iowa.

Dorosh, P., Pradesha, A., Raihan, S., Thurlow, J., 2020. COVID-19 in Bangladesh: Impacts on Production, Poverty and Food Systems presentaion Slide deck. https://www2.slideshare.net/ifpri/covid19-in-bangladesh-impacts-on-production-poverty-and-food-systems.

Doss, C.R., 2001. Designing agricultural technology for African women farmers: lessons from 25 years of experience. World Dev. 29 (12), 2075−2092.

Doss, C.R., Morris, M.L., 2001. How does gender affect the adoption of agricultural innovations? The case of improved maize technology in Ghana. Agric. Econ. 25, 27−39.

Duffy, P., Yamazaki, F., Zizza, A., 2012. Can dietary guidelines for Americans 2010 help trim America's waistline? Choice 27 (1).

Duflo, E., Dec. 2012. Women empowerment and economic development. J. Econ. Lit. 50 (4), 1051−1079.

Duflo, E., Rachel, Glennerster, Michael, Kremer, 2007. Using randomization in development economics research: a toolkit. In: Paul Schults, T., Strauss, J. (Eds.), Handbook of Development Economics, vol. 4. Elsevier Science Ltd., North Holland, pp. 3862–3895.

Duflo, E., Kremer, M., Robinson, J., 2011. Nudging farmers to use fertilizer: theory and experimental evidence from Kenya. Am. Econ. Rev. 101 (6), 2350–2390.

Dunn, W., 1994. Public Policy Analysis: An Introduction, second ed. Prentice Hall, Englewood Cliffs, NJ.

Dunn, R., 2010. The effect of fast-food availability on obesity: an analysis by gender, race, and residential location. Am. J. Agric. Econ. 92 (4), 1149–1164.

Dutko, P., Ver Ploeg, M., Farrigan, T., 2012. Characteristics and Influential Factors of Food Deserts. United States Department of Agriculture, Economic Research Service, Economic Research Report Number 140.

Dutta, P., Baruah, A., Deka, N., 2018. A study on post harvest losses in pulses in Nagaon district of Assam. Indian J. Econ. Dev. 14 (1), 191–194.

D'Ambrosio, C., Deutsch, J., Silber, J., 2011. Multidimensional approaches to poverty measurement: an empirical analysis of poverty in Belgium, France, Germany, Italy and Spain, based on the European panel. Appl. Econ. 43 (7–9), 951–961.

D'Souza, A., Jolliffe, D., 2012. Food security and wheat prices in Afghanistan: a distribution-sensitive analysis of household-level impacts. In: The World Bank, Policy Research Working Paper Series: 6024, Washington DC.

Economic Research Service, 2004. Rural America at a Glance, 2004. USDA. https://www.ers.usda.gov/publications/pub-details/?pubid=42579.

Elbers, C., Lanjouw, J.O., Lanjouw, P., 2001. Welfare in Villages and Towns: Microlevel Estimation of Poverty and Inequality. Mimeo. DECRG. World Bank, Washington, DC.

Engle, P.L., Zeitlen, M.F., 1996. Active feeding behavior compensates for low interest in food among young Nicaraguan children. J. Nutr. 126, 1808–1816.

Engle, P.L., Menon, P., Haddad, L., 1999. Care and nutrition: concepts and measurement. World Dev. 27 (8), 1309–1337.

Engle, P.L., Bentley, M., Pelto, G., 2000. The role of care in nutrition programmes: current research and a research agenda. Proc. Nutr. Soc. 59 (1), 25–35.

Erdman, L., Runge, C.F., 1990. American agricultural policy and the 1990 farm bill. Rev. Market. Agric. Econ. 58 (2–3), 109–126.

Ervin Paul, A., Vit, B., January 2019. Closing the rural-urban gap in child malnutrition: evidence from Paraguay, 1997–2012. Econ. Hum. Biol. 32, 1–10.

Espinoza-Delgado, J., Klasen, S., 2018. Gender and multidimensional poverty in Nicaragua: an individual based approach. World Dev. 110, 466–491.

Estefania, C., 2010. The economic and nutrition transition in Equatorial Guinea coincided with a double burden of over- and under nutrition. Econ. Hum. Biol. 8 (1), 80–87.

Europa, 2013. The G8 Summit in Lough Erne (UK) on 17–18 June 2013: The European Union's Role and Actions (Evidence from Pakistan. Asian Vegetable Research and Development Center, Discussion Paper). http://europa.eu/rapid/press-release_MEMO-13-551_en.htm?locale=en.

Fan, S., 2016. A nexus approach to food, water, and energy: sustainably meeting Asia's future food and nutrition requirements. Pak. Dev. Rev. 55 (4), 297–311. Winter.

Fan, S., Brzeska, J., 2011. The Nexus between Agriculture and Nutrition: Do Growth Patterns and Conditional Factors Matter? IFPRI Paper, Washington DC.

Fang, Di, et al., July 2019. WIC participation and relative quality of household food purchases: evidence from FoodAPS. South. Econ. J. 86 (1), 83–105.

FAO, 2000. Gender and Development Plan of Action 2002–2007 (Rome).

FAO, 2010. The State of Food Insecurity in the World, 2010. http://www.fao.org/fileadmin/templates/publications/pdf/i1683e_flyer.pdf.

FAO, 2012. The State of Food Insecurity in the World.
FAO, 2013. Statistical Yearbook — World Food and Agriculture.
FAO, IFAD and WFP, 2013. The State of Food Insecurity in the World 2013. The Multiple Dimensions of Food Security. FAO, Rome.
Favara, M., November 2012. 'United We Stand Divided We Fall' maternal social participation and children's nutritional status in Peru. In: The World Bank Policy Research Working Paper 6264.
Favara, M., May 2018. Maternal group participation and child nutritional status in Peru. Rev. Dev. Econ. 22 (2), 459—483.
Feeding America, 2011. Map the Meal Gap 2011: Preliminary Findings. A Report on County Level Food Insecurity and Food Cost in the United States in 2009. Nielson Company Report for Feeding America Organization. http://feedingamerica.org/hunger-in-america/hunger-studies/map-the-meal-gap.aspx.
Ferguson, E.L., Darmon, N., Fahmida, U., Fitriyanti, S., Harper, T.B., Premachandra, I.M., 2006. Design of optimal food-based complementary feeding recommendations and identification of key 'problem nutrients' using goal programming. J. Nutr. 136, 2399—2404.
Fertig, A., Glomm, G., Tchernis, R., 2009. The connection between maternal employment and childhood obesity: inspecting the mechanisms. Rev. Econ. Househ. 7 (3), 227—255.
Filmer, D., Pritchett, L., 1999. The Effect of Household Wealth on Educational Attainment: Demographic and Health Survey Evidence. World Bank Policy Research Paper No. 1980, Washington, DC.
Fitzpatrick, K., Greenhalgh-Stanley, N., Ver Ploeg, M., August 2019. Food deserts and diet-related health outcomes of the elderly. Food Pol. 87.
Floro, M.S., Bali, S.R., 2013. Food security, gender, and occupational choice among urban low-income households. World Dev. 42 (1), 89—99.
Food and Agriculture Organization (FAO), 1996. The Sixth World Food Survey. Rome, Italy.
Food and Agriculture Organization (FAO), 2001. Agricultural and Food Engineering Technologies Service: Dimensions of the Post-harvest Sector. Background Paper presented at FAO-Global Forum for Agricultural Research (GFAR) Regional Workshops, Rome.
Food and Agriculture Organization (FAO), April 2008. Soaring Food Prices: Facts, Perspectives, Impacts and Actions Required. Rome.
Food and Nutrition Assessment Technical Assistance (FANTA) Project, 2003. Anthropometric indicators measurement guide. In: For Improving Nutrition and Health; February 10—12, 2011; New Delhi, India. The Academy for Educational Development, Washington, DC.
Foster, J., Greer, J., Thorbecke, E., 1984. A class of decomposable poverty measures. Econometrica 52 (3), 761—765.
Freund, J.E., Walpole, R.E., 1986. Mathematical Statistics, fourth ed. Prentice Hall Inc, NJ.
Frongillo, E.A., Nguyen, P.H., Tina, S., Mahmud, Z., Bachera, A., Silvia, A., Menon, P., 2019. Nutrition interventions integrated into an existing maternal, neonatal, and child health program reduce food insecurity among recently delivered and pregnant women in Bangladesh. J. Nutr. 149 (1), 159—166.
Gabriel, A.H., Hundie, B., 2004. Farmers' post-harvest grain management choices under liquidity constraints and impending risks: implications for achieving food security objectives in Ethiopia. In: Ethiopian Civil Service College Working Paper. IFPRI's 2020 Vision Network for East Africa.
Gai, Y., Feng, L., 2012. Effects of federal nutrition program on birth outcomes. Atl. Econ. J. 40, 61—83.
Gaiha, R., Azam, Shafiul, Md, Annim, S., Imai, K.S., 2012. Agriculture, Markets and Poverty—A Comparative Analysis of Laos and Cambodia. Research Institute for Economics & Business Administration, Kobe University. Discussion Paper Series: DP2012-28.

Gajanan, S., Chandramohan, B.P., Babu Suresh, N., 2014. Willingness to pay for reliable electricity, in the presence of groundthwater depletion: A case-study of Tamil Nadu farmers. Int. J. Public Policy 10 (1/2/3).

Garcia, K., 2018. What's Next for the Enabling Environment for Agricultural Markets? AGRILINKS, Jul 31. Feed The Future. https://agrilinks.org/post/what-next-enabling-environment-agricultural-markets.

Garcia-Herreroa, Hoehna, D., Margalloa, M., Lasoa, J., Balab, A., Batlle-Bayerb, L., Fullanab, P., Vazquez-Rowec, I., Gonzaleza, M.J., Duráa, M.J., Sarabiaa, C., Abajasa, R., Amo-Setiena, F.J., Quiñonesa, A., Irabiena, A., Aldacoa, R., October 2018. On the estimation of potential food waste reduction to support sustainable production and consumption policies. Food Pol. 80, 24–38.

Garille, S.G., Gass, S.L., 2001. Stigler's diet problem revisited. Oper. Res. 49 (1), 1–13.

Garrett, J.L., Ruel, M.T., 1999. Are determinants of rural and urban food security and nutritional status different? Some insights from Mozambique. World Dev. 27 (11), 1955–1975.

Garrett, J., Ruel, M.T., 2003. Stunted Child-Overweight Mother Pairs: An Emerging Policy Concern? IFPRI, Washington, DC. FCND Discussion Paper No. 148.

Gayathri, R., Rajagopal, N., 2019. Multidimensional poverty index: an analysis of Tamil Nadu. Indian J. Econ. Dev. 15 (2), 177–185.

Genti, K., Mykerezi, E., Tanellari, E., 2011. Viability of organic production in rural counties: county and state-level evidence from the United States. J. Agric. Appl. Econ. 43 (3), 443–451.

Glauben, T., Herzfeld, T., Rozelle, S., Wang, X., 2012. Persistent poverty in rural China: where, why, and how to escape? World Dev. 40 (4), 784–795.

Glewwe, P., 1999. Why does mother's schooling raise child health in developing countries: evidence from Morocco. J. Hum. Resour. 34 (1), 124–159.

Glewwe, P., Koch, S., Linh, N.B., 2003. Child Nutrition, Economic Growth, and the Provision of Health Care Services in Vietnam in the 1990s. World Bank Working Paper No. 2776. Washington, DC.

Global Monitoring Report, 2012. Food Prices, Nutrition, and the Millennium Development Goals. International Monetary Fund. World Bank, Washington, D.C.

Goetz, S., 1993. Interlinked markets and the cash crop-food crop debate in land abundant tropical agriculture. Econ. Dev. Cult. Change 41, 343–361.

Goletti, F., Wolff, C., 1999. The Impact of Postharvest Research. IFPRI, Washington, DC. MSS Discussion Paper No. 29.

Govereh, J., Jayne, T.S., 2003. Cash cropping and food crop productivity: synergies or trade-offs? Agric. Econ. 28 (1), 39–50.

Govereh, J., Jayne, T.S., Nyoro, J., 1999. Smallholder Commercialization, Interlinked Markets and Food Crop Productivity: Cross-Country Evidence in Eastern and Southern Africa. Department of Agricultural Economics Working Paper. Michigan State University, East Lansing, MI.

Gragnolati, M., 1999. Children's Growth and Poverty in Rural Guatemala. World Bank Working Paper No. 2193, Washington, DC.

Greene, C., Dimitri, C., Lin, B.H., McBride, W., Oberholtzer, L., Smith, T.A., 2009. Emerging Issues in the U.S. Organic Industry. Economic Information Bulletin No. (EIB-55). USDA, Washington DC.

Greenwood, D.C., Cade, J.E., Draper, A., Barrett, J.H., Calvert, C., Greenhalgh, A., 2000. Seven unique food consumption patterns identified among women in the UK Women's cohort study. Eur. J. Clin. Nutr. 54, 314–320.

Gregory Christian, A., Smith, T.A., 2019. Salience, food security, and SNAP receipt. J. Policy Anal. Manag. 38 (1), 124–154. Winter.

Greve, J., et al., Oct. 2017. Fetal malnutrition and academic success: evidence from Muslim immigrants in Denmark. Econ. Educ. Rev. 60, 20–35. EBSCOhost. search.ebscohost.com/login. aspx?direct=true&db=ecn&AN=1684882&site=ehost-live.

Griliches, Z., 1957. Hybrid corn: an exploration in the economics of technological change. Econometrica 25 (4), 501–522.

Grillos, T., Aug. 2018. Women's participation in environmental decision-making: Quasi-experimental evidence from Northern Kenya. World Dev. 108, 115–130 (EBSCOhost, search.ebsco).

Grimaccia, E., Naccarato, A., May 2019. Food insecurity individual experience: a comparison of economic and social characteristics of the most vulnerable groups in the world. Soc. Indicat. Res. 143 (1), 391–410.

Grootaert, C., 1997. The determinants of poverty in Cote d'Ivoire in the 1980s. J. Afr. Econ. 6 (2), 169–196.

Guarnieri, E., Rainer, H., 2018. Female Empowerment and Male Backlash. EBSCOhost. http://www.cesifo-group.de/DocDL/cesifo1_wp7009.pdf.

Guldan, G.S., Zeitlin, M.F., Beiser, A.S., Super, C.M., Gershoff, S.N., Datta, S., 1993. Maternal education and child feeding practices in rural Bangladesh. Soc. Sci. Med. 36 (7), 925–935.

Gundersen, C., 2019a. Food assistance programmes and food insecurity in the United States. EuroChoices 18 (1), 56–60.

Gundersen, C., October 2019b. The right to food in the United States: the role of the supplemental nutrition assistance program (SNAP). Am. J. Agric. Econ. 101 (5), 1328–1336.

Gundersen, C., Ribar, D., 2011. Food insecurity and insufficiency at low levels of food expenditures. Rev. Income Wealth 57 (4), 704–726.

Gundersen, C., Ziliak James, P., March 2018. Food insecurity research in the United States: where we have been and where we need to go. Appl. Econ. Perspect. Pol. 40 (1), 119–135.

Gundersen, C., Lohman, B.J., Garasky, S., Stewart, S., Eisenmann, J., 2008. Food security, maternal stressors, and overweight among low-income US children: results from the national health and nutrition examination survey (1999–2002). Pediatrics 122 (3), 529–540.

Gundersen, C., Kreider, B., Pepper, J., 2011. The economics of food insecurity in the United States. Appl. Econ. Perspect. Policy 33 (3), 281–303. Autumn.

Gundersen, C., Brent, K., John, P., January 2012. The impact of the national school lunch program on child health: a Nonparametric bounds analysis. J. Econ. 166 (1), 79–91.

Gundersen, C., Brent, K., Pepper, J.V., July 2017. Partial identification methods for evaluating food assistance programs: a case study of the causal impact of SNAP on food insecurity. Am. J. Agric. Econ. 99 (4), 875–893.

Gupta, M., Avinash, K., July 2020. How COVID-19 may affect household expenditures in India: Unemployment shock, Household Consumption, and Transient Poverty. https://southasia.ifpri.info/2020/07/02/how-covid-19-may-affect-household-expenditures-in-india-unemployment-shock-household-consumption-and-transient-poverty/.

Gupta, S., Prabhu, P., Pinstrup-Andersen Per, October 2019. Women's empowerment and nutrition status: the case of iron deficiency in India. Food Pol. 88.

Gustavsson, J., Cederberg, C., Sonesson, U., van Otterdijk, R., Meybeck, A., 2011. Global Food Losses and Food Waste: Extent, Causes and Prevention. Food Agric. Org. United Nations (FAO), Rome.

Guthrie, J.F., Lin, B.-H., Frazao, E., 2002. Role of food prepared away from home in the American diet, 1977–78 versus 1994–96: changes and consequences. J. Nutr. Educ. Behav. 34 (3), 140–150.

Haddad, L., 1999. Women's status: levels, determinants, consequences for malnutrition, interventions, and policy. Asian Dev. Rev. 17 (12), 96–131.

Haddad, L., Kennedy, E., Sullivan, J., 1994. Choice of indicators for food security and nutrition monitoring. Food Pol. 19 (3), 329–343.

Haddad, L., Bhattarai, S., Immink, M., Kumar, S., 1996. Managing Interactions between Household Food Security and Preschooler Health. IFPRI, Washington, DC, 2020 Discussion Paper No. 16.

Haddad, L., Hoddinott, J., Alderman, H., 1997. Intrahousehold Resource Allocation: Policy Issues and Research Methods. Johns Hopkins University Press, Baltimore, MD.

Haeck, C., Lefebvre, P., 2016. A simple recipe: the effect of a prenatal nutrition program on child health at birth. Lab. Econ. 41 (Aug), 77–89.

Haines, M.R., Steckel, R.H., 2000. Childhood Mortality & Nutritional Status as Indicators of Standard of Living: Evidence from World War I Recruits in the United States. National Bureau of Economic Research Inc. NBER Historical Working Paper: 121.

Haines, M.,R., Craig, L.A., Weiss, T., 2011. Did African Americans experience the 'antebellum puzzle'? Evidence from the United States colored troops during the civil war. Econ. Hum. Biol. 9 (1), 45–55.

Hair, J.F., Black, B., Babin, B., Anderson, R.E., Tatham, R.L., 1998. Multivariate Data Analysis. Prentice Hall, NJ.

Hamburg, M., Young, P., 1994. Statistical Analysis for Decision Making, sixth ed. Harcourt Brace Publishers, New York.

Hamilton, L.C., 2006. Statistics with STATA, Version 12. Cengage Learning, USA.

Hanawa, H.P., Barkley, A., Chacon-Cascante, A., Kastens, T.L., 2012. The motivation for organic grain farming in the United States: profits, lifestyle, or the environment? J. Agric. Appl. Econ. 44 (2), 137–155.

Handa, S., Amber, P., August 2016. Is there catch-up growth? Evidence from three continents. Oxf. Bull. Econ. Stat. 78 (4), 470–500.

Hanson, K., Oliveira, V., 2012. How Economic Conditions Affect Par- Ticipation in USDA Nutrition Assistance Programs. United States Department of Agriculture, Economic Research Service. Economic Information Bulletin Number 100, September. www.ers.usda.gov.

Haroon, J., 2018. Exploring the relationship between Mother's empowerment and child nutritional status: an evidence from Pakistan. Pak. J. Appl. Econ. 28 (2), 189–211. Winter.

Hasan, R., Singh, H.P., 2017. Changing consumption pattern of agricultural commodities in Uttar Pradesh: an inter regional analysis. Indian J. Agric. Econ. 72 (3), 326–334. July–September.

Hausman, J., 1978. Specification tests in econometrics. Econometrica 46 (6), 1251–1271.

Hawkes, C., Ruel, M.T., Salm, L., Sinclair, B., Branca, F., 2020. Double-duty actions: seizing programme and policy opportunities to address malnutrition in all its forms. Lancet 395 (10218), 142–155.

Hazell, P., Norton, R.D., 1986. Mathematical Programming for Economic Analysis in Agriculture. Macmillan Publishing Company, New York.

Headey, D.D., Palloni, G., 2019. Water, sanitation, and child health: evidence from subnational panel data in 59 countries. Demography 56 (2), 729–752.

Headey Derek, D., Hoddinott, J.F., Park, S., 2016. Drivers of nutritional change in four South Asian countries: a dynamic observational analysis. Matern. Child Nutr. 12 (Suppl. 1), 210–218.

Headey, D., Martin, W., 2016. The impact of food prices on poverty and food. Ann. Rev. Resour. Econ. 8, 329–351.

Heady, D.D., 2013. The Impact of Global Food Crisis on Self-Assessed Food Security, Policy Research Paper 6329. The World Bank, Washington DC.

Heckman, J.J., Sergio, U., 2009. Comparing IV with structural models: What simple IV can and cannot identify. National Bureau of Economic Research. Working Paper 14706.

Hentschel, J., Lanjouw, J.O., Lanjouw, P., Poggi, J., 2000. Combining census and survey data to trace spatial dimensions of poverty: a case study of Ecuador. World Bank Econ. Rev. 14 (1), 147–165.

Hidrobo, M., Palloni, G., Aker, J.C., Gilligan, D.O., Ledlie, N., 2020. Paying for Digital Information: Assessing Farmers' Willingness to Pay for a Digital Agriculture and Nutrition Service in Ghana. International Food Policy Research Institute (IFPRI), Washington, DC. https://doi.org/10.2499/p15738coll2.133591. IFPRI Discussion Paper 1906.

Hill, R.V., Vigneri, M., 2009. Mainstreaming gender sensitivity in cash crop market supply chains. In: Background Paper Prepared for the State of Food and Agriculture 2010–11. FAO, Rome.

Hiroyuki, T., Nagarajan, L., 2012. Minor millets in Tamil Nadu, India: local market participation, on-farm diversity and farmer welfare. Environ. Dev. Econ. 17 (Special Issue 05), 603–632.

Hirvonen, K., Bai, Y., Headey, D., Masters, W.A., 2019. Affordability of the EAT–Lancet reference diet: a global analysis. Lancet Glob. Health 8 (1), e59–e66.

Hobcraft, J.N., McDonald, J.W., Rutstein, S.O., 1984. Socioeconomic factors in infant and child mortality: a cross-national comparison. Popul. Stud. 38 (2), 193–223.

Hoddinott, J., Maluccio, J., Behrman, J.R., Martorell, R., Melgar, P., Quisumbing, A.R., Ramirez-Zea, M., Stein, A.D., Yount, K.M., 2011. The Consequences of Early Childhood Growth Failure over the Life Course. International Food Policy Research Institute, Washington, DC. Discussion Paper 01073.

Hoddinott, J., Berhane, G., Gilligan, D.O., Kumar, N., Taffesse, A.S., 2012. The impact of Ethiopia's productive safety net programme and related transfers on agricultural productivity. J. Afr. Econ. 21 (5), 761–786.

Hoddinott, J.F., Ahmed, A., Roy, S., 2018. Randomized control trials demonstrate that nutrition-sensitive social protection interventions increase the use of multiple-micronutrient powders and iron supplements in rural pre-school Bangladeshi children. Publ. Health Nutr. 21 (9), 1753–1761.

Horn, J.L., 1965. A rationale and test for the number of factors in factor analysis. Psychometrika 30, 179–185.

Huang, J., Bouis, H., 1996. Structural Changes in the Demand for Food in Asia. IFPRI, Washington, DC, 2020 Discussion Paper No. 11.

Huang, F., Browne, B., 2017. Mortality forecasting using a modified continuous mortality investigation mortality projections model for China I: methodology and country-level results. Ann. Actuar. Sci. 11 (1), 20–45. EBSCOhost. http://journals.cambridge.org.pitt.idm.oclc.org/action/displayBackIssues?jid=AAS.

Huberty, C.J., Wisenbaker, J.M., Smith, J.C., 1987. Assessing predictive accuracy in discriminant analysis. Multivariate Behav. Res. 22, 307–329.

Hulme, D., Shepherd, A., 2003. Conceptualizing chronic poverty. World Dev. 31 (3), 403–423.

Hyman, G., Larrea, C., Farrow, A., 2005. Methods, results and policy implications of poverty and food security mapping assessments. Food Pol. 30, 453–460.

Ianchovichina, E., Loening, J., Wood, C., 2012. How Vulnerable are Arab Countries to Global Food Price Shocks? The World Bank, Policy Research Working Paper Series: 6018, Washington DC.

ICN (International Conference on Nutrition), 1992. Plan of Action for Nutrition. Rome.

IFPRI, 1995. More than Food Is Needed to Achieve Good Nutrition by 2020. 2020 Brief No. 25, Washington, DC.

IFPRI and CIAT, 2002b. Biofortification: Harnessing Agricultural Technology to Improve the Health of the Poor – Plant Breeding to Combat Micronutrient Deficiency (Washington, DC).

IFPRI and CSR, 2003. An assessment of the impact of ADMARC on welfare on Malawian households. In: Reforming the Malawi Agricultural Development and Marketing Corporation (ADMARC): Synthesis Report of the Poverty and Social Impact Analysis. World Bank, Washington, DC.

IFPRI, May 2008. High Food Prices: The what, Who, and How of Proposed Policy Actions. Policy Brief.

IFPRI, 2013a. A 2007 Social Accounting Matrix for China, Report by Zhang Yumei and Diao Xinshen. April 17, 2013.

IFPRI, 2013b. Bangladesh Integrated Household Survey (BIHS) 2011—2012, Authored by Ahmed, Akhter (IFPRI). April 24, 2013.

IFPRI, 2020. Jemimah Njuki looks to an inclusive future for small-holder farming in Africa. Editorial, Future Agricultures. https://www.future-agricultures.org/blog/jemimah-njuki-looks-to-an-inclusive-future-for-small-holder-farming-in-africa/.

Imai Katsushi, S., Annim Samuel, K., Gaiha, R., Kulkarni Veena, S., 2012. Does Women's Empowerment Reduce Prevalence of Stunted and Underweight Children, in Rural India? Research Institute for Economics & Business Administration. Kobe University. Discussion Paper Series: DP2012-11.

Imbens, G.W., Joshua, D.A., 1994. Identification and estimation of local average treatment effects. Econometrica 62 (2), 467—475.

Imbens, G.W., Kalyanaraman, K., 2012. Optimal bandwidth choice for the regression discontinuity estimator. Rev. Econ. Stud. 79, 933—959.

Imbens, W.G., 2010. Better LATE than nothing: Some comments on Deaton (2009) and Heckman and Urzua (2009). J. Econ. Lit. 48, 399—423.

Ingersent, K., 2003. The 2002 US Farm bill: a more positive assessment. EuroChoices 2 (1), 42—45. Spring.

International Livestock Research Institute, 1999. ILRI. International Livestock Research Institute, Nairobi, Kenya.

Investigators, S.E.N.E.C.A., 1996. Food patterns of elderly Europeans. Eur. J. Clin. Nutr. 50 (Suppl. 2), S86—S100.

Iqbal, A., Siddiqui, A., Zafar, M., 2020. A geographically disaggregated analysis of multidimensional poverty in Punjab. Int. J. Soc. Econ. 47 (3), 365—383.

Ir, P., et al., October 2019. Exploring the determinants of distress health financing in Cambodia. Health Pol. Plann. 34 (Suppl. 1), i26—37.

Islam, Md N., Al-Amin, Md, Sept. 2019. Life behind leaves: capability, poverty and social vulnerability of tea garden workers in Bangladesh. Labor Hist. 60 (5), 571—587.

Israel-Akinbo, S.O., Snowball, J., Fraser, G., 2018. An investigation of multidimensional energy poverty among South African low-income households. S. Afr. J. Econ. 86 (4), 468—487.

Jacobs, K., Sumner, D.A., 2002. The Food Balance Sheets of the Food and Agriculture Organization: A Review of Potential Ways to Broaden the Appropriate Uses of the Data. A review sponsored by FAO.

Jain, S., Koch, J., Jan. 2020. Crafting markets and fostering entrepreneurship within underserved communities: social ventures and clean energy provision in Asia. Enterpren. Reg. Dev. 32 (1—2), 176—196.

Jansen, E.C., Herran Oscar, F., Eduardo, V., December 2015. Trends and correlates of age at menarche in Colombia: results from a nationally representative survey. Econ. Hum. Biol. 19, 138—144.

Jayne, T.S., Jones, S., 1997. Food marketing and pricing policy in Eastern and Southern Africa: a survey. World Dev. 25 (9), 1505—1527.

Jehn, M., Brewis, A., 2009. Paradoxical malnutrition in mother-child pairs: untangling the phenomenon of over- and under-nutrition in underdeveloped economies. Econ. Hum. Biol. 7 (1), 28—35.

Jensen, R.E., 1969. A dynamic programming algorithm for cluster analysis. Oper. Res. 12, 1034—1057.

Jha, R., Kang, W., Nagarajan, H.K., Pradhan, K.C., 2012. Vulnerability as expected poverty in rural India, Australian National University, Australia South Asia Research Centre. ASARC Work. Papers, pp. 36.

Jones, S., 1996. Food Markets in Developing Countries: What Do We Know? Food Studies Group Working Paper No. 8. University of Oxford, Oxford.

Jordan, C., Jayne, T.S., 2013. Unpacking the meaning of 'market access': evidence from rural Kenya. World Dev. 41 (1), 245–264.

Joshi, A.R., 1994. Maternal schooling and child health: preliminary analysis of the intervening mechanisms in rural Nepal. Health Transit. Rev. 4, 1–26.

Justino, P., Litchfield, J., 2002. Poverty Dynamics in Rural Vietnam: Winners and Losers during Reform. University of Sussex Working Paper No. 10. Brighton, UK.

Kabubo-Mariara, Jane, N., Godfrey, K., Mwabu Domisiano, K., 2009. Determinants of children's nutritional status in Kenya: evidence from demographic and health surveys. J. Afr. Econ. 18 (3), 363–387.

Kabunga, N.S., Ghosh, S., Webb, P., 2017. Does ownership of improved dairy cow breeds improve child nutrition? A pathway analysis for Uganda. PLoS One 12 (11).

Kader, A., 2003. A perspective on postharvest horticulture (1978–2003). Hortscience 38 (5), 1004–1008.

Kadzandira, J.M., 2003. African Food Crisis: The Relevance of Asian Models. Unpublished Master's Thesis. Lund University, Sweden.

Karpinski, A., 2003. Planned Contrasts and Post-hoc Tests for One-Way ANOVA. Temple University Manuscript, Philadelphia, PA. Available at: http://astro.temple.edu/~andykarp/psych 522524/06_contrasts2.pdf.

Kassie, M., Zikhali, P., Manjur, K., Edwards, S., 2009. Adoption of Organic Farming Techniques: Evidence from a Semi-arid Region of Ethiopia, Resources for the Future (Discussion Papers).

Katz, E., 1995. Gender and trade within the household: observations from rural Guatemala. World Dev. 23 (2), 327–342.

Kelebe, H.E., Ayimut, K.M., Berhe, G.H., Hintsa, K., August 2017. Determinants for adoption decision of small scale biogas technology by rural households in Tigray, Ethiopia. Energy Econ. 66, 272–278.

Kennedy, E., Garcia, M., 1993. Effects of selected policies and porgrams on women's health and nutritional status. IFPRI Report 45, Washington DC.

Kennedy, E., Haddad, L., 1994. Are pre-schoolers from female-headed households less malnourished? A comparative analysis of results from Ghana and Kenya. J. Dev. Stud. 30 (3), 680–695.

Kennedy. (Eds.), 1993. Johns Hopkins University Press and IFPRI, Washington, DC. Kennedy, E., Garcia, M. (Eds.). Effects of Selected Policies and Programs on Women's Health and Nutritional Status. IFPRI Report, Washington, DC.

Khan, R.E.A., Azid, T., 2011. Malnutrition in primary school-age children: a case of urban and slum areas of Bahawalpur, Pakistan. Int. J. Soc. Econ. 38 (9–10), 748–766.

Khanna, R.A., et al., 2019. Comprehensive energy poverty index: measuring energy poverty and identifying micro-level solutions in South and Southeast Asia. Energy Pol. 132 (Sept), 379–391.

Kherallah, M., Delgado, C., Gabre-Madhin, E., Minot, N., Johnson, M., 2002. Reforming Agricultural Markets in Africa. The Johns Hopkins University Press, Baltimore and London.

Kiiza, B., Pederson, G., 2012. ICT-based market information and adoption of agricultural seed technologies: insights from Uganda. Telecommun. Pol. 36 (4), 253–259.

Kim, H., July 2019. In the wake of conflict: the long-term effect on child nutrition in Uganda. Oxf. Dev. Stud. 47 (3), 336–355.

Kim, H., 2019. Beyond monetary poverty analysis: the dynamics of multidimensional child poverty in developing countries. Soc. Indicat. Res. 141 (3), 1107–1136.

Kim, S.S., Nguyen Phuong, H., Mai, T.L., Silvia, A., Menon, P., Frongillo, E.A., 2020. Different combination of behavior change interventions and frequency of interpersonal contacts are associated with infant and young child feeding practices in Bangladesh, Ethiopia, and Viet Nam. Curr. Dev. Nutr. 4 (2).

Kishor, S., 2000. Empowerment of women in Egypt and links to the survival and health of their infants. In: Presser, H., Sen, G. (Eds.), Women's Empowerment and Demographic Processes, Ch. 6. Oxford University Press, Oxford.

Kline, P., 1994. An Easy Guide to Factor Analysis. Sage Publications, Thousand Oaks.

Koomson, I., Villano, R.A., Hadley, D., 2020. Effect of financial inclusion on poverty and vulnerability to poverty: evidence using a multidimensional measure of financial inclusion. Soc. Indicat. Res. 149 (2), 613–639.

Kreider, B., Pepper, J.V., Gundersen, C., Jolliffe, D., September 2012. Identifying the effects of SNAP (food stamps) on child health outcomes when participation is endogenous and misreported. J. Am. Stat. Assoc. 107 (499), 958–975.

Kreider, B., et al., April 2016. Identifying the effects of WIC on food insecurity among infants and children. South. Econ. J. 82 (4), 1106–1122.

Kumar, S.K., 1994. Adoption of Hybrid Maize in Zambia: Effects on Gender Roles, Food Consumption, and Nutrition. Research Report No. 100. IFPRI, Washington, DC.

Kumar, P., 2006. Contract farming through agribusiness firms and state corporation: a case study in Punjab. Econ. Polit. Wkly. 52 (30), A5367–A5375.

Kumar, P., Joshi, P.K., 2013. Household consumption pattern and nutritional security among poor rural households: impact of MGNREGA. Agric. Econ. Res. Rev. 26 (1), 73–82.

Kumar, N., Quisumbing, A., 2012. Inheritance practices and gender differences in poverty and well-being in rural Ethiopia. Dev. Pol. Rev. 30 (5), 573–595.

Kumar, T.K., Holla, J., Guha, P., 2008. Engel curve method for measuring poverty. Econ. Polit. Wkly. 43 (30), 115–123.

Kumar, A., Parappurathu, S., Babu, S.C., Betne, R., 2016. Public distribution system in Bihar, India: implications for food security. J. Agric. Food Inf. 17 (4), 300–315.

Kumar, A., Thapa, G., Mishra, A.K., Joshi, P.K., January 2020. Assessing Food and Nutrition Security in Nepal: Evidence from Diet Diversity and Food Expenditure Patterns. Food Security, pp. 1–28.

Laderchi, C.R., Saith, R., Stewart, F., 2003. Does it Matter that We Don't Agree on the Definition of Poverty? A Comparison of Four Approaches. Queen Elizabeth House Working Paper No. 107. University of Oxford, Oxford.

Lambrecht, I., Vanlauwe, B., Maertens, M., December 2016. Agricultural extension in eastern democratic republic of Congo: does gender matter? Eur. Rev. Agric. Econ. 43 (5), 841–874.

Larson, D.F., Lampietti, J., Gouel, C., Cafiero, C., Roberts, J., 2012. Food Security and Storage in the Middle East and North Africa. The World Bank, Policy Research Working Paper Series: 6031, Washington DC.

Lee, D.S., Lemieux, T., 2010. Regression discontinuity designs in economics. J. Econ. Lit. 48, 281–355.

Lee, H., Munk, T., 2008. Using regression discontinuity design for program evaluation. American Statistical Association, Joint Statistical Meeting, Denver, CO, pp. 3–7.

Leguizamon, J.S., Leguizamon, S., Jan. 2018. Health insurance subsidies and the expansion of an implicit marriage penalty: a regional comparison of various means-tested programmes. Appl. Econ. Lett. 25 (2), 130–135.

Leite, M.L.C., Nicolosi, A., Cristina, S., Hauser, W.A., Pugliese, P., Nappi, G., 2003. Dietary and nutritional patterns in an elderly rural population in Northern and Southern Italy: a cluster analysis of food consumption. Eur. J. Clin. Nutr. 57 (12), 1514—1521.

Lentz Erin, C., April 2018. Complicating narratives of women's food and nutrition insecurity: domestic violence in rural Bangladesh. World Dev. 104, 271—280.

Leroy, J.L., Olney Deanna, K., Ruel Marie, T., 2019. PROCOMIDA, a food-assisted maternal and child health and nutrition program, contributes to postpartum weight retention in Guatemala: a cluster-randomized controlled intervention trial. J. Nutr. 149 (12), 2219—2227.

Levine, R.A., Dexter, E., Velasco, P., Sara, L.V., Arun, J., Kathleen, S., Medardo, T.-U., 1994. Maternal literacy and health care in three countries: a preliminary report. Health Trans. Rev. 4, 186—191.

Li, G., 2003. The Impact of the One-Child Policy on Child-Wellbeing and Gender Differential. *Center for Research on Families Working Paper No. 2*. Seattle, Washington.

Lilja, N., Sanders, J.H., 1998. Welfare impacts of technological change on women in Southern Mali. Agric. Econ. 19, 73—79.

Limaye, K., Dharmendra, C., Choudhury, N., July 2020. Lockdown and Rural Distress: Highlights from Phone Surveys of 5,000 Households in 12 Indian States. https://southasia.ifpri.info/2020/07/13/lockdown-and-rural-distress-highlights-from-phone-surveys-of-5000-households-in-12-indian-states/.

Liu, X., June 2018. Public Policy, Health Insurance, and Labor Markets and Demographic Outcomes Among Young Adults. University at Albany.

Lone, Ahmad Parvaze and Rather, Ahmad, N., 2012. Poverty and food security nexus in India. Int. J. Res. Commer. Econ. Manag. 2 (5), 129—132.

Losa, F.B., Soldini, E., 2011. The similar faces of Swiss working poor: an empirical analysis across Swiss regions using logistic regression and classification trees. Schweizerische Zeitschrift fur Volkswirtsch, aft und Statistik/Swiss J. Econ. Stat. 147 (1), 17—44.

Lowe, M., Roth, B., June 2020. India's Supply Chains Unchained. https://southasia.ifpri.info/2020/06/18/indias-supply-chains-unchained/.

Lowry, R., 2003. Concepts and Applications of Inferential Statistics. Vassar College, Poughkeepsie, NY, USA. Available at: http://faculty.vassar.edu/lowry/webtext.html.

Ludwig, J., Miller, D.L., 2007. Does Head Start improve children's life chances? Evidence from a regression discontinuity design. Q. J. Econ. 122, 159—208.

Lutafali, S., et al., 2016. Expanding role of microfinance institutions to combat multidimensional poverty. Int. J. Ecol. Econ. Stat. 37 (2), 1—10.

Mackinnon, J., 1995. Health as an Informational Good: The Determinants of Child Nutrition and Mortality during Political and Economic Recovery in Uganda. Centre for the Study of African Economies Working Paper No.WPS/95-9. University of Oxford, UK.

Maduekwe, E., de Vries, W.T., Buchenrieder, G., 2020. Measuring human recognition for women in Malawi using the Alkire Foster method of multidimensional poverty counting. Soc. Indicat. Res. 147 (3), 805—824.

Mahadevan, R., Jayasinghe, M., 2020. Examining multidimensional poverty in Sri Lanka: transitioning through post war conflict. Soc. Indicat. Res. 149 (1), 15—39.

Mahmood, R., Shah, A., 2017. Deprivation counts: an assessment of energy poverty in Pakistan. Lahore J. Econ. 22 (1), 109—132.

Mahmood, T., Yu, X., Klasen, S., 2019. Do the poor really feel poor? Comparing objective poverty with subjective poverty in Pakistan. Soc. Indicat. Res. 142 (2), 543—580.

Mahrt, K., Rossi, A., Salvucci, V., Tarp, F., 2018. Multidimensional poverty of children in Mozambique. World Institute for Development Economic Research (UNU-WIDER). WIDER Working Paper Series: 108.

Maitra, P., Rammohan, A., Ray, R., Robitaille, M.C., 2013. Food consumption patterns and malnourished Indian children: is there a link? Food Pol. 38, 70–81.

Malapit Hazel Jean, L., Quisumbing Agnes, R., April 2015. What dimensions of women's empowerment in agriculture matter for nutrition in Ghana? Food Pol. 52, 54–63.

Marco, S., 2011. Upward and onward: high-society American women eluded the antebellum puzzle. Econ. Hum. Biol. 9 (2), 165–171.

Marivoet, W., Ulimwengu, J., Sedanoc, F., 2019. Spatial typology for targeted food and nutrition security interventions. World Dev. 120, 62–75.

Masanjala, W.H., 2006. Cash crop liberalization and poverty alleviation in Africa: evidence from Malawi. Agric. Econ. 35, 231–240.

Masters, W.A., Rosettie Katherine, L., Sarah, K., Goodarz, D., Webb, P., Dariush, M., Namukolo, C., George, M., 2018. Designing programs to improve diets for maternal and child health: estimating costs and potential dietary impacts of nutrition-sensitive programs in Ethiopia, Nigeria, and India. Health Pol. Plann. 33 (4), 564–573.

Maxwell, D.G., 1996. Measuring food insecurity: the frequency and severity of coping strategies. Food Pol. 21 (3), 291–303.

Maxwell, S., Frankenberger, T., 1992. Household Food Security: Concepts, Indicators and Measurements: A Technical Review. UNICEF and IFAD, New York and Rome.

McBride, W.D., Greene, C., 2009. Characteristics, Costs, and Issues for Organic Dairy Farming, Economic Research Report No. (ERR-82), November 2009.

McEwan, P.J., Shapiro, J.S., 2008. The benefits of delayed primary school enrollment: Discontinuity estimates using exact birth dates. J. Human Resour. Winter 43 (1), 1–29.

Mcguire, J.W., 2006. Basic health care provision and under-5 mortality: a cross-national study of developing countries. World Dev. 34 (3), 405–425.

McMurray, C., 1996. Cross-sectional anthropometry: what can it tell us about the health of young children? Health Trans. Rev. 6, 147–168.

Megan, S., Barrett Christopher, B., July 2017. Review: food loss and waste in Sub-Saharan Africa. Food Pol. 70, 1–12.

Melesse, M.B., et al., October 2018. Joint land certification programmes and women's empowerment: evidence from Ethiopia. J. Dev. Stud. 54 (10), 1756–1774. EBSCOhost. http://www-tandfonline-com.pitt.idm.oclc.org/loi/fjds20.

Mendoza Jr., C.B., Cayonte, D.D.D., Leabres, M.S., Manaligod, L.R.A., 2019. Understanding multidimensional energy poverty in the Philippines. Energy Pol. 133.

Menon, P., 2012. Childhood undernutrition in South Asia: perspectives from the field of nutrition. CESifo Econ. Stud. 58 (2), 274–295.

Menon, P., et al., Mar. 2020. Lessons from using cluster-randomized evaluations to build evidence on large-scale nutrition behavior change interventions. World Dev. 127. EBSCOhost. search.ebscohost.com/login.aspx?direct=true&db=ecn&AN=1815566&site=ehost-live.

Mensch, B.S., Lentzner, H., Preston, S., 1985. Socio-economic Differentials in Child Mortality in Developing Countries. United Nations Department of International Economic and Social Affairs Monograph No. ST/ESA/SER.A/97.

Metz, M., 2000. Methods for Analysis and Assessment of Aggregate Food Deficits. Unpublished Background Papers for the Training Course on Food and Nutrition Security in the Context of Poverty Alleviation and Disaster and Crisis Mitigation and Response. InWEnt, Feldafing.

Meyer, D.F., Keyser, E., Oct. 2016. Validation and testing of the Lived Poverty Index Scale (LPI) in a poor South African community. Soc. Indicat. Res. 129 (1), 147–159.

Miguel, E., Michael, K., 2004. Worms: Identifying impacts on education and health in the presence of treatment externalities. Econometrica 72 (1), 159–217.

Millen, B.E., Quatromoni, P.A., Copenhafer, D.L., Demissie, S., O'Horo, C.E., D'Agostino, R.B., 2001. Validation of a dietary pattern approach for evaluating nutritional risk: the Framingham Nutrition Studies. J. Am. Diet Assoc. 101 (3), 187–194.

Miller Laurie, C., Sumanta, N., Joshi, N., Shrestha, M., Shailes, N., Mahendra, L., Thorne-Lyman, A.L., February 2020. Diet Quality over Time Is Associated with Better Development in Rural Nepali Children. Maternal and Child Nutrition.

Mills, B.F., Mykerezi, E., 2009. Chronic and transient poverty in the Russian federation. Post Commun. Econ. 21 (3), 283–306.

Minten, B., Barrett, C.B., 2008. Agricultural technology, productivity, and poverty in Madagascar. World Dev. 36 (5), 797–822.

Mitra, S., 2018. Re-assessing "trickle-down" using a multidimensional criteria: the case of India. Soc. Indicat. Res. 136 (2), 497–515.

Mittal, S., 2008. Demand-supply Trends and Projections of Food in India. Indian Council for Research on International Economic Relations Working Paper No. 209, New Delhi, India.

Moffitt, R.A., 2003. Means-tested transfer programs in the United States. In: NBER Conference Report Series. University of Chicago Press, Chicago and London.

Morton, L.W., Bitto, E.A., Oakland, Mary Jane, Sand, M., Spring 2008. Accessing food resources: rural and urban patterns of giving and getting food. Agric. Hum. Val. 25 (1), 107–119.

Motoyuki, G., 2008. Agricultural multifunctionality and village viability: a case study from Japan. In: OECD, Multifunctionality in Agriculture: Evaluating the Degree of Jointness, and Policy Implications. OECD Publishing.

Murugani, V.G., Thamaga-Chitja, J.M., July 2019. How does women's empowerment in agriculture affect household food security and dietary diversity? The case of rural irrigation schemes in Limpopo Province, South Africa. Agrekon 58 (3), 308–323.

Muthoni, F., Guo, Z., Mateete, B., Haroon, S., Kizito, F., Frederick, B., Irmgard, H.-Z., 2017. Sustainable recommendation domains for scaling agricultural technologies in Tanzania. Land Use Pol. 66, 34–48.

Mykerezi, E., Bradford, M., April 2009. On Intra-annual Poverty in the U.S: Preva-Lence, Causes and Response to Food Stamp Program Use. World Bank Staff Papers, Staff Paper P09-7, Washington DC.

Nagelkerke, N.J.D., 1991. A note on a general definition of the coefficient of determination. Biometrika 78 (3), 691–692.

Najera Catalan, H.E., Gordon, D., 2020. The importance of reliability and construct validity in multidimensional poverty measurement: an illustration using the Multidimensional Poverty Index for Latin America (MPI-LA). J. Dev. Stud. 56 (9), 1763–1783. http://www.tandfonline.com/loi/fjds20.

Narayanan, S., July 2020. How India's Agrifood Supply Chains Fared during the COVID-19 Lockdown, from Farm to Fork. https://www.ifpri.org/blog/how-indias-agrifood-supply-chains-fared-during-covid-19-lockdown-farm-fork.

Narayanan, S., Erin, L., Fontana, M., De Anuradha, Kulkarni, B., December 2019. Developing the women's empowerment in nutrition index in two states of India. Food Pol. 89.

National Academy of Sciences, 1995. Estimated Mean Per Capita Energy Requirements for Plan- Ning Emergency Food Aid Rations. Committee on Nutrition. US National Academy of Sciences, 1995. National Academy Press, Washington, DC.

Ndiritu, S.W., Ruhinduka, R.D., 2019. Climate variability and post-harvest food loss abatement technologies: evidence from Rural Tanzania. Stud. Agric. Econ. 121 (1), 30–40.

Nguyen, T.T., et al., Mar. 2020. Health shocks and Natural resource extraction: a Cambodian case study. Ecol. Econ. 169.

Nguyen Phuong, H., Kim, S.S., Nguyen, T.T., Nemat, H., Tran, L.M., Alayon, S., Ruel Marie, T., Rawat, R., Frongillo Edward, A., Menon, P., 2016. Exposure to mass media and interpersonal counseling has additive effects on exclusive breastfeeding and its psychosocial determinants among Vietnamese mothers. Matern. Child Nutr. 12 (4), 713−725.

Nguyen Phuong, H., Rasmi, A., Headey Derek, D., Mai, T.L., Ruel Marie, T., Menon, P., 2018. Progress and inequalities in infant and young child feeding practices in India between 2006 and 2016. Matern. Child Nutr. 14 (S4).

Nobuhiko, F., 2000. A note on the analysis of female headed households in developing countries. In: Technical Bulletin of Faculty of Horticulture, vol. 54. Chiba University, pp. 125−138.

Novak, L., August 2014. The impact of access to water on child health in Senegal. Rev. Dev. Econ. 18 (3), 431−444.

Novella, R., January 2019. Parental education, gender preferences and child nutritional status in Peru. Oxf. Dev. Stud. 47 (1), 29−47.

Nyariki, D.M., 2011. Farm size, modern technology adoption, and efficiency of small holdings in developing countries: evidence from Kenya. J. Develop. Area. 45 (1), 35−52.

OECD, 2013. Global Food Security: Challenges for the Food and Agricultural System. OECD Publishing. https://doi.org/10.1787/9789264195363-en.

Ogutu, Ochieng, S., Ochieng, D.O., Qaim, M., August 2020. Supermarket contracts and smallholder farmers: implications for income and multidimensional poverty. Food Pol. 95, 101940. https://doi.org/10.1016/j.foodpol.2020.101940.

Olshansky, S.J., Ault, A.B., 1986. The fourth stage of the epidemiologic transition: the age of delayed degenerative diseases. Milbank Mem. Fund. Q. 64, 355−391.

Olson, C.M., 2004. Factors protecting against and contributing to food insecurity among rural families. Fam. Econ. Nutr. Rev. 16 (1), 12−20.

Olusegun, F., Mulubrhan, A., George, M., Dare, A., Adebayo, O., 2019. Mother's nutrition-related knowledge and child nutrition outcomes: empirical evidence from Nigeria. PLoS One 14 (2).

Omotilewa, O.J., Ricker-Gilbert, J., Ainembabazi, J.H., Shively, G.E., November 2018. Does improved storage technology promote modern input use and food security? Evidence from a randomized trial in Uganda. J. Dev. Econ. 135, 176−198.

Oparinde, A., Tahirou, A., Babatima, M.D., Bamire Adebayo, S., 2017. Will farmers intend to cultivate Provitamin A genetically modified (GM) cassava in Nigeria? Evidence from a k-means segmentation analysis of beliefs and attitudes. PLoS One 12 (7).

Opsomer, J.D., Jensen, H.H., Nusser, S.M., Drignei, D., Amemiya, Y., 2002. Statistical Considerations for the USDA Food Insecurity Index. Iowa State University, Department of Economics, Staff General Research Papers.

Orr, A., 2000. Green gold? Burley tobacco, smallholder agriculture, and poverty alleviation in Malawi. World Dev. 28 (2), 347−363.

Oum, S., 2019. Energy poverty in the Lao PDR and its impacts on education and health. Energy Pol. 132 (Sept), 247−253.

Overseas Development Institute (ODI), 2008. Rising Food Prices: A Global Crisis. ODI Briefing Paper, UK.

Ozor, N., Igbokwe, E.M., 2007. Roles of agricultural biotechnology in ensuring adequate food security in developing societies. Afr. J. Biotechnol. 6 (14), 1597−1602.

Ozughalu, U.M., Ogwumike, F.O., 2019. Extreme energy poverty incidence and determinants in Nigeria: a multidimensional approach. Soc. Indicat. Res. 142 (3), 997−1014.

Palma, A.M., Knutson, R.D., 2012. Implementing dietary goals and guide- lines. Choice (16), 1−6.

Paolisso, M.J., Hallman, K., Haddad, L., Regmi, S., 2001. Does cash crop adoption detract from childcare provision? Evidence from rural Nepal. In: FCND Discussion Paper No. 109. IFPRI, Washington, DC. Patterns and Conditional Factors Matter? 2020 Conference. Leveraging Agriculture.

Parappurathu, S., Kumar, A., Bantilan, C., Joshi, P.K., 2019. Household-level food and nutrition insecurity and its determinants in eastern India. Curr. Sci. 117 (1), 71–79.

Parra-Lopez, C., Groot, J.C.J., Carmona-Torres, C., Rossing, W.A.H., 2008. Integrating public demands into model-based design for multifunctional agriculture: an application to intensive Dutch dairy landscapes. Ecol. Econ. 67 (4), 538–551.

Paul, D., June 2020. Enhancing Household Food Security in a Pandemic: Policy Options for Rice Markets and Safety Nets in Bangladesh. https://southasia.ifpri.info/2020/06/25/enhancing-household-food-security-in-a-pandemic-policy-options-for-rice-markets-and-safety-nets-in-bangladesh/.

Paulson, N.D., Schnitkey, G.D., 2011. Policy concerns of midwestern grain producers for the 2012 farm bill. Am. J. Agric. Econ. 94 (2), 515–521.

Pelletier, D.L., Frongillo, E.A., Habicht, J.P., 1993. Epidemiologic evidence for a potentiating effect of malnutrition on child mortality. Am. J. Publ. Health 83, 1130–1133.

Pelletier, D.L., Low, J.W., Johnson, F.C., Msukwa, L.A.H., 1994. Child anthropometry and mortality in Malawi: testing for effect modifications by age and length of follow- up and confounding socio-economic factors. J. Nutr. 124, 2082S–2105S.

Penders, C.L., Staatz, J.M., Tefft, J.F., 2000. How Does Agricultural Development Affect Child Nutrition in Mali? Policy Synthesis Brief No. 51. Michigan State University, East Lansing, MI, USA.

Peters, P.E., Herrera, M.G., 1994. Tobacco cultivation, food production, and nutrition among smallholders in Malawi. In: von Braun, J., Kennedy, E. (Eds.), Agricultural Commercialization, Economic Development, and Nutrition. Johns Hopkins University Press and IFPRI, Washington, DC.

Pham, Trang, April 2018. The capability approach and evaluation of community-driven development programs. J. Hum. Dev. Capab. 19 (2), 166–180.

Pham, A.T.Q., Mukhopadhaya, P., Vu, H., 2020. Targeting administrative regions for multi-dimensional poverty alleviation: a study on Vietnam. Soc. Indicat. Res. 150 (1), 143–189.

Phoumin, H., Kimura, F., September 2019. Cambodia's energy poverty and its effects on social wellbeing: empirical evidence and policy implications. Energy Pol. 132, 283–289.

Pingali, P.L., Rosegrant, M.W., 1998. Supplying wheat for Asia's increasingly westernized diets. Am. J. Agric. Econ. 80 (5), 954–959.

Pingali, P.L., Stringer, R., 2003. Food Security and Agriculture in the Low Income Food Deficit Countries: 10 Years after the Uruguay Round. ESA Working Paper No. 03–18. FAO, Rome.

Pinstrup-Andersen, P., 2012. The Food System and Its Interaction with Human Health and Nutrition.

Pongou, R., Ezzati, M., Salomon, J.A., 2006. Household and community socioeconomic and environmental determinants of child nutritional status in Cameroon. BMC Publ. Health 6, 98–117.

Popkin, B.M., 2003. The nutrition transition in the developing world. Dev. Pol. Rev. 21 (5–6), 581–597.

Posel, D., 2001. Who are the heads of household, what do they do, and is the concept of headship useful? An analysis of headship in South Africa. Dev. South Afr. 18 (5), 651–670.

Prakash, M., Jain, K., 2016. Inequalities among malnourished children in India: a decomposition analysis from 1992–2006. Int. J. Soc. Econ. 43 (6), 643–659.

Pritchett, L., Suryahadi, A., Sumarto, S., 2000. Quantifying Vulnerability to Poverty: A Proposed Measure with Application to Indonesia. SMERU Working Paper, Indonesia.

Qiao, Y., Martin, F., Cook, S., He, X., Halberg, N., Scott, S., 2018. Certified Organic Agriculture as an Alternative Livelihood Strategy for Small-Scale Farmers in China: A Case Study in Wanzai County, Jiangxi Province. Ecol. Econ. 145, 301—307.

Qiao, Y., Martin, F., He, X., Zhen, H., Pan, X., January 2019. The changing role of local government in organic agriculture development in Wanzai County, China. Can. J. Dev. Stud. 40 (1), 64—77.

Quattrochi, J., et al., Mar. 2020. Contributions of experimental approaches to development and poverty alleviation: field experiments and humanitarian assistance. World Dev. 127. EBSCOhost. search.ebscohost.com/login.aspx?direct=true&db=ecn&AN=1815574&site=ehost-live.

Quisumbing, A.R., Otsuka, K., 2001. Land, Trees and Women: Evolution of Land Tenure Institutions in Western Ghana and Sumatra. Research Report No. 121. IFPRI, Washington, DC.

Ramachandran, A., Snehalatha, C., 2010. Rising burden of obesity in Asia. J. Obes. https://doi.org/10.1155/2010/868573. Article ID 868573.

Ramezani, C.A., Roeder, C., 1995. Health knowledge and nutritional adequacy of female heads of households in the United States. J. Consum. Aff. 29 (2), 381—402 (Winter).

Rana Pooja, P., Kshetri Indra, D., July 2020. Adapting a Large Multi-Sectoral Nutrition Program during the COVID-19 Pandemic: The Story of Suaahara II from Nepal. https://southasia.ifpri.info/2020/07/20/adapting-a-large-multi-sectoral-nutrition-program-during-the-covid-19-pandemic-the-story-of-suaahara-ii-from-nepal/.

Rao, N.D., Jihoon, M., July 2018. Decent living standards: material prerequisites for human wellbeing. Soc. Indicat. Res. 138 (1), 225—244.

Rasaily, R., Saxena, N.C., Pandey, S., et al., 2020. Effect of home-based newborn care on neonatal and infant mortality: A cluster randomised trial in India. BMJ Glob. Health 5, e000680. https://doi.org/10.1136/bmjgh-2017-000680.

Raschke, C., 2012. Food stamps and the time cost of food preparation. Rev. Econ. Househ. 10 (2), 259—275.

Rashad, A.S., Mesbah, F.S., Jan. 2019. Does maternal employment affect child nutrition status? New evidence from Egypt. Oxf. Dev. Stud. 47 (1), 48—62. EBSCOhost. http://www-tandfonline-com.pitt.idm.oclc.org/loi/cods20.

Ravallion, M., 1994. Poverty Comparisons. Harwood Academic Publishers, Chur, Switzerland.

Ravallion, M., 1998. Poverty Lines in Theory and Practice. LSMS Working Paper No. 133. World Bank, Washington, DC.

Ravallion, M., Bidani, B., 1994. How robust is a poverty profile? World Bank Econ. Rev. 8, 75—102.

Ravallion, M., March 2012. Fighting poverty one experiment at a time: poor economics: a radical rethinking of the way to fight global poverty: review essay. J. Econ. Lit. 50 (1), 103—114. EBSCOhost. http://www.aeaweb.org.pitt.idm.oclc.org/jel/index.php.

Ray, R., 2007. Changes in food consumption and the implications for food security and undernourishment: India in the 1990s. Dev. Change 38 (2), 321—343.

Ray, D., Subramanian, S., 2020. India's Lockdown: An Interim Report. NBER Working Paper No. 27282, May 2020.

Reardon, T., Kelly, V., Crawford, E., et al., 1997. Promoting sustainable intensification and productivity growth in Sahel agriculture after macroeconomic policy reform. Food Pol. 22 (4), 317—328.

Reis, M., 2012. Food insecurity and the relationship between household income and children's health and nutrition in Brazil. Health Econ. 21 (4), 405—427.

Ren, Y., Zhang, Y., Jens-Peter, L., Thomas, G., 2018. Food consumption among income classes and its response to changes in income distribution in rural China, China Agric. Econ. Rev. 10 (3), 406—424.

Rencher, A.C., 2002. Methods of Multivariate Analysis. John Wiley and Sons Inc., New York.
Resnick, D., April 2020. COVID-19 Lockdowns Threaten Africa's Vital Informal Urban Food Trade. https://gssp.ifpri.info/2020/04/16/covid-19-lockdowns-threaten-africas-vital-informal-urban-food-trade/.
Resnick, D., 2020. COVID-19 lockdowns threaten Africa's vital informal urban food trade. In: J., Swinnen, J., McDermott (Eds.), COVID-19 and global food security, Part Four: Food trade. International Food Policy Research Institute (IFPRI), Washington, DC, pp. 73–74. https://doi.org/10.2499/p15738coll2.133762_16. Chapter 16.
Rhoe, V., Babu, S., Reidhead, W., 2008. An analysis of food security and poverty in Central Asia – case study from Kazakhstan. J. Int. Dev. 20, 452–465.
Ricker-Gilbert, J., Jones, M., January 2016. Does storage technology affect adoption of improved maize varieties in Africa? Insights from Malawi's input subsidy program. Food Pol. 50, 92–105.
Riely, F., January 1999. Food Security Indicators and Framework for Use in the Monitoring and Evaluation of Food Aid Programs. FANTA study.
Roberts, I., Jotzo, J., 2002. 2002 US farm bill: an Australian perspective on its impact. Aust. Commod. Forecast. Iss. 9 (2), 365–368.
Robinson, Christina, 2016. Family composition and the benefits of participating in the special supplemental nutrition program for women, infants, and children (WIC). E. Econ. J. 42 (2), 232–251. Spring.
Robinson, Christina, 2018. The gendered health benefits of WIC participation. J. Econ. Insight 44 (1), 21–43.
Robinson, C., Bouzarovski, S., Lindley, S., 2018. Underrepresenting Neighbourhood vulnerabilities? The measurement of fuel poverty in England. Environ. Plann. A 50 (5), 1109–1127. http://epn.sagepub.com/content/by/year.
Rocha, C., Burlandy, L., Maluf, R., 2012. Small farms and sustainable rural development for food security: the Brazilian experience. Dev. South Afr. 29 (4), 519–529.
Rodgers, Yana van der, M., et al., Mar. 2020. Experimental approaches in development and poverty alleviation. World Dev. vol. 127. EBSCOhost. search.ebscohost.com/login.aspx?direct=true&db=ecn&AN=1815529&site=ehost-live.
Rodriguez, A.G., Smith, S.M., 1994. A comparison of determinants of urban, rural and farm poverty in Costa Rica. World Dev. 22 (3), 381–397.
Rogers, B., Youssef, N., 1988. The importance of women's involvement in economic activities in the improvement of child nutrition and health. Food Nutr. Bull. 10 (3), 33–41.
Rosen, H., 2002. Public Finance, sixth ed. McGraw-Hill Irwin, New York.
Rosen, S., Shapouri, S., 2008. Rising Food Prices Intensify Food Insecurity in Developing Countries. USDA Policy Brief, Washington, DC.
Rosenbaum, P.R., Rubin, D.B., 1983. The central role of the propensity score in observational studies for causal effects. Biometrika 70 (1), 41–55.
Rosenzweig, M.R., 1995. Why are there returns to schooling? Am. Econ. Rev. 85 (2), 153–158.
Rosenzweig, M.R., Mar. 2012. Thinking small: poor economics: a radical rethinking of the way to fight global poverty: review essay. J. Econ. Lit. 50 (1), 115–127.
Ruel, M., 2001. Operational Evaluation of the Hogares Comunitarios Program of Guatemala. IFPRI, Washington, DC.
Ruel, M.T., Hawkes, C., 2019. Double duty actions to tackle all forms of malnutrition (P10-053-19). Curr. Dev. Nutr. 3 (Suppl. 1), 814.
Ruel, M.T., Menon, P., 2002. Child feeding practices are associated with child nutritional status in Latin America: innovative uses of the demographic and health surveys. J. Nutr. 132, 1180–1187.

Ruel, M., Habicht, J.P., Pinstrup Andersen, P., Gröhn, Y., 1992. The mediating effect of maternal nutrition knowledge on the association between maternal schooling and child nutritional status in Lesotho. Am. J. Epidemiol. 135, 904–914.

Ruel, M., Levin, C.E., Klemesu, M.A., Maxwell, D., Morris, S.S., 1999. Good care practices can mitigate the negative effects of poverty and low maternal schooling on children's nutritional status: evidence from Accra. World Dev. 27, 1993–2009.

Sadath, A.C., Acharya, R.H., 2017. Assessing the extent and intensity of energy poverty using multidimensional energy poverty index: empirical evidence from households in India. Energy Pol. 102, 540–550.

Saenz, M., 2013. Gender and policy effects on technology adoption in Zambia. In: Paper Presented at the Annual Meetings of the American Agricultural and Applied Economics Association. Washington D.C. August, 2013.

Sahn, D., Alderman, H., 1997. On the determinants of nutrition in Mozambique: the importance of age-specific effects. World Dev. 25 (4), 577–588.

Saitio, K.A., 1994. Raising the Productivity of Women Farmers in Sub-saharan Africa. World Bank Discussion Paper No. 230, Washington, DC.

Sam Anu, S., Azhar, A., Surendran Padmaja, S., Kaechele Harald Kumar, R., Muller, K., February 2019. Capacity and drought risk: implications for sustainable rural development. Soc. Indicat. Res. 142 (1), 363–385.

Samad, H.A., Fan, Z., 2019. Electrification and Women's Empowerment: Evidence from Rural India. EBSCOhost. search.ebscohost.com/login.aspx?direct=true&db=ecn&AN=1796392&site=ehost-live.

Samman, E., June 2013. Eradicating Global Poverty: A Noble Goal, but How Do We Measure it? ODI Report. Working Paper 2.

Sanders, M.R., Patel, R.K., Le Grice, B., Shepherd, R.W., 1993. Children with persistent feeding difficulties: an observational analysis of the feeding interactions of problem and nonproblem eaters. Health Psychol. 12 (1), 64–73.

Sano, Y., Steven, G., Greder Kimberly, A., Cook Christine, C., Dawn, E., Browder, 2011. Understanding food insecurity among latino immigrant families in rural America. J. Fam. Econ. Issues 32, 111–123.

Santhosh, R., July–September 2018. Food expenditure profile and north-south divide: an analysis. Indian J. Econ. Dev. 14 (3), 521–527.

SC UK, 2003. Thin on the Ground: Questioning the Evidence behind World-Bank Funded Community Nutrition Projects in Bangladesh, Ethiopia and Uganda. Save the Children, UK.

Scaramozzino, P., 2006. Measuring Vulnerability to Food Insecurity. ESA Working Paper No. 06-12. FAO, Rome.

Scarborough, V., 1990. Domestic Food Marketing Liberalization in Malawi: A Preliminary Assessment. ADU Occasional Paper No. 13. University of London, London.

Scarlato, M., et al., Mar. 2016. Evaluating CCTs from a gender perspective: the impact of Chile Solidario on women's employment prospect. J. Int. Dev. 28 (2), 177–197. EBSCOhost. http://onlinelibrary.wiley.com.pitt.idm.oclc.org/journal/10.1002/%28ISSN%291099-1328/issues.

Schimmelpfennig, D., Ebel, R., 2011. On the doorstep of the information age: recent adoption of precision agriculture. Econ. Inf. Bull. (EIB-80) 31.

Schott, W., Elisabetta, A., Penny Mary, E., Behrman Jere, R., August 2019. The double burden of malnutrition among youth: trajectories and inequalities in four emerging economies. Econ. Hum. Biol. 34, 80–91.

Schultz, T.P., 1984. Studying the impact of household economic and community variables on child mortality. Popul. Dev. Rev. 10 (Suppl. 1), 215–235.

Seber, George, A.F., 1984. Multivariate Observations. Wiley, New York.
Sekabira, H., Matin, Q., December 2017. Can mobile phones improve gender equality and nutrition? Panel data evidence from farm households in Uganda. Food Pol. 73, 95–103.
Sen, A., 1981. Poverty and Famines: An Essay on Entitlement and Deprivation. Clarendon Press, Oxford.
Sen, A., 1985. Commodities and Capabilities. Oxford University Press, New ork.
Sen, A., 1999. Development as Freedom. Random House Inc., New York.
Sethi, A.S., Pandhi, R., March 2014. Interstate divergences in nutritional expenditure in India: a cluster analysis approach. Poverty Public Pol. 6 (1), 80–97. EBSCOhost. http://onlinelibrary.wiley.com.pitt.idm.oclc.org/journal/10.1002/%28ISSN%291944-2858/issues.
Sevinc, D., 2020. How poor is poor? A Novel look at multidimensional poverty in the UK. Soc. Indicat. Res. 149 (3), 833–859.
Sharma, A., Chandrasekhar, S., Feb. 2016. Impact of commuting by workers on household dietary diversity in rural India. Food Pol. 59, 34–43. EBSCOhost. http://www.sciencedirect.com.pitt.idm.oclc.org/science/journal/03069192.
Sharma, H., Swain, M., Kalamkar, S.S., 2017. Assessment of economic losses due to inadequate post-harvest infrastructure facilities for marine fisheries in Gujarat. Econ. Aff. 62 (1), 1–10.
Sheereen, Z., 2012. Concerns of food security in India amidst economic crisis. Int. J. Res. Commer. Econ. Manag. 2 (4), 66–69.
Shen, Y., Alkire, S., Zhan, P., 2019. Measurement and decomposition of multi-dimensional poverty in China. China Econ. 14 (3), 12–28. http://www.chinaeconomist.com.cn.
Shetty, P., Henry, C.J.K., Black, A.E., Prentice, A.M., 1996. Energy requirements of adults: an update on basal metabolic rates (BMRs) and physical activity levels (PALs). Eur. J. Clin. Nutr. 50 (Suppl. l), S11–S23.
Shi, A., 2000. How Access to Urban Potable Water and Sewerage Connections Affects Child Mortality. World Bank Working Paper No. 2274, Washington, DC.
Shikuku, K.M., March 2019. Information exchange links, knowledge exposure, and adoption of agricultural technologies in Northern Uganda. World Dev. 115, 94–106.
Shinns, L.H., Lyne, M.C., 2003. Symptoms of Poverty within a Group of Land Reform Beneficiaries in the Midlands of Kwazulu-Natal: Analysis and Policy Recommendations. BASIS CRSP Working Paper. Department of Agricultural and Applied Economics, University of Wisconsin-Madison.
Shiratori, S., Kinsey, J., 2011. Media Impact of Nutrition Information on Food Choice. Selected Paper Presented at the Agricultural and Applied Economics Associations Annual Meeting. Pittsburgh, Pennsylvania, July 2011.
Shireen, M., 2012. Distance to Market and Search Costs in an African Maize Market. The World Bank, Policy Research Working Paper Series: 6172, Washington DC.
Silberberg, E., 1985. Nutrition and the demand for tastes. J. Polit. Econ. 93 (5), 881–900.
Simatele, D., Binns, T., Simatele, M., 2012. Sustaining livelihoods under a changing climate: the case of urban agriculture in Lusaka, Zambia. J. Environ. Plann. Manag. 55 (9), 1175–1191.
Simler, K., Mukherjee, S., Dava, G.L., Datt, G., 2004. Rebuilding after War: Micro- Level Determinants of Poverty Reduction in Mozambique. Research Report No. 132. IFPRI, Washington, DC.
Singh, P., June 2017. Learning and behavioural spillovers of nutritional information. J. Dev. Stud. 53 (6), 911–931. EBSCOhost. http://www-tandfonline-com.pitt.idm.oclc.org/loi/fjds20.
Singh, R., Prajapati, M.R., Mishra, S., Vahoniya, D.R., Lad, Y.A., Zala, Y.C., 2018. Physical distribution system of drumstick (Moringa oliefera) in vadodara district of Gujarat. Indian J. Econ. Dev. 14 (1), 139–144.
Skolnik, R.L., 2007. Essentials of Global Health. Jones & Bartlett Publishers Inc.

Smale, M., Jayne, T., 2003. Maize in Eastern and Southern Africa: Seeds of Success in Retrospect. IFPRI, Washington, DC. EPTD Discussion Paper No. 97.

Smith, V.E., 1959. Linear programming models for the determination of palatable human diets. J. Farm Econ. 41 (2), 272−283.

Smith, R.D., 2008. Food security and international fisheries policy in Japan's postwar planning. Soc. Sci. Jpn. J. 11 (2), 259−276.

Smith, T.G., 2011. Economic stressors and the demand for "fattening" foods. Am. J. Agric. Econ. 94 (2), 324−330.

Smith, L.C., Haddad, L., 2000. Explaining Child Malnutrition in Developing Countries: Across-Country Analysis. Research Report No. 111. IFPRI, Washington, DC.

Smith, L., Elobeid, A., Jensen, H., 2000. The geography and causes of food insecurity in developing countries. Agric. Econ. 22 (1), 199−215.

Smith, L., Ramakrishnan, U., Ndiaye, A., Haddad, L., Martorell, R., 2003. The Importance of Women's Status for Child Nutrition in Developing Countries. Research Report No. 131. IFPRI, Washington, DC.

Smith, L., Alderman, H., Aduayom, D., 2006. Food Insecurity in Sub-saharan Africa: New Estimates from Household Expenditure Surveys. Research Report No.146. IFPRI, Washington, DC.

Smith, M.D., et al., May 2017. Who are the world's food insecure? New evidence from the food and agriculture organization's food insecurity experience scale. World Dev. 93, 402−412.

Sokal, R.R., Rohlf, F.J., 1981. Biometry: The Principles and Practice of Statistics in Biological Research, second ed. W.H. Freeman and Company, New York, NY.

Spanos, A., 1998. Probability Theory and Statistical Inference: Econometric Modeling with Observa- Tional Data. Cambridge University Press, Cambridge.

Spring, A., 2000. Commercialization and women farmers: old paradigms and new themes. In: Spring, A. (Ed.), Women Farmers and Commercial Ventures: Increasing Food Security in Developing Countries. Lynne Rienner Publishers, Boulder, CO.

Staatz, J.M., D'Agostino, V.C., Sundberg, S., 1990. Measuring food security in Africa: conceptual, empirical and policy issues. Am. J. Agric. Econ. 72 (5), 1311−1317.

Stark, O., Micevska, M., Mycielski, J., 2009. Relative poverty as a determinant of migration: evidence from Poland. Econ. Lett. 103 (3), 119−122.

Stenmarck, C., Jensen, T., Quested, G., Moates, M., Buksti, B., Cseh, S., Juul, A., Parry, A., Politano, B., Redlingshofer, S., Scherhaufer, K., Silvennoinen, H., Soethoudt, C., Zübert, K., Östergren, 2016. "Estimates of European Food Waste Levels", Fusions Project. IVL Swedish Environmental Research Institute.

Stewart, K., Roberts, N., 2019. Child poverty measurement in the UK: assessing support for the downgrading of income-based poverty measures. Soc. Indicat. Res. 142 (2), 523−542.

Stifel, D.S., Minten, B., Dorosh, P., 2003. Transaction Costs and Agricultural Productiv- Ity: Implications of Isolation for Rural Poverty in Madagascar. MSSD Discussion Paper No. 56. IFPRI, Washington, DC.

Stigler, G.J., 1945. The cost of subsistence. J. Farm Econ. 27, 303−314.

Stock, J.H., Watson, M.W., 2011. Introduction to Econometrics, third ed. Addison- Wesley, New York.

Strasberg, P.J., 1997. Smallholder Cash-Cropping, Food Cropping and Food Security in Northern Mozambique. Unpublished PhD Dissertation. Michigan State University, USA.

Strauss, J., Thomas, D., 1995. Human resources: empirical modeling of household and family decisions. In: Behrman, J., Srinivasan, T.N. (Eds.), Handbook of Development Economics, vol. 3A. North Holland, New York, pp. 1883−2023.

Streeten, P., Burki, S.J., Haq, M.U., Hicks, N., Stewart, F., 1981. First Things First: Meeting Basic Human Needs in the Developing Countries. Oxford University Press, Oxford.

Streeter, J.L., 2017. Socioeconomic Factors Affecting Food Consumption and Nutrition in China: Empirical Evidence during the 1989-2009 Period, Chinese Economy. May 50 (3), 168−192.

Stringer, R., 2000. Food Security in Developing Countries. CIES Working Paper No. 11. University of Adelaide, Australia.

Subramanian, A., Dec. 2007. Harnessing ideas to idealism: arvind subramanian profiles Michael Kremer. Finance Dev. 44 (4), 6−9. EBSCOhost. http://www.imf.org.pitt.idm.oclc.org/external/pubs/ft/fandd/fda.htm.

Sumner, D.A., 2003. Implications of the US farm bill of 2002 for agricultural trade and trade negotiations. Aust. J. Agric. Resour. Econ. 47 (1), 99−122.

Sunil, S., Mamata, P., Roy, D., August 2020. Assessing India's Food Supply Situation Using a Food Balance Sheet and COVID-19-Sensitive Markers. https://southasia.ifpri.info/2020/08/06/assessing-indias-food-supply-situation-using-a-food-balance-sheet-and-covid-19-sensitive-markers/.

Suppa, N., 2018. Towards a multidimensional poverty index for Germany. Empirica 45 (4), 655−683.

Svedberg, P., 2004. Has the Relationship Between Undernutrition and Income Changed? Challenge Paper for Copenhagen Consensus.

Swaminathan, S., Menon, P., January 4 2020. The Double Burden of Malnutrition: Need for Urgent Policy Action. The Hindu.

Tabachnick, B.G., Fidell, L.S., 2001. Using Multivariate Statistics, fourth ed. Allyn and Bacon, Boston, MA.

Taylor, C.A., Spees, C.K., Markwordt, A.M., Watowicz, R.P., Clark, J.K., Hooker, N.H., 2017. Differences in US adult dietary patterns by food security status. J. Consum. Aff. Fall 51 (3), 549−565.

Theis, S., Lefore, N., Meinzen-Dick, R., Bryan, E., September 2018. What happens after technology adoption? Gendered aspects of small-scale irrigation technologies in Ethiopia, Ghana, and Tanzania. Agric. Hum. Values 35 (3), 671−684.

Thompson, W., April 2, 2012. India's Food Security Problem. The Diplomat.

Thomson, A., Metz, M., 1998. Implications of Economic Policy for Food Security: A Training Manual. Available at: http://www.fao.org/docrep/004/x3936e/X3936E00. HTM.

Timmer, C.P., 1997. Farmers and markets: the political economy of new paradigms. Am. J. Agric. Econ. 79, 621−627.

Timmer, C.P., Falcon, W.P., Pearson, S.R., 1983. Food Policy Analysis. Johns Hopkins University Press for the World Bank, Baltimore, MD.

Toke, D., Raghavan, S., 2010. Ecological modernisation as bureaucracy−organic food and its certification in the UK and India. Int. J. Green Econ. 4 (3), 313−326.

Toma, R.B., Fansler, L.T., Knipe, M.T., 1991. World food shortage: the third dimension. In: James, V. (Ed.), Urban and Rural Development in Third World Countries: Problems of Population in Developing Nations. McFarland, Jefferson, North Carolina, pp. 61−66.

Toothacker, L.E., 1993. Multiple Comparisons Procedures. Sage Publications, Thousand Oaks, CA.

Topolyan, I., Xu, X., May 2017. Differential effects of mother's and child's postnatal WIC participation on breastfeeding. Appl. Econ. 49 (22−24), 2216−2225.

Torun, B., 1996. Energy requirements and dietary energy recommendations for children and adolescents 1 to 18 years old. Eur. J. Clin. Nutr. 50, S37−S81.

Toshiaki, A., December 2019. Ex-Ante inequality of opportunity in child malnutrition: new evidence from ten developing countries in asia. Econ. Hum. Biol. 35, 144–161.

Tranchant, J.-P., et al., July 2019. The impact of food assistance on food insecure populations during conflict: evidence from a quasi-experiment in Mali. World Dev. vol. 119, 185–202. EBSCOhost. search.ebscohost.com/login.aspx?direct=true&db=ecn&AN=1767310&site=ehost-live.

Trauger, A., Sachs, C., Barbercheck, M., Brasier, K., Nancy, E.K., 2010. 'Our market is our community': women farmers and civic agriculture in Pennsylvania, USA. Agric. Hum. Val. 27, 43–55.

Tshirley, D., Thieraut, V., 2013. On the institutional details that mediate the impact of cash crops on food crop intensification: the case of cotton. In: Paper Presented at the American Association of Agricultural and Applied Economics Meetings, Washington D.C. August 2013.

Uematsu, H., Mishra, A.K., 2012. Organic farmers or conventional farmers: where's the money? Ecol. Econ. 78 (1), 55–62.

Ukwuani, F.A., Suchindran, C.M., 2003. Implications of women's work for child nutritional status in Sub-Saharan Africa: a case study of Nigeria. Soc. Sci. Med. 56 (10), 2109–2121.

UNDP, 1997. Human Development Report. Oxford University Press, Oxford.

UNEP, 2013. GEO-5 for Business: Impacts of a Changing Environment on the Corporate Sector.

UNICEF, 1990. Strategy for Improved Nutrition of Children and Women in Developing Countries. UNICEF, New York.

UNICEF, 1998. The State of the World's Children. Oxford University Press, New York.

United Nations Administrative Committee on Coordination (ACC/SCN), 2000. Fourth Report on the World Nutrition Situation: Nutrition throughout the Life Cycle (Geneva).

USDA, 2005. Commercialization of Food Consumption in Rural China. Economic Research Report No. ERR8., Washington, DC.

USDA, 2009. Access to Affordable and Nutritious Food: Measuring and Understanding Food Deserts and Their Consequences—Report to Congress.

Valdivia, M., 2004. Poverty, health infrastructure and the nutrition of Peruvian children. Econ. Hum. Biol. 2 (3), 489–510.

Van Phan, P., O'Brien, M., Aug. 2019. Multidimensional wellbeing inequality in a developing country: a case study of Vietnam. Soc. Indicat. Res. 145 (1), 157–183.

Vansteenkiste, J., Schuller, M., April 2018. The gendered space of capabilities and functionings: lessons from Haitian community-based organizations. J. Hum. Dev. Capab. 19 (2), 147–165.

Varshney, D., Kumar, A., Joshi, P.K., July 2020. Situation Assessment of the Rural Economy amid COVID-19 Crisis: Evidence from India. https://southasia.ifpri.info/2020/07/23/situation-assessment-of-the-rural-economy-amid-covid-19-crisis-evidence-from-india/.

Vartanian, T.P., Houser, L., 2012. The effects of childhood SNAP use and neighborhood conditions on adult body mass index. Demography 49 (3), 1127–1154.

Vaz, A., Pratley, P., Alkire, S., 2016. Measuring women's autonomy in Chad using the relative autonomy index. Fem. Econ. 22 (1), 264–294. http://www.tandfonline.com/loi/rfec20.

Vaz, A., Alkire, S., Quisumbing, A., Sraboni, E., 2018. Measuring autonomy: evidence from Bangladesh. Asia-Pacific Sustain. Dev. J. 25 (2), 21–51.

Venkatasubramanian, K., Ramnarain, S., November 2018. Gender and adaptation to climate change: perspectives from a pastoral community in Gujarat, India. Dev. Change 49 (6), 1580–1604.

Verwimp, P., Juan Carlos, M.-M., June 2018. Returning home after civil war: food security and nutrition among Burundian households. J. Dev. Stud. 54 (6), 1019–1040. EBSCOhost. http://www-tandfonline-com.pitt.idm.oclc.org/loi/fjds20.

References

von Braun, J., 1988. Effects of technological change in agriculture on food consumption and nutrition: rice in a West African setting. World Dev. 16 (9), 1083−1098.

von Braun, J., 2005. The world food situation: an overview. In: Paper Prepared for the CGIAR Annual General Meeting, Marrakech, Morocco, December 6, 2005. IFPRI, Washington D.C.

von Braun, J., Immink, M.D.C., 1994. Nontraditional vegetable crops and food security among smallholder farmers in Guatemala. In: von Braun, J., Kennedy, E. (Eds.), Agricultural Commercialization, Economic Development, and Nutrition. Johns Hopkins University Press and IFPRI, Washington, DC.

von Braun, J., Bouis, H., Kumar, S., Pandya-Lorch, R., 1992. Improving household food security. In: Theme Paper for the International Conference on Nutrition (ICN). IFPRI, Washington, DC.

von Braun, J., McComb, J., Fred-Mensah, B.K., Pandya-Lorch, R., 1993. Urban Food Insecurity and Malnutrition in Developing Countries: Trends, Policies, and Research Implications. IFPRI, Washington, DC.

von Braun, J., Johm, K.B., Puetz, D., 1994. Nutritional effects of commercialization of a woman's crop: irrigated rice in the Gambia. In: von Braun, J., Kennedy, E. (Eds.), Agricultural Commercialization, Economic Development, and Nutrition. Johns Hopkins University Press and IFPRI, Washington, DC.

Von Jacobi, N., 2018. Institutions as meso-factors of development: a human development perspective. Schmollers Jahrbuch: J. Context. Econ. 138 (1), 53−88.

Vorley, B., 2013. Meeting Small-Scale Farmers in Their Markets: Understanding and Improving the Institutions and Governance of Informal Agrifood Trade, IIED Report July 2013.

Vossenaar, M., Knight, F.A., Tumilowicz, A., Hotz, C., Chege, P., Ferguson, E.L., 2017. Context-specific complementary feeding recommendations developed using Optifood could improve the diets of breast-fed infants and young children from diverse livelihood groups in northern Kenya. Public Health Nutr. 20, 971−983.

Wang, Y., Gao, Q., Yang, S., 2019. Prioritising Health and Food: Social Assistance and Family Consumption in Rural China. China Int. J., February 17 (1), 48−75.

Warren, A.M., Frongillo, E.A., Nguyen, P.H., Menon, P., 2020. Nutrition intervention using behavioral change communication without additional material inputs increased expenditures on key food groups in Bangladesh. J. Nutr. 150 (5), 1284−1290.

Warsaw, P., Phaneuf, D.J., November 2019. The implicit price of food access in an urban area: evidence from milwaukee property markets. Land Econ. 95 (4), 515−530.

Webb, P., Block, S., 2003. Nutrition Knowledge and Parental Schooling as Inputs to Child Nutrition in the Long and Short Run. Food Policy and Applied Nutrition Program Work- Ing Paper No. 21. Tufts University, Boston, MA.

WFP, 2011. Fortified Wheat Flour Keeps Tribal Children Healthy. WFP report. Feb 25 2011. http://www.wfp.org/node/3485/4530/30705.

WFP, 2013a. Food is Main Priority for New Arrivals in Uganda. http://www.wfp.org/node/3608/3501/397795.

WFP, 2013b. http://www.wfp.org/countries/india.

White, H., 2014. Current challenges in impact evaluation. Eur. J. Dev. Res. 26, 18−30.

Whitehead, J.C., 2006. Improving willingness to pay estimates for quality improve- ments through joint estimation with quality perceptions. South. Econ. J. 73 (1), 100−111.

Winne, M., Joseph, H., Fisher, A., 1997. Community Food Security: Promoting Food Security and Building Healthy Food Systems. Community Food Security Coalition, Venice, CA.

Wirfalt, A.K., Jeffery, R.W., 1997. Using cluster analysis to examine dietary patterns: nutrient intakes, gender, and weight status differ across food pattern clusters. J. Am. Diet Assoc. 97 (3), 272−279.

Wooldridge, J.M., 2002. Econometric Analysis of Cross Section and Panel Data. MIT Press, Boston MA.

World Bank, 1981. Accelerated Development in Sub-saharan Africa: An Agenda for Action (Washington, DC).

World Bank, 1986. Poverty and Hunger — Issues and Options for Food Security in Developing Countries. The World Bank, Washington, DC.

World Bank, 2000. World Development Report (WDR) 2000/2001: Attacking Poverty (Washington, DC).

World Bank, 2002a. Reaching the Rural Poor: A Rural Development Strategy for the Latin American and Caribbean Region (Washington, DC).

World Bank, 2006. Repositioning Nutrition as Central to Development: A Strategy for Large-Scale Action (Washington, DC).

World Bank, 2007. Global Monitoring Report 2007. Ch. 3. Washington, DC.

World Bank, 2008. World Development Report (WDR) 2008: Agriculture for Development. Ch. 7. Washington, DC.

World Bank, Natural Resources Institute, & FAO., 2011. Missing Food: The Case of Postharvest Grain Losses in Sub-saharan Africa. Vol. 60371–AFR.

World Food Program (WFP), May 10, 2008. WFP's Response to Global Food Crisis. Rome.

World Health Organization (WHO), 1995. Physical Status: The Use and Interpretation of Anthropometry. Report of a WHO Expert Committee. Geneva. WHO (2005 and 2006). Data Tables on Deaths by Age, Sex and Cause for the Year 2002; 20 Leading Causes of Deaths and Burden of Disease at All Ages; 20 Leading Causes of DALYs Due to Selected Risk Factors for Each OECD Countries, BRIICS Countries and the World. Data provided by MHI/EIP/WHO between November 2005 and April 2006.

Wu, Q., et al., September 2017. Food access, food deserts, and the women, infants, and children program. J. Agric. Resour. Econ. 42 (3), 310–328.

Wu, P., et al., Jan. 2018. Research on fertility policy in China: the relative necessity for reform among the different provinces. Soc. Indicat. Res. 135 (2), 751–767.

You, J., Imai, K., Raghav, G., January 2016. Declining nutrient intake in a growing China: does household heterogeneity matter? World Dev. 77, 171–191.

Young, M.F., Nguyen, P.H., Gonzalez Casanova, I., Addo, O.Y., Tran, L.M., Nguyen, S., 2018. Role of maternal preconception nutrition on offspring growth and risk of stunting across the first 1000 days in Vietnam: a prospective cohort study. PLoS One 13 (8).

Yuan, M., Seale J. Jr., L., Thomas, W., Bai, J., 2019. The Changing Dietary Patterns and Health Issues in China. China Agric. Econ. Rev. 11 (1), 143–159.

Yuka, T., 2012. Garden plots as an informal safety net in rural Russia — recovering from an income shock (in Japanese. with English summary). Econ. Rev. 63 (4), 305–317.

Zehetmayer, M., 2011. The continuation of the Antebellum puzzle: Stature in the US, 1847-1894. Eur. Rev. Econ. Hist. 15 (2), 313–327.

Zereyesus Yacob, A., Amanor-Boadu, V., Ross, K.L., Aleksan, S., July 2017. Does women's empowerment in agriculture matter for children's health status? Insights from Northern Ghana. Soc. Indicat. Res. 132 (3), 1265–1280.

Zhai, F., Gao, Q., 2010. Center-based care in the context of one-child policy in China: do child gender and siblings matter? Popul. Res. Pol. Rev. 29, 745–774.

Zhang, D., Li, J., Han, P., 2019. A multidimensional measure of energy poverty in China and its impacts on health: an empirical study based on the China family panel studies. Energy Pol. 131, 72–81.

Zhao, F., Bishai, D., 2004. The Interaction of Community Factors and Individual Characteristics on Child Height in China. World Bank Working Paper, Washington, DC.

Zhen, L., 2010. Arable land requirements based on food consumptions patterns: case study in rural Guyuan district, Western China. Ecol. Econ. 69 (7), 1443−1453.
Ziliak James, P., Craig, G., April 2016. Multigenerational families and food insecurity. South. Econ. J. 82 (4), 1147−1166.
Zivin, J.G., Thirumurthy, H., Goldstein, M., 2009. AIDS treatment and intrahousehold resource allocation: children's nutrition and schooling in Kenya. J. Publ. Econ. 93 (7−8), 1008−1015.

Further reading

Acemoglu, D., Johnson, S., Robinson, J.A., 2001. The colonial origins of comparative development: an empirical investigation. Am. Econ. Rev. 91 (5), 1369−1401.
Acock, A.C., 2012. A Gentle Introduction to Stata, third ed. STATA Press, TX.
Agnes, S.C., 1999. Anthropometric, Health and Demographic Indicators in Assessing Nutritional Status and Food Consumption. FAO, Rome.
Agricultural Policy Reform and the WTO: Where are we Heading? Elgar, Cheltenham, U.K. and Northampton, Mass.
Ahmed, A., Hill, R.V., Smith, L., Wiesmann, D., Frankenberger, T., 2007. The World's Most Deprived: Characteristics and Causes of Extreme Poverty and Hunger. IFPRI Conference Taking Action for the World's Poor and Hungry People. Beijing, China, 17−19 October, 2007.
Aizer, A., Shari, E., Joseph, F., Adriana, L.-M., April 2016. The long-run impact of cash transfers to poor families. Am. Econ. Rev. 106 (4), 935−971.
Alexis, D., 2009. Implementing horn's parallel analysis for principal component analysis and factor analysis. STATA J. 9 (2), 291−298.
Alkire, S., Roche, J.M., Vaz, A., 2017b. Changes over time in multidimensional poverty: methodology and results for 34 countries. World Dev. 94, 232−249.
Allcott, H., Diamond, R., Dube, J.-P., Handbury, J., Rahkovsky, I., Schnell, M., November 2019. Food deserts and the causes of nutritional inequality. Q. J. Econ. 4, 1793−1844, 134.
Anania, G., Bohman, M.E., Carter, C.A., McCalla, A.F., 2004. Agricultural policy reform and the WTO. Where are we heading? Q. J. Int. Agric. 43 (2), 188−191.
Arndt, C., et al., September 2020. Covid-19 lockdowns, income distribution, and food security: an analysis for South Africa. Glob. Food Secur. 26, 1−5.
Alejandra, A., Samuel, S., Neha, K., Purnima, M., Agnes, Q., August 2020. Being Social is Important for Women's Mental Health: Insights from a Pre-pandemic Survey in Rural India. https://southasia.ifpri.info/2020/08/19/being-social-is-important-for-womens-mental-health-insights-from-a-pre-pandemic-survey-in-rural-india/.
Atkinson, A.B., Brandolini, A., 2004. I cambiamenti di lungo periodo nelle disuguaglianze di reddito nei paesi industrializzati. Riv Ital. Degli Econ. 3, 389−421.
Atsushi Iimi Mengesha, H.A., Markland, J., Asrat, Y., Kassahu, K., 2018. Heterogeneous Impacts of Main and Feeder Road Improvements: Evidence from Ethiopia. The World Bank, Policy Research Working Paper Series: 8548.
Babu, S.C., 1997c. Multi-disciplinary capacity strengthening for food security and nutrition policy analysis: lessons from Malawi. Food Nutr. Bull. 18 (4), 363−375.
Babu, S.C., 2000. Rural nutrition interventions with indigenous plant foods − a case study of Vitamin A deficiency in Malawi. Biotechnol. Agron. Soc. Environ. 4 (3), 169−179.
Babu, S.C., 2011. Developing multi-disciplinary capacity for agriculture, health, and nutrition − challenges and opportunities. Afr. J. Food Agric Nutr. Sci. 11 (6), 1−3.
Babu, S.C., Andersen, P.P., 1994. Food security and nutrition monitoring: a conceptual framework, issues and challenges. Food Pol. 19 (3), 218−233.

References

Babu, S.C., Mthindi, G.B., 1994. Household food security and nutrition monitoring: the Malawi approach to development planning and policy interventions. Food Pol. 19 (3), 272–284.

Babu, S.C., Subramanian, S.R., 1988. Nutritional poverty – distribution and measure- ment. Indian J. Nutr. Diet 25 (3), 75–81.

Bailey, K.W., 1989. The impact of the food security act of 1985 on U.S. wheat exports. South. J. Agric. Econ. 21 (2), 116–128.

Bali, S.R., Varghese, A., December 2014. Evaluating the impact of training in self-help groups in India. Eur. J. Dev. Res. 26 (5), 870–885.

Barrett, C.B., 2002. Food Aid Effectiveness: It's the Targeting, Stupid! Cornell University, Department of Applied Economics and Management. Working Papers: 14754.

Behrman, J.R., Wolfe, B.L., 1984. More evidence on nutrition demand: income seems overrated and women's schooling underemphasized. J. Dev. Econ. 14 (1–2), 105–128.

Benson, T., 2004. Africa's Food and Nutrition Security Situation – where are We and How Did We Get Here? Paper Prepared for the 2020 Africa Conference Assuring Food and Nutrition Security in Africa by 2020: Prioritizing Actions, Strengthening Actors, and Facilitating Partnerships. Uganda, Kampala. April 1–3, 2004.

Bóo, F.L., Canon, M.E., 2012. Richer but More Unequal? Nutrition and Caste Gaps. Federal Reserve Bank of St. Louis, St. Louis, MO. Working Paper 2012-051A, October. http://research.stlouisfed.org/wp/2012/2012-051.pdf.

Bourguignon, F., Chakravarty, S.R., 2003. The measurement of multidimensional poverty. J. Econ. Inequal. 1 (1), 25–49.

Byerlee, D., Fischer, K., 2001. Accessing Modern Science: Policy and Institutional Options for Agricultural Biotechnology in Developing Countries. IP Strategy Today, No. 1.

Cafiero Carlo, 2013. What Do We Really Know about Food Security? National Bureau of Economic Research, Inc. NBER Working Papers: 18861 (Washington, DC).

Cameron Colin, A., Trivedi, Pravin K., 2010. Microeconometrics Using Stata. STATA Press, College Station, Texas.

Cason, K., Nieto-Montenegro, S., Chavez-Martinez, A., Ly, N., Snyder, A., 2004. Dietary Intake and Food Security Among Migrant Farm Workers in Pennsylvania, Harris School of Public Policy Studies, University of Chicago. Working Papers: 0402.

Chamberlin, J., Jayne, T., 2009. Has Kenyan Farmers' Access to Markets and Services Improved? Panel Survey Evidence, 1997–2007. Michigan State University. Food Security Working Papers. East Lansing, Michigan.

Chen, S., Ravallion, M., 2004. How have the world's poorest fared since the early 1980s? World Bank Res. Obs. 19 (2), 141–169.

Clemens, B., Abla, A., Mariam, R., Manfred, W., 2020. COVID-19 and the Egyptian Economy: Estimating the Impacts of Expected Reductions in Tourism, Suez Canal Revenues, and Remittances. MENA Policy Note 4, Washington, DC. https://doi.org/10.2499/p15738coll2.133663. International Food Policy Research Institute (IFPRI).

Coleman-Jensen, A., Nord, M., Andrews, M., Carlson, S., 2012. Household Food Security in the United States in 2011, Economic Research Report No. (ERR-141). USDA.

Committee on World Food Security, 2012. Global Strategic Framework for Food Security.

Costa Dora, L., 1999. Unequal at Birth: A Long-Term Comparison of Income and Birth Weight. National Bureau of Economic Research Inc. NBER Working Papers, p. 6313.

Courtemanche, C., Augustine, D., Rusty, T., July 2019. Estimating the associations between SNAP and food insecurity, obesity, and food purchases with imperfect administrative measures of participation. South. Econ. J. 86 (1), 202–228.

Davis, B., 2003. Choosing a Method for Poverty Mapping. FAO, Rome.

de Onis, M., Blössner, M., Borghi, E., Morris, R., Frongillo, E.A., 2004. Methodology for estimating regional and global trends of child malnutrition. Int. J. Epidemiol. 33 (6), 1260–1270.

Dorward, A.R., Kydd, J.G., Morrison, J.A., Urey, I., 2004. A policy agenda for pro- poor agricultural growth. World Dev. 32 (1), 73–89.

Esrey, S.A., Habicht, J.P., 1988. Maternal literacy modifies the effects of toilets and piped water on infant survival in Malaysia. Am. J. Epidemiol. 127 (5), 1079–1087.

Falcon, W.P., Naylor, R.L., 2005. Rethinking food security for the twenty-first century. Am. J. Agric. Econ. 87 (5), 1113–1127.

Fan, M., Jin, Y., 2015. The supplemental nutrition assistance program and childhood obesity in the United States: evidence from the National Longitudinal Survey of Youth 1997. Am. J. Health Econ. Fall 1 (4), 432–460.

FAO, 2005. The State of Food and Agriculture 2005 (Rome).

FAO and WHO, 1992. Major issues for nutrition strategies. In: International Conference on Nutrition. Rome, Italy, 5–11 December.

FAO/WHO, 2002. Human Vitamin and Mineral Requirements. Report of a joint FAO/WHO expert consultation, Geneva, Switzerland.

FAO/WHO/UNU, 1985. Energy and Protein Requirements. Report of a Joint FAO/WHO/UN Ad Hoc Expert Consultation. WHO Technical Report Series No. 724. Geneva).

FEWS, 2000. Status of the Food Security and Vulnerability Profile. Available at: http://v4.fews.net/docs/Publications/1000009.pdf.

Friss-Hansen, E. (Ed.), 2000. Agricultural Policy in Africa after Adjustment. CDR Policy Paper, Copenhagen.

Gillespie, S., Haddad, L., 2001. Attacking the Double Burden of Malnutrition in Asia and the Pacific. ADB Nutrition and Development Series No. 4. Available at: http://www.adb.org/Documents/Books/Nutrition/Malnutrition/default.asp.

Graham, R.D., Welch, R.M., Bouis, H.E., 2001. Addressing micronutrient malnutrition through enhancing the nutritional quality of staple foods: principles, perspectives and knowledge gaps. Adv. Agron. 70, 77–142.

Graham, T., Mbzibain, A., Ali, S., 2012. A comparison of the drivers influencing farmers' adoption of enterprises associated with renewable energy. Energy Pol. 49 (1), 400–409.

Greene, W.H., 2002. Econometric Analysis, fifth ed. Prentice Hall, Upper Saddle River, NJ.

Gupta, M., Avinash, K., July 2020. How COVID-19 May Affect Household Expenditures in India: Unemployment Shock, Household Consumption, and Transient Poverty. https://southasia.ifpri.info/2020/07/02/how-covid-19-may-affect-household-expenditures-in-india-unemployment-shock-household-consumption-and-transient-poverty/.

Haddad, L., Westbrook, M.D., Driscoll, D., Rozen, J., Weeks, M., 1995. Strengthening Policy Analysis: Econometric Tests Using Microcomputer Software. IFPRI, Washington, DC.

Haider, S., Schoeni, R., Bao, Y., Danielson, C., 2001. Immigrants, Welfare Reform, and the Economy in the 1990s. RAND Corporation Publications Department. Working Papers: 01–13.

Hassan, R.M., Babu, S.C., 1991. Measurement and determinants of rural poverty: household consumption patterns and food poverty in rural Sudan. Food Pol. 16 (6), 451–460.

Headey, D., Hirvonen, K., Hoddinott, J., 2018. Animal sourced foods and child stunting. Am. J. Agric. Econ. 100 (5), 1302–1319.

Hippolyte, A., Mutungiab, C., Pascal, S., Borgemeisterad, C., February 2015. Unpacking post-harvest losses in Sub-Saharan Africa: a meta-analysis. World Dev. 66, 49–68.

Hoddinott, J., 1997. Water, Health, and Income: A Review. FCND Discussion Paper No. 25. IFPRI, Washington, DC.

Hosmer, D.W., Lemeshow, S., 1989. Applied Logistic Regression. Wiley, New York.

References

HTF (Hunger Task Force), 2003. Halving Hunger by 2015: A Framework for Action. Interim Report of the Millennium Project. UNDP, New York. http://thediplomat.com/indian-decade/2012/04/02/india%E2%80%99s-food-security-problem/.

Huberty, C.J., 1994. Applied Discriminant Analysis. John Wiley and Sons, New York.

Huffman, S.K., Jensen, H.H., 2003. Do Food Assistance Programs Improve Household Food Security? Recent Evidence from the United States. Iowa State University, Department of Economics, Staff General Research Papers.

IFPRI, 2002a. Commercial Vegetable and Polyculture Fish Production in bangladesh: Their Impacts on Income, Household Resource Allocation, and Nutrition (Washington, DC).

IFPRI Datasets, Various Years. Available at: http://www.ifpri.org/data/dataset.htm.

Ishdorj, A., Capps Jr., O., Dec. 2017. The impact of policy changes on milk and beverage consumption of Texas WIC children. Agric. Resour. Econ. Rev. 46 (3), 421–442.

Johnson, M., Akeem, A., 2016. Postharvest processing, marketing, and competitiveness of domestic rice. In: The Nigerian Rice Economy: Policy Options for Transforming Production, Marketing, and Trade. University of Pennsylvania Press, IFPRI-UPP, Washington, DC, pp. 111–138.

Joshi, P.K., Gulati, A., Birthal, P.S., Tewari, L., 2005. Agricultural diversification in South Asia: patterns, determinants and policy implications. In: Babu, S.C., Gulati, A. (Eds.), Economic Reforms and Food Security: The Impact of Trade and Technology in South Asia. Haworth Press Inc., New York.

Kemp, R., 1997. In: Lyme, N.H. (Ed.), Environmental Policy and Technical Change: A Comparison of the Technological Impact of Policy Instruments Cheltenham. Elgar, Williston, VT. distributed by American International Distribution Corporation.

Kennedy, E., 1994. Health and nutrition effects of commercialization of agriculture. In: Agricultural Commercialization, Economic Development, and Nutrition (J. von Braun and E).

Klasen, S., 2008. Economic growth and poverty reduction: measurement issues using income and non-income indicators. World Dev. 36 (3), 420–445.

Klaus, D., Liu, Y., 2009. Longer-term economic impacts of self-help groups in India. In: The World Bank, Policy Research Working Paper Series, p. 4886.

Kramer-LeBlanc, Carol, S., Basiotis, P.P., Kennedy, E.T., 1997. Maintaining food and nutrition security in the United States with welfare reform. Am. J. Agric. Econ. 79 (5), 1600–1607.

Lambrecht, I., Schuster, M., Asare Samwini, S., Pelleriaux, L., 2018. Changing gender roles in agriculture? Evidence from 20 years of data in Ghana. Agric. Econ. 49 (6), 691–710.

Le Bihan, G., Delpeuch, F., Maire, B., 2002. Food, Nutrition and Public Policies. Institut de Recherche pour le Développement, Montpellier, France.

Lecoutere, E., Spielman, D.J., Van Campenhout, B., 2019. Women's Empowerment, Agricultural Extension, and Digitalization: Disentangling Information and Role Model Effects in Rural Uganda. International Food Policy Research Institute (IFPRI), Washington, DC. https://doi.org/10.2499/p15738coll2.133523. IFPRI Discussion Paper 1889.

Luzzi, A.F., June 2002. Individual food intake survey methods. In: Measurement and Assessment of Food Deprivation and Undernutrition, Proceedings of the International Scientific Symposium, Rome, pp. 26–28.

McCalla, A.F., 1999. Prospects for food security in the 21st century with special emphasis on Africa. Agric. Econ. 20, 95–103.

Mrema, G.C., Rolle, R.S., 2003. Status of the postharvest sector and its contribution to agricultural development and economic growth. In: JIRCAS International Symposium, 11, pp. 13–20.

Nanak, K., Shi, L., Wang, X., Mengbing, Z., February 2019. Evaluating the effectiveness of the rural minimum living standard guarantee (Dibao) program in China. China Econ. Rev. 53, 1–14.

Narayan, N., Patel, R., Schafft, K., Rademacher, A., Koch-Schulte, S., 1999. Can Anyone Hear Us? Voices from 47 Countries. Oxford University Press, New York.

Nguyen Phuong, H., Frongillo, E.A., Kim Sunny, S., Zongrone, A., Jilani Amir, H., Tran Lan, M., Sanghvi, T., Menon, P., 2019. Information diffusion and social norms are associated with infant and young child feeding practices in Bangladesh. J. Nutr. 149 (11).

Nguyen, P.H., DiGirolamo Ann, M., Gonzalez-Casanova, I., Young, M., Kim, N., Nguyen, S., Martorell, R., Ramakrishnan, U., 2018. Influences of early child nutritional status and home learning environment on child development in Vietnam. Matern. Child Nutr. 14 (1).

Nord, M., Andrews, M., Carlson, S., 2006. Household Food Security in the United States, 2005. USDA, Economic Research Service. Available at: http://www.ers.usda.gov/Publications/ERR29/.

Norusis, M.J., 2005. SPSS 13.0 Advanced Statistical Procedures Companion. Prentice Hall, New Jersey.

Nutrition and health. Washington D.C.: IFPRI. In: Fan, S., Pandya-Lorch, R. (Eds.), Reshaping Agriculture for Nutrition.

Oniang'o, R., Mukudi, E., 2002. Nutrition and gender. In: Brief 07 — Nutrition: A Foundation for Development. IFPRI, Washington, DC.

Pacifico, D., Poege, F., 2017. Estimating measures of multidimensional poverty with STATA. STATA J. 17 (3), 687—703.

Philip, V., 2012. Undernutrition, subsequent risk of mortality and civil war in Burundi. Econ. Hum. Biol. 10 (3), 221—231.

Pingali, P.L., Stringer, R., 1995. Agricultural commercialization and diversification: processes and policies. Food Pol. 20 (3), 171—185.

Prusty, S.R., Tripathy, S., 2017. Marketing of pulses in Jagatsinghpur District of Odisha. Indian J. Econ. Dev. 13 (1), 159—164.

Quintano, Claudio, Castellano, Rosalia, Punzo, Gennaro, 2011. Measuring poverty and living conditions in Italy through a combined analysis at a sub-national level. J. Econ. Soc. Meas. 36 (1—2), 93—118.

Ram, R., Athalye, S., 2009. Drought resilience in agriculture: the role of technological options, land use dynamics, and risk perception. Nat. Resour. Model. 22 (3), 437—462.

Randolph, S., Gaye, I., Hathie, I., Perez-Escamilla, R., 2007. Monitoring the Realization of the Right to Food: Adaptation and Validation of the U.S. Department of Agriculture Food Insecurity Module to Rural Senegal, vol. 6. University of Connecticut, Human Rights Institute. Economic Rights Working Papers.

Reardon, T., Crawford, E., Kelly, V., Diagana, B., 1995. Promoting Farm Investment for Sustainable Intensification of African Agriculture. MSU International Development Paper No. 18. East Lansing, Michigan.

Riely, F., Mock, N., Cogill, B., Bailey, L., Kenefick, E., 1999. Food Security Indicators and Framework for Use in the Monitoring and Evaluation of Food Aid Programs. Academy for Educational Development, Washington, DC.

Rosegrant, M., Cline, S.A., Li, W., Sulser, T.B., Valmonte-Santos, R.A., 2005. Looking Ahead. Long-Term Prospects for Africa's Agricultural Development and Food Security. IFPRI, Washington, DC, 2020 Discussion Paper No. 41.

Sahu, G., December 2019. Primitive tribes and undernutrition: a study of Katkari Tribe from Maharashtra, India. J. Soc. Econ. Dev. 21 (2), 234—251.

Sarah, S., Ibrahim, K., December 2012. Causes of health inequalities in Uganda: evidence from the demographic and health surveys. Afr. Dev. Rev./Revue Africaine de Developpement, 24 (4), 327—341.

Shahidur, K.R., Gayatri, K.B., Hussain, S.A., 2010. Handbook on Impact Evaluation: Quantitative Methods and Practices. World Bank, Washington, D.C., p. 239

Sharma, S., Shukla, R., 2017. Economics of post harvest losses in onion in jhunjhunu district of Rajasthan. Int. J. Commer. Bus. Manag. 10 (1), 15−19.

Skinner, C., 2012. State immigration legislation and SNAP take-up among immigrant families with children. J. Econ. Iss. 46 (3), 661−681.

Smith, L., June 2002. The use of household expenditure surveys for the assessment of food insecurity. In Measurement and assessment of food deprivation and undernutrition. In: Proceedings of the International Scientific Symposium, Rome, pp. 26−28.

SPSS, 2003. SPSS 12.0 Statistical Procedures Companion. Prentice Hall, New Jersey.

Ssewanyana, S., Ibrahim, K., December 2012. Causes of Health Inequalities in Uganda: Evidence from the Demographic and Health Surveys. African Development Review/Revue Africaine de Developpement 24 (4), 327−341.

State University, Department of Economics, Staff General Research Papers, Dorward, A.R., 2013. Agricultural labour productivity, food prices and sustainable development impacts and indicators. Food Pol. 39, 40−50.

Steiner, F.R., 1990. Soil Conservation in the United States: Policy and Planning. Johns Hopkins University Press, Baltimore and London.

Sunder, M., 2011. Upward and Onward: High-Society American Women Eluded the Antebellum Puzzle, vol. 9, pp. 165−171, 2.

Supawat, R., Wang, X., 2012. Investigating agricultural productivity improvements in transition economies. China Agric. Econ. Rev. 4 (4), 450−467.

Takeda, Y., 2012. Structure of employment in rural areas and personal subsidiary farming during the period of economic growth in Russia. Graduate School of Economics, The University of Tokyo, pp. 250−274 (CIRJE Research Report Series; vol. CIRJE-R-9).

Takeshi, F., Norikazu, I., Yoshiro, K., 2010. The role of neighborhood effects on agricultural technology diffusion: the case of crossbred 'Aigamo' duck farming technology in Japan. Stud. Reg. Sci. 40 (2), 397−412.

The FAO Report, 2012. The State of Food and Agriculture: Women in Agriculture. Closing the Gender Gap for Development. FAO, United Nations, Rome.

Tweeten, L., Thompson, S.R., 2002. Agricultural Policy for the Twenty-First Century. Iowa University Press, Ames.

Udry, C., Duflo, E., 2010. John bates clark medalist. J. Econ. Perspect. 25 (3), 197−216. Summer 2011.

UN Millennium Project, 2005. Investing in Development: A Practical Plan to Achieve the MDGs. www.unmillenniumproject.org.

UN Summit, 2008. High-level Conference on World Food Security: The Challenges of Climate Change and Bioenergy. Held at FAO, Rome.

UNDP, 2003. Human Development Report. Oxford University Press, New York.

United Nations Millennium Declaration, 2000. Millennium Summit of the United Nations. 6-8 September (New York, USA).

UNON Publishing Services. UNON Nairobi.

Vandana, S., Jalees, K., 2009. Why is Every 4th Indian Hungry? The Causes and Cures for Food Insecurity. Navdanya Press, New Delhi.

Vickery Jr., R.E., 2011. The Eagle and the Elephant: Strategic Aspects of US-India Eco- Nomic Engagement. Woodrow Wilson Center Press, Washington, D.C. Baltimore (Johns Hopkins University Press).

von Braun, J., 1996. Food Security and Nutrition. Technical Background Document for the World Food Summit 13−17 Nov 1996, 1(5). FAO, Rome.

von Braun, J., 2008. Rising Food Prices: What Should be Done? IFPRI Policy Brief.

von Braun, J., Swaminathan, M.S., Rosegrant, M.W., 2004. Agriculture, Food Security, Nutrition and the Millennium Development Goals. 2003–04. IFPRI Annual Report Essay, Washington, DC.
WFP, 2019. Fill the nutrient gap: Analysis for decision-making towards sustainable food systems for healthy diets and improved human capital. World Food Programme, Rome, p. 30.
WHO, 2005. Water, Sanitation and Hygiene Links to Health. https://www.who.int/water_sanitation_health/publications/facts2004/en/.
Wondimagegn, T., Tirivayi, N., 2018. The impacts of postharvest storage innovations on food security and welfare in Ethiopia. Food Pol. 75, 52–67.
World Bank, 2002b. Reaching the Rural Poor; a Renewed Strategy for Rural Development. The World Bank, Washington, DC.
World Bank Living Standards Measurement Study, Various Years. Available at: http://econ.worldbank.org/WBSITE/EXTERNAL/EXTDEC/EXTRESEARCH/EXTLSMS/0,contentMDK:21485765~isCURL:Y~menuPK:4196952~pagePK:64168445~piPK:64168309~theSitePK:3358997,00.html.
World Health Organization (WHO), 1978. Declaration of Alma-Ata. In: International Conference on Primary Health Care, Alma-Ata, USSR, September, pp. 6–12.
Yamano, T., Rajendran, S., Malabayabas, M.L., October 2015. Farmers' self-perception toward agricultural technology adoption: evidence on adoption of submergence-tolerant rice in eastern India. J. Soc. Econ. Dev. 17 (2), 260–274.

Index

Note: 'Page numbers followed by "*f*" indicate figures and "*t*" indicate tables.'

A

Absolute poverty, 443
ADD. *See* Agricultural Development Districts
Addition of matrices, 670
Adequate diet, 6—7
ADMARC, 174
Adult equivalents (AE), OECD scale of, 17
Afghanistan, food security in, 181—182
Africa
　access, information, and food security in, 178—180
　AIDS and double burden in, 309—310
　food security and welfare in, 393—395
Agglomerative hierarchical algorithm, 502
AGGREGATE command, 507—508
Aggregation, levels of, 16
Agribusiness, 36
　high-tech, 36
Agricultural biotechnology
　adoption of, 29
　commercial applications of, 29
Agricultural commercialization, chi-square statistic application. *See* Chi-square statistic application, in agricultural commercialization
Agricultural Development Districts (ADD), 316
Agricultural innovations, 106—108
Agricultural market reform linkages, 171, 172f
Agricultural productivity, 31, 471
Agricultural technology, 6
Agriculture
　alternative, 36
　civic, 112
　gender gap in, 113—116
　and poverty, in Laos and Cambodia, 463—465
　precision. *See* Precision agriculture
　sustainable, 113
　technologies in, 115b—116b
　women in, 113—116
Agriculture in the middle, 36
Alley cropping, 108
Alternative agriculture, 36
Alternative hypothesis, 150
American Standard Code for Information Interchange (ASCII), 634
　importing data into SPSS, 634—635
Analysis of variance (ANOVA), 30—31, 159—160
　approach, 149
　assumptions in, 160
　technique, 159—160
Anderson—Rubin method, 199—200
ANOVA. *See* Analysis of variance
Antebellum paradox, 462b—463b
ANTHRO, 667
Anthropometric data
　percentage of median, 666—667
　software programs, 667—668
　uses, 668
　Z-scores computation, 665—666
Anthropometric indicators, 299—300
ASCII. *See* American Standard Code for Information Interchange
Asia, childhood undernutrition and climate change in, 396
Asset indicators, 189
Assets and technology adoption (ASSETTECH), 197—198
Asymmetric price transmission, 181
Average linkage method, 503

B

Bacille Calmette-Guérin (BCG), 301
Bangladesh, women empowerment and nutrition status, 271
Bartlett factor scores, 199—200
Bartlett's test, 157
Basal metabolic rate (BMR), 19
Basic capabilities, 436

Basic needs approach, to evaluate poverty, 436
Bayley scale, 227
BCG. *See* Bacille Calmette-Guérin
Behavioral rating method, 227
Bernoulli distribution, 486
Bernoulli observations, 487
BFEEDNEW. *See* Breastfeeding
Binary logistic regression analysis, 468
Binary variables, 607
Biofuels in China, 38—40
Bivariate scatterplot, 278—279
BMI. *See* Body mass index
BMR. *See* Basal metabolic rate
Body mass index (BMI), 143, 271, 298
Box's M test, 403, 403t
Brazil, food insecurity in, 269
Brazilian Demographic and Health Survey, 269
Breastfeeding (BFEEDNEW), 274, 317, 351, 353t
Brown-Forsythe test, 55
Bt-maize, 29
Bureau of Labor Statistics, 563
Burundi, malnutrition and mortality in, 309—310

C
CALADEQ, 83
Calorie distribution, 182
Calorie income function, 444
Calorie intake, 75, 83, 181—182
Calorie requirements met (CALREQ), 83—84, 91—92, 352, 360—361
  cashcrop and, 90—91
Canonical Discriminant Analysis, 412—416
Capability approach, in measuring poverty, 436—437, 437f
Care effect, 234
Care index by age group, 249
Care practices
  behavioral rating method, 227
  observation method, 226—227
  quantitative assessment method, 227
Care-giver practices, 225
CASHCROP, 33, 69, 83—85
CBN method. *See* Cost of basic needs method
Centers for Disease Control and Prevention (CDC), 667
Central Asian population, 12
Centroids, 409, 418
CFLs. *See* Compact fluorescent lamps

CGIAR, 108
Chi-square statistic application, in agricultural commercialization, 68—70
  data description and analysis, 83—84
  descriptive analysis: cross-tabulation results, 84—93
    chi-square tests using STATA, 89—93
  effects, 73
    on food consumption, 73—74
    on nutrition, 74
  empirical analysis, 82—83
  introduction, 70—74
  organic farms and, 79—81
  policy implications, 93—95
  recent cases, 71—72
  selected studies, 74—78
  technical appendices, 95—96
Chi-square test statistic, 87—89
Child health, 274
  conditions, 301
  estimating determinants of, 541—543
Child malnutrition, 222—223
Child mortality, 263, 533
Child-care
  conceptual and measurement issues on, 529
  extended model of, 226f
  practices, estimating, 540—541
  variables, 376
Child-care index (CARE), 235
Child's nutritional status, 274—275, 298
  age-specific effects on, 361
  causes of, 383
  determinants of, 392b—393b
  in Kenya, 345
  outcome variables of, 274
  role of maternal education and community characteristics on, 337—342
  in United States, 307—309
  and women's status
    direct linkages, 385—388
    indirect linkages, 388—390
Children's obesity level, 230b—231b
China
  biofuel production in, 38
  food consumption patterns in, 147—149
  poverty in, 467—468
China Health and Nutrition Survey (CHNS), 339—340, 389
Chinese agriculture, commercialization of, 71
CHNS. *See* China Health and Nutrition Survey

Chronic food shortages, 135–136
Chronic malnutrition, 296
Chronic poverty, 457–458
Chronic vulnerability approach in measuring food security, 20
CILSS. *See* Cote d'Ivoire Living Standards Survey
Civic agriculture, 112
Classical linear regression model, 679
Classification statistics, 409–412
Classification table, 475–476, 476t
Clinic feeding (CLINFEED), 274, 289
Clusters, 494–495
  analysis, 501, 503–506
  centers, 493–494
  of households, descriptive characteristics of, 508–510
  initial partitions and optimum number of, 507–508
Coefficient of determination, 321
Common factor model, 206
Common variance, 194t
Communalities, 196–197, 196t
Community characteristics, 336, 344
  variables, 274–275
  vector, 350
Community food security, 21
Community infrastructure
  variables, 341
Community level resources, 224–225
Community level variables, 305
Community-based programs, 438–439
Complete linkage, 502
Complex samples general linear model (CSGLM), 627
Component matrix, 195, 195t
Component plot, in rotated space, 199f
Component score coefficient matrix, 198–199
COMPOSIT, 471
Composite child-care index, 235
Comprehensive nutritional program, 277–278
Compute statement, 644
Concentrated poverty, 176
Conceptual framework, 297–303
Conditional probability, 457
Consumption expenditure, 440–441
Consumption patterns, study of, 136
Continuous variables, 59, 627
Contract farming, types of, 71–72
Conventional measures of poverty, 437

Coping strategy approach in measuring food security, 20
Correlation analysis concepts in, 275–276
  of outcome variables, 279–281
Cost of basic needs (CBN) method, 434
Cote d'Ivoire Living Standards Survey (CILSS), 455
CPS. *See* Current Population Survey
Cramer's V and phi coefficient application, in gender role in technology adoption
  data description and analysis, 119
  descriptive analysis: cross-tabulation results, 120
  empirical analysis, 118–119
  female farm operators, 109–112
  introduction, 105–107
  policy implications, 123–126
  selected studies, 107–109
  women in agriculture, 113–116
Crop insurance, 44
Cross-sectional surveys, 273
Cross-tabulation results, 120
  χ^2 test. *See also* Chi-square statistic application, in agricultural commercialization
  of cash crop growers
    and CALREQ, 85t
    and INSECURE, 85t
  descriptive analysis, 84–93, 97–99
  limitations, 96–101
  Pearson chi-square, 95–96
  student's *t*-test *vs.*, 96
Current Population Survey (CPS), 10–11

D

Data editor toolbar, 622
Data file, opening, 636f
Data generating process (DGP), 681
Data handling
  data structure, 635–637
  importing ASCII data, into SPSS, 634–635
  importing data from excel, 635
  useful terms, 637
Data selection statements, 645
Decomposition of total variation, 160
Degrees of freedom, 161, 347–348
Demographic and Health survey, 388
Demographic indicators, 300
Demographic variables, household level characteristics and, 470–471
Dependency ratio, 83, 454

738 Index

Determinants, 303—307
DGP. *See* Data generating process
Diagonal matrix, 671
Dialog recall box, 624f
Diarrhea, 222, 262, 273, 275, 301
Dichotomous logistic regression model, 469
Dichotomous variable, 58—59, 189
Diet, adequate, 6—7
Dietary diversity, 137—138
Dietary guidelines, 18
Dietary intake, 260
  of macronutrients, 6—7
Discrete variables, 637
Discriminant analysis (DA), 382—383, 403—405, 417—424
  decision process, 417—424
Discriminant functions
  analysis, 419—420
  correlation between predictor variables and, 418—419
  estimating, 421—422
  interpretation of, 423—424
Discriminant weights methods, 423
Distance to a health facility (HEALTDST), 275
DPD regressions. *See* Dynamic panel data regressions
Drinking distance (DRINKDST), 274—275, 318, 324, 352, 540
Dummy variable. *See* Dichotomous variable
Dynamic panel data (DPD) regressions, 619

E

Earned Income Tax Credit (EITC), 460
East Asia, 24t
  financial crisis and child nutrition in, 345, 465
Eastern and Southern African (ESA) countries, 173
Ecological theory of human development, 342b—343b
Econometric methodology, 170—171
Econometric simulations, 232
Economic conditions, food security and, 201
Education of the household head (EDUCHEAD), 317
Education of the spouse (EDUCSPOUS), 190, 235, 239, 241—242
Education variable, 223
Educational level effect, 234
EDUCHEAD. *See* Education of the household head

EDUCSPOUS. *See* Education of the spouse
EDUCSPOUS* NCARE, 239—241
EEP. *See* Export Enhancement Program
Efficient marketing system, 170
Eigenvalues, 186
  examination, 193—201
  properties, 187—188
Eigenvectors, 186
EITC. *See* Earned Income Tax Credit
Empirical analysis, 150, 185—186, 234
  basic univariate approach, 45—46
  chi-square test, 82—83
  of Cramer's V and phi coefficient, 118—119
Employment effects, 456—457
Employment targeting, 439
Endogeneity/omitted variable bias, 684
Energy intake, 18—19
Engel curve, 449—454
Epidemiological transition, 137—138
Equal prior probabilities, classification function based on, 410—412
Equality of matrices, 670
Equivalence scales, 441
Equivalent expenditure method, 442
Ethanol, 38
  fuel production, 40
Ethiopia, female farm operators in, 109—112
Ethiopian Rural Household Survey, 394
Europe, poverty in, 465—467
Excel, importing data from, 635
Expectation Maximization (EM) algorithm, 627—628
Expenditure function, 439—440
Expenditure quartiles, 150
Export Enhancement Program (EEP), 40—41
Export-producing cooperative in Guatemala, 32

F

F ratio test, 54
F-statistic, 239
F-test, 161—167, 321
  and distribution, 161
F&V. *See* Fruits and vegetable
Factor analysis, 170—171
  application, market access impact in, 170—171
  data description and methodology, 188—190
  empirical analysis, 185—186
  food security. *See* Food security
  principal components, 190—193

selected studies, 173—175
technical appendices, 206—211
technical concepts, 186—188
decision process, 206—211
using principal components, 170—171, 193
Factor loadings, 187, 196
Factor score coefficients, 192
Factor scores computation, 198—201
Factors estimation, 192
Family Entitlement Program, 340
Family planning policy, 340
FAO report, 22—23, 113, 115b—116b
Farm-gate buying, 178
Farmers, improving market access for, 170
Federal food and nutrition assistance programs, 7
Feeding America system, 267
FEI method. *See* Food energy intake method
Female farm operators
  in Kenya and Ethiopia, 109—112
  in United States, 112—113
Female-headed households (FEMHHH), 107—108, 116, 118, 387
Fertilizer use (FERTILIZ), 189, 199—200
FGT measure, 451
FICRCD. *See* Food Intakes Converted to Retail Commodities Databases
Financial crisis and poverty, in Russian Federation, 465
Fisher's test, 249—250
Fisheries, Japan, 12
Food access, 17, 175—176
  measuring, 17
Food aid policies, 141
Food and nutrition program evaluation
  difference-in-difference, 583—585
  instrumental variables, 579—583
  randomization, 577—579
  regression discontinuity design, 585—587
Food assistance programs, 145b—147b
Food availability, 16—17, 137—138, 546
  measuring, 17
  stability of, 19—20
Food balance sheets, 16
Food component (FOOD), 197—198
Food consumption
  commercialization of agriculture on, 73—74
  patterns, 141—144
    determinants of, 141
  substantial increase in, 135—136
Food Corporation of India, 499

Food deserts in United States, 175—178
Food energy intake (FEI) method, 434, 444—446
Food Expenditure Data, 24t
Food grains, 143
Food group shares of household calorie, 151t
Food indicators, 189
Food insecurity, 3—4
  data, 109—110
  and malnutrition, 11—12
  measuring, 23—26
Food Intakes Converted to Retail Commodities Databases (FICRCD), 153
Food poverty line, 447—448
Food prices, level of, 17
Food sector reforms, 12
Food security, 3—4, 136—137, 139—140, 259—260, 307
  access and, India, 147—149
  in Afghanistan, 181—182
  in Africa, 393—395
  alternative approaches in measuring, 19—20
  assessment, 19, 23
  chronic vulnerability approach, 20
  conceptual framework of, 4—7
  concerns in other countries, 11—16
  coping strategy approach, 20
  core original determinants of, 16
  definition of, 3—4
  determinants of, 16
  in developed world, 7—11
  and economic conditions, 201
  food availability, 16—17
  and food production in India, 14
  food utilization, 18—19
  and hunger, 7
  impact on, 139—140
  index, components of, 47—51
  issues and technology, 36—38
  issues in MENA, 171
  measurement of determinants of, 16
  measures of household, 47—53
    consumption components of food security index, 49—51
    descriptive statistics, 51
    tests for equality of variances, 52—53
  policy issues in US, 11
  post-harvest technology, 33—36

740 Index

Food security (*Continued*)
  and productivity, market reforms on, 171–185
    role of, 180
    scaling approach, 20
    stability of availability, 19–20
    technology adoption impact on, 107
    threshold of, 52
    US Farm Policy and, 40–44
    USDA to assess household, 8b–10b
Food Security Act of 1985, 40
Food Security and Nutrition Monitoring Survey (FSNM), 634
Food Stamp and Temporary Assistance for Needy Families Programs, 11
Food Stamp Program (FSP), 11, 342, 497
Food stamps, 392b–393b
Food utilization, measuring, 18–19
FOODSEC, 49–50, 507
  group distribution of, 51t
"Foot-loose activities," low rate of employment in, 466
Formal education system, 222–223
Free market option, 145b–147b
Fruits and vegetable (F&V), 144
FSNM. *See* Food Security and Nutrition Monitoring Survey

G

Gender gap in agriculture, 113–116
Gender role in technology adoption. *See* Cramer's V and phi coefficient application, in gender role in technology adoption
Generalized coefficient of determination, 474–475
Generalized linear models (GLM), 621–622
GEO-5 and coping mechanisms, 44–45
German agency for Technical Cooperation (GTZ), 541
Get File
  command, 643
  statement, 647
Ghana, 532
GLM. *See* Generalized linear models
Global Hunger Index, 499–500
Global Positioning System (GPS), 42b–44b
  guidance systems, 42b–44b
Global undernourishment, 15t
Goodness-of-fit test, 473–474
Graphics, SPSS programming basics
  histograms, 652–653

  scatterplots, 651–652
Green Revolution, 28–29
Group centroids, 409
  functions at, 410t
Group differences, evaluating, 422
Growth Monitoring and Promotion (Gmp) Programs, percent of children in, 665
GTZ. *See* German agency for Technical Cooperation
Guatemala
  export-producing cooperative in, 32
  food security and nutritional status in, 74
  malnutrition in, 272–273
Guttman scale, 226–227

H

H/A. *See* Height for age
Head Start program, 460
Headcount measure, 449–450
HEALTDST, 275, 352
Health clinic data for STATA, 479t–480t
Health indicators, 300
Health service indicators, 301–302
Health status indicators, 300–301
Hedgerow intercropping. *See* Alley cropping
Height for age (H/A), 296, 299–300
  model for determinants, 356t
  Z-Scores
    analysis of variance table for, 358t
    individual coefficients for determinants, 358–361
Herbicide-tolerant crops, 29
Heteroskedasticity, 357
HHLDQTY. *See* Household size
Hierarchical clustering method, 501–503
High weight for height, 664
High-tech agribusiness, 36–37
Highly indebted poor countries (HIPC), 439
Hindustan Lever Limited (HLL), 71
HIPC. *See* Highly indebted poor countries
Histograms, 652–653
HLL. *See* Hindustan Lever Limited
Homogeneity of variance, 364
  checking for, 364
Homoscedasticity, 166, 357
Honduran Social Investment Fund, 340
Hosmer–Lemeshow goodness of fit test, 473–474, 481
  contingency table for, 474t
HOSPITAL, 401
Household Asset Building Program (HABP), 393–394

Index **741**

Household assets, 232
Household calorie, food group shares of, 151t
Household characteristics, gender of head and, 303
Household composition, 441
Household food security, 20
Household income, 108, 135–136
Household level characteristics, 190
  and demographic variables, 470–471
Household production function approach, 261
Household size (HHLDQTY), 317
Household survey consumption modules, 182
Household utility maximization framework, 297–298
Household welfare, expected determinants of, 470–472
Households with access to sanitation, percentage of, 303
Human poverty index (HPI), 436
Hunger and malnutrition, 14
Hybrid maize adoption (HYBRID), 47, 51, 59, 119–120, 189, 199–200
Hybrid maize variety, 151
Hygiene education, 302, 324
Hypotheses testing, 357–361

I

ICRAF. *See* International Center for Research in Agroforestry
ICT. *See* Information and Communication Technology
Identification problem, 443–444
Identity matrix, 671
If statement, 644–645
IFPRI. *See* International Food Policy Research Institute
IITA. *See* International Institute for Tropical Agriculture
ILCA. *See* International Livestock Center for Africa
ILRI. *See* International Livestock Research Institute
IMCI. *See* Integrated Management of Childhood Illness
Immunization coverage for children, 301
Inadequate food, 260
Income measure, 440
Independent group t-test, 59–60, 62–63
Independent sample t-test, 60–63
Independent variables, descriptive summary of, 352–353
India, 34

access and food security in, 182–185
  caste and factors affecting nutrition in, 296–297
  food consumption patterns in, 147–149
  food security
    economic crisis, and poverty in, 499–501
    and food production in, 28
    malnutrition and chronic disease in, 271–272
  poverty in, 499–501
Indicators, in measuring poverty, 439–440
Indirect utility function, 442
Indonesia, 28–29
Infant mortality, 263
  rate, 301
Influence, 298
Informal sector, role of, 180
Information age and precision agriculture, 42b–44b
Information and Communication Technology (ICT), 179
  market information, 179
INSECURE, 47, 412
Instrumental variables (IV), 527–529, 579–583
  estimation, 532
    empirical analysis, 539–545
    endogeneity/omitted variable bias, 684
    federal nutrition programs and children's health in US, 534–536
    food security using Gallup World Poll, 538–539
  Hausman test, 553
  least squares estimate, problem of bias in, 679–680
  measurement error, 683
  overidentification test, 682
  parental unemployment and children's health in Germany, 536–538
  properties of, 680
  simultaneous equations, 683
Integrated Child Development Scheme, 271
Integrated Management of Childhood Illness (IMCI), 351
Interaction approach, 20
  food security, 20
Interaction effect, 223, 234
  interpretation, 241–248
International Agricultural Research Centers, 28–29

742 Index

International Center for Research in
    Agroforestry (ICRAF), 108
International Center for Tropical Agriculture
    (CIAT), 32, 35
International Food Policy Research Institute
    (IFPRI), 32
International Institute for Tropical
    Agriculture (IITA), 108
International Livestock Center for Africa
    (ILCA), 108
International Livestock Research Institute
    (ILRI), 108
International Maize and Wheat Improvement
    Center (CIMMYT), 28–29
International Rice Research Institute (IRRI),
    28–29, 35
Interval variable. *See* Continuous variables
Intra-annual poverty, 497
Inverse matrix computing, 672–673
    properties of, 673
IRRI. *See* International Rice Research
    Institute
Item Response Theory, 496

J

Japan, collaborative farming system,
    11–12

K

K-mean cluster analysis application,
    classifying households on
    cluster analysis, 501
    cluster centers, 510–513
    cluster of households, descriptive
        characteristics of, 508–510
    data description, 507
    empirical analysis, 506–507
    food hardships and economic status in
        United States, 495–499
    food security, economic crisis, and poverty
        in India, 499–501
    hierarchical clustering method
        average linkage method, 503
        complete linkage, 502
        single linkage, 502
    initial partitions and optimum number of
        clusters, 507–508
    K-mean method, 503
    review of studies using, 503–506
    STATA, cluster analysis in, 513–516
K-means cluster analysis, 495

Kaiser-Meyer-Olkin (KMO) measure, 171,
    190, 192t, 199–200, 202
Kenya, 108–109, 173
    children's nutrition and maternal education
        in, 232
    community characteristics and child
        nutrition in, 345
    female farm operators in, 109–112
KMO measure. *See* Kaiser-Meyer-Olkin
    measure
KMO–Bartlett test, 192t

L

LAC. *See* Latin American and Caribbean
Land owned (LANDO), 189, 191t, 195t, 469
Land reform, 505
Latin America, adaptive strategies and
    sustainability lessons from, 397–398
Latin American and Caribbean (LAC), 144,
    387–388
Least significant difference test (LSD),
    249–250
Least squares method, 353–354
Leave-one-out (L-O-O)
    method, 410
    principle, 424
Levels of aggregation, 16
Levene's test, 30–31, 52–53, 53t
    equality of variances, 52
Leverage, 362
Liberalization process, 171–172
Life expectancy at birth, 301
Life-cycle approach, 4–5
Lightness, 299
Likelihood function, 487
Linear programming model, in achieving
    ideal diet, 557–559, 563–565
    graphical solution approach, 565–571
    review of literature, 559–563
    solution procedures, 565
Linear regression, 319
Linear systems of equations, solutions of, 673
Living standard measurement surveys
    (LSMS), 382
LIVSTOCKSCALE, 47–50, 507
LNPRODLMAIZ, 189, 195t
LNXFD, 317
LNXTOTAL. *See* Total expenditure
Log-likelihood ratio, 472–473
Logistic coefficients, interpreting,
    476–478

Logistic regression models application in
    poverty measurement and
    determinants, 387–388, 434
  empirical results, 472
  exercises, 489–490
  selected review, 454–458
  in STATA, 478–485
  technical appendices, 486–487
Logistic regression results, 477t
Logit, 477
Low height for age, 664
Low weight for age, 664
Low weight for height, 663–664
Lower bound of non-food poverty line, 448
LSD. *See* Least significant difference test

M

Main effect, definition of, 237
Malawi, 30, 151t
  dataset, 469–470
    example with, 470f
Malawi, measures for, 452–453
  measuring. *See* Measuring poverty
  monetary approach, 435–436
  multiple dimensions of, 485–486
  participatory poverty approach (PPA),
    437–439
  relative, 466
  squared poverty gap index, 489
  for STATA, 490–492
  and welfare in United States, 458–463
Malnutrition, 4, 222, 259–260
  food insecurity and, 12
  hunger and, 11–12
Manova command, 412
Manovatest command, 413
Market access indicators, 189
Marketing loan, 40
Maternal education
  in Kenya, 223
  in United States, 229–231
Maternal schooling, child-care and nutritional
    status linkages between, 223–229
Matrix
  algebra, 669
  command, 413
  definition of, 669
  equality of, 670
  functions of, 671–673
  operations with, 670–671
  transpose of, 669–670
Maximum likelihood (ML), 186

estimation, 472
MDGs. *See* Millennium Development Goals
Means-tested programs, 458–459
Measuring poverty, 435, 439
  basic needs approach, 436
  capability approach, 436–437
  consumption expenditure, 440–441
  defining and, 435
  income measure, 440
  indicators in, 439–440
  monetary approach, 435–436
  PPA, 437–439
  rationale for, 439
Median, percentage of, 666–667
Medicaid, 266
  program, 459–460
MENA. *See* Middle East and North Africa
Mexico, 28–29
Micronutrient deficiencies, 6–7
  of woman, 388
Middle East and North Africa (MENA), 171
  region, food security issues in, 180–181
Millennium Development Goals (MDGs),
    269–270
  target, 22–23
ML. *See* Maximum likelihood
Model fit, measuring
  classification table, 475–476
  generalized coefficient of determination,
    474–475
  Hosmer-Lemeshow goodness-of-fit test,
    473–474
  log-likelihood ratio, 472–473
Mother's educational levels
  cross-tabulation of weight for height with,
    235–241
  prevalence of
    stunting by, 236t
    wasting by, 236t
Multinomial logit model, 456
Multiple comparison test, 241t
Multiple dimensions of poverty, 485–486
Multiple regression, parameters in, 367t
Multiplication of matrices, 670–671
Multivariate regression model, 174, 328

N

National Advisory Council (NAC), 13
National Health and Nutrition Examination
    Survey (NHANES), 342
National Longitudinal Surveys, 229–231
National Sample Survey (NSS), 142

National School Lunch Program (NSLP), 11, 265, 392
Natural causal analysis, 230b—231b
Natural resources, 6
NCHS. *See* US National Center for Health Statistics
Nearest neighbor method, 502
"Nested," multivariate equations, 473
Newly industrializing countries (NIC), 24t
NHANES. *See* National Health and Nutrition Examination Survey
Nigeria, 108—109, 389
Nigerian Demographic and Health Survey, 388
Nominal/categorical variable, 96
Non-food poverty lines, construction of, 448
NSLP. *See* National School Lunch Program
NSS. *See* National Sample Survey
Null hypothesis, 52—53, 321—323, 541
Number of times child fed during sickness (SICKFEED), 318
Nutrient deficiencies, 500
Nutrient intake, 18
Nutrient requirements, 18
Nutrition, 222
 commercialization of agriculture on, 74
 global monitoring report on, 269—270
 technology adoption impact on, 110—112
Nutrition policy-focused conceptual framework, 4
Nutrition security, 4
 causes of, 4
Nutrition-sensitive approaches in multisectoral interventions, 285
Nutritional status, 222
 of children, 227
 conceptual framework and indicators of, 297—303
 core indicators of, 299—303
 gender of head and, 303
 in United States, 229—231

O

OASI program. *See* Old-Age Social Security program
"Obesogenic environment," 145b—147b
Oblique, 197, 206
Observation method, 226—227
Observed correlation coefficients matrix, 190
OECD scale of adult equivalents, 441
Off-farm activity, 458
Old-Age Social Security (OASI) program, 460
OLS. *See* Ordinary least squares
One-way ANOVA approach, 139, 149
Order, definition of, 669
Ordinary least squares (OLS), 475, 529, 541
Organic farming, 80b—81b
 and commercialization, 79—81
 in global context, 81—82
Other Food Security Program (OFSP), 393—394
Output market participation, 180

P

P-values, 246, 252, 412, 414, 428—429
PAF. *See* Principal axis factoring
PAFC. *See* Punjab Agro Foodgrains Corporation
Pakistan, 70, 181
 malnutrition and mortality in, 310—312
Panel Study of Income Dynamics (PSID), 268b—269b
 data, 344
Parameter estimates, standard errors of, 347—348
Parliamentary Standing Committee on Food, 13
Partial correlation coefficients, 361
Partial F-values methods, 423
Participatory poverty approach (PPA), 437—439
Participatory surveys, 438
Partitioning sum of squares, 237—241
Patchwork system, 460—461
PBTRADER, 472
PC. *See* Principal components
Pearson's chi-square test. *See* Chi-square statistic application, in agricultural commercialization
Pepsi, 66—67
Per capita expenditure (PXTOTAL), 150, 507
Per capita expenditure on food (PXFOOD), 189, 196t, 198t, 199—200
Per capita food consumption, 136—137
Per capita health expenditure, 302
Phi coefficient, 128
Philippines, 28—29
Physical activity levels, 19
Pipeline Comparison method, 587—588
Policy makers, 296
Policy measures, 457

Poor households, characteristics of, 453—454, 453t—454t
Population parameters in correlation, inference about, 276—277
Possible linkages, 223—224
Post-harvest crop loss, 33
Post-harvest technology, 28, 33—36
　grain management practices, 34
Post-hoc procedures, 249—252
Post-hoc tests, interpretation, 241—248
Potatoes, 29, 39—40
Poverty, 12, 338, 346—347
　absolute, 443
　agriculture and, Laos and Cambodia, 463—465
　basic needs approach, 436
　capability approach, 436—437
　determinants of, 434, 454—458
　in developing countries, 467—468
　Engel curve, measures based on, 449—454
　in Europe, 465—467
　financial crisis and, Russian Federation, 465
　gap index, 450—451
　headcount measure, 449—450
　indicators
　　consumption expenditure, 440—441
　　income measure, 440
Poverty line
　absolute and relative poverty, 443
　deriving, 444
　objective poverty line
　　cost of basic needs method, 446—448
　　food energy intake method, 444—446
　referencing and identification problems, 443—444
　in theory, 442—443
Poverty mapping exercises, 503—504
Poverty maps, 504
Poverty reduction strategies paper (PRSP), 439
PPA. *See* Participatory poverty approach
Precision agriculture, 42b—44b
　information age and, 42b—44b
　use of, 42b—44b
Predictor variables and discriminant function, correlation between, 408—409
Preliminary statistical concepts, specification tests
　differentiate between good and bad, 417—418
　principles of, 675—677

Preschoolers' nutrition, maternal education and care impact. *See* Two-way ANOVA application, in preschoolers' nutrition, maternal education and care impact
Principal axis factoring (PAF), 186
Principal components (PC) analysis, 170—171, 193
　in STATA, 201—204
Private food assistance programs, effects of, 267
Private traders, 170
Probit analysis, 77
Probit model, 455
PRODLMAIZ, 471, 507
Production function approach, 8—10
Productive Safety Net Program (PSNP), 393—394
Propensit Score Matching (PSM) method, 577, 587—588
Provision of sanitation (LATERINE), 275, 352
PRSP. *See* Poverty reduction strategies paper
PSID. *See* Panel Study of Income Dynamics
PSM method. *See* Propensit Score Matching method
Psychosocial care, 226
　practices, 225
Punjab Agro Foodgrains Corporation (PAFC), 71—72
PXFD. *See* Per capita expenditure on food
PXFOOD. *See* Per capita expenditure on food

Q

q-statistic tests, 249—250
QR estimator. *See* Quantile Regression estimator
Quality of care approach, 222, 225
Quantile Regression (QR) estimator, 182
Quantitative analysis of poverty, 439—440
Quantitative assessment method, 227

R

Randomization, 577—579
Ravallion, 442
Recode statement, 644
Recommended intakes, nutrients, 18
Reference group, 447—448
Referencing problem, 443—444
Regional differences in poverty, 466

Regression assumptions, checking for violations of, 362—368
Regression diagnostics, 362
Regression discontinuity design, 577, 585—587, 587f
Regression framework, 297
Regression method factor scores, 199—200
Regression models, 456—457
Regression standardized residuals, normal P-P plot of, 364f
Regression techniques, 316
Relative poverty, 443
Rice, 35
Right to Food Campaign, 13
Rotated component matrix, 197—198
Rotation procedure, 197, 198t
Rule of thumb, 280
RUNDUM, 507
Rural Bangladesh, impact of food price spike on, 270—271
Russian Household Budget Survey, 398

S

SADMDIST, 401
Salima add-code, 452—453
Sanitation facility, 303
Save Outfile statement, 646
Scaling approach in measuring food security, 20
Scatterplots, 278—279
SCHIP. *See* State Children's Health Insurance Program
School Breakfast Program (SBP), 392
Scoring system to create care index, 249
Scree plot, 193—197, 201—203
  eigenvalues, 202—203
Scree test criterion, 187
Security issues, 458
Selection bias, 576, 578
Selling point (SELPOINT), 189, 196, 199—200
Showorder command, 413
SICKFEED. *See* Number of times child fed during sickness
Single linkage, 502
Size of land owned (LANDO), 453—454
Smallholder agricultural sector, 174
SNAP. *See* Supplemental Nutrition Assistance Program
Social insurance programs, 458—460
Social sciences, statistical programs in, 619

Social Security Disability Insurance program, 459—460
  recipients, 459
Social-desirability bias, 497
Socioeconomic factors, 330—331
Sod-swamp-buster programs, 40
South Asia, 181—185
Specification tests
  differentiate between good and bad, 417—418
  principles of, 675—677
Split file box, 625f
SPSS. *See* Statistical Package for the Social Sciences
Square matrix, definition of, 669
Squared poverty gap index, 451—452
SSB. *See* Sum of squares between
SSI. *See* Supplemental Security Income
SSW. *See* Sum of squares within
Standard multiple linear regression model, 486
Standardized canonical discriminant coefficients, 432
  function, 415
Standardized discriminant functions, 408, 421—422
Staple left over (STAPLEFT), 189, 199—200
STATA, 513
  canonical discriminant analysis using, 412—416
  chi-square tests using, 89—93
  cluster analysis in, 513—516
  command, 513
  logistic regression models in, 478—485
  multiple regression in, 364—368
  one-way ANOVA in, 153—158
  PC analysis in, 201—204
  statistical data analysis, 621—622
  for t-tests, 201—204
State Children's Health Insurance Program (SCHIP), 460—461
Statistical Package for the Social Sciences (SPSS), 186, 198, 502
  categories, 623
  complex samples, 627
  conjoint, 627
  data transformation techniques, 643—653
  environment and commands
    add-ons menu, 627—628
    analyze menu, 626
    data definition and manipulation, 629
    data editor toolbar, 624

data menu, 626
edit menu, 625
file menu, 625
graphs menu, 626
help menu, 628
main SPSS window, 623
menus, 623
operations, 630–631
output window, 628
procedure, 629
syntax window, 629
transform menu, 626
utilities menu, 627
view menu, 626
window menu, 628
exact tests, 627
importing ASCII data into, 634–635
maps, 627
missing value analysis, 627–628
statistical procedures, submitting, 643
syntax, 641–643
Structure correlations methods, 423
Structure matrix, 408–409
Student's t-test. *See* T-statistic test application, in technological change implications and post-harvest technology
Stunting, 259–260, 296, 318–319
and wasting, incidence of, 232
Sub-Saharan Africa (SSA), 105–106, 135–136, 387–388
Subjective poverty line construction, 444
Subtraction of matrices, 670
Sum of squares between (SSB), 239, 406
partition, 237–241
Sum of squares within (SSW), 249, 349
Supplemental Nutrition Assistance Program (SNAP), 11, 145b–147b, 264, 267–269, 342, 459
Supplemental Security Income (SSI), 459–460
Sustainable agriculture, 113
Sweet corn, 29
Symmetric matrix, 671

T

T-distribution, relation of F to, 161–167
T-statistic test application, in technological change implications and post-harvest technology
biofuels, 38–40
data description and analysis, 46–47
empirical analysis, 45–46
equality of means, 53–56
food security. *See* Food security
food-security index, 58
GEO-5 and coping mechanisms, 44–45
introduction, 28–30
vs. Pearson's chi-square (χ^2) test, 56–58
policy implications, 56–58
selected studies, 30–36
using STATA for, 59–63
independent group *t*-test, 59–60
independent sample *t*-test, 60–63
technical appendices, 58–59
US Farm Policy and food security, 44
variable definitions, 58–59
TANF program. *See* Temporary Assistance for Needy Families program
Tanzaniás maize market, 178
Targeting interventions, 386
Technological change implications, t-statistic test application in. *See* T-statistic test application, in technological change implications and post-harvest technology
Technology adoption
effect of, 31
impact of, 56
Cramer's V and phi coefficient application, in gender role in. *See* Cramer's V and phi coefficient application, in gender role in technology adoption
food security and, 30–31, 55–56
implementation of, 56
Temporary Assistance for Needy Families (TANF) program, 11, 459
Terciles (NCARE), 235
Text files
delimited format, 634
fixed format, 634
TLUs. *See* Tropical livestock units
Total expenditure (LNXTOTAL), 317
Total food expenditure (LNXFD), 317
Trade reforms, 457
Trade-related reforms on poverty, impact of, 455–456
Transaction cost, 178–179
Transient poverty, 465
Transitory poverty, 458
Transpose of matrix, 664
Tropical livestock units (TLUs), 47–48
Tukey honestly significant difference (HSD) test, 249–250

Tukey's method, 244
Turkish Household Budget Survey, 466
Two-stage least squares, 530
Two-way ANOVA application, in preschoolers' nutrition, maternal education and care impact
   conceptual framework
      conceptual and measurement issues, on child-care, 224–225
      possible linkages, 223–224
      review of selected studies, 227–229
   cross-tabulation of W/H, with mother's educational levels, 235–241
      data description, 234–235
      empirical analysis, 234
      interaction effect and post-hoc tests, interpreting, 241–248
      technical appendices, 249
Two-way ANOVA model, 249
Type I error, 676
Type II error, 676

U

Uganda's coffee market, 116–118
UI program. *See* Unemployment Insurance program
Unbiasedness, 677
UNDP. *See* United Nations Development Programme
Unemployment Insurance (UI) program, 460
UNEP, 44–45
Unequal prior probabilities
   classification function based on, 410–412
   classification results based on, 411t
Unhealthy foods
   reducing/limiting advertising of, 145b–147b
   tax on, 145b–147b
UNICEF, 528
   nutrition conceptual framework, 248
United Nations Development Programme (UNDP), 436
United States
   community characteristics and children's nutrition in, 342–344
   female farm operators in, 112–113
   food consumption patterns in, 145–147
   food deserts in, 175–178
   food hardships and economic status in, 495–499
   food insecurity and nutrition in, 264–269
   food security issues and technology, 36–38
   maternal education and nutrition status in, 229–231
   obesity in, 504
   organic farms and commercialization in, 79–81
   poverty and welfare in, 458–463
   types of agriculture, 36
Urbanization, 138–139
US Department of Agriculture (USDA), 390
   Agricultural Research Service (ARS), 153
   Economic Research Service, 175–176
US Farm Policy, 40–44
US food assistance programs, 268b–269b
US National Center for Health Statistics (NCHS), 665
USAID, 28–29
Utilization rate, 301

V

Value labels, 624
Variable-rate technology (VRT), 42b–44b
Variables, 150, 505
   definitions, 58–59
   dichotomous, 58–59
   of household welfare, 434
   insecure, 351
   interval, 59
   labels, 624
   normality tests and transformation of, 319–320
   vector of, 483
   for WTP analysis, 482t–483t
Variance partitioning for two-way ANOVA, 238f
Variation in ANOVA framework, 160
VARIMAX method, 187
   orthogonal rotation, 197–198, 206
Vector, definition of, 663–664
Vegetable and fruit cash crop (VFC) program, 76–77
   goal of, 76–77
VFC program. *See* Vegetable and fruit cash crop program
Vietnam Living Standards Survey (VLSS), 143, 456
VRT. *See* Variable-rate technology
Vulnerability, 434, 436
Vulnerability and access (VA), 197–198

W

W/A. *See* Weight for age
W/H. *See* Weight for height
Wald statistics, 476
Ward's method, 510
Wasting, 318–319
    incidence of stunting and, 284
    index, 300
Water and sanitation indicators, 302–303
Water for STATA, 479t–480t
Water source (WATER), 318
    access to improve, 303
Weight for age (W/A), 299
    Z-Scores
        analysis of variance table for, 358t
        model for determinants of, 356t
Weight for height (W/H), 296, 299
Welfare in United States, poverty and, 458–463
Welfare ratio approach, 442
WFP. *See* World Food Programme
Wheat prices in Afghanistan, 181–182
WIC. *See* Women, Infants, and Children
Wilks' lambda coefficient, 412
Wilks' lambda test, 406–407, 406t
Within sum of squares (WSS), 419–420
Women, Infants, and Children (WIC), 145b–147b, 459, 534
    food package, 391
Women empowerment and nutrition status, 312–315
Women in agriculture, 113–116
    and techniques, 115b–116b
Women's status, 383
    and child nutrition
        direct linkages between, 385–388
        indirect linkages between, 388–390
        linkages between, 385
World Bank, 434
    funded projects, 296–297
World Bank's Global Monitoring Report, 269–270
World Fertility Survey (WFS) program, 222
World Food Programme (WFP), 13–15
World Food Summit, 16
WTP analysis, variables for, 482t–483t

Y

Yield monitoring, precision agriculture, 42b–44b
Yield/technology indicators, 189
Young Lives project (YL), 233
Younger Cohort, 310–311

Z

Z transformation, 277
Z-score weight for height (ZWH), 300
Z-scores, 223
    computation, 665–666
    scale, 198–199
Zambia, 107, 173
ZHANEW, 84
Zimbabwe, 77–78, 173
ZWANEW, 84
ZWHNEW, 84, 399–400, 405
    effect of child-care on, 238t
    dependent variable, 240t
    effect of mothers' education on, 238t
    predictor variables on, 407–408

CPI Antony Rowe
Eastbourne, UK
December 28, 2023